# Mechanics of Fluids
## Fourth Edition, SI

### Merle C. Potter
*Michigan State University*

### David C. Wiggert
*Michigan State University*

### Bassem Ramadan
*Kettering University*

with

### Tom I-P. Shih
*Purdue University*

*SI Edition prepared by:*

### Shaligram Tiwari
*Indian Institute of Technology Madras*

CENGAGE
Learning™

Australia • Brazil • Japan • Korea • Mexico • Singapore • Spain • United Kingdom • United States

*Mechanics of Fluids, Fourth Edition, SI*
Merle C. Potter, David C. Wiggert,
Bassem Ramadan, and Tom I-P. Shih
SI Edition Prepared by: Shaligram Tiwari

Publisher, Global Engineering:
   Christopher M. Shortt

Acquisitions Editor, SI Edition:
   Swati Meherishi

Assistant Developmental Editor:
   Debarati Roy

Senior Developmental Editor: Hilda Gowans

Assistent Development Editor: Debarati Roy

Editorial Assistant: Tanya Altieri

Team Assistant: Carly Rizzo

Marketing Manager: Lauren Betsos

Media Editor: Chris Valentine

Content Project Manager: D Jean Buttrom

Production Service: RPK Editorial Services

Copyeditor: Harlan James

Proofreader: Martha McMaster

Indexer: Merle Potter

Compositor: Integra Software Solutions

Senior Art Director: Michelle Kunkler

Internal Designer: Carmela Periera

Cover Designer: Andrew Adams

Cover Image: © NASA

Text and Image Permissions Researcher:
   Kristiina Paul

First Print Buyer: Arethea L. Thomas

Library of Congress Control Number: 2011920709

ISBN-13: 978-1-4390-6203-6
ISBN-10: 1-4390-6203-X

**Cengage Learning**
200 First Stamford Place, Suite 400
Stamford, CT 06902
USA

Cengage Learning is a leading provider of customized learning solutions with office locations around the globe, including Singapore, the United Kingdom, Australia, Mexico, Brazil, and Japan. Locate your local office at **international.cengage.com/region**.

Cengage Learning products are represented in Canada by Nelson Education, Ltd.

For your course and learning solutions, visit **www.cengage.com/engineering**.

Purchase any of our products at your local college store or at our preferred online store **www.cengagebrain.com**.

Printed in the United States of America
1 2 3 4 5 6 7 13 12 11

# Contents

# Preface to the SI Edition

This edition of *Mechanics of Fluids* has been adapted to incorporate the International System of Units (*Le Système International d'Unités* or SI) throughout the book.

## Le Système International d'Unités

The United States Customary System (USCS) of units uses FPS (foot-pound-second) units (also called English or Imperial units). SI units are primarily the units of the MKS (meter-kilogram-second) system. However, CGS (centimeter-gram-second) units are often accepted as SI units, especially in textbooks.

## Using SI Units in this Book

In this book, we have used both MKS and CGS units. USCS units or FPS units used in the US Edition of the book have been converted to SI units throughout the text and problems. However, in case of data sourced from handbooks, government standards, and product manuals, it is not only extremely difficult to convert all values to SI, it also encroaches upon the intellectual property of the source. Some data in figures, tables, and references, therefore, remains in FPS units. For readers unfamiliar with the relationship between the FPS and the SI systems, a conversion table has been provided inside the front cover.

To solve problems that require the use of sourced data, the sourced values can be converted from FPS units to SI units just before they are to be used in a calculation. To obtain standardized quantities and manufacturers' data in SI units, the readers may contact the appropriate government agencies or authorities in their countries/regions.

## Instructor Resources

The Instructors' Solution Manual in SI units is available through your Sales Representative or online through the book website at www.cengage.com/engineering.

The readers' feedback on this SI Edition will be highly appreciated and will go a long way in helping us improve subsequent editions.

*The Publishers*

# Preface

The motivation to write a book is difficult to describe. Most often the authors suggest that the other texts on the subject have certain deficiencies that they will correct, such as an accurate description of entrance flows and flows around blunt objects, the difference between a one-dimensional flow and a uniform flow, the proper presentation of the control volume derivation, or a definition of laminar flow that makes sense. New authors, of course, introduce other deficiencies that future authors hope to correct! And life goes on. This is another fluids book that has been written in hopes of presenting an improved view of fluid mechanics so that the undergraduate can understand the physical concepts and follow the mathematics. This is not an easy task: Fluid mechanics is a subject that contains many difficult-to-understand phenomena. For example, how would you explain the hole scooped out in the sand by the water on the upstream side of an abutment? Or the high concentration of smog contained in the Los Angeles area (it doesn't exist to the same level in New York)? Or the unexpected strong wind around the corner of a tall building in Chicago? Or the vibration and subsequent collapse of a large concrete-steel bridge due to the wind? Or the trailing vortices observed from a large aircraft? We have attempted to present fluid mechanics so that the student can understand and analyze many of the important phenomena encountered by the engineer.

The mathematical level of this book is based on previous mathematics courses required in all engineering curricula. We use solutions to differential equations and vector algebra. Some use is made of vector calculus with the use of the gradient operator, but this is kept to a minimum since it tends to obscure the physics involved.

Many popular texts in fluid mechanics have not presented fluid flows as fields. That is, they have presented primarily those flows that can be approximated as one-dimensional flows and have treated other flows using experimental data. We must recognize that when a fluid flows around an object, such as a building or an abutment, its velocity possesses all three components which depend on all three space variables and often, time. If we present the equations that describe such a general flow, the equations are referred to as field equations, and velocity and pressure fields become of interest. This is quite analogous to electrical and magnetic fields in electrical engineering. In order for the difficult problems of the future, such as large-scale environmental pollution, to be analyzed by engineers, it is imperative that we understand fluid fields. Thus in Chapter 5 we introduce the field equations and discuss several solutions for some relatively simple geometries. The more conventional manner of treating the flows individually is provided as an alternate route for those who wish this more standard approach. The field equations can then be included in a subsequent course.

Perhaps a listing of the additions made in this fourth edition would be of interest. We have:

- Included inside the back cover the second edition of a DVD entitled "Multimedia Fluid Mechanics," by G.M. Homsy, distributed by Cambridge University Press. A number of virtual labs and demonstrations have been identified at various places in the text.
- Added many examples and problems, most of which have real-life applications.
- Collected all the multiple-choice problems to the front of the problem sets. They can be used to review the subject of Fluid Mechanics for the Fundamentals of Engineering and the GRE/Engineering exams.
- Deleted the chapter on Environmental Fluid Mechanics in an attempt to shorten the book.
- Simplified the chapter on Computational Fluid Mechanics.
- Made numerous changes to clarify the presentation.

The introductory material included in Chapters 1 through 9 has been selected carefully to introduce students to all fundamental areas of fluid mechanics. Not all of the material in each chapter need be covered in an introductory course. The instructor can fit the material to a selected course outline. Some sections at the end of each chapter may be omitted without loss of continuity in later chapters. In fact, Chapter 5 can be omitted in its entirety if it is decided to exclude field equations in the introductory course, a relatively common decision. That chapter can then be included in an intermediate fluid mechanics course. After the introductory material has been presented, there is sufficient material to present in one or two additional courses. This additional course or courses could include material that had been omitted in the introductory course and combinations of material from the more specialized Chapters 9 through 14. Much of the material is of interest to all engineers, although several of the chapters are of interest only to particular disciplines.

We have included examples worked out in detail to illustrate each important concept presented in the text material. Numerous home problems, many having multiple parts for better homework assignments, provide the student with ample opportunity to gain experience solving problems of various levels of difficulty. Answers to selected home problems are presented just prior to the Index. We have also included design-type problems in several of the chapters. After studying the material, reviewing the examples, and working several of the home problems, students should gain the needed capability to work many of the problems encountered in actual engineering situations. Of course, there are numerous classes of problems that are extremely difficult to solve, even for an experienced engineer. To solve these more difficult problems, the engineer must gain considerably more information than is included in this introductory text. There are, however, many problems that can be solved successfully using the material and concepts presented herein.

Many students take the FE/EIT exam at the end of their senior year, the first step in becoming a professional engineer. The problems in the FE/EIT exam are all four-part, multiple choice. Consequently, we have included this type of problem at the beginning of the chapters. Multiple-choice problems will be presented using SI units since the FE/EIT exam uses SI units exclusively. Additional information on the FE/EIT exam can be obtained from a website at **ppi2pass.com.**

The book is written emphasizing SI units;

The authors are very much indebted to both their former professors and to their present colleagues. Chapter 10 was written with inspiration from F.M. Henderson's book titled *Open Channel Flow* (1996), and D. Wood of the University of Kentucky encouraged us to incorporate comprehensive material on pipe network analysis in Chapter 11. Several illustrations in Chapter 11 relating to the water hammer phenomenon were provided by C.S. Martin of the Georgia Institute of Technology. R.D. Thorley provided some of the problems at the end of Chapter 12. Tom Shih assisted in the writing of Chapter 14 on Computational Fluid Dynamics. Thanks to Richard Prevost for writing the MATLAB® solutions. We would also like to thank our reviewers: Sajjed Ahmed, University of Nevada; Mohamed Alawardy, Louisiana State University; John R. Biddle, California State Polytechnic University; Nancy Ma, North Carolina State University; Saeed Moaveni, Minnesota State University; Nikos J. Mourtos, San Jose (CA) State University; Julia Muccino, Arizona State University; Emmanuel U. Nzewi, North Carolina A&T State University; and Yiannis Ventikos, Swiss Federal Institute of Technology.

*Merle C. Potter*
*David C. Wiggert*
*Bassem Ramadan*

# Multimedia Fluid Mechanics

The following entries from the Multimedia Fluid Mechanics DVD have been added to the text. They are experiments and demonstrations that are included on the DVD inside the back cover. The instructor may assign additional files from the DVD at appropriate locations in the text. Also, the reader may find selected entries that are not listed of particular interest. To view a file on a specified page, simply open any of the eight major headings, then enter a page number in the box at the top and click on "go to page." The page numbers after the descriptors refer to the pages on the DVD.

**Example 8.17a:** Laminar Boundary-Layer Growth, 625–627

**Problem 8.14:** Flow Past a Sphere, 5

**Below Fig. 3.13:** Taylor Cells, 24

**In margin, p.105:** Reynolds Number, 524

**In margin, p.105:** Pipe Flow, 202

**In margin, p.380:** A Doublet, 281

**In margin, p.239:** Dimensional Analysis, 588

**In margin, p.242:** Dimensional Analysis, 521–523

**In margin, p.246:** Dimensionless Numbers, 524–528

**In margin, p.249:** Dynamic Similarity, 195

**In margin, p.250:** Similarity and Scaling, 494, 534, 535, 568

**In margin, p.274:** Laminar Flow in a Pipe, 686

**In margin, p.288:** Flow between Cylinders, 735

**In margin, p.293:** Reynolds Decompostion, 689

**In margin, p.295:** Turbulent Flow in a Pipe, 687

**In margin, p.347:** Flow Past a Cylinder, 116, 131, 190

**In margin, p.347:** Flow over an Airfoil, 649

**In margin, p.349:** Separated Flow over an Airfoil, 167

**In margin, p.350:** Separated Flow past Sharp Edges, 662, 664, 666

**In margin, p.353:** Flow around Immersed Bodies, 652, 657, 694

**In margin, p.355:** Drag Curve for a Golf Ball, 265

**In margin, p.359:** Vortex Shedding, 81, 216

**In margin, p.362:** Streamlining, 651

**In margin, p.370:** Trailing Vortex, 725, 577

**In margin, p.373:** Potential Flows, 123, 271

**In margin, p.378:** Simple Potential Flows, 277

**In margin, p.387:** Boundary Layers, 163, 260, 602, 622, 677, 852

**In margin, p.426:** Compressible Flows, 50

**In margin, p.347:** Speed of Sound, 57

**In margin, p.754:** CFD Solutions, 669–672, 807, 821

# Nomenclature

$A$ - area

$A_2, A_3$ - profile type

$a$ - acceleration, speed of a pressure wave

$\mathbf{a}$ - acceleration vector

$a_x, a_y, a_z$ - acceleration components

$B$ - bulk modulus of elasticity, free surface width

$b$ - channel bottom width

$C$ - centroid, Chezy coefficient, Hazen-Williams coefficient

$C_1, C_3$ - profile type

$C_D$ - drag coefficient

$C_d$ - discharge coefficient

$C_f$ - skin-friction coefficient

$C_H$ - head coefficient

$C_L$ - lift coefficient

$C_P$ - pressure recovery factor, pressure coefficient

$C_{NPSH}$ - net positive suction head coefficient

$C_Q$ - flow-rate coefficient

$C_V$ - velocity coefficient

$C_{\dot{W}}$ - power coefficient

$c$ - specific heat, speed of sound, chord length, celerity

$c_f$ - local skin-friction coefficient

$c_p$ - constant pressure specific heat

$c_v$ - constant volume specific heat

c.s. - control surface

c.v. - control volume

$D$ - diameter

$\dfrac{D}{Dt}$ - substantial derivative

$d$ - diameter

$dx$ - differential distance

$d\theta$ - differential angle

$E$ - energy, specific energy, coefficient

$E_c$ - critical energy

EGL - energy grade line

Eu - Euler number

$e$ - the exponential, specific energy, wall roughness height, pipe wall thickness

exp - the exponential $e$

$\mathbf{F}$ - force vector

$F$ - force

$F_B$ - buoyant force

$F_H$ - horizontal force component

$F_V$ - vertical force component

$F_W$ - body force equal to the weight

$f$ - friction factor, frequency

$G$ - center of gravity

$\overline{GM}$ - metacentric height

$\mathbf{g}$ - gravity vector

$g$ - gravity

$H$ - enthalpy, height, total energy

$H_2, H_3$ - profile type

$H_D$ - design head

$H_P$ - pump head

$H_T$ - turbine head

HGL - hydraulic grade line

$h$ - distance, height, specific enthalpy

$h_j$ - head loss across a hydraulic jump

$I$ - second moment of an area

$\bar{I}$ - second moment about the centroidal axis

$I_{xy}$ - product of inertia

$\hat{\mathbf{i}}$ - unit vector in the $x$-direction

$\hat{\mathbf{j}}$ - unit vector in the $y$-direction

$\mathbf{k}$ - unit vector in the $z$-direction

$K$ - thermal conductivity, flow coefficient

$K_c$ - contraction coefficient

$K_e$ - expansion coefficient

$K_{uv}$ - correlation coefficient

$k$ - ratio of specific heats

$L$ - length

$L_E$ - entrance length

$L_e$ - equivalent length

$\ell$ - length

$\ell_m$ - mixing length

$M$ - molar mass, Mach number, momentum function

M - Mach number

$M_1, M_2, M_3$ - profile type

$m$ - mass, side-wall slope, constant for curve fit

$\dot{m}$ - mass flux

$\dot{m}_r$ - relative mass flux

$m_a$ - added mass

$m_1, m_2$ - side-wall slopes

$\dot{m}om$ - momentum flux

$N$ - general extensive property, an integer, number of jets

$NPSH$ - net positive suction head

$n$ - normal direction, number of moles, power-law exponent, Manning number

$\hat{\mathbf{n}}$ - unit normal vector

$P$ - power, force, wetted perimeter

$p$ - pressure

$Q$ - flow rate (discharge), heat transfer

$Q_D$ - design discharge

$\dot{Q}$ - rate of heat transfer

$q$ - source strength, specific discharge, heat flux

$R$ - radius, gas constant, hydraulic radius, radius of curvature

Re - Reynolds number

$\text{Re}_{\text{crit}}$ - critical Reynolds number

$R_u$ - universal gas constant

$R_x, R_y$ - force components

$r$ - radius, coordinate variable

$\mathbf{r}$ - position vector

$S$ - specific gravity, entropy, distance, slope of chennel, slope of EGL

$S_1, S_2, S_3$ - profile type

$S_c$ - critical slope

St - Strouhal number

$\mathbf{S}$ - position vector

$S_0$ - slope of channel bottom

$s$ - specific entropy, streamline coordinate

$\hat{\mathbf{s}}$ - unit vector tangent to streamline

sys - system

$T$ - temperature, torque, tension

$t$ - time, tangential direction

$U$ - average velocity

$U_\infty$ - free-stream velocity away from a body

$u$ - $x$-component velocity, circumferential blade speed

$u'$ - velocity perturbation

$\tilde{u}$ - specific internal energy

$\bar{u}$ - time average velocity

$u_\tau$ - shear velocity

$V$ - velocity

$V_c$ - critical velocity

$V_{ss}$ - steady-state velocity

$\mathbf{V}$ - velocity vector

$\overline{V}$ - spatial average velocity

$\mathahead{V}$ - volume

$V_B$ - blade velocity

$V_n$ - normal component of velocity

$V_r$ - relative speed

$V_t$ - tangential velocity

$\nu$ - velocity, $y$-component velocity

$\nu'$ - velocity perturbation

$\nu_r, \nu_z, \nu_\theta, \nu_\phi$ - velocity components

$W$ - work, weight, change in hydraulic grade line

$\dot{W}$ - work rate (power)

$\dot{W}_f$ - actual power

We - Weber number

$\dot{W}_S$ - shaft work (power)

$\omega$ - $z$-component velocity, velocity of a hydraulic bore

$X_T$ - distance where transition begins

$x$ - coordinate variable

$x_m$ - origin of moving reference frame

$\tilde{x}$ - distance relative to a moving reference frame

$\overline{x}$ - $x$-coordinate of centroid

$Y$ - upstream water height above top of wier

$y$ - coordinate variable, flow energy head

$y_p$ - distance to center of pressure

$\overline{y}$ - $y$-coordinate of centroid

$y_c$ - critical depth

$z$ - coordinate variable

$\alpha$ - angle, angle of attack, lapse rate, thermal diffusivity, kinetic-energy correction factor, blade angle

$\beta$ - angle, momentum correction factor, fixed jet angle, blade angle

$\Delta$ - a small increment

$\nabla$ - gradient operator

$\nabla^2$ - Laplacian

$\delta$ - boundary layer thickness

$\delta(x)$ - Dirac-delta function

$\delta_d$ - displacement thickness

$\delta_\nu$ - viscous wall layer thickness

$\varepsilon$ - a small volume

$\varepsilon_{xx}, \varepsilon_{xy}, \varepsilon_{xz}$ - rate-of-strain components

$\phi$ - angle, coordinate variable, velocity potential function, speed factor

$\Gamma$ - circulation, vortex strength

$\gamma$ - specific weight

$\eta$ - a general intensive property, eddy viscosity, efficiency, a position variable

$\eta_P$ - pump efficiency

$\eta_T$ - turbine efficiency

$\lambda$ - mean free path, a constant, wave length

$\mu$ - viscosity, doublet strength

$\nu$ - kinematic viscosity

$\pi$ - a pi term

$\theta$ - angle, momentum thickness, laser beam angle

$\rho$ - density

$\Omega$ - angular velocity

$\Omega_P$ - specific speed of a pump

$\Omega_T$ - specific speed of a turbine

$\boldsymbol{\Omega}$ - angular velocity vector

$\sigma$ - surface tension, cavitation number, circumferential stress

$\sigma_{xx}, \sigma_{yy}, \sigma_{zz}$ - normal stress components

$\tau$ - stress vector

$\bar{\tau}$ - time average stress

$\tau_{xy}, \tau_{xz}, \tau_{yz}$ - shear stress components

$\omega$ - angular velocity, vorticity

$\boldsymbol{\omega}$ - vorticity vector

$\psi$ - stream function

$\dfrac{\partial}{\partial x}$ - partial derivative

# Mechanics of Fluids

**Left**: Contemporary windmills are used to generate electricity at many locations in the United States. They are located in areas where consistent prevailing winds exist. (IRC/Shutterstock) **Top right**: Hurricane Bonnie, Atlantic Ocean about 800 km from Bermuda. At this stage in its development, the storm has a well-developed center, or "eye," where air currents are relatively calm. Vortex-like motion occurs away from the eye. (U.S. National Aeronautics and Space Administration) **Bottom right**: The Space Shuttle Discovery leaves the Kennedy Space Center on October 29, 1998. In 6 seconds the vehicle cleared the launch tower with a speed of 160 km/h, and in about two minutes it was 250 km down range from the Space Center, 47 km above the ocean with a speed of 6150 km/h. The wings and rudder on the tail are necessary for successful reentry as it enters the earth's atmosphere upon completion of its mission. (U.S. National Aeronautics and Space Administration)

# 1

# Basic Considerations

## Outline

## Chapter Objectives

The objectives of this chapter are to:

▲ Introduce many of the quantities encountered in fluid mechanics including their dimensions and units.
▲ Identify the liquids to be considered in this text.
▲ Introduce the fluid properties of interest.
▲ Present the thermodynamic laws and associated quantities.

## 1.1 INTRODUCTION

A proper understanding of fluid mechanics is extremely important in many areas of engineering. In biomechanics the flow of blood and cerebral fluid are of particular interest; in meteorology and ocean engineering an understanding of the motions of air movements and ocean currents requires a knowledge of the mechanics of fluids, chemical engineers must understand fluid mechanics to design the many different kinds of chemical-processing equipment; aeronautical engineers use their knowledge of fluids to maximize lift and minimize drag on aircraft and to design fan-jet engines; mechanical engineers design pumps, turbines, internal combustion engines, air compressors, air-conditioning equipment, pollution-control equipment, and power plants using a proper understanding of fluid mechanics; and civil engineers must also utilize the results obtained from a study of the mechanics of fluids to understand the transport of river sediment and erosion, the pollution of the air and water, and to design piping systems, sewage treatment plants, irrigation channels, flood control systems, dams, and domed athletic stadiums.

It is not possible to present fluid mechanics in such a way that all of the foregoing subjects can be treated specifically; it is possible, however, to present the fundamentals of the mechanics of fluids so that engineers are able to understand the role that the fluid plays in a particular application. This role may involve the proper sizing of a pump (the horsepower and flow rate) or the calculation of a force acting on a structure.

**KEY CONCEPT** *We will present the fundamentals of fluids so that engineers are able to understand the role that fluid plays in particular applications.*

In this book we present the general equations, both integral and differential, that result from the conservation of mass principle, Newton's second law, and the first law of thermodynamics. From these a number of particular situations will be considered that are of special interest. After studying this book the engineer should be able to apply the basic principles of the mechanics of fluids to new and different situations.

In this chapter topics are presented that are directly or indirectly relevant to all subsequent chapters. We include a macroscopic description of fluids, fluid properties, physical laws dominating fluid mechanics, and a summary of units and dimensions of important physical quantities. Before we can discuss quantities of interest, we must present the units and dimensions that will be used in our study of fluid mechanics.

## 1.2 DIMENSIONS, UNITS, AND PHYSICAL QUANTITIES

Before we begin the more detailed studies of fluid mechanics, let us discuss the dimensions and units that will be used throughout the book. Physical quantities require quantitative descriptions when solving an engineering problem. Density is one such physical quantity. It is a measure of the mass contained in a unit volume. Density does not, however, represent a fundamental dimension. There are nine quantities that are considered to be fundamental dimensions: length, mass, time, temperature, amount of a substance, electric current, luminous intensity, plane angle, and solid angle. The dimensions of all other quantities can be expressed in terms of the fundamental dimensions. For example, the quantity "force" can be related to the fundamental dimensions of mass, length, and time. To do this, we use Newton's second law, named after Sir Isaac Newton (1642–1727), expressed in simplified form in one direction as

$$F = ma \qquad (1.2.1)$$

Using brackets to denote "the dimension of," this is written dimensionally as

$$[F] = [m][a]$$
$$F = M \frac{L}{T^2} \qquad (1.2.2)$$

where $F$, $M$, $L$, and $T$ are the dimensions of force, mass, length, and time, respectively. If force had been selected as a fundamental dimension rather than mass, a common alternative, mass would have dimensions of

$$[m] = \frac{[F]}{[a]}$$
$$M = \frac{FT^2}{L} \qquad (1.2.3)$$

where $F$ is the dimension[1] of force.

There are also systems of dimensions in which both mass and force are selected as fundamental dimensions. In such systems conversion factors, such as a gravitational constant, are required; we do not consider these types of systems in this book, so they will not be discussed.

To give the dimensions of a quantity a numerical value, a set of units must be selected. In the United States, two primary systems of units are presently being used, the British Gravitational System, which we will refer to as English units, and the International System, which is referred to as SI (Système International) units. SI units are preferred and are used internationally; the United States is the only major country not requiring the use of SI units, but there is now a program of conversion in most industries to the predominant use of SI units. Following this trend, we have used SI units in this SI Edition of the text.

**KEY CONCEPT** *SI units are preferred and are used internationally.*

The fundamental dimensions and their units are presented in Table 1.1; some derived units appropriate to fluid mechanics are given in Table 1.2. Other units that are acceptable are the hectare (ha), which is 10 000 m², used for large areas; the metric ton (t), which is 1000 kg, used for large masses; and the liter (L), which is 0.001 m³. Also, density is occasionally expressed as grams per liter (g/L).

In chemical calculations the mole is often a more convenient unit than the kilogram. In some cases it is also useful in fluid mechanics. For gases the kilogram-mole (kmol) is the quantity that fills the same volume as 32 kilograms

---

[1]Unfortunately, the quantity force $F$ and the dimension of force $[F]$ use the same symbol.

**Table 1.1** Fundamental Dimensions and Their Units

| Quantity | Dimensions | SI units | |
|---|---|---|---|
| Length $\ell$ | $L$ | meter | m |
| Mass $m$ | $M$ | kilogram | kg |
| Time $t$ | $T$ | second | s |
| Electric current $i$ | | ampere | A |
| Temperature $T$ | $\Theta$ | kelvin | K |
| Amount of substance | $M$ | kg-mole | kmol |
| Luminous intensity | | candela | cd |
| Plane angle | | radian | rad |
| Solid angle | | steradian | sr |

of oxygen at the same temperature and pressure. The mass (in kilograms) of a gas filling that volume is equal to the molecular weight of the gas; for example, the mass of 1 kmol of nitrogen is 28 kilograms.

When expressing a quantity with a numerical value and a unit, prefixes have been defined so that the numerical value may be between 0.1 and 1000. These

**Table 1.2** Derived Units

| Quantity | Dimensions | SI units |
|---|---|---|
| Area $A$ | $L^2$ | $m^2$ |
| Volume $V$ | $L^3$ | $m^3$ |
| | | L (liter) |
| Velocity $V$ | $L/T$ | m/s |
| Acceleration $a$ | $L/T^2$ | $m/s^2$ |
| Angular velocity $\omega$ | $T^{-1}$ | rad/s |
| Force $F$ | $ML/T^2$ | $kg \cdot m/s^2$ |
| | | N (newton) |
| Density $\rho$ | $M/L^3$ | $kg/m^3$ |
| Specific weight $\gamma$ | $M/L^2 T^2$ | $N/m^3$ |
| Frequency $f$ | $T^{-1}$ | $s^{-1}$ |
| Pressure $p$ | $M/LT^2$ | $N/m^2$ |
| | | Pa (pascal) |
| Stress $\tau$ | $M/LT^2$ | $N/m^2$ |
| | | Pa (pascal) |
| Surface tension $\sigma$ | $M/T^2$ | N/m |
| Work $W$ | $ML^2/T^2$ | $N \cdot m$ |
| | | J (joule) |
| Energy $E$ | $ML^2/T^2$ | $N \cdot m$ |
| | | J (joule) |
| Heat rate $\dot{Q}$ | $ML^2/T^3$ | J/s |
| Torque $T$ | $ML^2/T^2$ | $N \cdot m$ |
| Power $P$ | $ML^2/T^3$ | J/s |
| $\dot{W}$ | | W (watt) |
| Viscosity $\mu$ | $M/LT$ | $N \cdot s/m^2$ |
| Mass flux $\dot{m}$ | $M/T$ | kg/s |
| Flow rate $Q$ | $L^3/T$ | $m^3/s$ |
| Specific heat $c$ | $L^2/T^2\Theta$ | $J/kg \cdot K$ |
| Conductivity $K$ | $ML/T^3\Theta$ | $W/m \cdot K$ |

**Table 1.3**   SI Prefixes

| Multiplication factor | Prefix | Symbol |
|---|---|---|
| $10^{12}$ | tera | T |
| $10^{9}$ | giga | G |
| $10^{6}$ | mega | M |
| $10^{3}$ | kilo | k |
| $10^{-2}$ | centi[a] | c |
| $10^{-3}$ | milli | m |
| $10^{-6}$ | micro | $\mu$ |
| $10^{-9}$ | nano | n |
| $10^{-12}$ | pico | p |

[a]Permissible if used alone as cm, $cm^2$, or $cm^3$.

prefixes are presented in Table 1.3. Using scientific notation, however, we use powers of 10 rather than prefixes (e.g., $2 \times 10^6$ N rather than 2 MN). If larger numbers are written the comma is not used; twenty thousand would be written as 20 000 with a space and no comma.[2]

Newton's second law relates a net force acting on a rigid body to its mass and acceleration. This is expressed as

$$\Sigma \mathbf{F} = m\mathbf{a} \qquad (1.2.4)$$

> **KEY CONCEPT**   *When using SI units, if larger numbers are written (5 digits or more), the comma is not used. The comma is replaced by a space (i.e., 20 000).*

Consequently, the force needed to accelerate a mass of 1 kilogram at 1 meter per second squared in the direction of the net force is 1 newton. This allows us to relate the units by

$$N = kg \cdot m/s^2 \qquad (1.2.5)$$

> **KEY CONCEPT**   *The relationship* $N = kg \cdot m/s^2$ *is often used in the conversion of units.*

which are included in Table 1.2. These relationships between units are often used in the conversion of units. In the SI system, weight is always expressed in newtons, never in kilograms. To relate weight to mass, we use

$$W = mg \qquad (1.2.6)$$

where $g$ is the local gravity. The standard value for gravity is 9.80665 $m/s^2$ and it varies from a minimum of 9.77 $m/s^2$ at the top of Mt. Everest to a maximum of 9.83 $m/s^2$ in the deepest ocean trench. A nominal value of 9.81 $m/s^2$ will be used unless otherwise stated.

Finally, a note on significant figures. In engineering calculations we often do not have confidence in a calculation beyond three significant digits since the information given in the problem statement is often not known to more than three significant

---

[2]In many countries commas represent decimal points, so they are not used where confusion may occur.

digits; in fact, viscosity and other fluid properties may not be known to even three significant digits. The diameter of a pipe may be stated as 2 cm; this would, in general, not be as precise as 2.000 cm would imply. If information used in the solution of a problem is known to only two significant digits, it is incorrect to express a result to more than two significant digits. In the examples and problems we will assume that all information given is known to three significant digits, and the results will be expressed accordingly. If the numeral 1 begins a number, it is not counted in the number of significant digits, i.e., the number 1.210 has three significant digits.

**KEY CONCEPT**  *We will assume that all information given is known to three significant digits.*

## Example 1.1

A mass of 100 kg is acted on by a 400-N force acting vertically upward and a 600-N force acting upward at a 45° angle. Calculate the vertical component of the acceleration. The local acceleration of gravity is 9.81 m/s².

### Solution
The first step in solving a problem involving forces is to draw a free-body diagram with all forces acting on it, as shown in Fig. E1.1.

**Fig. E1.1**

Next, apply Newton's second law (Eq. 1.2.4). It relates the net force acting on a mass to the acceleration and is expressed as

$$\Sigma F_y = ma_y$$

Using the appropriate components in the $y$-direction, with $W = mg$, we have

$$400 + 600 \sin 45° - 100 \times 9.81 = 100a_y$$

$$a_y = -1.567 \text{ m/s}^2$$

The negative sign indicates that the acceleration is in the negative $y$-direction, i.e., down. *Note:* We have used only three significant digits in the answer since the information given in the problem is assumed known to three significant digits. (The number 1.567 has three significant digits. A leading "1" is not counted as a significant digit.)

## 1.3 CONTINUUM VIEW OF GASES AND LIQUIDS

Substances referred to as fluids may be **liquids** or **gases**. In our study of the fluid mechanics we restrict the liquids that are studied. Before we state the restriction, we must define a shearing stress. A force $\Delta F$ that acts on an area $\Delta A$ can be decomposed into a normal component $\Delta F_n$ and a tangential component $\Delta F_t$, as shown in Fig. 1.1. The force divided by the area upon which it acts is called a *stress*. The force vector divided by the area is a **stress vector**[3], the normal component of force divided by the area is a **normal stress**, and the tangential force divided by the area is a **shear stress**. In this discussion we are interested in the shear stress $\tau$. Mathematically, it is defined as

$$\tau = \lim_{\Delta A \to 0} \frac{\Delta F_t}{\Delta A} \qquad (1.3.1)$$

Our restricted family of fluids may now be identified; the fluids considered in this book are *those liquids and gases that move under the action of a shear stress, no matter how small that shear stress may be*. This means that even a very small shear stress results in motion in the fluid. Gases obviously fall within this category of fluids, as do water and tar. Some substances, such as plastics and catsup, may resist small shear stresses without moving; a study of these substances is included in the subject of *rheology* and is not included in this book.

It is worthwhile to consider the microscopic behavior of fluids in more detail. Consider the molecules of a gas in a container. These molecules are not stationary but move about in space with very high velocities. They collide with each other and strike the walls of the container in which they are confined, giving rise to the pressure exerted by the gas. If the volume of the container is increased while the temperature is maintained constant, the number of molecules impacting on a given area is decreased, and as a result, the pressure decreases. If the temperature of a gas in a given volume increases (i.e., the velocities of the molecules increase), the pressure increases due to increased molecular activity.

Molecular forces in liquids are relatively high, as can be inferred from the following example. The pressure necessary to compress 20 grams of water vapor at 20 °C into 20 $cm^3$, assuming that no molecular forces exist, can be shown by the ideal gas law to be approximately 1340 times the atmospheric pressure. Of course, this pressure is not required, because 20 g of water occupies 20 $cm^3$. It follows that the cohesive forces in the liquid phase must be very large.

Despite the high molecular attractive forces in a liquid, some of the molecules at the surface escape into the space above. If the liquid is contained, an equilibrium is established between outgoing and incoming molecules. The presence of molecules above the liquid surface leads to a so-called *vapor pressure*.

**Stress vector:**   *The force vector divided by the area.*

**Normal stress:**   *The normal component of force divided by the area.*

**Shear stress:**   *The tangential force divided by the area.*

**Liquid:**   *A state of matter in which the molecules are relatively free to change their positions with respect to each other but restricted by cohesive forces so as to maintain a relatively fixed volume.*[4]

**Gas:**   *A state of matter in which the molecules are practically unrestricted by cohesive forces. A gas has neither definite shape nor volume.*

**KEY CONCEPT**   *Fluids considered in this text are those that move under the action of a shear stress, no matter how small that stress may be.*

Components

**Fig. 1.1**   Normal and tangential components of a force.

---

[3] *A quantity that is defined in the margin is in bold face whereas a quantity not defined in the margin is italic.*

[4] *Handbook of Chemistry and Physics*, 40th ed. CRC Press, Boca Raton, Fla.

This pressure increases with temperature. For water at 20°C this pressure is approximately 0.02 times the atmospheric pressure.

In our study of fluid mechanics it is convenient to assume that both gases and liquids are continuously distributed throughout a region of interest, that is, the fluid is treated as a **continuum**. The primary property used to determine if the continuum assumption is appropriate is the *density* $\rho$, defined by

$$\rho = \lim_{\Delta v \to 0} \frac{\Delta m}{\Delta V} \qquad (1.3.2)$$

where $\Delta m$ is the incremental mass contained in the incremental volume $\Delta V$. The density for air at **standard atmospheric conditions**, that is, at a pressure of 101.3 kPa and a temperature of 15°C, is 1.23 kg/m³. For water, the nominal value of density is 1000 kg/m³.

Physically, we cannot let $\Delta V \to 0$ since, as $\Delta V$ gets extremely small, the mass contained in $\Delta V$ would vary discontinuously depending on the number of molecules in $\Delta V$; this is shown graphically in Fig. 1.2. Actually, the zero in the definition of density should be replaced by some small volume $\varepsilon$, below which the continuum assumption fails. For most engineering applications, the small volume $\varepsilon$ shown in Fig. 1.2 is extremely small. For example, there are $2.7 \times 10^{16}$ molecules contained in a cubic millimeter of air at standard conditions; hence, $\varepsilon$ is much smaller than a cubic millimeter. An appropriate way to determine if the continuum model is acceptable is to compare a characteristic length $l$ (e.g., the diameter of a rocket) of the device or object of interest with the **mean free path** $\lambda$, the average distance a molecule travels before it collides with another molecule; if $l \gg \lambda$, the continuum model is acceptable. The mean free path is derived in molecular theory. It is

$$\lambda = 0.225 \frac{m}{\rho d^2} \qquad (1.3.3)$$

where $m$ is the mass (kg) of a molecule, $\rho$ the density (kg/m³) and $d$ the diameter (m) of a molecule. For air $m = 4.8 \times 10^{-26}$ kg and $d = 3.7 \times 10^{-10}$ m. At standard atmospheric conditions the mean free path is approximately $6.4 \times 10^{-6}$ cm, at an elevation of 100 km it is 10 cm, and at 160 km it is 5000 cm. Obviously, at

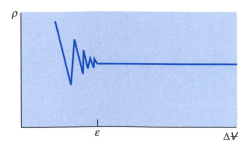

**Fig. 1.2**   Density at a point in a continuum.

higher elevations the continuum assumption is not acceptable and the theory of rarefied gas dynamics (or free molecular flow) must be utilized. Satellites are able to orbit the earth if the primary dimension of the satellite is of the same order of magnitude as the mean free path.

With the continuum assumption, fluid properties can be assumed to apply uniformly at all points in a region at any particular instant in time. For example, the density $\rho$ can be defined at all points in the fluid; it may vary from point to point and from instant to instant; that is, in Cartesian coordinates $\rho$ is a continuous function of $x, y, z$, and $t$, written as $\rho(x, y, z, t)$.

## 1.4 PRESSURE AND TEMPERATURE SCALES

In fluid mechanics pressure results from a normal compressive force acting on an area. The *pressure p* is defined as (see Fig. 1.3)

$$p = \lim_{\Delta A \to 0} \frac{\Delta F_n}{\Delta A} \qquad (1.4.1)$$

where $\Delta F_n$ is the incremental normal compressive force acting on the incremental area $\Delta A$. The metric units to be used on pressure are newtons per square meter ($N/m^2$) or pascal (Pa). Since the pascal is a very small unit of pressure, it is more conventional to express pressure in units of kilopascal (kPa). For example, standard atmospheric pressure at sea level is 101.3 kPa. Atmospheric pressure is often expressed as millimeters (mm) of mercury, as shown in Fig. 1.4; such a column of fluid creates the pressure at the bottom of the column, providing the column is open to atmospheric pressure at the top.

Both pressure and temperature are physical quantities that can be measured using different scales. There exist absolute scales for pressure and temperature, and there are scales that measure these quantities relative to selected reference points. In many thermodynamic relationships (see Section 1.7) absolute scales must be used for pressure and temperature. Figures 1.4 and 1.5 summarize the commonly used scales.

The **absolute pressure** reaches zero when an ideal vacuum is achieved, that is, when no molecules are left in a space; consequently, a negative absolute pressure is an impossibility. A second scale is defined by measuring pressures relative

**KEY CONCEPT** *In many relationships, absolute scales must be used for pressure and temperature.*

**Absolute pressure:** *The scale measuring pressure, where zero is reached when an ideal vacuum is achieved.*

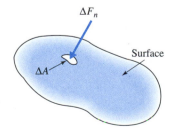

**Fig. 1.3**  Definition of pressure.

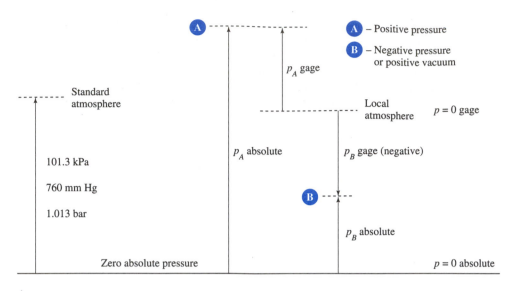

**Fig. 1.4**   Gage pressure and absolute pressure.

**Gage pressure:** *The scale measuring pressure relative to the local atmospheric pressure.*

to the local atmospheric pressure. This pressure is called **gage pressure**. A conversion from gage pressure to absolute pressure can be carried out using

$$p_{\text{absolute}} = p_{\text{atmospheric}} + p_{\text{gage}} \qquad (1.4.2)$$

**KEY CONCEPT**
*Whenever the absolute pressure is less than the atmospheric pressure, it may be called a vacuum.*

Note that the atmospheric pressure in Eq. 1.4.2 is the local atmospheric pressure, which may change with time, particularly when a weather "front" moves through. However, if the local atmospheric pressure is not given, we use the value given for a particular elevation, as given in Table B.3 of Appendix B, and assume zero elevation if the elevation is unknown. The gage pressure is negative whenever the absolute pressure is less than atmospheric pressure; it may then be called a **vacuum**. In this book the word "absolute" will generally follow the pressure value if the pressure is given as an absolute pressure (e.g., $p = 50$ kPa absolute). If it were stated as $p = 50$ kPa, the pressure would be taken as a gage pressure, except that atmospheric pressure is always an absolute pressure. Most often in fluid mechanics gage pressure is used.

**Vacuum:** *Whenever the absolute pressure is less than the atmospheric pressure.*

|  | °C | K |
|---|---|---|
| Steam point | 100° | 373 |
| Ice point | 0° | 273 |
| Special point | −18° | 255 |
|  |  |  |
| Zero absolute temperature |  |  |

**Fig. 1.5**   Temperature scales.

Two temperature scales are commonly used, the Celsius (C) and Fahrenheit (F) scales. Both scales are based on the ice point and steam point of water at an atmospheric pressure of 101.3 kPa. In this SI Edition, we will not use the Fahrenheit scale. Figure 1.5 shows that the ice and steam point are 0 and 100°C on the Celsius scale. There are two corresponding absolute temperature scales. The absolute scale corresponding to the Celsius scale is the kelvin (K) scale. The relation between these scales is

$$K = °C + 273.15 \qquad (1.4.3)$$

The absolute scale corresponding to the Fahrenheit scale is the Rankine scale (°R). The relation between these scales is

$$°R = °F + 459.67 \qquad (1.4.4)$$

Note that in the SI system we do not write 100°K but simply 100 K, which is read "100 kelvins," similar to other SI units.

Reference will often be made to "standard atmospheric conditions" or "standard temperature and pressure." This refers to sea-level conditions at 40° latitude, which are taken to be 101.3 kPa for pressure and 15°C for temperature. Actually, The standard pressure is usually taken as 100 kPa, sufficiently accurate for engineering calculations.

## Example 1.2

A pressure gage attached to a rigid tank measures a vacuum of 42 kPa inside the tank shown in Fig. E1.2, which is situated at a site in Colorado where the elevation is 2000 m. Determine the absolute pressure inside the tank.

air    — −42 kPa

**Fig. E1.2**

### Solution

To determine the absolute pressure, the atmospheric pressure must be known. If the elevation were not given, we would assume a standard atmospheric pressure of 100 kPa. However, with the elevation given, the atmospheric pressure is found from Table B.3 in Appendix B to be 79.5 kPa. Thus

$$p = -42 + 79.5 = 37.5 \text{ kPa absolute}$$

*Note:* A vacuum is always a negative gage pressure. Also, using standard atmospheric pressure of 100 kPa is acceptable, rather than 101.3 kPa, since it is within 1%, which is acceptable engineering accuracy.

## 1.5 FLUID PROPERTIES

In this section we present several of the more common fluid properties. If density variation or heat transfer is significant, several additional properties, not presented here, become important.

### 1.5.1 Density and Specific Weight

**Specific weight:** *Weight per unit volume (γ = ρg).*

Fluid density was defined in Eq. 1.3.2 as mass per unit volume. A fluid property directly related to density is the **specific weight** γ or weight per unit volume. It is defined by

$$\gamma = \frac{W}{V} = \frac{mg}{V} = \rho g \qquad (1.5.1)$$

where $g$ is the local gravity. The units of specific weight are N/m³. For water we use the nominal value of 9800 N/m³.

**Specific gravity:** *The ratio of the density of a substance to the density of water.*

The **specific gravity** $S$ is often used to determine the specific weight or density of a fluid (usually a liquid). It is defined as the ratio of the density of a substance to the density of water at a reference temperature of 4°C:

$$S = \frac{\rho}{\rho_{\text{water}}} = \frac{\gamma}{\gamma_{\text{water}}} \qquad (1.5.2)$$

**KEY CONCEPT** *Specific gravity is often used to determine the density of a fluid.*

For example, the specific gravity of mercury is 13.6, a dimensionless number; that is, the mass of mercury is 13.6 times that of water for the same volume. The density, specific weight, and specific gravity of air and water at standard conditions are given in Table 1.4.

The density and specific weight of water do vary slightly with temperature; the approximate relationships are

$$\rho_{\text{H}_2\text{O}} = 1000 - \frac{(T-4)^2}{180}$$

$$\gamma_{\text{H}_2\text{O}} = 9800 - \frac{(T-4)^2}{18} \qquad (1.5.3)$$

**TABLE 1.4** Density, Specific Weight, and Specific Gravity of Air and Water at Standard Conditions

|  | Density ρ kg/m³ | Specific weight γ N/m³ | Specific gravity S |
|---|---|---|---|
| Air | 1.23 | 12.1 | 0.00123 |
| Water | 1000 | 9810 | 1 |

For mercury the specific gravity relates to temperature by

$$S_{Hg} = 13.6 - 0.0024T \qquad (1.5.4)$$

Temperature in the three equations above is measured in degrees Celsius. For temperatures under 50°C, using the nominal values stated earlier for water and mercury, the error is less than 1%, certainly within engineering limits for most design problems. Note that the density of water at 0°C is less than that at 4°C; consequently, the lighter water at 0°C rises to the top of a lake so that ice forms on the surface. For most other liquids the density at freezing is greater than the density just above freezing.

### 1.5.2  Viscosity

**Viscosity** can be thought of as the internal stickiness of a fluid. It is one of the properties that influences the power needed to move an airfoil through the atmosphere. It accounts for the energy losses associated with the transport of fluids in ducts, channels, and pipes. Further, viscosity plays a primary role in the generation of turbulence. Needless to say, viscosity is an extremely important fluid property in our study of fluid flows.

**Viscosity:** *The internal stickiness of a fluid.*

**KEY CONCEPT** *Viscosity plays a primary role in the generation of turbulence.*

The rate of deformation of a fluid is directly linked to the viscosity of the fluid. For a given stress, a highly viscous fluid deforms at a slower rate than a fluid with a low viscosity. Consider the flow of Fig. 1.6 in which the fluid particles move in the *x*-direction at different speeds, so that particle velocities *u* vary with the *y*-coordinate. Two particle positions are shown at different times; observe how the particles move relative to one another. For such a simple flow field, in which $u = u(y)$, we can define the **viscosity** $\mu$ of the fluid by the relationship

$$\tau = \mu \frac{du}{dy} \qquad (1.5.5)$$

where $\tau$ is the shear stress of Eq. 1.3.1 and *u* is the velocity in the *x*-direction. The units of $\tau$ are N/m$^2$ or Pa, and of $\mu$ are N·s/m$^2$. The quantity $du/dy$ is a velocity gradient and can be interpreted as a **strain rate**. Stress velocity-gradient relationships for more complicated flow situations are presented in Chapter 5.

**Strain rate:** *The rate at which a fluid element deforms.*

The concept of viscosity and velocity gradients can also be illustrated by considering a fluid within the small gap between two concentric cylinders, as shown

**Fig. 1.6**  Relative movement of two fluid particles in the presence of shear stresses.

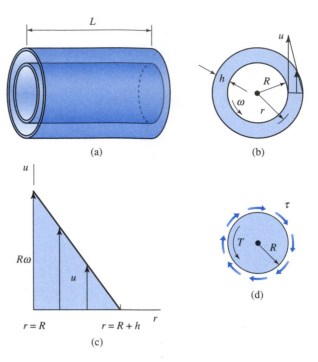

**Fig. 1.7**   Fluid being sheared between cylinders with a small gap: (a) the two cylinders; (b) rotating inner cylinder; (c) velocity distribution; (d) the inner cylinder. The outer cylinder is fixed and the inner cylinder is rotating.

in Fig. 1.7. A torque is necessary to rotate the inner cylinder at constant rotational speed $\omega$ while the outer cylinder remains stationary. This resistance to the rotation of the cylinder is due to viscosity. The only stress that exists to resist the applied torque for this simple flow is a shear stress, which is observed to depend directly on the velocity gradient; that is,

$$\tau = \mu \left| \frac{du}{dr} \right| \tag{1.5.6}$$

where $du/dr$ is the velocity gradient and $u$ is the tangential velocity component, which depends only on $r$. For a small gap ($h \ll R$), this gradient can be approximated by assuming a linear velocity distribution[5] in the gap. Thus

$$\left| \frac{du}{dr} \right| = \frac{\omega R}{h} \tag{1.5.7}$$

where $h$ is the gap width. We can thus relate the applied torque $T$ to the viscosity and other parameters by the equation

$$T = \text{stress} \times \text{area} \times \text{moment arm}$$
$$= \tau \times 2\pi RL \times R$$
$$= \mu \frac{\omega R}{h} \times 2\pi RL \times R = \frac{2\pi R^3 \omega L \mu}{h} \tag{1.5.8}$$

---

[5]If the gap is not small relative to $R$, the velocity distribution will not be linear (see Section 7.5). The distribution will also not be linear for relatively small values of $\omega$.

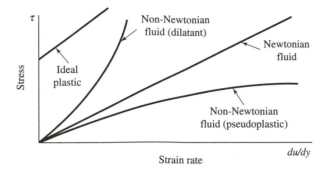

**Fig. 1.8**    Newtonian and non-Newtonian fluids.

where the shearing stress acting on the ends of the cylinder is negligible; $L$ represents the length of the rotating cylinder. Note that the torque depends directly on the viscosity; thus the cylinders could be used as a *viscometer*, a device that measures the viscosity of a fluid.

 Viscosity[6], 454

If the shear stress of a fluid is directly proportional to the velocity gradient, as was assumed in Eqs. 1.5.5 and 1.5.6, the fluid is said to be a **Newtonian fluid**. Fortunately, many common fluids, such as air, water, and oil, are Newtonian. *Non-Newtonian fluids*, with shear stress versus strain rate relationships as shown in Fig. 1.8, often have a complex molecular composition.

**Newtonian fluid:** *the shear stress of the fluid is directly proportional to the velocity gradient.*

*Dilatants* (quicksand, slurries) become more resistant to motion as the strain rate increases, and *pseudoplastics* (paint and catsup) become less resistant to motion with increased strain rate. *Ideal plastics* (or *Bingham fluids*) require a minimum shear stress to cause motion. Clay suspensions and toothpaste are examples that also require a minimum shear to cause motion, but they do not have a linear stress-strain rate relationship.

**KEY CONCEPT** *Viscosity causes fluid to adhere to a surface.*

An extremely important effect of viscosity is to cause the fluid to adhere to the surface; this is known as the **no-slip condition**. This was assumed in the example of Fig. 1.7. The velocity of the fluid at the rotating cylinder was taken to be $\omega R$, and the velocity of the fluid at the stationary cylinder was set equal to zero, as shown in Fig. 1.7b. When a space vehicle reenters the atmosphere, the high speed creates very large velocity gradients at the surface of the vehicle, resulting in large stresses that heat up the surface; the high temperatures can cause the vehicle to disintegrate if not properly protected.

**No-slip condition:** *Condition where viscosity causes fluid to adhere to the surface.*

The viscosity is very dependent on temperature in liquids in which cohesive forces play a dominant role; note that the viscosity of a liquid decreases with increased temperature, as shown in Fig. B.1 in Appendix B. The curves are often approximated by the equation

$$\mu = Ae^{Bt} \tag{1.5.9}$$

known as *Andrade's equation*; the constants $A$ and $B$ would be determined from measured data. For a gas it is molecular collisions that provide the internal stresses, so that as the temperature increases, resulting in increased molecular

---

[6]To view a file on a specified page, simply open any of the eight major headings, then enter a page number in the box at the top and click on "go to page." The numbers after the descriptors refer to the pages on the DVD.

activity, the viscosity increases. This can be observed in the bottom curve for a gas of Fig. B.1 in Appendix B. Note, however, that the percentage change of viscosity in a liquid is much greater than in a gas for the same temperature difference. Also, one can show that cohesive forces and molecular activity are quite insensitive to pressure, so that $\mu = \mu(T)$ only for both liquids and gases.

Since the viscosity is often divided by the density in the derivation of equations, it has become useful and customary to define *kinematic viscosity* to be

$$v = \frac{\mu}{\rho} \tag{1.5.10}$$

where the units of $v$ are m²/s. Note that for a gas, the kinematic viscosity will also depend on the pressure since density is pressure sensitive. The kinematic viscosity is shown, at atmospheric pressure, in Fig. B.2 in Appendix B.

## Example 1.3

A viscometer is constructed with two 30-cm-long concentric cylinders, one 20.0 cm in diameter and the other 20.2 cm in diameter. A torque of 0.13 N·m is required to rotate the inner cylinder at 400 rpm (revolutions per minute). Calculate the viscosity.

### Solution

The applied torque is just balanced by a resisting torque due to the shear stresses (see Fig. 1.7c). This is expressed by the small gap equation, Eq. 1.5.8.

The radius is $R = d/2 = 10$ cm; the gap $h = (d_2 - d_1)/2 = 0.1$ cm; the rotational speed, expressed as rad/s, is $\omega = 400 \times 2\pi/60 = 41.89$ rad/s.

Equation 1.5.8 provides:

$$\mu = \frac{Th}{2\pi R^3 \omega L}$$

$$= \frac{0.13(0.001)}{2\pi(0.1)^3(41.89)(0.3)} = 0.001646 \text{ N·s/m}^2$$

*Note:* All lengths are in meters so that the desired units on $\mu$ are obtained. The units can be checked by substitution:

$$[\mu] = \frac{\text{N·m·m}}{\text{m}^3(\text{rad/s})\text{m}} = \frac{\text{N·s}}{\text{m}^2}$$

### 1.5.3   Compressibility

**Bulk modulus of elasticity:**
*The ratio of change in pressure to relative change in density.*

In the preceding section we discussed the deformation of fluids that results from shear stresses. In this section we discuss the deformation that results from pressure changes. All fluids compress if the pressure increases, resulting in a decrease in volume or an increase in density. A common way to describe the compressibility of a fluid is by the following definition of the **bulk modulus of elasticity** $B$:

$$B = \lim_{\Delta V \to 0} \left[ -\frac{\Delta p}{\Delta V / V} \right]_T = \lim_{\Delta \rho \to 0} \frac{\Delta p}{\Delta \rho / \rho} \bigg|_T$$

$$= -V \frac{\partial p}{\partial V} \bigg|_T = \rho \frac{\partial p}{\partial \rho} \bigg|_T \qquad (1.5.11)$$

In words, the bulk modulus, also called the *coefficient of compressibility*, is defined as the ratio of the change in pressure ($\Delta p$) to relative change in density ($\Delta \rho / \rho$) while the temperature remains constant. The bulk modulus has the same units as pressure.

The bulk modulus for water at standard conditions is approximately 2100 MPa, or 21 000 times the atmospheric pressure. For air at standard conditions, $B$ is equal to 1 atm. In general, $B$ for a gas is equal to the pressure of the gas. To cause a 1% change in the density of water a pressure of 21 MPa (210 atm) is required. This is an extremely large pressure needed to cause such a small change; thus liquids are often assumed to be incompressible. For gases, if significant changes in density occur, say 4%, they should be considered as compressible; for small density changes under 3% they may also be treated as incompressible. This occurs for atmospheric airspeeds under about 100 m/s, which includes many airflows of engineering interest: air flow around automobiles, landing and take-off of aircraft, and air flow in and around buildings.

**KEY CONCEPT** *Gases with small density changes under 3% may be treated as incompressible.*

Small density changes in liquids can be very significant when large pressure changes are present. For example, they account for "water hammer," which can be heard shortly after the sudden closing of a valve in a pipeline; when the valve is closed an internal pressure wave propagates down the pipe, producing a hammering sound due to pipe motion when the wave reflects from the closed valve or pipe elbows. Water hammer is considered in detail in Section 11.5.

The bulk modulus can also be used to calculate the speed of sound in a liquid; in Section 9.2 it will be shown to be given by

$$c = \sqrt{\frac{\Delta p}{\Delta \rho} \bigg|_T} = \sqrt{\frac{B}{\rho}} \qquad (1.5.12)$$

This yields approximately 1450 m/s for the speed of sound in water at standard conditions. The speed of sound in a gas will be presented in Section 1.7.3.

## 1.5.4 Surface Tension

**Surface tension** is a property that results from the attractive forces between molecules. As such, it manifests itself only in liquids at an interface, usually a liquid-gas interface. The forces between molecules in the bulk of a liquid are equal in all directions, and as a result, no net force is exerted on the molecules. However, at an interface the molecules exert a force that has a resultant in the interface layer. This force holds a drop of water suspended on a rod and limits the size of the drop that may be held. It also causes the small drops from a

*Surface tension: A property resulting from the attractive forces between molecules.*

**Fig. 1.9** Internal forces in (a) a droplet and (b) a bubble.

sprayer or atomizer to assume spherical shapes. It may also play a significant role when two immiscible liquids (e.g., oil and water) are in contact with each other.

Surface tension has units of force per unit length, N/m. The force due to surface tension results from a length multiplied by the surface tension; the length to use is the length of fluid in contact with a solid, or the circumference in the case of a bubble. A surface tension effect can be illustrated by considering the free-body diagrams of half a droplet and half a bubble, as shown in Fig. 1.9. The droplet has one surface, and the bubble is composed of a thin film of liquid with an inside surface and an outside surface. An expression for the pressure inside the droplet and bubble can now be derived.

The pressure force $p\pi R^2$ in the droplet balances the surface tension force around the circumference. Hence

$$p\pi R^2 = 2\pi R\sigma$$

$$\therefore p = \frac{2\sigma}{R} \tag{1.5.13}$$

Similarly, the pressure force in the bubble is balanced by the surface tension forces on the two circumferences assuming the bubble thickness is small. Therefore,

$$p\pi R^2 = 2(2\pi R\sigma)$$

$$\therefore p = \frac{4\sigma}{R} \tag{1.5.14}$$

From Eqs. 1.5.13 and 1.5.14 we can conclude that the internal pressure in a bubble is twice as large as that in a droplet of the same size.

Figure 1.10 shows the rise of a liquid in a clean glass capillary tube due to surface tension. The liquid makes a contact angle $\beta$ with the glass tube. Experiments have shown that this angle for water and most liquids in a clean glass tube is zero. There are also cases for which this angle is greater than 90° (e.g., mercury); such liquids have a capillary drop. If $h$ is the capillary rise, $D$ the diameter, $\rho$ the density, and $\sigma$ the surface tension, $h$ can be determined from

**KEY CONCEPT** *Force due to surface tension results from a length multiplied by the surface tension.*

Drop Formation, 453

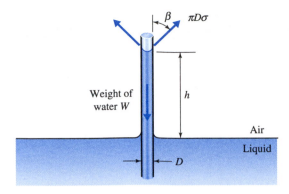

**Fig. 1.10**  Rise in a capillary tube.

equating the vertical component of the surface tension force to the weight of the liquid column:

$$\sigma \pi D \cos \beta = \gamma \frac{\pi D^2}{4} h \qquad (1.5.15)$$

or, rearranged,

$$h = \frac{4\sigma \cos \beta}{\gamma D} \qquad (1.5.16)$$

Surface tension may influence engineering problems when, for example, laboratory modeling of waves is conducted at a scale that surface tension forces are of the same order of magnitude as gravitational forces.

## Example 1.4

A 2-mm-diameter clean glass tube is inserted in water at 15°C (Fig. E1.4). Determine the height that the water will climb up the tube. The water makes a contact angle of 0° with the clean glass.

Capillary, 346

**Fig. E1.4**

**Solution**

A free-body diagram of the water shows that the upward surface-tension force is equal and opposite to the weight. Writing the surface-tension force as surface tension times distance, we have

$$\sigma\pi D = \gamma\frac{\pi D^2}{4}h$$

or

$$h = \frac{4\sigma}{\gamma D} = \frac{4 \times 0.0741 \text{ N/m}}{9800 \text{ N/m}^3 \times 0.002 \text{ m}} = 0.01512 \text{ m}\quad\text{or}\quad 15.12 \text{ mm}$$

The numerical values for $\sigma$ and $\rho$ were obtained from Table B.1 in Appendix B. Note that the nominal value used for the specific weight of water is $\gamma = \rho g = 9800 \text{ N/m}^3$.

**Example 1.4a**   on the DVD, Similarity and Scaling, Capillary Rise 512

### 1.5.5  Vapor Pressure

When a small quantity of liquid is placed in a closed container, a certain fraction of the liquid will vaporize. Vaporization will terminate when equilibrium is reached between the liquid and gaseous states of the substance in the container—in other words, when the number of molecules escaping from the water surface is equal to the number of incoming molecules. The pressure resulting from molecules in the gaseous state is the **vapor pressure**.

The vapor pressure is different from one liquid to another. For example, the vapor pressure of water at 15°C is 1.70 kPa absolute, and for ammonia it is 33.8 kPa absolute.

The vapor pressure is highly dependent on temperature; it increases significantly when the temperature increases. For example, the vapor pressure of water increases to 101.3 kPa if the temperature reaches 100°C. Water vapor pressures for other temperatures are given in Appendix B.

It is, of course, no coincidence that the water vapor pressure at 100°C is equal to the standard atmospheric pressure. At that temperature the water is **boiling**; that is, the liquid state of the water can no longer be sustained because the attractive forces are not sufficient to contain the molecules in a liquid phase. In general, a transition from the liquid state to the gaseous state occurs if the local absolute pressure is less than the vapor pressure of the liquid. At high elevations where the atmospheric pressure is relatively low, boiling occurs at temperatures less than 100°C; see Fig. 1.11. At an elevation of 3000 m, boiling would occur at approximately 90°C; see Tables B.3 and B.1.

In liquid flows, conditions can be created that lead to a pressure below the vapor pressure of the liquid. When this happens, bubbles are formed locally. This phenomenon, called **cavitation**, can be very damaging when these bubbles are transported by the flow to higher-pressure regions. What happens is that the bubbles collapse upon entering the higher-pressure region, and this collapse produces local pressure spikes which have the potential of damaging a pipe wall or a ship's propeller. Cavitation on a propeller is shown in Fig. 1.12. Additional information on cavitation is included in Section 8.3.4.

**Fig. 1.11**   Cooking food in boiling water takes a longer amount of time at a high altitude. It would take longer to boil these eggs in Denver than in New York City. (Thomas Firak Photography/FoodPix/ Getty Images)

**Vapor pressure:**  *The pressure resulting from molecules in a gaseous state.*

**KEY CONCEPT**  *Cavitation can be very damaging.*

**Boiling:**  *The point where vapor pressure is equal to the atmospheric pressure.*

**Cavitation:**  *Bubbles form in a liquid when the local pressure falls below the vapor pressure of the liquid.*

**Fig. 1.12**    A photograph of a cavitating propeller inside MIT's water tunnel. (Courtesy of Prof. S. A. Kinnas, Ocean Engineering Group, University of Texas - Austin.)

## Example 1.5

Calculate the vacuum necessary to cause cavitation in a water flow at a temperature of 80°C in Colorado where the elevation is 2500 m.

### Solution
The vapor pressure of water at 80°C is given in Table B.1. It is 47.3 kPa absolute. The atmospheric pressure is found by interpolation using Table B.3 to be 79.48 − (79.48 − 61.64)500\2000 ≅ 75.0. The required pressure is then

$$p = 47.3 - 75.0 = -27.7 \text{ kPa} \quad \text{or} \quad 27.7 \text{ kPa vacuum}$$

## 1.6  CONSERVATION LAWS

From experience it has been found that fundamental laws exist that appear exact; that is, if experiments are conducted with the utmost precision and care, deviations from these laws are very small and in fact, the deviations would be even smaller if improved experimental techniques were employed. Three such laws form the basis for our study of fluid mechanics. The first is the **conservation of mass**, which states that matter is indestructible. Even though Einstein's theory of relativity postulates that under certain conditions, matter is convertible into energy and leads to the statement that the extraordinary quantities of radiation from the sun are associated with a conversion of $3.3 \times 10^{14}$ kg of matter per day into energy, the destructibility of matter under typical engineering conditions is not measurable and does not violate the conservation of mass principle.

For the second and third laws it is necessary to introduce the concept of a system. A **system** is defined as a fixed quantity of matter upon which attention is focused. Everything external to the system is separated by the system boundaries.

**Conservation of mass:** *Matter is indestructible.*

**System:**  *A fixed quantity of matter.*

**Newton's second law:** *The sum of all external forces acting on a system is equal to the time rate of change of linear momentum of the system.*

These boundaries may be fixed or movable, real, or imagined. With this definition we can now present our second fundamental law, the *conservation of momentum*: The momentum of a system remains constant if no external forces are acting on the system. A more specific law based on this principle is **Newton's second law**: The sum of all external forces acting on a system is equal to the time rate of change of linear momentum of the system. A parallel law exists for the moment of momentum: The rate of change of angular momentum is equal to the sum of all torques acting on the system.

**Conservation of energy:** *The total energy of an isolated system remains constant. Also known as the* first law of thermodynamics.

The third fundamental law is the **conservation of energy**, which is also known as the *first law of thermodynamics*: The total energy of an isolated system remains constant. If a system is in contact with the surroundings, its energy increases only if the energy of the surroundings experiences a corresponding decrease. It is noted that the total energy consists of potential, kinetic, and internal energy, the latter being the energy content due to the temperature of the system. Other forms of energy[7] are not considered in fluid mechanics. The first law of thermodynamics and other thermodynamic relationships are presented in the following section.

## 1.7 THERMODYNAMIC PROPERTIES AND RELATIONSHIPS

For incompressible fluids, the three laws mentioned in the preceding section suffice. This is usually true for liquids but also for gases if relatively small pressure, density, and temperature changes occur. However, for a compressible fluid, it may be necessary to introduce other relationships, so that density, temperature, and pressure changes are properly taken into account. An example is the prediction of changes in density, pressure, and temperature when compressed gas is released from a rocket through a nozzle.

Thermodynamic properties, quantities that define the state of a system, either depend on the system's mass or are independent of the mass. The former is called an **extensive property**, and the latter is called an **intensive property**. An intensive property can be obtained by dividing the extensive property by the mass of the system. Temperature and pressure are intensive properties; momentum and energy are extensive properties.

**Extensive property:** *A property that depends on the system's mass.*

**Intensive property:** *A property that is independent of the system's mass.*

### 1.7.1 Properties of an Ideal Gas

The behavior of gases in most engineering applications can be described by the ideal-gas law, also called the perfect-gas law. When the temperature is relatively low and/or the pressure relatively high, caution should be exercised and real-gas laws should be applied. For air with temperatures higher than $-50°C$ the ideal-gas law approximates the behavior of air to an acceptable degree provided that the pressure is not extremely high.

---

[7]Other forms of energy include electric and magnetic field energy, the energy associated with atoms, and energy released during combustion.

The *ideal-gas law* is given by

$$p = \rho RT \qquad (1.7.1)$$

where $p$ is the absolute pressure, $\rho$ the density, $T$ the absolute temperature, and $R$ the gas constant. The gas constant is related to the universal gas constant $R_u$ by the relationship

$$R = \frac{R_u}{M} \qquad (1.7.2)$$

where $M$ is the molar mass. Values of $M$ and $R$ are tabulated in Table B.4 in Appendix B. The value of $R_u$ is

$$R_u = 8.314 \text{ kJ/kmol·K} \qquad (1.7.3)$$

For air $M = 28.97$ kg/kmol, so that for air $R = 0.287$ kJ/kg·K, a value used extensively in calculations involving air.

Other forms that the ideal-gas law takes are

$$p\mathcal{V} = mRT \qquad (1.7.4)$$

and

$$p\mathcal{V} = nR_u T \qquad (1.7.5)$$

where $n$ is the number of moles.

## Example 1.6

A tank with a volume of 0.2 m$^3$ contains 0.5 kg of nitrogen. The temperature is 20°C. What is the pressure?

#### Solution

Assume this is an ideal gas. Apply Eq. 1.7.1 ($R$ can be found in Table B.4). Solving the equation, $p = \rho RT$, we obtain, using $\rho = m/\mathcal{V}$,

$$p = \frac{0.5 \text{ kg}}{0.2 \text{ m}^3} \times 0.2968 \frac{\text{kJ}}{\text{kg·K}} (20 + 273) \text{ K} = 218 \text{ kPa absolute}$$

*Note:* The resulting units are kJ/m$^3$ = kN·m/m$^3$ = kN/m$^2$ = kPa. The ideal-gas law requires that pressure and temperature be in absolute units.

### 1.7.2   First Law of Thermodynamics

In the study of incompressible fluids, the first law of thermodynamics is particularly important. The first law of thermodynamics states that when a system, which is a fixed quantity of fluid, changes from state 1 to state 2, its energy content changes from $E_1$ to $E_2$ by energy exchange with its surroundings. The energy exchange is in the form of heat transfer or work. If we define heat transfer to the system as positive and work done by the system as positive,[8] the first law of thermodynamics can be expressed as

$$Q_{1\text{-}2} - W_{1\text{-}2} = E_2 - E_1 \qquad (1.7.6)$$

where $Q_{1\text{-}2}$ is the amount of heat transfer to the system and $W_{1\text{-}2}$ is the amount of work done by the system. The energy $E$ represents the total energy, which consists of kinetic energy ($mV^2/2$), potential energy ($mgz$), and internal energy ($m\tilde{u}$), where $\tilde{u}$ is the internal energy per unit mass; hence

$$E = m\left(\frac{V^2}{2} + gz + \tilde{u}\right) \qquad (1.7.7)$$

Note that $V^2/2$, $gz$, and $\tilde{u}$ are all intensive properties and $E$ is an extensive property.

For an isolated system, one that is thermodynamically disconnected from the surroundings (i.e., $Q_{1\text{-}2} = W_{1\text{-}2} = 0$), Eq. 1.7.6 becomes

$$E_1 = E_2 \qquad (1.7.8)$$

This equation represents the conservation of energy.

The work term in Eq. 1.7.6 results from a force $F$ moving through a distance as it acts on the system's boundary; if the force is due to pressure, it is given by

$$W_{1\text{-}2} = \int_{l_1}^{l_2} F\,dl$$
$$= \int_{l_1}^{l_2} pA\,dl = \int_{V_1}^{V_2} p\,dV \qquad (1.7.9)$$

where $A\,dl = dV$. An example that demonstrates an application of the first law of thermodynamics follows.

---

[8]In some presentations the work done on the system is positive, so that Eq. 1.7.6 would appear as $Q + W = \Delta E$. Either choice is acceptable.

## Example 1.7

A cart with a mass of 30 kg is pushed up a ramp with an initial force of 450 N (Fig. E1.7). The force decreases according to

$$F = 75(6 - l) \text{ N}$$

If the cart starts from rest at $l = 0$, determine its velocity after it has traveled 6 m up the ramp. Neglect friction.

**Fig. E1.7**

### Solution

The energy equation (Eq. 1.7.6) allows us to relate the quantities of interest. Since there is no heat transfer, we have

$$-W_{1\text{-}2} = E_2 - E_1$$

Recognizing that the force is doing work on the system, the work is negative. Hence the energy equation becomes

$$-\left[-\int_0^6 75(6 - l)\,dl\right] = m\left(\frac{V_2^2}{2} + gz_2\right) - m\left(\cancel{\frac{V_1^2}{2}}^{0} + \cancel{gz_1}^{0}\right)$$

Taking the datum as $z_1 = 0$, we have $z_2 = 6 \sin 30° = 3$ m. Thus

$$450 \times 6 - 75 \times \frac{6^2}{2} = 30\left(\frac{V_2^2}{2} + 9.8 \times 3\right)$$

$$\therefore V_2 = 5.58 \text{ m/s}$$

*Note:* We have assumed no internal energy change and no heat transfer.

### 1.7.3    Other Thermodynamic Quantities

In compressible fluids it is sometimes useful to define thermodynamic quantities that are combinations of other thermodynamic quantities. One such combination is the sum $(m\tilde{u} + p\mathcal{V})$, which can be considered a system property; it is encountered in numerous thermodynamic processes. This property is defined as **enthalpy** $H$:

**Enthalpy:** *A property created to aid in thermodynamic calculations.*

$$H = m\tilde{u} + p\mathcal{V} \qquad\qquad (1.7.10)$$

The corresponding intensive property ($H/m$) is

$$h = \tilde{u} + \frac{p}{\rho} \tag{1.7.11}$$

**KEY CONCEPT** *Constant-volume specific heat and constant-pressure specific heat are used to calculate the enthalpy and internal energy changes.*

Other useful thermodynamic quantities are the *constant-pressure specific heat $c_p$* and the *constant-volume specific heat $c_v$*; they are used to calculate the enthalpy and the internal energy changes in an ideal gas as follows:

$$\Delta h = \int c_p \, dT \tag{1.7.12}$$

and

$$\Delta \tilde{u} = \int c_v \, dT \tag{1.7.13}$$

For many situations we can assume constant specific heats in the foregoing relationships. Specific heats for common gases are listed in Table B.4. For an ideal gas $c_p$ is related to $c_v$ by using Eq. 1.7.11 in differential form:

$$dh = d\tilde{u} + RdT \qquad c_p = c_v + R \tag{1.7.14}$$

**Ratio of specific heats:** *The ratio of $c_p$ to $c_v$.*

where we used $p/\rho = RT$. The **ratio of specific heats** $k$ is often of use for an ideal gas; it is expressed as

$$k = \frac{c_p}{c_v} \tag{1.7.15}$$

For liquids and solids we use $\Delta u = c \, \Delta T$ where $c$ is the specific heat of the substance. For water $c \cong 4.18$ kJ/kg·°C.

A process in which pressure, temperature, and other properties are essentially constant at any instant throughout the system is called a **quasi-equilibrium** or *quasi-static* process. An example of such a process is the compression and expansion in the cylinder of an internal combustion engine.[9] If, in addition, no heat is transferred ($Q_{1\text{-}2} = 0$), the process is called an *adiabatic*, quasi-equilibrium process or an *isentropic* process. For such an isentropic[10] process the following relationships may be used:

**Quasi-equilibrium process:** *A process in which properties are essentially constant at any instant throughout a system.*

$$\frac{p_1}{p_2} = \left(\frac{\rho_1}{\rho_2}\right)^k \qquad \frac{T_1}{T_2} = \left(\frac{p_1}{p_2}\right)^{(k-1)/k} \qquad \frac{T_1}{T_2} = \left(\frac{\rho_1}{\rho_2}\right)^{k-1} \tag{1.7.16}$$

[9]Even though these processes may seem fast, they are thermodynamically slow. Molecules move very fast.
[10]An isentropic process occurs when the entropy is constant. We will not define or calculate entropy here; it is discussed in Section 9.1.

For a small pressure wave traveling in a gas at relatively low frequency, the wave speed is given by an isentropic process so that

$$c = \sqrt{\left.\frac{dp}{d\rho}\right|_s} = \sqrt{kRT} \qquad (1.7.17)$$

If the frequency is relatively high, entropy is not constant and we use

$$c = \sqrt{\left.\frac{dp}{d\rho}\right|_T} = \sqrt{RT} \qquad (1.7.18)$$

These are the primary thermodynamic relationships that will be used when considering compressible fluids.

## Example 1.8

A cylinder fitted with a piston has an initial volume of 0.5 m$^3$. It contains 2.0 kg of air at 400 kPa absolute. Heat is transferred to the air while the pressure remains constant until the temperature is 300°C. Calculate the heat transfer and the work done. Assume constant specific heats.

### Solution

Using the first law, Eq. 1.7.9, and the definition of enthalpy, we see that

$$Q_{1\text{-}2} = p_2 V_2 - p_1 V_1 + m\tilde{u}_2 - m\tilde{u}_1$$
$$= m\tilde{u}_2 + p_2 V_2 - (m\tilde{u}_1 + p_1 V_1)$$
$$= H_2 - H_1 = m(h_2 - h_1) = mc_p(T_2 - T_1)$$

where Eq. 1.7.12 is used assuming $c_p$ to be constant. The initial temperature is

$$T_1 = \frac{p_1 V_1}{mR} = \frac{400 \text{ kN/m}^2 \times 0.5 \text{ m}^3}{2.0 \text{ kg} \times 0.287 \text{ kJ/kg·K}} = 348.4 \text{ K}$$

(Use kJ = kN·m to check the units.) Thus the heat transfer is ($c_p$ is found in Table B.4)

$$Q_{1\text{-}2} = 2.0 \times 1.0[(300 + 273) - 348.4] = 449 \text{ kJ}$$

The final volume is found using the ideal-gas law:

$$V_2 = \frac{mRT_2}{p_2} = \frac{2 \text{ kg} \times (0.287 \text{ kJ/kg·K}) \times 573 \text{ K}}{400 \text{ kN/m}^2} = 0.822 \text{ m}^3$$

The work done for the constant-pressure process is, using Eq. 1.7.9 with $p = $ const,

$$W_{1\text{-}2} = p(V_2 - V_1)$$

$$= 400 \text{ kN/m}^2(0.822 - 0.5) \text{ m}^3 = 129 \text{ kN·m or } 129 \text{ kJ}$$

## Example 1.9

The temperature on a cold winter day in the mountains of Wyoming is $-30°C$ at an elevation of 3000 m. Calculate the density of the air assuming the same pressure as in the local atmosphere; also find the speed of sound.

**Solution**
From Table B.3 we find the atmospheric pressure at an elevation of 3000 m to be 70 KPa. The absolute temperature is found to be

$$T = -30 + 273 = 243 \text{ K}$$

Using the ideal-gas law, the density is calculated as

$$\rho = \frac{p}{RT}$$
$$= \frac{70000 \text{ N/m}^2}{287 \text{ J/kg K} \times 243 \text{ K}} = 1.0 \text{ kg/m}^3$$

The speed of sound, using Eq. 1.7.17, is determined to be

$$c = \sqrt{kRT}$$
$$= \sqrt{1.4 \times 287 \text{ J/kg K} \times 243 \text{ K}} = 312 \text{ m/s}$$

## 1.8 SUMMARY

To relate units we often use Newton's second law, which allows us to write

$$N = kg \cdot m/s^2 \tag{1.8.1}$$

When making engineering calculations, an answer should have the same number of significant digits as the least accurate number used in the calculations. Most fluid properties are known to at most four significant digits. Hence, answers should be expressed to at most four significant digits, and often to only three significant digits.

In fluid mechanics pressure is expressed as gage pressure unless stated otherwise. This is unlike thermodynamics, in which pressure is assumed to be absolute. If absolute pressure is needed, add 101 kPa if the atmospheric pressure is not given in the problem statement.

The density, or specific weight, of a fluid is known if the specific gravity is known:

$$\rho_x = S_x \rho_{\text{water}} \qquad \gamma_x = S_x \gamma_{\text{water}} \qquad\qquad (1.8.2)$$

The shear stress due to viscous effects in a simple flow where $u = u(y)$ is given by

$$\tau = \mu \frac{du}{dy} \qquad\qquad (1.8.3)$$

This stress can be used to calculate the torque needed to rotate a shaft in a bearing.

Many air flows, and other gases too, are assumed to be incompressible at low speeds, speeds under about 100 m/s for atmospheric air.

The three fundamental laws used in our study of fluid mechanics are the conservation of mass, Newton's second law, and the first law of thermodynamics. These will take on various forms depending on the problem being studied. Much of our study of fluid mechanics will be expressing these laws in mathematical forms so that quantities of interest can be calculated.

## FUNDAMENTALS OF ENGINEERING EXAM REVIEW PROBLEMS

**1.1**  If force, length, and time are selected as the three fundamental dimensions, the units of mass in the SI system could be written as:
(A)  $FT^2/L$        (B)  $FL/T^2$
(C)  $N \cdot s^2/m$        (D)  $N \cdot m/s^2$

**1.2**  Select the dimensions of viscosity using the $F$-$L$-$T$ system:
(A)  $FT^2/L$
(B)  $FT/L^2$
(C)  $N \cdot s/m^2$
(D)  $N \cdot s^2/m$

**1.3**  The quantity $2.36 \times 10^{-8}$ Pa can be written as:
(A)  23.6 nPa        (B)  236 μPa
(C)  $236 \times 10^{-3}$ mPa        (D)  236 nPa

**1.4**  A body that weighs 250 N on earth would weigh how much on the moon where $g \cong 1.6$ m/s²?
(A)  5030 N        (B)  250 N
(C)  40.77 N        (D)  6.2 N

**1.5**  A 4200-N force acts on a 250-cm area at an angle of 30° to the normal. The shear stress acting on the area is:
(A)  84 Pa        (B)  84 mPa
(C)  84 kPa        (D)  84 MPa

**1.6**  The temperature at 11 000 m in the standard atmosphere, using a parabolic interpolation of the entries in Table B.3, is nearest:
(A)  −62.4°C        (B)  −53.6°C
(C)  −32.8°C        (D)  −17.3°C

**1.7**  Using an equation, estimate the density of water at 80°C:
(A)  980 kg/m³        (B)  972 kg/m³
(C)  976 kg/m³        (D)  968 kg/m³

**1.8**  The velocity distribution in a 4-cm-diameter pipe transporting 20°C water is given by $u(r) = 10(1 - 2500r^2)$ m/s. The shearing stress at the wall is nearest:
(A)  1.0 Pa        (B)  0.1 Pa
(C)  0.01 Pa        (D)  0.001 Pa

**1.9**  The distance 20°C water would climb in a long 10-μm-diameter, clean glass tube is nearest:
(A)  50 cm
(B)  100 cm
(C)  200 cm
(D)  300 cm

**1.10**  Which of the following is an intensive property?
(A)  Kinetic energy        (B)  Enthalpy
(C)  Density        (D)  Momentum

**1.11** The mass of propane contained in a 4-m³ tank maintained at 800 kPa and 10°C is nearest:
(A) 100 kg (B) 80 kg
(C) 60 kg (D) 20 kg

**1.12** Five 40-cm³ ice cubes completely melt in 2 liters of warm water (it takes 320 kJ to melt a kilogram of ice). The temperature drop in the water is nearest:
(A) 10°C (B) 8°C
(C) 6°C (D) 4°C

**1.13** The speed of sound of a dog whistle in the atmosphere at a location where the temperature is 50°C is nearest:
(A) 396 m/s (B) 360 m/s
(C) 332 m/s (D) 304 m/s

## PROBLEMS

### Dimensions, Units, and Physical Quantities

**1.14** State the three basic laws that are used in the study of the mechanics of fluids. State at least one global (integral) quantity that occurs in each. State at least one quantity that may be defined at a point that occurs in each.

**1.15** Verify the dimensions given in Table 1.2 for the following quantities:
(a) Density (b) Pressure
(c) Power (d) Energy
(e) Mass (f) Flow rate

**1.16** Express the dimensions of the following quantities using the $F$-$L$-$T$ system:
(a) Density (b) Pressure
(c) Power (d) Energy
(e) Mass flux (f) Flow rate

**1.17** Recognizing that all terms in an equation must have the same dimensions, determine the dimensions on the constants in the following equations:
(a) $d = 4.9\,t^2$ where $d$ is distance and $t$ is time.
(b) $F = 9.8\,m$ where $F$ is a force and $m$ is mass.
(c) $Q = 80AR^{2/3}\,S_0^{1/2}$ where $A$ is area, $R$ is a radius, $S_0$ is a slope and $Q$ is a flow rate with dimensions of $L^3/T$.

**1.18** Determine the units on each of the constants in the following equations, recognizing that all terms in an equation have the same dimensions:
(a) $d = 4.9\,t^2$ where $d$ is in meters and $t$ is in seconds.
(b) $F = 9.8\,m$ where $F$ is in newtons and $m$ is in kilograms.
(c) $Q = 80AR^{2/3}S_0^{1/2}$ where $A$ is in meters squared, $R$ is in meters, $S_0$ is the slope, and $Q$ has units of meters cubed per second.

**1.19** State the SI units of Table 1.1 on each of the following:
(a) Pressure (b) Energy
(c) Power (d) Viscosity
(e) Heat flux (f) Specific heat

**1.20** Determine the units on $c$, $k$ and $f(t)$ in
$$m\,\frac{d^2y}{dt^2} + c\,\frac{dy}{dt} + ky = f(t)$$
if $m$ is in kilograms, $y$ is in meters, and $t$ is in seconds.

**1.21** Write the following with the use of prefixes:
(a) $2.5 \times 10^5$ N (b) $5.72 \times 10^{11}$ Pa
(c) $4.2 \times 10^{-8}$ Pa (d) $1.76 \times 10^{-5}$ m³
(e) $1.2 \times 10^{-4}$ m² (f) $7.6 \times 10^{-8}$ m³

**1.22** Write the following with the use of powers; do not use a prefix:
(a) 125 MN (b) 32.1 μs
(c) 0.67 GPa (d) 0.0056 mm³
(e) 520 cm² (f) 7.8 km³

**1.23** Rewrite Eq. 1.3.3 using the units of grams for mass and cm for the molecular diameter.

**1.24** Using the table of conversions on the inside front cover, express each of the following in the SI units of Table 1.2:
(a) 20 cm/hr (b) 2000 rpm
(c) 500 hp (d) 1000 cm³/min
(e) 2000 kN/cm² (f) 300 g/min
(g) 500 g/L (h) 500 kWh

**1.25** What net force is needed to accelerate a 10-kg mass at the rate of 40 m/s² (neglect all friction):
(a) Horizontally?
(b) Vertically upward?
(c) On an upward slope of 30°?

**1.26** A particular body has a mass of 27 kg on earth. Calculate its weight on the moon, where $g \simeq 1.63$ m/s².

**1.27** Calculate the mean free path in the atmosphere using Eq. 1.3.3 and Table B.3 in the Appendix at an elevation of:
(a) 30 000 m (b) 50 000 m (c) 80 000 m

## Pressure and Temperature

**1.28** A gage pressure of 52.3 kPa is read on a gage. Find the absolute pressure if the elevation is:
   (a)   At sea level          (b)   1000 m
   (c)   5000 m                (d)   10 000 m
   (e)   30 000 m

**1.29** A vacuum of 31 kPa is measured in an airflow at sea level. Find the absolute pressure in:
   (a)   kPa
   (b)   mm Hg
   (c)   $mN/cm^2$
   (d)   $N/mm^2$
   (e)   cm Hg

**1.30** For a constant-temperature atmosphere, the pressure as a function of elevation is given by $p(z) = p_0 e^{-gz/RT}$, where $g$ is gravity, $R = 287$ J/kg·K, and $T$ is the absolute temperature. Use this equation and estimate the pressure at 4000 m assuming that $p_0 = 101$ kPa and $T = 15°C$. What is the error?

**1.31** Estimate the pressure and temperature at an elevation of 6876 m using Table B.3. Employ:
   (a)   A linear interpolation: $f \simeq f_0 + n(f_1 - f_0)$.
   (b)   A parabolic interpolation: $f \simeq f_0 + n(f_1 - f_0) + (n/2)(n - 1)(f_2 - 2f_1 + f_0)$.

**1.32** Estimate the temperature in °C at 10,000 m, an elevation at which many commercial airplanes fly. Use Table B.3.

**1.33** An applied force of 26.5 MN is distributed uniformly over a 152-$cm^2$ area; however, it acts at an angle of 42° with respect to a normal vector (see Fig. P1.33). If it produces a compressive stress, calculate the resulting pressure.

**Fig. P1.33**

**1.34** The force on an area of 0.2 $cm^2$ is due to a pressure of 120 kPa and a shear stress of 20 Pa, as shown in Fig. P1.34. Calculate the magnitude of the force acting on the area and the angle of the force with respect to a normal coordinate.

**Fig. P1.34**

## Density and Specific Weight

**1.35** Calculate the density and specific weight of water if 2.92 kg occupies 2950 $cm^3$.

**1.36** Use Eq. 1.5.3 to determine the density and specific gravity of water at 70°C. What is the error in the calculation for density? Use Table B.1.

**1.37** The specific gravity of mercury is usually taken as 13.6. What is the percent error in using a value of 13.6 at 50°C?

**1.38** The specific weight of an unknown liquid is 12 400 $N/m^3$. What mass of the liquid is contained in a volume of 500 $cm^3$? Use:
   (a)   The standard value of gravity.
   (b)   The minimum value of gravity on the earth.
   (c)   The maximum value of gravity on the earth.

**1.39** A liquid with a specific gravity of 1.2 fills a volume. If the mass in the volume is 146 kg, what is the magnitude of the volume?

## Viscosity

**1.40** In combustion systems that burn hydrocarbon fuels, the carbon dioxide gas that is produced eventually escapes to the atmosphere thereby contributing to global warming. Calculate the density, specific weight, viscosity, and kinematic viscosity of carbon dioxide at a pressure of 200 kPa absolute and 90°C.

**1.41** In a single cylinder engine a piston without rings is designed to slide freely inside the vertical cylinder. Lubrication between the piston and cylinder is maintained by a thin oil film. Determine the velocity with which the 120-mm-diameter piston will fall inside the 120.5-mm-diameter cylinder. The 350-g piston is 10 cm long. The lubricant is SAE 10W-30 oil at 60°C.

**1.42** Consider a fluid flow between two parallel fixed plates 5 cm apart, as shown in Fig. P1.42. The velocity distribution for the flow is given by $u(y) = 120(0.05y - y^2)$ m/s where $y$ is in meters. The fluid is water at 10°C. Calculate the magnitude of the shear stress acting on each of the plates.

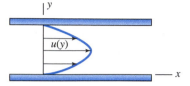

**Fig. P1.42**

**1.43** A velocity distribution in a 5-cm-diameter pipe is measured to be $u(r) = 9(1 - r^2/r_0^2)$ m/s, where $r_0$ is the radius of the pipe. Calculate the shear stress at the wall if water at 24°C is flowing.

**1.44** The velocity distribution in a 1.0-cm-diameter pipe is given by $u(r) = 16(1 - r^2/r_0^2)$ m/s, where $r_0$ is the pipe radius. Calculate the shearing stress at the centerline, at $r = 0.25$ cm, and at the wall if water at 20°C is flowing.

**1.45** For two 0.2-m-long rotating concentric cylinders, the velocity distribution is given by $u(r) = 0.4/r - 1000r$ m/s. If the diameters of the cylinders are 2 cm and 4 cm, respectively, calculate the fluid viscosity if the torque on the inner cylinder is measured to be 0.0026 N·m.

**1.46** A 1.2-m-long, 2.54-cm-diameter shaft rotates inside an equally long cylinder that is 2.6 cm in diameter. Calculate the torque required to rotate the inner shaft at 2000 rpm if SAE-30 oil at 21°C fills the gap. Also, calculate the horsepower required. Assume concentric cylinders.

**1.47** A 60-cm-wide belt moves at 10 m/s, as shown in Fig. P1.47. Calculate the horsepower requirement assuming a linear velocity profile in the 10°C water.

**Fig. P1.47**

**1.48** A 15-cm-diameter horizontal disk rotates a distance of 2 mm above a solid surface. Water at 16°C fills the gap. Estimate the torque required to rotate the disk at 400 rpm.

**1.49** Calculate the torque needed to rotate the cone shown in Fig. P1.49 at 2000 rpm if SAE-30 oil at 40°C fills the gap. Assume a linear velocity profile between the cone and the fixed wall.

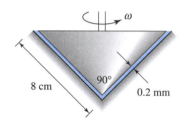

**Fig. P1.49**

**1.50** A free-body diagram of the liquid between a moving belt and a fixed wall shows that the shear stress in the liquid is constant. If the temperature varies according to $T(y) = K/y$, where $y$ is measured from the wall (the temperature at the wall is very large), what would be the shape of the velocity profile if the viscosity varies according to Andrade's equation $\mu = Ae^{B/T}$?

**1.51** The viscosity of water at 20°C is 0.001 N·s/m² and at 80°C it is $3.57 \times 10^{-4}$ N·s/m². Using Andrade's equation $\mu = Ae^{B/T}$ estimate the viscosity of water at 40°C. Determine the percent error.

## Compressibility

**1.52** Show that $d\rho/\rho = -d\mathcal{V}/\mathcal{V}$, as was assumed in Eq. 1.5.11.

**1.53** What is the volume change of 2 m³ of water at 20°C due to an applied pressure of 10 MPa?

**1.54** Two engineers wish to estimate the distance across a lake. One pounds two rocks together under water on one side of the lake and the other submerges his head and hears a small sound 0.62 s later, as indicated by a very

accurate stopwatch. What is the distance between the two engineers?

**1.55** A pressure is applied to 20 L of water. The volume is observed to decrease to 18.7 L. Calculate the applied pressure.

**1.56** Calculate the speed of propagation of a small-amplitude wave through water at:
   (a)   4°C
   (b)   37°C
   (c)   97°C

**1.57** The change in volume of a liquid with temperature is given by $\Delta V = \alpha_T V \Delta T$, where $\alpha_T$ is the *coefficient of thermal expansion*. For water at 40°C, $\alpha_T = 3.8 \times 10^{-4}$ K$^{-1}$. What is the volume change of 1 m$^3$ of 40°C water if $\Delta T = -20$°C? What pressure change would be needed to cause that same volume change?

## Surface Tension

**1.58** Calculate the pressure in the small 10-μm-diameter droplets that are formed by spray machines. Assume the properties to be the same as water at 15°C. Calculate the pressure for bubbles of the same size.

**1.59** A small 1.5-mm-diameter bubble is formed by a stream of 16°C water. Estimate the pressure inside the bubble.

**1.60** In diesel engines diesel fuel is injected directly into the engine cylinder during the compression stroke where the average air pressure could reach 8000 kPa. Assuming that liquid fuel droplets are formed as the fuel flows from the injector, determine the interior pressure in a 5-μm diameter spherical droplet. The surface tension for diesel fuel in air is 0.025 N/m.

**1.61** Determine the height that 20°C water would climb in a vertical 0.02-cm-diameter tube if it attaches to the wall with an angle β of 30° to the vertical.

**1.62** Mercury makes an angle of 130° (β in Fig. 1.10) when in contact with clean glass. What distance will mercury depress in a vertical 2-cm-diameter glass tube? Use $\sigma = 0.47$ N/m.

**1.63** Find an expression for the rise of liquid between two parallel plates a distance $t$ apart. Use a contact angle β and surface tension σ.

**1.64** Write an expression for the maximum diameter $d$ of a needle of length $L$ that can float in a liquid with surface tension σ. The density of the needle is ρ.

**1.65** Would a 7-cm-long 4-mm-diameter steel needle be able to float in 15°C water? Use $\rho_{steel} = 7850$ kg/m$^3$.

**1.66** Find an expression for the maximum vertical force $F$ needed to lift a thin wire ring of diameter $D$ slowly from a liquid with surface tension σ.

**1.67** Two flat plates are positioned as shown in Fig. P1.67 with a small angle $\alpha$ in an open container with a small amount of liquid. The plates are vertical, and the liquid rises between the plates. Find an expression for the location $h(x)$ of the surface of the liquid assuming that β = 0.

**Fig. P1.67**

## Vapor Pressure

**1.68** Water is transported through the pipe of Fig. P1.68 such that a vacuum of 80 kPa exists at a particular location. What is the maximum possible temperature of the water? Use $p_{atm} = 92$ kPa.

−80 kPa

Water

**Fig. P1.68**

**1.69** A group of explorers desired their elevation. An engineer boiled water and measured the temperature to be 82°C. They found a fluid mechanics book in a backpack and the engineer told the group their elevation! What elevation should the engineer have quoted?

**1.70** A tank half-filled with 40°C water is to be evacuated. What is the minimum pressure that can be expected in the space above the water?

**1.71** Water is forced through a contraction causing low pressure. The water is observed to "boil" at a pressure of −79.3 KPa gauge. If atmospheric pressure is 101.3 KPa, what is the temperature of the water?

**1.72** Oil is transported through a pipeline by a series of pumps that can produce a pressure of 10 MPa in the oil leaving each pump. The losses in the pipeline cause a pressure drop of 600 kPa each kilometer. What is the maximum possible spacing of the pumps?

## Ideal Gas

**1.73** Determine the density and the specific gravity of air at standard conditions (i.e., 15°C and 101.3 kPa absolute).

**1.74** Calculate the density of air inside a house and outside a house using 20°C inside and −25°C outside. Use an atmospheric pressure of 85 kPa. Do you think there would be a movement of air from the inside to the outside (infiltration), even without a wind? Explain.

**1.75** A 0.41 m³ air tank is pressurized to 5.17 MPa. When the temperature reaches −13°C, calculate the density and the air mass.

**1.76** Estimate the weight of air contained in a classroom that measures 10 m × 20 m × 4 m. Assume reasonable values for the variables.

**1.77** A car tire is pressurized to 240 KPa in Michigan when the temperature is −13°C. The car is driven to Arizona, where the temperature on the highway, and in the tire, reaches 66°C. Estimate the maximum pressure in the tire.

**1.78** The mass of all the air in the atmosphere contained above a 1-m² area is to be contained in a spherical volume. Estimate the diameter of the sphere if the air is at standard conditions.

## First Law

**1.79** A body falls from rest. Determine its velocity after 3 m and 6 m, using the energy equation.

**1.80** Determine the final velocity of the 15-kg mass of Fig. P1.80 moving horizontally if it starts at 10 m/s and moves a distance of 10 m while the following net force acts in the direction of motion (where $s$ is the distance in the direction of motion):
**(a)** 200 N    **(b)** $20s$ N
**(c)** $200 \cos (s\pi/20)$ N

**Fig. P1.80**

**1.81** The 10-kg mass shown in Fig. P1.81 is traveling at 40 m/s and strikes a plunger that is attached to a piston. The piston compresses 0.2 kg of air contained in a cylinder. If the mass is brought to rest, calculate the maximum rise of temperature in the air. What effects could lead to a lower temperature rise?

**Fig. P1.81**

**1.82** A 1500-kg automobile traveling at 100 km/h is suddenly grabbed by a hook and all of its kinetic energy is dissipated in a hydraulic absorber containing 2000 cm³ of water. Calculate the maximum temperature rise in the water.

**1.83** A fuel mass of 0.2 kg contains 40 MJ/kg of energy. Calculate the temperature rise of 100 kg of water if complete combustion occurs and the water, which surrounds the fuel, is completely insulated from the surroundings.

**1.84** 2 kg of air is compressed in a cylinder–piston arrangement while the temperature remains constant at 21°C. If the initial pressure is 207 KPa absolute, calculate the work needed to compress the air such that the absolute pressure doubles. Also calculate the heat transfer.

**1.85** Determine the heat transfer needed to double the absolute pressure in a fixed 2-m³ volume containing air at 200 kPa absolute if the initial temperature is:
**(a)** 20°C
**(b)** 100°C
**(c)** 200°C

**1.86** Heat is transferred to 2 kg of air in a cylinder so that the temperature doubles while the pressure remains constant. What work is required if the initial temperature is:
**(a)** 60°C?
**(b)** 150°C?
**(c)** 200°C?

## Isentropic Flow

**1.87** Air flows from a tank maintained at 5 MPa absolute and 20°C. It exits a hole and reaches a pressure of 500 kPa absolute. Assuming an adiabatic, quasi-equilibrium process, calculate the exiting temperature.

**1.88** An airstream flows with no heat transfer such that the temperature changes from 20°C to 150°C. If the initial pressure is measured to be 150 kPa, estimate the maximum final pressure.

**1.89** Air is compressed in an insulated cylinder from 20°C to 200°C. If the initial pressure is 100 kPa absolute, what is the maximum final pressure? What work is required?

## Speed of Sound

**1.90** Calculate the speed of sound at 20°C in:
- **(a)** Air
- **(b)** Carbon dioxide
- **(c)** Nitrogen
- **(d)** Hydrogen
- **(e)** Steam

**1.91** Compare the speed of sound in the atmosphere at an elevation of 10 000 m with that at sea level by calculating a percentage decrease.

**1.92** A lumberman, off in the distance, is chopping with an axe. An observer, using her digital stopwatch, measures a time of 8.32 s from the instant the axe strikes the tree until the sound is heard. How far is the observer from the lumberman if:
- **(a)** $T = -20°C$?
- **(b)** $T = 20°C$?
- **(c)** $T = 45°C$?

The Morrow Point Dam is an example of an arch-type dam. The curved wall enables large hydrostatic loads to exist on the upstream face while minimizing the required thickness of the structure. (U.S. Bureau of Reclamation)

# 2

# Fluid Statics

## Outline

## Chapter Objectives

The objectives of this chapter are to:

▲ Establish the variation of pressure in a fluid at rest.
▲ Learn how to use manometers to measure pressure.
▲ Calculate forces on plane and curved surfaces including buoyant forces.
▲ Determine the stability of submerged and floating objects.
▲ Calculate pressures and forces in accelerating and rotating containers.
▲ Present numerous examples and problems that demonstrate how pressures and forces are calculated in fluids at rest.

## 2.1 INTRODUCTION

**Fluid statics:** *The study of fluids in which there is no relative motion between fluid particles.*

KEY CONCEPT *The only stress that exists where there is no motion is a normal stress, the pressure.*

**Fluid statics** is the study of fluids in which there is no relative motion between fluid particles. If there is no relative motion, no shearing stresses exist, since velocity gradients, such as $du/dy$, are required for shearing stresses to be present. The only stress that exists is a normal stress, the pressure, so it is the pressure that is of primary interest in fluid statics.

Three situations, depicted in Fig. 2.1, involving fluid statics will be investigated. These include fluids at rest, such as water pushing against a dam, fluids contained in devices that undergo linear acceleration, and fluids contained in rotating cylinders. In each of these three situations the fluid is in static equilibrium with respect to a reference frame attached to the boundary surrounding the fluid. In addition to the examples shown for fluids at rest, we consider instruments called manometers and investigate the forces of buoyancy. Finally, the stability of floating bodies such as ships will also be presented.

## 2.2 PRESSURE AT A POINT

We have defined pressure as being the infinitesimal normal compressive force divided by the infinitesimal area over which it acts. This defines the pressure at a point. One might question whether the pressure at a given point varies as the normal to the area changes direction. To show that this is not the case, even for fluids in motion with no shear, consider the wedge-shaped element of unit depth (in the $z$-direction) shown in Fig. 2.2. Assume that a pressure $p$ acts on the hypotenuse and that a different pressure acts on each of the other areas, as shown. Since the forces on the two end faces are in the $z$-direction, we have not included them on the element. Now, let us apply Newton's second law to the element, for both the $x$- and $y$-directions:

$$\Sigma F_x = ma_x: \qquad p_x \Delta y - p\Delta s \sin\theta = \rho \frac{\Delta x \Delta y}{2} a_x$$

$$\Sigma F_y = ma_y: \qquad p_y \Delta x - \rho g \frac{\Delta x \Delta y}{2} - p\Delta s \cos\theta = \rho \frac{\Delta x \Delta y}{2} a_y$$

(2.2.1)

where we have used $\Delta V = \Delta x \, \Delta y/2$ (we could include $\Delta z$ in each term to account

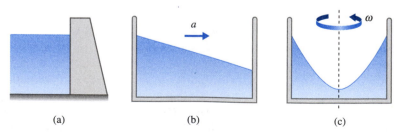

(a)        (b)        (c)

**Fig. 2.1** Examples included in fluid statics: (a) liquids at rest; (b) linear acceleration; (c) angular rotation.

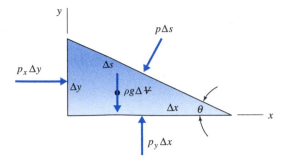

**Fig. 2.2**    Pressure at a point in a fluid.

for the depth). The pressures shown are due to the surrounding fluid and are the average pressure on the areas. Substituting

$$\Delta s \sin \theta = \Delta y \qquad \Delta s \cos \theta = \Delta x \qquad (2.2.2)$$

we see that Eqs. 2.2.1 take the forms

$$p_x - p = \frac{\rho a_x}{2} \Delta x$$

$$p_y - p = \frac{\rho(a_y + g)}{2} \Delta y \qquad (2.2.3)$$

Note that in the limit as the element shrinks to a point, $\Delta x \to 0$ and $\Delta y \to 0$. Hence the right-hand sides in the equations above go to zero, even for fluids in motion, providing us with the result that, at a point,

$$p_x = p_y = p \qquad (2.2.4)$$

Since $\theta$ is arbitrary, this relationship holds for all angles at a point. We could have analyzed an element in the $xz$-plane and concluded that $p_x = p_z = p$. Thus we conclude that the pressure in a fluid is constant at a point; that is, pressure is a scalar function. It acts equally in all directions at a given point for both a static fluid and a fluid that is in motion in the absence of shear stress.

**KEY CONCEPT**    *Pressure in a fluid acts equally in all directions at a given point.*

## 2.3 PRESSURE VARIATION

A general equation is derived to predict the pressure variation of fluids at rest or fluids undergoing an acceleration while the relative position of fluid elements to one another remains the same (this eliminates shear stress). To determine the pressure variation in such fluids, consider the infinitesimal element displayed in Fig. 2.3, where the $z$-axis is in the vertical direction. The pressure variation from one point to another will be determined by applying Newton's second law; that is, the sum of the forces acting on the fluid element is equal to the mass times the acceleration of the element.

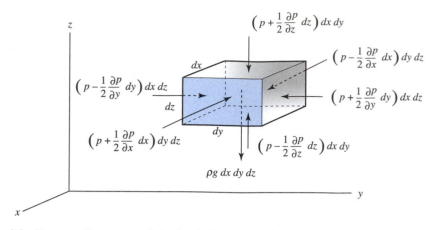

**Fig. 2.3** Forces acting on an infinitesimal element that is at rest in the *xyz*-reference frame. The reference frame may be accelerating or rotating.

If we assume that a pressure $p$ exists at the center of this element, the pressures at each of the sides can be expressed by using the chain rule from calculus with $p(x, y, z)$:

$$dp = \frac{\partial p}{\partial x} dx + \frac{\partial p}{\partial y} dy + \frac{\partial p}{\partial z} dz \tag{2.3.1}$$

If we move from the center to a face a distance $(dx/2)$ away, we see that the pressure is

$$p\left(x + \frac{dx}{2}, y, z\right) = p(x, y, z) + \frac{\partial p}{\partial x}\frac{dx}{2} \tag{2.3.2}$$

The pressures at all faces are expressed in this manner, as shown in Fig. 2.3. Newton's second law is written in vector form for a constant-mass system as

$$\Sigma \mathbf{F} = m\mathbf{a} \tag{2.3.3}$$

This results in the three component equations, assuming $z$ to be vertical and using the mass as $\rho\, dx\, dy\, dz$,

$$-\frac{\partial p}{\partial x} dx\, dy\, dz = \rho a_x\, dx\, dy\, dz$$

$$-\frac{\partial p}{\partial y} dx\, dy\, dz = \rho a_y\, dx\, dy\, dz \tag{2.3.4}$$

$$-\frac{\partial p}{\partial z} dx\, dy\, dz = \rho(a_z + g)\, dx\, dy\, dz$$

where $a_x$, $a_y$, and $a_z$ are the components of the acceleration of the element. Division by the element's volume $dx\,dy\,dz$ yields

$$\frac{\partial p}{\partial x} = -\rho a_x$$

$$\frac{\partial p}{\partial y} = -\rho a_y \qquad (2.3.5)$$

$$\frac{\partial p}{\partial z} = -\rho(a_z + g)$$

The pressure differential in any direction can now be determined from Eq. 2.3.1 as

$$dp = -\rho a_x dx - \rho a_y dy - \rho(a_z + g)dz \qquad (2.3.6)$$

where $z$ is always vertical. Pressure differences between specified points can be found by integrating Eq. 2.3.6. This equation is useful in a variety of problems, as will be demonstrated in the remaining sections of this chapter.

## 2.4  FLUIDS AT REST

A fluid at rest does not undergo any acceleration. Therefore, set $a_x = a_y = a_z = 0$ and Eq. 2.3.6 reduces to

$$dp = -\rho g\,dz \qquad (2.4.1)$$

or

$$\frac{dp}{dz} = -\gamma \qquad (2.4.2)$$

This equation implies that there is no pressure variation in the $x$- and $y$-directions, that is, in the horizontal plane. The pressure varies in the $z$-direction, the vertical direction, only. Also note that $dp$ is negative if $dz$ is positive; that is, the pressure decreases as we move up and increases as we move down.

### 2.4.1  Pressures in Liquids at Rest

If the density can be assumed constant, Eq. 2.4.2 is integrated to yield

$$\Delta p = -\gamma \Delta z \quad \text{or} \quad p + \gamma z = \text{constant} \quad \text{or} \quad \frac{p}{\gamma} + z = \text{constant} \qquad (2.4.3)$$

so that pressure increases with depth. Note that $z$ is positive in the upward direction. The quantity $(p/\gamma + z)$ is often referred to as the *piezometric head*. If the point of interest were a distance $h$ below a **free surface** (a surface separating a gas

**Free surface:**  *A surface separating a gas from a liquid.*

**Fig. 2.4** Pressure below a free surface.

from a liquid), as shown in Fig. 2.4, Eq. 2.4.3 would result in

$$p = \gamma h \tag{2.4.4}$$

**KEY CONCEPT** *The equation $p = \gamma h$ is used to convert pressure to a height of liquid.*

where $p = 0$ at $h = 0$. This equation will be quite useful in converting pressure to an equivalent height of liquid. For example, atmospheric pressure is often expressed as millimeters of mercury; that is, the atmospheric pressure is equal to the pressure at a certain depth in a mercury column, and by knowing the specific weight of mercury, we can then determine that depth using Eq. 2.4.4.

### 2.4.2 Pressures in the Atmosphere

**Standard atmosphere:**
*Position at 40° latitude where calculations are standardized.*

For the atmosphere where the density depends on height [i.e., $\rho = \rho(z)$], we must integrate Eq. 2.4.1 along a vertical path. The atmosphere is divided into four layers: the *troposphere* (nearest Earth), the *stratosphere*, the *mesosphere*, and the *ionosphere*.[1] Because conditions change with time and latitude in the atmosphere with the layers being thicker at the equator and thinner at the poles, we base calculations on the **standard atmosphere**, which is at 40° latitude. In the standard atmosphere the temperature in the troposphere varies linearly with elevation, $T(z) = T_0 - \alpha z$, where the *lapse rate* $\alpha = 0.0065$ K/m and $T_0$ is 288 K. In the part of the stratosphere between 11 and 20 km the temperature is constant at −56.5°C. (Commercial aircraft usually fly in the lower part of this constant-temperature region.) The temperature then increases again and reaches a maximum near 50 km; it then decreases to the edge of the ionosphere. The standard atmosphere is sketched in Fig. 2.5. Because the density of the air in the ionosphere is so low, it is possible for satellites to orbit the earth in this layer.

Figure 2.6 shows how atmospheric pressure varies with altitude on three mountains. A column of air from the outer atmosphere to a given point on Earth contains gases that exert a force equal to 101,352 N on each square meter. This pressure is 1 atm or 760 mm Hg. At a higher altitude the pressure is less because the mass of the column of air from the outer atmosphere to that point is less. Examples of pressure on the three mountains are given on the right of Fig. 2.6.

---

[1] The ionosphere is composed of the thermosphere and exosphere.

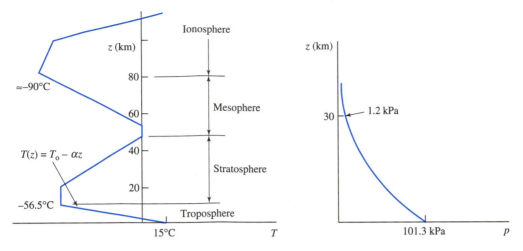

**Fig. 2.5**   Standard atmosphere.

To determine the pressure variation of the troposphere, we can use the ideal-gas law $p = \rho RT$ and Eq. 2.4.1; there results

$$dp = -\frac{pg}{RT}\,dz$$

or, collecting the pressure $p$ on the left-hand side,

$$\frac{dp}{p} = -\frac{g}{RT}\,dz \tag{2.4.5}$$

This can be integrated, between sea level and an elevation $z$ in the troposphere:

$$\int_{P_{atm}}^{p} \frac{dp}{p} = -\frac{g}{R}\int_{0}^{z}\frac{dz}{T_0 - \alpha z} \tag{2.4.6}$$

Upon integration this gives

$$\ln\frac{p}{P_{atm}} = \frac{g}{\alpha R}\ln\frac{T_0 - \alpha z}{T_0} \tag{2.4.7}$$

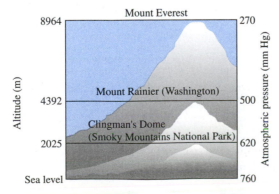

**Fig. 2.6**   Atmospheric pressure and altitude on three mountains.

which can be put in the form

$$p = p_{atm}\left(\frac{T_0 - \alpha z}{T_0}\right)^{g/\alpha R}$$   (2.4.8)

If we use standard atmospheric conditions in Eq. 2.4.8, we find that $p/p_{atm} = 0.999$ at $z = 10$ m. Consequently, we ignore changes in pressure in a gas such as air unless $z$ is relatively large. At $z = 1000$ m, the pressure decreases by about 2%.

In the lower part of the stratosphere, where the temperature is constant, Eq. 2.4.5 is integrated again as follows:

$$\int_{p_s}^{p} \frac{dp}{p} = -\frac{g}{RT_s}\int_{z_s}^{z} dz$$   (2.4.9)

$$\ln\frac{p}{p_s} = -\frac{g}{RT_s}(z - z_s)$$   (2.4.10)

or

$$p = p_s \exp\left[\frac{g}{RT_s}(z_s - z)\right]$$   (2.4.11)

The subscript $s$ denotes conditions at the troposphere–stratosphere interface. Properties of the standard atmosphere up to 80 km are listed in Appendix B.3.

### Example 2.1

The atmospheric pressure is given as 680 mm Hg at a mountain location. Convert this to kilopascals and meters of water. Also, calculate the pressure decrease due to 500-m elevation increase, starting at 2000 m elevation, assuming constant density.

**Solution**
Use Eq. 2.4.4 and find, using $S_{Hg} = 13.6$ with Eq. 1.5.2,

$$p = \gamma_{Hg}h$$
$$= (9.81 \text{ kN/m}^3 \times 13.6) \times 0.680 \text{ m} = 90.7 \text{ kPa}$$

To convert this to meters of water, we have

$$h = \frac{p}{\gamma_{H_2O}}$$
$$= \frac{90.7}{9.810} = 9.25 \text{ m of water}$$

To find the pressure decrease, we use Eq. 2.4.3 and find the density in Table B.3:

$$\Delta p = -\gamma\Delta z = -\rho g\Delta z$$
$$= -1.007 \text{ kg/m}^3 \times 9.81 \text{ m/s}^2 \times 500 \text{ m} = -4940 \text{ Pa}$$

where we used kg = N·s²/m.

*Note:* Since gravity is known to three significant digits, we express the answer to three significant digits.

## Example 2.2

Assume an isothermal atmosphere and approximate the pressure at 10 000 m. Calculate the percent error when compared with the values using Eq. 2.4.8 and from Appendix B.3. Use a temperature of 256 K, the temperature at 5000 m.

### Solution

Integrate Eq. 2.4.5 assuming that $T$ is constant, as follows:

$$\int_{101}^{p} \frac{dp}{p} = -\frac{g}{RT}\int_{0}^{z} dz$$

$$\ln\frac{p}{101} = -\frac{gz}{RT} \quad \text{or} \quad p = 101e^{-gz/RT}$$

Substituting $z = 10\,000$ m and $T = 256$ K, there results

$$p = 101e^{-9.81\times10\,000/(287\times256)}$$

$$= 26.57 \text{ kPa}$$

Using Eq. 2.4.8 we have

$$p = p_{\text{atm}}\left(\frac{T_0 - \alpha z}{T_0}\right)^{g/\alpha R}$$

$$= 101\left(\frac{288 - 0.0065 \times 10\,000}{288}\right)^{9.81/0.0065\times287} = 26.3 \text{ kPa}$$

The actual pressure at 10 000 m is found from Table B.3 to be 26.50 kPa. Hence the percent errors are

$$\% \text{ error} = \left(\frac{26.57 - 26.3}{26.3}\right) \times 100 = 1.03\%$$

$$\% \text{ error} = \left(\frac{26.57 - 26.50}{26.50}\right) \times 100 = 0.26\%$$

Because the error is so small, we often assume the atmosphere to be isothermal. *Note:* When evaluating $gz/RT$ we use $R = 287$ J/kg·K, not 0.287 kJ/kg·K. To observe that $gz/R$ is dimensionless, which it must be since it is an exponent, use $N = kg \cdot m/s^2$ so that

$$\left[\frac{gz}{RT}\right] = \frac{(m/s^2)m}{(J/kg\cdot K)K} = \frac{m^2/s^2}{N\cdot m/kg} = \frac{m^2/s^2}{(kg\cdot m^2/s^2)/kg} = \frac{m^2/s}{m^2/s}$$

### 2.4.3 Manometers

Hydrostatic Pressure, 517

Manometers are instruments that use columns of liquids to measure pressures. Three such instruments, shown in Fig. 2.7, are discussed to illustrate their use. Part (a) displays a U-tube manometer, used to measure relatively small pressures. In this case the pressure in the pipe can be determined by defining a point 1 at the center of the pipe and a point 2 at the surface of the right column. Then, using Eq. 2.4.3,

$$p_1 + \gamma z_1 = p_2 + \gamma z_2$$

where the datum from which $z_1$ and $z_2$ are measured is located at any desired position, such as through point 1. Since $p_2 = 0$ (gage pressure is selected; if absolute pressure is desired, we would select $p_2 = p_{\text{atm}}$) and $z_2 - z_1 = h$,

$$p_1 = \gamma h \tag{2.4.12}$$

Figure 2.7b shows a manometer used to measure relatively large pressures since we can select $\gamma_2$ to be quite large; for example, we could select $\gamma_2$ to be that of mercury so that $\gamma_2 = 13.6 \, \gamma_{\text{water}}$. The pressure can be determined by

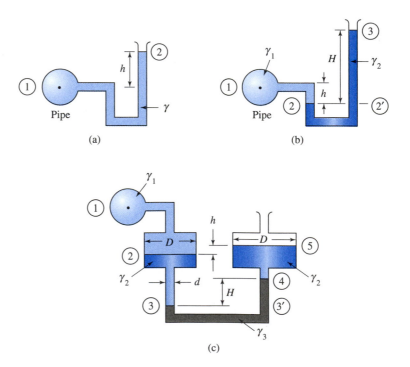

**Fig. 2.7** Manometers: (a) U-tube manometer (small pressures); (b) U-tube manometer (large pressures); (c) micromanometer (very small pressure changes).

introducing the points indicated. This is necessary because Eq. 2.4.3 applies throughout one fluid; $\gamma$ must be constant. The value of $\gamma$ changes abruptly at point 2. The pressure at point 2 and at point 2' is the same since the points are at the same elevation in the same fluid. Hence

$$p_2 = p_2{}'$$
$$p_1 + \gamma_1 h = p_3 + \gamma_2 H \qquad (2.4.13)$$

Setting $p_3 = 0$ (gage pressure is used) results in

$$p_1 = -\gamma_1 h + \gamma_2 H \qquad (2.4.14)$$

Figure 2.6c shows a micromanometer that is used to measure very small pressure changes. Introducing the points indicated, requiring that $p_3 = p_3{}'$, we can write

$$p_1 + \gamma_1(z_1 - z_2) + \gamma_2(z_2 - z_3) = p_5 + \gamma_2(z_5 - z_4) + \gamma_3(z_4 - z_3) \qquad (2.4.15)$$

Observe that $z_2 - z_3 + h = H + z_5 - z_4$ and set $p_5 = 0$; then

$$p_1 = \gamma_1(z_2 - z_1) + \gamma_2(h - H) + \gamma_3 H$$
$$= \gamma_1(z_2 - z_1) + \gamma_2 h + (\gamma_3 - \gamma_2)H \qquad (2.4.16)$$

Note that in all of the equations above for all three manometers, we have identified all interfaces with a point. This is always necessary when analyzing a manometer.

The micromanometer is capable of measuring small pressure changes because a small pressure change in $p_1$ results in a relatively large deflection $H$. The change in $H$ due to a change in $p_1$ can be determined using Eq. 2.4.16. Suppose that $p_1$ increases by $\Delta p_1$ and, as a result, $z_2$ decreases by $\Delta z$; then $h$ and $H$ also change. Using the fact that a decrease in $z_2$ is accompanied by an increase in $z_5$ leads to an increase in $h$ of $2\Delta z$ and, similarly, assuming that the volumes are conserved, it can be shown that $H$ increases by $2\Delta z D^2/d^2$. Hence a pressure change $\Delta p_1$ can be evaluated from changes in deflections as follows:

$$\Delta p_1 = \gamma_1(-\Delta z) + \gamma_2(2\Delta z) + \frac{(\gamma_3 - \gamma_2)2\Delta z D^2}{d^2} \qquad (2.4.17)$$

The rate of change in $H$ with $p_1$ is

$$\frac{\Delta H}{\Delta p_1} = \frac{2\Delta z D^2/d^2}{\Delta p_1} \qquad (2.4.18)$$

Using Eq. 2.4.17 we have

$$\frac{\Delta H}{\Delta p_1} = \frac{2D^2/d^2}{-\gamma_1 + 2\gamma_2 + 2(\gamma_3 - \gamma_2)D^2/d^2}$$ (2.4.19)

An example of this type of manometer is given in Example 2.4.

## Example 2.3

Water and oil flow in horizontal pipelines. A double U-tube manometer is connected between the pipelines, as shown in Fig. E2.3. Calculate the pressure difference between the water pipe and the oil pipe.

### Solution

We first identify the relevant points as shown in the figure. Begin at point ① and add pressure when the elevation decreases and subtract pressure when the elevation increases until point ⑤ is reached:

$$p_1 + \gamma(z_1 - z_2) - \gamma S_1(z_3 - z_2) - \gamma S_{\text{air}}(z_4 - z_3) + \gamma S_2(z_4 - z_5) = p_5$$

**Fig. E2.3**

where $\gamma = 9810 \text{ N/m}^3$, $S_1 = 1.6$, $S_2 = 0.9$, and $S_{\text{air}} \approx 0$. Thus

$$p_1 - p_5 = 9810\left(-\frac{250}{1000} + 1.6 \times \frac{275}{1000} + 0 \times \frac{150}{1000} - 0.9 \times \frac{150}{1000}\right)$$

$$= 539.6 \text{ Pa} \quad \text{or} \quad 0.5396 \text{ kPa}$$

Note that by neglecting the weight of the air, the pressure at point 3 is equal to the pressure at point 4.

## Example 2.4

For a given condition the liquid levels in Fig. 2.7c are $z_1 = 0.95$ m, $z_2 = 0.70$ m, $z_3 = 0.52$ m, $z_4 = 0.65$ m, and $z_5 = 0.72$ m. Further, $\gamma_1 = 9810$ N/m$^3$, $\gamma_2 = 11\ 500$ N/m$^3$, and $\gamma_3 = 14\ 000$ N/m$^3$. The diameters are $D = 0.2$ m and $d = 0.01$ m. (a) Calculate the pressure $p_1$ in the pipe, (b) calculate the change in $H$ if $p_1$ increases by 100 Pa, and (c) calculate the change in $h$ of the manometer of Fig. 2.7a if $h = 0.5$ m of water and $\Delta p_1 = 100$ Pa.

### Solution

(a) Referring to Fig. 2.7c, we have

$$h = 0.72 - 0.70 = 0.02 \text{ m}$$

$$H = 0.65 - 0.52 = 0.13 \text{ m}$$

Substituting the given values into Eq. (2.4.16) leads to

$$
\begin{aligned}
p_1 &= \gamma_1(z_2 - z_1) + \gamma_2 h + (\gamma_3 - \gamma_2)H \\
&= 9810(0.70 - 0.95) + 11\ 500(0.02) + (14\ 000 - 11\ 500)(0.13) \\
&= -1898 \text{ Pa}
\end{aligned}
$$

(b) If the pressure $p_1$ is increased by 100 Pa to $p_1 = -1798$ Pa, the change in $H$ is, using Eq. 2.4.19,

$$\Delta H = \Delta p_1 \frac{2D^2/d^2}{-\gamma_1 + 2\gamma_2 + 2(\gamma_3 - \gamma_2)D^2/d^2}$$

$$\Delta H = 100 \frac{2(20^2)}{-9810 + 2(11\ 500) + 2(14\ 000 - 11\ 500) \times 20^2} = 0.0397 \text{ m}$$

Thus $H$ increases by 3.97 cm as a result of increasing the pressure by 100 Pa.

(c) For the manometer in Fig. 2.7a, the pressure $p_1$ is given by $p = \gamma h$. Assume that initially $h = 0.50$ m. Thus the pressure initially is

$$p_1 = 9810 \times 0.50 = 4905 \text{ Pa}$$

Now if $p_1$ is increased by 100 Pa, $h$ can be found:

$$p_1 = \gamma h$$

$$h = \frac{p_1}{\gamma} = \frac{5005}{9810} = 0.510 \text{ m.} \quad \therefore \Delta h = 0.510 - 0.5 = 0.01 \text{ m}$$

Thus an increase of 100 Pa increases $h$ by 1 cm in the manometer shown in part (a), 25% of the change in the micromanometer.

### 2.4.4    Forces on Plane Areas

In the design of devices and objects that are submerged, such as dams, flow obstructions, surfaces on ships, and holding tanks, it is necessary to calculate the magnitudes and locations of forces that act on both plane and curved surfaces. In this section we consider only plane surfaces, such as the plane surface of general shape shown in Fig. 2.8. Note that a side view is given as well as a view showing

**Fig. 2.8** Force on an inclined plane area.

the shape of the plane. The total force of the liquid on the plane surface is found by integrating the pressure over the area, that is,

$$F = \int_A p \, dA \qquad (2.4.20)$$

where we usually use gage pressure. (Atmospheric pressure cancels out since it acts on both sides of the area.) The $x$ and $y$ coordinates are in the plane of the plane surface, as shown. Assuming that $p = 0$ at $h = 0$, we know that

$$\begin{aligned} p &= \gamma h \\ &= \gamma y \sin \alpha \end{aligned} \qquad (2.4.21)$$

where $h$ is measured vertically down from the free surface to the elemental area $dA$ and $y$ is measured from point $O$ on the free surface. The force may then be expressed as

$$\begin{aligned} F &= \int_A \gamma h \, dA \\ &= \gamma \sin \alpha \int_A y \, dA \end{aligned} \qquad (2.4.22)$$

The distance to a centroid is defined as

$$\bar{y} = \frac{1}{A} \int_A y \, dA \qquad (2.4.23)$$

The expression for the force then becomes

$$\begin{aligned} F &= \gamma \bar{y} A \sin \alpha \\ &= \gamma \bar{h} A = p_C A \end{aligned} \qquad (2.4.24)$$

where $\bar{h}$ is the vertical distance from the free surface to the centroid of the area and $p_C$ is the pressure at the centroid. Thus we see that the magnitude of the force on a plane surface is the pressure at the centroid multiplied by the area. The force does not, in general, act at the centroid.

To find the location of the resultant force $F$, we note that the sum of the moments of all the infinitesimal pressure forces acting on the area $A$ must equal the moment of the resultant force. Let the force $F$ act at the point $(x_p, y_p)$, the **center of pressure** (c.p.). The value of $y_p$ can be obtained by equating moments about the $x$-axis:

$$y_p F = \int_A yp\, dA$$

$$= \gamma \sin \alpha \int_A y^2\, dA = \gamma I_x \sin \alpha \qquad (2.4.25)$$

where the second moment of the area about the $x$-axis is

$$I_x = \int_A y^2\, dA \qquad (2.4.26)$$

The second moment of an area is related to the second moment of an area $\bar{I}$ about the centroidal axis by the parallel-axis-transfer theorem,

$$I_x = \bar{I} + A\bar{y}^2 \qquad (2.4.27)$$

Substitute Eqs. 2.4.24 and 2.4.27 into Eq. 2.4.25, and obtain

$$y_p = \frac{\gamma(\bar{I} + A\bar{y}^2)\sin \alpha}{\gamma \bar{y} A \sin \alpha}$$

$$= \bar{y} + \frac{\bar{I}}{A\bar{y}} \qquad (2.4.28)$$

where $\bar{y}$ is measured parallel to the plane area to the free surface.

Centroids and moments for several areas are presented in Appendix C. Using the expression above, we can show that the force on a rectangular gate, with the top edge even with the liquid surface, as shown in Fig. 2.9, acts two-thirds of the way down. This is also obvious considering the triangular pressure distribution acting on the gate. Note that Eq. 2.4.28 shows that $y_p$ is always greater

**KEY CONCEPT**   *The force on a plane surface is the pressure at the centroid multiplied by the area.*

**Center of pressure:**   *The point where the resultant force acts.*

**KEY CONCEPT**   *The force on a rectangular gate, with the top edge even with the liquid surface, acts two-thirds of the way down.*

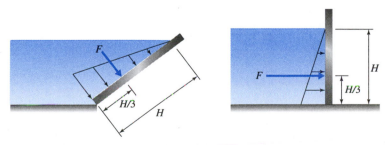

**Fig. 2.9**   Force on a plane area with top edge in a free surface.

than $\bar{y}$; that is, the resultant force of the liquid on a plane surface always acts below the centroid of the area, except on a horizontal area for which $\bar{y} = \infty$; then the center of pressure and the centroid coincide.

Similarly, to locate the $x$-coordinate $x_p$ of the c.p., we write

$$x_p F = \int_A xp \, dA$$

$$= \gamma \sin \alpha \int_A xy \, dA = \gamma I_{xy} \sin \alpha \qquad (2.4.29)$$

where the product of inertia of the area $A$ is

$$I_{xy} = \int_A xy \, dA \qquad (2.4.30)$$

Using the transfer theorem for the product of inertia,

$$I_{xy} = \bar{I}_{xy} + A\bar{x}\bar{y} \qquad (2.4.31)$$

Equation 2.4.29 becomes

$$x_p = \bar{x} + \frac{\bar{I}_{xy}}{A\bar{y}}$$

$$(2.4.32)$$

We now have expressions for the coordinates locating the center of pressure.

Finally, we should note that the force $F$ in Fig. 2.8 is the result of a *pressure prism* acting on the area. For the rectangular area shown in Fig. 2.10, the pressure increases, as shown by the pressure distribution in Fig. 2.10b. If we form the integral $\int p \, dA$, we obtain the volume of the pressure prism, which equals the force $F$ acting on the area, shown in Fig. 2.10c. The force acts through the centroid of the volume. For the rectangular area shown in Fig. 2.10a, the volume could be divided into two volumes: a rectangular volume with centroid at its center, and a triangular volume with centroid one-third the distance from the appropriate base. The location of the force is then found by locating the centroid of the composite volume.

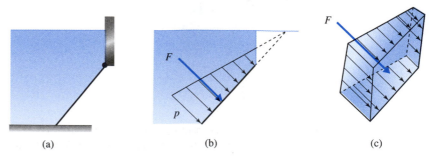

(a)  (b)  (c)

**Fig. 2.10** Pressure prism: (a) rectangular area; (b) pressure distribution on the area; (c) pressure prism.

# Example 2.5

A plane area of 80 cm × 80 cm acts as an escape hatch on a submersible in the Great Lakes. If it is on a 45° angle with the horizontal, what force applied normal to the hatch at the bottom edge is needed to just open the hatch, if it is hinged at the top edge when the top edge is 10 m below the surface? The pressure inside the submersible is assumed to be atmospheric.

**Fig. E2.5**

## Solution

First, a sketch of the hatch would be very helpful, as in Fig. E2.5. The force of the water acting on the hatch is

$$F = \gamma \bar{h} A$$

$$= 9810(10 + 0.4 \times \sin 45°)(0.8 \times 0.8) = 64\,560 \text{ N}$$

The distance $\bar{y}$ is

$$\bar{y} = \frac{\bar{h}}{\sin 45°} = \frac{10 + 0.4 \times \sin 45°}{\sin 45°} = 14.542 \text{ m}$$

so that

$$y_p = \bar{y} + \frac{\bar{I}}{A\bar{y}}$$

$$= 14.542 + \frac{0.8 \times 0.8^3/12}{(0.8 \times 0.8) \times 14.542} = 14.546 \text{ m}$$

Taking moments about the hinge provides the needed force $P$ to open the hatch:

$$0.8P = (y_p - \bar{y} + 0.4)F$$

$$\therefore \quad P = \frac{14.546 - 14.542 + 0.4}{0.8} \, 64\,560 = 32\,610 \text{ N}$$

Alternatively, we could have sketched the pressure prism, composed of a rectangular volume and a triangular volume. Moments about the top hinge would provide the desired force.

## Example 2.6

Find the location of the resultant force $F$ of the water on the triangular gate and the force $P$ necessary to hold the gate in the position shown in Fig. E2.6a. Neglect the weight of the gate, as usual.

**Fig. E2.6**

### Solution

First we draw a free-body diagram of the gate, including all the forces acting on the gate (Fig. E2.6c). The centroid of the gate is shown in Fig. E2.6b. The $y$-coordinate of the location of the resultant $F$ can be found using Eq. 2.4.28 as follows:

$$\bar{y} = 2 + 5 = 7$$

$$y_p = \bar{y} + \frac{\bar{I}}{A\bar{y}}$$

$$= 7 + \frac{2 \times 3^3/36}{3 \times 7} = 7.071 \text{ m}$$

To find $x_p$ we could use Eq. 2.4.32. Rather than that, we recognize that the resultant force must act on a line connecting the vertex and the midpoint of the opposite side since each infinitesimal force acts on this line (the moment of the resultant must equal the moment of its components). Thus, using similar triangles we have

$$\frac{x_p}{1} = \frac{2.071}{3}$$

$$\therefore \quad x_p = 0.690 \text{ m}$$

The coordinates $x_p$ and $y_p$ locate where the force due to the water acts on the gate.

If we take moments about the hinge, assumed to be frictionless, we can determine the force $P$ necessary to hold the gate in the position shown:

$$\Sigma M_{\text{hinge}} = 0$$

$$\therefore \quad 3 \times P = (3 - 2.071)F$$

$$= 0.929 \times \gamma \bar{h} A$$

$$= 0.929 \times 9810 \times (7 \sin 53°) \times 3$$

where $\bar{h}$ is the vertical distance from the centroid to the free surface. Hence

$$P = 50\,900 \text{ N} \quad \text{or} \quad 50.9 \text{ kN}$$

## 2.4.5   Forces on Curved Surfaces

We do not use a direct method of integration to find the force due to the hydro-static pressure on a curved surface. Rather, a free-body diagram that contains the curved surface and the liquids directly above or below the curved surface is identified. Such a free-body diagram contains only plane surfaces upon which unknown fluid forces act; these unknown forces can be found as in the preceding section.

As an example, let us determine the force of the curved gate on the stop, shown in Fig. 2.11a. The free-body diagram, which includes the water contained directly above the gate, is shown in Fig. 2.11b; $F_1$ and $F_2$ are due to the surrounding water and are the resultant forces of the pressure distributions shown; the body force $F_W$ is due to the weight of the water shown. In Fig. 2.11c the gate is the free body; the forces $F_x$ and $F_y$ are the horizontal and vertical components, respectively, of the force acting on the hinge. By summing moments about an axis passing through the hinge, we can determine the force $P$ acting on the stop.

If the curved surface is a quarter circle, the problem can be greatly simplified. This is observed by considering the free-body diagram of the gate only (Fig. 2.11c). The horizontal force $F_H$ acting on the gate is equal to $F_1$ of Fig. 2.11b, and the component $F_V$ is equal to the combined force $F_2 + F_W$ of Fig. 2.11b. Now, $F_H$ and $F_V$ are due to the differential pressure forces acting on the circular arc; each differential pressure force acts through the center of the circular arc. Hence the resultant force $\mathbf{F}_H + \mathbf{F}_V$ (this is a vector addition) must act through the center. Consequently, we can locate the components $F_H$ and $F_V$ at the center of the quarter circle, resulting in a much simpler problem. Example 2.7 will illustrate.

**KEY CONCEPT**   *The resultant force* $\mathbf{F}_H + \mathbf{F}_V$ *must act through the center of the circular arc.*

If the pressure on the free surface is $p_0$, we can simply add a depth of liquid necessary to provide $p_0$ at the location of the free surface and then work the resulting problem with a fictitious free surface located the appropriate distance above the original free surface. Or, the pressure force $p_0 A$ is added to the force $F_2$ of Fig. 2.11b.

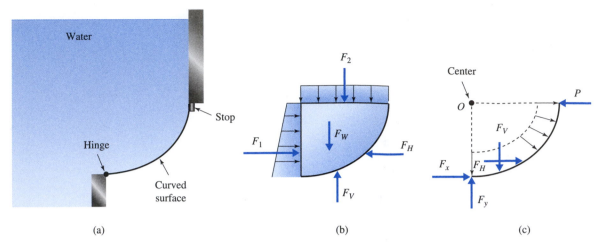

**Fig. 2.11** Forces acting on a curved surface: (a) curved surface; (b) free-body diagram of water and gate; (c) free-body diagram of gate only.

## Example 2.7

Calculate the force $P$ necessary to hold the 4-m-wide gate in the position shown in Fig. E2.7a. Neglect the weight of the gate.

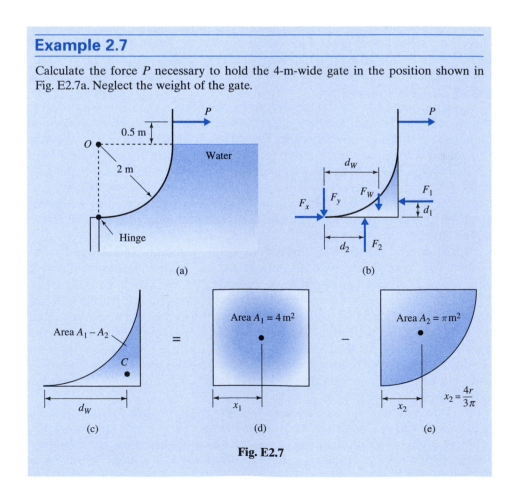

**Fig. E2.7**

**Solution**

The first step is to draw a free-body diagram. One choice is to select the gate and the water directly below the gate, as shown in Fig. E2.7b. To calculate $P$, we must determine $F_1$, $F_2$, $F_W$, $d_1$, $d_2$, and $d_W$; then moments about the hinge will allow us to find $P$. The force components are given by

$$F_1 = \gamma \overline{h}_1 A_1$$
$$= 9810 \times 1 \times (2 \times 4) = 78\,480 \text{ N}$$
$$F_2 = \gamma \overline{h}_2 A_2$$
$$= 9810 \times 2 \times (2 \times 4) = 156\,960 \text{ N}$$
$$F_W = \gamma V_{\text{water}}$$
$$= 9810 \times 4\left(4 - \frac{\pi \times 2^2}{4}\right) = 33\,700 \text{ N}$$

The distance $d_W$ is the distance to the centroid of the volume. It can be determined by considering the area as the difference of a square and a quarter circle as shown in Fig. E2.7c–e. Moments of areas yield

$$d_W(A_1 - A_2) = x_1 A_1 - x_2 A_2$$
$$d_W = \frac{x_1 A_1 - x_2 A_2}{A_1 - A_2}$$
$$= \frac{1 \times 4 - (4 \times 2/3\pi) \times \pi}{4 - \pi} = 1.553 \text{ m}$$

The distance $d_2 = 1$ m. Because $F_1$ is due to a triangular pressure distribution (see Fig. 2.9), $d_1$ is given by

$$d_1 = \frac{1}{3}(2) = 0.667 \text{ m}$$

Summing moments about the frictionless hinge gives

$$2.5P = d_1 F_1 + d_2 F_2 - d_W F_W$$
$$P = \frac{0.667 \times 78.5 + 1 \times 157.0 - 1.553 \times 33.7}{2.5} = 62.8 \text{ kN}$$

Rather than the somewhat tedious procedure above, we could observe that all the infinitesimal forces that make up the resultant force $(\mathbf{F}_H + \mathbf{F}_V)$ acting on the circular arc pass through the center $O$, as noted in Fig. 2.11c. Since each infinitesimal force passes through the center, the resultant force must also pass through the center. Hence we could have located the resultant force $(\mathbf{F}_H + \mathbf{F}_V)$ at point $O$. If $F_V$ and $F_H$ were located at $O$, $F_V$ would pass through the hinge, producing no moment about the hinge. Then, realizing that $F_H = F_1$ and summing moments about the hinge gives

$$2.5P = 2F_H$$

Therefore,

$$P = 2 \times \frac{78.48}{2.5} = 62.8 \text{ kN}$$

This was obviously much simpler. All we needed to do was calculate $F_H$ and then sum moments!

## Example 2.8

Find the force $P$ needed to hold the gate in the position shown in Fig. E2.8a if $P$ acts 3 m from the $y$-axis. The parabolic gate is 150 cm wide.

(a)　　　　　　　　　　　(b)

**Fig. E2.8**

### Solution

A free-body diagram of the gate and the water directly above the gate is shown in Fig. E2.8b. The forces are found to be

$$F_1 = \gamma \bar{h} A$$
$$= 9810 \times 1 \times (2 \times 1.5) = 29\,430\text{ N}$$

$$F_W = \gamma V$$
$$= 9810 \int_0^2 1.5x\,dy = 14\,715 \int_0^2 \frac{y^2}{2}\,dy = 14\,715\,\frac{2^3}{6} = 19\,620\text{ N}$$

The distance $d_1$ is $\frac{1}{3}(2) = 0.667$ m since the top edge is in the free surface. The distance $d_W$ through the centroid is found using a horizontal strip:

$$d_W = \frac{\displaystyle\int_0^2 x(x/2)\,dy}{\displaystyle\int_0^2 x\,dy} = \frac{\dfrac{1}{8}\displaystyle\int_0^2 y^4\,dy}{\dfrac{1}{2}\displaystyle\int_0^2 y^2\,dy} = \frac{1}{4}\frac{2^5/5}{2^3/3} = 0.6\text{ m}$$

Sum moments about the hinge and find $P$ as follows:

$$3P = d_1 F_1 + d_W F_W$$
$$= 0.667 \times 29\,430 + 0.6 \times 19\,620 \qquad \therefore P = 10\,470\text{ N}$$

## 2.4.6  Buoyancy

The law of buoyancy, known as Archimedes' principle, dates back some 2200 years to the Greek philosopher Archimedes. Legend has it that Hiero, king of Syracuse, suspected that his new gold crown may have been constructed of materials other than pure gold, so he asked Archimedes to test it. Archimedes probably made a lump of pure gold that weighed the same as the crown. The lump was discovered to weigh more in water than the crown weighed in water, thereby proving to Archimedes that the crown was not pure gold. The fake material possessed a larger volume to have the same weight as gold, hence it displaced more water. *Archimedes' principle* is: There is a buoyancy force on an object equal to the weight of displaced liquid.

To prove the law of buoyancy, consider the submerged body shown in Fig. 2.12a. In part (b) a cylindrical free-body diagram is shown that includes the submerged body with weight $W$ and liquid having a weight $F_W$; the cross-sectional area $A$ is the maximum cross-sectional area of the body. From the diagram we see that the resultant vertical force acting on the free-body diagram due to the water only (do not include $W$) is equal to

**KEY CONCEPT**
*Archimedes' principle states the buoyancy force on an object equals the weight of displaced liquid.*

$$\Sigma F = F_2 - F_1 - F_W \qquad (2.4.33)$$

This resultant force is by definition the *buoyant force* $F_B$. It can be expressed as

$$F_B = \gamma(h_2 A - h_1 A - \forall_W) \qquad (2.4.34)$$

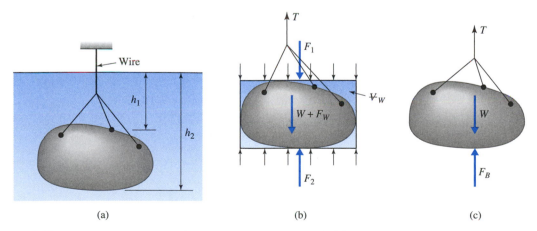

(a)                    (b)                    (c)

**Fig. 2.12**  Forces on a submerged body: (a) submerged body; (b) free-body diagram; (c) free body showing the buoyant force $F_B$.

where $V_W$ is the liquid volume included in the free-body diagram. Recognizing that the volume of the submerged body is

$$V_B = (h_2 - h_1)A - V_W \tag{2.4.35}$$

we see from Eq. 2.4.34 that

$$F_B = \gamma V_{\text{displaced liquid}} \tag{2.4.36}$$

thereby proving the law of buoyancy.

The force necessary to hold the submerged body in place (see Fig. 2.12c) is equal to

$$T = W - F_B \tag{2.4.37}$$

where $W$ is the weight of the submerged body.

For a floating object, as in Fig. 2.13, the buoyant force is

$$F_B = \gamma V_{\text{displaced liquid}} \tag{2.4.38}$$

Obviously, $T = 0$, so that Eq. 2.4.36 gives

$$F_B = W \tag{2.4.39}$$

where $W$ is the weight of the floating object.

**KEY CONCEPT** *The buoyant force acts through the centroid of the displaced liquid volume.*

From the foregoing analysis it is apparent that the buoyant force $F_B$ acts through the centroid of the displaced liquid volume. For the floating object, the weight of the object acts through its center of gravity, so the center of gravity of the object must lie on the same vertical line as the centroid of the liquid volume.

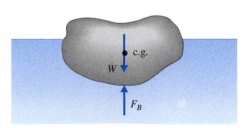

**Fig. 2.13** Forces on a floating object.

A **hydrometer**, an instrument used to measure the specific gravity of liquids, operates on the principle of buoyancy. A sketch is shown in Fig. 2.14. The upper part, the stem, has a constant diameter. When placed in pure water the specific gravity is marked to read 1.0. The force balance is

$$W = \gamma_{water}\mathcal{V} \tag{2.4.40}$$

where $W$ is the weight of the hydrometer and $\mathcal{V}$ is the submerged volume below the $S = 1.0$ line. In an unknown liquid of specific weight $\gamma_x$, a force balance would be

$$W = \gamma_x(\mathcal{V} - A\Delta h) \tag{2.4.41}$$

where $A$ is the cross-sectional area of the stem. Equating these two expressions gives

$$\Delta h = \frac{\mathcal{V}}{A}\left(1 - \frac{1}{S_x}\right) \tag{2.4.42}$$

where $S_x = \gamma_x/\gamma_{water}$. For a given hydrometer, $\mathcal{V}$ and $A$ are fixed so that the quantity $\Delta h$ is dependent only on the specific gravity $S_x$. Thus the stem can be calibrated to read $S_x$ directly. Hydrometers are used to measure the amount of antifreeze in the radiator of an automobile, or the charge in a battery since the density of the fluid changes as $H_2SO_4$ is consumed or produced.

**Hydrometer:** *An instrument used to measure the specific gravity of liquids.*

**Fig. 2.14** Hydrometer: (a) in water; (b) in an unknown liquid.

## Example 2.9

The specific weight and the specific gravity of a body of unknown composition are desired. Its weight in air is found to be 1000 N, and in water it weighs 750 N.

### Solution

The volume is found from a force balance when submerged as follows (see Fig. 2.12c):

$$T = W - F_B$$

$$750 = 1000 - 9810 V \quad \therefore V = 0.02548 \text{ m}^3$$

The specific weight is then

$$\gamma = \frac{W}{V} = \frac{1000}{0.02548} = 39246 \text{ N/m}^3$$

The specific gravity is found to be

$$S = \frac{\gamma}{\gamma_{\text{water}}} = \frac{39246}{9810} = 4.00$$

## 2.4.7 Stability

**KEY CONCEPT** *A floating object has vertical stability.*

The notion of stability can be demonstrated by considering the vertical stability of a floating object. If the object is raised a small distance, the buoyant force decreases and the object's weight returns the object to its original position. Conversely, if a floating object is lowered slightly, the buoyant force increases and the larger buoyant force returns the object to its original position. Thus a floating object has vertical stability, since a small departure from equilibrium results in a restoring force.

**Center of buoyancy:** *Centroid of a floating body.*

Consider now the rotational stability of a submerged body, shown in Fig. 2.15. In part (a) the center of gravity $G$ of the body is above the centroid $C$ (also referred to as the **center of buoyancy**) of the displaced volume, and a small angular rotation results in a moment that will continue to increase the rotation; hence the body is unstable and overturning would result. If the center of gravity is below the centroid, as in part (c), a small angular rotation provides a restoring moment and the body is stable. Part (b) shows neutral stability for a body in which the center of gravity and the centroid coincide, a situation that is encountered whenever the density is constant throughout the submerged body.

Next, consider the rotational stability of a floating body. If the center of gravity is below the centroid, the body is always stable, as with the submerged body of Fig. 2.15c. The body may be stable, though, even if the center of gravity is above the centroid, as sketched in Fig. 2.16a. When the body rotates the centroid of the volume of displaced liquid moves to the new location $C'$, shown in part (b). If the centroid $C'$ moves sufficiently far, a restoring moment develops and the body is stable, as shown. This is determined by the *metacentric height* $\overline{GM}$ defined as the distance from $G$ to the point of intersection of the buoyant force before rotation

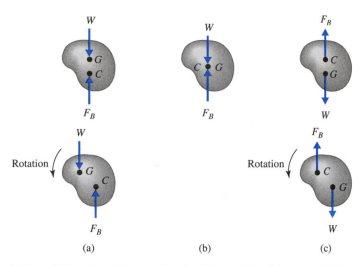

**Fig. 2.15** Stability of a submerged body: (a) unstable; (b) neutral; (c) stable.

with the buoyant force after rotation. If $\overline{GM}$ is positive, as shown, the body is stable; if $\overline{GM}$ is negative ($M$ lies below $G$), the body is unstable.

To determine a quantitative relationship for the distance $\overline{GM}$ refer to the sketch of Fig. 2.17, which shows a uniform cross section. Let us find an expression for $\bar{x}$, the $x$-coordinate of the centroid of the displaced liquid volume. It can be found by considering the volume to be the original volume plus the added wedge with cross-sectional area $DOE$ minus the subtracted wedge with cross-sectional area $AOB$; to locate the centroid of a composite volume, we take moments as follows:

$$\bar{x}\mathcal{V} = \bar{x}_0\mathcal{V}_0 + \bar{x}_1\mathcal{V}_1 - \bar{x}_2\mathcal{V}_2 \qquad (2.4.43)$$

where $\mathcal{V}_0$ is the original volume below the water line, $\mathcal{V}_1$ is the area $DOE$ times the length $\bar{x}_1$, and $\mathcal{V}_2$ is the area $AOB$ times the length $\bar{x}_2$; the cross section is assumed to be uniform so that the length $l$ is constant for the body. The quantity $\bar{x}_0$, the $x$-coordinate of point $C$, is zero. The remaining two terms can best be represented by integrals so that

> **KEY CONCEPT**   *If GM is positive, the body is stable.*

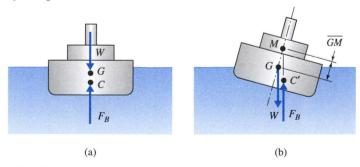

**Fig. 2.16** Stability of a floating body: (a) equilibrium position; (b) rotated position.

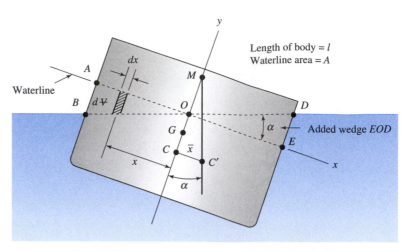

**Fig. 2.17** Uniform cross section of a floating body.

$$\overline{x}\,V = \int_{V_1} x\,dV - \int_{V_2} x\,dV \tag{2.4.44}$$

Then $dV = x \tan \alpha\, dA$ in volume 1 and $dV = -x \tan \alpha\, dA$ in volume 2, where $dA = l\,dx$, $l$ being the constant length of the body. The equation above becomes

$$\overline{x}\,V = \tan \alpha \int_{A_1} x^2\,dA + \tan \alpha \int_{A_2} x^2\,dA$$

$$= \tan \alpha \int_A x^2\,dA$$

$$= I_O \tan \alpha \tag{2.4.45}$$

where $I_O$ is the second moment (moment of inertia) of the waterline area about an axis passing through the origin $O$. The waterline area would be the length $\overline{AE}$ times the length $l$ of the body if $l$ were of constant length. Using $\overline{x} = \overline{CM} \tan \alpha$, we can write

$$\overline{CM}\,V = I_O \tag{2.4.46}$$

or, with $\overline{CG} + \overline{GM} = \overline{CM}$, we have

$$\overline{GM} = \frac{I_O}{V} - \overline{CG} \tag{2.4.47}$$

For a given body orientation, if $\overline{GM}$ is positive, the body is stable. Even though this relationship (2.4.47) is derived for a floating body with uniform cross section, it is applicable for floating bodies in general. We will apply it to a floating cylinder in the following example.

## Example 2.10

A 0.25-m-diameter cylinder is 0.25 m long and composed of material with specific weight 8000 N/m³. Will it float in water with the ends horizontal?

### Solution

With the ends horizontal, $I_O$ will be the second moment of the circular cross section,

$$I_O = \frac{\pi d^4}{64} = \frac{\pi \times 0.25^4}{64} = 0.000192 \text{ m}^4$$

The displaced volume will be

$$V = \frac{W}{\gamma_{water}} = \frac{8000 \times \pi/4 \times 0.25^2 \times 0.25}{9810} = 0.0100 \text{ m}^3$$

The depth the cylinder sinks in the water is

$$depth = \frac{V}{A} = \frac{0.01}{\pi \times 0.25^2/4} = 0.204 \text{ m}$$

Cylinder

0.204 m

0.102 m

G
C

0.125 m

**Fig. E2.10**

Hence, the distance $\overline{CG}$, as shown in Fig. E2.10, is

$$\overline{CG} = 0.125 - \frac{0.204}{2} = 0.023 \text{ m}$$

Finally,

$$\overline{GM} = \frac{0.000192}{0.01} - 0.023 = -0.004 \text{ m}$$

This is a negative value showing that the cylinder will not float with ends horizontal. It would undoubtedly float on its side.

## 2.5 LINEARLY ACCELERATING CONTAINERS

In this section the fluid will be at rest relative to a reference frame that is linearly accelerating with a horizontal component $a_x$ and a vertical component $a_z$. Then Eq. 2.3.6 simplifies to

$$dp = -\rho a_x \, dx - \rho(g + a_z) \, dz \tag{2.5.1}$$

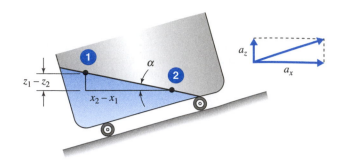

**Fig. 2.18** Linearly accelerating tank.

Integrating between two arbitrary points 1 and 2 results in

$$p_2 - p_1 = -\rho a_x(x_2 - x_1) - \rho(g + a_z)(z_2 - z_1) \qquad (2.5.2)$$

If points 1 and 2 lie on a constant-pressure line, such as the free surface in Fig. 2.18, then $p_2 - p_1 = 0$ and we have

$$\frac{z_1 - z_2}{x_2 - x_1} = \tan \alpha = \frac{a_x}{g + a_z} \qquad (2.5.3)$$

where $\alpha$ is the angle that the constant-pressure line makes with the horizontal.

**KEY CONCEPT** *We often utilize the conservation of mass and equate the volumes before and after the acceleration is applied.*

In the solution of problems involving liquids, we must often utilize the conservation of mass and equate the volumes before and after the acceleration is applied. After the acceleration is initially applied, sloshing may occur. Our analysis will assume that sloshing is not present; either sufficient time passes to dampen out time-dependent motions, or the acceleration is applied in such a way that such motions are minimal.

## Example 2.11

The tank shown in Fig. E2.11a is accelerated to the right. Calculate the acceleration $a_x$ needed to cause the free surface, shown in Fig. E2.11b, to touch point $A$. Also, find $p_B$ and the total force acting on the bottom of the tank if the tank width is 1 m.

**Fig. E2.11**

**Solution**

The angle the free surface takes is found by equating the air volume (actually, areas since the width is constant) before and after since no water spills out:

$$0.2 \times 2 = \tfrac{1}{2}(1.2x)$$

$$x = 0.667 \text{ m}$$

The quantity $\tan \alpha$ is now known. It is

$$\tan \alpha = \frac{1.2}{0.667} = 1.8$$

Using Eq. 2.5.3, we find $a_x$ to be, letting $a_z = 0$,

$$a_x = g \tan \alpha$$

$$= 9.81 \times 1.8 = 17.66 \text{ m/s}^2$$

We can find the pressure at $B$ by noting the pressure dependence on $x$. At $A$, the pressure is zero. Hence, Eq. 2.5.2 yields

$$p_B - \cancelto{0}{p_A} = -\rho a_x(x_B - x_A)$$

$$p_B = -1000 \times 17.66(-2)$$

$$= 35\,300 \text{ Pa}\quad \text{or}\quad 35.3 \text{ kPa}$$

To find the total force acting on the bottom of the tank, we realize that the pressure distribution is decreasing linearly from $p = 35.3$ kPa at $B$ to $p = 0$ kPa at $A$. Hence, we can use the average pressure over the bottom of the tank:

$$F = \frac{p_B + p_A}{2} \times \text{area}$$

$$= \frac{35\,300 + 0}{2} \times 2 \times 1 = 35\,300 \text{ N}$$

## 2.6 ROTATING CONTAINERS

In this section we consider the situation of a liquid contained in a rotating container, such as that shown in Fig. 2.19. After a relatively short time the liquid reaches static equilibrium with respect to the container and the rotating $rz$-reference frame. The horizontal rotation will not alter the pressure distribution in the vertical direction. There will be no variation of pressure with respect to the $\theta$-coordinate. Applying Newton's second law ($\Sigma F_r = ma_r$) in the $r$-direction to the element shown, using $\sin d\theta/2 \cong d\theta/2$, yields

> **KEY CONCEPT** *Horizontal rotation will not alter the pressure distribution in the vertical direction.*

$$-\frac{\partial p}{\partial r} dr\, rd\theta\, dz - pr d\theta\, dz - p\, dr\, d\theta\, dz - \frac{\partial p}{\partial r}(dr)^2 d\theta\, dz$$

$$+ 2p\frac{d\theta}{2} dr\, dz + pr d\theta\, dz = -\rho\, rd\theta\, dr\, dz\, r\omega^2 \qquad (2.6.1)$$

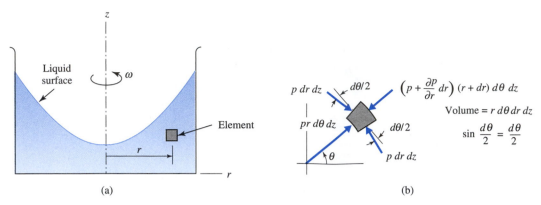

**Fig. 2.19** Rotating container: (a) liquid cross section; (b) top view of element.

where the acceleration is $r\omega^2$ toward the center of rotation. Simplify and divide by the volume $rd\theta\, dr\, dz$; then

$$\frac{\partial p}{\partial r} = \rho r \omega^2 \tag{2.6.2}$$

where we have neglected the higher-order term that contains the differential $dr$. The pressure differential then becomes

$$dp = \frac{\partial p}{\partial r}\, dr + \frac{\partial p}{\partial z}\, dz$$

$$= \rho r \omega^2\, dr - \rho g\, dz \tag{2.6.3}$$

where we have used the static pressure variation given by Eq. 2.3.5 with $a_z = 0$. We can now integrate between any two points $(r_1, z_1)$ and $(r_2, z_2)$ to obtain

$$p_2 - p_1 = \frac{\rho \omega^2}{2}(r_2^2 - r_1^2) - \rho g(z_2 - z_1) \tag{2.6.4}$$

If the two points are on a constant-pressure surface, such as the free surface, locating point 1 on the $z$-axis so that $r_1 = 0$, there results

$$\frac{\omega^2 r_2^2}{2} = g(z_2 - z_1) \tag{2.6.5}$$

**KEY CONCEPT** *The free surface is a paraboloid of revolution.*

which is the equation of a parabola. Hence the free surface is a paraboloid of revolution. The equations above can now, with the conservation of mass, be used to solve problems of interest.

## Example 2.12

The cylinder shown in Fig. E2.12 is rotated about its centerline. Calculate the rotational speed that is necessary for the water to just touch the origin $O$. Also, find the pressures at $A$ and $B$.

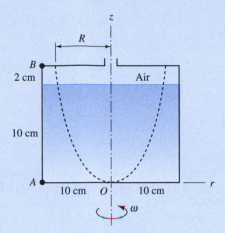

**Fig. E2.12**

### Solution

Since no water spills from the container, the air volume remains constant, that is,

$$\pi \times 10^2 \times 2 = \tfrac{1}{2}\pi R^2 \times 12$$

where we have used the fact that the volume of a paraboloid of revolution is one-half that of a circular cylinder with the same height and radius. This gives the value

$$R = 5.77 \text{ cm}$$

Using Eq. 2.6.5 with $r_2 = R$, we have

$$\frac{\omega^2 \times 0.0577^2}{2} = 9.81 \times 0.12$$

$$\therefore \omega = 26.6 \text{ rad/s}$$

To find the pressure at point $A$, we simply calculate the pressure difference between $A$ and $O$. Using Eq. 2.6.4 with $r_2 = r_A = 0.1$ m, $r_1 = r_0 = 0$, and $p_1 = p_0 = 0$, there results

$$p_A = \frac{\rho\omega^2}{2}(r_A^2 - r_0^2) = \frac{1000 \text{ kg/m}^3 \times 26.6^2 \text{ rad/s}^2}{2} \times 0.1^2 \text{ m}^2 = 3540 \text{ Pa} \quad \text{or} \quad 3.54 \text{ kPa}$$

using kg = N·s²/m. The pressure at $B$ can be found by applying Eq. 2.6.4 to points $A$ and $B$. This equation simplifies to

$$p_B - p_A = -\rho g(z_B - z_A)$$

Hence

$$p_B = 3540 - 1000 \text{ kg/m}^3 \times 9.81 \text{ m/s}^2 \times 0.12 \text{ m} = 2360 \text{ Pa} \quad \text{or} \quad 2.36 \text{ kPa}$$

## 2.7 SUMMARY

The pressure variation in the vertical $z$-direction in a constant density fluid is found using

$$\Delta p = -\gamma \Delta z \tag{2.7.1}$$

This is used to interpret manometers and to establish the force on a plane as

$$F = \gamma \bar{h} A \tag{2.7.2}$$

where $\bar{h}$ is the vertical distance to the centroid of the area. The force is located a distance from the free surface to the center of pressure parallel to the area given by

$$y_p = \bar{y} + \frac{\bar{I}}{A\bar{y}} \tag{2.7.3}$$

where $\bar{I}$ is about the centroidal axis. Forces on curved surfaces are found using the above relationships and the weight of liquid contained above the surface.

   Pressures and forces in linearly accelerating containers are determined using the angle $\alpha$ of a constant-pressure line:

$$\tan \alpha = \frac{a_x}{g + a_z} \tag{2.7.4}$$

Quite often the acceleration $a_z$ in the vertical direction is zero.

   In a container rotating with angular velocity $\omega$, a constant-pressure surface is described by

$$\frac{1}{2} \omega^2 r_2^2 = g(z_2 - z_1) \tag{2.7.5}$$

where point 1 is on the axis of rotation and point 2 is anywhere on the constant-pressure surface.

## FUNDAMENTALS OF ENGINEERING EXAM REVIEW PROBLEMS

**2.1**  A meteorologist states that the barometric pressure is 28.5 inches of mercury. Convert this pressure to kilopascals.

(A)  98.6 kPa      (B)  97.2 kPa
(C)  96.5 kPa      (D)  95.6 kPa

**2.2**  The pressure in the foothills of the Rockies near Boulder, Colorado is 84 kPa. The pressure, assuming a constant density of $1.00 \text{ kg/m}^3$, at the top of a nearby 4000-m-high mountain is closest to:

(A)  60 kPa      (B)  55 kPa
(C)  50 kPa      (D)  45 kPa

**2.3**  Estimate the pressure in the water pipe shown in Fig. P2.3. The manometer is open to the atmosphere.

(A)  10 kPa      (B)  9 kPa
(C)  8 kPa      (D)  7 kPa

$\gamma = 30 \text{ kN/m}^3$

**Fig. P2.3**

**2.4**  If the pressure in the air shown in Fig. P2.4, is increased by 10 kPa, the magnitude of $H$ will be nearest (initially $H = 16$ cm):

(A)  8.5 cm      (B)  10.5 cm
(C)  16 cm      (D)  24.5 cm

**Fig. P2.4**

**2.5**  The rectangular gate shown in Fig. P2.5 is 3 m wide. The force $P$ needed to hold the gate in the position shown is nearest:

(A)  24.5 kN      (B)  32.7 kN
(C)  98 kN      (D)  147 kN

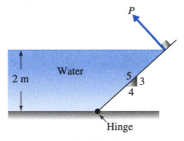

**Fig. P2.5**

**2.6**  The rigid gate hinged at a central point as shown in Fig. P2.6 opens when $H = 5$ m. How far is the hinge above the water bottom?

(A)  1.08 m      (B)  1.10 m
(C)  1.12 m      (D)  1.14 m

**Fig. P2.6**

**2.7**  A force $P = 300$ kN is needed to just open the gate of Fig. P2.7 with $R = 1.2$ m and $H = 4$ m. How wide is the gate?

(A)  2.98 m      (B)  3.67 m
(C)  4.32 m      (D)  5.16 m

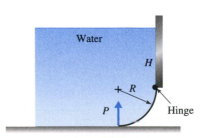

**Fig. P2.7**

**2.8** The rectangular barge of Fig. P2.8 is known to be 15 m long. A load having a mass of 900 kg is added to the barge causing it to sink 10 mm. How wide is the barge?

(A)  6 m          (B)  9.2 m
(C)  7.5 m        (D)  0.62 m

15 m

**Fig. P2.8**

**2.9** The tank, with an initial pressure of $p = 20$ kPa, is accelerated as shown in Fig. P2.9 at the rate of 5 m/s$^2$. The force on the 4-cm-diameter plug is nearest:

(A)  30 N         (B)  50 N
(C)  130 N       (D)  420 N

Plug              $p$

Gasoline        $a$

1.2 m

**Fig. P2.9**

## PROBLEMS

### Pressure

**2.10** Assume the element of Fig. 2.2 to be in the $yz$-plane with unit depth in the $x$-direction. Find a result similar to that of Eq. 2.2.4. Assume gravity to be in the $z$-direction.

**2.11** Calculate the pressure at a depth of 10 m in a liquid with specific gravity of:
(a)  1.0           (b)  0.8
(c)  13.6         (d)  1.59
(e)  0.68

**2.12** What depth is necessary in a liquid to produce a pressure of 250 kPa if the specific gravity is:
(a)  1.0?      (b)  0.8?      (c)  13.6?
(d)  1.59?     (e)  0.68?

**2.13** A pressure of 140 kPa is measured at a depth of 6 m. Calculate the specific gravity and the density of the liquid if $p = 0$ on the surface.

**2.14** How many meters of water are equivalent to:
(a)  760 mm Hg?      (b)  75 cm Hg?
(c)  10 mm Hg?

**2.15** Determine the pressure at the bottom of an open tank if it contains layers of:
(a)  20 cm of water and 2 cm of mercury
(b)  52 mm of water and 26 mm of carbon tetrachloride
(c)  3 m of oil, 2 m of water, and 10 cm of mercury

**2.16** Assuming the density of air to be constant at 1.27 kg/m$^3$, calculate the pressure change from the top of a mountain to its base if the elevation change is 3000 m.

**2.17** Assume that the air pressure is 100 kPa absolute at the top of a 3-m-high wall. Assuming constant density, estimate the difference in pressure at the base of the wall if outside the wall the temperature is $-20°$C and inside the wall it is 20°C. This pressure difference induces an infiltration even though no wind is present.

**2.18** The specific gravity of a liquid varies linearly from 1.0 at the surface to 1.1 at a depth of 10 m. Calculate the pressure at $h = 10$ m.

**2.19** If the gradient of $p(x, y, z)$ in rectangular coordinates is $\nabla p = \dfrac{\partial p}{\partial x} \hat{\mathbf{i}} + \dfrac{\partial p}{\partial y} \hat{\mathbf{j}} + \dfrac{\partial p}{\partial z} \hat{\mathbf{k}}$, write the simplest expression for $\nabla p$ using Eq. 2.3.5, recognizing that $\mathbf{a} = a_x\hat{\mathbf{i}} + a_y\hat{\mathbf{j}} + a_z\hat{\mathbf{k}}$.

**2.20** Use Eq. 2.4.8 to determine the pressure at the top of a building 300 m tall. Then assume that the density is constant at the $z = 0$ value and estimate $p$ at 300 m; also calculate the percent error in this second calculation. Use standard conditions at $z = 0$. Comment as to the advisability of an engineer assuming the atmosphere to be incompressible over heights of up to 300 m.

**2.21** Calculate the pressure change of air over a height of 20 m assuming standard conditions and using Eq. 2.4.8. Comment as to the advisability of an engineer to ignore completely changes in pressure up to heights of 20 m or so, in a gas such as air.

**2.22** Assume the bulk modulus to be constant and find an expression for the pressure as a function of depth $h$ in the ocean. Use this expression and estimate the pressure assuming $\rho_0 = 1064$ kg/m$^3$. Now, assume a constant density of 1064 kg/m$^3$ and calculate the pressure and the percent error, assuming the estimate in the first calculation to be correct. Use depths of (a) 450 m, (b) 1500 m, and (c) 4500 m.

**2.23** Estimate the pressure at 10 000 m assuming an isothermal atmosphere with temperature:
(a)  0°C       (b)  15°C      (c)  $-15°$C

**2.24** The temperature in the atmosphere is approximated by $T(z) = 15 - 0.0065z$°C for elevations less than 11 000 m. Calculate the pressure at elevations of:
(a)  3000 m        (b)  6000 m
(c)  9000 m        (d)  11 000 m

**2.25** Determine the elevation where $p = 7$ Pa assuming an isothermal atmosphere with $T = -20°$C.

## Manometers

**2.26** Calculate the pressure in a pipe transporting air if a U-tube manometer measures 25 cm Hg. Note that the weight of air in the manometer is negligible.

**2.27** If the pressure of air in a pipe is 450 kPa, what will a U-tube manometer with mercury read? Use $h = 1.5$ cm in Fig. 2.7b.
   **(a)** Neglect the weight of the air column.
   **(b)** Include the weight of the air column, assuming that $T_{air} = 20°C$, and calculate the percent error of part (a).

**2.28** A U-tube manometer is attached to a pipe transporting a liquid. It is known that the pressure in the pipe at the location of the manometer is 2.4 kPa. Select the liquid from Table B.5 that is most likely being transported if the manometer indicates the following height of liquid above the pipe:
   **(a)** 36.0 cm          **(b)** 27.2 cm
   **(c)** 24.5 cm          **(d)** 15.4 cm

**2.29** It is known that the pressure at the nose of an airplane traveling at relatively low speed is related to its speed by $p = \frac{1}{2}\rho V^2$, where $\rho$ is the density of air. Determine the speed of an airplane traveling near the earth's surface if a U-tube manometer, that measures the nose pressure, reads:
   **(a)** 6 cm of water    **(b)** 7.5 cm of water
   **(c)** 10 cm of water   **(d)** 12.5 cm of water

**2.30** Oil with $S = 0.86$ is being transported in a pipe. Calculate the pressure if a U-tube manometer reads 24 cm Hg. The oil in the manometer is depressed 12.5 cm below the pipe centerline.

**2.31** Several liquids are layered inside a tank with pressurized air at the top. If the air pressure is 3.2 kPa, calculate the pressure at the bottom of the tank if the layers include 20 cm of SAE 10 oil, 10 cm of water, 15 cm of glycerin, and 18 cm of carbon tetrachloride.

**2.32** For the setup shown in Fig. P2.32, calculate the manometer reading $H$.

**Fig. P2.32**

**2.33** For the setup shown in Fig. P2.33, find the difference in pressure between the oil pipe and the water pipe.

**Fig. P2.33**

**2.34** What is the pressure in the water pipe shown in Fig. P2.34?

**Fig. P2.34**

**2.35** Determine the pressure difference between the water pipe and the oil pipe shown in Fig. P2.35.

**Fig. P2.35**

**2.36**  What is the pressure in the oil pipe shown in Fig. P2.36 if the pressure in the water pipe is 15 kPa?

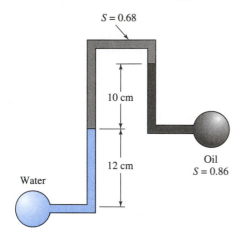

**Fig. P2.36**

**2.37**  For the tank shown in Fig. P2.37, determine the reading of the pressure gage if:
(a)  $H = 2$ m, $h = 10$ cm.
(b)  $H = 0.8$ m, $h = 20$ cm.
(c)  $H = 1.8$ m, $h = 10$ cm.
(d)  $H = 0.6$ m, $h = 20$ cm.

**Fig. P2.37**

**2.38**  For the tank shown in Fig. P2.38, if $H = 16$ cm, what will the pressure gage read?

**Fig. P2.38**

**2.39**  The pressure in the water pipe of Fig. 2.7b is 8.2 kPa, with $h = 25$ cm and $S_2 = 1.59$. Find the

pressure in the water pipe if the $H$ reading increases by 27.3 cm.

**2.40**  Find the pressure in the water pipe shown in Fig. P2.40.

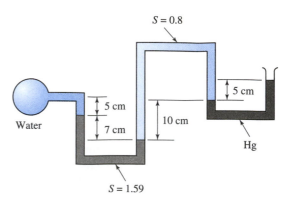

**Fig. P2.40**

**2.41**  For the inclined manometer containing mercury, shown in Fig. P2.41, determine the pressure in pipe $B$ if the pressure in pipe $A$ is 10 kPa. Pipe $A$ has water flowing through it, and oil is flowing in pipe $B$.

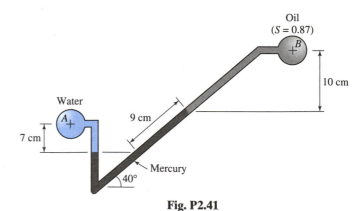

**Fig. P2.41**

**2.42**  The pressure in pipe $B$ in the previous problem is reduced slightly. Determine the new pressure in pipe $B$ if the pressure in pipe $A$ remains the same and the reading along the inclined leg of the manometer is 11 cm.

**2.43**  In Fig. P2.43, with the manometer top open the mercury level is 20 cm below the air pipe; there is no pressure in the air pipe. The manometer top is then sealed. Calculate the manometer reading $H$

for a pressure of 270 kPa in the air pipe. Assume an isothermal process for the air in the sealed tube.

**Fig. P2.43**

**2.44** Referring to Fig. 2.7c, determine the manometer reading $H$ for the following conditions:
$z_i = (22, 16, 10, z_4, 17)$ cm, $p_1 = 4$ kPa,
$\gamma_i = (9800, 15\,600, 133\,400)$ N/m$^3$,
$d = 5$ mm, $D = 100$ mm.

**2.45** Calculate the percentage increase in the manometer reading if the pressure $p_1$ is increased by 10% in:
**(a)** Problem 2.44a.
**(b)** Problem 2.44b.

**2.46** The pressure in the water pipe of Problem 2.36 is increased to 15.5 kPa, while the pressure in the oil pipe remains constant. What will be the new manometer reading?

**2.47** Determine the new manometer reading $h$ if the air pressure is increased by 10% in:
**(a)** Problem 2.37a
**(b)** Problem 2.37b
**(c)** Problem 2.37c
**(d)** Problem 2.37d

---

## Forces on Plane Areas

**2.48** Calculate the force acting on a 30-cm-diameter porthole of a ship if the center of the porthole is 10 m below water level.

**2.49** A swimming pool is filled with 2 m of water. Its bottom is square and measures 4 m on a side. Two opposite sides are vertical; one end is at 45° and the other makes an angle of 60° with the horizontal. Calculate the force of the water on:
**(a)** The bottom
**(b)** A vertical side
**(c)** The 45° end
**(d)** The 60° end

**2.50** An enclosed rectangular concrete vault with outside dimensions 2 m × 1 m × 1.5 m and wall thickness 10 cm is buried with the top surface even with the ground surface. Will the vault tend to rise out of the ground if the ground becomes completely saturated with water? Use $S_{concrete} = 2.4$.

**2.51** Gasoline fills a tank 4 m in diameter and 6 m long. Calculate the force that the gasoline exerts on one end of the tank. Assume that the tank is not pressurized and the ends are vertical.

**2.52** The sides of a triangular area measure 2 m, 3 m, and 3 m, respectively. Calculate the force of water on one side of the area if the 2-m side is

horizontal and 10 m below the surface and the triangle is:
**(a)** Vertical          **(b)** Horizontal
**(c)** On a 60° angle sloped upward

**2.53** The triangular gate shown in Fig. P2.53 has its 1.8-m side parallel and 9 m below the water surface. Calculate the magnitude and location of the force acting on the gate if it is:
**(a)** Vertical
**(b)** Horizontal
**(c)** On a 45° angle sloped upward

**Fig. P2.53**

**2.54** The top of each gate shown in Fig. P2.54 lies 4 m below the water surface. Find the location and magnitude of the force acting on one side assuming a vertical orientation.

(a)                          (b)

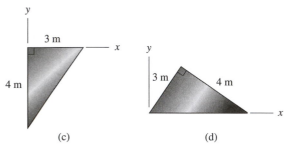

(c)                          (d)

**Fig. P2.54**

**2.55**   A 1.8-m-wide, 3-m-high vertical rectangular gate
has its top edge 6 ft below the water level. It is
hinged along its lower edge. What force, acting on
the top edge, is necessary to hold the gate shut?

**2.56**   Determine the force $P$ needed to hold the 4-m-
wide gate in the position shown in Fig. P2.56.

**Fig. P2.56**

**2.57**   Calculate the force $P$ necessary to hold the 4-m-
wide gate in the position shown in Fig. P2.57 if:
(a)   $H = 6$ m          (b)   $H = 8$ m
(c)   $H = 10$ m

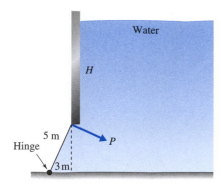

**Fig. P2.57**

**2.58**   Use Eq. 2.4.28 and show that the force $F$ in
Fig. 2.9 acts one-third the way up on a vertical rectangular area and also on a sloped rectangular area.
Assume the sloped gate to be on an angle $\alpha$ with
the horizontal.

**2.59**   Find the force $P$ needed to hold the 3-m-wide rectangular gate as shown in Fig. P2.59 if:
(a)   $l = 2$ m      (b)   $l = 4$ m      (c)   $l = 5$ m

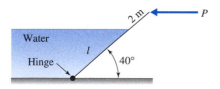

**Fig. P2.59**

**2.60**   A trapezoidal channel, with cross section shown in
Fig. P2.60, is gated at one end. What is the minimum force $P$ needed to hold the vertical gate
closed if it is hinged at the bottom? The gate has
the same dimensions as the channel and the force $P$
acts at the water surface.

**Fig. P2.60**

**2.61**   A vertical gate at the end of a channel (Fig. P2.61)
opens when the water above the hinge produces a
moment greater than the moment of the water

below the hinge. What height $h$ of water is needed to open the gate if:

**(a)** $H = 0.9$ m      **(b)** $H = 1.2$ m

**(c)** $H = 1.5$ m

**Fig. P2.61**

**2.62** At what height $H$ will the rigid gate, hinged at a central point as shown in Fig. P2.62, open if $h$ is:

**(a)** 0.6 m?    **(b)** 0.8 m?    **(c)** 1.0 m?

**Fig. P2.62**

**2.63** For the gate shown in Fig. P2.63, calculate the height $H$ that will result in the gate opening automatically if (neglect the weight of the gate):

**(a)** $l = 2$ m   **(b)** $l = 1$ m   **(c)** $l = 1.8$ m

**(d)** $l = 0.9$ m

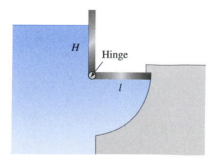

**Fig. P2.63**

**2.64** A 2-m-wide side rectangular gate is hinged at the bottom, as shown in Fig.P2.64. It is attached to a 70-kN, 1-m-diameter cylindrical block by a cable. Determime the height $H$ needed if the gate barely touches the stop.

**Fig. P2.64**

**2.65** The pressure distribution over the base of a concrete ($S = 2.4$) dam varies linearly, as shown in Fig. P2.65, producing an *uplift*. Will the dam topple over (sum moments of all forces about the bottom, right-hand corner)? Use:

**(a)** $H = 45$ m

**(b)** $H = 60$ m

**(c)** $H = 75$ m

**Fig. P2.65**

**2.66**  Assume a linear pressure distribution over the base of the concrete ($S = 2.4$) dam shown in Fig. P2.66. Will the dam topple over (sum moments about the bottom, right-hand corner)? Use:
(a)  $H = 12$ m
(b)  $H = 18$ m
(c)  $H = 24$ m

**Fig. P2.66**

## Forces on Curved Surfaces

**2.67**  In Example 2.7 assume that the water is above the gate rather than below the gate. Water above the gate will produce the same pressure distribution on the gate and hence the same forces (except the forces will be in opposite directions). Consequently, the force $P$ will be numerically the same (it will act to the left). With water above the gate, draw a free-body diagram and calculate $P$. Compare with the details of the first method of Example 2.7.

**2.68**  Find the force $P$ needed to hold the 10-m-long cylindrical object in position as shown in Fig. P2.68.

**Fig. P2.68**

**2.69**  Find the force $P$ needed to just open the gate shown in Fig. P2.69 if:
$H = 6$ m, $R = 2$ m, and the gate is 4 m wide.

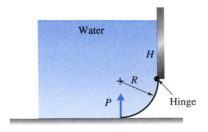

**Fig. P2.69**

**2.70  (a)** Determine the magnitude, direction, and line of action of the horizontal and vertical components of the hydrostatic force acting on curved surface $AB$ shown in Fig. P2.70, which has a radius of 2 m and a width of 4 m.
**(b)**  Assume that the water exists on the opposite side of the vertical barrier (the surface $AB$ remains as shown with point $A$ 8 m below the surface level). Determine the information requested in part (a).

**Fig. P2.70**

**2.71**  What $P$ is needed to hold the 4-m-wide gate shown in Fig. P2.71 closed?

**Fig. P2.71**

**2.72** Find the force $P$ required to hold the gate in the position shown in Fig. P2.72. The gate is 5 m wide.

**Fig. P2.72**

**2.73** The 3-m-wide circular gate shown in Fig. P2.73 weighs 400 N with cesnter of gravity 0.9 m to the left of the hinge. Estimate the force $P$ needed to open the gate.

**Fig. P2.73**

**2.74** The quarter-circle cylindrical gate (Fig. P2.74, $S = 0.2$) is in equilibrium, as shown. Calculate the value of $\gamma_x$.

**Fig. P2.74**

**2.75** A log is in equilibrium, as shown in Fig. P2.75. Calculate the force pushing it against the dam and the specific gravity of the log if its length is 6 m and $R = 0.6$ m

**Fig. P2.75**

**2.76** Find the force on the weld shown in Fig. P2.76 if:
**(a)** Air fills the hemisphere
**(b)** Oil fills the hemisphere

**Fig. P2.76**

**2.77** Find the force $P$ if the parabolic gate shown in Fig. P2.77 is:
**(a)** 2 m wide and $H = 2$ m
**(b)** 1.2 m wide and $H = 2.4$ m

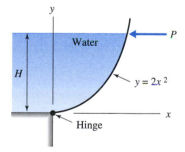

**Fig. P2.77**

**Buoyancy**

**2.78** The 3-m-wide barge shown in Fig. P2.78 weighs 20 kN empty. It is proposed that it carry a 250-kN load. Predict the draft in:
**(a)** Fresh water
**(b)** Salt water ($S = 1.03$)

**Fig. P2.78**

**2.79**   An object weighs 100 N in the air and 25 N when submerged in water. Calculate its volume and specific weight.

**2.80**   A car ferry is essentially rectangular with dimensions 7.5 m wide and 90 m long. If 60 cars, with an average weight per car of 15,000 N, are loaded on the ferry, how much farther will it sink into the water?

**2.81**   A 30-m-long vessel, with cross section shown in Fig. P2.81, carries a load of 6000 kN. How far will the water level be from the top of the vessel if its mass is 100 000 kg?

**Fig. P2.81**

**2.82**   A body, with a volume of 2 m³, weighs 40 kN. Determine its weight when submerged in a liquid with $S = 1.59$.

**2.83**   A hot-air balloon carries a load of 1000 N, including its own weight. If it is 10 m in diameter, estimate the average temperature of the air inside if the air outside is at 20°C.

**2.84**   A large blimp is proposed to travel near the Earth's surface. If the blimp resembles a large cylinder 1500 m long with a diameter of 300 m, estimate the payload if its own weight is 10% of the payload. How many 800-N people could it carry? The blimp is filled with helium and standard conditions prevail. (This vehicle will not cause sea sickness and sunsets are spectacular!)

**2.85**   An object is constructed of a material lighter than water. It weighs 50 N in air and a force of 10 N is required to hold it under water. What is its density, specific weight, and specific gravity?

**2.86**   The plug and empty cylinder shown in Fig. P2.86 weigh 6670 N. Calculate the height $h$ needed to lift the plug if the radius $R$ of the 3-m-long cylinder is:
    **(a)**   0.3 m    **(b)**   0.4 m    **(c)**   0.5 m

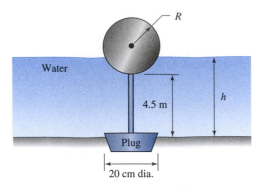

**Fig. P2.86**

**2.87**   The hydrometer shown in Fig. P2.87 with no mercury has a mass of 0.01 kg. It is designed to float at the midpoint of the 12-cm-long stem in pure water.
    **(a)**   Calculate the mass of mercury needed.
    **(b)**   What is the specific gravity of the liquid if the hydrometer is just submerged?
    **(c)**   What is the specific gravity of the liquid if the stem of the hydrometer is completely exposed?

**Fig. P2.87**

**2.88**   The hydrometer of Problem 2.87 is weighted so that in fresh water the stem is just submerged.
    **(a)**   What is the maximum specific gravity that can be read?
    **(b)**   What mass of mercury is required?

## Stability

**2.89** A 25-cm-diameter cylinder is composed of material with specific gravity 0.8. Will it float in water with the ends horizontal if its length is:
(a)  30 cm?  (b)  25 cm?  (c)  20 cm?

**2.90** Over what range of specific weights will a circular cylinder with uniform specific weight $\gamma_x$ float in water with ends horizontal if its height equals its diameter?

**2.91** Over what range of specific weights will a homogeneous cube float with sides horizontal and vertical?

**2.92** For the object shown in Fig. P2.92, calculate $S_A$ for neutral stability when submerged.

**Fig. P2.92**

**2.93** Orient the object shown in Fig. P2.93 for rotational stability when submerged if:
(a)  $t = 2$ cm  (b)  $t = 2.5$ cm

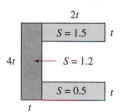

**Fig. P2.93**

**2.94** The barge shown in Fig. P2.94 is loaded such that the center of gravity of the barge and the load is at the waterline. Is the barge stable?

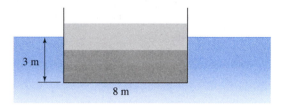

**Fig. P2.94**

**2.95** Is the symmetrically loaded barge shown in Fig. P2.95 stable? The center of gravity of the barge and load is located as shown.

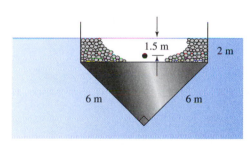

**Fig. P2.95**

## Linearly Accelerating Containers

**2.96** The tank shown in Fig. P2.96 is completely filled with water and accelerated. Calculate the maximum pressure in the tank if:
(a)  $a_x = 20$ m/s$^2$, $a_z = 0$, $L = 2$ m
(b)  $a_x = 0$, $a_z = 20$ m/s$^2$, $L = 2$ m
(c)  $a_x = 18$ m/s$^2$, $a_z = 18$ m/s$^2$, $L = 1.8$ m
(d)  $a_x = 0$, $a_z = 18$ m/s$^2$, $L = 1.8$ m

**Fig. P2.96**

**2.97** The tank shown in Fig. P2.97 is accelerated to the right at 10 m/s². Find:
    **(a)** $p_A$     **(b)** $p_B$     **(c)** $p_C$

**Fig. P2.97**

**2.98** The tank of Problem 2.97 is accelerated so that $p_B$ = 60 kPa. Find $a_x$ assuming that:
    **(a)** $a_z = 0$
    **(b)** $a_z = 10$ m/s²
    **(c)** $a_z = 5$ m/s²

**2.99** The tank shown in Fig. P2.99 is filled with water and accelerated. Find the pressure at $A$ if:
    **(a)** $a = 20$ m/s², $L = 1$ m
    **(b)** $a = 10$ m/s², $L = 1.5$ m

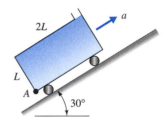

**Fig. P2.99**

**2.100** The tank of Problem 2.97 is 4 m wide. Find the force acting on:
    **(a)** The end $AB$
    **(b)** The bottom
    **(c)** The top

**2.101** The tank of Problem 2.98(a) is 1.5 m wide. Calculate the force on:
    **(a)** The bottom
    **(b)** The top
    **(c)** The left end

**2.102** For the U-tube shown in Fig. P2.102, determine the pressure at points $A$, $B$, and $C$ if:
    **(a)** $a_x = 0$, $a_z = 10$ m/s², $L = 60$ cm
    **(b)** $a_x = 10$ m/s², $a_z = 0$, $L = 60$ cm
    **(c)** $a_x = 20$ m/s², $a_z = 10$ m/s², $L = 60$ cm
    **(d)** $a_x = 0$, $a_z = -18$ m/s², $L = 62.5$ cm
    **(e)** $a_x = 18$ m/s², $a_z = 0$, $L = 62.5$ cm
    **(f)** $a_x = -9$ m/s², $a_z = 9$ m/s², $L = 62.5$ cm

**Fig. P2.102**

## Rotating Containers

**2.103** The U-tube of Problem 2.102 is rotated about the left leg at 50 rpm. Find $p_A$, $p_B$, and $p_C$ if:
    **(a)** $L = 60$ cm     **(b)** $L = 40$ cm

**2.104** The U-tube of Problem 2.102 is rotated about the right leg at 10 rad/s. Find the pressures at points $A$, $B$, and $C$ if:
    **(a)** $L = 60$ cm     **(b)** $L = 40$ cm

**2.105** The U-tube of Problem 2.102 is rotated about a vertical axis passing through the center of the

horizontal leg so that the pressure at the center of that leg is zero. Calculate $\omega$ if:
    **(a)** $L = 60$ cm     **(b)** $L = 40$ cm

**2.106** For the cylinder shown in Fig. P2.106, determine the pressure at point $A$ for a rotational speed of:
    **(a)** 5 rad/s     **(b)** 7 rad/s
    **(c)** 10 rad/s     **(d)** 20 rad/s

Air

Water

20 cm

60 cm

*A*

60 cm

$\omega$

**Fig. P2.106**

**2.107** The hole in the cylinder of Problem 2.106 is closed and the air pressurized to 25 kPa. Find the pressure at point *A* if the rotational speed is:
**(a)**  5 rad/s  **(b)**  7 rad/s
**(c)**  10 rad/s  **(d)**  20 rad/s

**2.108** Find the force on the bottom of the cylinder of:
**(a)**  Problem 2.106a
**(b)**  Problem 2.106b
**(c)**  Problem 2.106c
**(d)**  Problem 2.106d

White water rafting is a popular sport in North America. It embodies the thrill of experiencing rapidly moving water in a craft that demands quick thinking and manipulative skills. (ArtmannWitte/Shutterstock)

# 3

# Introduction to Fluids in Motion

## Outline

## Chapter Objectives

The objectives of this chapter are to:

▲ Mathematically describe the motion of a fluid.
▲ Express the acceleration and vorticity of a fluid particle given the velocity components.
▲ Describe the deformation of a fluid particle.
▲ Classify various fluid flows. Is a flow viscous, is it turbulent, is it incompressible, is it a uniform flow?
▲ Derive the Bernoulli equation and identify its restrictions.
▲ Present several examples and numerous problems that demonstrate how fluid flows are described, how flows are classified, and how Bernoulli's equation is used to estimate flow variables.

## 3.1 INTRODUCTION

This chapter serves as an introduction to all the following chapters that deal with fluid motions. Fluid motions manifest themselves in many different ways. Some can be described very easily, while others require a thorough understanding of physical laws. In engineering applications, it is important to describe the fluid motions as simply as can be justified. This usually depends on the required accuracy. Often, accuracies of $\pm$ 10% are acceptable, although in some applications higher accuracies have to be achieved. The general equations of motion are very difficult to solve; consequently, it is the engineer's responsibility to know which simplifying assumptions can be made. This, of course, requires experience and, more importantly, an understanding of the physics involved.

> **KEY CONCEPT**  *Under certain conditions, viscous effects can be neglected.*

Some common assumptions used to simplify a flow situation are related to fluid properties. For example, under certain conditions, the viscosity can affect the flow significantly; in others, viscous effects can be neglected, greatly simplifying the equations without significantly altering the predictions. It is well known that the compressibility of a gas in motion should be taken into account if the velocities are very high. But compressibility effects do not have to be taken into account to predict wind forces on buildings or to predict any other physical quantity that is a direct effect of wind. Wind speeds are simply not high enough. Numerous examples could be cited. After our study of fluid motions, the appropriate assumptions used should become more obvious.

This chapter has three sections. In the first section we introduce the reader to some important general approaches used to analyze fluid mechanics problems. In the second section we give a brief overview of different types of flow, such as compressible and incompressible flows, and viscous and inviscid flows. Detailed discussions of each of these flow types follow in later chapters. The third section introduces the reader to the commonly used Bernoulli equation, an equation that establishes how pressures and velocities vary in a flow field. The use of this equation, however, requires many simplifying assumptions, and its application is, therefore, limited.

## 3.2 DESCRIPTION OF FLUID MOTION

The analysis of complex fluid flow problems is often aided by the visualization of flow patterns, which permit the development of a better intuitive understanding and help in formulating the mathematical problem. The flow in a washing machine is a good example. An easier, yet difficult problem is the flow in the vicinity of where a wing attaches to a fuselage, or where a bridge support interacts with the water at the bottom of a river. In Section 3.2.1 we discuss the description of physical quantities as a function of space and time coordinates. The second topic in this section introduces the different flow lines that are useful in our objective of describing a fluid flow. Finally, the mathematical description of motion is presented.

### 3.2.1   Lagrangian and Eulerian Descriptions of Motion

In the description of a flow field, it is convenient to think of individual particles each of which is considered to be a small mass of fluid, consisting of a large number of molecules, that occupies a small volume $\Delta V$ that moves with the flow. If the fluid is incompressible, the volume does not change in magnitude but may deform. If the fluid is compressible, as the volume deforms, it also changes its magnitude. In both cases the particles are considered to move through a flow field as an entity.

In the study of particle mechanics, where attention is focused on individual particles, motion is observed as a function of time. The position, velocity, and acceleration of each particle are listed as $\mathbf{s}(x_0, y_0, z_0, t)$, $\mathbf{V}(x_0, y_0, z_0, t)$, and $\mathbf{a}(x_0, y_0, z_0, t)$, and quantities of interest can be calculated. The point $(x_0, y_0, z_0)$ locates the starting point—the name—of each particle. This is the **Lagrangian** description, named after Joseph L. Lagrange (1736–1813), of motion that is used in a course on dynamics. In the Lagrangian description many particles can be followed and their influence on one another noted. This becomes, however, a difficult task as the number of particles becomes extremely large in even the simplest fluid flow.

An alternative to following each fluid particle separately is to identify points in space and then observe the velocity of particles passing each point; we can observe the rate of change of velocity as the particles pass each point, that is, $\partial\mathbf{V}/\partial x$, $\partial\mathbf{V}/\partial y$, and $\partial\mathbf{V}/\partial z$, and we can observe if the velocity is changing with time at each particular point, that is, $\partial\mathbf{V}/\partial t$. In this **Eulerian** description, named after Leonhard Euler (1707–1783), of motion, the flow properties, such as velocity, are functions of both space and time. In Cartesian coordinates the velocity is expressed as $\mathbf{V} = \mathbf{V}(x, y, z, t)$. The region of flow that is being considered is called a **flow field**.

An example may clarify these two ways of describing motion. An engineering firm is hired to make recommendations that would improve the traffic flow in a large city. The engineering firm has two alternatives: Hire college students to travel in automobiles throughout the city recording the appropriate observations (the Lagrangian approach), or hire college students to stand at the intersections and record the required information (the Eulerian approach). A correct interpretation of each set of data would lead to the same set of recommendations, that is, the same solution. In this example it may not be obvious which approach would be preferred; in an introductory course in fluids, however, the Eulerian description is used exclusively since the physical laws using the Eulerian description are easier to apply to actual situations. Yet, there are examples where a Lagrangian description is needed, such as drifting buoys used to study ocean currents.

If the quantities of interest do not depend on time, that is, $\mathbf{V} = \mathbf{V}(x, y, z)$, the flow is said to be a **steady flow**. Most of the flows of interest in this introductory textbook are steady flows. For a steady flow, all flow quantities at a particular point are independent of time, that is,

$$\frac{\partial\mathbf{V}}{\partial t} = 0 \qquad \frac{\partial p}{\partial t} = 0 \qquad \frac{\partial\rho}{\partial t} = 0 \qquad (3.2.1)$$

to list a few. It is implied that $x$, $y$, and $z$ are held fixed in the above. Note that the properties of a fluid particle do, in general, vary with time; the velocity and pressure vary with time as a particular fluid particle progresses along its path in a flow, even in a steady flow. In a steady flow, however, properties do not vary with time at a fixed point.

**Lagrangian:** *Description of motion where individual particles are observed as a function of time.*

**Eulerian:** *Description of motion where the flow properties are functions of both space and time.*

**Flow field:** *The region of interest in a flow.*

 Eulerian vs. Lagrangian, 31–33

**Steady flow:** *Where flow quantities do not depend on time.*

### 3.2.2 Pathlines, Streaklines, and Streamlines

**Pathline:** *The history of a particle's locations.*

Pathlines, 91

Three different lines help us in describing a flow field. A **pathline** is the locus of points traversed by a given particle as it travels in a field of flow; the pathline provides us with a "history" of the particle's locations. A photograph of a pathline would require a time exposure of an illuminated particle. A photograph showing pathlines of particles below a water surface with waves is given in Fig. 3.1.

**Streakline:** *An instantaneous line.*

Streamlines, 122

A **streakline** is defined as an instantaneous line whose points are occupied by all particles originating from some specified point in the flow field. Streaklines tell us where the particles are "right now." A photograph of a streakline would be a snapshot of the set of illuminated particles that passed a certain point. Figure 3.2 shows streaklines produced by the continuous release of a small-diameter stream of smoke as it moves around a cylinder.

**Fig. 3.1** Pathlines underneath a wave in a tank of water. (Photograph by A. Wallet and F. Ruellan. Courtesy of M. C. Vasseur.)

Streaklines, 122

**Fig. 3.2** Streaklines in the unsteady flow around a cylinder. (Photography by Sadatoshi Taneda. From Album of Fluid Motion, 1982, The Parabolic Press, Stanford, California.)

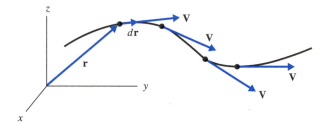

**Fig. 3.3**   Streamline in a flow field.

A **streamline** is a line in the flow possessing the following property: the velocity vector of each particle occupying a point on the streamline is tangent to the streamline. This is shown graphically in Fig. 3.3. An equation that expresses that the velocity vector is tangent to a streamline is

*Streamline:   The velocity vector is tangent to the streamline.*

$$\mathbf{V} \times d\mathbf{r} = 0 \qquad (3.2.2)$$

since $\mathbf{V}$ and $d\mathbf{r}$ are in the same direction, as shown in the figure; recall that the cross product of two vectors in the same direction is zero. This equation will be used in future chapters as the mathematical expression of a streamline. A photograph of a streamline cannot be made directly. For a general unsteady flow the streamlines can be inferred from photographs of short pathlines of a large number of particles.

A **streamtube** is a tube whose walls are streamlines. Since the velocity is tangent to a streamline, no fluid can cross the walls of a streamtube. The streamtube is of particular interest in fluid mechanics. A pipe is a streamtube since its walls are streamlines; an open channel is a streamtube since no fluid crosses the walls of the channel. We often sketch a streamtube with a small cross section in the interior of a flow for demonstration purposes.

*Streamtube:   A tube whose walls are streamlines.*

In a steady flow, pathlines, streaklines, and streamlines are all coincident. All particles passing a given point will continue to trace out the same path since the velocity in our Eulerian system does not change with time; hence the pathlines and streaklines coincide. In addition, the velocity vector of a particle at a given point will be tangent to the line that the particle is moving along; thus the line is also a streamline. Since the flows that we observe in laboratories are invariably steady flows, we call the lines that we observe streamlines even though they may actually be streaklines, or for the case of time exposures, pathlines.

**KEY CONCEPT**   *In a steady flow, pathlines, streaklines, and streamlines are all coincident.*

### 3.2.3   Acceleration

The acceleration of a fluid particle is found by considering a particular particle shown in Fig. 3.4. Its velocity changes from $\mathbf{V}(t)$ at time $t$ to $\mathbf{V}(t + dt)$ at time $t + dt$. The acceleration is, by definition,

**Fig. 3.4** Velocity of a fluid particle.

$$\mathbf{a} = \frac{d\mathbf{V}}{dt} \tag{3.2.3}$$

where $d\mathbf{V}$ is shown in Fig. 3.4. The velocity vector $\mathbf{V}$ is given in component form as

$$\mathbf{V} = u\hat{\mathbf{i}} + v\hat{\mathbf{j}} + w\hat{\mathbf{k}} \tag{3.2.4}$$

where $(u, v, w)$ are the velocity components in the $x$-, $y$-, and $z$- directions, respectively, and $\hat{\mathbf{i}}$, $\hat{\mathbf{j}}$ and $\hat{\mathbf{k}}$ are the unit vectors. The quantity $d\mathbf{V}$ is, using the chain rule from calculus with $\mathbf{V} = \mathbf{V}(x, y, z, t)$,

$$d\mathbf{V} = \frac{\partial \mathbf{V}}{\partial x} dx + \frac{\partial \mathbf{V}}{\partial y} dy + \frac{\partial \mathbf{V}}{\partial z} dz + \frac{\partial \mathbf{V}}{\partial t} dt \tag{3.2.5}$$

This gives the acceleration using Eg. 3.2.3 as

$$\mathbf{a} = \frac{\partial \mathbf{V}}{\partial x} \frac{dx}{dt} + \frac{\partial \mathbf{V}}{\partial y} \frac{dy}{dt} + \frac{\partial \mathbf{V}}{\partial z} \frac{dz}{dt} + \frac{\partial \mathbf{V}}{\partial t} \tag{3.2.6}$$

Since we have followed a particular particle, as in Fig. 3.4, we recognize that

$$\frac{dx}{dt} = u \qquad \frac{dy}{dt} = v \qquad \frac{dz}{dt} = w \tag{3.2.7}$$

The acceleration is then expressed as

$$\mathbf{a} = u \frac{\partial \mathbf{V}}{\partial x} + v \frac{\partial \mathbf{V}}{\partial y} + w \frac{\partial \mathbf{V}}{\partial z} + \frac{\partial \mathbf{V}}{\partial t} \tag{3.2.8}$$

The scalar component equations of the above vector equation for Cartesian coordinates are written as

$$a_x = \frac{\partial u}{\partial t} + u\frac{\partial u}{\partial x} + v\frac{\partial u}{\partial y} + w\frac{\partial u}{\partial z}$$

$$a_y = \frac{\partial v}{\partial t} + u\frac{\partial v}{\partial x} + v\frac{\partial v}{\partial y} + w\frac{\partial v}{\partial z} \qquad (3.2.9)$$

$$a_z = \frac{\partial w}{\partial t} + u\frac{\partial w}{\partial x} + v\frac{\partial w}{\partial y} + w\frac{\partial w}{\partial z}$$

We often return to Eq. 3.2.3 and write Eq. 3.2.8 in a simplified form as

$$\mathbf{a} = \frac{D\mathbf{V}}{Dt} \qquad (3.2.10)$$

where, in Cartesian coordinates,

$$\frac{D}{Dt} = u\frac{\partial}{\partial x} + v\frac{\partial}{\partial y} + w\frac{\partial}{\partial z} + \frac{\partial}{\partial t} \qquad (3.2.11)$$

This derivative is called the **substantial derivative**, or **material derivative**. It is given a special name and special symbol ($D/Dt$ instead of $d/dt$) because we followed a particular fluid particle, that is, we followed the substance (or material). It represents the relationship between a Lagrangian derivative in which a quantity depends on time $t$ and an Eulerian derivative in which a quantity depends on position $(x, y, z)$ and time $t$. The substantial derivative can be used with other dependent variables; for example, $DT/Dt$ would represent the rate of change of the temperature of a fluid particle as we followed the particle along.

The substantial derivative and acceleration components in cylindrical and spherical coordinates are presented in Table 3.1 on page 96.

The time-derivative term on the right side of Eqs. 3.2.8 and 3.2.9 for the acceleration is called the **local acceleration** and the remaining terms on the right side in each equation form the **convective acceleration**. Hence the acceleration of a fluid particle is the sum of the local acceleration and convective acceleration. In a pipe, local acceleration results if, for example, a valve is being opened or closed; and convective acceleration occurs in the vicinity of a change in the pipe geometry, such as a pipe contraction or an elbow. In both cases fluid particles change speed, but for very different reasons.

We must note that the foregoing expressions for acceleration give the acceleration relative to an observer in the observer's reference frame only. In certain situations the observer's reference frame may be accelerating; then

**Substantial or material derivative:**  *The derivative $D/Dt$.*

**Local acceleration:**  *The time-derivative term $\partial V/\partial t$ for acceleration.*

**Convective acceleration:** *All terms other than the local acceleration term.*

**KEY CONCEPT**  *Convective acceleration occurs in the vicinity of a change in the geometry.*

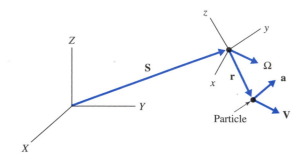

**Fig. 3.5**    Motion relative to a noninertial reference frame.

the acceleration of a particle relative to a fixed reference frame may be needed. It is given by

$$\mathbf{A} = \mathbf{a} \; + \; \frac{d^2\mathbf{S}}{dt^2} \; + \; 2\boldsymbol{\Omega} \times \mathbf{V} \; + \; \boldsymbol{\Omega} \times (\boldsymbol{\Omega} \times \mathbf{r}) + \frac{d\boldsymbol{\Omega}}{dt} \times \mathbf{r} \qquad (3.2.12)$$

| | acceleration of reference frame | Coriolis acceleration | normal acceleration | angular acceleration |

where $\mathbf{a}$ is given by Eq. 3.2.8, $d^2\mathbf{S}/dt^2$ is the acceleration of the observer's reference frame, $\mathbf{V}$ and $\mathbf{r}$ are the velocity and position vectors of the particle, respectively, in the observer's reference frame, and $\boldsymbol{\Omega}$ is the angular velocity of the observer's reference frame (see Fig. 3.5). Note that all vectors are written using the unit vectors of the $XYZ$-reference frame. For most engineering applications, reference frames attached to the earth yield $\mathbf{A} = \mathbf{a}$, since the other terms in Eq. 3.2.12 are often negligible with respect to $\mathbf{a}$. We may decide, however, to attach the $xyz$-reference frame to an accelerating device (a rocket), or to a rotating device (a sprinkler arm); then certain terms of Eq. 3.2.12 must be included along with $\mathbf{a}$ of Eq. 3.2.8.

If the acceleration of all fluid particles is given by $\mathbf{A} = \mathbf{a}$ in a selected reference frame, it is an *inertial* reference frame. If $\mathbf{A} \neq \mathbf{a}$, it is a *noninertial* reference frame. A reference frame that moves with constant velocity without rotating is an inertial reference frame. When analyzing flow about, for example, an airfoil moving at a constant speed, we attach the reference frame to the airfoil so that a steady flow is observed in that reference frame.

### 3.2.4    Angular Velocity and Vorticity

A fluid flow may be thought of as the motion of a collection of fluid particles. As a particle travels along it may rotate or deform. The rotation and deformation of the fluid particles are of particular interest in our study of fluid mechanics. There are certain flows, or regions of a flow, in which the fluid particles do not rotate; such flows are of special importance, particularly in flows around objects, and are referred to as **irrotational flows**. Flow outside a thin boundary layer on airfoils, outside the separated flow region around autos and other moving vehicles, in the

**Irrotational flows:** *Flows where the fluid particles do not rotate.*

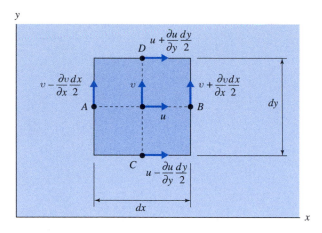

**Fig. 3.6**  Fluid particle occupying an infinitesimal parallelepiped at a particular instant.

Vorticity, 134

flow around submerged objects, and many other flows are examples of irrotational flows. Irrotational flows are extremely important.

Let us consider a small fluid particle that occupies an infinitesimal volume that has the $xy$-face as shown in Fig. 3.6. The **angular velocity** $\Omega_z$ about the $z$-axis is the average of the angular velocity of line segment $AB$ and line segment $CD$. The two angular velocities, counterclockwise being positive, are

**Angular velocity:** *The average velocity of two perpendicular line segments of a fluid particle.*

$$\Omega_{AB} = \frac{v_B - v_A}{dx}$$

$$= \left[ v + \frac{\partial v}{\partial x} \frac{dx}{2} - \left( v - \frac{\partial v}{\partial x} \frac{dx}{2} \right) \right] \bigg/ dx = \frac{\partial v}{\partial x} \tag{3.2.13}$$

$$\Omega_{CD} = -\frac{u_D - u_C}{dy}$$

$$= -\left[ u + \frac{\partial u}{\partial y} \frac{dy}{2} - \left( u - \frac{\partial u}{\partial y} \frac{dy}{2} \right) \right] \bigg/ dy = -\frac{\partial u}{\partial y} \tag{3.2.14}$$

Consequently, the angular velocity $\Omega_z$ of the fluid particle is

$$\Omega_z = \frac{1}{2} (\Omega_{AB} + \Omega_{CD})$$

$$= \frac{1}{2} \left( \frac{\partial v}{\partial x} - \frac{\partial u}{\partial y} \right) \tag{3.2.15}$$

If we had considered the $xz$-face, we would have found the angular velocity about the $y$-axis to be

$$\Omega_y = \frac{1}{2} \left( \frac{\partial u}{\partial z} - \frac{\partial w}{\partial x} \right) \tag{3.2.16}$$

**Table 3.1** The Substantial Derivative, Acceleration, and Vorticity in Cartesian, Cylindrical, and Spherical Coordinates

| *Substantial Derivative* | *Vorticity* |
|---|---|
| **Cartesian** | **Cartesian** |
| $\dfrac{D}{Dt} = u\dfrac{\partial}{\partial x} + v\dfrac{\partial}{\partial y} + w\dfrac{\partial}{\partial z} + \dfrac{\partial}{\partial t}$ | $\omega_x = \dfrac{\partial w}{\partial y} - \dfrac{\partial v}{\partial z}$  $\omega_y = \dfrac{\partial u}{\partial z} - \dfrac{\partial w}{\partial x}$  $\omega_z = \dfrac{\partial v}{\partial x} - \dfrac{\partial u}{\partial y}$ |
| **Cylindrical** | **Cylindrical** |
| $\dfrac{D}{Dt} = v_r\dfrac{\partial}{\partial r} + \dfrac{v_\theta}{r}\dfrac{\partial}{\partial \theta} + v_z\dfrac{\partial}{\partial z} + \dfrac{\partial}{\partial t}$ | $\omega_r = \dfrac{1}{r}\left(\dfrac{\partial v_z}{\partial \theta}\right) - \dfrac{\partial v_\theta}{\partial z}$  $\omega_\theta = \dfrac{\partial v_r}{\partial z} - \dfrac{\partial v_z}{\partial r}$  $\omega_z = \dfrac{1}{r}\left(\dfrac{\partial(rv_\theta)}{\partial r} - \dfrac{\partial v_r}{\partial \theta}\right)$ |
| **Spherical** | **Spherical** |
| $\dfrac{D}{Dt} = v_r\dfrac{\partial}{\partial r} + \dfrac{v_\theta}{r}\dfrac{\partial}{\partial \theta} + \dfrac{v_\phi}{r\sin\theta}\dfrac{\partial}{\partial \phi} + \dfrac{\partial}{\partial t}$ | $\omega_r = \dfrac{1}{r\sin\theta}\left[\dfrac{\partial}{\partial \theta}(v_\phi \sin\theta) - \dfrac{\partial v_\theta}{\partial \phi}\right]$  $\omega_\phi = \dfrac{1}{r}\left[\dfrac{\partial}{\partial r}(rv_\theta) - \dfrac{\partial v_r}{\partial \theta}\right]$ |
| *Acceleration* | $\omega_\theta = \dfrac{1}{r}\left[\dfrac{1}{\sin\theta}\dfrac{\partial v_r}{\partial \phi} - \dfrac{\partial}{\partial r}(rv_\phi)\right]$ |

**Cartesian**

$$a_x = \frac{\partial u}{\partial t} + u\frac{\partial u}{\partial x} + v\frac{\partial u}{\partial y} + w\frac{\partial u}{\partial z}$$

$$a_y = \frac{\partial v}{\partial t} + u\frac{\partial v}{\partial x} + v\frac{\partial v}{\partial y} + w\frac{\partial v}{\partial z}$$

$$a_z = \frac{\partial w}{\partial t} + u\frac{\partial w}{\partial x} + v\frac{\partial w}{\partial y} + w\frac{\partial w}{\partial z}$$

**Cylindrical**

$$a_r = \frac{\partial v_r}{\partial t} + v_r\frac{\partial v_r}{\partial r} + \frac{v_\theta}{r}\frac{\partial v_r}{\partial \theta} + v_z\frac{\partial v_r}{\partial z} - \frac{v_\theta^2}{r}$$

$$a_\theta = \frac{\partial v_\theta}{\partial t} + v_r\frac{\partial v_\theta}{\partial r} + \frac{v_\theta}{r}\frac{\partial v_\theta}{\partial \theta} + v_z\frac{\partial v_\theta}{\partial z} + \frac{v_r v_\theta}{r}$$

$$a_z = \frac{\partial v_z}{\partial t} + v_r\frac{\partial v_z}{\partial r} + \frac{v_\theta}{r}\frac{\partial v_z}{\partial \theta} + v_z\frac{\partial v_z}{\partial z}$$

**Spherical**

$$a_r = \frac{\partial v_r}{\partial t} + v_r\frac{\partial v_r}{\partial r} + \frac{v_\theta}{r}\frac{\partial v_r}{\partial \theta} + \frac{v_\phi}{r\sin\theta}\frac{\partial v_r}{\partial \phi} - \frac{v_\phi^2 + v_\theta^2}{r}$$

$$a_\theta = \frac{\partial v_\theta}{\partial t} + v_r\frac{\partial v_\theta}{\partial t} + \frac{v_\theta}{r}\frac{\partial v_\theta}{\partial \theta} + \frac{v_\phi}{r\sin\phi}\frac{\partial v_\theta}{\partial \phi} + \frac{v_r v_\theta - v_\phi^2 \cot\theta}{r}$$

$$a_\phi = \frac{\partial v_\phi}{\partial t} + v_r\frac{\partial v_\phi}{\partial r} + \frac{v_\theta}{r}\frac{\partial v_\phi}{\partial \theta} + \frac{v_\phi}{r\sin\theta}\frac{\partial v_\phi}{\partial \phi} + \frac{v_r v_\phi + v_\theta v_\phi \cot\theta}{r}$$

and the *yz*-face would provide us with the angular velocity about the *x*-axis:

$$\Omega_x = \frac{1}{2}\left(\frac{\partial w}{\partial y} - \frac{\partial v}{\partial z}\right) \tag{3.2.17}$$

These are the three components of the angular velocity vector. A cork placed in a water flow in a wide channel (the *xy*-plane) would rotate with an angular velocity about the *z*-axis, given by Eq. 3.2.15.

**Vorticity:** *Twice the angular velocity.*

It is common to define the **vorticity** $\omega$ to be twice the angular velocity; its three components are then

$$\omega_x = \frac{\partial w}{\partial y} - \frac{\partial v}{\partial z} \qquad \omega_y = \frac{\partial u}{\partial z} - \frac{\partial w}{\partial x} \qquad \omega_z = \frac{\partial v}{\partial x} - \frac{\partial u}{\partial y} \tag{3.2.18}$$

The vorticity components in cylindrical and spherical coordinates are included in Table 3.1 above.

An irrotational flow possesses no vorticity; the cork mentioned above would not rotate in an irrotational flow. We consider this special flow in Section 8.5.

The deformation of the particle of Fig. 3.6 is the rate of change of the angle that line segment $AB$ makes with line segment $CD$. If $AB$ is rotating with an angular velocity different from that of $CD$, the particle is deforming. The deformation is represented by the **rate-of-strain tensor**; its component $\epsilon_{xy}$ in the $xy$-plane is given by

**Rate-of-strain tensor:** *The rate at which deformation occurs.*

$$\epsilon_{xy} = \frac{1}{2}\left(\Omega_{AB} - \Omega_{CD}\right)$$
$$= \frac{1}{2}\left(\frac{\partial v}{\partial x} + \frac{\partial u}{\partial y}\right) \tag{3.2.19}$$

For the $xz$-plane and the $yz$-plane we have

$$\epsilon_{xz} = \frac{1}{2}\left(\frac{\partial w}{\partial x} + \frac{\partial u}{\partial z}\right) \qquad \epsilon_{yz} = \frac{1}{2}\left(\frac{\partial w}{\partial y} + \frac{\partial v}{\partial z}\right) \tag{3.2.20}$$

Observe that $\epsilon_{xy} = \epsilon_{yx}$, $\epsilon_{xz} = \epsilon_{zx}$, and $\epsilon_{yz} = \epsilon_{zy}$. By observation, we see that the rate-of-strain tensor is symmetric.

The fluid particle could also deform by being stretched or compressed in a particular direction. For example, if point $B$ of Fig. 3.6 is moving faster than point $A$, the particle would be stretching in the $x$-direction. This normal rate of strain is measured by

$$\epsilon_{xx} = \frac{u_B - u_A}{dx}$$
$$= \left[u + \frac{\partial u}{\partial x}\frac{dx}{2} - \left(u - \frac{\partial u}{\partial x}\frac{dx}{2}\right)\right]\bigg/dx = \frac{\partial u}{\partial x} \tag{3.2.21}$$

Similarly, in the $y$- and $z$-directions we would find that

$$\epsilon_{yy} = \frac{\partial v}{\partial y} \qquad \epsilon_{zz} = \frac{\partial w}{\partial z} \tag{3.2.22}$$

The symmetric rate-of-strain tensor can be displayed as

$$\epsilon_{ij} = \begin{pmatrix} \epsilon_{xx} & \epsilon_{xy} & \epsilon_{xz} \\ \epsilon_{xy} & \epsilon_{yy} & \epsilon_{yz} \\ \epsilon_{xz} & \epsilon_{yz} & \epsilon_{zz} \end{pmatrix} \tag{3.2.23}$$

where the subscripts $i$ and $j$ take on numerical values 1, 2, or 3. Then $\epsilon_{12}$ represents $\epsilon_{xy}$ in row 1 column 2.

We will see in Chapter 5 that the normal and shear stress components in a flow are related to the foregoing rate-of-strain components. In fact, in the one-dimensional flow of Fig. 1.6, the shear stress was related to $\partial u/\partial y$ with Eq. 1.5.5; note that $\partial u/\partial y$ is twice the rate-of-strain component given by Eq. 3.2.19 with $v = 0$.

## Example 3.1

The velocity field is given by $\mathbf{V} = 2x\hat{\mathbf{i}} - yt\hat{\mathbf{j}}$ m/s, where $x$ and $y$ are in meters and $t$ is in seconds. Find the equation of the streamline passing through $(2, -1)$ and a unit vector normal to the streamline at $(2, -1)$ at $t = 4$ s.

### Solution

The velocity vector is tangent to a streamline so that $\mathbf{V} \times d\mathbf{r} = 0$ (the cross product of two parallel vectors is zero). For the given velocity vector we have, at $t = 4$ s,

$$(2x\hat{\mathbf{i}} - 4y\hat{\mathbf{j}}) \times (dx\hat{\mathbf{i}} + dy\hat{\mathbf{j}}) = (2x\,dy + 4y\,dx)\hat{\mathbf{k}} = 0$$

where we have used $\hat{\mathbf{i}} \times \hat{\mathbf{j}} = \mathbf{k}, \hat{\mathbf{j}} \times \hat{\mathbf{i}} = -\mathbf{k}$, and $\hat{\mathbf{i}} \times \hat{\mathbf{i}} = 0$. Consequently,

$$2x\,dy = -4y\,dx \qquad \text{or} \qquad \frac{dy}{y} = -2\frac{dx}{x}$$

Integrate both sides:

$$\ln y = -2\ln x + \ln C$$

where we used $\ln C$ for convenience. This is written as

$$\ln y = \ln x^{-2} + \ln C = \ln(Cx^{-2})$$

Hence

$$x^2 y = C$$

At $(2, -1)$ $C = -4$, so that the streamline passing through $(2, -1)$ has the equation

$$x^2 y = -4$$

A normal vector is perpendicular to the streamline, hence the velocity vector, so that using $\hat{\mathbf{n}} = n_x\hat{\mathbf{i}} + n_y\hat{\mathbf{j}}$ we have at $(2, -1)$ and $t = 4$ s

$$\mathbf{V} \cdot \hat{\mathbf{n}} = (4\hat{\mathbf{i}} + 4\hat{\mathbf{j}}) \cdot (n_x\hat{\mathbf{i}} + n_y\hat{\mathbf{j}}) = 0$$

Using $\hat{\mathbf{i}} \cdot \hat{\mathbf{i}} = 1$ and $\hat{\mathbf{i}} \cdot \hat{\mathbf{j}} = 0$, this becomes

$$4n_x + 4n_y = 0 \qquad \therefore n_x = -n_y$$

Then, because $\hat{\mathbf{n}}$ is a unit vector, $n_x^2 + n_y^2 = 1$ and we find that

$$n_x^2 = 1 - n_x^2 \qquad \therefore n_x = \frac{\sqrt{2}}{2}$$

The unit vector normal to the streamline is written as

$$\hat{\mathbf{n}} = \frac{\sqrt{2}}{2}(\hat{\mathbf{i}} - \hat{\mathbf{j}})$$

# Example 3.2

A velocity field in a particular flow is given by $\mathbf{V} = 20y^2\hat{\mathbf{i}} - 20xy\hat{\mathbf{j}}$ m/s. Calculate the acceleration, the angular velocity, the vorticity vector, and any nonzero rate-of-strain components at the point $(1, -1, 2)$.

## Solution

We could use Eq. 3.2.9 and find each component of the acceleration, or we could use Eq. 3.2.8 and find a vector expression. Using Eq. 3.2.8, we have

$$\mathbf{a} = u\frac{\partial \mathbf{V}}{\partial x} + v\frac{\partial \mathbf{V}}{\partial y} + w\frac{\partial \mathbf{V}}{\partial z}^{\,0} + \frac{\partial \mathbf{V}}{\partial t}^{\,0}$$

$$= 20y^2(-20y\hat{\mathbf{j}}) - 20xy(40y\hat{\mathbf{i}} - 20x\hat{\mathbf{j}})$$

$$= -800xy^2\hat{\mathbf{i}} - 400(y^3 - x^2y)\hat{\mathbf{j}}$$

where we have used $u = 20y^2$ and $v = -20xy$, as given by the velocity vector. All particles passing through the point $(1, -1, 2)$ have the acceleration

$$\mathbf{a} = -800\hat{\mathbf{i}} \text{ m/s}^2$$

The angular velocity has two zero components:

$$\Omega_x = \frac{1}{2}\left(\frac{\partial w}{\partial y}^{\,0} - \frac{\partial v}{\partial z}^{\,0}\right) = 0, \quad \Omega_y = \frac{1}{2}\left(\frac{\partial u}{\partial z}^{\,0} - \frac{\partial w}{\partial x}^{\,0}\right) = 0$$

The non-zero $z$-component is, at the point $(1, -1, 2)$,

$$\Omega_z = \frac{1}{2}\left(\frac{\partial v}{\partial x} - \frac{\partial u}{\partial y}\right)$$

$$= \frac{1}{2}(-20y - 40y) = 30 \text{ rad/s}$$

The vorticity vector is twice the angular velocity vector:

$$\boldsymbol{\omega} = 2\Omega_z\hat{\mathbf{k}} = 60\hat{\mathbf{k}} \text{ rad/s}$$

The nonzero rate-of-strain components are

$$\epsilon_{xy} = \frac{1}{2}\left(\frac{\partial v}{\partial x} + \frac{\partial u}{\partial y}\right)$$

$$= \frac{1}{2}(-20y + 40y) = -10 \text{ rad/s}$$

$$\epsilon_{yy} = \frac{\partial v}{\partial y}$$

$$= -20x = -20 \text{ rad/s}$$

All other rate-of-strain components are zero.

## 3.3  CLASSIFICATION OF FLUID FLOWS

In this section we provide an overview of some of the aspects of fluid mechanics that are considered in more depth in subsequent sections and chapters. Although most of the notions presented here are redefined and discussed in detail later, it will be helpful at this point to introduce the general classification of fluid flows.

### 3.3.1   One-, Two-, and Three-Dimensional Flows

**Three-dimensional flow:**  *The velocity vector depends on three space variables.*

**Stagnation point:**  *The point where the fluid comes to rest.*

In the Eulerian description of motion the velocity vector, in general, depends on three space variables and time, that is, $\mathbf{V} = \mathbf{V}(x, y, z, t)$. Such a flow is a **three-dimensional flow**, because the velocity vector depends on three space coordinates. The solutions to problems in such a flow are very difficult and are beyond the scope of an introductory course. Even if the flow could be assumed to be steady [i.e., $\mathbf{V} = \mathbf{V}(x, y, z)$], it would remain a three-dimensional flow. A particular flow is shown in Fig. 3.7. This flow is normal to a plane surface; the fluid decelerates and comes to rest at the **stagnation point**. The velocity components, $u$, $v$, and $w$ depend on $x$, $y$ and $z$; that is, $u = u(x, y, z)$, $v = v(x, y, z)$ and $w = w(x, y, z)$.

**Two-dimensional flow:**  *The velocity vector is dependent on only two space variables.*

**Plane flow:**  *The velocity vector depends on the two coordinates x and y.*

Often a three-dimensional flow can be approximated as a two-dimensional flow. For example, the flow over a wide dam is three-dimensional because of the end conditions, but the flow in the central portion away from the ends can be treated as two-dimensional. In general, a **two-dimensional flow** is a flow in which the velocity vector depends on only two space variables. An example is a **plane flow**, in which the velocity vector depends on two spatial coordinates, $x$ and $y$, but not $z$, i.e., $\mathbf{V} = \mathbf{V}(x, y)$. In an axisymmetric flow, the velocity vector would depend in $r$ and $\theta$, i.e., $\mathbf{V} = \mathbf{V}(r, \theta)$; the flow in Fig. 3.7 will be considered two dimensional if described in a cylindrical coordinate system.

**One-dimensional flow:**  *The velocity vector depends on only one space variable.*

A **one-dimensional flow** is a flow in which the velocity vector depends on only one space variable. Such flows occur away from geometry changes in

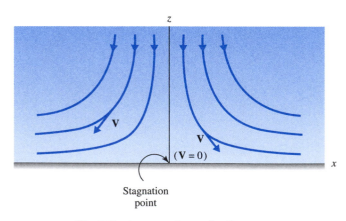

**Fig. 3.7**   A stagnation point flow.

long, straight pipes or between parallel plates, as shown in Fig. 3.8. The velocity in the pipe varies only with $r$, i.e., $u = u(r)$. The velocity between parallel plates varies only with the coordinate $y$, i.e., $u = u(y)$. Even if the flow is unsteady so that $u = u(y, t)$, as would be the situation during startup, the flow is one-dimensional.

The flows shown in Fig. 3.8 may also be referred to as **developed flows**; that is, the velocity profile does not vary with respect to the space coordinate in the direction of flow. This demands that the region of interest be a substantial distance from an entrance or a sudden change in geometry.

There are many engineering problems in fluid mechanics in which a flow field is simplified to a **uniform flow**: the velocity, and other fluid properties, are constant over the area, as in Fig. 3.9. This simplification is made when the velocity is essentially constant over the area, a rather common occurrence. Examples of such flows are relatively high speed flow in a pipe section, and flow in a stream. The average velocity may change from one section to another; the flow conditions depend only on the space variable in the flow direction. For large conduits, however, it may be necessary to consider hydrostatic variation in the pressure normal to the streamlines.

*Developed flows: The velocity profile does not vary with respect to the space coordinate in the direction of flow.*

*Uniform flow: The fluid properties are constant over the area.*

## 3.3.2 Viscous and Inviscid Flows

A fluid flow may be broadly classified as either a viscous flow or an inviscid flow. An **inviscid flow** is one in which viscous effects do not significantly influence the flow and are thus neglected. In a **viscous flow** the effects of viscosity are important and cannot be ignored.

To model an inviscid flow analytically, we can simply let the viscosity be zero; this will obviously make all viscous effects zero. It is more difficult to create an inviscid flow experimentally, because all fluids of interest (such as water and air)

*Inviscid flow: Viscous effects do not significantly influence the flow.*

*Viscous flow: The effects of viscosity are significant.*

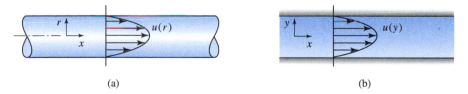

(a)  (b)

**Fig. 3.8** One-dimensional flow: (a) flow in a pipe; (b) flow between parallel plates.

**Fig. 3.9** Uniform velocity profiles.

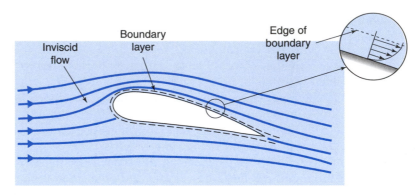

**Fig. 3.10**  Flow around an airfoil.

have viscosity. The question then becomes: Are there flows of interest in which the viscous effects are negligibly small? The answer is "yes, if the shear stresses in the flow are small and act over such small areas that they do not significantly affect the flow field." This statement is very general, of course, and it will take considerable analysis to justify the inviscid flow assumption.

Based on experience, it has been found that the primary class of flows, which can be modeled as inviscid flows, is **external flows**, that is, a flow which exists exterior to bodies. Inviscid flows are of primary importance in flows around streamlined bodies, such as flow around an airfoil or a hydrofoil. Any viscous effects that may exist are confined to a thin layer, called a **boundary layer**, which is attached to the boundary, such as that shown in Fig. 3.10; the velocity in a boundary layer is always zero at a fixed wall, a result of viscosity. For many flow situations, boundary layers are so thin that they can simply be ignored when studying the gross features of a flow around a streamlined body. For example, the inviscid flow solution provides an excellent prediction to the flow around the airfoil, except inside the boundary layer and possibly near the trailing edge. Inviscid flow is also encountered in contractions inside piping systems and in short regions of internal flows where viscous effects are negligible.

Viscous flows include the broad class of internal flows, such as flows in pipes and conduits and in open channels. In such flows viscous effects cause substantial "losses" and account for the huge amounts of energy that must be used to transport oil and gas in pipelines. The no-slip condition resulting in zero velocity at the wall and the resulting shear stresses lead directly to these losses.

**External flows:**  *Flows that exist exterior to bodies.*

**Boundary layer:**  *A thin layer attached to the boundary in which viscous effects are concentrated.*

**KEY CONCEPT**  *Inviscid flow provides an excellent prediction to the flow around an airfoil.*

### 3.3.3  Laminar and Turbulent Flows

A viscous flow can be classified as either a laminar flow or a turbulent flow. In a **laminar flow** the fluid flows with no significant mixing of neighboring fluid particles. If dye were injected into the flow, it would not mix with the neighboring fluid except by molecular activity; it would retain its identity for a relatively long period of time. Viscous shear stresses always influence a laminar flow. The flow may be highly time dependent, due to the erratic motion of a piston as shown by the output of a velocity probe in Fig. 3.11a, or it may be steady, as shown in Fig. 3.11b.

**Laminar flow:**  *A flow with no significant mixing of particles but with significant viscous shear stresses.*

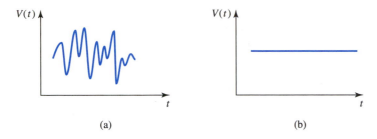

**Fig. 3.11**   Velocity as a function of time in a laminar flow: (a) unsteady flow; (b) steady flow.

In a **turbulent flow** fluid motions vary irregularly so that quantities such as velocity and pressure show a random variation with time and space coordinates. The physical quantities are often described by statistical averages. In this sense we can define a "steady" turbulent flow as a flow in which the time-average physical quantities do not change in time. Figure 3.12 shows instantaneous velocity measurements in an unsteady and a steady turbulent flow. A dye injected into a turbulent flow would mix immediately by the action of the randomly moving fluid particles; it would quickly lose its identity in this diffusion process.

A laminar flow and a turbulent flow can be observed by performing a simple experiment with a water faucet. Turn the faucet on so the water flows out very slowly as a silent stream. This is laminar flow. Open the faucet slowly and observe the flow becoming turbulent. Note that a turbulent flow develops with a relatively small flow rate.

The reason why a flow can be laminar or turbulent has to do with what happens to a small flow disturbance, a perturbation to the velocity components. A flow disturbance can either increase or decrease in size. If a flow disturbance in a laminar flow increases (i.e., the flow is unstable), the flow may become turbulent; if the disturbance decreases, the flow remains laminar. In certain situations the flow may develop into a different laminar flow, as is the case between concentrically rotating cylinders shown in Fig. 3.13. At low rotational speed the flow would be in simple circles. But at a sufficiently high speed the flow becomes unstable and the vortices suddenly appear; it is a much more complex laminar flow called Taylor-Conette flow.

**Turbulent flow:** *Flow varies irregularly so that flow quantities show random variation.*

**KEY CONCEPT**   *A dye injected into a turbulent flow would mix immediately.*

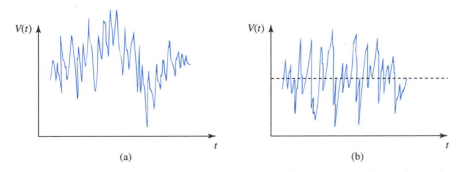

**Fig. 3.12**   Velocity as a function of time in a turbulent flow: (a) unsteady flow; (b) "steady" flow.

Whether a flow is laminar or turbulent depends on three physical parameters describing the flow conditions. The first parameter is a length scale of the flow field, such as the thickness of a boundary layer or the diameter of a pipe. If this length scale is sufficiently large, a flow may be turbulent. The second parameter is a velocity scale such as a spatial average of the velocity; for a large enough velocity the flow may be turbulent. The third parameter is the kinematic viscosity; for a small enough viscosity the flow may be turbulent.

The three parameters can be combined into a single parameter that can serve as a tool to predict the flow regime. This quantity is the **Reynolds number**, named after Osborne Reynolds (1842–1912), a dimensionless parameter, defined as

$$\mathrm{Re} = \frac{VL}{\nu} \tag{3.3.1}$$

where $L$ and $V$ are a characteristic length and velocity, respectively, and $\nu$ is the kinematic viscosity; for example, in a pipe flow $L$ would be the pipe diameter and $V$ would be the average velocity. If the Reynolds number is relatively small, the flow is laminar as shown in Figs. 3.13 and 3.14; if it is large, the flow is turbulent. This is more precisely stated by defining a **critical Reynolds number**, $\mathrm{Re}_{\mathrm{crit}}$, so that the flow is laminar if $\mathrm{Re} < \mathrm{Re}_{\mathrm{crit}}$. For example, in a flow inside a rough-walled pipe it is found that $\mathrm{Re}_{\mathrm{crit}} \approx 2000$. This is the minimum critical Reynolds number and is used for most engineering applications. If the pipe wall is extremely smooth and free of vibration, the critical Reynolds number can be

**Fig. 3.13** Laminar flow between rotating cylinders. A secondary flow occurs as regularly spaced toroidal vortices. ("Steady supercritical Taylor vortex flow," by Burkhalter and Koschmieder. From Journal of Fluid Mechanics vol. 58, pp. 547–560 (1973). 2006 © Cambridge Journals, reproduced with permission.)

 Taylor cells, 24

**Reynolds number:**
*Parameter combining a length scale, a velocity scale, and the kinematic viscosity into*

$$\mathrm{Re} = VL/\nu$$

**Critical Reynolds number:**
*The number above which a primary laminar flow ceases to exist.*

**Fig. 3.14  Streamlines around a semicircular arc.** At this Reynolds number of 0.031 the centers of the pair of eddies in the cavity are separated by 0.52 diameter, in good agreement with an analytical solution. Aluminum powder dispersed in glycerine is illuminated by a slit of light. (Photograph by Sadatoshi Taneda. From Album of Fluid Motion, 1982, The Parabolic Press, Stanford, California.)

**Fig. 3.15**   Velocity versus time signal from a velocity probe in an intermittent flow.

increased as the fluctuation level in the flow is decreased; values in excess of 40 000 have been measured. The critical Reynolds number is different for every geometry, e.g., it is 1500 for flow between parallel plates using the average velocity and the distance between the plates.

The flow can also be intermittently turbulent and laminar; this is called an *intermittent flow*. This phenomenon can occur when the Reynolds number is close to $\mathrm{Re_{crit}}$. Figure 3.15 shows the output of a velocity probe for such a flow.

In a boundary layer that exists on a flat plate, due to a constant-velocity fluid stream, as shown in Fig. 3.16, the length scale changes with distance from the upstream edge. A Reynolds number is calculated using the length $x$ as the characteristic length. For a certain $x_T$, Re becomes $\mathrm{Re_{crit}}$ and the flow undergoes transition from laminar to turbulent. For a smooth rigid plate in a uniform flow with a low free-stream fluctuation level, values as high as $\mathrm{Re_{crit}} = 10^6$ have been observed. In most engineering applications we assume a rough wall, or high free-stream fluctuation level, with an associated critical Reynolds number of approximately $3 \times 10^5$.

It is not appropriate to refer to an inviscid flow as laminar or as turbulent. The inviscid flow of Fig. 3.10 is often called the **free stream**. The free stream can be irrotational or it can possess vorticity; most often it is irrotational.

 Reynolds number, 524

 Pipe flow, 202

**KEY CONCEPT**   *In most applications, we assume a critical Reynolds number of $3 \times 10^5$ in flow on a flat plate.*

**Free stream:**   *The inviscid flow outside the boundary layer in an external flow.*

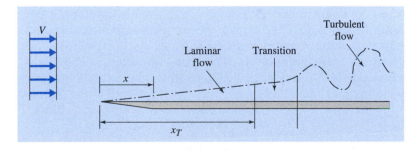

**Fig. 3.16**   Boundary layer flow on a flat plate.

## Example 3.3

The 2-cm-diameter pipe of Fig. E3.3 is used to transport water at 20°C. What is the maximum average velocity that may exist in the pipe for which laminar flow is guaranteed?

Water @ 20°C

**Fig. E3.3**

### Solution

The kinematic viscosity is found in Appendix B to be $\nu = 10^{-6}$ m$^2$/s. Using a Reynolds number of 2000 so that a laminar flow is guaranteed, we find that

$$V = \frac{2000\nu}{D}$$

$$= \frac{2000 \times 10^{-6}}{0.02} = 0.1 \text{ m/s}$$

This average velocity is quite small. Velocities this small are not usually encountered in actual situations; hence laminar flow is seldom of engineering interest except for specialized topics such as lubrication. Most internal flows are turbulent flows, and thus the study of turbulence gains much attention.

### 3.3.4   Incompressible and Compressible Flows

**Incompressible flow:** *The density of each fluid particle remains constant.*

The last major classification of fluid flows to be considered in this chapter separates flows into incompressible and compressible flows. An **incompressible flow** exists if the density of each fluid particle remains relatively constant as it moves through the flow field, that is,

$$\frac{D\rho}{Dt} = 0 \qquad (3.3.2)$$

**KEY CONCEPT** *Constant density is more restrictive than incompressibility.*

This does not demand that the density is everywhere constant. If the density is constant, then obviously, the flow is incompressible, but that would be a more restrictive condition. Atmospheric flow, in which $\rho = \rho(z)$, where $z$ is vertical, and flows that involve adjacent layers of fresh and salt water, as happens when rivers enter the ocean, are examples of incompressible flows in which the density varies.

**Mach number:** *A parameter in a gas flow defined as M = V/c.*

In addition to liquid flows, low-speed gas flows, such as the atmospheric flow referred to above, are also considered to be incompressible flows. The **Mach number**, named after Ernst Mach (1838–1916), is defined as

$$M = \frac{V}{c} \qquad (3.3.3)$$

where $V$ is the gas speed and the wave speed $c = \sqrt{kRT}$. Equation 3.3.3 is useful in deciding whether a particular gas flow can be studied as an incompressible flow. If M < 0.3, density variations are at most 3% and the flow is assumed to be incompressible; for standard air this corresponds to a velocity below about 100 m/s. If M > 0.3, the density variations influence the flow and compressibility effects should be accounted for; such flows are **compressible flows** and are considered in Chapter 9.

*Compressible flow:  Density variations influence the flow.*

Incompressible gas flows include atmospheric flows, the aerodynamics of landing and takeoff of commercial aircraft, heating and air-conditioning airflows, flow around automobiles and through radiators, and the flow of air around buildings, to name a few. Compressible flows include the aerodynamics of high-speed aircraft, airflow through jet engines, steam flow through the turbine in a power plant, airflow in a compressor, and the flow of the air-gas mixture in an internal combustion engine.

## 3.4  THE BERNOULLI EQUATION

In this section we present an equation that is probably used more often in fluid flow applications than any other equation. It is also often misused; it is thus important to understand its limitations. Its limitations are a result of several assumptions made in the derivation. One of the assumptions is that viscous effects are neglected. In other words, in view of Eq. 1.5.5, shear stresses introduced by velocity gradients are not taken into consideration. These stresses are often very small compared with pressure differences in the flow field. Locally, these stresses have little effect on the flow field and the assumption is justified. However, over long distances or in regions of high-velocity gradients, these stresses may affect the flow conditions so that viscous effects must be included.

**KEY CONCEPT**  *Shear stresses are often very small compared with pressure differences.*

The derivation of this important equation, the Bernoulli equation, starts with the application of Newton's second law to a fluid particle. Let us use an infinitesimal cylindrical particle positioned as shown in Fig. 3.17, with length $ds$ and cross-sectional area $dA$. The forces acting on the particle are the pressure forces and the weight, as shown. Summing forces in the direction of motion, the $s$-direction, there results

$$p \, dA - \left( p + \frac{\partial p}{\partial s} \, ds \right) dA - \rho g \, ds \, dA \cos \theta = \rho \, ds \, dA \, a_s \qquad (3.4.1)$$

where $a_s$ is the acceleration of the particle in the $s$-direction. It is given by[1]

$$a_s = V \frac{\partial V}{\partial s} + \frac{\partial V}{\partial t} \qquad (3.4.2)$$

where $\partial V / \partial t = 0$ since we will assume steady flow. Also, we see that

$$dh = ds \cos \theta = \frac{\partial h}{\partial s} \, ds \qquad (3.4.3)$$

so that

$$\cos \theta = \frac{\partial h}{\partial s} \qquad (3.4.4)$$

_____

[1] This can be verified by considering Eq. 3.2.9a, assuming that $v = w = 0$. Think of the $x$-direction being tangent to the streamline at the instant shown, so that $u = V$.

**Fig. 3.17**    Particle moving along a streamline.

Then, after dividing by $ds\ dA$, and using the above equations for $a_s$ and $\cos\theta$, Eq 3.4.1 takes the form

$$-\frac{\partial p}{\partial s} - \rho g \frac{\partial h}{\partial s} = \rho V \frac{\partial V}{\partial s} \tag{3.4.5}$$

Now, we assume constant density and note that $V\partial V/\partial s = \partial(V^2/2)/\partial s$; then we can write Eq. 3.4.5 as

$$\frac{\partial}{\partial s}\left(\frac{V^2}{2} + \frac{p}{\rho} + gh\right) = 0 \tag{3.4.6}$$

Bernoulli's equation, 910      This is satisfied if, along the streamline,

$$\frac{V^2}{2} + \frac{p}{\rho} + gh = \text{const} \tag{3.4.7}$$

where the constant may have a different value on a different streamline. Between two points on the same streamline,

$$\frac{V_1^2}{2} + \frac{p_1}{\rho} + gh_1 = \frac{V_2^2}{2} + \frac{p_2}{\rho} + gh_2 \tag{3.4.8}$$

This is the well-known *Bernoulli equation*, named after Daniel Bernoulli (1700–1782). Note the five assumptions:

-Inviscid flow (no shear stresses)
-Steady flow ($\partial V/\partial t = 0$)
-Along a streamline ($a_s = V\partial V/\partial s$)
-Constant density ($\partial\rho/\partial s = 0$)
-Inertial reference frame ($\mathbf{A} = \mathbf{a}$ as in Eq. 3.2.12)

If Eq. 3.4.8 is divided by $g$, the dimension of each term is length and Bernoulli's equation takes the alternate form

$$\frac{V_1^2}{2g} + \frac{p_1}{\gamma} + h_1 = \frac{V_2^2}{2g} + \frac{p_2}{\gamma} + h_2 \qquad (3.4.9)$$

**KEY CONCEPT** *The total pressure is* $p + \rho V^2/2$.

The sum of the two terms $(p/\gamma + h)$ is called the *piezometric head* and the sum of the three terms the *total head*. The pressure $p$ is often referred to as **static pressure**, and the sum of the two terms

$$p + \rho \frac{V^2}{2} = p_T \qquad (3.4.10)$$

**Static pressure:** *The pressure* $p$, *usually expressed as gage pressure.*

is called the *total pressure* $p_T$ or **stagnation pressure**, the pressure at a stagnation point (see Fig. 3.7) in the flow.

The static pressure in a pipe can be measured simply by installing a so-called **piezometer**, shown[2] in Fig. 3.18a. A device, known as a **pitot probe**, sketched in Fig. 3.18b, is used to measure the total pressure in a fluid flow. Point 2 just inside the pitot tube is a stagnation point; the velocity there is zero. The difference between the readouts can be used to determine the velocity at point 1. A **pitot-static probe** is also used to measure the difference between total and static pressure with one probe (Fig. 3.18c). The velocity at point 1 (using the readings of the piezometer and pitot probes, or the reading from the pitot-static probe) can be determined by applying the Bernoulli equation between points 1 and 2:

**Stagnation pressure:** *The pressure that exists at a stagnation point.*

**Piezometer:** *Gauge designed to measure static pressure.*

**Pitot probe:** *Gauge designed to measure total pressure.*

**Pitot-static probe:** *Gauge designed to measure the difference between total and static pressure.*

$$\frac{V_1^2}{2g} + \frac{p_1}{\gamma} = \frac{p_2}{\gamma} \qquad (3.4.11)$$

where we have assumed point 2 to be a stagnation point so that $V_2 = 0$. This gives

$$V_1 = \sqrt{\frac{2}{\rho}(p_2 - p_1)} \qquad (3.4.12)$$

We will find many uses for Bernoulli's equation in our study of fluids. We must be careful, however, never to use it in an unsteady flow or if viscous effects are significant (the primary reasons for making Bernoulli's equation inapplicable). We must also never confuse Bernoulli's equation with the energy equation; they are independent equations as Example 3.6 illustrates.

**KEY CONCEPT** *Never confuse Bernoulli's equations with the energy equation.*

**Fig. 3.18**   Pressure probes: (a) piezometer; (b) pitot probe; (c) pitot-static probe.

---

[2] When drilling the hole in the wall necessary for the piezometer, burrs are often formed on the inner surface. It is important that such burrs be removed since they may cause errors as high as 30% in the pressure reading.

**Fig. 3.19**   Internal inviscid flows: (a) flow through a contraction; (b) flow from a plenum.

The Bernoulli equation can be used to determine how high the water from a fireman's hose will reach, to find the pressure on the surface of a low-speed air-foil,[3] and to find the wind force on the window in a house. These examples are all external flows, flows around objects submerged in the fluid.

Another class of problems where inviscid flow can be assumed and where the Bernoulli equation finds frequent application involves internal flows over relatively short distances, for example, flow through a contraction, as shown in Fig. 3.19a, or flow from a plenum, as shown in Fig. 3.19b. For a given velocity profile entering the short contraction, the pressure drop ($p_1 - p_2$) and the velocity profile at section 2 can be approximated assuming an inviscid flow. Viscous effects are typically very small and require substantial distances and areas over which to operate in order to become significant; so in situations such as those shown in Fig. 3.19, viscous effects can often be neglected.

Inviscid flow does not always give a good approximation to the actual flow that exists around a body. Consider the inviscid flow around the sphere (or cylinder), shown in Fig. 3.20. A stagnation point where $V = 0$ exists at both the front and the back of the sphere. Bernoulli's equation predicts a maximum pressure at the stagnation points $A$ and $C$ because the velocity is zero at such points. A maximum velocity, and thus a minimum pressure, would exist at point $B$. In

Flow over a cylinder, 95

Flow over a cylinder, 190

**Fig. 3.20**   Flow around a sphere: (a) inviscid flow; (b) actual flow.

---

[3] To consider the flow around an aircraft as a steady flow, we simply make the aircraft stationary and move the air, as is done in model studies using a wind tunnel. The pressures and forces remain unchanged.

the inviscid flow of part (a), the fluid flowing from $B$ to $C$ must flow from the low-pressure region near $B$ to the high-pressure region near $C$. In the actual flow there exists a thin boundary layer in which the velocity goes to zero at the surface of the sphere. This slow-moving fluid near the boundary does not have sufficient momentum to move into the higher-pressure region near $C$; the result is that the fluid *separates* from the boundary—the boundary streamline leaves the boundary—creating a **separated region**, a region of recirculating flow, as shown in the actual flow sketched in part (b). The pressure does not increase but remains relatively low over the rear part of the sphere. The high pressure that exists near the front stagnation point is never recovered on the rear of the sphere, resulting in a relatively large drag force in the direction of flow. A similar situation occurs in the flow around an automobile.

The flow on the front of the sphere is well approximated by an inviscid flow, but it is obvious that the flow over the rear of the sphere deviates radically from an inviscid flow. Viscous effects in the boundary layer have led to a separated flow, a phenomenon that is often undesirable. For example, separated flow on an airfoil is called **stall** and must never occur, except on the wings of special stunt planes. On the blades of a turbine separated flows lead to substantially reduced efficiency. The air deflector on the roof of the cab of a semitruck reduces the separated region, thereby reducing drag and fuel consumption.

If viscous effects are negligible in a steady liquid flow, we can use Bernoulli's equation to locate points of possible *cavitation*. This condition occurs when the local pressure becomes equal to the vapor pressure of the liquid. It is to be avoided, if at all possible, because of damage to solid surfaces or because the vaporized liquid may cause devices to not operate effectively. Figure 3.21 shows cavitating flow just downstream of a contraction in a pipe. At the point where cavitation occurs, small vapor bubbles are generated, and these bubbles collapse when they enter a higher-pressure region. The collapse is accompanied by very large local pressures that last for only a small fraction of a second. These pressure spikes may reach a wall, where they can, after repeated applications, result in significant damage.

An important observation regarding pressure changes in a fluid needs to be made regarding entrances and exits to a pipe or conduit. Consider a flow from a reservoir through a pipe, as sketched in Fig. 3.22. At the entrance the streamlines are curved and the pressure is not constant across section 1, so the pressure at section 1 cannot be assumed to be uniform. At the exit, however, the streamlines

 Flow past a cylinder, 166

**Separated region:** *A region of recirculating flow due to the fluid separating from the boundary.*

**KEY CONCEPT** *Pressure remains relatively low over the rear part of a sphere.*

**KEY CONCEPT** *The flow on the front of a sphere is approximated by an inviscid flow.*

**Stall:** *Separated flow on an airfoil.*

**KEY CONCEPT** *Cavitation occurs when the local pressure equals the vapor pressure.*

**KEY CONCEPT** *At a pipe entrance, the streamlines are curved and the pressure is not constant across the inlet area.*

(a)

(b)

**Fig. 3.21**   Cavitation in a nozzle, with water flowing at a velocity of 15 m/s: (a) incandescent lamp, exposure time $\frac{1}{30}$ s; (b) strobe light exposure time 5 $\mu$s. (Photograph courtesy of the Japan Society of Mechanical Engineers and Pergamon Press.)

**Fig. 3.22**    Exit flow into the atmosphere.

are straight, so that no acceleration exists normal to the streamlines; hence the pressure forces acting on the ends of the small cylindrical control volume must be equal. We write this as $p_2 = p_{\text{atm}}$ or $p_2 = 0$ gage pressure.

In Fig. 3.17 we summed forces on the fluid element along the streamline and derived Bernoulli's equation. We can gain additional insight into the pressure field if we sum forces normal to the streamline. Let us consider the fluid particle to be a parallelepiped with thickness $dn$ in the $n$-direction and area $dA_s$ on the side with length $ds$. Applying Newton's second law in the $n$-direction provides

$$p\,dA_s - \left(p + \frac{\partial p}{\partial n}\, dn\right)dA_s = \rho\,dA_s dn \frac{V^2}{R} \tag{3.4.13}$$

where we have neglected the weight since we are not intending to integrate over significant distances. We have assumed the acceleration in the normal direction to be $V^2/R$, where $R$ is the radius of curvature in this plane flow (in a three-dimensional flow there would be a principal radius of curvature and a binormal radius of curvature). Equation (3.4.13) reduces to

$$-\frac{\partial p}{\partial n} = \rho \frac{V^2}{R} \tag{3.4.14}$$

From this equation we can qualitatively describe how the pressure changes normal to a streamline (Bernoulli's equation predicts the pressure changes along a streamline). If we replace $\partial p/\partial n$ with $\Delta p/\Delta n$, the incremental pressure change $\Delta p$ over the short distance $\Delta n$ normal to the streamline is given by

$$\Delta p = -\rho \frac{V^2}{R} \Delta n \tag{3.4.15}$$

**KEY CONCEPT** *The pressure decreases in the n-direction.*

This says that the pressure decreases in the $n$-direction; this decrease is directly proportional to $\rho$ and $V^2$ and inversely proportional to $R$. Consequently, a tornado, with $p = 0$ outside the tornado, will have a very low pressure at its center where $R$ is relatively small and $V$ is quite large.

In Fig. 3.22 the pressure would be relatively low at the corner of section 1 and relatively high at the center of section 1. Such qualitative descriptions can be quite helpful in understanding fluid flow behavior.

## Example 3.4

The wind reaches a speed of 144 km/h in a storm. Calculate the force acting on the 0.9 m × 1.8 m window of Fig. E3.4 facing the storm. The window is in a high-rise building, so the wind speed is not reduced due to ground effects. Use $\rho = 1.27$ kg/m$^3$.

V = 65 mph

Window

**Fig. E3.4**

### Solution

The window facing the storm will be in a stagnation region where the wind speed is brought to zero. Working with gage pressures, the pressure $p$ upstream in the wind is zero. The velocity $V$ must have units of ft/sec. It is

$$V = 144\,\frac{\text{km}}{\text{hr}} \times \frac{1\ \text{hr}}{3600\ \text{s}} \times \frac{1000\ \text{m}}{1\ \text{km}} = 40\ \text{m/s}$$

Bernoulli's equation can be used in this situation since we can neglect viscous effects, and steady flow occurs along a streamline at constant density (air is incompressible at speeds below about 360 km/h). We calculate the pressure on the window selecting state 1 in the free stream and state 2 on the window, as follows:

$$\frac{V_1^2}{2g} + \frac{p_1}{\gamma}_{\,0} + h_1 = \frac{V_2^2}{2g}_{\,0} + \frac{p_2}{\gamma} + h_2$$

$$\therefore p_2 = \frac{\rho V_1^2}{2}$$

$$= \frac{1.27\ \text{kg/m}^3 \times (40)^2\,\text{m}^2/\text{s}^2}{2000} = 1.016\ \text{kPa}$$

where we have used $\gamma = \rho g$, $h_2 = h_1$, $p_1 = 0$, and $V_2 = 0$. Multiply by the area and find the force to be

$$F = pA$$

$$= 1.016 \times 0.9 \times 1.8 = 1.646\ \text{kN}$$

## Example 3.5

The static pressure head in an air pipe (Fig. E3.5) is measured with a piezometer as 16 mm of water. A pitot probe indicates 24 mm of water. Calculate the velocity of the 20°C air. Also, calculate the Mach number and comment as to the compressibility of the flow.

**Fig. E3.5**

### Solution

Bernoulli's equation is applied between two points on the streamline that terminates at the stagnation point of the pitot probe. Point 1 is upstream and $p_2$ is the total pressure at point 2; then, with no elevation change,

$$\frac{V_1^2}{2g} + \frac{p_1}{\gamma} = \frac{p_T}{\gamma}$$

The pressure measured with the piezometer is $p_1 = \gamma h = 9810 \times 0.016 = 157$ Pa. We use the ideal gas law to calculate the density:

$$\rho = \frac{p}{RT}$$

$$= \frac{(157 + 101\,000)\text{Pa}}{287\ \text{kJ/kg·K} \times (273 + 20)\text{K}} = 1.203\ \text{kg/m}^3$$

where standard atmospheric pressure, which is 101 000 Pa (if no elevation is given, assume standard conditions), is added since absolute pressure is needed in the ideal gas law. The units are checked by using Pa = N/m² and J = N·m. The velocity is then

$$V_1 = \sqrt{\frac{2}{\rho}(p_T - p_1)}$$

$$= \sqrt{\frac{2(0.024 \times 9810 - 157)\ \text{Pa}}{1.203\ \text{kg/m}^3}} = 11.42\ \text{m/s}$$

where the units can be verified by using kg = N·s²/m. To find the Mach number, we must calculate the speed of sound. From Eq. 1.7.17 it is

$$c = \sqrt{kRT}$$

$$= \sqrt{1.4 \times 287\ \text{kJ/kg·K} \times 293\ \text{K}} = 343\ \text{m/s}$$

The Mach number is then

$$M = \frac{V}{c} = \frac{11.44}{343} = 0.0334$$

Obviously, the flow can be assumed to be incompressible since M < 0.3. The velocity would have to be much higher before compressibility would be significant.

# Example 3.6

Bernoulli's equation, in the form of Eq. 3.4.8, looks very much like the energy equation developed in thermodynamics for a control volume. Discuss the differences between the two equations.

### Solution

From thermodynamics we recall that the steady-flow energy equation for a control volume with one inlet and one outlet takes the form

$$\dot{Q} - \dot{W}_s = \dot{m}\left(\frac{V_2^2}{2} + \frac{p_2}{\rho_2} + \tilde{u}_2 + gz_2\right) - \dot{m}\left(\frac{V_1^2}{2} + \frac{p_1}{\rho_1} + \tilde{u}_1 + gz_1\right)$$

This becomes, after dividing through by $g$,

$$\frac{V_2^2}{2g} + \frac{p_2}{\gamma} + z_2 = \frac{V_1^2}{2g} + \frac{p_1}{\gamma} + z_1$$

where we have made the following assumptions:

No heat transfer ($\dot{Q} = 0$)

No shaft work ($\dot{W}_s = 0$)

No temperature change ($\tilde{u}_2 = \tilde{u}_1$, i.e., no losses due to shear stresses)

Uniform velocity profiles at the two sections

Steady flow

Constant density ($\gamma_2 = \gamma_1$)

Even though several of these assumptions are the same as those made in the derivation of the Bernoulli equation (steady flow, constant density, and no shear stress), we must not confuse the two equations; the Bernoulli equation is derived from Newton's second law and is valid along a streamline, whereas the energy equation is derived from the first law of thermodynamics and is valid between two sections in a fluid flow. The energy equation can be used across a pump to determine the horsepower required to provide a particular pressure rise; the Bernoulli equation can be used along a stagnation streamline to determine the pressure at a stagnation point, a point where the velocity is zero. The equations are quite different, and just because the energy equation degenerates to the Bernoulli equation for particular situations, the two should not be used out of context.

# Example 3.7

Explain why a burr on the upstream side of the piezometer opening of Fig. 3.18a will result in a low reading of the pressure.

$p_{high}$   $p_{low}$

Flow

**Fig. E3.7**

*(continued)*

**Solution**

A burr on the upstream side of the piezometer opening would result in a flow in the vicinity of the burr somewhat like that shown in Fig. E3.7. A streamline pattern would develop so that a relatively high pressure would occur on the upstream side of the burr and a relatively low pressure on the downstream side at the opening of the piezometer tube. Consequently, since the center of the curvature of the streamline is in the vicinity of the opening, a lower reading of the pressure would be recorded. If the burr were on the downstream side of the opening, a high pressure reading would be recorded.

## 3.5 SUMMARY

The Eulerian description of motion was used to express the acceleration as

$$\mathbf{a} = u\,\frac{\partial \mathbf{V}}{\partial x} + v\,\frac{\partial \mathbf{V}}{\partial y} + w\,\frac{\partial \mathbf{V}}{\partial z} + \frac{\partial \mathbf{V}}{\partial t} \tag{3.5.1}$$

Fluid motion may result in fluid particles being rotated and/or deformed. For a flow in the $xy$-plane a particle would rotate with angular velocity

$$\Omega_z = \frac{1}{2}\left(\frac{\partial v}{\partial x} - \frac{\partial u}{\partial y}\right) \tag{3.5.2}$$

and deform with

$$\epsilon_{xx} = \frac{\partial u}{\partial x}, \qquad \epsilon_{yy} = \frac{\partial v}{\partial y}, \qquad \epsilon_{xy} = \frac{1}{2}\left(\frac{\partial v}{\partial x} + \frac{\partial u}{\partial y}\right) \tag{3.5.3}$$

Fluid flows are classified as steady or unsteady; viscous or inviscid; laminar, turbulent, or free stream; and incompressible or compressible. Any of these can be uniform, one-, two-, or three-dimensional flows. Experience and practice are needed to appropriately classify a particular flow of interest. It is only for the simpler flows (e.g., a steady, laminar, incompressible, one-dimensional flow) that we hope to obtain a relatively simple solution.

Finally, the famous Bernoulli equation

$$\frac{V_1^2}{2g} + \frac{p_1}{\gamma} + h_1 = \frac{V_2^2}{2g} + \frac{p_2}{\gamma} + h_2 \tag{3.5.4}$$

was presented for a steady, inviscid, constant-density flow along a streamline in an inertial reference frame. The estimate of the pressure change normal to a streamline,

$$\Delta p = -\rho\,\frac{V^2}{R}\,\Delta n \tag{3.5.5}$$

was also obtained.

# FUNDAMENTALS OF ENGINEERING EXAM REVIEW PROBLEMS

**3.1**  Determine the unit vector normal to the streamline at a point where $\mathbf{V} = 3\hat{\mathbf{i}} - 4\hat{\mathbf{j}}$ in a plane flow.
   (A) $0.6\hat{\mathbf{i}} + 0.8\hat{\mathbf{j}}$   (B) $-0.6\hat{\mathbf{i}} + 0.8\hat{\mathbf{j}}$
   (C) $0.8\hat{\mathbf{i}} - 0.6\hat{\mathbf{j}}$   (D) $0.8\hat{\mathbf{i}} + 0.6\hat{\mathbf{j}}$

**3.2**  A velocity field is given by $\mathbf{V} = 2xy\hat{\mathbf{i}} - y^2\hat{\mathbf{j}}$ m/s. The magnitude of the acceleration at $(-1\text{ m}, 2\text{ m})$ is nearest:
   (A) $11.21$ m/s$^2$   (B) $14.69$ m/s$^2$
   (C) $17.89$ m/s$^2$   (D) $1.2$ m/s$^2$

**3.3**  The velocity shown in Fig. P3.3 is given by $V(x) = 10/(4 - x)^2$ m/s. The acceleration at $x = 2$ m is nearest:
   (A) $52.5$ m/s$^2$   (B) $42.5$ m/s$^2$
   (C) $25$ m/s$^2$   (D) $6.25$ m/s$^2$

**Fig. P3.3**

**3.4**  The flow in the section of a conduit shown in Fig. P3.3 is a:
   (A) developed flow
   (B) uniform flow
   (C) one-dimensional flow
   (D) two-dimensional flow

**3.5**  The velocity of an airplane is measured with a pitot tube. If the pitot tube measures 800 mm of water, estimate the speed of the airplane. Use $\rho_{air} = 1.23$ kg/m$^3$.
   (A) $125$ m/s   (B) $113$ m/s
   (C) $80$ m/s   (D) $36$ m/s

**3.6**  A pitot tube measures 600 mm of water in a pipe transporting water. A static pressure probe at the same location measures 200 mm of water. The velocity of water in the pipe is nearest:
   (A) $1.10$ m/s   (B) $1.98$ m/s
   (C) $2.8$ m/s   (D) $3.43$ m/s

**3.7**  A manometer, utilizing a pitot probe, measures 10 mm of mercury. If the velocity is desired in a pipe transporting water to which the manometer is attached, what additional information in the following list is needed?

   I.  The temperature of the water   (A) I and II
   II.  The pressure in the pipe   (B) II and III
   III.  The density of the mercury   (C) III and IV
   IV.  The diameter of the pipe   (D) III and IV

**3.8**  A water hose is pressurized to 800 kPa with a nozzle in the off position. If the nozzle is opened a small amount, as shown in Fig. P3.8, estimate the exiting velocity of the water. Assume the velocity inside the hose to be negligible.
   (A) $40$ m/s   (B) $30$ m/s
   (C) $20$ m/s   (D) $10$ m/s

**Fig. P3.8**

**3.9**  Benzene flows out through the disks shown in Fig. P3.9. If $V_2 = 30$ m/s, the pressure $p_1$, is nearest:
   (A) $150$ kPa   (B) $200$ kPa
   (C) $250$ kPa   (D) $300$ kPa

**Fig. P3.9**

# PROBLEMS

## Flow Fields

**3.10**  A fire is started and the smoke from the chimney goes straight up; no wind is present. After a few minutes a wind arises but the smoke continues to rise slowly. Sketch the streakline of the smoke, the pathline of the first particles leaving the chimney, and a few streamlines, assuming that the wind blows parallel to the ground in a constant direction.

**3.11**  A researcher has a large number of small flotation devices each equipped with a battery and light bulb. Explain how she would determine the pathlines and streaklines near the surface of a stream with some unknown currents that vary with time.

**3.12**  A little boy chases his dad around the yard with the water hose of Fig. P3.12. Sketch a pathline and a streakline if the boy is running perpendicular to the water jet.

**Fig. P3.12**

**3.13** The hot-air balloon of Fig. P3.13 travels with the wind. The wind velocity vector is $\mathbf{V} = 2\hat{\mathbf{i}} + 3\hat{\mathbf{j}}$ m/s for the first hour and then it is $3\hat{\mathbf{i}} + 1.5\hat{\mathbf{j}}$ m/s for 2 hr. On $xy$-coordinates, sketch the pathline of the balloon and streamlines at $t = 2$ hr. If a number of hot-air balloons started from the same location, sketch the streakline formed by the balloons at $t = 3$ hr. The balloons start at the origin.

**Fig. P3.13**

**3.14** A velocity field is given by $\mathbf{V} = (2t + 2)\hat{\mathbf{i}} + 2t\hat{\mathbf{j}}$ m/s. Sketch the pathlines of two particles up to $t = 5$ s, one that originates from the origin at $t = 0$, and another that originates from the origin at $t = 2$ s. Also, sketch the streamlines at $t = 5$ s.

**3.15** Using rectangular coordinates, express the $z$-component of Eq. 3.2.2.

**3.16** The traffic situation on Mackinac Island, Michigan, where no automobiles are allowed (bikes are permitted), is to be studied. Comment on how such a study could be performed using a Lagrangian approach and an Eulerian approach.

**3.17** Determine the speed of a fluid particle at the origin and at the point $(1, -2, 0)$ for each of the following

velocity fields when $t = 2$ s. All distances are in meters and $t$ is in seconds.
(a) $\mathbf{V} = (x + 2)\hat{\mathbf{i}} + xt\hat{\mathbf{j}} - z\hat{\mathbf{k}}$ m/s
(b) $\mathbf{V} = xy\hat{\mathbf{i}} - 2y^2\hat{\mathbf{j}} + tyz\hat{\mathbf{k}}$ m/s
(c) $\mathbf{V} = x^2t\hat{\mathbf{i}} - (xz + 2t)\hat{\mathbf{j}} + xyt\hat{\mathbf{k}}$ m/s

**3.18** Calculate the angle that the velocity vector makes with the $x$-axis and a unit vector normal to the streamline at $(1, -2)$ for the following velocity fields when $t = 2$ s. All distances are in meters and $t$ is in seconds.
(a) $\mathbf{V} = (x + 2)\hat{\mathbf{i}} + xt\hat{\mathbf{j}}$ m/s
(b) $\mathbf{V} = xy\hat{\mathbf{i}} - 2y^2\hat{\mathbf{j}}$ m/s
(c) $\mathbf{V} = (x^2 + 4)\hat{\mathbf{i}} - y^2t\hat{\mathbf{j}}$ m/s

**3.19** Find the equation of the streamline that passes through $(1, -2)$ at $t = 2$ s for the flow of:
(a) Problem 3.18a      (b) Problem 3.18b
(c) Problem 3.18c

**3.20** Find the acceleration vector field for a fluid flow that possesses the following velocity field where $x, y, z$ are in meters. Evaluate the acceleration at $(2, -1, 3)$ at $t = 2$ s.
(a) $\mathbf{V} = 20 (1 - y^2)\hat{\mathbf{i}}$ m/s
(b) $\mathbf{V} = 2x\hat{\mathbf{i}} + 2y\hat{\mathbf{j}}$ m/s
(c) $\mathbf{V} = x^2\, t\hat{\mathbf{i}} + 2xyt\hat{\mathbf{j}} + 2yzt\hat{\mathbf{k}}$ m/s
(d) $\mathbf{V} = x\hat{\mathbf{i}} - 2xyz\hat{\mathbf{j}} + tz\hat{\mathbf{k}}$ m/s

**3.21** Find the angular velocity vector for the following flow fields. Evaluate the angular velocity at $(2, -1, 3)$ at $t = 2$ s.
(a) Problem 3.20a
(b) Problem 3.20b
(c) Problem 3.20c
(d) Problem 3.20d

**3.22** Find the vorticity vector for the following flow fields. Evaluate the vorticity at $(2, -1, 3)$ at $t = 2$ s.
(a) Problem 3.20a
(b) Problem 3.20b
(c) Problem 3.20c
(d) Problem 3.20d

**3.23** Determine the components of the rate-of-strain tensor for the velocity field of the following at $(2, -1, 3)$ at $t = 2$ s.
(a) Problem 3.20a
(b) Problem 3.20b
(c) Problem 3.20c
(d) Problem 3.20d

**3.24** The velocity components, in m/s, in cylindrical coordinates are given by

$$v_r = \left(10 - \frac{40}{r^2}\right) \cos \theta, \qquad v_\theta = -\left(10 + \frac{40}{r^2}\right) \sin \theta$$

(a) Calculate the acceleration of a fluid particle occupying the point (4 m, 180°).

(b) Calculate the vorticity component at (4 m, 180°).

**3.25** The velocity components, in m/s, in spherical coordinates are given by

$$v_r = \left(10 - \frac{80}{r^3}\right) \cos\theta, \qquad v_\theta = -\left(10 + \frac{80}{r^3}\right) \sin\theta$$

(a) Calculate the acceleration of a fluid particle occupying the point (4 m, 180°).

(b) Calculate the vorticity component at (4 m, 180°).

**3.26** An unsteady flow occurs between parallel plates such that $u = u(y, t), v = 0$, and $w = 0$. Write an expression for the acceleration. What is the acceleration if the flow is steady, that is, $u = u(y), v = 0$, and $w = 0$?

**3.27** Consider a symmetrical steady flow in a pipe with the axial and radial velocity components designated $u(r, x)$ and $v(r, x)$, respectively. Write the equations for the two acceleration components $a_r$ and $a_x$. Use equations from Table 3.1. See Fig. P3.27 for the coordinates.

(a)

**Fig. P3.27**

**3.28** The velocity in the 2-cm-diameter pipe of Fig. P3.28 has only one nonzero velocity component given by $u(r, t) = 2(1 - r^2/r_0^2)(1 - e^{-t/10})$ m/s, where $r_0$ is the radius of the pipe and $t$ is in seconds. Calculate the maximum velocity and the maximum acceleration:

(a) Along the centerline of the pipe

(b) Along a streamline at $r = 0.5$ cm

(c) Along a streamline just next to the pipe wall

[*Hint:* Let $v_z = u(r, t), v_r = 0$, and $v_\theta = 0$ in the appropriate equations in Table 3.1.]

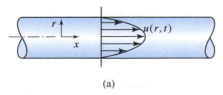

(a)

**Fig. P3.28**

**3.29** The temperature changes periodically in a flow according to $T(y, t) = 20(1 - y^2) \cos \pi t/100°$C. If the velocity is given by $u = 2(1 - y^2)$ m/s, determine the rate of change of the temperature of a fluid particle located at $y = 0$ if $t = 20$ s.

**3.30** The density of air in the atmosphere varies according to $\rho(z) = 1.23e^{-10^{-4}z}$ kg/m³. Air flowing over the mountain of Fig. P3.30 has the velocity vector $\mathbf{V} = 20\hat{\mathbf{i}} + 10\hat{\mathbf{k}}$ m/s at a location of interest where $z = 3000$ m. Find the rate at which a particle's density is changing at that location.

**Fig. P3.30**

**3.31** The density variation with elevation is given by $\rho(z) = 1000(1 - z/4)$ kg/m³. At a location where $\mathbf{V} = 10\hat{\mathbf{i}} + 10\hat{\mathbf{k}}$ m/s, find $D\rho/Dt$.

**3.32** Salt is slowly added to water in a pipe so that $\partial\rho/\partial x = 0.01$ kg/m⁴. Determine $D\rho/Dt$ if the velocity is uniform at 4 m/s.

**3.33** Express the substantial derivative in terms of the gradient $\nabla$ and the velocity vector $\mathbf{V}$. Recall from calculus that in rectangular coordinates

$$\nabla = \frac{\partial}{\partial x}\hat{\mathbf{i}} + \frac{\partial}{\partial y}\hat{\mathbf{j}} + \frac{\partial}{\partial z}\hat{\mathbf{k}}$$

**3.34** We can write Eqs. 3.2.9 in a simplified vector form. The gradient is a vector operator expressed in rectangular coordinates as $\nabla = \frac{\partial}{\partial x}\hat{\mathbf{i}} + \frac{\partial}{\partial y}\hat{\mathbf{j}} + \frac{\partial}{\partial z}\hat{\mathbf{k}}$.

Write the dot product of the velocity vector $\mathbf{V}$ and the gradient $\nabla$ and then write Eqs. 3.2.9 as one vector equation expressing the acceleration $\mathbf{a}$ as the sum of the local acceleration and the convective acceleration.

**3.35** For the flow shown in Fig. P3.35, relative to a fixed reference frame, find the acceleration of a fluid particle at:

(a) Point A

(b) Point B

The water at B makes an angle of 30° with respect to the ground and the sprinkler arm is horizontal.

20 rad/s

A ● → 3.6 m/s

B

1.36 m      1.36 m

18 m/s

**Fig. P3.35**

**3.36** A river is flowing due south at 5 m/s at a latitude of 45°. Calculate the acceleration of a particle floating with the river relative to a fixed reference frame. The radius of the Earth is 6400 km.

## Classification of Fluid Flows

**3.37** Consider each of the following flows and state whether it could be approximated as a one-, two-, or three-dimensional flow or as a uniform flow:
  **(a)** Flow from a vertical pipe striking a horizontal wall
  **(b)** Flow in the waves of the ocean near a beach
  **(c)** Flow near the entrance of a pipe
  **(d)** Flow around a rocket with a blunt nose
  **(e)** Flow around an automobile
  **(f)** Flow in an irrigation channel
  **(g)** Flow through an artery
  **(h)** Flow through a vein

**3.38** Which of the flows in Problem 3.37 could be assumed to be a steady flow? Which must be modeled as an unsteady flow?

**3.39** Which flow in Problem 3.37 could best be modeled as a plane flow?

**3.40** Select the flows in Problem 3.37 that would possess a stagnation point. Sketch each of the selected flows indicating the location of the stagnation point.

**3.41** Which of the flows in Problem 3.37 could be modeled as a developed flow?

**3.42** State whether each of the flows in Problem 3.37 could be considered as primarily an inviscid flow or a viscous flow.

**3.43** Select the flows in Problem 3.37 that are external flows. Does each of the external flows possess a stagnation point?

**3.44** Sketch the flow around a razor blade positioned parallel to the flow showing the boundary layers.

**3.45** The 32°C water exiting the 1.5-cm-diameter faucet of Fig. P3.45 has an average velocity of 2 m/s. Would you expect the flow to be laminar or turbulent?

**Fig. P3.45**

**3.46** The Red Cedar River flows placidly through Michigan State University's campus. In a certain section the depth is 0.75 m and the average velocity is 0.18 m/s. Is the flow laminar or turbulent?

**3.47** Air at 40°C flows in a rectangular 30 cm × 6 cm heating duct at an average velocity of 4 m/s. Is the flow laminar or turbulent?

**3.48** The sphere of diameter $D$ of Fig. P3.48 is traveling with a velocity $V$ of 1.2 m/s in 20°C atmospheric air. If Re $= VD/\nu$ is less than $4 \times 10^4$, the boundary layer around the front of the sphere is completely laminar. Determine if the boundary layer is completely laminar on a sphere of diameter:
  **(a)** 1 cm          **(b)** 1 m

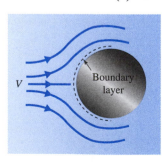

$V$                Boundary layer

**Fig. P3.48**

**3.49**  The airfoil on a commercial airliner is approximated as a flat plate sketched in Fig. P3.49. How long would you expect the laminar portion of the boundary layer to be if it is flying:

At an altitude of 10 000 m and a speed of 900 km/h?

**Fig. P3.49**

**3.50**  A leaf keeps cool by transpiration, a process in which water flows from the leaf to the atmosphere. An experimenter wonders if the boundary layer on a leaf influences the transpiration, so an "experimental" leaf is set up in the laboratory and air is blown parallel to it at 6 m/s. Comment as to whether the boundary layer is expected to be laminar or turbulent.

**3.51**  For the following situations state whether a compressible flow is required or if the flow can be approximated with an incompressible flow:

(a)  An aircraft flying at 100 m/s at an elevation of 8000 m

(b)  A golf ball traveling at 72 m/s

(c)  Flow around an object being studied in a high-temperature wind tunnel if the temperature is 100°C and the air velocity is 100 m/s

**3.52**  Write out Eq. 3.3.2 using Eq. 3.2.11. For a steady, plane flow what relationship must exist for an incompressible flow in which the density is allowed to vary?

**3.53**  If $\rho = \rho_0 (1 + cz)$ models the density variation in a channel (there is heavy salt water at the bottom and fresh water at the top) in which $u(y, z)$ is the only velocity component, is the flow incompressible?

## Bernoulli's Equation

**3.54**  A pitot tube is used to measure the velocity of a small aircraft flying at 900 m. Calculate its velocity if the pitot tube measures:
(a)  2 kPa     (b)  6 kPa     (c)  0.6 kPa

**3.55**  Approximate the force acting on the 15-cm-diameter headlight shown in Fig. P3.55 of an automobile traveling at 120 km/h.

**Fig. P3.55**

**3.56**  A vacuum cleaner is capable of creating a vacuum of 2 kPa just inside the hose of Fig. P3.56. What maximum average velocity would be expected in the hose?

**Fig. P3.56**

**3.57**  The inviscid, incompressible flow in the vicinity of a stagnation point (Fig. P3.57) is approximated by $u = -10x$, $v = 10y$. If the pressure at the origin is $p_0$, find an expression for the pressure neglecting gravity effects:
(a)  Along the negative $x$-axis
(b)  Along the positive $y$-axis

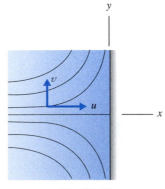

**Fig. P3.57**

**3.58**  The inviscid, incompressible flow field exterior to the cylinder shown in Fig. P3.58 is given by

$$v_r = U_\infty\left(1 - \frac{r_c^2}{r^2}\right)\cos\theta, \qquad v_\theta = U_\infty\left(1 + \frac{r_c^2}{r^2}\right)\sin\theta$$

If the pressure at $r = \infty$ is zero (i.e., $p_\infty = 0$), find an expression for the pressure neglecting gravity effects:

(a)  Along the negative $x$-axis
(b)  At the stagnation point
(c)  On the surface of the cylinder
(d)  On the surface of the cylinder at $\theta = 90°$

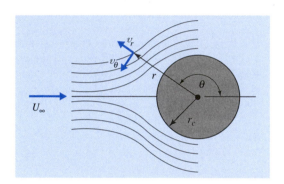

**Fig. P3.58**

**3.59**  The inviscid, incompressible flow field exterior to a sphere (see Fig. P3.58) is given by

$$v_r = U_\infty\left(1 - \frac{r_c^3}{r^3}\right)\cos\theta, \qquad v_\theta = U_\infty\left(1 + \frac{r_c^3}{r^3}\right)\sin\theta$$

If the pressure at $r = \infty$ is zero, find an expression for the pressure neglecting gravity effects:

(a)  Along the negative $x$-axis
(b)  At the stagnation point
(c)  On the surface of the sphere
(d)  On the surface of the sphere at $\theta = 90°$

**3.60**  The velocity along the negative $x$-axis in the inviscid, incompressible flow field exterior to the body shown in Fig. P3.60 is given by $u(x) = U_\infty + q/2\pi x$. If the pressure at $x = -\infty$ is zero, find an expression for the pressure neglecting gravity effects:

(a)  Along the negative $x$-axis if $U_\infty = 10$ m/s and $q = 20\pi$ m²/s
(b)  At the stagnation point if $U_\infty = 10$ m/s and $q = 20\pi$ m²/s

(c)  Along the negative $x$-axis if $U_\infty = 9$ m/s and $q = 6\pi$ m²/s
(d)  At the stagnation point if $U_\infty = 9$ m/s and $q = 6\pi$ m²/s

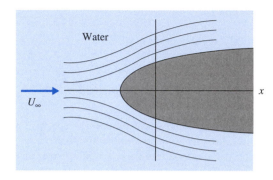

**Fig. P3.60**

**3.61**  The incompressible flow of water through the short contraction of Fig. 3.19a is assumed to be inviscid. If a pressure drop of 20 kPa is measured, estimate the velocity at the wall at section 2 just downstream of the contraction. (Actually, a boundary layer would develop, and the velocity calculated at the wall would be the velocity at the edge of the boundary layer; see the inset of Fig. 3.10.)

**3.62**  Air flows from a relatively large plenum in a furnace out a relatively small rectangular duct. If the pressure in the plenum measures 60 Pa and in the duct 10.2 Pa, estimate the velocity of the 40°C air in the duct.

**3.63**  A short contraction is followed by an expansion, as shown in Fig. P3.63. The manometer is used to determine the velocity of the fluid providing the viscous effects are negligible. If water is flowing steadily, determine the velocity $V_1$ if $H = 12$ cm. Losses through the contraction are negligible.

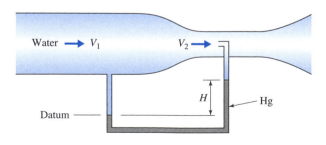

**Fig. P3.63**

**3.64** What is the velocity of the water in the pipe if the manometer shown in Fig. P3.64 reads:
**(a)** 4 cm?
**(b)** 10 cm?

**Fig. P3.64**

**3.65** Air at 120 kPa absolute and 30°C flows vertically upward in a pipe, as shown in Fig. P3.65. If the water manometer deflection $H = 5$ cm, determine the velocity in the smaller pipe. Assume the air to be incompressible.

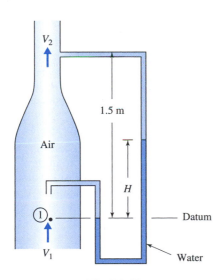

**Fig. P3.65**

**3.66** A manometer, positioned inside a cylinder as shown in Fig. P3.66, reads 4 cm of water. Estimate $U_\infty$ assuming an inviscid flow. Refer to the velocity field in Problem 3.58.

**Fig. P3.66**

**3.67** A manometer, positioned as shown in Figure P3.66 inside a sphere, reads 4 cm of water. Estimate $U_\infty$ assuming an inviscid flow. Refer to the velocity field in Problem 3.59.

**3.68** For the flow shown in Fig. P3.68, estimate the pressure $p_1$ and velocity $V_1$ if $V_2 = 20$ m/s and:
**(a)** $H = 1$ cm
**(b)** $H = 5$ cm
**(c)** $H = 10$ cm

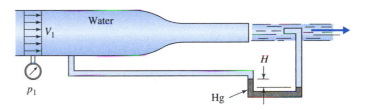

**Fig. P3.68**

**3.69** Water at 15°C flows steadily through the contraction shown in Fig. P3.69 such that $V_2 = 4V_1$. If the gage reading is maintained at 120 kPa, determine the maximum velocity $V_1$ possible before cavitation occurs.

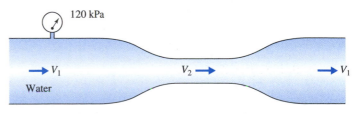

**Fig. P3.69**

**3.70** In the pipe contraction shown in Fig. P3.70, water flows steadily with a velocity of $V_1 = 0.5$ m/s and $V_2 = 1.125$ m/s. Two piezometer tubes are attached to the pipe at sections 1 and 2. Determine the height $H$. Neglect any losses through the contraction.

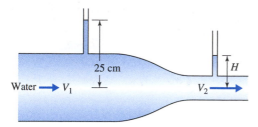

25 cm

Water $\longrightarrow V_1$     $V_2 \longrightarrow$    $H$

**Fig. P3.70**

**3.71** Air at 20°C is drawn into the hose of a vacuum cleaner through a head that is relatively free of obstructions (the flow can be assumed to be inviscid). Estimate the velocity in the hose if the vacuum in the hose measures:
**(a)** 2 cm of water     **(b)** 8 cm of water

**3.72** A wind tunnel is designed to draw in air from the atmosphere and produce a velocity of 100 m/s in the test section. The fan is located downstream of the test section. What pressure is to be expected in the test section if the atmospheric temperature and pressure are:
**(a)** −20°C, 90 kPa?    **(b)** 0°C, 95 kPa?
**(c)** 20°C, 92 kPa?    **(d)** 40°C, 100 kPa?

**3.73** The pump shown in Fig. P3.73 creates a flow such that $V = 14$ m/s. Predict the pressure at the gage shown assuming an inviscid flow in the entrance and a uniform flow at the gage. Use a streamline starting at:
**(a)** Point $A$        **(b)** Point $B$

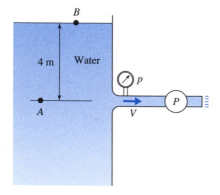

$B$

4 m   Water

$p$

$A$    $V$   $P$

**Fig. P3.73**

**3.74** A fireman reduces the exit area on a nozzle so that the velocity inside the hose is quite small relative to the exiting velocity. What is the maximum exiting velocity and what is the maximum height the water can reach if the pressure inside the hose is:
**(a)** 700 kPa?      **(b)** 1400 kPa?

**3.75** The velocity downstream of a sluice gate is assumed to be uniform (Fig. P3.75). Express $V$ in terms of $H$ and $h$ for this inviscid flow. Use a streamline:
**(a)** Along the top downstream
**(b)** Along the bottom downstream

Water    Sluice gate

$H$    $V_1 \cong 0$      $h$

**Fig. P3.75**

**3.76** To what maximum velocity can water be accelerated before it reaches the turbine blades of a hydro-turbine if it enters with relatively low velocity at:
**(a)** 600 kPa?      **(b)** 300 kPa?

**3.77** Water exists in a city's water system at a pressure of 500 kPa at a particular location. The water pipe must traverse over a hill. How high could the hill be, above that location, for the system to possibly supply water to the other side of the hill?

**3.78** Fluid flows between the radial disks shown in Fig. P3.78. Estimate the pressure in the 2-cm-diameter pipe if the fluid exits to the atmosphere. Neglect viscous effects. The fluid is:
**(a)** Water        **(b)** Benzene
**(c)** Gasoline       **(d)** Air

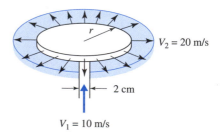

$r$     $V_2 = 20$ m/s

2 cm

$V_1 = 10$ m/s

**Fig. P3.78**

**3.79** Estimate the pressure at $r = 10$ cm if the velocity there is 30 m/s in Problem 3.78d.

**3.80** It is proposed that air blown through a tube attached to a metal disk be used to pick up envelopes, as shown in Fig. P3.80. Would such a setup actually pick up an envelope? Explain. Assume an inviscid flow with the air slowing down as it moves radially outward.

**Fig. P3.80**

**3.81** From which of the following objects would you expect the flow to separate and form a substantial separated region?
(a)    A golf ball
(b)    A telephone wire
(c)    A wind machine blade
(d)    A 2-mm-diameter wire in a low-speed wind tunnel
(e)    An automobile
(f)    An aircraft
(*Note:* Separation occurs whenever the Reynolds number exceeds a value around 20 over a blunt object.)

**3.82** Explain, with the use of a sketch, why a burr on the downstream side of a piezometer opening in a pipe (see Fig. 3.18a) will result in a reading of the pressure that is too high.

**3.83** An incompressible, inviscid flow of water enters a bend with a uniform velocity of $V_1 = 10$ m/s (Fig. P3.83). Estimate the pressure difference between points $A$ and $B$ if the average radius of curvature at the bend is 5 cm. Sketch an anticipated velocity profile along $AB$. Assume that $p_A < p_1$ and $p_B > p_1$.

**Fig. P3.83**

**3.84** Viscosity causes a fluid to stick to a surface. If the fluid in Problem 3.83 sticks to the surface, explain why, in a viscous flow, a secondary flow is caused by the bend. Sketch such a flow on a circular cross section at section 2.

**3.85** In Fig. P3.85, assuming an inviscid flow, insert one of these signs between the pressures: $>, <, \cong$ .

$$p_A \qquad p_B$$
$$p_C \qquad p_D$$
$$p_B \qquad p_D$$

**Fig. P3.85**

Collegiate Rowing Regatta on the Schuylkill River, Philadelphia, Pennsylvania. With each stroke, the work performed by the crew is transferred to the hull to overcome drag forces. (John Kropewnicki/Shutterstock)

# 4

# The Integral Forms of the Fundamental Laws

## Outline

## Chapter Objectives

The objectives of this chapter are to:

▲ Derive an equation that will allow us to convert the three basic laws that are stated for a system to a form that is applicable to a control volume.

▲ Apply the conservation of mass to control volumes of interest.

▲ Analyze the work-rate term of the energy equation.

▲ Apply the energy equation to numerous engineering situations.

▲ Apply Newton's second law to control volumes of interest.

▲ Apply the moment-of-momentum equation to rotating devices.

▲ Present numerous examples of the basic laws applied to control volumes so that students will be able to correctly solve fluid flow problems involving many of the control volumes of interest to engineers.

▲ Express the basic laws in their most general control-volume form so that complex problems encountered in engineering applications can be properly analyzed and hopefully solved.

## 4.1 INTRODUCTION

Quantities of interest to engineers can often be expressed in terms of integrals. For example, volume flow rate is the integral of the velocity over an area; heat transfer is the integral of the heat flux over an area; force is the integral of a stress over an area; mass is the integral of the density over a volume; and kinetic energy is the integral of $V^2/2$ over each mass element in a volume. There are, of course, many other integral quantities. To determine an integral quantity, the integrand must be known, or information must be available so that a good approximation to the integrand can be made. If the integrand is not known or cannot be approximated with any degree of certainty, appropriate differential equations (see Chapter 5) must be solved yielding the needed integrand; the integration is then performed giving the engineer the desired integral quantity.

> **KEY CONCEPT**  *To determine an integral quantity the integrand must be known, or information must be available so that a good approximation to the integrand can be made.*

In this chapter we present the integral quantities of interest, develop equations that relate the integral quantities, and work a number of problems for which the integrands are given or can be approximated. This includes a surprisingly large variety of problems. There are, however, many integral quantities that cannot be determined since the integrands are unknown. These would include the lift and drag on an airfoil, the torque on the blades of a wind machine, and the kinetic energy in the wake of a submarine. To determine such integrands, it would be necessary to solve the appropriate differential equations, a task that is often quite difficult; some relatively simple flow geometries are considered in subsequent chapters.

Also, there are many quantities of interest that are not integral in nature. Included would be the point of separation of the flow around a body, the concentration of a pollutant in a stream at a certain location, the pressure distribution on the side of a building, and the wave-shore interaction along a lake. To study subjects such as these, it is necessary to consider the differential equations that describe the flow. Most of the topics mentioned are relegated to specialized graduate courses; however, some topics that require the solution of the more easily solvable differential equations are included in this book.

## 4.2 THE THREE BASIC LAWS

> **System:**  *A fixed collection of material particles.*

The integral quantities of primary interest in fluid mechanics are contained in three basic laws: conservation of mass, first law of thermodynamics, and Newton's second law. These basic laws are expressed using a Lagrangian description in terms of a **system**, a fixed collection of material particles. For example, if we consider flow through a pipe, we could identify a fixed quantity of fluid at time $t$ as the system (Fig. 4.1); this system would then move due to velocity to a downstream

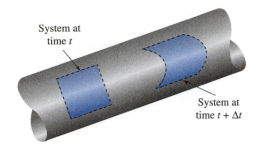

System at
time $t$

System at
time $t + \Delta t$

**Fig. 4.1**   Example of a system in fluid mechanics.

location at time $t + \Delta t$. Any of the three basic laws could be applied to this system. This is, however, not an easy task. First let us state the basic laws in their general form.

   **Conservation of Mass:**   The law stating that mass must be conserved is:

> The mass of a system remains constant.

The mass of a fluid particle is $\rho \, d\mathcal{V}$, where $d\mathcal{V}$ is the volume occupied by the particle and $\rho$ is its density. Knowing that the density can change from point to point in the system, the conservation of mass can be expressed in integral form as

$$\frac{D}{Dt} \int_{\text{sys}} \rho \, d\mathcal{V} = 0 \qquad (4.2.1)$$

where $D/Dt$ is used since we are following a specified group of material particles, a system.

   **First Law of Thermodynamics:**   The law that relates heat transfer, work, and energy change is the first law of thermodynamics; it states:

> The rate of heat transfer to a system minus the rate at which the system does work equals the rate at which the energy of the system is changing.

Recognizing that both density and specific energy may change from point to point in the system, it may be expressed as

$$\dot{Q} - \dot{W} = \frac{D}{Dt} \int_{\text{sys}} e\rho \, d\mathcal{V} \qquad (4.2.2)$$

*Specific energy: Accounts for kinetic energy, potential energy, and internal energy per unit mass.*

where the **specific energy** $e$ accounts for kinetic energy, potential energy, and internal energy per unit mass. Other forms of energy—chemical, electrical, nuclear—are not included in an introductory course in fluid mechanics. Equation 4.2.2 is often referred to as the energy equation.

In its basic form stated here, the first law of thermodynamics applies only to a system, a collection of fluid particles; therefore, $D/Dt$ is used. We study $\dot{Q}$ and $\dot{W}$ in Section 4.5, where we consider the energy equation in detail.

**Newton's Second Law:** Newton's second law, also called the momentum equation, states:

> The resultant force acting on a system equals the rate at which the momentum of the system is changing.

The momentum of a fluid particle of mass is a vector quantity given by $\mathbf{V}\rho\, d\forall$; consequently, Newton's second law may be expressed in an inertial reference frame as

$$\Sigma\, \mathbf{F} = \frac{D}{Dt} \int_{\text{sys}} \mathbf{V}\rho\, d\forall \tag{4.2.3}$$

recognizing that both density and velocity may change from point to point in the system. This equation reduces to $\Sigma\, \mathbf{F} = m\mathbf{a}$ if $\mathbf{V}$ and $\rho$ are constant throughout the entire system; $\rho$ is often a constant, but in fluid mechanics the velocity vector invariably changes from point to point. Again, $D/Dt$ is used to provide the rate-of-change since Newton's second law is applied to a system.

**Moment-of-Momentum Equation:** The moment-of-momentum equation results from Newton's second law; it states:

> The resultant moment acting on a system equals the rate of change of the angular momentum of the system.

In equation form this becomes, relative to an inertial reference frame,

$$\Sigma\, \mathbf{M} = \frac{D}{Dt} \int_{\text{sys}} \mathbf{r} \times \mathbf{V}\rho\, d\forall \tag{4.2.4}$$

where $\mathbf{r} \times \mathbf{V}\rho\, d\forall$ represents the angular momentum of a fluid particle with mass $\rho\, d\forall$. The vector $\mathbf{r}$ locates the volume element $d\forall$ and is measured from the origin of the coordinate axes, the point relative to which the resultant moment is measured.

Note that in each of the basic laws the integral quantity is an extensive property of the system (see Section 1.7). We will use the symbol $N_{sys}$ to denote this extensive property; for example, $N_{sys}$ could represent the mass, the momentum, or the energy of the system. The left-hand side of Eq. 4.2.1 and the right-hand sides of Eqs. 4.2.2, 4.2.3, and 4.2.4 may all be expressed as

$$\frac{DN_{sys}}{Dt} \qquad (4.2.5)$$

where $N_{sys}$ represents an integral quantity, either a scalar quantity or a vector quantity.

It is also useful to introduce the variable $\eta$ for the intensive property, the property of the system per unit mass. The relation between $N_{sys}$ and $\eta$ is given by

$$N_{sys} = \int_{sys} \eta \rho \, d\mathcal{V} \qquad (4.2.6)$$

As an example, the extensive property of Newton's second law is the momentum

$$\textbf{momentum}_{\textbf{system}} = \int_{sys} \textbf{V} \rho \, d\mathcal{V} \qquad (4.2.7)$$

which is a vector quantity. The corresponding intensive property would be the velocity vector $\textbf{V}$. Note that the density and velocity, which may vary from point to point within the system, may also be functions of time, as in unsteady flow.

Our interest is most often focused on a device, or a region of space, into which fluid enters and/or from which fluid leaves; we identify this region as a **control volume**. An example of a fixed control volume is shown in Fig. 4.2a. A control volume need not be fixed; it could deform as in a piston-cylinder during exhaust or in a balloon as it deflates. We will, however, consider only fixed control volumes in this book. This will not limit us in most problem situations.

The difference between a control volume and a system is illustrated in Fig. 4.2b. The figure indicates that the system occupies the control volume at time $t$ and has partially moved out of it at time $t + \Delta t$. Since it is often more convenient to focus on a control volume (e.g., a pump) rather than on a system, the first

**Control volume:**    *A region of space, into which fluid enters and/or from which fluid leaves.*

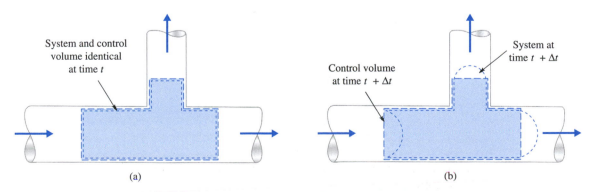

Fig. 4.2    Example of a fixed control volume and a system: (a) time $t$; (b) time $t + \Delta t$.

order of business is to find a transformation that will allow us to express the substantial derivative of a system (a Lagrangian description) in terms of quantities associated with a control volume (an Eulerian description) so that the basic laws can be applied directly to a control volume. This will be done in general and then applied to the specific laws.

## 4.3 SYSTEM-TO-CONTROL-VOLUME TRANSFORMATION

We are interested in the time rate of change of the extensive property $N_{sys}$ as we follow the system along, that is, $DN_{sys}/Dt$, and we would like to express this in terms of quantities that pertain to the control volume. In this section we present the derivation of such a transformation.

The derivation involves fluxes of the extensive property in and out of the control volume. A flux is a measure of the rate at which an extensive property crosses an area; for example, a mass flux is the rate at which mass crosses an area. It is useful to introduce vector notation to describe these fluxes. Consider an area element $dA$ of the **control surface**, the surface area that completely encloses a control volume. The property flux across an elemental area $dA$ (see Fig. 4.3) may be expressed by

**Control surface:** *The surface area that completely encloses the control volume.*

$$\text{flux across } dA = \eta \rho \hat{\mathbf{n}} \cdot \mathbf{V} \, dA \qquad (4.3.1)$$

where $\hat{\mathbf{n}}$, a unit vector normal to the area element $dA$, always points out of the control volume, and $\eta$ represents the intensive property associated with $N_{sys}$. Note that this expression yields a negative value if it concerns a property influx. Only the normal component $\hat{\mathbf{n}} \cdot \mathbf{V}$ of the velocity vector contributes to this flux term. If there is no normal component of velocity on a particular area, such as the wall of a pipe, no flux occurs across that area. A positive $\hat{\mathbf{n}} \cdot \mathbf{V}$ indicates a flux out of the volume; a negative $\hat{\mathbf{n}} \cdot \mathbf{V}$, that is, $\mathbf{V}$ has a component in the opposite direction of $\hat{\mathbf{n}}$, indicates a flux into the volume. We must always use $\hat{\mathbf{n}}$ pointing out of the volume. The velocity vector $\mathbf{V}$ may be at some angle to the unit vector $\hat{\mathbf{n}}$; the

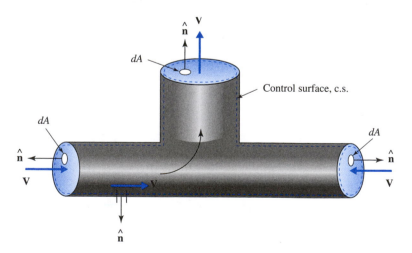

**Fig. 4.3** Illustration showing the flux of an extensive property.

dot product $\hat{\mathbf{n}} \cdot \mathbf{V}$ accounts for the appropriate component of $\mathbf{V}$ that produces a flux through the area.

The net property flux out of the control surface is then obtained by integrating over the entire control surface:

$$\text{net flux of property} = \int_{\text{c.s.}} \eta \rho \hat{\mathbf{n}} \cdot \mathbf{V} \, dA \qquad (4.3.2)$$

If the net flux is positive, the flux out is larger than the flux in.

Let us return now to the derivative $DN_{\text{sys}}/Dt$. The definition of a derivative from calculus allows us to write

$$\frac{DN_{\text{sys}}}{Dt} = \lim_{\Delta t \to 0} \frac{N_{\text{sys}}(t + \Delta t) - N_{\text{sys}}(t)}{\Delta t} \qquad (4.3.3)$$

The system is shown in Fig. 4.4 at times $t$ and $t + \Delta t$. Assume that the system occupies the full control volume at time $t$; if we were considering a device, such as a pump, the particles of the system would just fill the device at time $t$. Since the device, the control volume shown in Fig. 4.4, is assumed to be fixed in space, the system will move through the device. Equation 4.3.3 can then be written

> **KEY CONCEPT**  *The system occupies the full control volume at time $t$.*

$$\frac{DN_{\text{sys}}}{Dt} = \lim_{\Delta t \to 0} \frac{N_3(t + \Delta t) + N_2(t + \Delta t) - N_2(t) - N_1(t)}{\Delta t} \qquad (4.3.4a)$$

$$= \lim_{\Delta t \to 0} \frac{N_2(t + \Delta t) + N_1(t + \Delta t) - N_2(t) - N_1(t)}{\Delta t}$$

$$+ \lim_{\Delta t \to 0} \frac{N_3(t + \Delta t) - N_1(t + \Delta t)}{\Delta t} \qquad (4.3.4b)$$

where, in this second expression, we have simply added and subtracted $N_1(t + \Delta t)$ in the numerator. In the equations above, the numerical subscript denotes the region; for example, $N_2(t)$ signifies the extensive property in region 2 at time $t$. Now, we observe that the first limit on the right-hand side of Eq. 4.3.4b refers to the control volume, so we can write

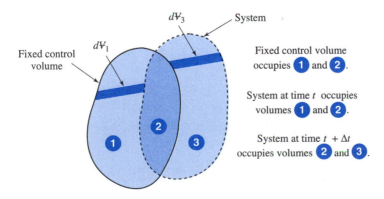

**Fig. 4.4**   System and fixed control volume.

$$\frac{DN_{\text{sys}}}{Dt} = \lim_{\Delta t \to 0} \frac{N_{\text{c.v.}}(t + \Delta t) - N_{\text{c.v.}}(t)}{\Delta t} + \lim_{\Delta t \to 0} \frac{N_3(t + \Delta t) - N_1(t + \Delta t)}{\Delta t} \qquad (4.3.5)$$

The first ratio on the right-hand side is $dN_{\text{c.v.}}/dt$, where we use an ordinary derivative since we are not following specific fluid particles. Thus there results

$$\frac{DN_{\text{sys}}}{Dt} = \frac{dN_{\text{c.v.}}}{dt} + \lim_{\Delta t \to 0} \frac{N_3(t + \Delta t) - N_1(t + \Delta t)}{\Delta t} \qquad (4.3.6)$$

Now, we must find expressions for the extensive quantities $N_3(t + \Delta t)$ and $N_1(t + \Delta t)$. They, of course, depend on the mass contained in the volume elements $d\mathcal{V}_1$ and $d\mathcal{V}_3$, shown in Fig. 4.4 and enlarged in Fig. 4.5. Note that the unit vector $\hat{\mathbf{n}}$ always points out of the volume, and hence to obtain a positive differential volume a negative sign is required for region 1. Also, note that the cosine of the angle between the velocity vector and the normal vector is required,[1] thus the presence of the dot product. Referring to Figs. 4.4 and 4.5, we have

$$N_3(t + \Delta t) = \int_{A_3} \eta \rho \hat{\mathbf{n}} \cdot \mathbf{V} \, \Delta t \, dA_3$$

$$\qquad (4.3.7)$$

$$N_1(t + \Delta t) = -\int_{A_1} \eta \rho \hat{\mathbf{n}} \cdot \mathbf{V} \, \Delta t \, dA_1$$

Recognizing that $A_3$ plus $A_1$ completely surrounds the control volume, we combine the two integrals into one integral. That is,

$$N_3(t + \Delta t) - N_1(t + \Delta t) = \int_{\text{c.s.}} \eta \rho \hat{\mathbf{n}} \cdot \mathbf{V} \, \Delta t \, dA \qquad (4.3.8)$$

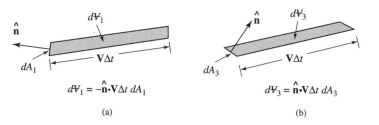

$$d\mathcal{V}_1 = -\hat{\mathbf{n}} \cdot \mathbf{V} \Delta t \, dA_1 \qquad\qquad d\mathcal{V}_3 = \hat{\mathbf{n}} \cdot \mathbf{V} \Delta t \, dA_3$$

(a)                                                                 (b)

**Fig. 4.5**   Differential volume elements.

---

[1]To obtain the volume of a box, we multiply the height by the area of the base, provided that the box is upright. If it is completely collapsed, its volume is zero. Hence for some intermediate position, the volume is the height times the area of the base times the cosine of the appropriate angle.

where the control surface, denoted by c.s., is the area that completely surrounds the control volume. Substituting Eq. 4.3.8 back into Eq. 4.3.6 yields the desired result, the system-to-control-volume transformation, or equivalently, the **Reynolds transport theorem**:

**Reynolds transport theorem:** *The system-to-control-volume transformation.*

$$\frac{DN_{\text{sys}}}{Dt} = \frac{d}{dt} \int_{\text{c.v.}} \eta\rho \, d\mathcal{V} + \int_{\text{c.s.}} \eta\rho\hat{\mathbf{n}} \cdot \mathbf{V} \, dA \qquad (4.3.9)$$

This is a Lagrangian-to-Eulerian transformation of the rate of change of an extensive integral quantity.

The first integral represents the rate of change of the extensive property in the control volume. The second integral represents the flux of the extensive property across the control surface; it is nonzero only where fluid crosses the control surface. We study this flux term in considerable detail in the following sections. Thus we can now express the basic laws in terms of a fixed volume in space. We will do this in subsequent sections for each of the basic laws.

We can move the time derivative of the control volume term inside the integral since, for a fixed control volume, the limits on the volume integral are independent of time; we then write the Reynolds transport theorem as

**KEY CONCEPT** *The time derivative of the control volume term can be moved inside the integral for a fixed control volume.*

$$\frac{DN_{\text{sys}}}{Dt} = \int_{\text{c.v.}} \frac{\partial}{\partial t}(\rho\eta) \, d\mathcal{V} + \int_{\text{c.s.}} \eta\rho\hat{\mathbf{n}} \cdot \mathbf{V} \, dA \qquad (4.3.10)$$

In this form we have used $\partial/\partial t$ since $\rho$ and $\eta$ are, in general, dependent on the position variables.

### 4.3.1   Simplifications of the Reynolds Transport Theorem

Many flows of interest are steady flows, so that $\partial(\eta\rho)/\partial t = 0$. Our system-to-control-volume transformation then takes the form

$$\frac{DN_{\text{sys}}}{Dt} = \int_{\text{c.s.}} \eta\rho\hat{\mathbf{n}} \cdot \mathbf{V} \, dA \qquad (4.3.11)$$

Furthermore, there is often only one area $A_1$ across which fluid enters the control volume and one area $A_2$ across which fluid leaves the control volume; assuming that the velocity vector is normal to the area (Fig. 4.6), we can write $\hat{\mathbf{n}} \cdot \mathbf{V}_1 = -V_1$ over area $A_1$ and $\hat{\mathbf{n}} \cdot \mathbf{V}_2 = V_2$ over area $A_2$. Then Eq. 4.3.11 becomes

$$\frac{DN_{\text{sys}}}{Dt} = \int_{A_2} \eta_2\rho_2 V_2 \, dA - \int_{A_1} \eta_1\rho_1 V_1 \, dA \qquad (4.3.12)$$

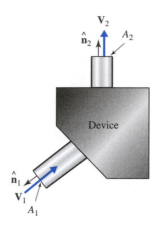

**Fig. 4.6**  Flow into and from a device.

Finally, there are many situations that are modeled acceptably by assuming uniform properties over each plane area (see Fig. 3.9); then the equation simplifies to

$$\frac{DN_{\text{sys}}}{Dt} = \eta_2 \rho_2 V_2 A_2 - \eta_1 \rho_1 V_1 A_1 \tag{4.3.13}$$

We will find that the system-to-control-volume transformation in this simplified form is most often used in the application of the basic laws to problems of interest in an introductory course in fluid mechanics. Some applications will, however, be included that will illustrate nonuniform distributions and unsteady flows.

   If we generalize Eq. 4.3.13 to include several areas across which the fluid flows, we could write

$$\frac{DN_{\text{sys}}}{Dt} = \sum_{i=1}^{N} \eta_i \rho_i \mathbf{V}_i \cdot \hat{\mathbf{n}}_i A_i \tag{4.3.14}$$

where $N$ is the number of areas. The dot product $\hat{\mathbf{n}} \cdot \mathbf{V}$ would provide us with the appropriate sign at each area; for an inlet area, $\hat{\mathbf{n}} \cdot \mathbf{V}$ introduces a negative sign, and for an exit area, $\hat{\mathbf{n}} \cdot \mathbf{V}$ introduces a positive sign.

   For an unsteady flow in which flow properties are assumed to be uniform throughout the control volume, the system-to-control-volume equation takes the form

$$\frac{DN_{\text{sys}}}{Dt} = V_{\text{c.v.}} \frac{d(\eta\rho)}{dt} + \eta_2 \rho_2 V_2 A_2 - \eta_1 \rho_1 V_1 A_1 \tag{4.3.15}$$

for one inlet and one outlet with uniform properties.

## 4.4 CONSERVATION OF MASS

A system is a given collection of fluid particles; hence its mass remains fixed. This is expressed as

$$\frac{Dm_{\text{sys}}}{Dt} = \frac{D}{Dt} \int_{\text{sys}} \rho \, d\mathcal{V} = 0 \qquad (4.4.1)$$

In Eq. 4.2.6, $N_{\text{sys}}$ represents the mass of the system, so we simply let $\eta = 1$. Thus the conservation of mass, referring to Eq. 4.3.9, becomes

$$0 = \frac{d}{dt} \int_{\text{c.v.}} \rho \, d\mathcal{V} + \int_{\text{c.s.}} \rho \hat{\mathbf{n}} \cdot \mathbf{V} \, dA \qquad (4.4.2)$$

or, if we prefer, it takes the equivalent form for a fixed control volume,

$$0 = \int_{\text{c.v.}} \frac{\partial \rho}{\partial t} \, d\mathcal{V} + \int_{\text{c.s.}} \rho \hat{\mathbf{n}} \cdot \mathbf{V} \, dA \qquad (4.4.3)$$

Conservation of mass, 881–883

If the flow is steady, there results (see Eq. 4.3.11)

$$\int_{\text{c.s.}} \rho \hat{\mathbf{n}} \cdot \mathbf{V} \, dA = 0 \qquad (4.4.4)$$

which, for a uniform flow with one entrance and one exit, takes the form (see Eq. 4.3.13)

$$\rho_2 A_2 V_2 = \rho_1 A_1 V_1 \qquad (4.4.5)$$

where for an inlet we have used $\hat{\mathbf{n}}_1 \cdot \mathbf{V}_1 = -V_1$ and for an exit $\hat{\mathbf{n}} \cdot \mathbf{V}_2 = V_2$. Recall $\hat{\mathbf{n}}$ always points out of the control volume.

**KEY CONCEPT** $\hat{\mathbf{n}}$ *always points out of the control volume.*

If the density is constant in the control volume, the derivative $\partial \rho / \partial t = 0$ even if the flow is unsteady; the continuity equation (4.4.3) then reduces to

$$A_1 V_1 = A_2 V_2 \qquad (4.4.6)$$

This form of the continuity equation is used quite often, particularly with liquids and low-speed gas flows.

At this point we wish to discuss again the use of uniform velocity profiles (see also Section 3.3.1). Suppose that the velocity profiles at the entrance and the

**Fig. 4.7** Nonuniform velocity profiles.

exit are not uniform, such as sketched in Fig. 4.7. Furthermore, suppose that the density is uniform over each area. Then the continuity equation takes the form

$$\rho_1 \int_{A_1} V_1\, dA = \rho_2 \int_{A_2} V_2\, dA \qquad (4.4.7)$$

or, letting an overbar denote an average, we can write

$$\rho_1 \overline{V}_1 A_1 = \rho_2 \overline{V}_2 A_2 \qquad (4.4.8)$$

where $\overline{V}_1$ and $\overline{V}_2$ are the *average velocities* over the areas at sections 1 and 2, respectively. In examples and problems the overbar is often omitted. It should be kept in mind, however, that actual velocity profiles are usually not uniform; Eqs. 4.4.5 and 4.4.6 are used with the velocities representing average velocities.

 Mass flux, 888–889

Any one of the above equations (4.4.2) through (4.4.8) is referred to as the *continuity equation.*

Before presenting some examples applying the continuity equation, two fluxes are defined that will be useful in specifying the quantity of flow. The *mass flux $\dot{m}$*, or the mass rate of flow, is

$$\dot{m} = \int_A \rho V_n\, dA \qquad (4.4.9)$$

and has units of kg/s; $V_n$ is the normal component of velocity. The *flow rate $Q$*, or the volume rate of flow, is

$$Q = \int_A V_n\, dA \qquad (4.4.10)$$

and has units of m³/s or sometimes L/s. The mass flux is usually used in specifying the quantity of flow for a compressible flow and the flow rate for an incompressible flow. We often refer to the flow rate as **discharge**.

**Discharge:** *Another term for flow rate.*

In terms of average velocity, we have

$$Q = A\overline{V} \qquad (4.4.11)$$

$$\dot{m} = \rho A \overline{V} \qquad (4.4.12)$$

where for the mass flux we assume a uniform density profile; we also assume that the velocity is normal to the area.

The following examples are solved by first selecting a control volume. If you study the examples carefully, you will notice that often there is only one proper choice for the control volume. We must position the inlet and exit areas at locations where the integrands are either known or where they can be approximated; also, the quantity being sought is often included at an inlet or exit area. In a few cases there may be more freedom in the selection of the control volume (Example 4.5).

This first example represents the primary use of the continuity equation. It allows us to calculate the velocity at one section if it is known at another section.

## Example 4.1

Water flows at a uniform velocity of 3 m/s into a nozzle that reduces the diameter from 10 cm to 2 cm (Fig. E4.1). Calculate the water's velocity leaving the nozzle and the flow rate.

**Fig. E4.1**

### Solution

The control volume is selected to be the inside of the nozzle as shown. Flow enters the control volume at section 1 and leaves at section 2. The simplified continuity equation (4.4.6) is used since the density of water is assumed constant and the velocity profiles are uniform:

$$A_1 V_1 = A_2 V_2$$

$$\therefore V_2 = V_1 \frac{A_1}{A_2} = 3 \, \frac{\pi \times 0.1^2/4}{\pi \times 0.02^2/4} = 75 \text{ m/s}$$

The flow rate, or discharge, is found to be

$$Q = V_1 A_1$$

$$= 3 \times \pi \times 0.1^2/4 = 0.0236 \text{ m}^3/\text{s}$$

## Example 4.2

Water flows in and out of a device as shown in Fig. E4.2a. Calculate the rate of change of the mass of water ($dm/dt$) in the device.

(a)

(b)

**Fig. E4.2**

### Solution

The control surface of the control volume selected is shown in Fig. E4.2b. The continuity equation (4.4.2), with three surfaces across which water flows, takes the following form:

$$0 = \frac{d}{dt}\int_{c.v.} \rho \, d\mathcal{V} + \int_{c.s.} \rho \hat{\mathbf{n}} \cdot \mathbf{V} \, dA$$

$$= \frac{dm}{dt} - \rho_1 A_1 V_1 + \rho_2 A_2 V_2 + \rho_3 A_3 V_3$$

where we have assumed the density to be constant over the volume and we have used $\mathbf{V}_1 \cdot \hat{\mathbf{n}} = -V_1$, since $\hat{\mathbf{n}}_1$ points out of the volume, opposite to the direction of $\mathbf{V}_1$. The last three terms come from the area integral. In terms of the quantities given, the above can be expressed as

$$0 = \frac{dm}{dt} - \rho_1 A_1 V_1 + \dot{m}_2 + \rho_3 Q_3$$

$$= \frac{dm}{dt} - 1028 \text{ kg/m}^3 \times \left(\pi \times \frac{3.75^2}{10000}\right) \text{m}^2 \times 9 \text{ m/s} + 4.4 \text{ kg/s}$$

$$+ 1028 \text{ kg/m}^3 \times 0.0083 \text{ m}^3/\text{s}$$

This is solved to yield

$$\frac{dm}{dt} = 28.9 \text{ kg/s}$$

Hence the mass is increasing at the rate of 28.9 kg/s. To accomplish this, the device could have a spongelike material that absorbs water.

## Example 4.3

A uniform flow of air approaches a cylinder as shown in Fig. E4.3a. The symmetrical velocity distribution at the location shown downstream in the wake of the cylinder is approximated by

$$u(y) = 1.25 + \frac{y^2}{4} \qquad -1 < y < 1$$

where $u(y)$ is in m/s and $y$ is in meters. Determine the mass flux across the surface $AB$ per meter of depth (into the paper). Use $\rho = 1.23 \text{ kg/m}^3$.

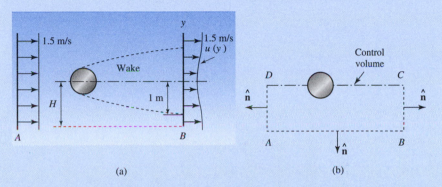

(a)                                        (b)

**Fig. E4.3**

### Solution

Select $ABCD$ as the control volume (Fig. E4.3b). Outside the wake (a region of retarded flow) the velocity is constant at 1.5 m/s. Hence the velocity normal to plane $AD$ is 1.5 m/s. No mass flux crosses the surface $CD$ because of symmetry. Assuming a steady flow, the continuity equation (4.4.3) becomes

$$0 = \int_{c.s.} \rho \mathbf{V} \cdot \hat{\mathbf{n}} \, dA$$

Mass flux occurs across three surfaces: $AB$, $BC$, and $AD$. Thus the equation above takes the form

$$0 = \int_{A_{AB}} \rho \mathbf{V} \cdot \hat{\mathbf{n}} \, dA + \int_{A_{BC}} \rho \mathbf{V} \cdot \hat{\mathbf{n}} \, dA + \int_{A_{AD}} \rho \mathbf{V} \cdot \hat{\mathbf{n}} \, dA$$

$$= \dot{m}_{AB} + \int_0^H \rho u(y) \, 1 \times dy - \rho \text{ kg/m}^3 \times H \text{ m} \times 1 \text{ m} \times 1.5 \text{ m/s}$$

where the negative sign for surface $AD$ results from the fact that the unit vector points out of the volume to the left while the velocity vector points to the right. Recall that a negative sign in the steady-flow continuity equation is always associated with an influx and a positive sign with an outflux. Now, we integrate out to 1 m instead of $H$, since the mass that enters on the left beyond 1 m simply leaves on the right with no net gain or loss. So, letting $H = 1$ m, we have

$$0 = \dot{m}_{AB} + \int_0^1 1.23\left(1.25 + \frac{y^2}{4}\right) dy - 1.23 \times 1 \times 1.5 \text{ kg/s}$$

Perform the integration and there results

$$\dot{m}_{AB} = 0.205 \text{ kg/s per meter}$$

## Example 4.4

A balloon is being inflated with a water supply of $0.6 \text{ m}^3/\text{s}$ (Fig. E4.4a). Find the rate of growth of the radius at the instant when $R = 0.5$ m.

(a)                                             (b)

Fig. E4.4

### Solution

The objective is to find $dR/dt$ when the radius $R = 0.5$ m. This growth rate $V_R = dR/dt$ is the same as the water velocity normal to the wall of the balloon. Therefore, we select as our fixed control volume a sphere with a constant radius of 0.5 m (see Fig. E4.4b) so that we can calculate the velocity of the water at the surface at the instant shown moving radially out at $R = 0.5$ m. The continuity equation is written as

$$0 = \int_{c.v.} \underbrace{\frac{\partial \rho}{\partial t}}_{0} dV + \int_{c.s.} \rho \mathbf{V} \cdot \hat{\mathbf{n}} \, dA$$

The first term is zero because the density of water inside the control volume does not change in time. Further, the water crosses two areas: the inlet area $A_1$ with a velocity $V_1$ and the remainder of the sphere surface $A_R$ with a velocity $V_R$. We will assume that $A_1 \ll A_R$. The continuity equation then takes the form

$$0 = -\rho A_1 V_1 + \rho A_R V_R$$

Since the flow rate into the volume is $A_1 V_1 = 0.6 \text{ m}^3/\text{s}$ and $A_R \simeq 4\pi R^2$ assuming that $A_1$ is quite small, we can solve for $V_R$. At $R = 0.5$ m

$$V_R = \frac{A_1 V_1}{4\pi R^2} = \frac{0.6 \text{ m}^3/\text{s}}{4\pi \times 0.5^2 \text{ m}^2} = 0.191 \text{ m/s}$$

$$\therefore \frac{dR}{dt} = 0.191 \text{ m/s}$$

We have used a fixed control volume and allowed the moving surface of the balloon to pass through it at the instant considered. With this approach it is possible to model situations in which surfaces, such as a piston, are allowed to move.

# Example 4.5

This example shows that there may be more than one good choice for a control volume. We want to determine the rate at which the water level rises in an open container if the water coming in through a 0.10-m$^2$ pipe has a velocity of 0.5 m/s and the flow rate going out is 0.2 m$^3$/s (Fig. E4.5a). The container has a circular cross section with a diameter of 0.5 m.

(a)                              (b)

**Fig. E4.5**

## Solution

First we select a control volume that extends above the water surface as shown in Fig. E4.5a. Apply the continuity equation (Eq. 4.3.2):

$$\frac{d}{dt}\int_{c.v.} \rho \, dV + \rho(-V_1)A_1 + \rho V_2 A_2 = 0$$

in which the first term describes the rate of change of mass in the control volume. Hence, neglecting the airmass above the water, we have

$$\frac{d(\rho h \pi D^2/4)}{dt} - \rho V_1 A_1 + \rho Q_2 = 0$$

Divide by the constant $\rho$,

$$\frac{\pi D^2}{4}\frac{dh}{dt} - V_1 A_1 + Q_2 = 0$$

The rate at which the water level is rising is then

$$\frac{dh}{dt} = \frac{V_1 A_1 - Q_2}{\pi D^2/4}$$

Thus

$$\frac{dh}{dt} = \frac{0.5 \times 0.1 - 0.2}{\pi \times 0.5^2/4} = -0.764 \text{ m/s}$$

The negative sign indicates that the water level is actually decreasing.

Let's solve this problem again but with another choice for the control volume, one with its top surface below the water level (Fig. E4.5b). The velocity at the top surface is then equal to the rate at which the surface rises, i.e., $dh/dt$. The flow condition inside the control volume is steady. Hence we can apply Eq. 4.3.4. There are three areas across which fluid flows. On the third area, the velocity is $dh/dt$; hence the continuity equation takes the form

$$\rho(-V_1)A_1 + \rho Q_2 + \rho \frac{dh}{dt}\frac{\pi}{4}D^2 = 0$$

so that

$$\frac{dh}{dt} = \frac{V_1 A_1 - Q_2}{\pi D^2/4}$$

This is the same result as given above.

## 4.5 ENERGY EQUATION

Many problems involving fluid motion demand that the first law of thermodynamics, often referred to as the *energy equation*, be used to relate quantities of interest. If the heat transferred to a device (a boiler or compressor), or the work done by a device (a pump or turbine), is desired, the energy equation is obviously needed. It is also used to relate pressures and velocities when Bernoulli's equation is not applicable; this is the case whenever viscous effects cannot be neglected, such as flow through a piping system in an industrial plant or a golf course, or in an open channel bringing water to Los Angeles. Let us express the energy equation in control volume form. For a system it is

$$\dot{Q} - \dot{W} = \frac{D}{Dt} \int_{\text{sys}} e\rho \, d\mathcal{V} \tag{4.5.1}$$

where the specific energy $e$ includes specific kinetic energy $V^2/2$, specific potential energy $gz$, and specific internal energy $\tilde{u}$; that is,

$$e = \frac{V^2}{2} + gz + \tilde{u} \tag{4.5.2}$$

We will not include other forms of energy, such as energy due to magnetic or electric field-flow field interactions or those due to chemical reactions. In terms of a control volume, Eq. 4.5.1 becomes

$$\dot{Q} - \dot{W} = \frac{d}{dt} \int_{\text{c.v.}} e\rho \, d\mathcal{V} + \int_{\text{c.s.}} \rho e \mathbf{V} \cdot \hat{\mathbf{n}} \, dA \tag{4.5.3}$$

This can be put in simplified forms for certain restricted flows, but first let us discuss the rate-of-heat transfer term $\dot{Q}$ and the work-rate term $\dot{W}$.

The term $\dot{Q}$ represents the rate-of-energy transfer across the control surface due to a temperature difference. (Do not confuse this term with the flow rate $Q$.) The rate-of-heat transfer term is either given or results from using Eq. 4.5.3. The calculation of $\dot{Q}$ from given temperatures is the objective of a course in heat transfer and is quite difficult, in general, to calculate. An entire course in Heat Transfer is often devoted to the determination of $\dot{Q}$ from given temperatures. The work-rate term is discussed in detail in the following section.

> **KEY CONCEPT**  $\dot{Q}$ represents the rate-of-energy transfer across the control surface due to a temperature difference.

### 4.5.1   Work-Rate Term

The work-rate term results from work being done by the system. Or, since we consider the instant that the system occupies the control volume, we can also state that the work-rate term results from work being done by the control volume. Work of interest in fluid mechanics is due to a force moving through a distance while it acts on the control volume. The rate of doing work $\dot{W}$, or power, is given by the dot product of a force $\mathbf{F}$ with its velocity:

> **KEY CONCEPT**  *The rate of doing work is given by the dot product of a force* **F** *with its velocity.*

$$\dot{W} = -\mathbf{F} \cdot \mathbf{V}_I \tag{4.5.4}$$

where $\mathbf{V}_I$ is the velocity measured with respect to a fixed inertial reference frame. The negative sign results because we have selected the convention that work done on the control volume is negative.

If the force results from a variable stress acting over the control surface, we must integrate,

$$\dot{W} = -\int_{\text{c.s.}} \boldsymbol{\tau} \cdot \mathbf{V}_I \, dA \tag{4.5.5}$$

where $\boldsymbol{\tau}$ is the stress vector acting on the elemental area $dA$, the differential force being represented by $d\mathbf{F} = \boldsymbol{\tau} \, dA$, as shown in Fig. 4.8.

For a moving control volume, such as a car, we have to evaluate the velocity with respect to a fixed reference frame. For example, let us consider an automobile traveling at constant speed (see Example 4.10). If we want to apply the energy equation, we could make the car the control volume. In that case, the velocity in Eq. 4.5.4 would be measured relative to a fixed reference and not relative to the car. If the velocity relative to the car were used, the drag force would have a zero velocity which would result in no work done; but we know that at high speed, energy from the gasoline goes primarily to overcome the drag. Thus the stationary reference frame is needed.

In general, for moving control volumes the velocity vector $\mathbf{V}_I$ is related to a relative velocity $\mathbf{V}$, observed in a reference frame attached to the control volume by

$$\mathbf{V}_I = \mathbf{V} + \dot{\mathbf{S}} + \boldsymbol{\Omega} \times \mathbf{r} \tag{4.5.6}$$

where $\dot{\mathbf{S}}$ is the velocity of the control volume (see Fig. 3.5). We can now write the work rate of Eq. 4.5.5 as

$$\dot{W} = -\int \boldsymbol{\tau} \cdot \mathbf{V} \, dA + \dot{W}_I \tag{4.5.7}$$

where the "inertial work-rate" term is given by

$$\dot{W}_I = -\int_{\text{c.s.}} \boldsymbol{\tau} \cdot (\dot{\mathbf{S}} + \boldsymbol{\Omega} \times \mathbf{r}) \, dA \tag{4.5.8}$$

**Fig. 4.8**   Stress vector acting on the control surface.

Next, express the stress vector as the sum of a normal component and a shear component, as shown in Fig. 4.8, that is,

$$\boldsymbol{\tau} = -p\hat{\mathbf{n}} + \boldsymbol{\tau}_s \tag{4.5.9}$$

where the normal stress component is assumed to be negative the pressure $p$. Then

$$\dot{W} = \int_{\text{c.s.}} p\hat{\mathbf{n}} \cdot \mathbf{V}\, dA - \int_{\text{c.s.}} \boldsymbol{\tau}_s \cdot \mathbf{V}\, dA + \dot{W}_I \tag{4.5.10}$$

**KEY CONCEPT**   *Shaft work $W_S$ is transmitted by a rotating shaft that is cut by the control surface.*

We will allow the shear stress term to consist of two parts. One part accounts for work called *shaft work $\dot{W}_S$* transmitted by a rotating shaft that is cut by the control surface; this term is important when we deal with flows in pumps and turbines. The other part will be denoted *shear work $\dot{W}_{\text{shear}}$* and results from moving boundaries; this term is required if the control surface itself moves relative to the control volume, as occurs with a moving belt.

Hence the work-rate term becomes

$$\dot{W} = \int_{\text{c.s.}} p\hat{\mathbf{n}} \cdot \mathbf{V}\, dA + \dot{W}_S + \dot{W}_{\text{shear}} + \dot{W}_I \tag{4.5.11}$$

The terms are summarized as follows:

$\int p\hat{\mathbf{n}} \cdot \mathbf{V}\, dA$   Work rate resulting from the force due to pressure moving at the control surface. It is often referred to as **flow work**.

**Flow work:**   *Work rate resulting from the force due to pressure moving at the control surface.*

$\dot{W}_S$   Work rate resulting from rotating shafts such as that of a pump or turbine, or the equivalent electric power.

$\dot{W}_{\text{shear}}$   Work rate due to the shear acting on a moving boundary, such as a moving belt.

$\dot{W}_I$   Work rate that occurs when the control volume moves relative to a fixed reference frame.

We should note that the work-rate terms $\dot{W}_{\text{shear}}$ and $\dot{W}_I$ are seldom encountered in problems in an introductory course and are often omitted from textbooks. They are included here for completeness.

### 4.5.2   General Energy Equation

When the work-rate term of Eq. 4.5.11 is substituted into Eq. 4.5.3, we obtain the energy equation in the form

$$\dot{Q} - \dot{W}_S - \dot{W}_{\text{shear}} - \dot{W}_I = \frac{d}{dt}\int_{\text{c.v.}} e\rho\, d\mathcal{V} + \int_{\text{c.s.}} \left(e + \frac{p}{\rho}\right)\rho\hat{\mathbf{n}} \cdot \mathbf{V}\, dA \tag{4.5.12}$$

The work-rate term needed to move the pressure force has been moved to the right-hand side, as is common, and is treated like an energy flux term.

Substitution of Eq. 4.5.2 into Eq. 4.5.12 results in

$$\dot{Q} - \dot{W}_S - \dot{W}_{\text{shear}} - \dot{W}_I = \frac{d}{dt} \int_{\text{c.v.}} \left( \frac{V_I^2}{2} + gz + \tilde{u} \right) \rho \, dV$$

$$+ \int_{\text{c.s.}} \left( \frac{V_I^2}{2} + gz + \tilde{u} + \frac{p}{\rho} \right) \rho \mathbf{V} \cdot \hat{\mathbf{n}} \, dA \qquad (4.5.13)$$

This general form of the energy equation is useful in analyzing fluid flow problems that may include time-dependent effects and nonuniform profiles. Before we simplify the equation for steady flow and uniform profiles, let us introduce the notion of "losses."

In many fluid flows, useful forms of energy (kinetic energy and potential energy) and flow work are converted into unusable energy forms (internal energy or heat transfer). If we assume that the temperature of the control volume remains unchanged, the internal energy does not change and the losses are balanced by heat transfer across the control surface. This heat transfer can be the result of convection, radiation, or conduction at the control surfaces. The theory of heat transfer presents a detailed description of these effects. However, in an introductory fluid mechanics course the sum of these effects is lumped together and denoted $\dot{Q}$. Thus we define **losses** as the sum of all the terms representing unusable forms of energy:

**Losses:** *The sum of all the terms representing unusable forms of energy.*

$$\text{losses} = -\dot{Q} + \frac{d}{dt} \int_{\text{c.v.}} \tilde{u} \rho dV + \int_{\text{c.s.}} \tilde{u} \rho \, \mathbf{V} \cdot \hat{\mathbf{n}} \, dA \qquad (4.5.14)$$

We can now rewrite the energy equation as

$$-\dot{W}_S - \dot{W}_{\text{shear}} - \dot{W}_I = \frac{d}{dt} \int_{\text{c.v.}} \left( \frac{V_I^2}{2} + gz \right) \rho \, dV$$

$$+ \int_{\text{c.s.}} \left( \frac{V_I^2}{2} + gz + \frac{p}{\rho} \right) \rho \, \mathbf{V} \cdot \hat{\mathbf{n}} \, dA + \text{losses} \qquad (4.5.15)$$

Losses are due to two primary effects:

1. Viscosity causes internal friction that results in increased internal energy (temperature increase) or heat transfer.
2. Changes in geometry result in separated flows that require useful energy to maintain the resulting secondary motions in which viscous dissipation occurs.

**KEY CONCEPT** *Losses are due primarily to internal friction and separated flows.*

In a conduit, the losses due to viscous effects are distributed over the entire length, whereas the loss due to a geometry change (a valve, an elbow, an enlargement) is concentrated in the vicinity of the geometry change.

The analytical calculation of losses is rather difficult, particularly when the flow is turbulent. In general, the prediction of losses is based on empirical formulas. Such formulas will be given in subsequent chapters. In this chapter we discuss losses qualitatively, and, in examples and problems, losses will be given. For a pump or a turbine the losses are expressed in terms of the efficiency. For example, if the efficiency of a pump is 80%, the losses would be 20% of the energy input to the pump.

It may be that the objective in a particular fluid flow is to change the internal energy of the fluid, such as in the steam generator (boiler) of a power plant, by the transfer of heat; then the definition of losses above must be altered so that the loss term includes only the dissipative effects of the viscosity of the fluid. Generally, for problems of interest in fluid mechanics, Eq. 4.5.15 is acceptable.

### 4.5.3   Steady Uniform Flow

Consider a steady-flow situation in which there is one entrance and one exit across which uniform profiles can be assumed. Also, assume that $\dot{W}_{shear} = \dot{W}_I = 0$ with $V_I = V$. For such a flow the term $(V^2/2 + gz + p/\rho)$ in Eq. 4.5.15 is constant across the cross section because $V$ is constant (we assume a uniform velocity profile) and the sum of $p/\rho + gz$ is constant if the streamlines at each section are parallel. The energy equation (Eq. 4.5.15) then simplifies to

$$-\dot{W}_S = \rho_2 V_2 A_2\left(\frac{V_2^2}{2} + \frac{p_2}{\rho_2} + gz_2\right) - \rho_1 V_1 A_1\left(\frac{V_1^2}{2} + \frac{p_1}{\rho_1} + gz_1\right) + \text{losses}   \quad (4.5.16)$$

where the subscripts 1 and 2 refer to the entrance and exit, respectively. The mass flux is given by $\dot{m} = \rho_1 A_1 V_1 = \rho_2 A_2 V_2$. After dividing by $\dot{m}g$ we have

$$-\frac{\dot{W}_S}{\dot{m}g} = \frac{V_2^2 - V_1^2}{2g} + \frac{p_2}{\gamma_2} - \frac{p_1}{\gamma_1} + z_2 - z_1 + h_L   \quad (4.5.17)$$

where we have introduced the *head loss* $h_L$, defined to be

$$h_L = \frac{\tilde{u}_2 - \tilde{u}_1}{g} - \frac{\dot{Q}}{\dot{m}g}   \quad (4.5.18)$$

It is often written in terms of a *loss coefficient K* as

$$h_L = K\frac{V^2}{2g}   \quad (4.5.19)$$

where $V$ is most often $V_1$ or $V_2$; if it is not obvious, it will be specified. Loss coefficients are discussed in some detail in Chapter 7 and are tabulated in Table 7.2.

The head loss is referred to as a "head" since it has dimensions of length. We may also refer to $V^2/2g$ as the *velocity head* and $p/\gamma$ as the *pressure head* since those terms also have dimensions of length. Also recall from Chapter 3 that $p/\gamma + z$ is called the *piezometric head*. Further, the sum of the piezometric head and the velocity head is called the *total head*.

The energy equation, in the form of Eq. 4.5.17, is useful in many applications and is, perhaps, the most often used form of the energy equation. If the losses are negligible and if there is no shaft work, we note that the energy equation takes the form

$$\frac{V_2^2}{2g} + \frac{p_2}{\gamma_2} + z_2 = \frac{V_1^2}{2g} + \frac{p_1}{\gamma_1} + z_1 \tag{4.5.20}$$

Observe that the energy equation has been reduced to a form identical with Bernoulli's equation when $\gamma_2 = \gamma_1$ (a constant density flow). We must remember, however, that Bernoulli's equation is a momentum equation applicable along a streamline and the equation above is an energy equation applied between two sections of a flow. It is not surprising that both should predict identical results from the conditions stated because the velocity head is constant over a cross section and the sum of pressure head and elevation remains constant over a cross section.

> **KEY CONCEPT**
> *Bernoulli's equation is applicable along a streamline and the energy equation is applied between two sections.*

The energy equation (4.5.17) may be applied to any steady, uniform flow with one entrance and one exit. The control volume is usually selected such that the entrance and exit sections have a uniform total head. For example, it may be applied to water flow through a long pipeline; the total head at the entrance and exit may then be evaluated conveniently at the center of the pipe entrance and exit. The energy equation may be applied to the flow passing a gate (Fig. 4.9). An appropriate control volume is shown. The total head at the entrance and exit can be evaluated at any point at the entrance and exit, respectively. However, a convenient choice would be the points at the water surface. Thus the energy equation becomes

$$\frac{V_1^2}{2g} + \overset{0}{\cancel{\frac{p_1}{\gamma}}} + h_1 = \frac{V_2^2}{2g} + \overset{0}{\cancel{\frac{p_2}{\gamma}}} + h_2 + h_L \tag{4.5.21}$$

where the pressures have been set to zero. If we had picked the centroids of the entrance and exit, as shown in Fig. 4.9, we would have obtained

$$\frac{V_1^2}{2g} + \frac{p_1}{\gamma} + \frac{h_1}{2} = \frac{V_2^2}{2g} + \frac{p_2}{\gamma} + \frac{h_2}{2} + h_L \tag{4.5.22}$$

We see that this result is the same as in Eq. 4.5.21 if we substitute $p_1 = \gamma h_1/2$ and $p_2 = \gamma h_2/2$. For completeness, it is noted that the losses between 1 and 2 in

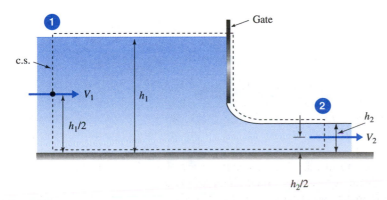

**Fig. 4.9**   Application of the energy equation to a gate in an open channel.

Fig. 4.9 could be neglected because internal viscous effects occur only over a relatively short distance and no significant secondary flows are generated.

The energy equation (4.5.15) can be applied to any control volume. For example, consider steady, uniform incompressible flow through a T-section in a pipe (Fig. 4.10) in which there is one entrance and two exits. The energy equation can be applied to each of two control volumes, one for the mass flux that exits section 2 and the other for the mass flux that exits section 3:

$$
\begin{aligned}
\frac{V_1^2}{2g} + \frac{p_1}{\gamma} + z_1 &= \frac{V_2^2}{2g} + \frac{p_2}{\gamma} + z_2 + h_{L_{1-2}} \\[2mm]
\frac{V_1^2}{2g} + \frac{p_1}{\gamma} + z_1 &= \frac{V_3^2}{2g} + \frac{p_3}{\gamma} + z_3 + h_{L_{1-3}}
\end{aligned}
\tag{4.5.23}
$$

The loss terms in Eq. 4.5.23 include the losses between the inlet and the respective exits. If the losses are negligible, the energy equation reduces to a form similar to the Bernoulli equation being applied along a streamline going from 1 to 2 or a streamline going from 1 to 3.

A final note to this section regards nomenclature for pumps and turbines in a flow system. It is often conventional to call the energy term ($\dot{W}_S/\dot{m}g$) associated with a pump the *pump head* $H_P$, and the term ($\dot{W}_S/\dot{m}g$) associated with a turbine the *turbine head* $H_T$. Then the energy equation, for an incompressible flow, takes the form

$$
H_P + \frac{V_1^2}{2g} + \frac{p_1}{\gamma} + z_1 = H_T + \frac{V_2^2}{2g} + \frac{p_2}{\gamma} + z_2 + h_L
\tag{4.5.24}
$$

**KEY CONCEPT** *$H_P$ and $H_T$ represent the energy that is transferred to and from the fluid, respectively.*

In this form we have equated the energy at the inlet plus added energy to the energy at the exit plus extracted energy (energy per unit weight, of course). If any of the quantities is zero (e.g., there is no pump), the appropriate term is simply omitted. The terms $H_P$ and $H_T$ above represent the energy that is transferred to and from the fluid, respectively. If the energy delivered by the turbine or required by the pump is desired, the efficiency of each device must be used.

**Fig. 4.10**   Application of the energy equation to a T-section.

The power generated by the turbine with an efficiency of $\eta_T$ is simply

$$\dot{W}_T = \dot{m}gH_T\eta_T = \gamma QH_T\eta_T \qquad (4.5.25)$$

The power requirement by a pump with an efficiency of $\eta_P$ would be

$$\dot{W}_P = \frac{\dot{m}gH_P}{\eta_P} = \frac{\gamma QH_P}{\eta_P} \qquad (4.5.26)$$

We will calculate power in watts, ft-lb/sec, or horsepower. Recall that one horsepower is equivalent to 746 W.

### 4.5.4  Steady Nonuniform Flow

If the assumption of uniform velocity profiles is not acceptable for a problem of interest, as is sometimes the situation, we have to consider the control-surface integral in Eq. 4.5.15 with the proper expression for the velocity distribution. In practice, a velocity distribution can be accounted for by introducing the *kinetic-energy correction factor* $\alpha$, defined by

$$\alpha = \frac{\int V^3 dA}{\overline{V}^3 A} \qquad (4.5.27)$$

where $\overline{V}$ is the average velocity over the area $A$, given by Eq. 4.4.11. Then the term that accounts for the flux of kinetic energy in Eq. 4.5.15 is

$$\frac{1}{2}\rho \int_A V^3\, dA = \frac{1}{2}\alpha\rho\, \overline{V}^3\, A \qquad (4.5.28)$$

where we have used $\mathbf{V}\cdot\hat{\mathbf{n}} = V$ and $V_I = V$. Using this factor, we can account for nonuniform velocity distributions by modifying Eq. 4.5.24 to read

$$H_P + \alpha_1\frac{\overline{V}_1^2}{2g} + \frac{p_1}{\gamma} + z_1 = H_T + \alpha_2\frac{\overline{V}_2^2}{2g} + \frac{p_2}{\gamma} + z_2 + h_L \qquad (4.5.29)$$

where $\overline{V}_1$ and $\overline{V}_2$ are the average velocities at sections 1 and 2, respectively. For a flow with a parabolic profile in a pipe we can calculate $\alpha = 2.0$ (see Example 4.9). For most internal turbulent flows, however, the profile is nearly uniform with $\alpha \simeq 1.05$. Hence we simply let $\alpha = 1$ since it is so close to unity; this will always be done unless otherwise stated, since most of the internal flows that we encounter are, in fact, turbulent flows.

**KEY CONCEPT**  *For most internal turbulent flows, we let $\alpha = 1$.*

## Example 4.6

The pump of Fig. E4.6 is to increase the pressure of 0.2 m³/s of water from 200 kPa to 600 kPa. If the pump is 85% efficient, how much electrical power will the pump require? The exit area is 20 cm above the inlet area. Assume inlet and exit areas are equal.

**Fig. E4.6**

### Solution
Equation (4.5.24) across the pump provides

$$H_P = \frac{p_2 - p_1}{\gamma} + z_2 - z_1$$

$$= \frac{(600\,000 - 200\,000)\ \text{N/m}^2}{9810\ \text{N/m}^3} + 0.2\ \text{m} = 41.0\ \text{m}$$

where $V_2 = V_1$ since the inlet and exit areas are equal, and any losses are accounted for with the efficiency of Eq. 4.5.26. That equation provides the power:

$$\dot{W}_P = \frac{\gamma Q H_P}{\eta_P}$$

$$= \frac{9810\ \text{N/m}^3 \times 0.2\ \text{m}^3/\text{s} \times 41.0\ \text{m}}{0.85} = 94\,600\ \text{J/s} \quad \text{or} \quad 94.6\ \text{kW}$$

## Example 4.7

Water flows from a reservoir through a 0.75-m-diameter pipeline to a turbine-generator unit and exits to a river that is 30 m below the reservoir surface. If the flow rate is 2.5 m³/s, and the turbine-generator efficiency is 88%, calculate the power output. Assume the loss coefficient in the pipeline (including the exit) to be $K = 2$.

**Fig. E4.7**

### Solution

Referring to Fig. E4.7, we select the control volume to extend from section 1 to section 2 on the reservoir and river surfaces, where we know the velocities, pressures, and elevations; we consider the water surface of the left reservoir to be the entrance and the water surface of the river to be the exit. The velocity in the pipe is

$$V = \frac{Q}{A} = \frac{2.5}{\pi \times 0.75^2/4} = 5.66 \text{ m/s}$$

Now, consider the energy equation. We will use gage pressures so that $p_1 = p_2 = 0$; the datum is placed through the lower section 2 so that $z_2 = 0$; the velocities $V_1$ and $V_2$ on the reservoir surfaces are negligibly small; $K$ is assumed to be based on the 0.75-m-diameter pipe velocity. The energy equation (4.5.24) then becomes

$$\cancel{H_P}^0 + \cancel{\frac{V_1^2}{2g}}^0 + \cancel{\frac{p_1}{\gamma}}^0 + z_1 = H_T + \cancel{\frac{V_2^2}{2g}}^0 + \cancel{\frac{p_2}{\gamma}}^0 + \cancel{z_2}^0 + K\frac{V^2}{2g}$$

$$30 = H_T + 2\,\frac{5.66^2 \text{ m}^2/\text{s}^2}{2 \times 9.81 \text{ m/s}^2}$$

$$\therefore H_T = 26.73 \text{ m}$$

From this the power output is found using Eq. 4.5.25 to be

$$\dot{W}_T = Q\gamma H_T \eta_T$$
$$= 2.5 \text{ m}^3/\text{s} \times 9807 \text{ N/m}^3 \times 26.73 \text{ m} \times 0.88 = 576.7 \text{ kW} \quad \text{or} \quad 773 \text{ hp}$$

In this example we have used gage pressure; the potential-energy datum was assumed to be placed through section 2, $V_1$ and $V_2$ were assumed to be insignificantly small, and $K$ was assumed to be based on the 0.75-m-diameter pipe velocity.

---

 **Example 4.7a**   Control Volumes, Pipe Flow Virtual Lab, 947–948

## Example 4.8

The venturi meter shown reduces the pipe diameter from 10 cm to a minimum of 5 cm (Fig. E4.8). Calculate the flow rate and the mass flux assuming ideal conditions.

**Fig. E4.8**

### Solution

The control volume is selected as shown such that the entrance and exit correspond to the sections where the pressure information of the manometer can be applied. The manometer's reading is interpreted as follows:

$$p_a = p_b$$
$$p_1 + \gamma(z + 1.2) = p_2 + \gamma z + 13.6\gamma \times 1.2$$

where $z$ is the distance from the pipe centerline to the top of the mercury column. The manometer then gives

$$\frac{p_1 - p_2}{\gamma} = (13.6 - 1) \times 1.2 = 15.12 \text{ m}$$

Continuity (4.4.6) allows us to relate $V_2$ to $V_1$ by

$$V_1 A_1 = V_2 A_2$$
$$\therefore V_2 = \frac{A_1}{A_2} V_1 = \frac{\pi \times 10^3/4}{\pi \times 5^2/4} V_1 = 4V_1$$

The energy equation (4.5.17) assuming ideal conditions (no losses and uniform flow) with $h_L = \dot{W}_S = 0$ takes the form

$$0 = \frac{V_2^2 - V_1^2}{2g} + \frac{p_2 - p_1}{\gamma} + \overset{0}{(z_2 - z_1)}$$
$$= \frac{16V_1^2 - V_1^2}{2g} - 15.12$$
$$\therefore V_1 = 4.45 \text{ m/s}$$

The flow rate is

$$Q = A_1 V_1 = \pi \times 0.05^2 \times 4.45 = 0.0350 \text{ m}^3/\text{s}$$

The mass flux is

$$\dot{m} = \rho Q = 1000 \times 0.035 = 35.0 \text{ kg/s}$$

# Example 4.9

The velocity distribution for a certain flow in a pipe is $V(r) = V_{max}(1 - r^2/r_0^2)$, where $r_0$ is the pipe radius (Fig. E4.9). Determine the kinetic-energy correction factor.

$dA = 2\pi r \, dr$

**Fig. E4.9**

## Solution

To find the kinetic-energy correction factor $\alpha$, we must know the average velocity. It is (combine Eqs. 4.4.10 and 4.4.11)

$$\overline{V} = \frac{\int V \, dA}{A}$$

$$= \frac{1}{\pi r_0^2} \int_0^{r_0} V_{max}\left(1 - \frac{r^2}{r_0^2}\right) 2\pi r \, dr = \frac{2\pi V_{max}}{\pi r_0^2} \int_0^{r_0}\left(r - \frac{r^3}{r_0^2}\right) dr$$

$$= \frac{2V_{max}}{r_0^2}\left(\frac{r_0^2}{2} - \frac{r_0^4}{4r_0^2}\right) = \frac{1}{2} V_{max}$$

Using Eq. 4.5.27, there results

$$\alpha = \frac{\int V^3 \, dA}{\overline{V}^3 A}$$

$$= \frac{\int_0^{r_0} V_{max}^3 (1 - r^2/r_0^2)^3 \, 2\pi r \, dr}{\left(\frac{1}{2} V_{max}\right)^3 \pi r_0^2} = \frac{16}{r_0^2} \int_0^{r_0}\left(1 - \frac{3r^2}{r_0^2} + \frac{3r^4}{r_0^4} - \frac{r^6}{r_0^6}\right) r \, dr$$

$$= \frac{16}{r_0^2}\left(\frac{r_0^2}{2} - \frac{3r_0^2}{4} + \frac{3r_0^2}{6} - \frac{r_0^2}{8}\right) = 2$$

Consequently, the kinetic energy flux associated with a parabolic velocity distribution across a circular area is given by

$$\int \rho \mathbf{V} \cdot \hat{\mathbf{n}} \frac{V^2}{2} \, dA = 2 \times \frac{\dot{m}\overline{V}^2}{2}$$

Parabolic velocity distributions are encountered in laminar flows in pipes and between parallel plates, downstream of inlets and geometry changes (valves, elbows, etc.). The Reynolds number must be quite small, usually less than about 2000.

## Example 4.10

The drag force on an automobile (Fig. E4.10) is approximated by the expression $0.15\rho V_\infty^2 A$, where $A$ is the projected cross-sectional area and $V_\infty$ is the automobile's speed. If $A = 1.2$ m$^2$, calculate the efficiency $\eta$ of the engine if the rate of fuel consumption $f$ (the gas mileage) is 15 km/L and the automobile travels at 90 km/h. Assume that the fuel releases 44 000 kJ/kg during combustion. Neglect the energy lost due to the exhaust gases and coolant and assume that the only resistance to motion is the drag force. Use $\rho_{\text{air}} = 1.12$ kg/m$^3$ and $\rho_{\text{fuel}} = 0.68$ kg/L.

**Fig. E4.10**

## Solution

If the car is taken as the moving control volume (note that the control volume is fixed), as shown, we can simplify the energy equation (Eq. 4.5.3 in combination with 4.5.11) to

$$\dot{Q} - \dot{W}_I = 0$$

since all other terms are negligible; there is no velocity crossing the control volume, so $\mathbf{V} \cdot \hat{\mathbf{n}} = 0$ (neglect the energy of the exhaust gases); there is no shear or shaft work; the energy of the c.v. remains constant. The energy input $\dot{Q}$ which accomplishes useful work is $\eta$ times the energy released during combustion; that is,

$$\dot{Q} = \dot{m}_f \times 44\,000\eta \text{ kJ/s}$$

where $\dot{m}_f$ is the mass flux of the fuel. The mass flux of fuel is determined knowing the rate of fuel consumption $\dot{f}$ and the density of fuel as 0.68 kg/L, as follows:

$$\dot{f} = \frac{\text{distance}}{\text{volume}} = \frac{V_\infty \times \text{time}}{Q \times \text{time}} = \frac{V_\infty}{\dot{m}_f/\rho_f} = \frac{\rho_f V_\infty}{\dot{m}_f}$$

with $V_\infty = 90\,000/3600 = 25$ m/s, we have, using $\dot{f} = 15 \times 1000$ m/L,

$$15 \times 1000 = \frac{0.68 \times 25}{\dot{m}_f}$$

$$\therefore \dot{m}_f = 0.001133 \text{ kg/s}$$

The inertial work-rate term is

$$\dot{W}_I = V_\infty \times \text{drag}$$
$$= 0.15\rho V_\infty^3 A = 0.15 \times 1.12 \times 25^3 \times 1.2 = 3150 \text{ J/s}$$

Equating $\dot{Q} = \dot{W}_I$, we have

$$44\,000\eta \times 0.001133 = 3.15$$

$$\therefore \eta = 0.0632 \quad \text{or} \quad 6.32\%$$

This is obviously a very low percentage, perhaps surprisingly low to the reader. Very little power (3.15 kJ/s = 4.22 hp) is actually needed to propel the automobile at 90 km/h. The relatively large engine, needed primarily for acceleration, is quite inefficient when simply propelling the automobile.

Note the importance of using a stationary reference frame. The reference frame attached to the automobile is an inertial reference frame since it is moving at constant velocity. Yet the energy equation demands a stationary reference frame allowing the energy required by the drag force to be properly included.

## 4.6 MOMENTUM EQUATION

### 4.6.1  General Momentum Equation

Newton's second law, often called the *momentum equation*, states that the resultant force acting on a system equals the rate of change of momentum of the system when measured in an inertial reference frame; that is,

$$\Sigma \mathbf{F} = \frac{D}{Dt} \int_{\text{sys}} \rho \mathbf{V} \, d\mathcal{V} \tag{4.6.1}$$

Using Eq. 4.3.9, with $\eta$ replaced by $\mathbf{V}$, this is written for a control volume as

$$\Sigma \mathbf{F} = \frac{d}{dt} \int_{\text{c.v.}} \rho \mathbf{V} \, d\mathcal{V} + \int_{\text{c.s.}} \rho \mathbf{V}(\mathbf{V} \cdot \hat{\mathbf{n}}) \, dA \tag{4.6.2}$$

where $\mathbf{V} \cdot \hat{\mathbf{n}}$ is simply a scalar for each differential area $dA$. The control surface integral on the right represents the net momentum flux across the control surface of the fluid entering and/or leaving the control volume.

When applying Newton's second law the quantity $\Sigma \mathbf{F}$ represents all forces acting on the control volume. The forces include the surface forces resulting from the surroundings acting on the control surface and body forces that result from gravity and magnetic fields. The momentum equation is often used to determine the forces induced by the flow. For example, the equation allows us to calculate the force on the support of an elbow in a pipeline or the force on a submerged body in a free-surface flow.

When we apply the momentum equation the surrounding fluid and sometimes the entire conduit or container is separated from the control volume. For example, in the horizontal nozzle of Fig. 4.11a, the nozzle and the fluid in the nozzle are isolated. Thus, care must be taken to include the pressure forces shown and the force $\mathbf{F}_{\text{joint}}$. It is convenient to use gage pressures so that the pressure acting on the exterior of the pipe is then zero. Alternatively, we could have selected a control volume that includes only the fluid in the nozzle (Fig. 4.11b). In that case we have to consider the pressure forces at the entrance and exit and the resultant pressure force $\mathbf{F}_{\text{nozzle}}$ of the interior wall of the nozzle on the fluid. A free body of the nozzle excluding the fluid shows that the force $\mathbf{F}_{\text{joint}}$ and $\mathbf{F}_{\text{nozzle}}$ are equal in magnitude. If the problem is to determine the forces exerted by the flow on the nozzle (Fig. 4.11b), we have to reverse the direction of the calculated force $\mathbf{F}_{\text{nozzle}}$. Examples at the end of this section illustrate this.

 Forces, 899–904

> **KEY CONCEPT**  *The momentum equation is used primarily to determine the forces induced by the flow.*

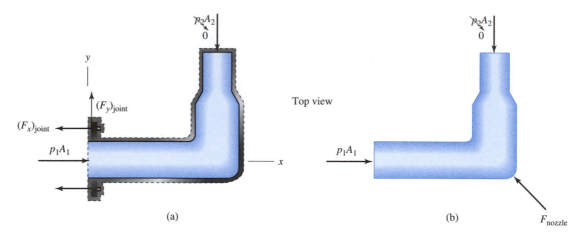

**Fig. 4.11** Forces acting on the control volume of a horizontal nozzle: (a) control volume includes nozzle and fluid in nozzle; (b) control volume includes fluid in nozzle only. We have neglected body forces.

### 4.6.2 Steady Uniform Flow

Equation 4.6.2 can be simplified considerably if a device has entrances and exits across which the flow may be assumed to be uniform and if the flow is steady. Then there results

🔘 Momentum flux, 896–897

$$\Sigma \mathbf{F} = \sum_{i=1}^{N} \rho_i A_i \mathbf{V}_i (\mathbf{V}_i \cdot \hat{\mathbf{n}}) \tag{4.6.3}$$

where $N$ is the number of flow exit/entrance areas.

At an entrance $\mathbf{V} \cdot \hat{\mathbf{n}} = -V$ since the unit vector points out of the volume and at the exit $\mathbf{V} \cdot \hat{\mathbf{n}} = V$. If there is only one entrance and one exit, as in Fig. 4.11, the momentum equation becomes

$$\Sigma \mathbf{F} = \rho_2 A_2 V_2 \mathbf{V}_2 - \rho_1 A_1 V_1 \mathbf{V}_1 \tag{4.6.4}$$

Using continuity,

$$\dot{m} = \rho_1 A_1 V_1 = \rho_2 A_2 V_2$$

the momentum equation takes the simplified form

$$\Sigma \mathbf{F} = \dot{m} \, (\mathbf{V}_2 - \mathbf{V}_1) \tag{4.6.5}$$

Note that the momentum equation is a vector equation which represents the three scalar equations

$$\Sigma \, F_x = \dot{m}(V_{2x} - V_{1x})$$

$$\Sigma \, F_y = \dot{m}(V_{2y} - V_{1y}) \tag{4.6.6}$$

$$\Sigma \, F_z = \dot{m}(V_{2z} - V_{1z})$$

If we consider the nozzle of Fig. 4.11a and we want to determine the $x$-component of the force of the joint on the nozzle, $(\mathbf{V}_1)_x = V_1$ and $(\mathbf{V}_2)_x = 0$ so that the momentum equation for the $x$-direction becomes

$$\Sigma \, F_x = -(F_x)_{\text{joint}} + p_1 A_1 = -\dot{m} \, V_1 \tag{4.6.7}$$

Similarly, we could write the $y$-component equation that would contain the term $(F_y)_{\text{joint}}$.

An example of a free-surface flow in a rectangular channel is shown in Fig. 4.12. If the force of the gate on the flow is desired, the following expression can be derived from the momentum equation:

$$\Sigma \, F_x = -F_{\text{gate}} + F_1 - F_2 = \dot{m}(V_2 - V_1) \tag{4.6.8}$$

where $F_1$ and $F_2$ are pressure forces (see Fig. 4.12).

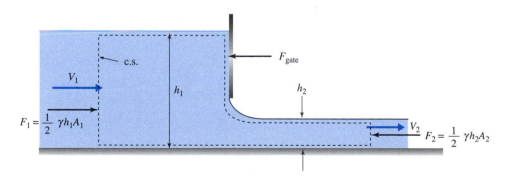

**Fig. 4.12**    Force of the flow on a gate in a free-surface flow.

## Example 4.11

Water flows through a horizontal pipe bend and exits into the atmosphere (Fig. E4.11a). The flow rate is 0.0083 m³/s. Calculate the force in each of the rods holding the pipe bend in position. Neglect body forces and viscous effects and shear force in the rods.

**Fig. E4.11**

### Solution

We have selected a control volume that surrounds the bend, as shown in Fig. E4.11b. Since the rods have been cut, the forces that the rods exert on the control volume are included. The pressure force at the entrance of the control volume is also shown. The flexible section is capable of resisting the interior pressure, but it transmits no axial force or moment. The body force (weight of the control volume) does not act in the $x$- or $y$-direction but normal to it. Therefore, no other forces are shown. The average velocities are found to be

$$V_1 = \frac{Q}{A_1} = \frac{0.0083}{\pi \times (7.5/100)^2/4} = 1.88 \text{ m/s}; \quad V_2 = \frac{Q}{A_2} = \frac{0.0083}{\pi \times (3.75/100)^2/4} = 7.52 \text{ m/s}$$

Before we can calculate the forces $R_x$ and $R_y$ we need to find the pressures $p_1$ and $p_2$. The pressure $p_2$ is zero because the flow exits into the atmosphere. The pressure at section 1 can be determined using the energy equation or the Bernoulli equation. Neglecting losses between sections 1 and 2, the energy equation gives

$$\frac{V_1^2}{2g} + \frac{p_1}{\gamma} = \frac{V_2^2}{2g} + \frac{\cancel{p_2}^{\,0}}{\gamma}$$

$$\therefore p_1 = \frac{\gamma}{2g}(V_2^2 - V_1^2) = \frac{9807 \text{ N/m}^3}{2 \times 9.81 \text{ m/s}^2}(7.52^2 - 1.88^2) \text{ m}^2/\text{s}^2 = 26.5 \text{ kPa}$$

Now we can apply the momentum equation (4.6.5) in the $x$-direction to find $R_x$ and in the $y$-direction to find $R_y$:

$x$-direction:
$$p_1 A_1 - R_x = \dot{m}(\cancel{V_{2x}}^{\,0} - V_{1x})$$

$$26500 \times \frac{\pi}{4}\left(\frac{7.5}{100}\right)^2 - R_x = 1000 \text{ kg/m}^3 \times 0.0083 \text{ m}^3/\text{s} \times (-1.88) \text{ m/s}$$

$$\therefore R_x = 132.6 \text{ N}$$

$y$-direction:
$$R_y = \dot{m}(V_{2y} - \cancel{V_{1y}}^{\,0})$$

$$= 1000 \times 0.0083 \times 7.52 = 62.4 \text{ N}$$

Note that we have assumed uniform profiles and steady flow and used $\dot{m} = \rho Q$. These are the usual assumptions if information is not given otherwise.

---

**Example 4.11a**   Control Volumes, Pipe Elbow Example, 918–923

## Example 4.12

When the velocity of a flow in an open rectangular channel of width $w$ is relatively large, it is possible for the flow to "jump" from a depth $y_1$ to a depth $y_2$ over a relatively short distance, as shown in Fig. E4.12a; this is referred to as a *hydraulic jump*. Express $y_2$ in terms of $y_1$ and $V_1$; assume a horizontal uniform flow.

(a)

(b)

**Fig. E4.12**

### Solution

A control volume is selected as shown in Fig. E4.12b with inlet and exit areas upstream and downstream of the "jump" sufficiently far that the streamlines are parallel to the wall with hydrostatic pressure distributions. Neglecting the drag that is present on the walls (if the distance between the sections is relatively small, the drag force should be negligible), the momentum equation can be manipulated as follows:

$$\Sigma F_x = \dot{m}(V_{2x} - V_{1x})$$

$$F_1 - F_2 = \rho A_1 V_1 (V_2 - V_1)$$

$$\gamma \frac{y_1}{2}(y_1 w) - \gamma \frac{y_2}{2}(y_2 w) = \rho y_1 w V_1 \left( V_1 \frac{y_1}{y_2} - V_1 \right)$$

where we have expressed $F_1$ and $F_2$ using Eq. 2.4.24, and continuity in the form of Eq. 4.4.6, so that

$$V_2 = \frac{y_1}{y_2} V_1$$

The above momentum equation can be simplified to

$$\frac{\gamma}{2}(y_1^2 - y_2^2) = \rho y_1 V_1^2 \frac{y_1 - y_2}{y_2}$$

or

$$\frac{g}{2}(y_1 - y_2)(y_1 + y_2) = \frac{y_1}{y_2} V_1^2 (y_1 - y_2)$$

The factor $(y_1 - y_2)$ is divided out and $y_2$ is found assuming $y_1$ and $V_1$ are known as follows:

$$\frac{g}{2}(y_1 + y_2) = \frac{y_1}{y_2} V_1^2$$

$$y_2^2 + y_1 y_2 - \frac{2}{g} y_1 V_1^2 = 0$$

$$\therefore y_2 = \frac{1}{2} \left( -y_1 + \sqrt{y_1^2 + \frac{8}{g} y_1 V_1^2} \right)$$

where the quadratic formula has been used. The energy equation could now be used to provide an expression for the losses in the hydraulic jump.

## Example 4.12a    Control Volumes, Fire Hose Example, 911–917

## Example 4.13

Consider the symmetrical flow of air around the cylinder. The control volume, excluding the cylinder, is shown in Fig. E4.13. The velocity distribution downstream of the cylinder is approximated with the parabola, as shown. Determine the drag force per meter of length acting on the cylinder. Use $\rho = 1.23$ kg/m$^3$.

**Fig. E4.13**

### Solution

First, we must recognize that not all of the mass flux entering through $AB$ exits through $CD$; consequently, some mass flux must exit $AD$ and $BC$, as shown. The momentum equation (4.6.2) for the steady flow, applied to the control volume $ABCD$, takes the form

$$-F = \int_{c.s.} \rho V_x \mathbf{V} \cdot \hat{\mathbf{n}} \, dA = \int_{A_{CD}} \rho u \mathbf{V} \cdot \hat{\mathbf{n}} \, dA + \int_{A_{AD}} \rho u \mathbf{V} \cdot \hat{\mathbf{n}} \, dA + \int_{A_{BC}} \rho u \mathbf{V} \cdot \hat{\mathbf{n}} \, dA$$

$$+ \int_{A_{AB}} \rho u \mathbf{V} \cdot \hat{\mathbf{n}} \, dA$$

$$= \int_{A_{CD}} \rho u^2 \, dA + U_\infty \dot{m}_{AD} + U_\infty \dot{m}_{BC} - \int_{A_{AB}} \rho u^2 \, dA$$

$$= 2 \int_0^{10} 1.23 \left( 29 + \frac{y^2}{100} \right)^2 dy + 2 \times 30 \dot{m}_{AD} - 1.23 \times 30^2 \times 20$$

where $\dot{m}_{BC} = \dot{m}_{AD}$ is the mass flux crossing $BC$ and $AD$ with the $x$-component velocity equal to 30 m/s. The limit of 10 m was used since at $y = 10$ m, the parabola gives $u(10) = 30$ m/s. We have used Eq. 4.4.9 for $\dot{m}_{AD}$ and $\dot{m}_{BC}$ recognizing that $\mathbf{V} \cdot \hat{\mathbf{n}} = V_n$, which would be the small $y$-component velocity. We now use continuity to find $\dot{m}_{AD}$:

$$0 = \int \rho \hat{\mathbf{n}} \cdot \mathbf{V} \, dA = \int_{A_{AD}} \rho \hat{\mathbf{n}} \cdot \mathbf{V} \, dA + \int_{A_{BC}} \rho \hat{\mathbf{n}} \cdot \mathbf{V} \, dA + \int_{A_{CD}} \rho \hat{\mathbf{n}} \cdot \mathbf{V} \, dA + \int_{A_{AB}} \rho \hat{\mathbf{n}} \cdot \mathbf{V} \, dA$$

$$= \dot{m}_{AD} + \dot{m}_{BC} + 2 \int_0^{10} \rho u(y) \, dy - \rho \times 20 \times 30$$

$$= 2\dot{m}_{AD} + 2 \int_0^{10} 1.23 \times \left( 29 + \frac{y^2}{100} \right) dy - 1.23 \times 20 \times 30$$

$$\therefore \dot{m}_{AD} = 8.2 \text{ kg/s per meter of length}$$

Evaluating the terms in the momentum equation above gives us

$$F = -21\,170 - 492 + 22\,140$$

$$= 478 \text{ N/m}$$

# Example 4.14

Find an expression for the head loss in a sudden expansion in a pipe in terms of $V_1$ and the area ratio (Fig. E4.14a). Assume uniform velocity profiles and assume that the pressure at the sudden enlargement is $p_1$.

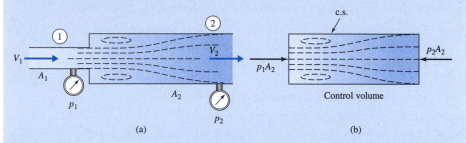

**Fig. E4.14**

## Solution

Figure E4.14a shows a sudden expansion with the diameter changing from $d_1$ to $d_2$. The pressure at the sudden enlargement is closest to $p_1$ since the streamlines are approximately parallel as shown (there is no pressure variation normal to parallel streamlines); they take some distance to again fill the pipe. Hence the force acting on the left end of the control volume shown in Fig. E4.14b is $p_1 A_2$. Newton's second law applied to the control volume yields, assuming uniform profiles,

$$\Sigma F_x = \dot{m}(V_2 - V_1)$$

$$(p_1 - p_2)A_2 = \rho A_2 V_2 (V_2 - V_1)$$

$$\therefore \frac{p_1 - p_2}{\rho} = V_2(V_2 - V_1)$$

The energy equation (4.5.17) provides

$$0 = \frac{V_2^2 - V_1^2}{2g} + \frac{p_2 - p_1}{\gamma} + z_2 \cancel{\nearrow}^{0} z_1 + h_L$$

$$\therefore h_L = \frac{p_1 - p_2}{\gamma} - \frac{V_2^2 - V_1^2}{2g}$$

$$= \frac{V_2(V_2 - V_1)}{g} - \frac{(V_2 + V_1)(V_2 - V_1)}{2g} = \frac{(V_1 - V_2)^2}{2g}$$

To express this in terms of only $V_1$, we can use continuity and relate

$$V_2 = \frac{A_1}{A_2} V_1$$

Then the expression above for the head loss becomes

$$h_L = \left(1 - \frac{A_1}{A_2}\right)^2 \frac{V_1^2}{2g}$$

### 4.6.3 Momentum Equation Applied to Deflectors

The application of the momentum equation to deflectors forms an integral part of the analysis of many turbomachines, such as turbines, pumps, and compressors. In this section we illustrate the steps in such an analysis. It will be separated into two parts: fluid jets deflected by stationary deflectors and fluid jets deflected by moving deflectors. For both problems we will assume the following:

- The pressure external to the fluid jets is everywhere constant so that the pressure in the fluid as it moves over a deflector remains constant.
- The frictional resistance due to the fluid-deflector interaction is negligible so that the relative speed between the deflector surface and the jet stream remains unchanged, a result of Bernoulli's equation.

- Lateral spreading of a plane jet is neglected.
- The body force, the weight of the control volume, is small and will be neglected.

**Stationary Deflector.** Let us first consider the stationary deflector, illustrated in Fig. 4.13. Bernoulli's equation allows us to conclude that the magnitudes of the velocity vectors are equal (i.e., $V_2 = V_1$), since the pressure is assumed to be constant external to the fluid jet and elevation changes are negligible (see Eq. 3.4.9). Assuming steady, uniform flow, the momentum equation takes the form of Eq. 4.6.5, which for the $x$- and $y$-directions becomes

$$-R_x = \dot{m}(V_2 \cos \alpha - V_1) = \dot{m}V_1(\cos \alpha - 1)$$
$$R_y = \dot{m}V_2 \sin \alpha = \dot{m}V_1 \sin \alpha \tag{4.6.9}$$

For given jet conditions the reactive force components can be calculated.

**Moving Deflectors.** The situation involving a moving deflector depends on whether a single deflector is moving (the blade on a snowplow or a water scoop used to slow a high-speed train) or whether a series of deflectors is moving (the vanes on a turbine). Let us first consider the single deflector shown in Fig. 4.14 to be moving in the positive $x$-direction with the speed $V_B$. In a reference frame attached to the stationary nozzle, from which the fluid jet issues, the flow is unsteady[2]; that is, at a particular point in space, the flow situation varies

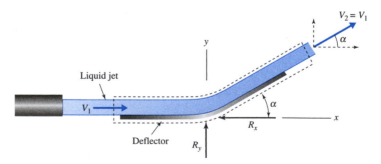

**Fig. 4.13** Stationary deflector.

[2]This is a rather subtle point. To determine whether a flow is steady, we observe the flow at a given point in space. If a flow property changes with time at that point, the flow is unsteady. In this situation, if we focus our attention on a particular point just before the blade, such as point $A$ in Fig. 4.14, first there is no flow, then the blade and the jet pass through the point, then there is again no flow. This is an unsteady flow.

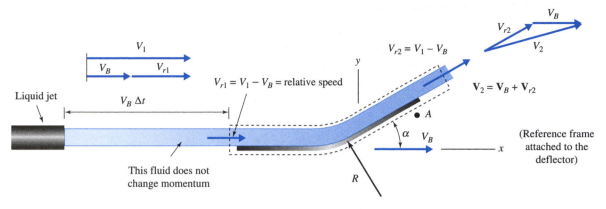

**Fig. 4.14**  Moving deflector.

with time. A steady flow is observed, however, from a reference frame attached to the deflector. From this inertial reference frame, moving with the constant velocity $V_B$, we observe the relative speed $V_{r1}$ entering the control volume to be $V_1 - V_B$, as shown. It is this relative speed that remains constant as the fluid flows relative to the deflector; it does not change since the pressure does not change. Hence, from this moving frame, the momentum equation (4.6.5) takes the forms

$$-R_x = \dot{m}_r(V_1 - V_B)(\cos \alpha - 1)$$
$$R_y = \dot{m}_r(V_1 - V_B) \sin \alpha \qquad (4.6.10)$$

where $\dot{m}_r$ represents only that part of the mass flux exiting the fixed jet that has its momentum changed. Since the deflector moves away from the fixed jet some of the fluid that exits the fixed jet never experiences a momentum change; this fluid is represented by the distance $V_B \Delta t$, shown in Fig. 4.14. Hence

$$\dot{m}_r = \rho A(V_1 - V_B) \qquad (4.6.11)$$

where the relative speed $(V_1 - V_B)$ is used in the calculation; the mass flux $\rho A V_B$ is subtracted from the exiting mass flux $\rho A V_1$ to provide the mass flux $\dot{m}_r$ that experiences a momentum change.

For a series of vanes (a cascade), the jets may be oriented to the side, as shown in Fig. 4.15. The actual force on a particular vane would be zero until the jet strikes the vane; then the force would increase to a maximum and decrease to zero as the vane leaves the jet. We will idealize the situation as follows: Assume that, on the average, the jet is deflected by the vanes as shown in Figs. 4.15 and 4.16a as viewed from a stationary reference frame; the fluid jet enters the vanes with an angle $\beta_1$ and exits with an angle $\beta_2$. What is desired, however, is that the relative velocity enter the vanes tangent to the leading edge of the vanes, that is, $V_{r1}$ in Fig. 4.16b is at the angle $\alpha_1$. The relative speed then remains constant as the fluid travels over the vane with the exiting relative velocity $V_{r2}$ leaving with the vane angle $\alpha_2$. The relative and absolute velocities are related with the velocity equations which are displayed by the velocity polygons of Figs. 4.16b and 4.16c.

**KEY CONCEPT**   *The relative speed remains constant as the fluid travels over a moving vane, i.e., $V_{r2} = V_{r1}$.*

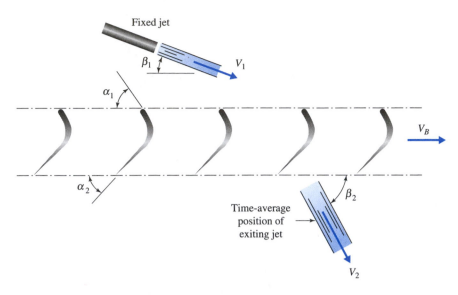

**Fig. 4.15** Fluid striking a series of vanes.

Assuming that all of the mass exiting the fixed jet has its momentum changed, we can write the momentum equation as

$$-R_x = \dot{m}\,(V_{2x} - V_{1x}) \tag{4.6.12}$$

Example 4.17 will illustrate the details.

Interest is usually focused on the x-component of force since it is this component that is related to the power output (or requirement). The power would be found by multiplying the x-component force by the blade speed for each jet; this takes the form

$$\dot{W} = NR_x V_B \tag{4.6.13}$$

where $N$ represents the number of jets. The y-component force does not move in the y-direction so it produces no power.

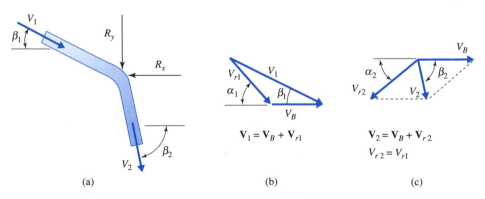

(a)    (b)    (c)

**Fig. 4.16** Detail of the flow situation involving a series of vanes: (a) average position of jet; (b) entrance velocity polygon; (c) exit velocity polygon.

## Example 4.15

A deflector turns a sheet of water through an angle of 30° as shown in Fig. E4.15. What force components are necessary to hold the deflector in place if $\dot{m} = 32$ kg/s?

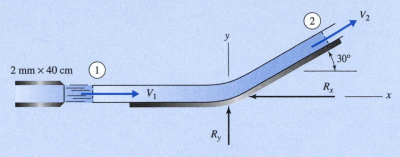

**Fig. E4.15**

### Solution

The control volume we have selected includes the deflector and the water adjacent to it. The only force that is acting on the control volume is due to a support needed to hold the deflector. This force has been decomposed into $R_x$ and $R_y$.

The velocity $V_1$ is found to be

$$V_1 = \frac{\dot{m}}{\rho A_1}$$

$$= \frac{32}{1000 \times 0.002 \times 0.4} = 40 \text{ m/s}$$

Bernoulli's equation (3.4.8) shows that if the pressure does not change, then the magnitude of the velocity does not change, provided that there is no significant change in elevation and that viscous effects are negligible; thus we can conclude that $V_2 = V_1$ since $p_2 = p_1$. Next, the momentum equation is applied in the $x$-direction to find $R_x$ and then in the $y$-direction for $R_y$:

$x$-direction:    $-R_x = \dot{m}(V_{2x} - V_{1x})$

$$= 32 \text{ kg/s } (40 \cos 30° - 40) \text{m/s}$$

$$\therefore R_x = 172 \text{ N}$$

$y$-direction:    $R_y = \dot{m}(V_{2y} - \overset{0}{\cancel{V_{1y}}})$

$$= 32(40 \sin 30°) = 640 \text{ N}$$

## Example 4.16

The deflector shown in Fig. E4.16 moves to the right at 30 m/s while the nozzle remains stationary. Determine (a) the force components needed to support the deflector, (b) $\mathbf{V}_2$ as observed from a fixed observer, and (c) the power generated by the vane. The jet velocity is 80 m/s.

**Fig. E4.16**

### Solution

(a) To solve the problem of a moving deflector, we observe the flow from a reference frame attached to the deflector. In this moving reference frame the flow is steady and Bernoulli's equation with $p_1 = p_2$ can then be used to show that $V_{r1} = V_{r2} = 50$ m/s, the velocity of the sheet of water as observed from the deflector. Note that we cannot apply Bernoulli's equation in a fixed reference frame since the flow would not be steady. Applying the momentum equation to the moving control volume, which is indicated again by the dashed line, we obtain the following:

$$x\text{-direction:} \quad -R_x = \dot{m}_r[(V_{r2})_x - (V_{r1})_x]$$
$$= 1000 \text{ kg/m}^3 \times 0.002 \text{ m} \times 0.4 \text{ m} \times 50 \text{ m/s}\,(50 \cos 30° - 50)\text{ m/s}$$
$$\therefore R_x = 268 \text{ N}$$

$$y\text{-direction:} \quad R_y = \dot{m}_r[(V_{r2})_y - \overset{0}{\cancel{(V_{r1})_y}}]$$
$$= 1000 \times 0.002 \times 0.4 \times 50(50 \sin 30°) = 1000 \text{ N}$$

When calculating $\dot{m}_r$ we must use only that water which has its momentum changed; hence the velocity used is 50 m/s.

(b) Observed from a fixed observer the velocity $\mathbf{V}_2$ of the fluid after the deflection is $\mathbf{V}_2 = \mathbf{V}_{r2} + \mathbf{V}_B$, where $\mathbf{V}_{r2}$ is directed tangential to the deflector at the exit and has a magnitude equal to $V_{r1}$ (see the velocity diagram above). Thus

$$(V_2)_x = V_{r2} \cos 30° + V_B$$
$$= 50 \times 0.866 + 30 = 73.3 \text{ m/s}$$
$$(V_2)_y = V_{r2} \sin 30°$$
$$= 50 \times 0.5 = 25 \text{ m/s}$$

Finally,

$$\mathbf{V}_2 = 73.3\,\hat{\mathbf{i}} + 25\,\hat{\mathbf{j}} \text{ m/s}$$

(c) The power generated by the moving vane is equal to the velocity of the vane times the force the vane exerts in the direction of the motion. Therefore,

$$\dot{W} = V_B \times R_x = 30 \text{ m/s} \times 268 \text{ N} = 8040 \text{ W}$$

# Example 4.17

High-speed air jets strike the blades of a turbine rotor tangentially while the 1.5-m-diameter rotor rotates at 140 rad/s (Fig. E4.17a). There are 10 such 4-cm-diameter jets. Calculate the maximum power output. The air density is 2.4 kg/m³.

**Fig. E4.17**

## Solution

The blade angle $\alpha_1$ is set by demanding that the air jet enter the blades tangentially, as observed from the moving blade; that is, the relative velocity vector $\mathbf{V}_r$ must make the angle $\alpha_1$ with respect to the blade velocity $\mathbf{V}_B$. This is shown in Fig. E4.17b. The relative entrance velocity is $\mathbf{V}_{r1}$ (Fig. E4.17b), and the relative exit velocity is $\mathbf{V}_{r2}$ (Fig. E4.17c). Both velocity polygons are presented by the vector equation

$$\mathbf{V} = \mathbf{V}_r + \mathbf{V}_B$$

which states that the absolute velocity equals the relative velocity plus the blade velocity. From the polygon at the entrance we have

$$V_1 \sin \beta_1 = V_{r1} \sin \alpha_1$$

$$V_1 \cos \beta_1 = V_{r1} \cos \alpha_1 + V_B$$

$$\therefore 200 \sin 30° = V_{r1} \sin \alpha_1$$

$$200 \cos 30° = V_{r1} \cos \alpha_1 + 0.75 \times 140$$

where $V_B$ is the radius multiplied by the angular velocity. A simultaneous solution yields

$$V_{r1} = 121 \text{ m/s} \qquad \alpha_1 = 55.7°$$

The friction between the air and the blade is quite small and can be neglected when calculating the maximum output. This allows us to assume $V_{r2} = V_{r1}$. From the exiting velocity polygon we can write

$$V_B - V_{r2} \cos \alpha_2 = V_2 \cos \beta_2$$

$$V_{r2} \sin \alpha_2 = V_2 \sin \beta_2$$

$$\therefore 0.75 \times 140 - 121 \cos 30° = V_2 \cos \beta_2$$

$$121 \sin 30° = V_2 \sin \beta_2 \qquad \textit{(continued)}$$

A simultaneous solution results in

$$V_2 = 60.5 \text{ m/s} \qquad \beta_2 = 89.8°$$

The momentum equation applied to the control volume, shown in Fig. E4.17d, gives

$$-R_x = \dot{m}(V_{2x} - V_{1x})$$
$$= 2.4 \text{ kg/m}^3 \times \pi \times 0.02^2 \text{ m}^2 \times 200 \text{ m/s}(60.5 \cos 89.8° - 200 \cos 30°) \text{ m/s}$$
$$\therefore R_x = 104.3 \text{ N}$$

There are 10 jets, each producing the force above. The maximum power output is then

$$\text{power} = 10 \times R_x \times V_B$$
$$= 10 \times 104.3 \text{ N} \times (0.75 \times 140) \text{ m/s} = 109\,600 \text{ W} \quad \text{or} \quad 109.6 \text{ kW}$$

### 4.6.4   Momentum Equation Applied to Propellers

The application of the momentum equation to propellers is also of sufficient interest that this section will be devoted to illustrating the procedure. Consider the propeller of Fig. 4.17 with the streamlines shown forming the surface of a control volume in which the fluid enters with a uniform velocity $V_1$ and exits with a uniform velocity $V_2$. The outer streamlines just touch the tips of the propeller. This flow situation can be seen to be identical to that of a propeller moving with velocity $V_1$ in a stagnant fluid by adding $V_1$ to the left in Fig. 4.17. The momentum equation, applied to the large control volume shown, gives

$$F = \dot{m}(V_2 - V_1) \tag{4.6.14}$$

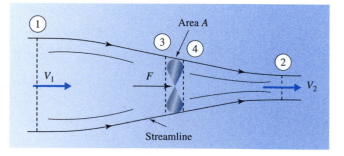

**Fig. 4.17**   Propeller in a fluid flow.

This control volume is not sufficient, however, since the areas $A_1$ and $A_2$ are unknown. We know the flow area $A$ of the propeller. So a control volume is drawn close to the propeller such that $V_3 \cong V_4$ and $A_3 \cong A_4 = A$. The momentum equation (4.6.5) in the $x$-direction gives

$$F + p_3 A - p_4 A = 0 \qquad (4.6.15)$$

or

$$F = (p_4 - p_3)A \qquad (4.6.16)$$

Now, since viscous effects would be quite small in this flow situation the energy equation up to the propeller and then downstream from the propeller is used to obtain

$$\frac{V_1^2 - V_3^2}{2} + \frac{p_1 - p_3}{\rho} = 0 \qquad \text{and} \qquad \frac{V_4^2 - V_2^2}{2} + \frac{p_4 - p_2}{\rho} = 0 \quad (4.6.17)$$

Adding these equations together, recognizing that $p_1 = p_2 = p_{\text{atm}}$, we have

$$(V_2^2 - V_1^2)\frac{\rho}{2} = p_4 - p_3 \qquad (4.6.18)$$

Inserting this and Eq. 4.6.16 into Eq. 4.6.14 results in

$$V_3 = \frac{1}{2}(V_2 + V_1) \qquad (4.6.19)$$

where we have used $\dot{m} = \rho A V_3$ since the propeller area is the only area known. This result shows that the velocity of the fluid moving through the propeller is the average of the upstream and downstream velocities.

The input power needed to produce this effect is found by applying the energy equation between sections 1 and 2, where the pressures are atmospheric; neglecting losses, Eq. 4.5.17 takes the form

$$\dot{W}_{\text{fluid}} = \frac{V_2^2 - V_1^2}{2}\dot{m} \qquad (4.6.20)$$

**KEY CONCEPT** *The velocity of the fluid moving through the propeller is the average of the upstream and downstream velocities.*

where $\dot{W}_{\text{fluid}}$ is the energy input between the two sections. The moving propeller requires power given by

$$\begin{aligned} \dot{W}_{\text{prop}} &= F \times V_1 \\ &= \dot{m} V_1 (V_2 - V_1) \end{aligned}$$

(4.6.21)

The theoretical propeller efficiency is then

$$\eta_P = \frac{\dot{W}_{\text{prop}}}{\dot{W}_{\text{fluid}}} = \frac{V_1}{V_3}$$

(4.6.22)

**KEY CONCEPT**   *In a wind machine, the downstream velocity is reduced and the diameter is increased.*

In contrast to the propeller, a wind machine extracts energy from the airflow; the downstream velocity is reduced and the diameter is increased.

### 4.6.5   Steady Nonuniform Flow

If we cannot assume uniform velocity profiles, we can let the momentum flux be expressed as

$$\int_A V^2 \, dA = \beta \overline{V}^2 A$$

(4.6.23)

where we have introduced the *momentum-correction factor* $\beta$, expressed explicitly as

$$\beta = \frac{\int V^2 \, dA}{\overline{V}^2 A}$$

(4.6.24)

The momentum equation (4.6.5), for a steady flow with one inlet and one exit, can then be written as

$$\Sigma \mathbf{F} = \dot{m}(\beta_2 \mathbf{V}_2 - \beta_1 \mathbf{V}_1)$$

(4.6.25)

For a laminar flow with a parabolic profile in a circular pipe, $\beta = 4/3$. If a profile is given, however, the integral is usually integrated and Eq. 4.6.2 is used.

# Example 4.18

Calculate the momentum correction factor for a parabolic profile (a) between parallel plates and (b) in a circular pipe. The parabolic profiles are shown in Fig. E4.18.

(a) A wide channel          (b) A circular pipe

**Fig. E4.18**

## Solution

(a) A parabolic profile between parallel plates can be expressed as

$$V(y) = V_{max}\left(1 - \frac{y^2}{h^2}\right)$$

where $y$ is measured from the centerline, the velocity is zero at the walls where $y = \pm h$, and $V_{max}$ is the centerline velocity at $y = 0$. First, let us find the average velocity. It is

$$\overline{V} = \frac{1}{A}\int V \, dA$$

$$= \frac{1}{hw}\int_0^h V_{max}\left(1 - \frac{y^2}{h^2}\right)w \, dy = \frac{V_{max}}{h}\left(h - \frac{1}{3}h\right) = \frac{2}{3}V_{max}$$

where we have integrated over the top half of the cross-section. Then

$$\beta = \frac{\int V^2 \, dA}{\overline{V}^2 A} = \frac{2}{\frac{4}{9}V_{max}^2 \times 2hw}\int_0^h V_{max}^2\left(1 - \frac{y^2}{h^2}\right)^2 w \, dy = \frac{6}{5}$$

where the factor "2" in the numerator accounts for the bottom half of the channel.

(b) For a circular pipe a parabolic profile can be written as

$$V(r) = V_{max}\left(1 - \frac{r^2}{R^2}\right)$$

where $R$ is the pipe radius and $V = 0$ at $r = R$. The average velocity is found to be

$$\overline{V} = \frac{1}{A}\int V \, dA = \frac{1}{\pi R^2}\int_0^R V_{max}\left(1 - \frac{r^2}{R^2}\right)2\pi r \, dr = \frac{1}{2}V_{max}$$

The momentum correction factor is then

$$\beta = \frac{\int V^2 dA}{\overline{V}^2 A} = \frac{1}{\frac{1}{4}V_{max}^2 \pi R^2}\int_0^R V_{max}^2\left(1 - \frac{r^2}{R^2}\right)^2 2\pi r \, dr = \frac{4}{3}$$

The correction factors above can be used to express the momentum flux across a cross-sectional area as $\beta\rho A\overline{V}^2$.

### 4.6.6 Noninertial Reference Frames

In certain situations it may be necessary to choose a noninertial reference frame in which the velocity is measured. This would be the case if we were to study the flow through a dishwasher arm, around a turbine blade, or from a rocket. Relative to a noninertial reference frame, Newton's second law takes the form (refer to Eq. 3.2.15)

$$\Sigma \mathbf{F} = \frac{D}{Dt} \int_{\text{sys}} \rho \mathbf{V} \, d\mathcal{V}$$

$$+ \int_{\text{sys}} \left[ \frac{d^2 \mathbf{S}}{dt^2} + 2\boldsymbol{\Omega} \times \mathbf{V} + \boldsymbol{\Omega} \times (\boldsymbol{\Omega} \times \mathbf{r}) + \frac{d\boldsymbol{\Omega}}{dt} \times \mathbf{r} \right] \rho \, d\mathcal{V} \quad (4.6.26)$$

where $\mathbf{V}$ is the velocity relative to the noninertial frame; the acceleration $\mathbf{a}$ of each particle in the system is already accounted for in the first integral. Equation 4.6.26 is often written as

$$\Sigma \mathbf{F} - \mathbf{F}_I = \frac{D}{Dt} \int_{\text{sys}} \rho \mathbf{V} \, d\mathcal{V}$$

$$= \frac{d}{dt} \int_{\text{c.v.}} \rho \mathbf{V} \, d\mathcal{V} + \int_{\text{c.s.}} \rho \mathbf{V} (\mathbf{V} \cdot \hat{\mathbf{n}}) \, dA \quad (4.6.27)$$

where $\mathbf{F}_I$ is called the "inertial body force," given by

$$\mathbf{F}_I = \int_{\text{sys}} \left[ \frac{d^2 \mathbf{S}}{dt^2} + 2\boldsymbol{\Omega} \times \mathbf{V} + \boldsymbol{\Omega} \times (\boldsymbol{\Omega} \times \mathbf{r}) + \frac{d\boldsymbol{\Omega}}{dt} \times \mathbf{r} \right] \rho \, d\mathcal{V} \quad (4.6.28)$$

Since the system and control volume are identical at time $t$ the system integration can be replaced with a control volume integration in the integral of Eq. 4.6.28. Example 4.19 will illustrate the use of a noninertial reference frame.

## Example 4.19

The rocket shown in Fig. E4.19, with an initial mass of 150 kg, burns fuel at the rate of 10 kg/s with a constant exhaust velocity of 700 m/s. What is the initial acceleration of the rocket and the velocity after 1 s? Neglect the drag on the rocket.

**Fig. E4.19**

### Solution

The control volume is sketched and includes the entire rocket. The reference frame attached to the rocket is accelerating upward at $d^2H/dt^2$. Newton's second law is written as, using $z$ upward,

$$\Sigma F_z - (F_I)_z = \frac{d}{dt} \int_{c.v.} \rho V_z \, d\Psi + \int_{c.s.} \rho V_z \mathbf{V} \cdot \hat{\mathbf{n}} \, dA$$

$$\therefore -W - \frac{d^2H}{dt^2} m_{c.v.} = \rho_e(-V_e)V_e A_e$$

where

$$\frac{d}{dt} \int_{c.v.} \rho V_z d\Psi \approx 0$$

since $V_z$ is the velocity of each mass element $\rho \, d\Psi$ relative to the reference frame attached to the control volume; the only vertical force is the weight $W$; and $m_{c.v.}$ is the mass of the control volume. From continuity we see that

$$m_{c.v.} = 150 - \dot{m}t = 150 - 10t$$

$$\therefore W = (150 - 10t) \times 9.81$$

The momentum equation becomes

$$-(150 - 10t) \times 9.81 - \frac{d^2H}{dt^2}(150 - 10t) = -\dot{m}_e V_e = -10 \times 700 = -7000$$

This is written as

$$\frac{d^2H}{dt^2} = \frac{700}{15 - t} - 9.81$$

The initial acceleration is found by letting $t = 0$:

$$\left.\frac{d^2H}{dt^2}\right|_{t=0} = \frac{700}{15} - 9.81 = 36.9 \text{ m/s}^2$$

Integrate the expression for $d^2H/dt^2$ and obtain

$$\frac{dH}{dt} = -700 \ln(15 - t) - 9.81t + C$$

The constant $C = 700 \ln 15$ since $dH/dt = 0$ at $t = 0$. Thus at $t = 1$ s the velocity is

$$\frac{dH}{dt} = 700 \ln \frac{15}{14} - 9.81 \times 1 = 38.5 \text{ m/s}$$

**Example 4.19a**   Pressure-Jet Virtual Lab, 932–935

## 4.7 MOMENT-OF-MOMENTUM EQUATION

In the preceding section we determined the magnitude of force components in a variety of flow situations. To determine the line of action of a given force component, it is often necessary to apply the moment-of-momentum equation. Also, in analyzing the flow situation in devices that have rotating components the moment-of-momentum equation is needed to relate the rotational speed to the other flow parameters. Since it may be advisable to attach the reference frame to the rotating component, we will write the general equation with the inertial forces included. It is (see Eq. 4.2.4)

$$\Sigma \mathbf{M} - \mathbf{M}_I = \frac{D}{Dt} \int_{\text{sys}} \mathbf{r} \times \mathbf{V} \, \rho \, d\mathcal{V} \qquad (4.7.1)$$

where

$$\mathbf{M}_I = \int \mathbf{r} \times \left[ \frac{d^2 \mathbf{S}}{dt^2} + 2\boldsymbol{\Omega} \times \mathbf{V} + \boldsymbol{\Omega} \times (\boldsymbol{\Omega} \times \mathbf{r}) + \frac{d\boldsymbol{\Omega}}{dt} \times \mathbf{r} \right] \rho \, d\mathcal{V} \qquad (4.7.2)$$

**KEY CONCEPT** *The inertial moment $\mathbf{M}_I$ accounts for the fact that a noninertial reference frame was selected.*

This inertial moment $\mathbf{M}_I$ accounts for the fact that a noninertial reference frame was selected; it is simply the moment of $\mathbf{F}_I$ (see Eq. 4.6.28). Applying the system-to-control volume transformation, the moment-of-momentum equation for a control volume becomes

$$\Sigma \mathbf{M} - \mathbf{M}_I = \frac{d}{dt} \int_{\text{c.v.}} \mathbf{r} \times \mathbf{V} \, \rho \, d\mathcal{V} + \int_{\text{c.s.}} \mathbf{r} \times \mathbf{V}(\mathbf{V} \cdot \hat{\mathbf{n}}) \, \rho \, dA \qquad (4.7.3)$$

Examples will illustrate the application of this equation.

## Example 4.20

A sprinkler has four 50-cm-long arms with nozzles at right angles with the arms and 45° with the ground (Fig. E4.20). If the total flow rate is 0.01 m³/s and a nozzle exit diameter is 12 mm, find the rotational speed of the sprinkler. Neglect friction.

**Fig. E4.20**

### Solution

The velocity exiting a nozzle as shown is

$$V_e = \frac{Q}{A}$$

$$= \frac{0.01/4}{\pi \times 0.006^2} = 22.1 \text{ m/s}$$

where the factor 4 accounts for the four exit areas. Attach the reference frame to the rotating arms as shown. Then, recognizing that $\mathbf{r} \times [\boldsymbol{\Omega} \times (\boldsymbol{\Omega} \times \mathbf{r})] = 0$ and assuming a stationary sprinkler so that $d^2\mathbf{S}/dt^2 = 0$ and constant angular velocity so that $d\boldsymbol{\Omega}/dt = 0$, we have

$$\mathbf{M}_I = \int_{c.v.} \mathbf{r} \times (2\boldsymbol{\Omega} \times \mathbf{V})\rho \, d\mathcal{V}$$

$$= 4 \int_0^{0.5} r\hat{\mathbf{i}} \times (2\Omega\hat{\mathbf{k}} \times V\hat{\mathbf{i}})\rho A \, dr$$

$$= 8\rho A V \Omega \hat{\mathbf{k}} \int_0^{0.5} r \, dr = \rho A V \Omega \hat{\mathbf{k}}$$

where the small mass of water in the end nozzles is neglected compared to that in the long arms; the factor 4 again accounts for the four arms (each arm would provide the unit vector $\hat{\mathbf{k}}$). Since there are no external moments applied to the sprinkler about the vertical z-axis, $\Sigma M_z = 0$. For the steady flow Eq. 4.7.3 provides

$$(\Sigma \overset{0}{\mathbf{M}})_z - (\mathbf{M}_I)_z = \int_{c.s.} (\mathbf{r} \times \mathbf{V})_z \, \mathbf{V} \cdot \hat{\mathbf{n}} \, \rho \, dA$$

$$-\rho A V \Omega = 4 \int_{A_{\text{exit}}} [0.5\hat{\mathbf{i}} \times (0.707V_e\hat{\mathbf{k}} - 0.707V_e\hat{\mathbf{j}})]_z \, V_e\rho \, dA$$

$$-V A \Omega = -4 \times 0.5 \times 0.707V_e^2 A_e$$

$$\therefore \Omega = 4 \times 0.5 \times 0.707 \times 22.1 = 31.25 \text{ rad/s}$$

where we have used $AV = A_eV_e$ from continuity considerations.

## Example 4.21

The nozzles of Example 4.20 make an angle of 0° with the ground and 90° with the arms. The water is suddenly turned on at $t = 0$ with the sprinkler motionless. Determine the resulting $\Omega(t)$ if the arm diameter is 24 mm. Neglect friction.

### Solution

The reference frame is again attached to the rotating arms, as sketched in Example 4.20. Referring to the control volume integral of Eq. 4.7.3, we observe that $\mathbf{r} \times \mathbf{V} = 0$ since $\mathbf{r}$ is in the same direction as $\mathbf{V}$ along an arm. Thus Eq. 4.7.3, along with Eq. 4.7.2, takes the form

$$\sum \overset{0}{\cancel{\mathbf{M}}} - 4 \int_0^{0.5} r\hat{\mathbf{i}} \times \left[ 2\Omega\hat{\mathbf{k}} \times V\hat{\mathbf{i}} + \Omega\hat{\mathbf{k}} \times (\Omega\hat{\mathbf{k}} \times r\hat{\mathbf{i}}) + \frac{d\Omega}{dt}\hat{\mathbf{k}} \times r\hat{\mathbf{i}} \right] \rho A \, dr$$

$$= \frac{d}{dt} \int_{\text{c.v.}} r\hat{\mathbf{i}} \times V\hat{\mathbf{i}} \, \rho \, d\Psi + 4 \int_{A_{\text{exit}}} 0.5\hat{\mathbf{i}} \times V_e(-\hat{\mathbf{j}})V_e\rho \, dA$$

Perform the vector operations and divide by $4\rho$,

$$-2AV\Omega \int_0^{0.5} r \, dr - \frac{d\Omega}{dt} A \int_0^{0.5} r^2 \, dr = -0.5V_e^2 A_e$$

The required integration, using $AV = A_e V_e = 0.01 \text{ m}^3/\text{s}$ and $V_e = 2.21$ m/s gives

$$\frac{d\Omega}{dt} + 132.6\Omega = 5862$$

This linear, first-order differential equation is solved by adding the homogeneous solution (suppress the right-hand side) to the particular solution to obtain

$$\Omega(t) = Ce^{-132.6t} + 44.2$$

Using the initial condition $\Omega(0) = 0$, we find that $C = -44.2$. Then

$$\Omega(t) = 44.2(1 - e^{-132.6t}) \text{ rad/s}$$

Observe that as time becomes large, the angular velocity is limited to 44.2 rad/s. If friction were included, this value would be reduced. If 44.2 is multiplied by 0.707 to account for the 45° angle, we obtain the value of Example 4.20.

## 4.8 SUMMARY

In this chapter we have presented the control-volume formulation of the fundamental laws. This formulation is useful when the integrands (the velocities and pressure) are known or can be approximated with an acceptable degree of accuracy. If this is not the case, the differential equations of Chapter 5 must be solved (numerically as in Chapter 14 or analytically as in Chapter 7), or experimental methods must be used to obtain the desired information; much of the remainder of this book is devoted to this task. After unknown velocities and pressures are determined, we often return to the control-volume formulation and calculate integral quantities of interest. Examples would include the lift and drag on an airfoil, the torque on a row of turbine blades, and the oscillating force on a suspension cable of a bridge.

As we have observed in the examples and problems of this chapter, the task of applying the control-volume equations is highly dependent on the appropriate selection of the boundaries of the control volume. Such boundaries are selected at locations either where information is known or where the unknowns appear. Experience is often required in the selection of a control volume, as it is in the selection of a free-body diagram in statics, dynamics, and solid mechanics. The student has undoubtedly gained some of this experience while working through the sections of this chapter. Table 4.1 presents the various forms of the fundamental laws to aid the user in selecting an appropriate form for a particular problem.

**TABLE 4.1**   Integral Forms of the Fundamental Laws

| *Continuity* | *Energy* | *Momentum* |
|---|---|---|

**General Form**

$$0 = \frac{d}{dt}\int_{c.v.} \rho \, d\Psi + \int_{c.s.} \rho \mathbf{V} \cdot \hat{\mathbf{n}} \, dA$$

$$-\Sigma \dot{W} = \frac{d}{dt}\int_{c.v.} \left(\frac{V^2}{2} + gz\right)\rho \, d\Psi$$

$$+ \int_{c.s.} \left(\frac{V^2}{2} + \frac{p}{\rho} + gz\right)\rho \mathbf{V} \cdot \hat{\mathbf{n}} \, dA + \text{losses}$$

$$\Sigma \mathbf{F} = \frac{d}{dt}\int_{c.v.} \rho \mathbf{V} \, d\Psi + \int_{c.s.} \rho \mathbf{V}(\mathbf{V} \cdot \hat{\mathbf{n}}) \, dA$$

**Steady Flow**

$$0 = \int_{c.s.} \rho \mathbf{V} \cdot \hat{\mathbf{n}} \, dA$$

$$-\Sigma \dot{W} = \int_{c.s.} \left(\frac{V^2}{2} + \frac{p}{\rho} + gz\right)\rho \mathbf{V} \cdot \hat{\mathbf{n}} \, dA + \text{losses}$$

$$\Sigma \mathbf{F} = \int_{c.s.} \rho \mathbf{V}(\mathbf{V} \cdot \hat{\mathbf{n}}) dA$$

**Steady Nonuniform Flow[a]**

$$\dot{m} = \rho_1 A_1 \overline{V}_1 = \rho_2 A_2 \overline{V}_2$$

$$\frac{-\Sigma \dot{W}}{\dot{m}g} = \alpha_2 \frac{\overline{V}_2^2}{2g} + \frac{p_2}{\gamma_2} + z_2 - \alpha_1 \frac{\overline{V}_1^2}{2g} - \frac{p_1}{\gamma_1} - z_1 + h_L$$

$$\Sigma F_x = \dot{m}(\beta_2 \overline{V}_{2x} - \beta_1 \overline{V}_{1x})$$

$$\Sigma F_y = \dot{m}(\beta_2 \overline{V}_{2y} - \beta_1 \overline{V}_{1y})$$

**Steady Uniform Form[a]**

$$\dot{m} = \rho_1 A_1 V_1 = \rho_2 A_2 V_2$$

$$-\frac{\Sigma \dot{W}}{\dot{m}g} = \frac{V_2^2}{2g} + \frac{p_2}{\gamma_2} + z_2 - \frac{V_1^2}{2g} - \frac{p_1}{\gamma_1} - z_1 + h_L$$

$$\Sigma \mathbf{F} = \dot{m}(\mathbf{V}_2 - \mathbf{V}_1)$$

**Steady Uniform Incompressible Flow[a]**

$$Q = A_1 V_1 = A_2 V_2$$

$$-\frac{\Sigma \dot{W}}{\dot{m}g} = \frac{V_2^2}{2g} + \frac{p_2}{\gamma} + z_2 - \frac{V_1^2}{2g} - \frac{p_1}{\gamma} - z_1 + h_L$$

$$\Sigma \mathbf{F} = \dot{m}(\mathbf{V}_2 - \mathbf{V}_1)$$

or

$$H_P + \frac{V_1^2}{2g} + \frac{p_1}{\gamma} + z_1 = H_T + \frac{V_2^2}{2g} + \frac{p_2}{\gamma} + z_2 + h_L$$

| | | |
|---|---|---|
| $\dot{m}$ = mass flux | $\alpha$ = kinetic energy correction factor | $h_L$ = head loss |
| $Q$ = flow rate | $= \dfrac{\int V^3 dA}{\overline{V}^3 A}$ | $\Sigma \dot{W} = \dot{W}_S + \dot{W}_{shear} + \dot{W}_I$ |
| $\overline{V}$ = average velocity | $\beta$ = momentum correction factor | $H_P$ = pump head = $\dot{W}_P / \dot{m}g$ |
| $= \dfrac{\int V dA}{A}$ | $= \dfrac{\int V^2 dA}{\overline{V}^2 A}$ | $H_T$ = turbine head = $\dot{W}_T / \dot{m}g$ |

[a]The control volume has one entrance (section 1) and one exit (section 2).

## Fundamentals of Engineering Exam Review Problems

**4.1**  Select the extensive property from the following:
  (A) Temperature     (B) Volume
  (C) Pressure        (D) Density

**4.2**  Air flows through an 8-cm-diameter pipe with an average velocity of 70 m/s with a temperature of 20°C and a pressure of 200 kPa. The mass flux is nearest:
  (A) 3.7 kg/s     (B) 2.37 kg/s
  (C) 1.26 kg/s    (D) 0.84 kg/s

**4.3**  Water flows with a velocity of 3 m/s in an 80-cm-diameter sewer pipe at a depth of 30 cm. The flow rate is nearest:
  (A) 516 L/s
  (B) 721 L/s
  (C) 938 L/s
  (D) 1262 L/s

**4.4**  What is the energy requirement of an 85% efficient pump that transports 40 L/s of water if the pressure increases from 200 kPa to 1200 kPa?

  (A) 4.8 kW

  (B) 14.2 kW

  (C) 34.0 kW

  (D) 47.1 kW

**4.5**  A high-speed water jet is used to cut a material. If the velocity issuing from the 2-mm-diameter jet is 120 m/s, the maximum pressure on the material at the point of impact is nearest:
  (A) 7200 kPa     (B) 3600 kPa
  (C) 735 kPa      (D) 452 kPa

**4.6**  Estimate $V_1$ in Fig. P4.6. Assume the air to be incompressible with $\rho = 1.2$ kg/m$^3$.
  (A) 62 m/s     (B) 40 m/s
  (C) 18 m/s     (D) 10 m/s

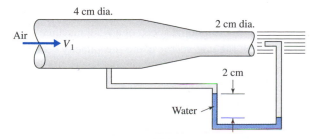

**Fig. P4.6**

**4.7**  The pressure drop across a valve, through which 40 L/s of water flows is measured to be 100 kPa. Estimate the loss coefficient if the nominal diameter of the valve is 8 cm.
  (A) 0.79     (B) 3.2
  (C) 8.7      (D) 31

**4.8**  An 89%-efficient pump is inserted in a 4-cm-diameter line transporting 40 L/s of water. A pressure rise of 400 kPa is desired. The power required by the pump is nearest:
  (A) 12 kW
  (B) 16 kW
  (C) 18 kW
  (D) 22 kW

**4.9**  A hydroturbine generates power by transporting 0.2 m$^3$/s of water from a dam. The water surface is 10 m above the turbine outlet. the overall loss coefficient for the 24-cm-diameter connecting pipe is 3.2. The maximum turbine output is nearest:
  (A) 42 kW
  (B) 21 kW
  (C) 18 kW
  (D) 13 kW

**4.10**  A 75%-efficient pump delivers 0.1 m$^3$/s of water from a reservoir to a device at an elevation of 50 m above the reservoir. The pressure at the 8-cm-diameter entrance to the device is 180 kPa. If the piping loss coefficient is 5.6, the necessary power input to the pump is nearest:
  (A) 263 kW     (B) 203 kW
  (C) 121 kW     (D) 91.3 kW

**4.11**  A strong wind blows directly against a window on a building. The force on the window can be approximated using:
  (A) Bernoulli's equation
  (B) The continuity equation
  (C) The momentum equation
  (D) All of the above

**4.12**  A nozzle with an exit diameter of 4 cm is attached to a 10-cm-diameter pipe transporting 0.1 m$^3$/s of water. The force the water exerts on the nozzle is nearest:
  (A) 6.7 kN     (B) 12.2 kN
  (C) 17.5 kN    (D) 24.2 kN

**4.13**  A 1 cm × 20 cm sheet of water is deflected as shown in Fig. P4.13. The magnitude of the total force acting on the stationary deflector is nearest:
  (A) 6830 N     (B) 5000 N
  (C) 4330 N     (D) 2500 N

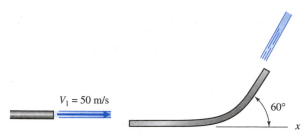

$V_1 = 50$ m/s

$60°$

$x$

**Fig. P4.13**

**4.14**   The water impacts one of the turbine blades as shown in Fig. P4.14. For a blade speed of 20 m/s, the maximum power output for a single jet is nearest:
(A) 18 kW          (B) 154 kW
(C) 206 kW        (D) 309 kW

$V_1 = 60$ m/s

4 cm dia.

$\overline{V_B = 20}$ m/s

$45°$

**Fig. P4.14**

**4.15**   A large vehicle in Fig. P4.15 is slowed by lowering a 2-m-wide scoop into a reservoir of water. Estimate the force exerted on the scoop if the vehicle is traveling at 60 m/s and it scoops off 5 cm of water? The scoop diverts the water through 180°.
(A) 720 kN          (B) 360 kN
(C) 12 kN            (D) 7.2 kN

Vehicle

5 cm

Water

**Fig. P4.15**

## PROBLEMS

### Basic Laws

**4.16**   (a)   State the necessary conditions for momentum of a system to remain constant.
(b)   State the necessary conditions for energy of a system to remain constant.
(c)   Show the detailed steps and state the assumptions that allow Eq. 4.2.3 to be reduced to $\Sigma\,\mathbf{F} = m\mathbf{a}$.

**4.17**   Make a list of five extensive properties that are of interest in fluid mechanics. Also, list their associated intensive properties. In addition, list five other intensive properties.

**4.18**   A control volume is identified as the volume interior to a balloon. At an instant the system is also identified as the air inside the balloon. Air escapes during a short time increment $\Delta t$. Sketch the system and the control volume at $t$ and at $t + \Delta t$.

**4.19**   At an instant in time the control volume and the system occupy the volume inside the pump shown in Fig. P4.19 and a few diameters of the pipe at the inlet side. Sketch the system and the control volume at times $t$ and $t + \Delta t$.

Pump

**Fig. P4.19**

**4.20**   Indicate which fundamental equation would be most useful in determining the following quantity:
(a)   The horsepower output of a pump
(b)   The mass flux from closing baffles
(c)   The drag force on an airfoil
(d)   The head loss in a pipeline
(e)   The rotational speed of a wind machine

**4.21**   Sketch the unit vector $\hat{\mathbf{n}}$ and the velocity vector $\mathbf{V}$ on each of the areas listed:
(a)   The exit area of a fireman's nozzle
(b)   The inlet area of a pump
(c)   The wall area of a pipe
(d)   The porous bottom area of a river into which a small amount of water flows
(e)   The cylindrical exit area of a rotating impeller

**4.22** Fluid flows through the enlargement shown in Fig. P4.22 with a velocity distribution $v_1(r)$ at the inlet and $v_2(r)$ at the exit. Sketch a control volume showing **V** and $\hat{\mathbf{n}}$ at selected locations on the control volume. Include locations on the lateral sides as well as the ends.

$v_1(r)$ $\qquad$ $v_2(r)$

**Fig. P4.22**

**4.23** Sketch the unit vector $\hat{\mathbf{n}}$ and the velocity vector **V** at several positions on a rectangular box surrounding the airfoil shown in Fig. P4.23.

**Fig. P4.23**

## System-to-Control-Volume Transformation

**4.24** Assume that $V_1 = V_2 = V_3 = 10$ m/s for the control volume shown in Fig. P4.24. Write $\hat{\mathbf{n}}_1$, $\hat{\mathbf{n}}_2$, and $\hat{\mathbf{n}}_3$ in terms of $\hat{\mathbf{i}}$, $\hat{\mathbf{j}}$, and $\hat{\mathbf{k}}$, and calculate the normal component of the velocity vector on each of the three plane areas. The volume is of uniform depth in the $z$-direction.

**Fig. P4.24**

**4.25** Write an expression for the flux of a property across each of the three areas of the control volume of Problem 4.24 if $\eta$ and $\rho$ are both constant over the entire control volume. Let the cross sectional area (normal to the $xy$-plane) be $A$. Use $V_1 = V_2 = V_3 = 10$ m/s.

**4.26** Show that $(\mathbf{B} \cdot \hat{\mathbf{n}})A$ is the volume of the parallelepiped of depth 12 cm (Fig. P4.26). Note that $\hat{\mathbf{n}}$ is normal to the area $A$.

**Fig. P4.26**

**4.27** We recognize that

$$\frac{d}{dt}\int_{\text{c.v.}} \rho\eta \, d\mathcal{V} = \int_{\text{c.v.}} \frac{\partial}{\partial t}\rho\eta \, d\mathcal{V}$$

What condition allows this equivalence? Why is an ordinary derivative used on the left and a partial derivative on the right?

**4.28** A spray can for mosquitos is activated at $t = 0$. Select the chemical inside the can as the system and sketch the system at $t = \Delta t$. Select a control volume and sketch the control volume at $t = \Delta t$.

**4.29** The air inside the lungs at the end of an inhaled breath is identified as the system at $t = 0$. Select a control volume and sketch both the system and control volume at $t = \Delta t$ if air is exhaled through the nose only.

**4.30** The control volume selected to analyze flow around an airfoil is the rectangular box as sketched in the figure of Problem 4.23. The system occupies the box at $t$. Sketch the system at $t + \Delta t$.

## Conservation of Mass

**4.31** Show that Eq. 4.4.5 results from Eq. 4.4.4 by assuming one inlet and one exit and uniform flow (constant properties).

**4.32** An incompressible fluid enters a volume filled with an absorbing material with a mass flux $\dot{m}$ and leaves the volume with a flow rate $Q$. Determine an expression for the rate of change of mass in the volume.

**4.33** A liquid of density $\rho$ flows into a volume filled with a sponge at a flow rate $Q_1$. It exits one area with a mass flux $\dot{m}_2$ and a second area $A_3$ with an average velocity $V_3$ as shown in Fig. P4.33. Write an expression for $dm_{\text{sponge}}/dt$, the rate of change of mass in the sponge.

**Fig. P4.33**

**4.34** Water is flowing in a 6-cm-diameter pipe at 18 m/s. If the pipe enlarges to a diameter of 12 cm, calculate the reduced velocity. Also, calculate the mass flux and the flow rate.

**4.35** Water flows in the 5-cm-diameter pipe shown in Fig. P4.35 with an average velocity of 10 m/s. It turns a 90° angle and flows radially between two parallel plates. What is the velocity at a radius of 60 cm? What are the mass flux and the discharge?

**Fig. P4.35**

**4.36** A pipe transports 200 kg/s of water. The pipe tees into a 5-cm-diameter pipe and a 7-cm-diameter pipe (Fig. P4.36). If the average velocity in the smaller-diameter pipe is 25 m/s, calculate the flow rate in the larger pipe.

**Fig. P4.36**

**4.37** Air at 16°C and 276 kPa (abs.) flows in a 10-cm-diameter pipe with a mass flux of 3 kg/s. The pipe undergoes a conversion to a 5 cm by 7.5 cm rectangular duct in which $T = 65°C$ and $p = 48$ kPa (abs.). Calculate the velocity in each section.

**4.38** Air at 120°C and 500 kPa absolute is flowing in a pipe at 600 m/s and suddenly undergoes an abrupt change to 249°C and 1246 kPa absolute at a location where the diameter is 10 cm. Calculate the velocity after the sudden change (a shock wave) illustrated in Fig P4.38. Also, calculate the mass flux and the flow rates before and after the abrupt change.

**Fig. P4.38**

**4.39** A laser velocimeter is used to measure velocities of 40 m/s and 120 m/s before and after an abrupt change in the diameter of a pipe from 10 cm to 6 cm, respectively. The pressure in the air before and after the change is measured to be 200 kPa and 120 kPa, respectively. If the temperature before the change is 20°C, what is the temperature after the change?

**4.40** Water flows with a velocity of 3 m/s at a depth of 1.5 m in a trapezoidal channel with a 2-m base and sides sloping at 45°. It empties into a circular pipe and flows at 2 m/s. What is the diameter if:
**(a)** The pipe flows full?
**(b)** The pipe flows half full?
**(c)** The water in the pipe flows at a depth of one-half the radius?

**4.41** Water flows in a 8-cm-diameter pipe with the profiles shown in Fig. P4.41. Find the average velocity, the mass flux, and the flow rate for each.

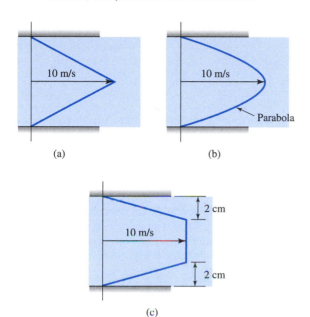

(a)

(b)

(c)

**Fig. P4.41**

**4.42** The profiles in Fig. P4.41 are assumed to exist in a wide rectangular channel, 8 cm high and 80 cm wide. Find the average velocity, the mass flux, and the flow rate for each.

**4.43** A constant-density fluid flows as shown in Fig. P4.43. Find the equation of the parabola if the conduit is:
**(a)** A pipe with $d = 2.5$ cm and $V = 1.8$ m/s
**(b)** A wide rectangular channel with $d = 2.5$ cm and $V = 1.8$ m/s
**(c)** A pipe with $d = 2$ cm and $V = 2$ m/s
**(d)** A wide rectangular channel with $d = 2$ cm and $V = 2$ m/s

**Fig. P4.43**

**4.44** In Example 4.2 let $\dot{m}_2$ be unknown and $V_1$ and $Q_3$ be as shown in the figure. Calculate $\dot{m}_2$ so that $dm/dt$ of the device is zero.

**4.45** A parabolic profile exists in a 10-mm-diameter pipe. The pipe contracts to a 5-mm-diameter pipe in which the velocity profile is essentially uniform at 2 m/s. Write the equation for the parabola. Assume an incompressible flow.

**4.46** As air flows as shown in Fig. P4.46 over a flat plate, the velocity is reduced to zero at the wall. If $u(y) = 10(20y - 100y^2)$ m/s, find the mass flux $\dot{m}$ through a surface parallel to the plate and 0.2 m above the plate. The plate is 2 m wide and $\rho = 1.23$ kg/m$^3$.

**4.47** A streamline is 5 cm above the plate shown in Fig. P4.46 at the leading edge. How far from the plate is that same streamline at the location of the profile $u(y) = 10(20y - 100y^2)$?

**4.48** Stratified salt water flows at a depth of 10 cm in a channel with a velocity distribution $12(y - 5y^2)$ m/s, where $y$ is measured in meters. If the density varies linearly from 1165 kg/m$^3$ at the bottom to clear water at the top, find $\dot{m}$. Also, show that $\dot{m} \neq \bar{\rho}\,\overline{V}A$. The channel is 1.5 m wide.

**4.49** Water flows as shown in Fig. P4.49. Calculate $V_2$.

**Fig. P4.46**

Fig. P4.49

Fig. P4.54

**4.50** Rain is falling vertically downward with an average velocity of 5.0 m/s onto a 9000-m² parking lot. All of the water flows from the lot in an open rectangular ditch with an average velocity of 1.5 m/s. Estimate the depth of flow in the 1.5-m-wide ditch if two thousand 3-mm-diameter drops of water are contained in each cubic meter of rain.

**4.51** Air at 255 kPa gage and 16°C is being forced into a tire, which has a volume of 0.47 m³, at a velocity of 54 m/s through a 6.25-mm-diameter opening. Determine the rate of change of density in the tire.

**4.52** In Fig. P4.52, if the mass of the control volume is not changing, find $\overline{V}_3$.

Fig. P4.52

**4.53** The average velocity $\overline{V}_3 = 10$ m/s in Problem 4.52. Find the rate at which the mass of the control volume is changing.

**4.54** Find the velocity of the gas-fuel interface shown in Fig. P4.54. Use $R_{gas} = 0.28$ kJ/kg·K and $d_e = 30$ cm.

**4.55** Conditioned air is delivered to a large conference room through four inlets, each transferring 40 m³/min. If the air is returned to the conditioner through a single 0.6 m × 1.2 m rectangular duct, estimate the average velocity in the duct. Make any needed assumptions.

**4.56** A rectangular scoop, 80 cm deep, collects air as shown in Fig. P4.56, and delivers it through a 30-cm-diameter pipe. Estimate the average velocity of air in the pipe if $u(y) = 20\, y^{1/5}$ m/s, where $y$ is in meters.

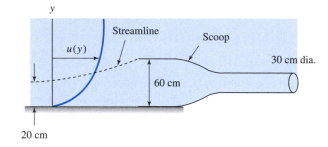

Fig. P4.56

**4.57** The jet pump operates by inducing a flow due to the high velocity in the 5-cm-diameter pipe as shown in Fig. P4.57. The velocity in the small pipe is $200[1 - (r/R)^2]$. Estimate the average velocity at the exit.

Fig. P4.57

**4.58** The experimental setup shown in Fig. P4.58 is used to provide liquid for the tissue. Derive an expression for the storage rate of liquid in the tissue in terms of the relevant information.

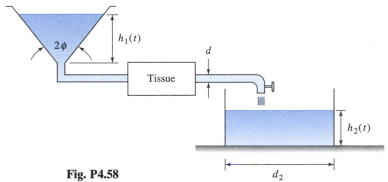

**Fig. P4.58**

**4.59** The water is 4 m deep behind a sluice gate in a rectangular channel that is suddenly opened (Fig. P4.59). Find the initial $dh/dt$ if $V_2 = 8$ m/s and $V_1 = 0.2$ m/s. The upstream channel length is 100 m.

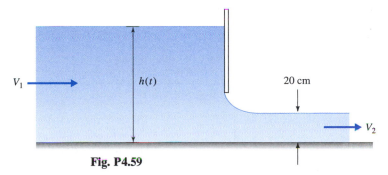

**Fig. P4.59**

**4.60** Water enters a kidney through a tube at 10 mL/min. It exits through a 6-mm-diameter tube at 20 mm/s. What is the rate of change of mass of water in the kidney?

**4.61** The solid fuel in a rocket is burning at the rate of $400 \, e^{-t/100}$ cm$^3$/s (Fig. P4.61). If the density of the fuel is 900 kg/m$^3$, estimate the exit velocity $V_e$ when $t = 10$ s assuming the density of the exiting gases to be 0.2 kg/m$^3$.

**Fig. P4.61**

**4.62** In Fig. P4.62, find the rate of change of $h(t)$ if water is the fluid at all locations:
  **(a)** $V_1 = 10$ m/s, $\dot{m}_2 = 10$ kg/s, $Q_3 = 600$ L/min
  **(b)** $V_1 = 0, \dot{m}_2 = 20$ kg/s, $Q_3 = 10$ L/s
  **(c)** $V_1 = 5$ m/s, $\dot{m}_2 = 10$ kg/s, $Q_3 = 1000$ L/min

**Fig. P4.62**

**4.63** A 1-m diameter cylindrical tank initially contains liquid fuel and has a 2-cm rubber plug at the bottom, as shown in Fig. P4.63. If the plug is removed, how long will it take to empty the tank. The initial height of liquid in the tank is 1.5 m. Bernoulli's equation is required.

**Fig. P4.63**

**4.64** A 1-m³-rigid tank initially contains compressed air at 15°C. Air is being removed from the tank through a 3-cm-diameter pipe, as shown in Fig. P4.64. If the velocity and density of air in the pipe are 200 m/s and 1.8 kg/m³, respectively, determine the initial rate at which the pressure inside the tank will vary assuming a constant temperature.

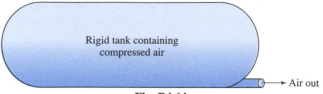

**Fig. P4.64**

**4.65** The velocity of the water entering the volume shown in Fig. P4.65 is $V(t) = 10e^{-t/10}$ m/s. Assuming $h(0) = 0$, find $h(t)$ if the volume is:
(a) A cone. The inlet is 4 cm in diameter.
(b) A 10-m-long trough. The inlet is 4 cm high.

**Fig. P4.65**

## Energy Equation

**4.66** In Fig. P4.66, determine the work rate by the air at the instant shown if $V_{piston} = 10$ m/s, the torque $T = 20$ N·m, the velocity gradient at the belt surface is $100$ s$^{-1}$, and the pressure acting on the piston is 400 Pa. The belt is 80 cm × 50 cm and the 40-cm-high rectangular piston is 50 cm deep (into the paper).

**Fig. P4.66**

**4.67** Assuming that the internal energy of natural gas depends on temperature only, what happens to the losses as natural gas is pumped from Texas to Michigan? The temperature remains essentially constant. Refer to Eq. 4.5.14.

**4.68** An insulated water pump requires 500 W while pumping 0.02 m³/s with an efficiency of 80%.

What is the temperature rise of the water from inlet to exit of the pump assuming equal inlet and exit areas? The specific heat of water is 4.18 kJ/kg°C.

**4.69** A water pump requires 5 hp to create a pump head of 20 m. If its efficiency is 87%, what is the flow rate of water?

**4.70** An 89% efficient hydro turbine operates with a turbine head of 40 m. What is the turbine output if the mass flux is:
(a)  200 kg/s?  (b)  90 000 kg/min?
(c)  $8 \times 10^6$ kg/h?

**4.71** The desired output from a set of 89% efficient turbines on a river is 10 MW. If the maximum turbine head attainable is 50 m, determine the average velocity at a location where the river is 60 m wide and 3 m deep.

**4.72** Water flows in an open rectangular channel at a depth of 1 m with a velocity of 4 m/s. The bottom of the channel drops over a short length a distance of 1 m. Calculate the two possible depths of flow after the drop. Neglect all losses.

**4.73** If the head loss in Problem 4.72 across the channel drop is 0.2 m, determine the two possible depths of flow.

**4.74** Find the velocity $V_1$ of the water in the vertical pipe shown in Fig. P4.74. Assume no losses.

10 cm dia.

$V_1$

2 m

5 cm dia.

40 cm

$V_2$

Hg

**Fig. P4.74**

**4.75** If the head-loss coefficient (based on $V_2$) between sections 1 and 2 of Problem 4.74 is 0.05, determine $V_1$ of the water.

**4.76** The flow rate of water in a 5-cm-diameter horizontal pipeline at a pressure of 414 kPa is 7 L/s. If the pipeline increases to 7.5 cm diameter, calculate the increased pressure after the expansion if the loss coefficient (based on $V_1$) is 0.37.

**4.77** Water flows at the rate of 600 L/min in a 4-cm-diameter horizontal pipeline with a pressure of 690 kPa. If the pressure after an enlargement to 6 cm diameter is measured to be 700 kPa, calculate the head loss across the enlargement.

**4.78** Calculate the pressure $p_1$ shown in Fig. P4.78 needed to maintain a flow rate of 0.08 m³/s of water in a 6-cm-diameter horizontal pipe leading to a nozzle if a loss coefficient based on $V_1$ is 0.2 between the pressure gage and the exit.

6 cm dia.

2 cm dia.

$V_1$

$V_2$

$p_1$

**Fig. P4.78**

**4.79** In Fig. P4.79, neglect all losses and predict the value of $H$ and $p$ if:
(a) $h = 15$ cm  (b) $h = 20$ cm

Water

$H$

14 cm dia.

10 cm dia.

$p$

14 cm dia.

5 cm

$h$

Hg

**Fig. P4.79**

**4.80** Water flows from the rectangular outlets shown in Fig. P4.80. Estimate the flow rate per unit width for each if $h = 80$ cm, $H = 2$ m. Neglect all losses.

$H$

$h$

(a)

$H$

$h$

(b)

**Fig. P4.80**

**4.81** The overall loss coefficient for the pipe shown in Fig. P4.81 is 5; up to $A$ it is 0.8, from $A$ to $B$ it is 1.2, from $B$ to $C$ it is 0.8, from $C$ to $D$ it is 2.2. Estimate the flow rate and the pressures at $A$, $B$, $C$, and $D$. The elevations are shown.

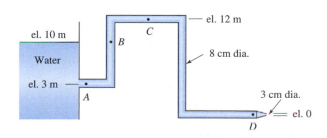

el. 12 m

$C$

el. 10 m

$B$

8 cm dia.

Water

el. 3 m

$A$

3 cm dia.

el. 0

$D$

**Fig. P4.81**

**4.82** Water exits from a pressurized reservoir as shown in Fig. P4.82. Calculate the flow rate if on section *A* we:
**(a)** Attach a nozzle with exit diameter 5 cm
**(b)** Attach a diffuser with exit diameter 18 cm
**(c)** Leave as an open pipe as shown
Neglect losses for all cases.

**Fig. P4.82**

**4.83** Rework Problem 4.82 assuming that $K_{pipe} = 1.5$, $K_{nozzle} = 0.04$ (based on $V_1$), and $K_{diffuser} = 0.8$ (based on $V_1$).

**4.84** Relate the flow rate of the water through the venturi meter shown in Fig. P4.84 to the diameter and the manometer reading. Assume no losses.

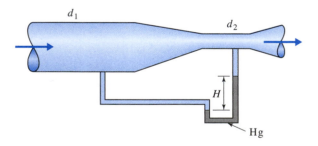

**Fig. P4.84**

**4.85** In the venturi meter of Fig. P4.84, calculate the flow rate if:
**(a)** $H = 20$ cm, $d_1 = 2d_2 = 16$ cm
**(b)** $H = 40$ cm, $d_1 = 3d_2 = 24$ cm

**4.86** In Fig. P4.86, determine the maximum possible height $H$ if cavitation is to be avoided. Let:
$d = 10$ cm and $T = 20°C$
Neglect all losses and assume $P_{atm} = 100$ kPa.

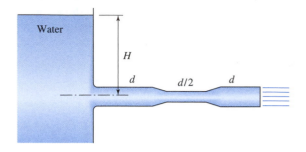

**Fig. P4.86**

**4.87** Cavitation is observed in the small section of pipe of Fig. 4.86 when $H = 65$ cm. Estimate the temperature of the water. Neglect all losses and assume $p_{atm} = 100$ kPa. Use:
**(A)** $d = 10$ cm　　　**(B)** $d = 12$ cm

**4.88** In Fig. P4.88, what is the minimum possible depth $H$ if cavitation is to be avoided? Assume a vapor pressure of 6 kPa absolute and an overall loss coefficient of 8 based on $V_2$ and including the exit loss. Neglect losses up to the enlargement.

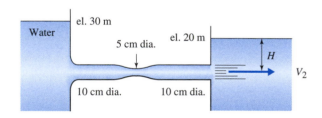

**Fig. P4.88**

**4.89** A contraction occurs in a 10-cm-diameter pipe to 6 cm followed by an enlargement back to 10 cm. The pressure upstream is measured to be 200 kPa when cavitation is first observed in the 20°C water. Calculate the flow rate. Neglect losses. Use $P_{atm} = 100$ kPa.

**4.90** In Fig. P4.90, calculate the maximum diameter $D$ such that cavitation will be avoided if:
$d = 20$ cm, $H = 5$ m, and $T_{water} = 20°C$
Neglect all losses and use $p_{atm} = 100$ kPa.

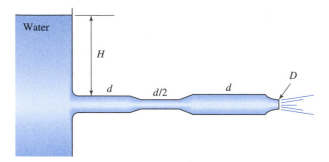

**Fig. P4.90**

**4.91**  The overall loss coefficient in the siphon shown in Fig. P4.91 is 4; up to section $A$ it is 1.5. At what height $H$ will the siphon cease to function?

**Fig. P4.91**

**4.92**  The pump shown in Fig. P4.92 is 85% efficient. If the pressure rise is 830 kPa, calculate the required energy input in horsepower.

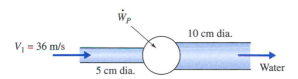

**Fig. P4.92**

**4.93**  The pump shown in Fig. P4.92 is powered by a 20-kW motor. If the pump is 82% efficient, determine the pressure rise.

**4.94**  A turbine, which is 87% efficient, accepts 2 m$^3$/s of water from a 50-cm-diameter pipe. The pressure drop is 600 kPa and the exit velocity is small. What is the turbine output?

**4.95**  A turbine receives 12.4 m$^3$/s of water from a 1.8 m diameter pipe at a pressure of 830 kPa and delivers 10 000 kW. The pressure in the 2.25-m-diameter exit pipe is 124 kPa. Calculate the turbine efficiency.

**4.96**  Air enters a compressor with negligible velocity at 85 kPa absolute and 20°C. It leaves with a velocity of 200 m/s at 600 kPa absolute. For a mass flux of 5 kg/s, calculate the exit temperature if the power required is 1500 kW and:

(a)  No heat transfer occurs

(b)  The heat transfer rate is 60 kW

**4.97**  Air enters a compressor at standard conditions with negligible velocity. At the 5-cm-diameter exit the pressure, temperature, and velocity are 420 kPa (abs.), 150°C, and 200 m/s, respectively. If the heat transfer is 24 kJ/kg of air, find the power required by the compressor.

**4.98**  A small river with a flow rate of 15 m$^3$/s feeds the reservoir shown in Fig. P4.98. Calculate the energy that is available continuously if the turbine is 80% efficient. The loss coefficient for the overall piping system is $K = 4.5$.

**Fig. P4.98**

**4.99**  For the system shown in Fig. P4.99 the average velocity in the pipe is 10 m/s. Up to point $A$, $K = 1.5$, from $B$ to $C$, $K = 6.2$, and the pump is 80% efficient. If $p_C = 200$ kPa find $p_A$ and $p_B$ and the power required by the pump.

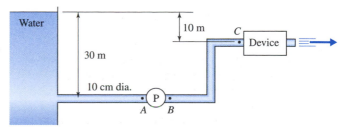

**Fig. P4.99**

**4.100** A gasoline-fueled engine uses a carburetor to mix fuel and air. The gasoline is pumped at a rate of $6.3 \text{ cm}^3/\text{s}$ through 5-mm-diameter tubing from a tank maintained at 100 kPa, as shown in Fig. P4. 100. The gasoline discharges through a 0.8-mm nozzle in the carburetor to a pressure of 95 kPa. The total head loss in the system is $210V^2/2g$, where $V$ is the average velocity in the pipe. If the pump has an efficiency of 75%, determine the power needed by the pump.

**Fig. P4.100**

**4.102** Find the power requirement of the 85%-efficient pump shown in Fig. P4.102 if the loss coefficient up to $A$ is 3.2, and from $B$ to $C$, $K = 1.5$. Neglect the losses through the exit nozzle. Also, calculate $p_A$ and $p_B$.

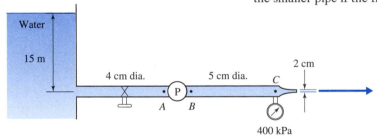

**Fig. P4.102**

**4.104** Neglect losses and find the depth of water on the raised section of the rectangular channel shown in Fig. P4.104. Assume uniform velocity profiles.

**Fig. P4.104**

**4.105** Determine the rate of kinetic energy loss, in watts, due to the cylinder shown in Fig. P4.105. Assume

**4.101** Determine the power output of the turbine shown in Fig. P4.101 for a water flow rate of $0.5 \text{ m}^3/\text{s}$. The turbine is 90% efficient.

**Fig. P4.101**

**4.103** Water leaves a reservoir with a head of 10 m and flows through a horizontal 10-cm-diameter pipe, then exits to the atmosphere. At the end of the pipe a short length of smaller-diameter pipe is added. The loss coefficient, including the reduction, is 2.2, and after the reduction the losses are negligible. Calculate the minimum diameter of the smaller pipe if the flow rate is $0.02 \text{ m}^3/\text{s}$.

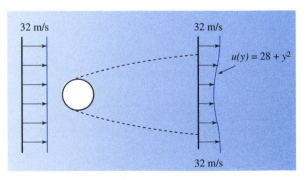

**Fig. P4.105**

a plane flow with $\rho = 1.23 \text{ kg/m}^3$ and make the calculation per meter of cylinder length.

**4.106** Calculate the head loss between the two sections shown in Fig. P4.106. Assume:
 (a) A pipe with $d = 1.2$ cm
 (b) A rectangular duct 1.2 cm $\times$ 8 cm

1.2 cm dia.     Parabola

$V_1 = 8$ m/s     Water

$p_1 = 150$ kPa     $p_2 = 110$ kPa

**Fig. P4.106**

**4.107** Calculate the kinetic-energy correction factor for the velocity profile at the downstream location in Fig. P4.105.

**4.108** Determine the kinetic-energy correction factor if:
 (a) $u(r) = 10(1 - r^2/R^2)$ in a 2-cm-diameter pipe.
 (b) $u(y) = 10(1 - y^2/h^2)$ in a 2-cm-high channel.

**4.109** A turbulent velocity profile in a pipe is often written as $u(r) = u_{max}(1 - r/R)^{1/n}$, where $n$ ranges between 5 and 9, 7 being the most common value. Calculate an expression for the kinetic energy passing a section of pipe and the kinetic-energy correction factor if:
 (a) $n = 5$
 (b) $n = 7$
 (c) $n = 9$

**4.110** A jet aircraft is flying with a velocity $V_\infty$. Use the energy equation to relate the fuel consumption $\dot{m}_f$ to other flow variables such as exhaust gas velocity $V_2$ and temperature $T_2$, inlet velocity $V_1$ and temperature $T_1$, drag $F_D$ acting on the aircraft, inlet air mass flux $\dot{m}$, and the fuel heating value $q_f$ (kJ/kg).

**4.111** An automobile moves at 100 km/h with a drag force of 1340 N. The gas consumption is noted to be 5 km/L. If the engine efficiency is 15%, determine the energy released per kilogram of fuel. The fuel density is 680 kg/m$^3$.

**4.112** A 180-m-long, 2-cm-diameter siphon provides water at 20°C from a reservoir to a field for irrigation. It exits 35 cm below the reservoir surface. The velocity distribution is assumed to be $u(r) = 2V(1 - r^2/r_0^2)$, where $V$ is the average velocity. Determine the flow rate if the head loss is given by $32\nu LV/(D^2 g)$, where $L$ is the siphon's length, $D$ its diameter, and $\nu$ the kinematic viscosity of the water.

**4.113** In the filtration system shown in Fig. P4.113, water is circulated steadily through a filter using a pump. The pump characteristic curve is given by $H_P = 15 + 11Q - 150Q^2$, where $H_P$ is in meters and $Q$ is in m$^3$/s. A constant diameter pipe of 10 cm is used in the system. The total loss coefficient for the system based on the average velocity in the pipe is $K = 51$. Determine the flow rate and power input to the pump.

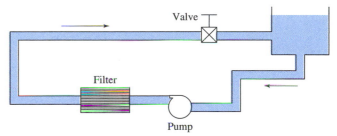

Valve

Filter

Pump

**Fig. P4.113**

**4.114** The manufacturer's pump curve for the pump in the flow system shown in Fig. P4.114a is provided in Fig. P4.114b. Estimate the flow rate. The overall loss coefficient is:

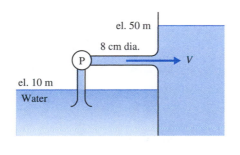

el. 50 m

8 cm dia.

P     $V$

el. 10 m

Water

(a)

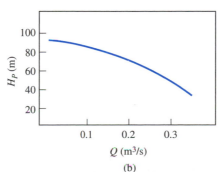

$H_P$ (m)

100

80

60

40

20

0.1    0.2    0.3

$Q$ (m$^3$/s)

(b)

**Fig. P4.114**

**(a)** $K = 5$       **(b)** $K = 20$

The solution involves a trial-and-error procedure, or the energy equation can be written as $H_p = H_p(Q)$ and plotted on the pump curve.

**4.115** A water pump has one inlet and two outlets as shown in Fig. P4.115, all at the same elevation. What pump power is required if the pump is 85% efficient? Neglect pipe losses.

**Fig. P4.115**

## Momentum Equation

**4.116** Water flows in a pipe of diameter $d$ at a pressure $p$. It flows out a nozzle of diameter $d/2$ to the atmosphere. Calculate the force of the water on the nozzle if:
**(a)** $d = 6$ cm, $p = 200$ kPa
**(b)** $d = 6$ cm, $p = 400$ kPa
**(c)** $d = 12$ cm, $p = 200$ kPa

**4.117** A nozzle and hose are attached to the ladder of a fire truck. What force is needed to hold a nozzle supplied by a 9-cm-diameter hose with a pressure of 2000 kPa? The nozzle outlet diameter is 3 cm.

**4.118** Water flows in a 10-cm-diameter pipe at a pressure of 400 kPa out a straight nozzle. Calculate the force of the water on the nozzle if the outlet diameter is:
**(a)** 8 cm       **(c)** 4 cm
**(b)** 6 cm       **(d)** 2 cm

**4.119** Find the horizontal force of the water on the horizontal bend shown in Fig. P4.119.

**Fig. P4.119**

**4.120** Find the horizontal force components of the water on the horizontal bend shown in Fig. P4.120 if $p_1$ is:
**(a)** 200 kPa
**(b)** 400 kPa
**(c)** 800 kPa

**Fig. P4.120**

**4.121** What is the net force needed to hold the orifice plate shown in Fig. P4.121 onto the pipe?

**Fig. P4.121**

**4.122** Assuming uniform velocity profiles, find $F$ needed to hold the plug in the pipe shown in Fig. P4.122. Neglect viscous effects.

Fig. P4.122

**4.123** Neglect viscous effects, assume uniform velocity profiles, and find the horizontal force component acting on the obstruction shown in Fig. P4.123.

Fig. P4.123

**4.124** Assuming hydrostatic pressure distributions, uniform velocity profiles, and negligible viscous effects, find the horizontal force needed to hold the sluice gate in the position shown in Fig. P4.124.

Fig. P4.124

**4.125** A sudden jump (a hydraulic jump) occurs in a rectangular channel as shown in Fig. P4.125. Find $y_2$ and $V_2$ if:
 (a) $V_1 = 8$ m/s, $y_1 = 60$ cm
 (b) $V_1 = 12$ m/s, $y_1 = 40$ cm

Fig. P4.125

**4.126** A hydraulic jump, as shown in Fig. P4.125, occurs so that $V_2 = \frac{1}{4}V_1$. Find $V_1$ and $y_2$ if:
 (a) $y_1 = 80$ cm
 (b) $y_1 = 0.5$ m

**4.127** For a flow rate of 9 m³/s, find $V_2$ and $y_2$ for the hydraulic jump shown in Fig. P4.127. The channel is 3 m wide. Neglect losses up to the jump.

Fig. P4.127

**4.128** The velocity is measured to be 3 m/s downstream of a hydraulic jump where the depth is 2 m. Calculate the velocity and depth prior to the jump.

**4.129** For the system shown in Fig. P4.129, estimate the downstream pressure $p_2$ if $p_1 = 60$ kPa and $V_1 = 20$ m/s. Neglect losses. (*Note:* The pressure immediately after the pipe expansion is $p_1$.)

Fig. P4.129

**4.130** Water flows at 15 m/s in a 10-cm-diameter stem of a horizontal T-section that branches into 5-cm-diameter pipes. Find the force of the water on the T-section if the branches exit to the atmosphere. Neglect viscous effects.

**4.131** Water flows steadily through the double elbow shown in Fig. P4.131. Water flows into the elbow from the top at 5 m/s, and from the left at 15 m/s. Determine the vertical and horizontal components of the force needed to hold the elbow in place.

**Fig. P4.131**

**4.132** A horizontal 10-cm-diameter jet of water with $\dot{m} = 300$ kg/s strikes a vertical plate. Calculate:
(a) The force needed to hold the plate stationary
(b) The force needed to move the plate away from the jet at 10 m/s
(c) The force needed to move the plate into the jet at 10 m/s

**4.133** A horizontal 6 cm-diameter jet of water strikes a vertical plate. Determine the velocity issuing from the jet if a force of 900 N is needed to:
(a) Hold the plate stationary
(b) Move the plate away from the jet at 9 m/s
(c) Move the plate into the jet at 9 m/s

**4.134** Determine the mass flux issuing from the jet shown in Fig. P4.134 if a force of 700 N is needed to:
(a) Hold the cone stationary
(b) Move the cone away from the jet at 8 m/s
(c) Move the cone into the jet at 8 m/s

**Fig. P4.134**

**4.135** Calculate the force components of the water acting on the deflecting blade shown in Fig. P4.135 if:
(a) The blade is stationary
(b) The blade moves to the right at 18 m/s
(c) The blade moves to the left at 18 m/s

**Fig. P4.135**

**4.136** The blade of Problem 4.135 is one of a series of blades that are attached to a 50-cm-radius rotor that has a rotational speed of 30 rad/s. If there are 10 such water jets, find the power output.

**4.137** Determine the force components of superheated steam acting on the blade shown in Fig. P4.137 if:
(a) The blade is stationary
(b) The blade moves to the right at 100 m/s
(c) The blade moves to the left at 100 m/s

**Fig. P4.137**

**4.138** The blade of Problem 4.137 is one of a series of blades that are attached to a 1.2-m-radius rotor that rotates at 150 rad/s. Calculate the power output if there are 15 such steam jets.

**4.139** Superheated steam jets strike the turbine blades shown in Fig. P4.139. Find the power output of the turbine if there are 15 jets and $\alpha_1$ is:
(a) 45°     (b) 60°     (c) 90°

**Fig. P4.139**

**4.140** Twelve high-speed water jets strike the blades as shown in Fig. P4.140. Find the power output and the blade angles if $V_B$ is:
(a) 20 m/s     (b) 40 m/s     (c) 50 m/s

**Fig. P4.140**

**4.141** Fifteen water jets strike the blades of a turbine as shown in Fig. P4.141. Calculate the power output and the blade angles if $\beta_2$ is:
(a) 60°
(b) 70°
(c) 80°

**Fig. P4.141**

**4.142** Water flows from the rectangular jet as shown in Fig. P4.142. Find the force $F$ and the mass fluxes $\dot{m}_2$ and $\dot{m}_3$ if
(a) $b = 20, h = 40$ cm, $V_1 = 40$ m/s

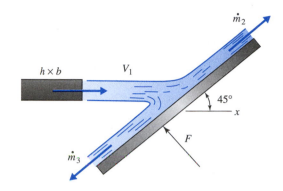

**Fig. P4.142**

**4.143** The plate of Problem 4.142A moves to the left at 20 m/s. Find the power required.

**4.144** Calculate the velocity that the plate of Problem 4.142A must move with (in the $x$-direction) to produce the maximum power output.

**4.145** A large vehicle with a mass of 100 000 kg is slowed by inserting a 180° deflector in a trough of water. If the 60-cm-wide deflector scoops off 10 cm of water, calculate the initial deceleration if the vehicle is traveling at 120 km/h. Also, find the time necessary to reach a speed of 60 km/h.

**4.146** A 2.5-m-wide snowplow travels at 50 km/h plowing snow at a depth of 0.8 m. The snow leaves the blade normal to the direction of motion of the snowplow. What power does the plowing operation require if the density of the snow is 90 kg/m³?

**4.147** A vehicle with a mass of 5000 kg is traveling at 900 km/h. It is decelerated by lowering a 20-cm-wide scoop into water a depth of 6 cm (Fig. P4.147). If the water is deflected through 180°, calculate the distance the vehicle must travel for the speed to be reduced to 100 km/h.

**Fig. P4.147**

**4.148** Modern jet engines have vanes that are deployed to produce a reverse thrust just after a plane touches down on a runway. The air flow rate through a particular engine is 100 kg/s, and the combustion gases exit the engine at a speed of 800 m/s relative to the engine. The fuel-air ratio is 1 to 40 in this case. Calculate the reverse thrust produced by the vanes for the configuration shown in Fig P4. 148.

**Fig. P4.148**

**4.149** A 300 kg vehicle is required to have an initial acceleration of 1.8 m/s². It is proposed that a 6 cm-diameter water jet strike a vane built into the rear of the vehicle that will divert the water through an angle of 180°. What jet velocity is necessary? What speed will be achieved after 2 sec?

**4.150** The swamp boat shown in Fig. P4.150 is propelled at 50 km/h by a 2-m-diameter propeller that requires a 20-kW engine. Calculate the thrust on the boat, the flow rate of air through the propeller, and the propeller efficiency.

**Fig. P4.150**

**4.151** An aircraft is propelled by a 2.2-m-diameter propeller at a speed of 200 km/h. The air velocity downstream of the propeller is 320 km/h relative to the aircraft. Determine the pressure difference across the propeller blades and the required power. Use $\rho = 1.2$ kg/m³.

**4.152** The 50-cm-diameter propeller on a boat moves at 30 km/h by causing a velocity of 60 km/h relative to the boat. Calculate the required power and the mass flux of water through the propeller.

**4.153** A jet boat takes in 0.2 m³/s of water and discharges it at a velocity of 20 m/s relative to the boat. If the boat travels at 10 m/s, calculate the thrust produced and the power required.

**4.154** Calculate the change in the momentum flux of the water flowing through the plane contraction shown in Fig. P4.154 if the flow rate is 0.2 m³/s. The slope of the two profiles is the same. The upstream profile is created by a plate containing slots of various widths.

**Fig. P4.154**

**4.155** Determine the momentum correction factor for the following profile shown in Problem 4.154:
(a) The entering profile
(b) The exiting profile

**4.156** Water at 16°C flows in a 3.8-cm-diameter horizontal pipe and experiences a pressure drop of 210 Pa over a 9-m length of pipe. Calculate the velocity gradient at the wall. Recall that $\tau = \mu \,|\, du/dr \,|$.

**4.157** Find the drag force on the walls between the two sections of the horizontal pipe shown in Fig. P4.157.

**Fig. P4.157**

**4.158** The velocity distribution downstream of a 10-m-long circular cylinder is as shown in Fig. P4.158. Determine the force of the air on the cylinder. Use $\rho = 1.23$ kg/m³.

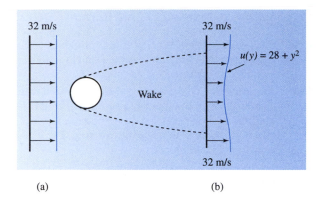

(a)                              (b)

**Fig. P4.158**

**4.159** Calculate the drag force acting on the flat 2-m-wide plate shown in Fig. P4.159. Outside the viscous region the velocity is uniform. Select:

**(a)** A rectangular control volume extending outside the viscous region (mass flux crosses the top).

**(b)** A control volume with the upper boundary a streamline (no mass flux crosses a streamline).

**Fig. P4.159**

---

## Momentum and Energy

**4.160** Determine the power lost in the hydraulic jump of:
**(a)** Problem 4.125a    **(b)** Problem 4.127
**(c)** Problem 4.128

**4.161** Calculate the loss coefficient for the expansion of Problem 4.129. Base the coefficient on the velocity $V_1$.

**4.162** Find a relationship between the acceleration of the cylindrical cart and the variables shown in Fig. P4.162. Neglect friction. The initial mass of the cart and water is $m_0$.

**Fig. P4.162**

**4.163** Set up the equations necessary to determine $H(t)$ for the air/water rocket shown in Fig. P4.163.

**Fig. P4.163**

## Moment of Momentum

**4.164** A four-armed water sprinkler has nozzles at right angles to the 30-cm-long arms and at 45° angles with the ground. If the outlet diameters are 8 mm and 4 kg/s of water exits the four nozzles, find the rotational speed.

**4.165** A four-armed rotor has 1.2 cm-diameter nozzles that exit water at 60 m/s relative to the arm. The nozzles are at right angles to the 25-cm-long arms and are parallel to the ground. If the rotational speed is 30 rad/sec, find the power output. The arms are 3.8 cm in diameter.

**4.166** Water flows out the 6-mm slots as shown in Fig. P4.166. Calculate $\Omega$ if 20 kg/s is delivered by the two arms.

**Fig. P4.166**

**4.167** A 1-kW motor drives the rotor shown in Fig. P4.167 at 500 rad/s. Determine the flow rate neglecting all losses. Use $\rho = 1.23$ kg/m$^3$.

**4.168** Find an expression for $\Omega(t)$ if the sprinkler of Problem 4.164 is suddenly turned on at $t = 0$. Assume the arms to be 2 cm in diameter.

**4.169** Air enters the centrifugal-type air pump of a leaf blower through the blue area shown in Fig. P4.169. The 10-cm-diameter 1.2-m-long tube has an attached nozzle with a 30-cm$^2$ exit area. The exit velocity is 240 km/h.

**(a)** Calculate the discharge.

**(b)** If the overall loss coefficient is 1.2, estimate the pump head.

**(c)** What power must the pump supply to the air?

**(d)** If the pump is 65% efficient, what is the required horsepower of the gasoline engine?

**(e)** Estimate the pressure at the tube entrance (just downstream of the pump).

**(f)** If the 10-kg blower hangs from a strap, what force must be applied at the handle located 30 cm above the nozzle? The center of gravity is 70 cm above and 120 cm to the left of the exit.

**Fig. P4.169**

**Fig. P4.167**

A commercial jetliner taking off. Current improvements in aircraft efficiency and speed are due in part to solving the differential equations of motion of air as particles move relative to the jet at subsonic and transonic speeds. (Mayskyphoto/Shutterstock)

# 5

# The Differential Forms of the Fundamental Laws

## Outline

## Chapter Objectives

The objectives of this chapter are to derive the differential equations and state the boundary and initial conditions needed to solve for the velocity and pressure fields in a fluid flow. The partial differential equations include:

▲ The continuity equation
▲ The momentum equations for inviscid flows (Euler's equations)
▲ The momentum equations for viscous flows (Navier–Stokes equations)
▲ The vorticity equations
▲ The energy equation

Numerous examples and problems will illustrate several relatively simple applications of the differential equations along with how to simplify the equations depending on the flow situation.

## 5.1 INTRODUCTION

The material in this chapter can be omitted in an introductory course. The subsequent chapters in this book have been designed to allow two possible routes: The general differential equations presented in this chapter may be used, or equations unique to a particular geometry may be derived without referring to these general equations.

In Chapter 4 the basic laws were expressed in terms of a fixed control volume, a finite volume in space. This is often described as the global approach to fluid mechanics. To find a solution using the control volume it was necessary to assume an approximation to the integrands (primarily the velocity and pressure distributions), or else expressions for the integrands were given. Suppose that we wish to find an integral quantity, such as the flow rate under a dam or the lift on an airfoil, and we cannot make a reasonable assumption for the velocity or pressure distribution. It is then necessary to determine the distributions that enter the integrands before the integral quantity can be found. This is done by solving the partial differential equations that express the basic laws.

> **KEY CONCEPT** *It is necessary to determine the distributions that enter the integrands before the integral quantity can be found.*

Not only do solutions of the differential forms of the basic laws allow us to determine integral quantities of interest, they often contain information in themselves. For example, we may wish to know the exact location of the minimum pressure on a body, or the region of separated flow from a surface may be of interest. So we often solve the differential equations to answer a specific question raised about a special flow.

There are two primary methods used in deriving the differential forms of the fundamental laws. One method involves the application of Gauss's theorem, which allows the area integrals of the basic equations of Chapter 4 to be transformed to volume integrals; the integrands are then collected under one integral, which can be set equal to zero. The integration is valid over any arbitrary control volume, and thus the integrand itself can be set equal to zero, providing us with the differential form of the basic law. The other approach, the one used in this book, is to identify an infinitesimal element in space and apply the basic laws directly to that element. Both methods result in the differential forms of the basic laws; the first method, however, demands the use of vector and tensor calculus, mathematics that is usually considered unnecessary in a first course in fluids. We introduce some vector calculus, but it will not be used at the operational level demanded by Gauss's theorem.

The conservation of mass, applied to an infinitesimal element, leads to the differential continuity equation; it relates the density and velocity fields.[1] Newton's second law (a vector relationship) results in three partial differential equations known as the Navier–Stokes equations; they relate the velocity, pressure, and density fields and include the viscosity and the gravity vector. The first law of thermodynamics provides us with the differential energy equation, which relates the temperature field to the velocity, density, and pressure fields and introduces the specific heat and thermal conductivity. Most problems considered in an introductory course are for isothermal, incompressible flows in which the

---

[1] When a dependent variable depends on more than one independent variable, it is referred to as a field; that is, $\mathbf{V}(x, y)$ is a velocity field and $p(x, y, z, t)$ is a pressure field. The partial differential equations describing the field quantities are often called *field equations*.

temperature field does not play a role; for such flows the three Navier-Stokes equations and the continuity equation provide four partial differential equations that relate the three velocity components and the pressure. Thus the energy equation would not be needed. We will, however, derive the differential energy equation for use in a limited number of situations.

Partial differential equations require conditions that specify certain values for the dependent variables at particular values of the independent variables. If the independent variable is time, the conditions are called **initial conditions**; if the independent variable is a space coordinate, the conditions are **boundary conditions**. The total problem is referred to as an *initial-value problem* or a *boundary-value problem*.

In fluid mechanics the boundary conditions result from:

- The no-slip conditions for a viscous flow. The viscosity causes the fluid to stick to the wall, and thus the velocity of the fluid at the wall assumes the velocity of the wall. Usually the velocity of the wall is zero.
- The normal component of velocity in an inviscid flow, a flow in which viscous effects are negligible. Near a non-porous wall the velocity vector must be tangent to the wall, demanding that the normal component be zero.
- The pressure in a flow involving a free surface. For problems with a free surface, such as a flow with a liquid–gas interface, the pressure is known on the interface. This would be the situation involving wave motion or in a separated flow involving cavitation.
- The temperature of the boundary or the temperature gradient at the boundary. If the temperature of the boundary is held constant, the temperature of the fluid next to the boundary will equal the boundary temperature. If the boundary is insulated, the temperature gradient will be zero at the boundary.

In an unsteady flow the initial condition demands that the three velocity components be specified at all points in the flow at a particular instant in time, usually taken as $t = 0$. This information would be very difficult to obtain in many situations, e.g., the atmosphere; to obtain the three velocity components everywhere in the atmosphere at an instant is obviously an improbable task.

Differential equations take very different forms depending on the coordinate system chosen. We derive the equations using Cartesian coordinates and then express the equations in vector form. Forms using both cylindrical and spherical coordinates are presented in Table 5.1 at the end of the chapter.

## 5.2 DIFFERENTIAL CONTINUITY EQUATION

Let us begin our quest for the partial differential equations that model the detailed motion of a fluid by applying the conservation of mass to a small volume in a fluid flow. Consider the mass flux through each face of the fixed infinitesimal control volume shown in Fig. 5.1. We set the net flux of mass entering the element equal to the rate of change of the mass of the element; that is,

$$\dot{m}_{\text{in}} - \dot{m}_{\text{out}} = \frac{\partial}{\partial t} m_{\text{element}} \qquad (5.2.1)$$

**Initial conditions:** *Conditions that depend on time.*

**Boundary conditions:** *Conditions that depend on a space coordinate.*

**KEY CONCEPT** *Viscosity causes the velocity of the fluid at the wall to assume the velocity of the wall.*

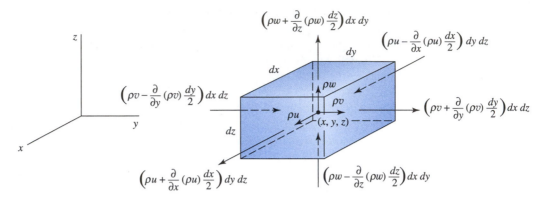

**Fig. 5.1** Infinitesimal control volume using Cartesian coordinates.

To perform this mass balance we identify $\rho u$, $\rho v$, and $\rho w$ at the center of the element and then treat each of these quantities as a single variable. See the analysis of pressure associated with Fig. 2.3. Refer to Fig. 5.1 showing the mass flux through each of the six faces; Eq. 5.2.1 takes the form

$$\left[\rho v - \frac{\partial(\rho v)}{\partial y}\frac{dx}{2}\right]dy\,dz - \left[\rho u + \frac{\partial(\rho u)}{ax}\frac{dx}{2}\right]dy\,dz$$

$$+ \left[\rho v - \frac{\partial(\rho v)}{\partial y}\frac{dx}{2}\right]dx\,dz - \left[\rho v + \frac{\partial(\rho v)}{ay}\frac{dx}{2}\right]dy\,dz$$

$$+ \left[\rho w - \frac{\partial(\rho w)}{\partial z}\frac{dz}{2}\right]dx\,dy - \left[\rho w + \frac{\partial(\rho w)}{\partial z}\frac{dz}{2}\right]dx\,dy$$

$$= \frac{\partial}{\partial t}(\rho\,dx\,dy\,dz) \tag{5.2.2}$$

Subtracting the appropriate terms and dividing by $dx\,dy\,dz$ yields

$$\frac{\partial}{\partial x}(\rho u) + \frac{\partial}{\partial y}(\rho v) + \frac{\partial}{\partial z}(\rho w) = -\frac{\partial\rho}{\partial t} \tag{5.2.3}$$

Since density is considered a variable, we differentiate the products and put Eq. 5.2.3 in the form

$$\frac{\partial\rho}{at} + u\frac{\partial\rho}{\partial x} + v\frac{\partial}{ax} + w\frac{\partial\rho}{\partial z} + \rho\left(\frac{\partial u}{\partial x} + \frac{\partial v}{\partial y} + \frac{\partial w}{\partial z}\right) = 0 \tag{5.2.4}$$

or, in terms of the substantial derivative (see Eq. 3.2.11),

**Differential continuity equation:** *The differential equation that results from the conservation of mass.*

$$\frac{D\rho}{Dt} + \rho\left(\frac{\partial u}{\partial x} + \frac{\partial v}{\partial y} + \frac{\partial w}{\partial z}\right) = 0 \tag{5.2.5}$$

This is the most general form of the **differential continuity equation** expressed in Cartesian coordinates.

We can introduce the *gradient operator*, called "del," which, in rectangular coordinates, is

$$\nabla = \frac{\partial}{\partial x}\,\hat{\mathbf{i}} + \frac{\partial}{\partial y}\,\hat{\mathbf{j}} + \frac{\partial}{\partial z}\,\hat{\mathbf{k}} \qquad (5.2.6)$$

The continuity equation can then be written in the form

$$\frac{D\rho}{Dt} + \rho\,\nabla\cdot\mathbf{V} = 0 \qquad (5.2.7)$$

where $\mathbf{V} = u\hat{\mathbf{i}} + v\hat{\mathbf{j}} + w\hat{\mathbf{k}}$; $\nabla \cdot \mathbf{V}$ is called the *divergence* of the velocity. This form of the continuity equation does not refer to any particular coordinate system, and thus is used to express the continuity equation using various coordinate systems.

For the case of *incompressible flow*, a flow in which the density of a fluid particle does not change as it travels along, we see that

$$\frac{D\rho}{Dt} = \frac{\partial \rho}{\partial t} + u\frac{\partial \rho}{\partial x} + v\frac{\partial \rho}{\partial y} + w\frac{\partial \rho}{\partial z} = 0 \qquad (5.2.8)$$

Note that this is less restrictive than the assumption of constant density, which would require each term in Eq. 5.2.8 to be zero. Incompressible flows that have density gradients are sometimes referred to as *stratified flows* or *nonhomogeneous flows*; atmospheric and oceanic flows are examples of such flows. Using Eq. 5.2.5, the continuity equation for an incompressible flow takes the form

$$\frac{\partial u}{\partial x} + \frac{\partial v}{\partial y} + \frac{\partial w}{\partial z} = 0 \qquad (5.2.9)$$

or, in vector form,

$$\nabla\cdot\mathbf{V} = 0 \qquad (5.2.10)$$

**KEY CONCEPT** *For an incompressible flow, $\nabla\cdot\mathbf{V} = 0$.*

The divergence of the velocity vector is zero for an incompressible flow.

In cylindrical and spherical coordinates the continuity equation for an incompressible flow is presented in Table 5.1. The expression for $D/Dt$ in cylindrical and spherical coordinates can also be found in Table 5.1.

## Example 5.1

The $x$-component velocity is given by $u(x, y) = Ay^2$ in an incompressible plane flow. Determine $v(x, y)$ if $v(x, 0) = 0$, as would be the case in flow between parallel plates.

### Solution

The differential continuity equation for an incompressible, plane flow is

$$\frac{\partial u}{\partial x} + \frac{\partial v}{\partial y} = 0$$

since in a plane flow the two velocity components depend only on $x$ and $y$. Using the given $u(x, y)$ we find that

$$\frac{\partial v}{\partial y} + \frac{\partial u}{\partial x} = \frac{\partial}{\partial y}(Ay^2) = 0$$

Since this is a partial differential equation, its solution is

$$v(x, y) = f(x)$$

But $v(x, 0) = 0$ requiring that $f(x) = 0$. Consequently,

$$v(x, y) = 0$$

is the $y$-component velocity demanded by the conservation of mass. In order for $v(x, y)$ to be non-zero, $u(x, y)$ would have to vary with $x$ or $v(x, 0)$ would have to be nonzero.

## Example 5.2

Air flows in a pipe and the velocity at three neighboring points $A$, $B$, and $C$, 10 cm apart, is measured to be 83, 86, and 88 m/s, respectively, as shown in Fig. E5.2. The temperature and pressure are 10°C and 345 kPa (abs.), respectively, at point $B$. Approximate $d\rho/dx$ at that point, assuming steady, uniform flow.

**Fig. E5.2**

### Solution

The continuity equation (5.2.4) for this steady $\left(\frac{\partial}{\partial t} = 0\right)$, uniform $\left(\frac{\partial}{\partial y} = \frac{\partial}{\partial z} = 0\right)$ flow reduces to

$$u\frac{d\rho}{dx} + \rho\frac{du}{dx} = 0$$

We used ordinary derivatives since $u$ and $\rho$ depend only on $x$. The velocity derivative is approximated by

$$\frac{du}{dx} \approx \frac{\Delta u}{\Delta x} = \frac{88 - 83}{0.2} = 25\frac{\text{m/s}}{\text{m}}$$

where the more accurate central difference has been used.[2] The density is

---

[2]Using a forward difference $\Delta u/\Delta x = (97 - 95)/0.05 = 40$; a backward difference provides $\Delta u/\Delta x = (95 - 91)/0.05 = 80$. These two approximations are less accurate than the central difference used in the example. A sketch of a general curve $u(x)$ could graphically show this by displaying three points at $x - \Delta x, x,$ and $x + \Delta x$ and sketching the slopes.

$$\rho = \frac{p}{RT} = \frac{345 \times 10^3 \text{ N/m}^2}{287 \text{ J/kg-K} \times 283 \text{ K}} = 4.25 \text{ kg/m}^3$$

where absolute pressure and temperature are used. The density derivative is then approximated to be

$$\frac{d\rho}{dx} = -\frac{\rho}{u}\frac{du}{dx}$$

$$= -\frac{4.25 \text{ kg/m}^3}{86 \text{ m/s}} \times 25 \text{ s}^{-1} = -1.23 \text{ kg/m}^4$$

## Example 5.3

The $x$-component of velocity at points $A$, $B$, $C$, and $D$, which are 10 mm apart, is measured to be 5.76, 6.72, 7.61, and 8.47 m/s, respectively, in the plane steady, symmetrical, incompressible flow shown in Fig. E5.3 in which $w = 0$. Approximate the $x$-component acceleration at $C$ and the $y$-component of velocity 6 mm above $B$.

**Fig. E5.3**

### Solution

The desired acceleration component on the centerline is found from Eq. 3.2.9 to be

$$a_x = \cancel{\frac{\partial u}{\partial t}}^{0} + u\cancel{\frac{\partial u}{\partial x}} + v\frac{\partial u}{\partial y}^{0} + \cancel{w}\frac{\partial u}{\partial z}^{0}$$

$$\cong u\frac{\Delta u}{\Delta x} = 7.61\frac{8.47 - 6.72}{0.02} = 666 \text{ m/s}^2$$

where we have assumed a symmetrical flow so that $v$ along the centerline is zero. We have used central differences to approximate $\partial u/\partial x$ at point $C$, as done in Example 5.2 (see footnote 2).

The $y$-component of velocity 6 mm above $B$ is found using the continuity equation (5.2.9) as follows:

$$\frac{\partial v}{\partial y} = -\frac{\partial u}{\partial x}$$

$$\frac{\Delta v}{\Delta y} \simeq -\frac{\Delta u}{\Delta x} = -\frac{7.61 - 5.76}{0.02} = -92.5$$

$$\therefore \Delta v = -92.5 \, \Delta y = -92.5 \times 0.006 = -0.555 \text{ m/s}$$

We know that $v = 0$ at $B$; hence at the desired location, with $\Delta v = v - v_B$, there results

$$v = -0.555 \text{ m/s}$$

## Example 5.4

The continuity equation can be used to change the form of an expression. Write the expression $\rho \, D\tilde{u}/Dt + p \, \boldsymbol{\nabla} \cdot \mathbf{V}$, which appears in the differential energy equation, in terms of enthalpy $h$ rather than internal energy $\tilde{u}$. Recall that $h = \tilde{u} + p/\rho$ (see Eq. 1.7.11).

**Solution**

Using the definition of enthalpy, we can write

$$\frac{D\tilde{u}}{Dt} = \frac{Dh}{Dt} - \frac{1}{\rho}\frac{Dp}{Dt} + \frac{p}{\rho^2}\frac{D\rho}{Dt}$$

where we used

$$\frac{D}{Dt}\left(\frac{p}{\rho}\right) = \frac{1}{\rho}\frac{Dp}{Dt} - \frac{p}{\rho^2}\frac{D\rho}{Dt}$$

The desired expression is then

$$\rho\frac{D\tilde{u}}{Dt} + p\,\boldsymbol{\nabla}\cdot\mathbf{V} = \rho\frac{Dh}{Dt} - \frac{Dp}{Dt} + \frac{p}{\rho}\frac{D\rho}{Dt} + p\,\boldsymbol{\nabla}\cdot\mathbf{V}$$

The continuity equation (5.2.7) is introduced resulting in

$$\rho\frac{D\tilde{u}}{Dt} + p\,\boldsymbol{\nabla}\cdot\mathbf{V} = \rho\frac{Dh}{Dt} - \frac{Dp}{Dt} + \frac{p}{\rho}\frac{D\rho}{Dt} + p\left(-\frac{1}{\rho}\frac{D\rho}{Dt}\right)$$

$$= \rho\frac{Dh}{Dt} - \frac{Dp}{Dt}$$

and enthalpy has been introduced.

## 5.3 DIFFERENTIAL MOMENTUM EQUATION

### 5.3.1  General Formulation

Suppose that we do not know the velocity field or the pressure field in an incompressible[3] flow of interest and we wish to solve differential equations to provide us with that information. The differential continuity equation is one differential equation to help us toward this end; however, it has three unknowns, the three velocity components. The differential momentum equation is a vector equation and thus provides us with three scalar equations. These component equations will aid us in our attempt to determine the velocity and pressure fields. There is a difficulty in deriving these equations, however, since we must use the stress components to determine the forces required in the momentum equation. Let us identify these stress components.

There are nine stress components that act at a particular point in a fluid flow. They are the nine components of the stress tensor $\tau_{ij}$. We will not study the properties of a stress tensor in detail in this study of fluid mechanics since we do not have to maximize or minimize the stress (as would be required in a solid

---

[3]An incompressible flow, when referred to in a general discussion such as in this section, will generally refer to a constant-density flow. This is true in most fluid mechanics literature, including textbooks on the subject.

mechanics course); we must, however, use the nine stress components in our derivations, then relate the stress components to the velocity and pressure fields with the appropriate equations. The stress components that act at a point are displayed on two- and three-dimensional rectangular elements in Fig. 5.2. These elements are considered to be an exaggerated point, a cubical point; the stress components act in the positive direction on a positive face (a normal vector points in the positive coordinate direction) and in the negative direction on a negative face (a normal vector points in the negative coordinate direction). The first subscript on a stress component denotes the face upon which the component acts, and the second subscript denotes the direction in which it acts; the component $\tau_{xy}$ acts in the positive $y$-direction on a positive $x$-face and in the negative $y$-direction on a negative $x$-face, as displayed in Fig. 5.2a. A stress component that acts perpendicular to a face is referred to as a **normal stress**; the components $\sigma_{xx}$, $\sigma_{yy}$, and $\sigma_{zz}$ are normal stresses. A stress component that acts tangential to a face is called a **shear stress**; the components $\tau_{xy}$, $\tau_{yx}$, $\tau_{xz}$, $\tau_{zx}$, $\tau_{yz}$, and $\tau_{zy}$ are the shear stress components.

**Normal stress:** *A stress component that acts perpendicular to an area.*

**Shear stress:** *A stress component that acts tangential to an area.*

There are nine stress components that act at a particular point in a fluid. To derive the differential momentum equation, consider the forces acting on the infinitesimal fluid particle shown in Fig. 5.3. Only the forces acting on positive faces are shown. The stress components are assumed to be function of $x, y, z$, and $t$, and hence the values of the stress components change from face to face since the location of each face is slightly different. The body force is assumed to act in an arbitrary direction.

Newton's second law applied to a fluid particle, for the $x$-component direction, is $\Sigma F_x = ma_x$. For the particle shown in Fig. 5.3, this takes the form

$$
\left( \sigma_{xx} + \frac{\partial \sigma_{xx}}{\partial x} \frac{dx}{2} \right) dy\, dz + \left( \tau_{yx} + \frac{\partial \tau_{yx}}{\partial y} \frac{dy}{2} \right) dx\, dz + \left( \tau_{zx} + \frac{\partial \tau_{zx}}{\partial z} \frac{dz}{2} \right) dx\, dy
$$

$$
- \left( \sigma_{xx} - \frac{\partial \sigma_{xx}}{\partial x} \frac{dx}{2} \right) dy\, dz - \left( \tau_{yx} - \frac{\partial \tau_{yx}}{\partial y} \frac{dy}{2} \right) dx\, dz
$$

$$
- \left( \tau_{zx} - \frac{\partial \tau_{zx}}{\partial z} \frac{dz}{2} \right) dx\, dy + \rho g_x\, dx\, dy\, dz = \rho\, dx\, dy\, dz \frac{Du}{Dt} \tag{5.3.1}
$$

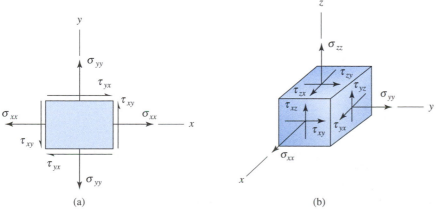

(a)                                    (b)

**Fig. 5.2** Stress components in Cartesian coordinates: (a) two-dimensional stress components; (b) three-dimensional stress components.

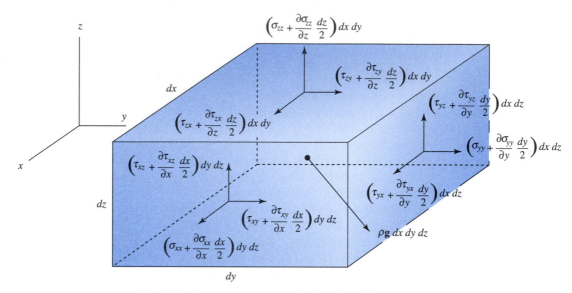

**Fig. 5.3**   Forces acting on an infinitesimal fluid particle.

where the component of the gravity vector **g** acting in the $x$-direction is $g_x$, and $Du/Dt$ is the $x$-component acceleration of the fluid particle (see Eq. 3.2.9). After we divide by the volume $dx\,dy\,dz$, the equation above can be simplified to

$$\rho\frac{Du}{Dt} = \frac{\sigma_{xx}}{\partial x} + \frac{\partial\tau_{yx}}{\partial y} + \frac{\partial\tau_{zx}}{\partial z} + \rho g_x \qquad (5.3.2)$$

Similarly, for the $y$- and $z$-directions we would have

$$\rho\frac{Dv}{Dt} = \frac{\partial\tau_{xy}}{\partial x} + \frac{\sigma_{yy}}{\partial y} + \frac{\partial\tau_{zy}}{\partial z} + \rho g_y$$

$$\rho\frac{Dw}{Dt} = \frac{\partial\tau_{xz}}{\partial x} + \frac{\partial\tau_{yz}}{\partial y} + \frac{\sigma_{zz}}{\partial z} + \rho g_z \qquad (5.3.3)$$

We can, by taking moments about axes passing through the center of the infinitesimal element, show that

$$\tau_{yx} = \tau_{xy} \qquad \tau_{yz} = \tau_{zy} \qquad \tau_{xz} = \tau_{zx} \qquad (5.3.4)$$

That is, the stress tensor is symmetric; so there are actually six independent stress components.

The stress tensor may be displayed in the usual way as

$$\tau_{ij} = \begin{pmatrix} \sigma_{xx} & \tau_{xy} & \tau_{xz} \\ \tau_{yx} & \sigma_{yy} & \tau_{yz} \\ \tau_{zx} & \tau_{zy} & \sigma_{zz} \end{pmatrix} \qquad (5.3.5)$$

The subscripts $i$ and $j$ take on numerical values 1, 2, or 3. Then $\tau_{12}$ represents the element $\tau_{xy}$ in the first row, second column.

## 5.3.2   Euler's Equations

Good approximations to the components of the stress tensor for many flows, especially for flow away from a boundary (flow around an airfoil) or in regions of sudden change (flow through a contraction) are displayed by the array

$$\tau_{ij} = \begin{pmatrix} -p & 0 & 0 \\ 0 & -p & 0 \\ 0 & 0 & -p \end{pmatrix} \tag{5.3.6}$$

For such flows, we have assumed the shear stress components that result from viscous effects to be negligibly small and the normal stress components to be equal to the negative of the pressure; this is precisely what we did in Fig. 3.16 when deriving Bernoulli's equation. If these stress components are introduced back into Eqs. 5.3.2 and 5.3.3 there results, for this frictionless flow,

**KEY CONCEPT**   *We often assume shear stress components to be negligibly small.*

$$\rho \frac{Du}{Dt} = -\frac{\partial p}{\partial x} + \rho g_x$$

$$\rho \frac{Dv}{Dt} = -\frac{\partial p}{\partial y} + \rho g_y \tag{5.3.7}$$

$$\rho \frac{Dw}{Dt} = -\frac{\partial p}{\partial z} + \rho g_z$$

The scalar equations above can then be written as the vector equation

$$\rho \frac{D}{Dt}(u\hat{\mathbf{i}} + v\hat{\mathbf{j}} + w\hat{\mathbf{k}}) = -\left(\frac{\partial p}{\partial x}\hat{\mathbf{i}} + \frac{\partial p}{\partial y}\hat{\mathbf{j}} + \frac{\partial p}{\partial z}\hat{\mathbf{k}}\right) - \rho \mathbf{g} \tag{5.3.8}$$

In vector form, we have the well-known **Euler's equation**

$$\rho \frac{D\mathbf{V}}{Dt} = -\nabla p - \rho \mathbf{g} \tag{5.3.9}$$

**Euler's equation:**   *The three differential equations that result from applying Newton's second law and neglecting viscous effects.*

If we assume a constant-density, steady flow, Eq. 5.3.9 can be integrated along a streamline to yield Bernoulli's equation, a result that does not surprise us since the same assumptions were imposed when deriving Bernoulli's equation in Chapter 3; this will be illustrated in Example 5.6.

With the differential momentum equations in the form of Eqs. 5.3.7, we have added three additional equations to the continuity equation to give four equations and four unknowns, $u, v, w$, and $p$. With the appropriate boundary and initial conditions, a solution, yielding the velocity and pressure fields for this inviscid, incompressible flow, would be possible.

## Example 5.5

A velocity field is proposed to be

$$u = \frac{10y}{x^2 + y^2} \qquad v = -\frac{10x}{x^2 + y^2} \qquad w = 0$$

(a) Is this a possible incompressible flow? (b) If so, find the pressure gradient $\nabla p$ assuming a frictionless air flow with the $z$-axis vertical. Use $\rho = 1.23$ kg/m³.

### Solution

(a) The continuity equation (5.2.9) is used to determine if the velocity field is possible. For this incompressible flow we have

$$\frac{\partial u}{\partial x} + \frac{\partial v}{\partial y} + \overset{0}{\cancel{\frac{\partial w}{\partial z}}} = 0$$

Substituting in the velocity components, we have

$$\frac{\partial}{\partial x}\left(\frac{10y}{x^2 + y^2}\right) + \frac{\partial}{\partial y}\left(-\frac{10x}{x^2 + y^2}\right) = \frac{-10y(2x)}{(x^2 + y^2)^2} - \frac{-10x(2y)}{(x^2 + y^2)^2} = \frac{1}{(x^2 + y^2)^2}[-20xy + 20xy] = 0$$

The quantity in brackets is obviously zero; hence the velocity field given is a possible incompressible flow.

  (b) The pressure gradient is found using Euler's equation. In component form we have the following:

$$\rho \frac{Du}{Dt} = -\frac{\partial p}{\partial x} + \overset{0}{\cancel{\rho g_x}}$$

$$\therefore \frac{\partial p}{\partial x} = -\rho\left[u\frac{\partial u}{\partial x} + v\frac{\partial u}{\partial y} + w\overset{0}{\cancel{\frac{\partial u}{\partial z}}} + \overset{0}{\cancel{\frac{\partial u}{\partial t}}}\right]$$

$$= -1.23\left[\frac{10y}{x^2 + y^2}\frac{-20xy}{(x^2 + y^2)^2} + \frac{-10x}{x^2 + y^2}\frac{(x^2 + y^2)10 - 10y(2y)}{(x^2 + y^2)^2}\right]$$

$$= \frac{123x}{(x^2 + y^2)^2}$$

$$\rho \frac{Dv}{Dt} = -\frac{\partial p}{\partial y} + \overset{0}{\cancel{\rho g_y}}$$

$$\therefore \frac{\partial p}{\partial y} = -\rho\left[u\frac{\partial v}{\partial x} + v\frac{\partial v}{\partial y} + w\overset{0}{\cancel{\frac{\partial v}{\partial z}}} + \overset{0}{\cancel{\frac{\partial v}{\partial t}}}\right]$$

$$= -1.23\left[\frac{10y}{x^2 + y^2}\frac{(x^2 + y^2)(-10) + 10x(2x)}{(x^2 + y^2)^2} + \frac{-10x}{x^2 + y^2}\frac{20xy}{(x^2 + y^2)^2}\right]$$

$$= \frac{123y}{(x^2 + y^2)^2}$$

$$\rho \overset{0}{\cancel{\frac{Dw}{Dt}}} = -\frac{\partial p}{\partial z} + \rho g_z$$

$$\therefore \frac{\partial p}{\partial z} = \rho g_z = 1.23 \text{ kg/m}^3 \times (-9.81) \text{ m/s}^2 = -12.07 \text{ N/m}^3$$

Thus
$$\nabla p = \frac{\partial p}{\partial x}\hat{\mathbf{i}} + \frac{\partial p}{\partial y}\hat{\mathbf{j}} + \frac{\partial p}{\partial z}\hat{\mathbf{k}} = \frac{123}{(x^2 + y^2)^2}(x\hat{\mathbf{i}} + y\hat{\mathbf{j}}) - 12.07\hat{\mathbf{k}} \text{ N/m}^3$$

# Example 5.6

Assume a steady, constant-density flow and integrate Euler's equation along a stream-line in a plane flow.

$$\frac{dz}{ds} = \sin\theta$$

$$(\hat{\mathbf{k}})_s = \sin\theta$$

**Fig. E5.6**

## Solution

First, let us express the substantial derivative in streamline coordinates. Since the veloc-ity vector is tangent to the streamline, we can write

$$\mathbf{V} = V\hat{\mathbf{s}}$$

where $\hat{\mathbf{s}}$ is the unit vector tangent to the streamline and $V$ is the magnitude of the veloc-ity, as shown in Fig. E5.6. The substantial derivative is then, for this plane flow,

$$\frac{D\mathbf{V}}{Dt} = \frac{\partial \mathbf{V}}{\partial t} + V\frac{\partial (V\hat{\mathbf{s}})}{\partial s} + \overset{0}{(\cancel{V})_n}\frac{\partial \mathbf{V}}{\partial n} = \frac{\partial \mathbf{V}}{\partial t} + V\frac{\partial V}{\partial s}\hat{\mathbf{s}} + V^2\frac{\partial \hat{\mathbf{s}}}{\partial s}$$

The quantity $\partial\hat{\mathbf{s}}/\partial s$ results from the change of the unit vector $\hat{\mathbf{s}}$; the unit vector cannot change magnitude (it must always have a magnitude of 1), it can only change direction. Hence the derivative $\partial\hat{\mathbf{s}}/\partial s$ is in a direction normal to the streamline and does not enter the streamwise component equation. For a steady flow $\partial\mathbf{V}/\partial t = 0$. Consequently, in the streamwise direction, Euler's equation (5.3.9) takes the form

$$\rho V\frac{\partial V}{\partial s} = -\frac{\partial p}{\partial s} - \rho g\frac{\partial z}{\partial s}$$

recognizing that the component of $\hat{\mathbf{k}}$ along the streamline can be expressed as $(\hat{\mathbf{k}})_s = \partial z/\partial s$ (see the sketch above). Note that we use partial derivatives in this equation since velocity and pressure also vary with the normal coordinate.

The equation above can be written, assuming constant density so that $\partial\rho/\partial s = 0$, as

$$\frac{\partial}{\partial s}\left(\rho\frac{V^2}{2} + p + \rho gz\right) = 0$$

Integrating along the streamline results in

$$\rho\frac{V^2}{2} + p + \rho gz = \text{const.}$$

or

$$\frac{V^2}{2} + \frac{p}{\rho} + gz = \text{const.}$$

This is, of course, Bernoulli's equation. We have integrated along a streamline assum-ing constant density, steady flow, negligible viscous effects, and an inertial reference frame, so it is to be expected that Bernoulli's equation will emerge.

### 5.3.3 Navier–Stokes Equations

Many fluids exhibit a linear relationship between the stress components and the velocity gradients. Such fluids are called **Newtonian fluids** and include common fluids such as water, oil, and air. If in addition to linearity, we require that the fluid be **isotropic**,[4] it is possible to relate the stress components and the velocity gradients using only two fluid properties, the *viscosity* $\mu$ and the *second coefficient of viscosity* $\lambda$. The stress–velocity gradient relations, often referred to as the *constitutive equations*,[5] are stated as follows:

$$\sigma_{xx} = -p + 2\mu \frac{\partial u}{\partial x} + \lambda\, \boldsymbol{\nabla}\!\cdot\!\mathbf{V} \qquad \tau_{xy} = \mu\!\left(\frac{\partial u}{\partial y} + \frac{\partial v}{\partial x}\right)$$

$$\sigma_{yy} = -p + 2\mu \frac{\partial v}{\partial y} + \lambda\, \boldsymbol{\nabla}\!\cdot\!\mathbf{V} \qquad \tau_{xz} = \mu\!\left(\frac{\partial u}{\partial z} + \frac{\partial w}{\partial x}\right) \qquad (5.3.10)$$

$$\sigma_{zz} = -p + 2\mu \frac{\partial w}{\partial z} + \lambda\, \boldsymbol{\nabla}\!\cdot\!\mathbf{V} \qquad \tau_{yz} = \mu\!\left(\frac{\partial v}{\partial z} + \frac{\partial w}{\partial y}\right)$$

For most gases, and for monatomic gases exactly, the second coefficient of viscosity is related to the viscosity by

$$\lambda = -\frac{2}{3}\mu \qquad (5.3.11)$$

a condition that is known as *Stokes's hypothesis*. With this relationship the negative average of the three normal stresses is equal to the pressure; that is,

$$-\frac{1}{3}(\sigma_{xx} + \sigma_{yy} + \sigma_{zz}) = p \qquad (5.3.12)$$

Using Eqs. 5.3.10, this can be shown to always be true for a liquid in which $\boldsymbol{\nabla}\!\cdot\!\mathbf{V} = 0$, and with Stokes's hypothesis it is also true for a gas.

If we substitute the constitutive equations into the differential momentum equations (5.3.2) and (5.3.3), there results, using Stokes's hypothesis,

$$\rho \frac{Du}{Dt} = -\frac{\partial p}{\partial x} + \rho g_x + \mu\!\left(\frac{\partial^2 u}{\partial x^2} + \frac{\partial^2 u}{\partial y^2} + \frac{\partial^2 u}{\partial z^2}\right) + \frac{\mu}{3}\frac{\partial}{\partial x}\!\left(\frac{\partial u}{\partial x} + \frac{\partial v}{\partial y} + \frac{\partial w}{\partial z}\right)$$

$$\rho \frac{Dv}{Dt} = -\frac{\partial p}{\partial y} + \rho g_y + \mu\!\left(\frac{\partial^2 v}{\partial x^2} + \frac{\partial^2 v}{\partial y^2} + \frac{\partial^2 v}{\partial z^2}\right) + \frac{\mu}{3}\frac{\partial}{\partial y}\!\left(\frac{\partial u}{\partial x} + \frac{\partial v}{\partial y} + \frac{\partial w}{\partial z}\right) \qquad (5.3.13)$$

$$\rho \frac{Dw}{Dt} = -\frac{\partial p}{\partial z} + \rho g_z + \mu\!\left(\frac{\partial^2 w}{\partial x^2} + \frac{\partial^2 w}{\partial y^2} + \frac{\partial^2 w}{\partial z^2}\right) + \frac{\mu}{3}\frac{\partial}{\partial z}\!\left(\frac{\partial u}{\partial x} + \frac{\partial v}{\partial y} + \frac{\partial w}{\partial z}\right)$$

where we have assumed a **homogeneous fluid**, that is, fluid properties (e.g., the viscosity) are independent of position.

For an incompressible flow the continuity equation allows the equations above to be reduced to

---

[4]The condition of isotropy exists if the fluid properties are independent of direction. Polymers are examples of anisotropic fluids.

[5]Details of the development of the constitutive equations can be found in any textbook on the subject of continuum mechanics.

$$\rho \frac{Du}{Dt} = -\frac{\partial p}{\partial x} + \rho g_x + \mu \left( \frac{\partial^2 u}{\partial x^2} + \frac{\partial^2 u}{\partial y^2} + \frac{\partial^2 u}{\partial z^2} \right)$$

$$\rho \frac{Dv}{Dt} = -\frac{\partial p}{\partial y} + \rho g_y + \mu \left( \frac{\partial^2 v}{\partial x^2} + \frac{\partial^2 v}{\partial y^2} + \frac{\partial^2 v}{\partial z^2} \right) \qquad (5.3.14)$$

$$\rho \frac{Dw}{Dt} = -\frac{\partial p}{\partial z} + \rho g_z + \mu \left( \frac{\partial^2 w}{\partial x^2} + \frac{\partial^2 w}{\partial y^2} + \frac{\partial^2 w}{\partial z^2} \right)$$

These are the **Navier–Stokes equations**, named after Louis M. H. Navier (1785–1836) and George Stokes (1819–1903); with these three differential equations and the differential continuity equation we have four equations and four unknowns, $u$, $v$, $w$, and $p$. The viscosity and density are fluid properties that are assumed to be known. With the appropriate boundary and initial conditions the equations can hopefully be solved. Several relatively simple geometries allow for analytical solutions; some of the solutions are presented in Chapter 7. Numerical solutions have also been determined for many flows of interest; computational methods are presented in Chapter 14. Because the equations are nonlinear partial differential equations (the acceleration terms cause the equations to be non-linear as observed in Eqs. 3.2.9), we cannot be assured that the solution we find will actually be realized in the laboratory; that is, the solutions are not unique. For example, a laminar flow and a turbulent flow may have the identical initial and boundary conditions, yet the two flows (the two solutions) are very different.

**Navier–Stokes equations:** *The three differential equations that result from applying Newton's second law to an incompressible, isotropic, homogeneous fluid.*

The Navier–Stokes equations have not been solved for a turbulent flow. All turbulent flows are unsteady and three-dimensional and hence the time-derivative terms must be retained. This requires an initial condition on all dependent variables; i.e., $u$, $v$, $w$, and $p$ must be known at all points in the flow field at $t = 0$. Such information would be extremely difficult, if not impossible, to obtain. To avoid this situation, time-averaged quantities are introduced for turbulent flows. This subject will be studied in a later chapter.

We can express the Navier–Stokes equations in vector form by multiplying Eqs. 5.3.14 by $\hat{\mathbf{i}}$, $\hat{\mathbf{j}}$, and $\hat{\mathbf{k}}$, respectively, and adding. We recognize that

$$\frac{Du}{Dt}\hat{\mathbf{i}} + \frac{Dv}{Dt}\hat{\mathbf{j}} + \frac{Dw}{Dt}\hat{\mathbf{k}} = \frac{D\mathbf{V}}{Dt}$$

$$\frac{\partial p}{\partial x}\hat{\mathbf{i}} + \frac{\partial p}{\partial y}\hat{\mathbf{j}} + \frac{\partial p}{\partial z}\hat{\mathbf{k}} = \nabla p \qquad (5.3.15)$$

$$\nabla^2 u \hat{\mathbf{i}} + \nabla^2 v \hat{\mathbf{j}} + \nabla^2 w \hat{\mathbf{k}} = \nabla^2 \mathbf{V}$$

where we have introduced the Laplacian

$$\nabla^2 = \frac{\partial^2}{\partial x^2} + \frac{\partial^2}{\partial y^2} + \frac{\partial^2}{\partial z^2} \qquad (5.3.16)$$

Combining the above, the Navier–Stokes equations (5.3.14) take the vector form

$$\rho \frac{D\mathbf{V}}{Dt} = -\nabla p + \rho \mathbf{g} + \mu \nabla^2 \mathbf{V} \qquad (5.3.17)$$

Using this vector form we can express the Navier–Stokes equations using other coordinate systems. The equations are listed for cylindrical and spherical coordinates in Table 5.1.

## Example 5.7

Simplify the $x$-component Navier–Stokes equation for steady flow in a horizontal, rectangular channel assuming all streamlines parallel to the walls. Let the $x$-direction be in the direction of flow (Fig. E5.7).

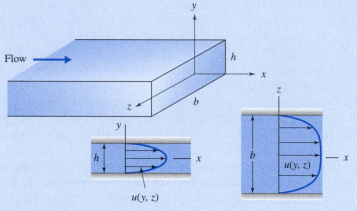

**Fig. E5.7**

**Solution**

If the streamlines are parallel to the walls, only the $x$-component of velocity will be nonzero. Letting $v = w = 0$ the continuity equation (5.2.9) for an incompressible flow becomes

$$\frac{\partial u}{\partial x} = 0$$

showing that $u = u(y, z)$. The acceleration is then

$$\frac{Du}{Dt} = \cancelto{0}{\frac{\partial u}{\partial t}} + u\cancelto{0}{\frac{\partial u}{\partial x}} + \cancelto{0}{v\frac{\partial u}{\partial y}} + \cancelto{0}{w\frac{\partial u}{\partial z}} = 0$$

The $x$-component momentum equation then simplifies to

$$0 = -\frac{\partial p}{\partial x} + \cancelto{0}{\rho g_x} + \mu\left(\cancelto{0}{\frac{\partial^2 u}{\partial x^2}} + \frac{\partial^2 u}{\partial y^2} + \frac{\partial^2 u}{\partial z^2}\right)$$

or

$$\frac{\partial p}{\partial x} = \mu\left(\frac{\partial^2 u}{\partial y^2} + \frac{\partial^2 u}{\partial z^2}\right)$$

With the appropriate boundary conditions (the no-slip conditions), a solution to the foregoing equation could be sought. It would provide the velocity profiles sketched in Fig. E5.7.

### 5.3.4   Vorticity Equations

There are certain fluid flow phenomena that cannot be explained or understood without reference to the vorticity equations, equations that are derived from the Navier–Stokes equations (Example 5.8 provides such a phenomenon). In addition to providing insight into such phenomena, the vorticity equations do not contain the pressure or gravity terms found in the Navier–Stokes equations but contain terms involving the velocity only. Since boundary conditions most often involve only the velocity, the vorticity equations are often the equations of choice for numerical solutions.

To derive the vorticity equations, we take the curl of Eq. 5.3.17, the vector form of the Navier–Stokes equations. This is a difficult task, so we will not display all the steps here but simply outline the process. First, let us define the vorticity of Eq. 3.2.18 in vector form using the del operator; using Eq. 5.2.6, we see that the three scalar Eq. 3.2.18 can be written as the single vector equation

$$\boldsymbol{\omega} = \nabla \times \mathbf{V} \tag{5.3.18}$$

where $\nabla \times \mathbf{V}$ is the curl of the velocity. The *curl* is the cross-product of the del operator and a vector function. Second, let us write the acceleration in vector form as

$$\mathbf{a} = \frac{D\mathbf{V}}{Dt} = \frac{\partial \mathbf{V}}{\partial t} + (\mathbf{V} \cdot \nabla) \, \mathbf{V} \tag{5.3.19}$$

where we have used Eqs. 3.2.8, 3.2.10, and 5.2.6. Finally, let us take the curl of the Navier–Stokes vector equation (5.3.17):

$$\nabla \times \left[ \rho \frac{\partial \mathbf{V}}{\partial t} + \rho(\mathbf{V} \cdot \nabla)\mathbf{V} \right] = -\nabla \times \nabla p + \rho \nabla \times \mathbf{g} + \mu \nabla \times \nabla^2 \mathbf{V} \tag{5.3.20}$$

The curl of the gradient of a scalar function and the curl of a constant are both zero. Also, since we can interchange differentiation, we can write

$$\nabla \times \frac{\partial \mathbf{V}}{\partial t} = \frac{\partial}{\partial t} \nabla \times \mathbf{V} = \frac{\partial \boldsymbol{\omega}}{\partial t}$$

$$\nabla \times \nabla^2 \mathbf{V} = \nabla^2 (\nabla \times \mathbf{V}) = \nabla^2 \boldsymbol{\omega} \tag{5.3.21}$$

The difficult step, which we will leave as a homework problem, comes in showing that

$$\nabla \times [(\mathbf{V} \cdot \nabla)\mathbf{V}] = (\mathbf{V} \cdot \nabla)\boldsymbol{\omega} - (\boldsymbol{\omega} \cdot \nabla)\mathbf{V} \tag{5.3.22}$$

**Vorticity equation:** *Equation derived by taking the curl of the Navier–Stokes equation. The vorticity equation does not contain terms involving pressure or gravity.*

Equation 5.3.20 then becomes, assuming that $\rho$ and $\mu$ are constants, the **vorticity equation**,

$$\frac{D\boldsymbol{\omega}}{Dt} = (\boldsymbol{\omega} \cdot \boldsymbol{\nabla}) \, \mathbf{V} + \nu \, \nabla^2 \boldsymbol{\omega} \tag{5.3.23}$$

The vorticity equation can be written as three scalar equations. Using Cartesian coordinates, the three vorticity equations are

$$\frac{D\omega_x}{Dt} = \omega_x \frac{\partial u}{\partial x} + \omega_y \frac{\partial u}{\partial y} + \omega_z \frac{\partial u}{\partial z} + \nu\nabla^2\omega_x$$

$$\frac{D\omega_y}{Dt} = \omega_x \frac{\partial v}{\partial x} + \omega_y \frac{\partial v}{\partial y} + \omega_z \frac{\partial v}{\partial z} + \nu\nabla^2\omega_y \tag{5.3.24}$$

$$\frac{D\omega_z}{Dt} = \omega_x \frac{\partial w}{\partial x} + \omega_y \frac{\partial w}{\partial y} + \omega_z \frac{\partial w}{\partial z} + \nu\nabla^2\omega_z$$

**KEY CONCEPT** *The vorticity equations involve only the velocity and its derivatives.*

Since the vorticity is the curl of the velocity, note that all of the terms in the vorticity equations involve only the velocity and its derivatives. Consequently, the vorticity equations often become the equations of choice when solving problems requiring the differential equations of motion. In fact, we refer to vortex lines and vortex tubes as being similar to streamlines and streamtubes. A **vortex line** is a line to which the vorticity vector is tangent. A **vortex tube**, or simply a *vortex*, is a tube whose walls contain vortex lines. A vortex is shown in Fig. 5.4.

**Vortex line:** *A line to which the vorticity vector is tangent.*

**Vortex tube:** *A bundle of vortex lines.*

An interesting conclusion can be made by considering the vorticity equation 5.3.23. If an inviscid flow is everywhere irrotational (i.e., $\boldsymbol{\omega} = 0$ at all points in the flow), it must remain irrotational since $D\boldsymbol{\omega}/Dt = 0$. This is referred to as the *persistence of irrotationality*. Also, if a uniform flow approaches an object, vorticity (rotation of fluid particles) is introduced to the flow only by the action of viscosity. Without viscous effects, vorticity cannot be created in an oncoming irrotational flow.

**KEY CONCEPT** *Vorticity cannot be created in an irrotational flow without viscous effects.*

Plane flows are often of particular interest. If $w = 0$, $u = u(x, y)$, and $v = v(x, y)$ the only nonzero vorticity component is $\omega_z$. For such a flow, the vorticity equation takes the simplified form

$$\frac{D\omega_z}{Dt} = \nu\nabla^2\omega_z \tag{5.3.25}$$

**KEY CONCEPT** *Viscous effects are needed to cause vorticity changes in a plane flow.*

We observe that viscous effects are needed to cause vorticity changes in a plane flow.

The following example illustrates a flow phenomenon that can be explained quite easily with the use of the vorticity equation.

$t=1.0$ s

$t=5.0$ s

$t=9.0$ s

$t=13.0$ s

**Fig. 5.4** Starting vortex on a wedge. A piston drives water normal to the axis of a wedge. Dye is injected into the water from small holes in the wedge surface. The characteristic Reynolds number is of order 1000. The piston stops at 12.5 s, producing a stopping vortex in the last photograph. (Photograph by D.I. Pullin and A.E. Perry.)

## Example 5.8

In a snowstorm, the snow is actually scooped out in front of a tree, or post, as shown in Fig. E5.8a. Explain this phenomenon by referring to the vorticity equations.

(a)

(b)

**Fig. E5.8**

### Solution

Let the velocity approaching the tree be in the $x$-direction with a velocity gradient $\partial u/\partial z$ near the ground. The vorticity components are then (refer to Eqs. 3.2.21)

$$\omega_x = 0 \qquad \omega_y = \frac{\partial u}{\partial z} \qquad \omega_z = 0$$

The vorticity equation (5.3.24) for $\omega_y$, ignoring viscous effects over the short flow length, reduces to

$$\frac{D\omega_y}{Dt} = \omega_y \frac{\partial v}{\partial y}$$

Observe from Fig. E5.8b that in the vicinity of the tree $\partial v/\partial y$ is positive since $v_C > v_B > v_A$. ($\partial v/\partial y$ can be shown to be positive for negative $y$ also.) Since $\omega_y$ and $\partial v/\partial y$ are both positive, $D\omega_y/Dt$ is positive and $\omega_y$ increases as the vortex tubes approach the tree. This increased vorticity creates a strong vortex in front of the tree resulting in the snow being scooped out as shown. This same phenomenon occurs in a sandstorm or in a water flow around a post in a riverbed.

## 5.4 DIFFERENTIAL ENERGY EQUATION

Most problems of interest in fluid mechanics do not involve temperature gradients; they involve flows in which the temperature is everywhere constant. For such flows it is not necessary to introduce the differential energy equation. There are situations, however, for both compressible and incompressible flows in which temperature gradients are important, and for such flows the differential energy equation may be needed. We will derive the differential energy equation assuming negligible viscous effects, an assumption that significantly simplifies the derivation. Since the shear stresses that result from viscosity are quite small for many applications, this assumption may be acceptable. These shear stresses do, however, account for the high temperatures that burn up satellites on reentry to the atmosphere; if they are significant, they must be included in any analysis.

Consider the infinitesimal fluid element, shown in Fig. 5.5. The heat transfer rate $\dot{Q}$ through an area $A$ is given by *Fourier's law of heat transfer*, named after Jean B. J. Fourier (1768–1830):

$$\dot{Q} = -KA\,\frac{\partial T}{\partial n} \tag{5.4.1}$$

where $n$ is the direction normal to the area, $T$ is the temperature, and $K$ is the *thermal conductivity*, assumed constant. The work rate done by a force is the magnitude of the force multiplied by the velocity in the direction of the force; for a pressure force $pA$ it is,

$$\dot{W} = pAV \tag{5.4.2}$$

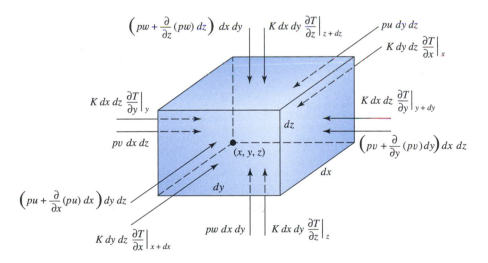

**Fig. 5.5**    Rate of heat transfer and work rate on an infinitesimal fluid element.

where $V$ is the velocity in the direction of the pressure force. The first law of thermodynamics (refer to Eq. 1.7.6) applied to a fluid particle is

$$\dot{Q} - \dot{W} = \frac{DE}{Dt} \tag{5.4.3}$$

where $D/Dt$ is used since we are following a fluid particle at the instant shown. For the particle occupying the infinitesimal element of Fig. 5.5, the relationships above allow us to write[6]

$$K \, dy \, dz\left(\frac{\partial T}{\partial x}\Big|_{x+dx} - \frac{\partial T}{\partial x}\Big|_x\right) - \frac{\partial}{\partial x}(pu) \, dx \, dy \, dz$$

$$+ K \, dx \, dz\left(\frac{\partial T}{\partial y}\Big|_{y+dy} - \frac{\partial T}{\partial y}\Big|_y\right) - \frac{\partial}{\partial y}(pv) \, dx \, dy \, dz$$

$$+ K \, dx \, dy\left(\frac{\partial T}{\partial z}\Big|_{z+dz} - \frac{\partial T}{\partial z}\Big|_z\right) - \frac{\partial}{\partial z}(pw) \, dx \, dy \, dz$$

$$= \rho \, dx \, dy \, dz \, \frac{D}{Dt}\left(\frac{u^2 + v^2 + w^2}{2} + gz + \tilde{u}\right) \tag{5.4.4}$$

where $E$ has included kinetic, potential, and internal energy, and the $z$-axis is assumed vertical. Also, since the mass of a fluid particle is constant, $\rho \, dx \, dy \, dz$ is outside the $D/Dt$-operator. Divide both sides by $dx \, dy \, dz$. The result is

$$K\left(\frac{\partial^2 T}{\partial x^2} + \frac{\partial^2 T}{\partial y^2} + \frac{\partial^2 T}{\partial z^2}\right) - \frac{\partial}{\partial x}(pu) - \frac{\partial}{\partial y}(pv) - \frac{\partial}{\partial z}(pw)$$

$$= \rho \frac{D}{Dt}\left(\frac{u^2 + v^2 + w^2}{2} + gz + \tilde{u}\right) \tag{5.4.5}$$

This can be rearranged as follows:

$$K\left(\frac{\partial^2 T}{\partial x^2} + \frac{\partial^2 T}{\partial y^2} + \frac{\partial^2 T}{\partial z^2}\right) - p\left(\frac{\partial u}{\partial x} + \frac{\partial v}{\partial y} + \frac{\partial w}{\partial z}\right) - u\frac{\partial p}{\partial x} - v\frac{\partial p}{\partial y} - w\frac{\partial p}{\partial z}$$

$$= \rho u \frac{Du}{Dt} + \rho v \frac{Dv}{Dt} + \rho w \frac{Dw}{Dt} + \rho g \frac{Dz}{Dt} + \rho \frac{D\tilde{u}}{Dt} \tag{5.4.6}$$

The Euler's scalar equations (5.3.7) are applicable for this inviscid flow; hence the last three terms on the left equal the first four terms on the right if we recognize that

$$\frac{Dz}{Dt} = \frac{\partial z}{\partial t}^{0} + u\frac{\partial z}{\partial x}^{0} + v\frac{\partial z}{\partial y}^{0} + w\frac{\partial z}{\partial z} = w \tag{5.4.7}$$

---

[6]We have used the lower corner at $(x, y, z)$ from which to measure the various quantities. Mathematically, we would let $\Delta x$ $\Delta y$, and $\Delta z$ approach zero so it matters not if we measure from a corner or from the center of the element.

since $x, y, z, t$ are all independent variables. The simplified energy equation then takes the form

$$\rho \frac{D\tilde{u}}{Dt} = K\left(\frac{\partial^2 T}{\partial x^2} + \frac{\partial^2 T}{\partial y^2} + \frac{\partial^2 T}{\partial z^2}\right) - p\left(\frac{\partial u}{\partial x} + \frac{\partial v}{\partial y} + \frac{\partial w}{\partial z}\right) \qquad (5.4.8)$$

In vector form this is expressed as

$$\rho \frac{D\tilde{u}}{Dt} = K\nabla^2 T - p\, \boldsymbol{\nabla} \cdot \mathbf{V} \qquad (5.4.9)$$

Before we simplify this equation for incompressible gas flow, let us write it in terms of enthalpy rather than internal energy. Using

$$\tilde{u} = h - \frac{p}{\rho} \qquad (5.4.10)$$

the energy equation becomes, using Eq. 5.2.7,

$$\rho \frac{Dh}{Dt} = K\nabla^2 T + \frac{Dp}{Dt} \qquad (5.4.11)$$

See Example 5.4 for the details of this conversion.

We have two special cases to be considered. First, for a liquid flow we can use $\boldsymbol{\nabla} \cdot \mathbf{V} = 0$ and with $\tilde{u} = c_p T$, $c_p$ being the specific heat,[7] Eq. 5.4.9 simplifies to

$$\frac{DT}{Dt} = \alpha \nabla^2 T \qquad (5.4.12)$$

where we have introduced the *thermal diffusivity* $\alpha$ defined by

$$\alpha = \frac{K}{\rho c_p} \qquad (5.4.13)$$

---

[7]The specific heat for a liquid is often listed in tables as $c_p$. The specific heat at constant volume for a liquid is approximately equal to $c_p$. Hence we often simply drop the subscript and let $c_p = c$ for a liquid. It is assumed to be a constant, but it does depend on temperature. Here we will use $c_p$.

For an incompressible gas flow, an interesting result occurs. In Example 5.10 we will show that

$$\left| \frac{Dp}{Dt} \right| \ll |p\, \mathbf{\nabla \cdot V}| \tag{5.4.14}$$

Thus, when comparing Eqs. 5.4.9 and 5.4.11, it is Eq. 5.4.11 that simplifies to

$$\rho c_p \frac{DT}{Dt} = K\nabla^2 T \tag{5.4.15}$$

for an incompressible gas flow if we make the ideal-gas assumption that

$$dh = c_p\, dT \tag{5.4.16}$$

If viscous effects are not negligible, the derivation would include the work input due to the shear stress components. This would add a term to the right-hand side of all of the differential energy equations above; this term is called the *dissipation function* $\Phi$, which, in Cartesian coordinates, is

$$\Phi = 2\mu\left[ \left(\frac{\partial u}{\partial x}\right)^2 + \left(\frac{\partial v}{\partial y}\right)^2 + \left(\frac{\partial w}{\partial z}\right)^2 + \frac{1}{2}\left(\frac{\partial u}{\partial y} + \frac{\partial v}{\partial x}\right)^2 + \frac{1}{2}\left(\frac{\partial v}{\partial z} + \frac{\partial w}{\partial y}\right)^2 \right.$$

$$\left. + \frac{1}{2}\left(\frac{\partial u}{\partial z} + \frac{\partial w}{\partial x}\right)^2 \right] \tag{5.4.17}$$

With the addition of the energy equation, problems involving temperature variations in a flow can now be considered. Such problems are present in compressible flows in which pressure and density are related to temperature by an equation of state. In incompressible gas flows (Mach number $< 0.3$) and liquid flows, temperature variations are often negligible so that the differential energy equation is not of interest. If, however, a temperature field does exist in a liquid flow or an incompressible gas flow (heat exchangers, atmospheric flows, lake inversions, lubrication flows, free convection flows), the energy equation provides an additional equation relating the quantities of interest. For liquid flows involving temperature gradients it is often necessary to assume that $\mu = \mu(T)$; in free convection flows we must assume that $\rho = \rho(T)$. In incompressible gas flows we can usually assume viscosity to be constant since the temperature variation is quite small.

# Example 5.9

A constant-density liquid flows into a wide, rectangular horizontal channel, the walls of which are maintained at a higher temperature than the liquid, as shown in Fig. E5.9. Assume a variable $\mu$, include viscous dissipation, and write the describing differential equations for a steady flow.

**Fig. E5.9**

## Solution

Let the $x$-axis coincide with centerline of the channel and the $y$-axis be vertical. The continuity equation would take the form

$$\nabla \cdot \mathbf{V} = \frac{\partial u}{\partial x} + \frac{\partial v}{\partial y} = 0$$

since $w = 0$ for the wide channel.

The flow will be primarily in the $x$-direction, but we must allow for variation of the $y$-component $v$. There will be no variation in the $z$-direction. The accelerations for this steady flow will be

$$\frac{Du}{Dt} = u\frac{\partial u}{\partial x} + v\frac{\partial u}{\partial y}$$

$$\frac{Dv}{Dt} = u\frac{\partial v}{\partial x} + v\frac{\partial v}{\partial y}$$

The stress terms of Eqs. 5.3.2 and 5.3.3 using Eqs. 5.3.10 with $\nabla \cdot \mathbf{V} = 0$, assuming a variable $\mu$, become

$$\frac{\partial \sigma_{xx}}{\partial x} + \frac{\partial \tau_{xy}}{\partial y} = -\frac{\partial p}{\partial x} + \mu\left(\frac{\partial^2 u}{\partial x^2} + \frac{\partial^2 u}{\partial y^2}\right) + 2\frac{\partial \mu}{\partial x}\frac{\partial u}{\partial x} + \frac{\partial \mu}{\partial y}\left(\frac{\partial u}{\partial y} + \frac{\partial v}{\partial x}\right)$$

$$\frac{\partial \tau_{xy}}{\partial x} + \frac{\partial \sigma_{yy}}{\partial y} = -\frac{\partial p}{\partial y} + \mu\left(\frac{\partial^2 v}{\partial x^2} + \frac{\partial^2 v}{\partial y^2}\right) + 2\frac{\partial \mu}{\partial y}\frac{\partial v}{\partial y} + \frac{\partial \mu}{\partial x}\left(\frac{\partial u}{\partial y} + \frac{\partial v}{\partial x}\right)$$

The momentum equations are then

$$\rho\left(u\frac{\partial u}{\partial x} + v\frac{\partial u}{\partial y}\right) = -\frac{\partial p}{\partial x} + \mu\left(\frac{\partial^2 u}{\partial x^2} + \frac{\partial^2 u}{\partial y^2}\right) + 2\frac{\partial \mu}{\partial x}\frac{\partial u}{\partial x} + \frac{\partial \mu}{\partial y}\left(\frac{\partial u}{\partial y} + \frac{\partial v}{\partial x}\right)$$

$$\rho\left(u\frac{\partial v}{\partial x} + v\frac{\partial v}{\partial y}\right) = -\frac{\partial p}{\partial y} + \mu\left(\frac{\partial^2 v}{\partial x^2} + \frac{\partial^2 v}{\partial y^2}\right) + 2\frac{\partial \mu}{\partial y}\frac{\partial v}{\partial y} + \frac{\partial \mu}{\partial x}\left(\frac{\partial u}{\partial y} + \frac{\partial v}{\partial x}\right)$$

The energy equation simplifies to

$$u\frac{\partial T}{\partial x} + v\frac{\partial T}{\partial y} = \alpha\left(\frac{\partial^2 T}{\partial x^2} + \frac{\partial^2 T}{\partial y^2}\right) + \frac{2\mu}{c_p}\left[\left(\frac{\partial u}{\partial x}\right)^2 + \left(\frac{\partial v}{\partial y}\right)^2 + \frac{1}{2}\left(\frac{\partial u}{\partial y} + \frac{\partial v}{\partial x}\right)^2\right]$$

where we have assumed $K$ to be constant. The nonlinear, partial differential equations above, although quite formidable when attempting an analytic solution, could be solved numerically with the appropriate boundary conditions, and for a sufficiently low flow rate so that laminar flow exists (a turbulent flow is always unsteady and three-dimensional).

## Example 5.10

Show that for an ideal gas, $|Dp/Dt| \ll |p\, \boldsymbol{\nabla} \cdot \mathbf{V}|$ in a low-speed flow, thereby concluding that Eq. 5.4.15 is the appropriate equation.

### Solution

Let us consider a steady, uniform flow in a pipe so that $|\mathbf{V}| = u$ and $Dp/Dt = u\, \partial p/\partial x$. Then the problem can be stated as follows: Show that

$$\left| u \frac{\partial p}{\partial x} \right| \ll \left| p \frac{\partial u}{\partial x} \right|$$

Viscous effects are small and would not change the conclusion, so we can ignore any possible viscous effects. Then Euler's equation (5.3.7) allows us to use

$$\frac{\partial p}{\partial x} = -\rho u \frac{\partial u}{\partial x}$$

Using the definition of the speed of sound (Eq. 1.7.17) and the equation of state, we see that

$$c = \sqrt{\frac{kp}{\rho}} \quad \text{or} \quad p = c^2 \frac{\rho}{k}$$

Thus

$$p \frac{\partial u}{\partial x} = \frac{c^2}{k} \rho \frac{\partial u}{\partial x}$$

Our problem can now be stated: Show that

$$\left| \rho u^2 \frac{\partial u}{\partial x} \right| \ll \left| \frac{c^2}{k} \rho \frac{\partial u}{\partial x} \right|$$

Or, more simply, is it true that

$$u^2 \ll \frac{c^2}{k}?$$

This can be seen to be true since we have assumed for a low-speed gas flow that the speed of the gas is much less than the speed of sound (e.g., $u < 0.3c$ or $M < 0.3$). We know that $k$ is of order unity ($k = 1.4$ for air), so it will not affect our conclusion that

$$\left| \frac{Dp}{Dt} \right| \ll |p\, \boldsymbol{\nabla} \cdot \mathbf{V}|$$

## 5.5 SUMMARY

We have now completed our derivation of the partial differential equations that are used in describing flows of interest. Let's summarize the equations in vector form for an incompressible flow:

Continuity:    $\nabla \cdot \mathbf{V} = 0$    (5.5.1)

Momentum:    $\rho \dfrac{D\mathbf{V}}{Dt} = -\nabla p + \rho \mathbf{g} + \mu \nabla^2 \mathbf{V}$    (5.5.2)

Energy:    $\rho c_p \dfrac{DT}{Dt} = K\nabla^2 T + \Phi$    Liquids    (5.5.3)

Energy:    $\rho c_p \dfrac{DT}{Dt} = K\nabla^2 T + \Phi$    Incompressible gases    (5.5.4)

To express these equations in the forms above, we have assumed:

- A Newtonian fluid (a linear relationship between the stress components and the velocity gradients).
- An isotropic fluid (the fluid properties are independent of direction).
- A homogeneous fluid (the fluid properties $\mu$, $c_p$, and $K$ do not depend on position).
- An incompressible flow (the density of a particle is constant, that is, $D\rho/Dt = 0$; we do not demand that $\rho = $ constant. For a gas flow we require that M < 0.3).
- An inertial reference frame.

The vorticity equation is also of interest, It is

$$\frac{D\boldsymbol{\omega}}{Dt} = (\boldsymbol{\omega} \cdot \nabla)\mathbf{V} + \nu \nabla^2 \boldsymbol{\omega}$$    (5.5.5)

Numerical methods often make use of this vorticity equation.

In the derivations of the differential equations in this chapter we have made no mention of laminar or turbulent flow. The equations are applicable to either class of flow. Some laminar flows in relatively simple geometries have been solved analytically, and many others have been solved numerically. Turbulent flows, however, have not been solved even for the simplest geometry. A turbulent flow is always an unsteady flow, and the presence of the time-derivative terms demands initial conditions; that is, at time $t = 0$ we must specify $u, v, w$ and $p$ at all points in the region of interest, information that is difficult, if not impossible, to obtain even in a simple pipe flow.

**KEY CONCEPT**  *The differential equations are applicable to laminar and turbulent flows.*

**Table 5.1** Fundamental Laws for Incompressible Flows.

*Continuity*

**Cartesian**
$$\frac{\partial u}{\partial x} + \frac{\partial v}{\partial y} + \frac{\partial w}{\partial z} = 0$$

**Cylindrical**
$$\frac{1}{r}\frac{\partial}{\partial r}(r v_r) + \frac{1}{r}\frac{\partial v_\theta}{\partial \theta} + \frac{\partial v_z}{\partial z} = 0$$

**Spherical**
$$\frac{1}{r^2}\frac{\partial}{\partial r}(r^2 v_r) + \frac{1}{r\sin\theta}\frac{\partial}{\partial \theta}(v_\theta \sin\theta) + \frac{1}{r\sin\theta}\frac{\partial v_\phi}{\partial \phi} = 0$$

*Momentum*

**Cartesian**
$$\frac{Du}{Dt} = -\frac{1}{\rho}\frac{\partial p}{\partial x} + g_x + \nu\nabla^2 u$$
$$\frac{Dv}{Dt} = -\frac{1}{\rho}\frac{\partial p}{\partial y} + g_y + \nu\nabla^2 v$$
$$\frac{Dw}{Dt} = -\frac{1}{\rho}\frac{\partial p}{\partial z} + g_z + \nu\nabla^2 w$$
$$\frac{D}{Dt} = \frac{\partial}{\partial t} + u\frac{\partial}{\partial x} + v\frac{\partial}{\partial y} + w\frac{\partial}{\partial z}$$
$$\nabla^2 = \frac{\partial^2}{\partial x^2} + \frac{\partial^2}{\partial y^2} + \frac{\partial^2}{\partial z^2}$$

**Cylindrical**
$$\frac{Dv_r}{Dt} - \frac{v_\theta^2}{r} = -\frac{1}{\rho}\frac{\partial p}{\partial r} + g_r + \nu\left(\nabla^2 v_r - \frac{v_r}{r^2} - \frac{2}{r^2}\frac{\partial v_\theta}{\partial \theta}\right)$$
$$\frac{Dv_\theta}{Dt} + \frac{v_r v_\theta}{r} = -\frac{1}{\rho r}\frac{\partial p}{\partial \theta} + g_\theta + \nu\left(\nabla^2 v_\theta + \frac{2}{r^2}\frac{\partial v_r}{\partial \theta} - \frac{v_\theta}{r^2}\right)$$
$$\frac{Dv_z}{Dt} = -\frac{1}{\rho}\frac{\partial p}{\partial z} + g_z + \nu\nabla^2 v_z$$
$$\frac{D}{Dt} = \frac{\partial}{\partial t} + v_r\frac{\partial}{\partial r} + \frac{v_\theta}{r}\frac{\partial}{\partial \theta} + v_z\frac{\partial}{\partial z}$$
$$\nabla^2 = \frac{\partial^2}{\partial r^2} + \frac{1}{r}\frac{\partial}{\partial r} + \frac{1}{r^2}\frac{\partial^2}{\partial \theta^2} + \frac{\partial^2}{\partial z^2}$$

**Spherical**
$$\frac{Dv_r}{Dt} - \frac{v_\theta^2 + v_\phi^2}{r} = -\frac{1}{\rho}\frac{\partial p}{\partial r} + g_r + \nu\left(\nabla^2 v_r - \frac{2v_r}{r^2} - \frac{2}{r^2}\frac{\partial v_\theta}{\partial \theta} - \frac{2v_\theta\cot\theta}{r^2} - \frac{2}{r^2\sin\theta}\frac{\partial v_\phi}{\partial \phi}\right)$$

$$\frac{Dv_\theta}{Dt} + \frac{v_r v_\theta - v_\phi^2\cot\theta}{r} = -\frac{1}{\rho r}\frac{\partial p}{\partial \theta} + g_\theta + \nu\left(\nabla^2 v_\theta + \frac{2}{r^2}\frac{\partial v_r}{\partial \theta} - \frac{v_\theta}{r^2\sin^2\theta} - \frac{2\cos\theta}{r^2\sin^2\theta}\frac{\partial v_\phi}{\partial \phi}\right)$$

$$\frac{Dv_\phi}{Dt} + \frac{v_\phi v_r + v_\theta v_\phi\cot\theta}{r} = -\frac{1}{\rho r\sin\theta}\frac{\partial p}{\partial \phi} + g_\phi + \nu\left(\nabla^2 v_\phi - \frac{v_\phi}{r^2\sin^2\theta} + \frac{2}{r^2\sin\theta}\frac{\partial v_r}{\partial \phi} + \frac{2\cos\theta}{r^2\sin^2\theta}\frac{\partial v_\theta}{\partial \phi}\right)$$

$$\frac{D}{Dt} = \frac{\partial}{\partial t} + v_r\frac{\partial}{\partial r} + \frac{v_\theta}{r}\frac{\partial}{\partial \theta} + \frac{v_\phi}{r\sin\theta}\frac{\partial}{\partial \phi}$$

$$\nabla^2 = \frac{1}{r^2}\frac{\partial}{\partial r}\left(r^2\frac{\partial}{\partial r}\right) + \frac{1}{r^2\sin^2\theta}\frac{\partial}{\partial \theta}\left(\sin\theta\frac{\partial}{\partial \theta}\right) + \frac{1}{r^2\sin^2\theta}\frac{\partial^2}{\partial \phi^2}$$

*Energy*

**Cartesian**
$$\rho\frac{Dh}{Dt} = K\nabla^2 T + 2\mu\left[\left(\frac{\partial u}{\partial x}\right)^2 + \left(\frac{\partial v}{\partial y}\right)^2 + \left(\frac{\partial w}{\partial z}\right)^2\right]$$
$$+ \frac{1}{2}\left(\frac{\partial u}{\partial y} + \frac{\partial v}{\partial x}\right)^2 + \frac{1}{2}\left(\frac{\partial v}{\partial z} + \frac{\partial w}{\partial y}\right)^2 + \frac{1}{2}\left(\frac{\partial u}{\partial z} + \frac{\partial w}{\partial x}\right)^2$$

**Cylindrical**
$$\rho\frac{Dh}{Dt} = K\nabla^2 T + 2\mu\left[\left(\frac{\partial v_r}{\partial r}\right)^2 + \left(\frac{1}{r}\frac{\partial v_\theta}{\partial \theta} + \frac{v_r}{r}\right)^2 + \left(\frac{\partial v_z}{\partial z}\right)^2\right]$$
$$+ \frac{1}{2}\left(\frac{1}{r}\frac{\partial v_z}{\partial \theta} + \frac{\partial v_\theta}{\partial z}\right)^2 + \frac{1}{2}\left(\frac{\partial v_r}{\partial z} + \frac{\partial v_z}{\partial r}\right)^2$$
$$+ \frac{1}{2}\left(\frac{1}{r}\frac{\partial v_r}{\partial \theta} + \frac{\partial v_\theta}{\partial r} - \frac{v_\theta}{r}\right)^2$$

**Spherical**
$$\rho\frac{Dh}{Dt} = K\nabla^2 T + 2\mu\left[\left(\frac{\partial v_r}{\partial r}\right)^2 + \left(\frac{1}{r}\frac{\partial v_\theta}{\partial \theta} + \frac{v_r}{r}\right)^2 + \left(\frac{1}{r}\frac{\partial v_\phi}{\partial \phi} + \frac{v_r}{r} + \frac{v_\theta\cot\theta}{r}\right)^2\right]$$
$$+ \mu\left[\left(\frac{1}{r\sin\theta}\frac{\partial v_\theta}{\partial \phi} + \frac{\sin\theta}{r}\frac{\partial}{\partial \theta}\left(\frac{v_\phi}{\sin\theta}\right)\right)^2\right]$$
$$+ \mu\left[\left(\frac{1}{r\sin\theta}\frac{\partial v_r}{\partial \phi} + r\frac{\partial}{\partial r}\left(\frac{v_\phi}{r}\right)\right)^2\right]$$
$$+ \mu\left[\left(r\frac{\partial}{\partial r}\left(\frac{v_\theta}{r}\right) + \frac{1}{r}\frac{\partial v_r}{\partial \theta}\right)^2\right]$$

*Stresses*

**Cartesian**
$$\sigma_{xx} = -p + 2\mu\frac{\partial u}{\partial x}$$
$$\sigma_{yy} = -p + 2\mu\frac{\partial v}{\partial y}$$
$$\sigma_{zz} = -p + 2\mu\frac{\partial w}{\partial z}$$
$$\tau_{xy} = \mu\left(\frac{\partial u}{\partial y} + \frac{\partial v}{\partial x}\right)$$
$$\tau_{yz} = \mu\left(\frac{\partial v}{\partial z} + \frac{\partial w}{\partial y}\right)$$
$$\tau_{xz} = \mu\left(\frac{\partial u}{\partial z} + \frac{\partial w}{\partial x}\right)$$

**Cylindrical**
$$\sigma_{rr} = -p + 2\mu\frac{\partial v_r}{\partial r}$$
$$\sigma_{\theta\theta} = -p + 2\mu\left(\frac{1}{r}\frac{\partial v_\theta}{\partial \theta} + \frac{v_r}{r}\right)$$
$$\sigma_{zz} = -p + 2\mu\frac{\partial v_z}{\partial z}$$
$$\tau_{r\theta} = \mu\left[r\frac{\partial}{\partial r}\left(\frac{v_\theta}{r}\right) + \frac{1}{r}\frac{\partial v_r}{\partial \theta}\right]$$
$$\tau_{\theta z} = \mu\left[\frac{\partial v_\theta}{\partial z} + \frac{1}{r}\frac{\partial v_z}{\partial \theta}\right]$$
$$\tau_{rz} = \mu\left[\frac{\partial v_r}{\partial z} + \frac{\partial v_z}{\partial r}\right]$$

**Spherical**
$$\sigma_{rr} = -p + 2\mu\frac{\partial v_r}{\partial r}$$
$$\sigma_{\theta\theta} = -p + 2\mu\left(\frac{1}{r}\frac{\partial v_\theta}{\partial \theta} + \frac{v_r}{r}\right)$$
$$\sigma_{\phi\phi} = -p + 2\mu\left(\frac{1}{r\sin\theta}\frac{\partial v_\phi}{\partial \phi} + \frac{v_r}{r} + \frac{v_\theta\cot\theta}{r}\right)$$
$$\tau_{r\theta} = \mu\left[r\frac{\partial}{\partial r}\left(\frac{v_\theta}{r}\right) + \frac{1}{r}\frac{\partial v_r}{\partial \theta}\right]$$
$$\tau_{\theta\phi} = \mu\left[\frac{\sin\theta}{r}\frac{\partial}{\partial \theta}\left(\frac{v_\phi}{\sin\theta}\right) + \frac{1}{r\sin\theta}\frac{\partial v_\theta}{\partial \phi}\right]$$
$$\tau_{r\phi} = \mu\left[\frac{1}{r\sin\theta}\frac{\partial v_r}{\partial \phi} + r\frac{\partial}{\partial r}\left(\frac{v_\phi}{r}\right)\right]$$

## PROBLEMS

### Differential Continuity Equation

**5.1** The divergence theorem (also knows as Gauss's theorem) states that

$$\int_A \mathbf{V} \cdot \hat{\mathbf{n}}\, dA = \int_V \nabla \cdot \mathbf{V}\, dV$$

where $\mathbf{V}$ represents any vector and $A$ completely surrounds the volume $V$. Apply this theorem to the integral continuity equation 4.3.3 and derive the differential continuity equation 5.2.5.

**5.2** Use the infinitesimal element shown in Fig. P5.2 and derive the differential continuity equation in cylindrical coordinates. The velocity vector is $\mathbf{V} = (v_r, v_\theta, v_z)$.

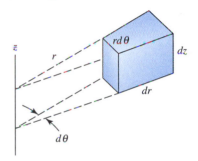

**Fig. P5.2**

**5.3** Use the infinitesimal elements shown in Fig. P5.3 and derive the differential continuity equation in spherical coordinates. The velocity vector is $\mathbf{V} = (v_r, v_\theta, v_\phi)$.

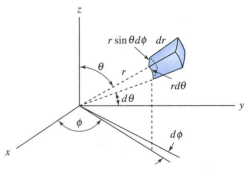

**Fig. P5.3**

**5.4** A uniform compressible flow occurs in a constant-diameter pipe. Write the simplified differential continuity equation for the steady flow.

**5.5** An incompressible flow of air over the mountain range shown in Fig. P5.5 can be approximated by a plane, steady flow. If the $z$-axis is vertical and density is allowed to vary, write the differential equations that result from conservation of mass considerations.

**Fig. P5.5**

**5.6** A stratified flow of salt water in which the density increases with depth occurs over an obstruction in the bottom of the channel of Fig. P5.6. Assuming a plane steady flow with the $z$-axis vertical, write the equations that result from the differential continuity equation.

**Fig. P5.6**

**5.7** Show that for an isothermal compressible flow,

$$\frac{1}{p}\frac{Dp}{Dt} = -\nabla \cdot \mathbf{V}$$

**5.8** An incompressible fluid is flowing radially into a sink (treated as a line or a point at the origin). Determine an expression for the radial velocity component if it is:

**(a)** A line sink          **(b)** A point sink

**5.9**    Consider steady, laminar, incompressible fluid flow in a two-dimensional diverging channel as shown in Fig. P5.9. The inclined walls of the channel are straight, and the fluid enters the diverging section with velocity $V_1 = 40$ m/s. Given $H = 1$ m, and assume unit width.

    **(a)**    Determine an expression for the velocity component $u$ as a function of position $x$ along the channel. ($u$ does not depend on y.)

    **(b)**    Determine an expression for the acceleration of the fluid in the $x$-direction

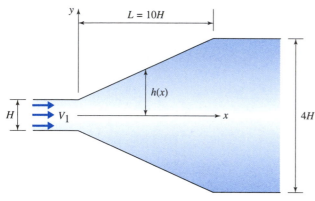

**Fig. P5.9**

**5.10**    For the same conditions given in Problem 5.9, determine:

    **(a)**    An expression for the velocity component $v$

    **(b)**    An expression for the acceleration in the $y$-direction

**5.11**    A compressible flow occurs such that

$$u = 200xy \qquad v = 200(x^2 + y^2) \qquad w = 0 \text{ m/s}$$

Find the rate at which the density is changing at the point (2 m, 1 m) where $\rho = 2.3$ kg/m$^3$.

**5.12**    If, in an incompressible plane flow, the velocity component $u = $ const, what can we say about the $y$-component of velocity? About the density?

**5.13**    In an incompressible flow we know that $u$ and $v$ are both nonzero but constant in magnitude. What can we infer about $w$ from the differential continuity equation? About the density?

**5.14**    In an incompressible plane flow $u = Ax$. Find $v(x, y)$ if $v(x, 0) = 0$.

**5.15**    If the velocity component $u$ is given by

$$u(x, y) = 10 + \frac{5x}{x^2 + y^2}$$

in an incompressible plane flow, determine $v(x, y)$. Let $v(x, 0) = 0$.

**5.16**    The $\theta$-component of velocity is given by

$$v_\theta = -\left(10 + \frac{0.4}{r^2}\right)\cos\theta$$

Find the $r$-component of velocity for the incompressible plane flow if $v_r(0.2, \theta) = 0$.

**5.17**    In an incompressible plane flow

$$v_\theta = 20\left(1 + \frac{1}{r^2}\right)\sin\theta - \frac{40}{r}$$

Find $v_r(r, \theta)$ if $v_r(1, \theta) = 0$.

**5.18**    In an incompressible axisymmetric flow ($v_\phi = 0$) the velocity component $v_\theta$ is given by

$$v_\theta = -\left(10 + \frac{40}{r^3}\right)\sin\theta$$

Find $v_r(r, \theta)$ if $v_r(2, \theta) = 0$.

**5.19**    The velocity of air in a pipe is measured at points 5 cm apart to be 137, 147, and 159 m/s, respectively. At the middle point the temperature is 5°C and the pressure is 124 kPa (abs.). Find $dp/dx$ at the middle point of this steady, uniform flow.

**5.20**    The $x$-component velocity on the $x$-axis (Fig. P5.20) is proposed to be $u(x) = -20\,(1 - e^{-x})$ m/s. Approximate the $y$-component velocity at the point $(2, 0.2)$ in this plane incompressible flow. The coordinates are in meters.

**Fig. P5.20**

**5.21**  Assume the flow in Problem 5.20 is axisymmetric and replace $y$ with $r$ and $x$ with $z$. Approximate the $r$-component velocity at $(2, 0.2)$. The coordinates are in meters.

**5.22**  The velocity component along the $x$-axis (Fig. P5.22) is $u(x) = 10 - 40/x^2$ m/s. What is the radius of the cylinder? Approximate the $y$-component velocity at $(-3, 0.1)$ assuming an incompressible flow. The coordinates are in meters.

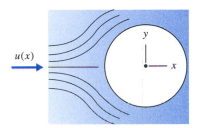

**Fig. P5.22**

**5.23**  Assume the flow in Problem 5.22 represents flow around a sphere and let $v_r(r) = (40/r^2) - 10$ m/s along the negative $x$-axis. What is the radius of the sphere? Approximate the $\theta$-component of velocity at $(-3, 0.1)$ assuming an incompressible flow. The coordinates are in meters.

**5.24**  The $x$-component of the velocity vector is measured, at points $A$, $B$, and $C$ 5 mm apart, as 11.3, 12.6, and 13.5 m/s, respectively, in the incompressible, steady plane flow shown in Fig. P5.24. Estimate:
  **(a)**  The $y$-component of velocity 4 mm above point $B$.
  **(b)**  The acceleration at point $B$.

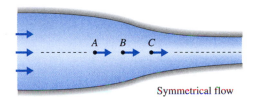

Symmetrical flow

**Fig. P5.24**

### Differential Momentum Equation

**5.25**  Sum forces on the element of Fig. 5.3 in the $y$-direction and show that Eq. 5.3.3a results.

**5.26**  Does the velocity field

$$u = \frac{10x}{x^2 + y^2} \qquad v = \frac{10y}{x^2 + y^2} \qquad w = 0$$

represent a possible incompressible flow? If so, find the pressure gradient $\nabla p$ assuming a frictionless flow with negligible body forces.

**5.27**  Does the velocity field

$$v_r = 10\left(1 - \frac{1}{r^2}\right)\cos\theta$$

$$v_\theta = -10\left(1 + \frac{1}{r^2}\right)\sin\theta$$

$$v_z = 0$$

represent a possible incompressible flow? If so, find the pressure gradient $\nabla p$ assuming a frictionless flow with negligible body forces.

**5.28**  Consider the velocity field

$$v_r = 10\left(1 - \frac{8}{r^3}\right)\cos\theta$$

$$v_\theta = -10\left(1 + \frac{4}{r^3}\right)\sin\theta$$

$$v_\phi = 0$$

Does it represent an incompressible flow? If so, find the pressure gradient $\nabla p$ assuming a frictionless flow, neglecting the body force.

**5.29**  For a steady plane flow shown in Fig. P5.29, find an expression for $D\mathbf{V}/Dt$ in terms of coordinates $(s, n)$ tangential and normal to a streamline. Let $R$ be the radius of curvature.

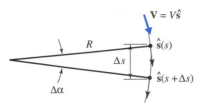

**Fig. P5.29**

**5.30** Write Euler's equation if the velocity is referred to a reference frame that is rotating with constant angular velocity.

**5.31** A velocity field is given by $u = 32\,(y - 96y^2)$ m/s, $v = 0$, and $w = 0$. Display the stress components at $y = 2.5$ mm using $\mu = 5 \times 10^{-4}$ kg-s/m$^2$ and $p = 210$ kPa. Find the ratio $\tau_{xy} / \sigma_{xx}$.

**5.32** The velocity field near a surface is approximated by $u = 10\,(2y/\delta - y^2/\delta^2)$, where $\delta = Cx^{4/5}$. If $\delta = 8$ m at $x = 1000$ m, find $v(x, y)$ assuming that $w = 0$ and $v(x, 0) = 0$. Also, display the stress components at $(1000, 0)$ using $\mu = 2 \times 10^{-5}$ N·s/m$^2$ and $p = 100$ kPa. Assume an incompressible flow.

**5.33** Show that for a steady flow $Du/Dt$ can be written as $(\mathbf{V} \cdot \boldsymbol{\nabla})\,u$, and that $D\mathbf{V}/Dt = (\mathbf{V} \cdot \boldsymbol{\nabla})\,\mathbf{V}$. Verify using rectangular coordinates.

**5.34** Write the compressible flow differential momentum equations (5.3.13) as one equation in vector form.

**5.35** Simplify the Navier–Stokes equations for incompressible steady flow between horizontal parallel plates assuming that $u = u(y)$; $w = 0$. Write all three equations.

**5.36** Consider two long, horizontal, parallel plates with a viscous incompressible fluid placed between them. The two plates move in two opposite directions with two different constant velocities, as shown in Fig. P5.36. There is no pressure gradient and the only body force is due to the weight. Starting with the Navier–Stokes equations, determine an expression for the velocity profile for laminar flow between the two plates.

**Fig. P5.36**

**5.37** Consider steady, laminar, and incompressible fluid flow between two fixed vertical parallel plates. If the fluid is moving downward between the two plates, use the Navier–Stokes equations to determine an expression for the velocity profile.

**5.38** Simplify the Navier–Stokes equations for incompressible steady flow in a horizontal pipe assuming that $v_z = v_z(r)$, $v_\theta = 0$. Write all three equations.

**5.39** An incompressible fluid flows steadily between two infinitely long concentric cylinders with the outer cylinder fixed and the inner cylinder moving with a velocity $V_c$, as shown in Fig. P5.39. For fully developed, laminar, axisymmetric flow between the two horizontal cylinders, determine an expression for the velocity $V_c$ needed to produce zero drag on the inner cylinder.

**Fig. P5.39**

**5.40** Fluid flows in the small gap between concentrically rotating spheres such that $v_\theta = v_\theta(r)$ and $v_\phi = 0$. Simplify the Navier–Stokes equations neglecting gravity for steady incompressible flow.

**5.41** Consider viscous fluid between two infinitely long concentric vertical cylinders with the inner cylinder, of radius $r_i$, fixed. The outer cylinder has a radius $r_o$ and rotates with angular velocity $\omega$. For steady, incompressible, axisymmetric laminar flow, determine an expression for the velocity profile using the Navier–Stokes equations.

**5.42** Substitute the constitutive equations (5.3.10) for incompressible flow into the differential momentum equations 5.3.2 and 5.3.3 and derive the Navier–Stokes equations 5.3.14.

**5.43** In Eqs. 5.3.14 the viscosity is assumed to be constant. If the temperature is not constant, as in a liquid flow with temperature gradients, we must let $\mu = \mu\,(T)$ so that $\mu = \mu\,(x, y, z)$ since $T = T(x, y, z)$. Modify Eqs. 5.3.14 to account for variable viscosity.

**5.44** A large flat plate oscillates beneath a liquid, as shown in Fig. P5.44. Write the differential equation that describes the motion if the plane laminar flow moves only parallel to the plate. Assume that $\mu$ = const.

$u_{\text{wall}} = U \sin \omega t$

**Fig. P5.44**

**5.45** For a gas flow in which Stokes's hypothesis is not applicable, the negative average of the three normal stresses, denoted $\bar{p}$, may be different from the pressure $p$. Find an expression for $(\bar{p} - p)$.

## Vorticity

**5.46** Show that the relationship (5.3.22) is indeed true by using rectangular coordinates.

**5.47** A uniform flow exists on the flat plate of Fig. P5.47 oriented parallel to the flow. The plate has a very sharp leading edge. Identify the term that is responsible for the creation of vorticity.

irrotational flow

$y$

region of vorticity

$x$

**Fig. P5.47**

**5.48** Neglect viscous effects and determine the velocity profile just downstream of the contraction at section 2 in the plane flow shown in Fig. P5.48. *Hint:* In an inviscid flow, the fluid does not stick to the wall.

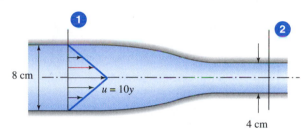

8 cm

$u = 10y$

4 cm

**Fig. P5.48**

**5.49** In Example 5.8, does $\omega_x$ remain zero as the vortex tube nears the tree? If not, explain why.

## Differential Energy Equation

**5.50** Derive the incompressible differential energy equation by applying Gauss's theorem (see Problem 5.1) to the integral energy equation 4.5.13 assuming that $\dot{W}_s = \dot{W}_{\text{shear}} = \dot{W}_I = 0$ and using

$\dot{Q}_s = \int_{\text{c.s.}} K\nabla T \cdot \hat{\mathbf{n}} \, dA$ assuming no viscous effects.

**5.51** Verify that Eq. 5.4.5 follows from Eq. 5.4.4. Recall the definition of a second derivative.

**5.52** Verify that Eq. 5.4.11 follows from Eq. 5.4.9.

**5.53** Simplify the differential energy equation for a liquid flow in which the temperature gradients are quite large and the velocity components are very small, such as in a lake heated from above.

**5.54** Explain which term in the differential energy equation accounts for the extremely high temperatures that exist on satellites during reentry.

**5.55** The velocity distribution in a 2.0-cm-diameter pipe is given by $u(r) = 10 \, (1 - 10\,000r^2)$ m/s. Find the magnitude of the dissipation function at the wall, at the centerline, and halfway between for air at 20°C.

**5.56** Let the plate in Problem 5.44 be heated. Write the simplified differential energy equation and the simplified Navier–Stokes equation assuming:
**(a)**  $\mu$ = const.
**(b)**  $\mu = \mu(T)$.

U.S. Olympic ski jumpers have used wind tunnels to improve their technique. The jumper is suspended by cables attached to his chest and thighs, enabling his lift and drag to be measured. The sport has recently been revolutionized by opening skis like a V into the wind. (© imagebroker/Alamy)

# 6

# Dimensional Analysis and Similitude

## Outline

## Chapter Objectives

The objectives of this chapter are to:

▲ Establish the parameters necessary to guide experimental studies.

▲ Present the technique used to apply the results of model studies to prototypes for a variety of flow situations.

▲ Extract the flow parameters from the differential equations and boundary conditions used to guide computational studies.

▲ Provide examples and problems that illustrate how dimensionless flow parameters are used, how model studies allow us to predict quantities of interest on a prototype, and how normalized differential equations are utilized.

## 6.1 INTRODUCTION

There are many problems of interest in the field of fluid mechanics in the real world of design that cannot be solved using the differential and integral equations only. It is often necessary to resort to experimental methods to establish relationships between the variables of interest. Since experimental studies are usually quite expensive, it is necessary to keep the required experimentation to a minimum. This is done using a technique called *dimensional analysis*, which is based on the notion of **dimensional homogeneity**—that all terms in an equation must have the same dimensions. For example, if we write Bernoulli's equation in the form

**Dimensional homogeneity:** *Condition where all terms in an equation have the same dimensions.*

$$\frac{V_1^2}{2g} + \frac{p_1}{\gamma} + z_1 = \frac{V_2^2}{2g} + \frac{p_2}{\gamma} + z_2 \tag{6.1.1}$$

we note that the dimension of each term is length. Furthermore, if we factored out $z_1$ from the left-hand side and $z_2$ from the right-hand side, we would have

$$\frac{V_1^2}{2gz_1} + \frac{p_1}{\gamma z_1} + 1 = \left( \frac{V_2^2}{2gz_2} + \frac{p_2}{\gamma z_2} + 1 \right) \frac{z_2}{z_1} \tag{6.1.2}$$

In this form of Bernoulli's equation, the terms are all dimensionless and we have written the equation as a combination of dimensionless parameters, the basic idea in dimensional analysis that will be presented in the next section.

Often in experimental work we are required to perform experiments on objects that are quite large, too large to experiment with for a reasonable cost. This would include flows over weirs and dams; wave interactions with piers and breakwaters; flows around submarines and ships; subsonic and supersonic flows around aircraft; flows around stadiums and buildings, as shown in Fig. 6.1; flows through large pumps and turbines; and flows around automobiles and trucks. Such flows are usually studied in laboratories using models that are smaller than the prototype, the actual device. This substantially reduces the costs when compared with full-scale studies and allows for the study of various configurations or flow conditions.

There are also flows of interest that involve rather small dimensions, such as flow around a turbine blade, flow into a capillary tube, flow around a microorganism, flow through a small control valve, and flow around and inside a falling droplet. These flows would require that the model be larger than the prototype so that observations could be made with an acceptable degree of accuracy.

**Similitude:** *The study of predicting prototype conditions from model observations.*

**Similitude** is the study of predicting prototype conditions from model observations. This will be presented following dimensional analysis. Similitude involves the use of the dimensionless parameters obtained in dimensional analysis.

**Buckingham $\pi$-theorem:** *A theory which organizes steps to ensure dimensional homogeneity.*

There are two approaches that can be used in the study of dimensional analysis. We will present both approaches. First, we use the **Buckingham $\pi$-theorem**, which organizes the steps of ensuring dimensional homogeneity; it requires some knowledge of the phenomenon being studied in order that the appropriate quantities of interest are included. Second, we extract the dimensionless parameters that influence a particular flow situation from the differential equations and boundary conditions that are needed to describe the phenomenon being investigated.

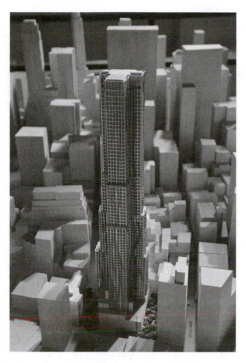

**Fig. 6.1**   Scale model of the large buildings in a city. Air flow around the buildings is studied. The roughening elements on the floor generate the desired wall turbulence. (© James Leynse/Corbis)

 Dimensional analysis, 588

## 6.2   DIMENSIONAL ANALYSIS

### 6.2.1   Motivation

In the study of phenomena involving fluid flows, either analytically or experimentally, there are invariably many flow and geometric parameters involved. In the interest of saving time and money, the fewest possible combinations of parameters should be utilized. For example, consider the pressure drop across the slider valve of Fig. 6.2. We may suspect that the pressure drop depends on such parameters as pipe mean velocity $V$, the density $\rho$ of the fluid, the fluid viscosity $\mu$, the pipe diameter $d$, and the gap height $h$. This could be expressed as

$$\Delta p = f(V, \rho, \mu, d, h) \qquad (6.2.1)$$

Now, if we attempt an experimental study of this problem, consider the strategy for finding the dependence of the pressure drop on the parameters involved. We could fix all parameters except the velocity and investigate the dependence of the pressure drop on the average velocity. Then the diameter could be changed and the experiment repeated. This would lead to the set of results shown in Fig. 6.3a. Following that set of experiments the gap height $h$ could be changed, leading to the curves of Fig. 6.3b. Again, different fluids could be studied, leading to curves with $\rho$ and $\mu$ changing values.

Consider next the notion that any equation that relates a certain set of variables, such as Eq. 6.2.1, can be written in terms of dimensionless parameters, as

> **KEY CONCEPT**   *The fewest possible combinations of parameters should be utilized.*

> **KEY CONCEPT**   *We deal with isotropic Newtonian fluids only. Non-isotropic fluids would possess additional parameters related to the appropriate stress-strain equations.*

> **KEY CONCEPT**   *Any equation can be written in terms of dimensionless parameters.*

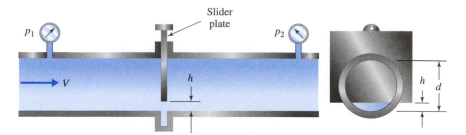

**Fig. 6.2** Flow around a slider valve.

was done with the Bernoulli equation (6.1.2). We can organize the variables of Eq. 6.2.1 into dimensionless parameters (the steps needed to do this will be presented in a subsequent section) as follows:

$$\frac{\Delta p}{\rho V^2} = f\left(\frac{V\rho d}{\mu}, \frac{h}{d}\right) \tag{6.2.2}$$

Obviously, this is a much simpler relationship. We could perform an experiment with a fixed $h/d$ (say, $h/d = 0.1$) by varying $V\rho d/\mu$ (this is done by simply varying $V$), resulting in the curve shown in Fig. 6.4. The quantity $h$ is changed so that $h/d = 0.5$ and the testing is repeated. Finally, the entire experiment is presented in one figure, as in Fig. 6.4. This has greatly reduced the effort and the cost in determining the actual form of $f(V\rho d/\mu, h/d)$; we would use only one pipe and one valve, and we would use only one fluid.

It is not always clear, however, which parameters should be included in an equation such as 6.2.1. The selection of these parameters requires a detailed understanding of the physics involved. In the selection of the parameters that affect the pressure drop across the slider valve, it was assumed that density and viscosity are important parameters, while parameters such as slider plate thickness, pipe pressure, and gravity are not. It should be kept in mind that the selection of the proper parameters is a first crucial step in the application of dimensional analysis.

## 6.2.2 Review of Dimensions

Before we present the dimensional analysis technique let's review the dimensions of the quantities of interest in an introductory course in fluid mechanics. All of the quantities have some combination of dimensions of length, time, mass, and

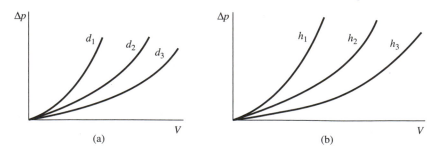

**Fig. 6.3** Pressure drop versus velocity curves: (a) $\rho, \mu, h$ fixed; (b) $\rho, \mu, d$ fixed.

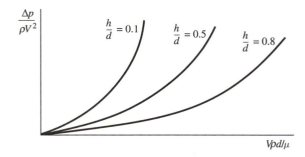

**Fig. 6.4**   Dimensionless pressure drop versus dimensionless velocity.

force which are related by Newton's second law,

$$\Sigma \mathbf{F} = m\mathbf{a} \tag{6.2.3}$$

In terms of dimensions, it is written as

$$F = \frac{ML}{T^2} \tag{6.2.4}$$

where $F$, $M$, $L$, and $T$ are the dimensions of force, mass, length, and time, respectively. Thus we see that it is sufficient to use only three basic dimensions. We will choose the $M$-$L$-$T$ system because we can eliminate the force dimension with Eq. 6.2.4.

> **KEY CONCEPT**  *Through-out this chapter, we will use the M-L-T system.*

If we were considering more complicated flow situations such as those involving electromagnetic field interactions, or those involving temperature gradients, we would need to include the appropriate additional dimensions. However, in this book such phenomena will not be introduced, except for the compressible flow of an ideal gas; for that case an equation of state relates the thermal effects to the dimensions above. That is,

$$p = \rho R T \tag{6.2.5}$$

where $T$ represents temperature. This allows us to write

$$[RT] = [p/\rho] = \frac{F}{L^2} \cdot \frac{L^3}{M} = \frac{ML/T^2}{L^2} \cdot \frac{L^3}{M} = \frac{L^2}{T^2} \tag{6.2.6}$$

where the brackets represent "the dimensions of." Notice that the equation of state does not introduce additional dimensions.

The quantities of interest in fluid mechanics are listed with their respective dimensions in Table 6.1. Reference to this table will simplify writing the dimensions of the quantities introduced in the problems.

### 6.2.3   Buckingham $\pi$-Theorem

In a given physical problem the dependent variable $x_1$ can be expressed in terms of the independent variables as

$$x_1 = f(x_2, x_3, x_4, \ldots, x_n) \tag{6.2.7}$$

**Table 6.1**  Symbols and Dimensions of Quantities Used in Fluid Mechanics

| Quantity | Symbol | Dimensions |
|---|---|---|
| Length | $l$ | $L$ |
| Time | $t$ | $T$ |
| Mass | $m$ | $M$ |
| Force | $F$ | $ML/T^2$ |
| Velocity | $V$ | $L/T$ |
| Acceleration | $a$ | $L/T^2$ |
| Frequency | $\omega$ | $T^{-1}$ |
| Gravity | $g$ | $L/T^2$ |
| Area | $A$ | $L^2$ |
| Flow rate | $Q$ | $L^3/T$ |
| Mass flux | $\dot{m}$ | $M/T$ |
| Pressure | $p$ | $M/LT^2$ |
| Stress | $\tau$ | $M/LT^2$ |
| Density | $\rho$ | $M/L^3$ |
| Specific weight | $\gamma$ | $M/L^2 T^2$ |
| Viscosity | $\mu$ | $M/LT$ |
| Kinematic viscosity | $\nu$ | $L^2/T$ |
| Work | $W$ | $ML^2/T^2$ |
| Power, heat flux | $\dot{W}, \dot{Q}$ | $ML^2/T^3$ |
| Surface tension | $\sigma$ | $M/T^2$ |
| Bulk modulus | $B$ | $M/LT^2$ |

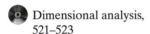

Dimensional analysis,
521–523

where $n$ represents the total number of variables. Referring to Eq. 6.2.1, $\Delta p$ is the dependent variable and $V$, $\rho$, $\mu$, $d$, and $h$ are the independent variables. The *Buckingham $\pi$-theorem*, named after Edgar Buckingham (1867–1940), states that $(n-m)$ dimensionless groups of variables, called $\pi$-terms, where $m$ is the number[1] of basic dimensions included in the variables, can be related by

$$\pi_1 = f_1(\pi_2, \pi_3, \ldots, \pi_{n-m}) \qquad (6.2.8)$$

where $\pi_1$ includes the dependent variable and the remaining $\pi$-terms include only independent variables, as in Eq. 6.2.2.

Further, it is noted that a requirement for a successful application of dimensional analysis is that a dimension must occur at least twice or not at all. For example, the equation $\Delta p = f(V, l, d)$ is ill-stated since pressure involves the dimensions of force and $V$, $l$, and $d$ do not contain such a dimension.

**KEY CONCEPT**  *A dimension must occur at least twice or not at all.*

The procedure used in applying the $\pi$-theorem is summarized as follows:

1.  Write the functional form of the dependent variable depending on the $(n - 1)$ independent variables. This step requires knowledge of the phenomenon being studied. All variables that effect the dependent variable must be included. These include geometric variables, fluid properties, and external effects that influence the variable being studied. Quantities that have no influence on the dependent variable must not be included. Also, do not include variables that depend on each other; e.g., both radius and diameter would not be included. The variables on the right-hand side of Eq. 6.2.7 should be independent.

---

[1] There are situations where $m$ is less than the number of basic dimensions. Example 6.2 will illustrate.

2. Identify $m$ **repeating variables**, variables that will be combined with each remaining variable to form the $\pi$-terms. The repeating variables selected from the independent variables must include all of the basic dimensions, but they must not form a $\pi$-term by themselves. An angle cannot be a repeating variable since it is dimensionless and forms a $\pi$-term itself.

3. Form the $\pi$-terms by combining the repeating variables with each of the remaining variables.

4. Write the functional form of the $(n - m)$ dimensionless $\pi$-terms.

Step 3 can be accomplished by an algebraic procedure; we will also illustrate a procedure in the examples that utilizes observation.

The algebraic procedure will now be illustrated with an example. Suppose that we desire to combine the variables surface tension $\sigma$, velocity $V$, density $\rho$, and length $l$ into a $\pi$-term; this can be written as

$$\pi = \sigma^a V^b \rho^c l^d \qquad (6.2.9)$$

The objective is to determine $a, b, c,$ and $d$ so that the grouping is dimensionless. In terms of dimensions (refer to Table 6.1), Eq. 6.2.9 is

$$M^0 L^0 T^0 = \left(\frac{M}{T^2}\right)^a \left(\frac{L}{T}\right)^b \left(\frac{M}{L^3}\right)^c L^d \qquad (6.2.10)$$

Equating exponents on each of the basic dimensions:

$$
\begin{aligned}
M: \quad & 0 = a + c \\
L: \quad & 0 = b - 3c + d \\
T: \quad & 0 = -2a - b
\end{aligned}
\qquad (6.2.11)
$$

The three algebraic equations are solved simultaneously to yield

$$a = -c \qquad b = 2c \qquad d = c \qquad (6.2.12)$$

so that the $\pi$-term becomes

$$\pi = \left(\frac{\rho l V^2}{\sigma}\right)^c \qquad (6.2.13)$$

A dimensionless parameter raised to any power remains dimensionless; consequently, we can select $c$ to be any number other than zero. It is usually selected as $c = 1$, depending on the ratio desired. Selecting $c = 1$, the $\pi$-term is

$$\pi = \frac{\rho l V^2}{\sigma} \qquad (6.2.14)$$

Actually, we could have selected $c = 1$ in Eq. 6.2.9 and proceeded with only three unknowns. Or if it were desired to have $\sigma$ in the numerator to the first power, we could have set $a = 1$ and let $b, c,$ and $d$ be unknowns.

A final note: If only one $\pi$-term results, the functional form would state that the $\pi$-term must be a constant since the right-hand side of Eq. 6.2.8 would contain no additional $\pi$-terms. This would result in an expression that includes an arbitrary constant that could be determined through analysis or experimentation.

## Example 6.1

The drag force $F_D$ on a cylinder of diameter $d$ and length $l$ is to be studied. What functional form relates the dimensionless variables if a fluid with velocity $V$ flows normal to the cylinder?

### Solution

First, we must determine the variables that have some influence on the drag force. If we include variables that do not influence the drag force, we would have additional $\pi$-terms that experimentation would show to be unimportant; if we do not include a variable that does influence the drag force, experimentation would also reveal that problem. Experience is essential in choosing the correct variables; in this example we will include as influential variables the free stream velocity $V$, the viscosity $\mu$, the density $\rho$ of the fluid, in addition to the diameter $d$ and the length $l$ of the cylinder, resulting in $n = 6$ variables. This is written as

$$F_D = f(d, l, V, \mu, \rho)$$

The variables are observed to include $m = 3$ dimensions. Refer to Table 6.1 and write

$$[F_D] = \frac{ML}{T^2} \quad [V] = \frac{L}{T} \quad [\mu] = \frac{M}{LT} \quad [d] = L \quad [l] = L \quad [\rho] = \frac{M}{L^3}$$

Consequently, we can expect $n - m = 6 - 3 = 3$ $\pi$-terms.

We choose repeating variables with the simplest combinations of dimensions such that they do not form a $\pi$-term by themselves (we could not include $d$ and $l$ as repeating variables); the repeating variables are chosen to be $d$, $V$, and $\rho$. These three variables are combined with each of the remaining variables to form the $\pi$-terms. Rather than writing equations similar to Eq. 6.2.9 for the $\pi$-terms, let us form the $\pi$-terms by inspection. When the repeating variables are combined with $F_D$ we observe that only $F_D$ and $\rho$ have the mass dimension; hence $F_D$ must be divided by $\rho$. Only $F_D$ and $V$ have the time dimension; hence, $F_D$ must be divided by $V^2$. Thus $F_D$ divided by $\rho$ has $L^4$ in the numerator; when divided by $V^2$ this results in $L^2$ remaining in the numerator. Hence we must have $d^2$ in the denominator resulting in

$$\pi_1 = \frac{F_D}{\rho V^2 d^2}$$

When $d$, $V$, and $\rho$ are combined with $l$ there results

$$\pi_2 = \frac{l}{d}$$

The last $\pi$-term results from combining $\mu$ with $d$, $V$, and $\rho$. The mass dimension disappears if we divide $\mu$ by $\rho$. The time dimension disappears if we divide $\mu$ by $V$. This leaves one length dimension in the numerator; hence $d$ is needed in the denominator resulting in

$$\pi_3 = \frac{\mu}{\rho V d}$$

The dimensionless, functional relationship relating the $\pi$-terms is

$$\pi_1 = f_1(\pi_2, \pi_3) \quad \text{or} \quad \frac{F_D}{\rho V^2 d^2} = f_1\left(\frac{l}{d}, \frac{\mu}{\rho V d}\right)$$

Rather than the original relationship of six variables we have reduced the relationship to one involving three $\pi$-terms, a much simpler expression. To determine the particular form of the functional relationship above, we would actually have to solve the problem; experimentation would be needed if analytical or numerical methods were not available. This is often the case in fluid mechanics.

Note that we could have included several additional variables in our original list, such as gravity $g$, the angle $\theta$ that the velocity makes with the cylinder, and the roughness $e$ of the cylinder surface. To not include variables that are significant, or to include variables that are not significant, is a matter of experience. The novice must learn how to identify significant variables; however, even the experienced researcher is often at a loss to correlate certain phenomena; much experimentation is often needed to discover the appropriate parameters.

## Example 6.2

The rise of liquid in a capillary tube is to be studied. It is anticipated that the rise $h$ will depend on surface tension $\sigma$, tube diameter $d$, liquid specific weight $\gamma$, and angle $\beta$ of attachment between the liquid and tube. Write the functional form of the dimensionless variables.

### Solution

The expression relating the variables is

$$h = f(\sigma, d, \gamma, \beta)$$

The dimensions of the variables are

$$[h] = L \qquad [\gamma] = \frac{M}{L^2 T^2} \qquad [\beta] = 1 \text{ (dimensionless)} \qquad [\sigma] = \frac{M}{T^2} \qquad [d] = L$$

By observation we see that $M/T^2$ occurs as that combination in both $\sigma$ and $\gamma$; hence $M$ and $T$ are not independent dimensions in this problem. There are only two independent groupings of basic dimensions, $L$ and $M/T^2$. Thus $m = 2$, and we choose $\sigma$ and $d$ as the repeating variables. When combined with $h$, the first $\pi$-term is

$$\pi_1 = \frac{h}{d}$$

When $\sigma$ and $d$ are combined with $\gamma$, the second $\pi$-term is

$$\pi_2 = \frac{\gamma d^2}{\sigma}$$

Finally, since the angle $\beta$ is dimensionless, it forms a $\pi$-term by itself; that is,

$$\pi_3 = \beta$$

The final functional form relating the $\pi$-terms is

$$\pi_1 = f_1(\pi_2, \pi_3) \qquad \text{or} \qquad \frac{h}{d} = f_1\left(\frac{\gamma d^2}{\sigma}, \beta\right)$$

Note: In this example we could not have chosen the angle $\beta$ as a repeating variable since it already is a dimensionless $\pi$-term. Also, we could not have chosen three repeating variables since $M$ and $T$ were not independent.

Also, note that we may have thought that gravity should have been included in the problem. If it had been included above, it would not have appeared in any of the $\pi$-terms, indicating that it should not have been included. If density and gravity, rather than specific weight, had been included, the relationship above would have resulted since $\gamma = \rho g$; this, by the way, would have avoided the necessity of observing that $M/T^2$ was a dimensional grouping.

A final note regarding the functional form of the $\pi$-terms: The relationship above could equally have been written as

$$\frac{h}{d} = f_1\left(\frac{\sigma}{\gamma d^2}, \beta\right)$$

Also, occasionally a different set of repeating variables could be selected. This simply expresses the final functional equation in a different but equivalent form. Actually, a second form can be shown to be a combination of the $\pi$-terms from an initial form.

### 6.2.4   Common Dimensionless Parameters

Consider a relatively general relationship between the pressure drop $\Delta p$, a characteristic length $l$, a characteristic velocity $V$, the density $\rho$, the viscosity $\mu$, the gravity $g$, the surface tension $\sigma$, the speed of sound $c$, and an angular frequency $\omega$, written as

$$\Delta p = f(l, V, \rho, \mu, g, c, \omega, \sigma) \tag{6.2.15}$$

The $\pi$-theorem applied to this problem, with $l$, $V$, and $\rho$ as repeating variables, results in

$$\frac{\Delta p}{\rho V^2} = f_1\left(\frac{V\rho l}{\mu}, \frac{V^2}{lg}, \frac{V}{c}, \frac{l\omega}{V}, \frac{V^2\rho l}{\sigma}\right) \tag{6.2.16}$$

Dimensionless numbers, 524–528

Each of the $\pi$-terms in this expression is a common dimensionless parameter that appears in numerous fluid flow situations. They are identified as follows:

$$\begin{aligned}
\text{Euler number, Eu} &= \frac{\Delta p}{\rho V^2} \\[2ex]
\text{Reynolds number, Re} &= \frac{V\rho l}{\mu} \\[2ex]
\text{Froude number}^2, \text{Fr} &= \frac{V}{\sqrt{lg}} \\[2ex]
\text{Mach number, M} &= \frac{V}{c} \\[2ex]
\text{Strouhal number}^2, \text{St} &= \frac{l\omega}{V} \\[2ex]
\text{Weber number}^2, \text{We} &= \frac{V^2 l\rho}{\sigma}
\end{aligned} \tag{6.2.17}$$

[2] Froude, Strouhal, and Weber number were named after William Froude (1810–1879), Vincenz Strouhal (1850–1922), and Moritz Weber (1871–1951), respectively.

The physical significance of each parameter can be determined by observing that each dimensionless number can be written as the ratio of two forces. The forces are observed to be

$$F_P = \text{pressure force} = \Delta p A \sim \Delta p l^2$$

$$F_I = \text{inertial force} = mV\frac{dV}{ds} \sim \rho l^3 V\frac{V}{l} = \rho l^2 V^2$$

$$F_\mu = \text{viscous force} = \tau A = \mu\frac{du}{dy}A \sim \mu\frac{V}{l}l^2 = \mu l V$$

$$F_g = \text{gravity force} = mg \sim \rho l^3 g \qquad (6.2.18)$$

$$F_B = \text{compressibility force} = BA \sim \rho\frac{dp}{d\rho}l^2 = \rho c^2 l^2$$

$$F_\omega = \text{centrifugal force} = mr\omega^2 \sim \rho l^3 l\omega^2 = \rho l^4 \omega^2$$

$$F_\sigma = \text{surface tension force} = \sigma l$$

Thus we see that

$$\text{Eu} \propto \frac{\text{pressure force}}{\text{inertial force}}$$

$$\text{Re} \propto \frac{\text{inertial force}}{\text{viscous force}}$$

$$\text{Fr} \propto \frac{\text{inertial force}}{\text{gravity force}} \qquad (6.2.19)$$

$$\text{M} \propto \frac{\text{inertial force}}{\text{compressibility force}}$$

$$\text{St} \propto \frac{\text{centrifugal force}}{\text{inertial force}}$$

$$\text{We} \propto \frac{\text{inertial force}}{\text{surface tension force}}$$

> **KEY CONCEPT**   *The ratios of forces allows us to anticipate the significant parameters in a flow.*

Thinking of the dimensionless parameters in terms of the ratios of forces allows us to anticipate the significant parameters in a particular flow of interest. If viscous forces are important, such as in the pipe flow of Fig. 3.8 or the boundary layer flow of Fig. 3.10, we know that the Reynolds number is a significant dimensionless parameter. If surface tension forces are instrumental in affecting the flow, as in droplet formation or flow over a weir with a small head, we expect the Weber number to be important. Similar analysis can be applied to other fluid flow situations.

Obviously, all of the effects included in the general relationship (6.2.16) would not be of interest in any one situation. It would be very unlikely that both compressibility effects and surface tension effects would influence a flow simultaneously. In addition, there is often more than one length of importance, thereby introducing additional geometric, dimensionless ratios. We have, however, introduced the more common dimensionless flow parameters of interest in fluid mechanics. Table 6.2 summarizes this section.

**Table 6.2**  Common Dimensionless Parameters in Fluid Mechanics

| Parameter | Expression | Flow situations where parameter is important |
|---|---|---|
| Euler number | $\dfrac{\Delta p}{\rho V^2}$ | Flows in which pressure drop is significant: most flow situations |
| Reynolds number | $\dfrac{\rho l V}{\mu}$ | Flows that are influenced by viscous effects: internal flows, boundary layer flows |
| Froude number | $\dfrac{V}{\sqrt{lg}}$ | Flows that are influenced by gravity: primarily free surface flows |
| Mach number | $\dfrac{V}{c}$ | Compressibility is important in these flows, usually if $V > 0.3\,c$ |
| Strouhal number | $\dfrac{l\omega}{V}$ | Flow with an unsteady component that repeats itself periodically |
| Weber number | $\dfrac{V^2 l\rho}{\sigma}$ | Surface tension influences the flow; flow with an interface may be such a flow |

## 6.3 SIMILITUDE

### 6.3.1 General Information

As stated in the introduction, *similitude* is the study of predicting prototype conditions from model observations. When an analytical or numerical solution is not practical, or when calculations are based on flow simplifications so that uncertainty is introduced, it is usually advisable to perform tests on a model if testing is not practical on a full-scale prototype, be it too large or too small.

If it is decided that a model study is to be performed, it is necessary to develop the means whereby a quantity measured on the model, signified with a subscript $m$, can be used to predict the associated quantity on the prototype, signified by a subscript $p$. We can develop such a means if we have **dynamic similarity** between model and prototype, that is, if the forces which act on corresponding masses in the model flow and the prototype flow are in the same ratio throughout the entire flow fields. Suppose that inertial forces, pressure forces, viscous forces, and gravity forces are present; then dynamic similarity requires that, at corresponding points in the flow fields,

**Dynamic similarity:** *Forces which act on corresponding masses in the model flow and prototype flow are in the same ratio throughout the entire flows.*

$$\frac{(F_I)_m}{(F_I)_p} = \frac{(F_P)_m}{(F_P)_p} = \frac{(F_\mu)_m}{(F_\mu)_p} = \frac{(F_g)_m}{(F_g)_p} = \text{const.} \tag{6.3.1}$$

These can be rearranged to read

$$\left(\frac{F_I}{F_P}\right)_m = \left(\frac{F_I}{F_P}\right)_p \qquad \left(\frac{F_I}{F_\mu}\right)_m = \left(\frac{F_I}{F_\mu}\right)_p \qquad \left(\frac{F_I}{F_g}\right)_m = \left(\frac{F_I}{F_g}\right)_p \tag{6.3.2}$$

which, in the preceding section, have been shown to be

$$\text{Eu}_m = \text{Eu}_p \qquad\qquad \text{Re}_m = \text{Re}_p \qquad\qquad \text{Fr}_m = \text{Fr}_p \tag{6.3.3}$$

If the forces above were the only ones present, we could write

$$F_I = f(F_P, F_\mu, F_g) \qquad (6.3.4)$$

Recognizing that there is only one basic dimension, namely force, dimensional analysis would allow us to write (see Eq. 6.2.8) the equation above in terms of force ratios or

$$\text{Eu} = f(\text{Re}, \text{Fr}) \qquad (6.3.5)$$

Hence we could conclude that if the Reynolds number and the Froude number are the same on the model and prototype, the Euler number must also be the same. Thus dynamic similarity between model and prototype is guaranteed by equating the Reynolds number and the Froude number of the model to those on the prototype, respectively. If compressibility forces were included here, the analysis above would result in the Mach number being included in Eq. 6.3.5.

Dynamic similarity, 195

We can write the inertial force ratio as

$$\frac{(F_I)_m}{(F_I)_p} = \frac{a_m m_m}{a_p m_p} = \text{const.} \qquad (6.3.6)$$

showing that the acceleration ratio between corresponding points on the model and prototype is a constant provided that the mass ratio of corresponding fluid elements is a constant. We can write the acceleration ratio as

$$\frac{a_m}{a_p} = \frac{V_m^2/l_m}{V_p^2/l_p} = \text{const.} \qquad (6.3.7)$$

showing that the velocity ratio between corresponding points is a constant providing the length ratio is a constant. The velocity ratio being a constant between all corresponding points in the flow fields is the statement of **kinematic similarity**. This would result in the streamline pattern around the model being the same as that around the prototype except for a scale factor. The length ratio being constant between all corresponding points in the flow fields is the demand of **geometric similarity**, which results in the model having the same shape as the prototype. Hence, to ensure complete similarity between model and prototype, we demand that:

**Kinematic similarity:** *Condition where the velocity ratio is a constant between all corresponding points in the flows.*

**KEY CONCEPT** *The streamline pattern around the model is the same as that around the prototype.*

- Geometric similarity be satisfied
- The mass ratio of corresponding fluid elements be a constant
- The appropriate dimensionless parameters of Eq. 6.2.17 be equal

Assuming that complete similarity between model and prototype exists, we can now predict quantities of interest on a prototype from measurements on a model. If we measure a drag force $F_D$ on a model and wish to predict the

**Geometric similarity:** *A condition where the model has the same shape as the prototype.*

corresponding drag on the prototype, we could equate the ratio of the drag forces to the ratio of the inertial forces (see Eq. 6.2.18) as

$$\frac{(F_D)_m}{(F_D)_p} = \frac{(F_I)_m}{(F_I)_p} = \frac{\rho_m V_m^2 l_m^2}{\rho_p V_p^2 l_p^2} \tag{6.3.8}$$

Similarity and Scaling, 494, 534, 535.

If we measure the power input to a model and wish to predict the power requirement of the prototype, we would recognize that power is force times velocity and write

$$\frac{\dot{W}_m}{\dot{W}_p} = \frac{(F_I)_m V_m}{(F_I)_p V_p} = \frac{\rho_m V_m^2 l_m^2 V_m}{\rho_p V_p^2 l_p^2 V_p} \tag{6.3.9}$$

Hence we can predict a prototype quantity if we select the model fluid (this provides $\rho_m/\rho_p$), the scale ratio (this provides $l_m/l_p$), and the appropriate dimensionless number from Table 6.2 (this provides $V_m/V_p$). Examples will illustrate.

### 6.3.2 Confined Flows

A confined flow is a flow that has no free surface (a liquid-gas surface) or interface (two different liquids forming an interface). It is confined to move within a specified region; such flows include external flows around objects, such as aircraft, buildings, and submarines, as well as internal flows in pipes and conduits.

**KEY CONCEPT** *Gravity does not influence the flow pattern in confined flows.*

Gravity does not influence the flow pattern in confined flows; that is, if gravity could be changed in magnitude, the flow pattern and associated flow quantities would not change. The dominant effect is that of viscosity in incompressible confined flows (all liquid flows and gas flows in which $M < 0.3$). Surface tension is obviously not a factor, as it would be in bubble formation, and for steady flows there would be no unsteady effects due to oscillations in the flow. The three relevant forces are pressure forces, inertial forces, and viscous forces. Therefore, in confined flows dynamic similarity is achieved if the ratios of viscous forces, inertial forces, and pressure forces between model and prototype are the same. This leads to the conclusion (see Eq. 6.3.5) that $Eu = f(Re)$, so that it is only necessary to consider the Reynolds number as the dominant dimensionless parameter in a confined incompressible flow. If compressibility effects are significant, the Mach number would also be important.

**KEY CONCEPT** *The Reynolds number is the dominant dimensionless parameter in a confined incompressible flow.*

### Example 6.3

A test is to be performed on a proposed design for a large pump that is to deliver 1.5 m³/s of water from a 40-cm-diameter impeller with a pressure rise of 400 kPa. A model with an 8-cm-diameter impeller is to be used. What flow rate should be used and what pressure rise is to be expected? The model fluid is water at the same temperature as the water in the prototype.

**Solution**

For similarity to exist in this confined incompressible flow problem, the Reynolds numbers must be equal; that is,

$$\text{Re}_m = \text{Re}_p$$

$$\frac{V_m d_m}{\nu_m} = \frac{V_p d_p}{\nu_p}$$

Recognizing that $\nu_m = \nu_p$ if the temperatures are equal, we see that

$$\frac{V_m}{V_p} = \frac{d_p}{d_m}$$

$$= \frac{0.4}{0.08} = 5$$

The ratio of flow rates is found recognizing that $Q = VA$:

$$\frac{Q_m}{Q_p} = \frac{V_m d_m^2}{V_p d_p^2}$$

$$= 5 \times \left(\frac{1}{5}\right)^2 = \frac{1}{5}$$

Thus we find that

$$Q_m = \frac{Q_p}{5} = \frac{1.5}{5} = 0.3 \text{ m}^3/\text{s}$$

The dimensionless pressure rise is found using the Euler number:

$$\left(\frac{\Delta p}{\rho V^2}\right)_m = \left(\frac{\Delta p}{\rho V^2}\right)_p$$

Hence the pressure rise for the model is

$$\Delta p_m = \Delta p_p \frac{\rho_m}{\rho_p} \frac{V_m^2}{V_p^2}$$

$$= 400 \times 1 \times 5^2 = 10\,000 \text{ kPa}$$

Note that in this example we see that the velocity in the model is equal to the velocity in the prototype multiplied by the length ratio, and the pressure rise in the model is equal to the pressure rise in the prototype multiplied by the length ratio squared. If the length ratio were very large, it is obvious that to maintain Reynolds number equivalence would be quite difficult, indeed. This observation is discussed in more detail in Section 6.3.4.

## 6.3.3   Free-Surface Flows

A free-surface flow is one in which part of the boundary involves a pressure boundary condition. This includes flows over weirs and dams, as shown in Fig. 6.5; flows in channels and spillways; flows involving two fluids separated by an interface; and flows around floating objects with waves and around submerged objects

**Fig. 6.5**   Model of the Bonneville lock and dam on the Columbia River. (Courtesy of the U.S. Army Corps of Engineers Waterways Experiment Station.)

with cavitation present. In all of these flows the location of the free surface is unknown and the velocity at the free surface is unknown; it is the pressure that must be the same[3] on either side of the interface. In free-surface flows, gravity controls both the location and the motion of the free surface. This introduces the Froude number because of the influence of the gravity forces. If we consider flows that do not exhibit periodic motions, which have negligible surface tension and compressibility effects, we can ignore the influence of St, M, and We. That leaves only the viscous effects to be considered. There are many free-surface flows in which viscous effects are significant. Consider, however, that in most model studies water is the only economical fluid to use; if the prototype fluid is also water, as it often is, we would find from the Froude numbers

> **KEY CONCEPT**   *We use the Froude number when modeling a free-surface flow.*

$$\frac{V_m^2}{l_m g_m} = \frac{V_p^2}{l_p g_p} \qquad \therefore \frac{V_m}{V_p} = \left(\frac{l_m}{l_p}\right)^{1/2} \qquad (6.3.10)$$

assuming that $g_m = g_p$. Matching the Reynolds numbers (using $\nu_m = \nu_p$):

$$\frac{V_m l_m}{\nu_m} = \frac{V_p l_p}{\nu_p} \qquad \therefore \frac{V_m}{V_p} = \frac{l_p}{l_m} \qquad (6.3.11)$$

---

[3] The surface tension, if significant, results in a pressure difference across the interface.

Thus we have a conflict. If we use the same fluid in the model study as in the prototype flow, we cannot satisfy both the Froude number criterion and the Reynolds number criterion. If we demand that both criteria be satisfied by using different fluids for model and prototype ($\nu_m \neq \nu_p$), we must select a model fluid with a viscosity $\nu_m = \nu_p(l_m/l_p)^{3/2}$ (this results from a matching of the Froude and Reynolds numbers). A fluid with this viscosity is probably either an impossibility or an impracticality. Hence, when modeling free-surface flows in which viscous effects are important, we equate the Froude numbers and include viscous effects by some other technique. For example, if we measured the total drag on the model of a ship, we would approximate the viscous drag (using some technique not included here) and subtract it from the total drag, leaving the drag due to wave resistance. The wave drag on the prototype would then be predicted using similitude, and the approximated viscous drag would be added to the wave drag, yielding the expected drag on the ship. For the better designs, the viscous drag on ship hulls can be of the same order of magnitude as the wave drag.

## Example 6.4

A 1:20 scale model of a surface vessel is used to test the influence of a proposed design on the wave drag. A wave drag of 27.5 N is measured at a model speed of 2.4 m/s. What speed does this correspond to on the prototype, and what wave drag is predicted for the prototype? Neglect viscous effects, and assume the same fluid for model and prototype.

### Solution

The Froude number must be equated for both model and prototype. Thus

$$\text{Fr}_m = \text{Fr}_p \qquad \frac{V_m}{\sqrt{l_m g}} = \frac{V_p}{\sqrt{l_p g}}$$

This yields, recognizing that $g$ does not vary significantly on the surface of the earth,

$$V_p = V_m \left(\frac{l_p}{l_m}\right)^{1/2} = 2.4\sqrt{20} = 10.7 \text{ m/s}$$

To find the wave drag on the prototype, we equate the drag ratio to the inertia force ratio:

$$\frac{(F_D)_m}{(F_D)_p} = \frac{\rho_m V_m^2 l_m^2}{\rho_p V_p^2 l_p^2}$$

This allows us to calculate the wave drag on the prototype as, using $\rho_p = \rho_m$,

$$(F_D)_p = (F_D)_m \frac{\rho_p V_p^2 l_p^2}{\rho_m V_m^2 l_m^2}$$

$$= 27.5 \times \frac{10.7^2}{2.4^2} \times 20^2 = 218.6 \text{ kN}$$

Note: We could have used the gravity force ratio rather than the inertial force ratio, but we could not have used the viscous force ratio since viscous forces were assumed negligible.

### 6.3.4 High-Reynolds-Number Flows

In a confined flow in which the Reynolds number is the dimensionless parameter that guarantees dynamic similarity, we note that if the same fluid is used in model and prototype, the velocity in the model study is $V_m = V_p l_p / l_m$; the velocity in the model is the velocity in the prototype multiplied by the scale factor. This often results in velocities that are prohibitively large in the model study. Also, the pressures encountered in the model study are large, as shown in Example 6.3, and the energy consumption is also very large. Because of these problems the Reynolds numbers may not be matched in studies involving large Reynolds numbers.

There is, however, some justification for not matching the Reynolds number in model studies. Consider a typical drag coefficient $C_D$ versus Reynolds number curve, as shown in Fig. 6.6 for a blunt object (the complete curve is presented in Fig. 8.8). The drag coefficient is a dimensionless drag, defined as $C_D = \text{drag}/\frac{1}{2}\rho V^2 A$. At a sufficiently high Reynolds number, between about $10^3$ and $10^5$, the flow is insensitive to changes in the Reynolds number; note that the drag coefficient is essentially constant and independent of Re. That implies that the flow field is similar at Re $= 10^3$ to that at Re $= 10^5$. Thus if Re$_p = 10^5$, it is only necessary that $10^3 < \text{Re}_m < 10^5$ for the viscous effects to have the same effect on model and prototype. This often allows another parameter of interest to be matched, such as the Froude number or the Mach number. There are, however, high-Reynolds-number flows in which compressibility effects and free-surface effects are negligible, so that neither the Froude number nor the Mach number are applicable. Examples include flow around automobiles, large smoke stacks, and dirigibles. For such flows we must only ensure that the Reynolds number be within the range where the drag coefficient is constant. We should note that for Reynolds numbers quite large (Re $> 5 \times 10^5$ in Fig. 6.6), the flow may also become Reynolds number independent. If that is the case, it is only necessary that Re$_m$ be sufficiently large.

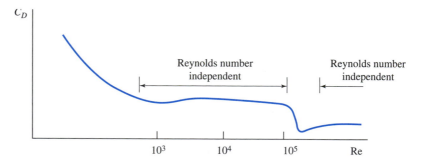

**Fig. 6.6**   Drag coefficient versus Reynolds number for typical blunt object.

### Example 6.5

A 1:10 scale model of an automobile is used to measure the drag on a proposed design. It is to simulate a prototype speed of 90 km/h. What speed should be used in the wind tunnel if Reynolds numbers are equated? For this condition, what is the ratio of drag forces?

## Solution

The same fluid exists on model and prototype; thus, equating the Reynolds numbers results in

$$\frac{V_m l_m}{\nu_m} = \frac{V_p l_p}{\nu_p} \qquad \therefore \quad V_m = V_p \frac{l_p}{l_m}$$

$$= 90 \times 10 = 900 \text{ km/h}$$

This speed would, of course, introduce compressibility effects, effects that do not exist in the prototype. Hence the proposed model study would be inappropriate.

If we did use this velocity in the model, the drag force ratio would be

$$\frac{(F_D)_p}{(F_D)_m} = \frac{\rho_p V_p^2 l_p^2}{\rho_m V_m^2 l_m^2} \qquad \therefore \quad \frac{(F_D)_p}{(F_D)_m} = 1$$

Thus we see that the drag force on the model is the same as the drag force on the prototype if the same fluids are used when we equate Reynolds numbers.

# Example 6.6

In Example 6.5, if the Reynolds numbers were equated, the velocity in the model study was observed to be in the compressible flow regime (i.e., $M > 0.3$ or $V_m > 360$ km/h). To conduct an acceptable model study, could we use a velocity of 90 km/h on a model with a characteristic length of 10 cm? Assume that the drag coefficient ($C_D = F_D / \frac{1}{2} \rho V^2 A$, where $A$ is the projected area) is independent of Re for Re $> 10^5$. If so, what drag force on the prototype would correspond to a drag force of 1.2 N measured on the model?

## Solution

The proposed model study in a wind tunnel is to be conducted with $V_m = 90$ km/h and $l_m = 0.1$ m. Using $\nu = 1.6 \times 10^{-5}$ m²/s, the Reynolds number is

$$\text{Re}_m = \frac{V_m l_m}{\nu_m} = \frac{(90 \times 1000/3600) \times 0.1}{1.6 \times 10^{-5}} = 1.56 \times 10^5$$

This Reynolds number is greater than $10^5$, so we will assume that similarity exists between model and prototype. The velocity of 90 km/h is sufficiently high.

The drag force on the prototype traveling at 90 km/h corresponding to 1.2 N on the model is found from

$$\frac{(F_D)_p}{(F_D)_m} = \frac{\rho_p V_p^2 l_p^2}{\rho_m V_m^2 l_m^2} \qquad \therefore (F_D)_p = (F_D)_m \frac{\overset{1}{\cancel{\rho_p}}}{\cancel{\rho_m}} \frac{\cancel{V_p^2}}{\overset{1}{\cancel{V_m^2}}} \frac{l_p^2}{l_m^2} = 1.2 \times 10^2 = 120 \text{ N}$$

Note that in this example we have assumed that the drag coefficient is independent of Re for Re $> 10^5$. If the drag coefficient continued to vary above Re $= 10^5$ (this would be evident from experimental data), the foregoing analysis would have to be modified accordingly.

### 6.3.5 Compressible Flows

For most compressible flow situations the Reynolds number is so large (refer to Fig. 6.6) that it is not a parameter of significance; the compressibility effects lead to the Mach number as the primary dimensionless parameter for model studies. Thus, for a particular model study we require

$$M_m = M_p \qquad \text{or} \qquad \frac{V_m}{c_m} = \frac{V_p}{c_p} \tag{6.3.12}$$

If the model study is carried out in a wind tunnel and the prototype fluid is air, we can assume that $c_m \simeq c_p$ if the temperature is the same in the two flows. For such a case, the velocity in the model study is equal to the velocity associated with the prototype. Of course, if the speeds of sound are different, the velocity ratio will be different from unity, accordingly.

---

### Example 6.7

The pressure rise from free stream to the nose of a fusilage section of an aircraft is measured in a wind tunnel at 20°C to be 34 kPa with a wind-tunnel airspeed of 900 km/h. If the test is to simulate flight at an elevation of 12 km, what is the prototype velocity and the expected nose pressure rise?

#### Solution

To find the prototype velocity corresponding to a wind-tunnel airspeed of 900 km/h, we equate the Mach numbers

$$M_m = M_p \qquad \text{or} \qquad \frac{V_m}{\sqrt{kRT_m}} = \frac{V_p}{\sqrt{kRT_p}}$$

Thus

$$V_p = V_m \left( \frac{kRT_p}{kRT_m} \right)^{1/2} = 900 \left( \frac{216.7}{293} \right)^{1/2} = 774 \text{ km/h}$$

The pressure at the nose of the prototype fusilage is found using the Euler number as follows:

$$\frac{\Delta p_m}{\rho_m V_m^2} = \frac{\Delta p_p}{\rho_p V_p^2}$$

$$\therefore \quad \Delta p_p = \Delta p_m \frac{\rho_p}{\rho_m} \frac{V_p^2}{V_m^2}$$

$$= 34 \times 0.2546 \times \frac{774^2}{900^2} = 6.4 \text{ kPa}$$

The density ratio and temperature $T_p$ were found in Appendix B.

---

## 6.3.6 Periodic Flows

In many flow situations there are regions of the flows in which periodic motions occur. Such flows include the periodic fluid motion (in Section 8.3.2 this is called vortex shedding) that takes place when a fluid flows past a cylindrical object such as a bridge, a TV tower, a cable, or a tall building; the flow past a wind machine; and the flow through turbomachinery. In flows such as these it is necessary to equate the Strouhal numbers, which can be written as

$$\frac{V_m}{\omega_m l_m} = \frac{V_p}{\omega_p l_p} \tag{6.3.13}$$

in order that the periodic motion be properly modeled.

In addition to the Strouhal number, there may be additional dimensionless parameters that must be equated: in viscous flows, the Reynolds number; in free-surface flows, the Froude number; and in compressible flows, the Mach number.

---

## Example 6.8

A large wind turbine, designed to operate at 50 km/h, is to be tested in a laboratory by constructing a 1:15 scale model. What airspeed should be used in the wind tunnel, what angular velocity should be used to simulate a prototype speed of 5 rpm, and what power output is expected from the model if the prototype output is designed to be 500 kW?

### Solution

The speed in the wind tunnel can be any speed above that needed to provide a sufficiently large Reynolds number. Let us select the same speed with which the prototype is to operate, namely, 50 km/h, and calculate the minimum characteristic length that a Reynolds number of $10^5$ would demand; this gives

$$\text{Re} = \frac{Vl}{\nu} \qquad 10^5 = \frac{(50 \times 1000/3600) \times l}{1.6 \times 10^{-5}} \qquad \therefore l = 0.12 \text{ m}$$

Obviously, in a reasonably large wind tunnel we can maintain a characteristic length (e.g., the blade length) that large.

The angular velocity is found by equating the Strouhal numbers. There results

$$\frac{V_m}{\omega_m l_m} = \frac{V_p}{\omega_p l_p} \qquad \therefore \omega_m = \omega_p \frac{\overset{1}{\cancel{V_m}}}{\cancel{V_p}} \frac{l_p}{l_m} = 5 \times 1 \times 15 = 75 \text{ rpm}$$

assuming that the wind velocities are equal.

The power is found by observing that power is force times velocity:

$$\frac{\dot{W}_m}{\dot{W}_p} = \frac{\rho_m V_m^3 l_m^2}{\rho_p V_p^3 l_p^2}$$

or

$$\dot{W}_m = \dot{W}_p \frac{\rho_m}{\rho_p} \frac{\overset{1}{\cancel{V_m^3}}}{\cancel{V_p^3}} \frac{\overset{1}{\cancel{l_m^2}}}{\cancel{l_p^2}} = 500 \times \left(\frac{1}{15}\right)^2 = 2.22 \text{ kW}$$

## 6.4 NORMALIZED DIFFERENTIAL EQUATIONS

In Chapter 5 we derived the partial differential equations used to describe flows of interest that involve isotropic, homogeneous, Newtonian fluids. Such flows could be laminar or turbulent, steady or unsteady, compressible or incompressible, confined flows or free-surface flows, flows with surface tension effects or ones in which surface tension is negligible. Most often, when utilizing the differential equations, we express them in dimensionless, or normalized, form. This form of the equations provides us with information not contained in the dimensional form, information similar to that provided by dimensional analysis. Let us normalize the differential equations that would describe the motion of an incompressible, homogeneous flow. The energy equation will not be needed.

Before we normalize the equations, let's review them. In vector form the continuity equation and the Navier–Stokes equation are

$$\nabla \cdot \mathbf{V} = 0$$

$$\rho \frac{D\mathbf{V}}{Dt} = - \nabla p - \rho g \nabla h + \mu \nabla^2 \mathbf{V} \tag{6.4.1}$$

where we have assumed $h$ to be vertical. Because the first two terms on the right are of the same form (the gradient of a scalar function) we can combine them as follows:

$$\nabla p + \rho g \nabla h = \nabla p_k \tag{6.4.2}$$

where the *kinetic pressure* $p_k$ is defined to be

$$p_k = p + \rho g h \tag{6.4.3}$$

**Kinetic pressure:**  *The pressure that results from fluid motion alone.*

In terms of the **kinetic pressure** the Navier–Stokes equation becomes

$$\rho \frac{D\mathbf{V}}{Dt} = - \nabla p_k + \mu \nabla^2 \mathbf{V} \tag{6.4.4}$$

So, as the fluid motion ceases, $p_k$ goes to zero.

If the pressure $p$ is never used in a boundary condition, we can retain the kinetic pressure in the equations, thereby "hiding" the gravity effects. If the pressure enters a boundary condition, we must return to Eq. 6.4.3, and gravity becomes important.

To normalize the differential equations and boundary conditions, we must select characteristic quantities that best describe the problem of interest. For example, consider a flow that has an average velocity $V$ and a primary dimension $l$. The dimensionless velocities and coordinate variables are

$$u* = \frac{u}{V} \qquad v* = \frac{v}{V} \qquad w* = \frac{w}{V} \qquad x* = \frac{x}{l} \qquad y* = \frac{y}{l} \qquad z* = \frac{z}{l} \qquad (6.4.5)$$

where asterisks are used to denote dimensionless quantities. The characteristic pressure will be twice the inviscid pressure rise between free stream and the stagnation point (i.e., $\rho V^2$); the characteristic time, provided that no characteristic time such as a period of oscillation exists in the problem, will be the time it takes a fluid particle to travel the distance $l$ at the velocity $V$ (i.e., $l/V$). Hence

$$p* = \frac{p}{\rho V^2} \qquad t* = \frac{t}{l/V} \qquad\qquad (6.4.6)$$

Using the dimensionless quantities of Eq. 6.4.5, we see that

$$\mathbf{V}* = u*\hat{\mathbf{i}} + v*\hat{\mathbf{j}} + w*\hat{\mathbf{k}}$$

$$= \frac{u}{V}\hat{\mathbf{i}} + \frac{v}{V}\hat{\mathbf{j}} + \frac{w}{V}\hat{\mathbf{k}} = \frac{\mathbf{V}}{V}$$

$$\boldsymbol{\nabla}* = \frac{\partial}{\partial x*}\hat{\mathbf{i}} + \frac{\partial}{\partial y*}\hat{\mathbf{j}} + \frac{\partial}{\partial z*}\hat{\mathbf{k}} \qquad\qquad (6.4.7)$$

$$= l\frac{\partial}{\partial x}\hat{\mathbf{i}} + l\frac{\partial}{\partial y}\hat{\mathbf{j}} + l\frac{\partial}{\partial z}\hat{\mathbf{k}} = l\boldsymbol{\nabla}$$

Introducing the above into the Navier–Stokes equation and the continuity equation, we have

$$\rho \frac{V}{l/V}\frac{D*\mathbf{V}*}{Dt*} = -\frac{\rho V^2}{l}\boldsymbol{\nabla}*p_k^* + \mu \frac{V}{l^2}\boldsymbol{\nabla}^{*2}\mathbf{V}*$$

$$\qquad\qquad (6.4.8)$$

$$\frac{V}{l}\boldsymbol{\nabla}*\cdot\mathbf{V}* = 0$$

where $D*/Dt*$ represents the dimensionless material derivative. Finally, in dimensionless form,

$$\frac{D^*\mathbf{V}^*}{Dt^*} = -\nabla^*p_k^* + \frac{\mu}{\rho Vl}\nabla^{*2}\mathbf{V}^*$$

$$\nabla^*\cdot\mathbf{V}^* = 0$$

(6.4.9)

Note that we have introduced the Reynolds number

$$\mathrm{Re} = \frac{\rho Vl}{\mu}$$

(6.4.10)

as a parameter in the normalized Navier–Stokes equation.

**KEY CONCEPT** *The boundary conditions introduce dimensionless parameters.*

Consider now the boundary conditions. The no-slip condition on a fixed boundary $\mathbf{V}^* = 0$ introduces geometric parameters that are necessary to specify the geometry of the fixed boundary. These could include, for example, a roughness parameter $e/l$, where $e$ is the average height of the roughness elements; a thickness parameter $t/l$, where $t$ is the maximum thickness of a turbine blade; and a nose radius parameter $r_0/l$, where $r_0$ is the nose radius of an airfoil.

Another boundary condition would be the entering velocity distribution, for example, $u(y)/V$. This may introduce parameters having to do with the profile and the turbulent structure of the entering flow. Such features could be extremely significant for a particular flow.

Also, a portion of the boundary may be oscillating, like that of a rotating component. On such a boundary we would require that the velocity of the fluid be the same as the velocity of the rotating part, that is, $v = r\omega$. This would introduce the Strouhal number St,

$$v^* = \mathrm{St}\,r^*$$

(6.4.11)

where

$$\mathrm{St} = \frac{\omega l}{V}$$

(6.4.12)

Another boundary condition that introduces an additional parameter is that of a free surface. It is required that, neglecting surface tension effects, the pressure be constant across a free surface. On an air–water surface this requires that $p = 0$ on the free surface. Returning to Eq. 6.4.3, and normalizing, there results

$$\rho V^2 p_k^* = \rho V^2 p^* + \rho g l h^* \qquad \text{or} \qquad p_k^* = p^* + \frac{gl}{V^2}h^*$$

(6.4.13)

where we have introduced the Froude number

$$\text{Fr} = \frac{V}{\sqrt{lg}} \qquad (6.4.14)$$

Finally, a boundary condition may involve surface tension, as in droplet formation problems. Such a boundary condition would involve the two radii of curvature leading to the normalized equation

$$\Delta p^* = \frac{\sigma}{l\rho V^2}\left(\frac{1}{r_1^*} + \frac{1}{r_2^*}\right) \qquad (6.4.15)$$

where we have introduced the Weber number

$$\text{We} = \frac{V^2 l\rho}{\sigma} \qquad (6.4.16)$$

Note that we have introduced all of the dimensionless parameters discussed in Section 6.3 with the exception of the Mach number. It could be introduced in a normalized energy equation, a task that is left to the home problems.

Obviously, if we can write the differential equations that describe a particular flow, those equations and the boundary conditions contain all the parameters of interest; hence the Buckingham $\pi$-theorem is not actually necessary for flow situations in which the equations and boundary conditions are known.

In addition, we can obtain the similitude conditions from the normalized equations and boundary conditions. The idea is to make the normalized differential equations and boundary conditions the same for both model and prototype. If they are the same, they should possess the same solution.[4] The normalized boundary conditions being identical would demand that geometric similarity be satisfied. In order that the equations and boundary conditions be identical on both model and prototype, it is necessary that the dimensionless parameters be the same for both model and prototype. Thus, similitude conditions are also included in the normalized problem.

> **KEY CONCEPT** *The differential equations and the boundary conditions contain all of the parameters of interest.*

> **KEY CONCEPT** *Similitude conditions are also included in the normalized problem.*

---

[4] Questions of uniqueness, which we will not consider, enter here. For example, it is possible to have identical conditions and equations, yet in one case a laminar flow could result and in the other it could be a turbulent flow or a second laminar flow different from the first; that is, different solutions would exist.

## 6.5 SUMMARY

Experimental studies are greatly simplified by reducing the number of variables that influence the phenomenon being studied. We demonstrated how the relationship

$$\Delta p = f(V, \rho, \mu, d, h) \tag{6.5.1}$$

could be written in the simplified form,

$$\frac{\Delta p}{\rho V^2} = f\left(\frac{V\rho d}{\mu}, \frac{h}{d}\right) \tag{6.5.2}$$

The most common flow parameters are, using $l$ as the characteristic length,

$$\mathrm{Re} = \frac{V\rho l}{\mu}, \quad \mathrm{Fr} = \frac{V}{\sqrt{lg}}, \quad \mathrm{M} = \frac{V}{c}, \quad \mathrm{Eu} = \frac{\Delta p}{\rho V^2}, \quad \mathrm{St} = \frac{l\omega}{V}, \quad \mathrm{We} = \frac{V^2 l \rho}{\sigma} \tag{6.5.3}$$

The parameters, used to guide model studies that are of primary significence in a particular flow are identified as follows:

$$\text{Confined flows:} \qquad \mathrm{Re} = \frac{V\rho l}{\mu} \tag{6.5.4}$$

$$\text{Free-surface flows:} \qquad \mathrm{Fr} = \frac{V}{\sqrt{lg}} \tag{6.5.5}$$

$$\text{High-Reynolds-number flows:} \quad \mathrm{Re} > (\mathrm{Re})_{\text{minimum}} \tag{6.5.6}$$

$$\text{Compressible flows:} \qquad \mathrm{M} = \frac{V}{c} \tag{6.5.7}$$

$$\text{Periodic flows:} \qquad \mathrm{St} = \frac{V}{l\omega} \tag{6.5.8}$$

The Navier–Stokes equation is written using dimensionless variables as

$$\frac{D\mathbf{V}}{Dt} = -\boldsymbol{\nabla} p_k + \frac{1}{\mathrm{Re}} \nabla^2 \mathbf{V} \tag{6.5.9}$$

where $p_k$ is the kinetic pressure. We know that all variables are dimensionless since the Reynolds number appears in the equation. We often write equations in dimensionless form without the asterisks for simplicity.

## Fundamentals of Engineering Exam Review Problems

**6.1** Combine power $W$, diameter $d$, pressure drop $\Delta p$, and average velocity $V$ into a dimensionless group.

(A) $\dfrac{\dot{W}}{d^2 V \Delta p}$  (B) $\dfrac{\dot{W}}{V \Delta p}$

(C) $\dfrac{\dot{W} d}{V \Delta p}$  (D) $\dfrac{\dot{W}}{V d \Delta p}$

**6.2** It is proposed that the velocity of a flow depends on a diameter $d$, a length $l$, gravity $g$, rotational speed $\omega$, and viscosity $\mu$, Select the variable that would not influence the velocity.

(A) $\mu$
(B) $\omega$
(C) $g$
(D) both $d$ and $l$

**6.3** What velocity should be selected in a wind tunnel where a 9:1 scale model of an automobile is to simulate a speed of 12 m/s? Neglect compressibility effects.

(A) 108 m/s  (B) 12 m/s
(C) 4 m/s  (D) 1.33 m/s

**6.4** Flow around an underwater structural component is to be studied in a 20°C-wind tunnel with a 10:1 scale model. What speed should be selected in the wind tunnel to simulate a 10°C-water speed of 4 m/s?

(A) 4.61 m/s
(B) 40 m/s
(C) 31.6 m/s
(D) 461 m/s

**6.5** What upstream velocity should be selected in a 16:1 scale model of a levee which has an upstream velocity of 2 m/s?

(A) 2 m/s  (B) 1 m/s
(C) 0.5 m/s  (D) 0.25 m/s

**6.6** A force of 10 N is measured on a 25:1 scale model of a ship tested in a water channel. What force should be expected on the protype ship? Neglect viscous effects.

(A) 156 kN  (B) 62.5 kN
(C) 6250 N  (D) 250 N

## PROBLEMS

**6.7** Write Bernoulli's equation in dimensionless form by dividing Eq. 6.1.1 by $V_1^2$ and multiplying by $g$. Express the equation in a form similar to that of Eq. 6.1.2.

**6.8** If the $F$-$L$-$T$ system of units were used, what would the dimensions be on each of the following?
(a) Mass flux  (b) Pressure
(c) Density  (d) Viscosity
(e) Work  (f) Power
(g) Surface tension

## Dimensional Analysis

**6.9** Verify that the dimensions of power are $ML^2/T^3$ as listed in Table 6.1.

**6.10** If the velocity $V$ in a fluid flow depends on a dimension $l$, the fluid density $\rho$, and the viscosity $\mu$, show that this implies that the Reynolds number $Vl\rho/\mu$ is constant.

**6.11** If the velocity $V$ in a fluid depends on the surface tension $\sigma$, the density $\rho$, and a diameter $d$, show that this implies that the Weber number $V^2 d\rho/\sigma$ is constant.

**6.12** Assume that the velocity $V$ of fall of an object depends on the height $H$ through which it falls, the gravity $g$, and the mass $m$ of the object. Find an expression for $V$.

**6.13** Include the density $\rho$ and viscosity $\mu$ of the surrounding fluid and repeat Problem 6.12. This would account for the resistance (drag) of the fluid.

**6.14** Select $l$, $V$, and $\rho$ as the repeating variables in Example 6.1 and find an expression for $F_D$. Show that this is an equivalent form to that of Example 6.1.

**6.15** Select $d$, $\mu$, and $V$ as the repeating variables in Example 6.1 and find an expression for $F_D$. Show that this is equivalent to the expression of Example 6.1.

**6.16** Include gravity $g$ in the list of variables of Example 6.2 and determine the final expression that results for $h$.

**6.17** Find an expression for the centrifugal force $F_c$ if it depends on the mass $m$, the angular velocity $\omega$, and the radius $R$ of an impeller.

**6.18** The normal stress $\sigma$ in a beam depends on the bending moment $M$, the distance $y$ from the neutral axis, and the moment of inertia $I$. Relate $\sigma$ to the other variables if we know that $\sigma$ varies linearly with $y$.

**6.19** Find an expression for the average velocity in a smooth pipe if it depends on the viscosity, the diameter, and the pressure gradient $\partial p/\partial x$.

**6.20** It is suggested that the velocity of the water flowing from a hole in the side of an open tank depends on the height $H$ of the hole from the surface, the gravity, and the density of the water. What expression relates the variables?

**6.21** Derive an expression for the velocity of liquid issuing from a hole in the side of an open tank if the velocity depends on the height $H$ of the hole from the free surface, the fluid viscosity and density, gravity, and the hole diameter.

**6.22** The pressure drop $\Delta p$ in the pipe of Fig. P6.22 depends on the average velocity, the pipe diameter, the kinematic viscosity, the length $L$ of pipe, the wall roughness height $e$, and the fluid density. Find an expression for $\Delta p$.

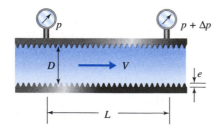

**Fig. P6.22**

**6.23** Select an appropriate set of variables that influence the drag force $F_D$ on an airfoil (Fig. P6.23) and write the final form in terms of dimensionless parameters.

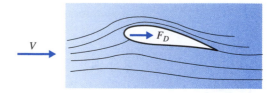

**Fig. P6.23**

**6.24** The flow rate $Q$ in an open channel depends on the hydraulic radius $R$, the cross-sectional area $A$, the wall roughness height $e$, gravity $g$, and the slope $S$. Relate $Q$ to the other variables using (a) the $M$-$L$-$T$ system, and (b) the $F$-$L$-$T$ system.

**6.25** The velocity of propagation of ripples in a shallow liquid depends on the liquid depth $h$, gravity $g$, surface tension $\sigma$, and the liquid density $\rho$. Find an expression for the propagation velocity $V$. See Fig. P6.25.

**Fig. P6.25**

**6.26** The drag force $F_D$ on a sphere depends on the velocity $V$, the viscosity $\mu$, the density $\rho$, the surface roughness height $e$, the freestream fluctuation intensity $I$ (a dimensionless quantity), and the diameter $D$. Find an expression for $F_D$.

**6.27** The drag force acting on a ship is considered to be a function of the fluid density $\rho$, viscosity $\mu$, gravity $g$, ship velocity $V$, and a characteristic length $l$. Determine a set of suitable dimensionless numbers to describe the relationship $F_D = f(\rho, \mu, g, V, l)$.

**6.28** The thrust force $T$, in newtons, from an airplane propeller is a function of propeller diameter $D$, air density $\rho$, air viscosity $\mu$, and airplane speed $V$. Determine a suitable set of dimensionless numbers to describe the relationship $T = f(\rho, D, \mu, V)$.

**6.29** The drag force $F_D$ on the smooth sphere of Fig. P6.29 falling in a liquid depends on the constant sphere speed $V$, the solid density $\rho_s$, the liquid density $\rho$ and viscosity $\mu$, the sphere diameter $D$, and gravity $g$. Find an expression for $F_D$ using (a) the $M$-$L$-$T$ system, and (b) the $F$-$L$-$T$ system.

**Fig. P6.29**

**6.30** The drag force $F_D$ on a golf ball depends on the velocity, the viscosity, the density, the diameter, the dimple depth, the dimple radius, and the concentration $C$ of dimples measured by the number of dimples per unit area. What expression relates $F_D$ to the other variables?

**6.31** The frequency $f$ with which vortices are shed from a cylinder depends on the viscosity, density, velocity, and diameter. Derive an expression for $f$.

**6.32** The lift $F_L$ on an airfoil depends on the aircraft speed $V$, the speed of sound $c$, the fluid density, the airfoil chord length $l_c$, the airfoil thickness $t$, and the angle of attack $\alpha$. Find an expression for $F_L$.

**6.33** Derive an expression for the torque $T$ needed to rotate a disk of diameter $d$ at the angular velocity $\omega$ in a fluid with density $\rho$ and viscosity $\mu$ if the disk is a distance $t$ from a wall. Also, find an expression for the power requirement.

**6.34** The cables holding a suspension bridge (Fig. P6.34) are observed to undergo large vibrations under certain wind conditions. Select a set of variables that influence the periodic force acting on a cable and write a simplified relationship of dimensionless parameters.

Cable

**Fig. P6.34**

**6.35** A vacuum cleaner creates a pressure drop $\Delta p$ across its fan. Relate this pressure drop to the impeller diameter $D$ and width $h$, its rotational speed $\omega$, the air density $\rho$, and the inlet and outlet diameters $d_i$ and $d_o$. Also, find an expression for the power requirement of the fan.

**6.36** Derive an expression for the maximum torque needed to rotate an agitator if it depends on the frequency of oscillation, the angular velocity with which the agitator rotates during an oscillation, the diameter of the agitator, the nominal height of the paddles, the length of the paddles, the number of paddles, the depth of liquid, and the liquid density.

**6.37** The flow rate of water over a weir depends on the head of water, the width of the weir, gravity, viscosity, density, and surface tension. Relate the flow rate to the other variables.

**6.38** The size of droplets in the spray of a fruit sprayer depends on the air velocity; the spray jet velocity, the spray jet exit diameter; the liquid spray surface tension, density, and viscosity; and the air density. Relate the droplet diameter to the other variables.

**6.39** Relate the torque $T$ to the other variables shown in Fig. P6.39.

**Fig. P6.39**

**6.40** A viscometer is composed of an open tank in which the liquid level is held constant. A small-diameter tube empties into a calibrated volume. Find an expression for the viscosity using the relevant parameters.

**6.41** Derive an expression for the torque $T$ necessary to rotate the shaft of Fig. P6.41.

Oil

**Fig. P6.41**

**6.42**  Derive an expression for the depth $y_2$ in the hydraulic jump of Fig. P6.42.

**Fig. P6.42**

**6.43**  Derive an expression for the frequency with which a cylinder, suspended by a string in a liquid flow, oscillates.

## Similitude

**6.44**  After a model study has been performed, it is necessary to predict quantities of interest that are to be expected on the prototype. Write expressions, in terms of density, velocity, and length, for the ratio of model to prototype of each of the following: flow rate $Q$, pressure drop $\Delta p$, pressure force $F_p$, shear stress $\tau_0$, torque $T$, and heat transfer rate $\dot{Q}$.

**6.45**  A 1:7 scale model simulates the operation of a large turbine that is to generate 200 kW with a flow rate of 1.5 m³/s. What flow rate should be used in the model, and what power output is expected?
  (a)  Water at the same temperature is used in both model and prototype.
  (b)  The model water is at 25°C and the prototype water is at 10°C.

**6.46**  A 1:5 scale model of a large pump is used to test a proposed change. The prototype pump produces a pressure rise of 600 kPa at a mass flux of 800 kg/s. Determine the mass flux to be used in the model and the expected pressure rise.
  (a)  Water at the same temperature is used in both model and prototype.
  (b)  The water in the model study is at 30°C and the water in the prototype is at 15°C.

**6.47**  A force on a component of a 1:10 scale model of a large pump is measured to be 45 N. What force is expected on the prototype component if water is used for both model and prototype:
  (a)  With the water at the same temperature?
  (b)  With the prototype water at 10°C and the model water at 20°C?

**6.48**  A 1:10 scale model study of an automobile is proposed. A prototype speed of 100 km/h is desired. What wind-tunnel speed should be selected for the model study? Comment on the advisability of such a speed selection.

**6.49**  A 1:10 scale model study of a torpedo is proposed. Prototype speeds of 90 km/h are to be studied. Should a wind tunnel or a water facility be used?

**6.50**  A 1:10 scale model of an airplane propeller is to be tested in a laboratory using a wind tunnel. The prototype on which we are to estimate the thrust moves at $V = 200$ km/h at 8000 m altitude. Assuming the prototype and model have geometric similarity, use the result from Problem 6.28 to achieve dynamic similarity. (a) What air speed should be used in the wind tunnel using standard air? (b) What is the corresponding thrust for the prototype, if a thrust of 10 N is measured on the model?

**6.51**  It is proposed that a model study be carried out on a proposed low-speed airfoil that is to fly at low altitudes at a speed of 50 m/s. If a 1:10 scale model is to be constructed, what velocity should be used in a wind tunnel? Comment as to the advisability of such a test. Would it be more advisable to perform the test in a 20°C water channel? If a water channel study were undertaken, calculate the drag ratio between the model and prototype.

**6.52**  A model study of oil (SAE-10W) at −2°C flowing in a 0.75-m-diameter pipe is performed by using water at 20°C. Pipe of what diameter should be used if the average velocities are to be the same? What pressure drop ratio is expected? $S_{oil} = 0.9$.

**6.53**  A 0.025-mm-long microorganism moves through water at the rate of 0.1 body length per second. Could a model study be performed for such a prototype in a water channel or a wind tunnel?

**6.54**  A 1:30 scale model of a ship is to be tested with complete similarity. What should be the viscosity of the model fluid? Is such a liquid possible? What conclusion can be stated?

**6.55**  A 1:60 scale model of a ship is used in a water tank to simulate a ship speed of 10 m/s. What should be the model speed? If a towing force of 10 N is measured on the model, what force is expected on the prototype? Neglect viscous effects.

**6.56** The flow rate over a weir is 2 m³/s of water. A 1:10 scale model of the weir is tested in a water channel.
**(a)** What flow rate should be used?
**(b)** If a force of 12 N is measured on the model, what force would be expected on the prototype?

**6.57** A 1:10 scale model of a hydrofoil is tested in a water channel with a force of 4 N measured at a speed of 6 m/s. Determine the speed and force predicted for the prototype. Neglect viscous effects.

**6.58** The propeller of a ship is to be studied with a 1:10 scale model.
**(a)** Assuming that the propeller operates close to the surface, select the propeller speed of the model if the prototype speed is 600 rpm.
**(b)** What torque would be expected if 1.2 N·m is measured on the model?

**6.59** The pontoon of a seaplane is to be studied in a water channel that has a channel speed capability of 6 m/s. If the seaplane is to lift off at 100 m/s, what model scale should be selected?

**6.60** A 1:30 scale model study of a submarine is proposed in an attempt to study the influence of a suggested shape modification. The prototype is 2 m in diameter, and it is designed to travel at 15 m/s. The model is towed in a water tank at 2 m/s, and a drag force of 2.15 N is measured. Does similitude exist for this test? If so, predict the power needed for the prototype.

**6.61** The smoke emanating from the stacks on a ship has a tendency to make its way to the deck down the outside of the stacks causing discomfort to the passengers. This problem is studied with a 1:20 scale model of a 4-m-diameter stack. The ship is traveling at 10 m/s. What range of wind-tunnel airspeeds could be used in this study?

**6.62** A model study of a dirigible (a large balloon that travels through the air) is to be performed. The 10-m-diameter dirigible is to travel at 20 m/s. If a 40-cm-diameter model is proposed for use in a wind tunnel, or a 10-cm-diameter model for use in a 20°C water channel, which should be selected? Suppose that the wind-tunnel model is used with a speed of 15 m/s and a drag force of 3.2 N is measured. What force would be expected on the water-channel model with a speed of 2.4 m/s in the water channel? What horsepower would be predicted to overcome

drag on the prototype? Assume that the flow is Reynolds number independent for $Re > 10^5$.

**6.63** A model study is to be performed of a 300-m-tall, 13.5-m-diameter smokestack of a power plant. It is known that the smokestack is submerged in a ground boundary layer that is 360 m thick. Could the study be performed in a wind tunnel that produces a 1.2-m-thick boundary layer?

**6.64** A 1:20 scale model of an aircraft is tested in a 23°C wind tunnel. A speed of 200 m/s is used in the model study. A drag force of 10 N is measured. What prototype speed and drag force does this simulate if the elevation is:
**(a)** Sea level?
**(b)** 5000 m?
**(c)** 10 000 m?

**6.65** A 1:10 scale model of an airfoil is tested in a wind tunnel that uses outside air at 0°C. The test is to simulate an aircraft speed of 250 m/s at an elevation of 10 000 m. What wind-tunnel speed should be used? If a velocity of 290 m/s and a pressure of 80 kPa absolute are measured on the model at a particular location at a 5° angle of attack, what speed and pressure are expected on the prototype at the corresponding location, and what is the angle of attack?

**6.66** A 1:10 scale model of a propeller on a ship is to be tested in a water channel. What should be the rotational speed of the model if the rotational speed of the propeller is 2000 rpm and:
**(a)** The Froude number governs the study?
**(b)** The Reynolds number governs the study?

**6.67** A torque of 12 N·m is measured on a 1:10 scale model of a large wind machine with a wind tunnel speed of 60 m/s. This is to simulate a wind velocity of 15 m/s since viscous effects are considered negligible. What torque is expected on the prototype? If the model rotates at 500 rpm, what prototype angular velocity is simulated?

**6.68** An underwater study of a porpoise is to be performed by using a 1:10 scale model. A porpoise swimming at 10 m/s and making one swimming motion each second is to be simulated. What speed could be used in the water channel, and for that speed, how many swimming motions per second should be used?

**6.69** A ship model test is used to predict the drag on a ship, which is very important in the design of ships. Model resistance tests are usually carried out in a towing tank. A ship model (1.2 to 2.2 m long for a small towing tank, 4 to 10 m for a large one) is towed at a constant velocity by a mechanically propelled towing carriage. The resistance of the model at constant velocity is recorded by the instruments on the carriage. Usually the test is carried out at a number of constant velocities, and a resistance curve is thus obtained. The drag coefficient $C_D = \rho V^2 A/2$, where $A$ is the wetted area, is proportional to $l^2$. From dimensional analysis we can write $C_D = f$ (Re, Fr). Since the drag force consists of two components, a friction drag and a wave drag, we write: $C_D = C_f + C_w$. Assuming that the two forces are independent, we can consider that the friction drag is only a function of Reynolds number, and the wave drag is only a function of Froude number; i.e., $C_f = f$ (Re) and $C_w = f$ (Fr). Since we cannot satisfy both Re and Fr similarity between model and prototype, we will use Fr to achieve dynamic similarity and use Re to determine the friction drag. Hence, from the model test data we can calculate

$$(C_D)_m = \left( \frac{F_D}{\frac{1}{2} \rho V^2 A} \right)_m = (C_f)_m + (C_w)_m.$$

Assuming friction resistance on the ship is similar to that of a flat plate, we calculate $(C_f)_m$ from $\text{Re}_m$ and an appropriate equation. The wave drag coefficient is determined using $(C_w)_m = (C_D)_m - (C_f)_m$. Using $(\text{Fr})_p = (\text{Fr})_m$ to determine $(C_w)_p = (C_w)_m$ we then calculate $(C_f)_p$ from $\text{Re}_p$. Finally, we calculate the drag coefficient for the prototype using $(C_D)_p = (C_f)_p + (C_w)_p$. Test data for the model are shown in Tables P6.69(1) and P6.69(2) below.

**Table 6.69(1)**

| Model Features | Model Data (1:20 scale) |
|---|---|
| Wetted area (m²) | 12.205 |
| Model length (m) | 7.11 |
| Water temperature (°C) | 12.6 |

**Table P6.69(2).** Test Data for Model

| V(m/s) | $F_D$ (N) |
|---|---|
| 0.805 | 17.189 |
| 0.995 | 25.49 |
| 1.199 | 35.80 |
| 1.399 | 49.450 |
| 1.599 | 69.66 |
| 1.759 | 92.18 |
| 1.899 | 132.98 |

If the prototype is used in seawater at 19.2°C, determine the following:
  **(a)** Using this data, calculate the drag coefficient for the prototype.
  **(b)** Plot the drag coefficient versus Re.
  **(c)** Plot the wave drag coefficient versus Fr.

## Normalized Differential Equations

**6.70** Normalize the continuity equation

$$\frac{\partial \rho}{\partial t} + \frac{\partial}{\partial x} (\rho u) + \frac{\partial}{\partial y} (\rho v) = 0$$

using a characteristic velocity $V$, length $l$, density $\rho_0$, and time $f^{-1}$, where $f$ is the frequency. What dimensionless parameter is introduced?

**6.71** Normalize Euler's equation

$$\frac{\partial \mathbf{V}}{\partial t} + u \frac{\partial \mathbf{V}}{\partial x} + v \frac{\partial \mathbf{V}}{\partial y} + w \frac{\partial \mathbf{V}}{\partial z} = -\frac{\nabla p}{\rho}$$

using a characteristic velocity $U$, length $l$, pressure $\rho U^2$, and time $f^{-1}$, where $f$ is the frequency. Find any dimensionless parameters that are introduced.

**6.72**   Normalize Euler's equation

$$\rho \frac{D\mathbf{V}}{Dt} = -\nabla p - \rho g \nabla h$$

using a characteristic velocity $U$, length $l$, pressure $\rho U^2$, and time $l/U$. Determine any dimensionless parameters introduced in the normalized equation.

**6.73**   A fluid is at rest between the large horizontal plates shown in Fig. P6.73. The top plate is suddenly given a velocity $U$. Show that the Navier–Stokes equation that describes the resulting motion simplifies to

$$\frac{\partial u}{\partial t} = \nu \frac{\partial^2 u}{\partial y^2}$$

Nondimensionalize this equation using characteristic velocity $U$, length $h$, and times **(a)** $h/U$ and **(b)** $h^2/\nu$. Identify any dimensionless parameters that result.

**Fig.P6.73**

**6.74**   A fluid flows in a horizontal pipe of diameter $d$ (Fig. P6.74). The flow is suddenly increased to an average velocity $V$. Show that the appropriate Navier–Stokes equation, using the coordinates shown, simplifies to

$$\rho \frac{\partial u}{\partial t} = -\frac{\partial p}{\partial x} + \mu \left( \frac{\partial^2 u}{\partial r^2} + \frac{1}{r} \frac{\partial u}{\partial r} \right)$$

Normalize this equation using characteristic velocity $V$, length $d$, and time **(a)** $d/V$ and **(b)** $d^2/\nu$. Identify any dimensionless parameters that result.

**Fig. P6.74**

**6.75**   A highly viscous liquid such as honey is flowing down a vertical flat surface. Its thickness decreases as it progresses down the surface (Fig. P6.75). Show that the steady flow is described by

$$\rho u \frac{\partial u}{\partial x} = \mu \left( \frac{\partial^2 u}{\partial x^2} + \frac{\partial^2 u}{\partial y^2} \right) + g$$

where we have neglected the $y$-component of velocity. Normalize this equation using characteristic length $h$ (measured at $x = 0$) and velocity $V$ (the average velocity). Identify any dimensionless parameters that result.

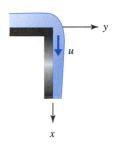

**Fig. P6.75**

**6.76**   Nondimensionalize the energy equation

$$\rho c_p \left( u \frac{\partial T}{\partial x} + v \frac{\partial T}{\partial y} \right) = K \nabla^2 T$$

using a characteristic velocity $U$, length $l$, and temperature $T_0$. Express the dimensionless parameter that results in terms of the Prandtl number

$$\text{Pr} = \mu \frac{c_p}{K}$$

**6.77**   Nondimensionalize the compressible flow differential momentum equation

$$\rho \frac{D\mathbf{V}}{Dt} = -\nabla p + \mu \nabla^2 \mathbf{V} + \frac{\mu}{3} \nabla (\nabla \cdot \mathbf{V})$$

and the energy equation

$$\rho c_v \frac{DT}{Dt} = K \nabla^2 T - p \nabla \cdot \mathbf{V}$$

using characteristic quantities $\rho_0, p_0, T_0, U,$ and $l$. The characteristic time is $l/U$. The speed of sound is $c = \sqrt{kRT_0}$. Find any dimensionless parameters that result if $\text{Pr} = \mu c_p / K$.

The Trans Alaska Pipeline conveys crude oil over long distances and through various types of terrain. Pumping stations are spaced along its length to overcome pressures lost due to viscous forces and elevation changes. (U.S. Bureau of Land Management)

# 7

# Internal Flows

## Outline

## Chapter Objectives

The objectives of this chapter are to:

▲ Establish the length of the entrance region for both laminar and turbulent flows.

▲ Determine the laminar flow solution for pipes, parallel plates, and rotating cylinders.

▲ Present the quantities of interest for turbulent pipe flow, with losses being of particular interest.

▲ Calculate the flow rate with a centrifugal pump in a single piping system.

▲ Determine flow rates in open channels.

▲ Provide numerous examples and problems that demonstrate entrance and developed flow solutions for both laminar and turbulent flows, including losses due to wall friction and various devices. Open channel flows are also analyzed.

## 7.1 INTRODUCTION

In this chapter the effects of viscosity on an incompressible, internal flow will be studied. Such flows are of particular importance to engineers. Flow in a circular pipe is undoubtedly the most common internal fluid flow. It is encountered in the veins and arteries in a body, in a city's water system, in a farmer's irrigation system, in the piping systems transporting fluids in a factory, in the hydraulic lines of an aircraft, and in the ink jet of a computer's printer. Flows in noncircular ducts and open channels must also be included in our study. In Chapter 6 we discovered that the viscous effects in a flow resulted in the introduction of the Reynolds number,

$$\text{Re} = \frac{V\rho l}{\mu} \tag{7.1.1}$$

**KEY CONCEPT** *When the surface areas are relatively large, viscous effects become important.*

The Reynolds number was observed to be a ratio of the inertial force to the viscous force. Hence, when this ratio becomes large, it is expected that the inertial forces may dominate the viscous forces. This is usually true when short, sudden geometric changes occur; for long reaches of pipe or open channels, this is not the situation. When the surface areas, such as the wall area of a pipe, are relatively large, viscous effects become quite important and must be included in our study.

Internal flow between parallel plates, in a pipe, between rotating cylinders, and in an open channel will be considered in detail. For a sufficiently low Reynolds number (Re < 2000 in a pipe and Re < 1500 in a wide channel) a laminar flow results, and at sufficiently high Reynolds number a turbulent flow occurs. Review Section 3.3.3 for a more detailed discussion. We consider laminar flow first and then turbulent flow.

## 7.2 ENTRANCE FLOW AND DEVELOPED FLOW

*Developed laminar flow: A flow where the velocity profile ceases to change in the flow direction.*

When considering internal flows we are interested primarily in developed flows in conduits. A **developed laminar flow** results when the velocity profile ceases to change in the flow direction. Let us first focus our attention on a laminar flow. In the *entrance region* of a laminar flow the velocity profile changes in the flow direction, as shown in Fig. 7.1. The idealized flow from a reservoir begins at the inlet as a uniform flow (in reality, there is a thin viscous layer on the wall, as shown); the viscous wall layer then grows over the *inviscid core length* $L_i$ until the viscous stresses dominate the entire cross section; the profile then continues to change in the *profile development region* due to viscous effects until a developed flow is achieved. The inviscid core length is one-fourth to one-third of the *entrance length* $L_E$, depending on the conduit geometry, shape of the inlet velocity profile, and the Reynolds number.

**KEY CONCEPT** *The inviscid core length in a laminar flow is one-fourth to one-third of the entrance length.*

For a laminar flow in a circular pipe with a uniform profile at the inlet, the entrance length is given by

$$\frac{L_E}{D} = 0.065\text{Re} \qquad \text{Re} = \frac{VD}{\nu} \tag{7.2.1}$$

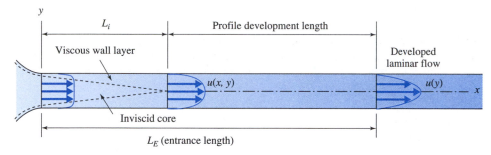

**Fig. 7.1**   Entrance region of a laminar flow in a pipe or a wide rectangular channel.

where the Reynolds number is based on the average velocity and the diameter. Laminar flow in a pipe has been observed at Reynolds numbers in excess of 40 000 for carefully controlled conditions in a smooth pipe. However, for engineering applications a value of about 2000 is the highest Reynolds number for which laminar flow is assured; this is due to vibrations of the pipe, fluctuations in the flow, and/or roughness elements on the pipe wall.

For a laminar flow in a high-aspect-ratio channel (the aspect ratio is the width divided by the distance between the top and bottom plates) with a uniform profile at the inlet the entrance length is

$$\frac{L_E}{h} = 0.04\text{Re} \qquad \text{Re} = \frac{Vh}{\nu} \qquad (7.2.2)$$

where the Reynolds number is based on the average velocity and the distance $h$ between the plates. The inviscid core length is approximately one-third of the entrance length. A laminar flow cannot exist over Re = 7700; for engineering situations a value of 1500 is often used as the upper limit for laminar flow.

For a turbulent flow the situation is slightly different, as shown in Fig. 7.2 for flow in a pipe. A developed flow results when all characteristics of the flow cease to change in the flow direction; this includes details of the turbulence that will be introduced later in this chapter. The inviscid core exists followed by the velocity profile development region, which terminates at $x = L_d$. An additional length is

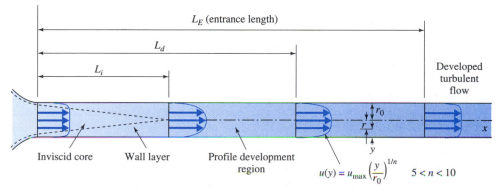

**Fig. 7.2**   Velocity profile development in a turbulent pipe flow.

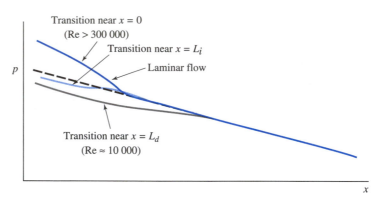

**Fig. 7.3** Pressure variation in a horizontal pipe flow for both laminar and turbulent flows. (From PhD Thesis of Dr. Jack Backus, Michigan State University)

needed, however, for the detailed structure of the turbulent flow to develop. The detailed structure is important in certain calculations such as accurate estimates of wall heat transfer. For large Reynolds number flow ($Re > 10^5$) in a pipe, where the turbulence fluctuations initiate near $x = 0$, tests have yielded

$$\frac{L_i}{D} \simeq 10 \qquad \frac{L_d}{D} \simeq 40 \qquad \frac{L_E}{D} \simeq 120 \qquad (7.2.3)$$

For turbulent flow with $Re = 4000$ the foregoing developmental lengths would be substantially higher, perhaps five times the values listed. This is true since, for low-Re turbulent flows, transition to a turbulent flow occurs in the profile development region; hence much of the entry region is laminar with relatively low wall shear. Experimental data are not available for low-Reynolds-number turbulent flow.

In Fig. 7.3 the pressure variation is sketched. In the flow beyond a sufficiently large $x$ it is noted that the pressure variation decreases linearly with $x$. If transition to turbulent flow occurs near the origin, the linear pressure variation begins near $L_i$ and the pressure gradient [the slope of the $p(x)$ curve] in the inlet region is higher than in the developed flow region; if transition occurs near $L_d$, as it does for low Re, the linear variation begins at the end of the transition process and the pressure gradient in the inlet region is less than that of developed flow.

For a laminar flow, the pressure variation qualitatively resembles that associated with a large Reynolds number. The pressure gradient is higher than in the developed flow region due to the higher shear stress at the wall and the increasing momentum flux.

## 7.3 LAMINAR FLOW IN A PIPE

Laminar flow in a pipe, 686

In this section we investigate incompressible, steady, developed laminar flow in a pipe, as sketched in Fig. 7.4. Two methods will be used: an elemental approach and a direct solution of the appropriate Navier–Stokes equation. Both develop the same equations so either can be used.

**Fig. 7.4**   Developed flow in a circular pipe.

## 7.3.1  Elemental Approach

An elemental volume of the fluid is shown in Fig. 7.4. This volume can be considered an infinitesimal control volume into which and from which fluid flows, or it can be considered an infinitesimal fluid mass upon which forces are acting. If it is considered a control volume, we would apply the momentum equation (4.5.6); if it is a fluid mass, we would apply Newton's second law. Since the velocity profile does not change in the $x$-direction, the momentum flux in equals the momentum flux out and the resultant force is zero; since there is no acceleration of the mass element, the resultant force must also be zero. Consequently, a force balance in the $x$-direction yields

> **KEY CONCEPT**   *Since the velocity profile does not change in the x-direction, the resultant force is zero.*

$$p\pi r^2 - (p + dp)\pi r^2 - \tau 2\pi r\, dx + \gamma\pi r^2\, dx \sin\theta = 0 \qquad (7.3.1)$$

which can be simplified to

$$\tau = -\frac{r}{2}\frac{d}{dx}(p + \gamma h) \qquad (7.3.2)$$

where we have used $\sin\theta = -dh/dx$, the vertical direction being denoted by $h$. Note that Eq. 7.3.2 can be applied to either a laminar or turbulent flow. The shear stress[1] in this laminar flow is related to the velocity gradient and the viscosity (see Eq. 1.5.5), giving

$$-\mu\frac{du}{dr} = -\frac{r}{2}\frac{d}{dx}(p + \gamma h) \qquad (7.3.3)$$

Since $u = u(r)$ for this developed flow, we can conclude from the above that $d/dx(p + \gamma h)$ must be at most a constant; it cannot depend on $x$, and from fluid statics $(p + \gamma h)$ does not depend on $x$ since there is no acceleration in the radial direction. Equation (7.3.3) can then be integrated to give the velocity distribution,

$$u(r) = \frac{r^2}{4\mu}\frac{d}{dx}(p + \gamma h) + A \qquad (7.3.4)$$

---

[1]The shear stress is considered a positive quantity, as displayed in Fig. 7.4. The minus sign in $\tau = -\mu\, du/dr$ is necessary since $du/dr$ is negative.

where $A$ is a constant of integration. Using $u = 0$ at $r = r_0$, we can evaluate $A$ and find the velocity distribution to be

$$u(r) = \frac{1}{4\mu} \frac{d(p + \gamma h)}{dx} (r^2 - r_0^2) \qquad (7.3.5)$$

**Poiseuille flow:** *A laminar flow with a parabolic profile in a pipe or between parallel plates.*

a parabolic profile. It is often referred to as **Poiseuille flow**, named after Jean L. Poiseuille (1799–1869).

The foregoing result can also be obtained by integrating the appropriate Navier–Stokes equation, as illustrated in the following section. If such an exercise is not of interest, go to Section 7.3.3.

### 7.3.2 Solving the Navier–Stokes Equations

For developed flow in a circular pipe the streamlines are parallel to the wall with no swirl, so that $v_r = v_\theta = 0$ and $u = u(r)$ only. Referring to the momentum equations in cylindrical coordinates of Table 5.1 (note that $u = v_z$ and the $z$-coordinate has been replaced by $x$) the $x$-component Navier–Stokes equation is

$$\rho\left(\underset{\text{steady}}{\cancel{\frac{\partial u}{\partial t}}} + \underset{\text{Streamlines}}{\cancel{v_r} \frac{\partial u}{\partial r}} + \underset{\text{no swirl}}{\cancel{\frac{v_\theta}{r}} \frac{\partial u}{\partial \theta}} + \underset{\substack{\text{developed} \\ \text{flow}}}{u \cancel{\frac{\partial u}{\partial x}}}\right)$$

$$= -\frac{\partial p}{\partial x} + \gamma \sin\theta + \mu\left(\frac{\partial^2 u}{\partial r^2} + \frac{1}{r}\frac{\partial u}{\partial r} + \underset{\substack{\text{symmetric} \\ \text{flow}}}{\frac{1}{r^2}\cancel{\frac{\partial^2 u}{\partial \theta^2}}} + \underset{\substack{\text{developed} \\ \text{flow}}}{\cancel{\frac{\partial^2 u}{\partial x^2}}}\right) \qquad (7.3.6)$$

**KEY CONCEPT** *An individual particle does not accelerate even though u = u(r).*

Note that there is no acceleration (the left-hand side is zero) of the fluid particles as they move in the pipe. Equation (7.3.6) simplifies to, using $\sin\theta = -dh/dx$ (see Fig. 7.4),

$$\frac{1}{\mu}\frac{\partial}{\partial x}(p + \gamma h) = \frac{1}{r}\frac{\partial}{\partial r}\left(r\frac{\partial u}{\partial r}\right) \qquad (7.3.7)$$

where the first two terms in the parentheses on the right-hand side of Eq. 7.3.6 have been combined. We see that the left-hand side is at most a function of $x$ and the right-hand side is at most a function of $r$. Since $x$ and $r$ can be varied independently, we must have

**KEY CONCEPT** *We use ordinary derivatives since u depends on only one variable.*

$$\frac{1}{r}\frac{d}{dr}\left(r\frac{du}{dr}\right) = \lambda \qquad (7.3.8)$$

where $\lambda$ is a constant[2] and we have used ordinary derivatives since $u$ depends on only one variable $r$. Multiply both sides by $r$ and integrate:

---

[2]If $\lambda$ were a function of $x$, $u$ would depend on $x$, which is not acceptable for this developed flow. If $\lambda$ were a function of $r$, $p + \gamma h$ would depend on $r$, also unacceptable. Hence it is at most a constant.

$$r \frac{du}{dr} = \frac{\lambda}{2} r^2 + A \qquad (7.3.9)$$

Divide both sides by $r$ and integrate:

$$u(r) = \frac{\lambda}{4} r^2 + A \ln r + B \qquad (7.3.10)$$

The velocity must remain finite at $r = 0$; hence $A = 0$. Also, at $r = r_0$, $u = 0$; thus $B$ can be evaluated and we have

$$u(r) = \frac{\lambda}{4} (r^2 - r_0^2)$$

$$= \frac{1}{4\mu} \frac{d}{dx} (p + \gamma h)(r^2 - r_0^2) \qquad (7.3.11)$$

where we have used $\lambda$ as the left-hand side of Eq. 7.3.7. This is the parabolic velocity distribution for flow in a pipe, often referred to as *Poiseuille flow*.

### 7.3.3  Pipe Flow Quantities

For steady, laminar, developed flow in a circular pipe, the velocity distribution has been shown to be

$$u(r) = \frac{1}{4\mu} \frac{d(p + \gamma h)}{dx} (r^2 - r_0^2) \qquad (7.3.12)$$

The average velocity $V$ is found to be

$$V = \frac{Q}{A} = \frac{1}{\pi r_0^2} \int_0^{r_0} u(r)\, 2\pi r \, dr$$

$$= \frac{2}{r_0^2} \int_0^{r_0} \frac{1}{4\mu} \frac{d(p + \gamma h)}{dx} (r^2 - r_0^2) r \, dr = - \frac{r_0^2}{8\mu} \frac{d(p + \gamma h)}{dx} \qquad (7.3.13)$$

Or, expressing the pressure drop $\Delta p$ in terms of the average velocity, we have, for a horizontal[3] pipe,

$$\qquad (7.3.14)$$

$$\Delta p = \frac{8\mu VL}{r_0^2}$$

where we have used $\Delta p/L = -dp/dx$ since $dp/dx$ is a constant for developed flow. Note that the pressure drop is a positive quantity, whereas the pressure gradient is negative.

**KEY CONCEPT**  *The pressure drop is a positive quantity, whereas the pressure gradient is negative.*

[3]For a pipe on an incline, simply replace $p$ with $(p + \gamma h)$.

The maximum velocity at $r = 0$ from Eq. 7.3.12 is

$$u_{max} = -\frac{r_0^2}{4\mu}\frac{d(p + \gamma h)}{dx} \tag{7.3.15}$$

so that (see Eq. 7.3.13)

$$V = \frac{1}{2}u_{max} \tag{7.3.16}$$

The shearing stress is determined to be

$$\tau = -\mu\frac{du}{dr}$$

$$= -\frac{r}{2}\frac{d(p + \gamma h)}{dx} \tag{7.3.17}$$

Letting $\tau = \tau_0$ at $r = r_0$, we see that the pressure drop $\Delta p$ over a length $L$ of a horizontal section of pipe is

$$\Delta p = \frac{2\tau_0 L}{r_0} \tag{7.3.18}$$

where we have again used $dp/dx = -\Delta p/L$.

**Friction factor:** *A dimensionless wall shear valid for both laminar and turbulent flow.*

If we introduce the **friction factor** $f$, a quantity of substantial interest in pipe flow, a dimensionless wall shear, defined by

$$f = \frac{\tau_0}{\frac{1}{8}\rho V^2} \tag{7.3.19}$$

we see that

$$\frac{\Delta p}{\gamma} = h_L = f\frac{L}{D}\frac{V^2}{2g} \tag{7.3.20}$$

where $h_L$ is the head loss (see Eq. 4.5.17) with dimension of length. This equation is often referred to as the *Darcy–Weisbach equation*,[4] named after Henri P. G. Darcy (1803–1858) and Julius Weisbach (1806–1871). Combining Eqs. 7.3.14 and 7.3.20, we find that

$$f = \frac{64}{Re} \tag{7.3.21}$$

for laminar flow in a pipe. Substituting this back into Eq. 7.3.20, we see that

$$h_L = \frac{32\mu LV}{\gamma D^2} \tag{7.3.22}$$

**KEY CONCEPT** *The head loss is directly proportional to the average velocity in a laminar flow.*

The head loss is directly proportional to the average velocity (and hence the discharge also) to the first power, a result that generally is applied to developed, laminar flows in conduits, including conduits of shape other than circular.

[4]This equation will be shown to be valid for both laminar and turbulent flows.

# Example 7.1

A small-diameter horizontal tube is connected to a supply reservoir as shown in Fig. 7.1. If 6600 mm$^3$ is captured at the outlet in 10 s, estimate the viscosity of the water.

**Fig. E7.1**

## Solution

The tube is very small, so we expect viscous effects to limit the velocity to a small value. Using Bernoulli's equation from the surface to the entrance to the tube, and neglecting the velocity head, we have, letting 0 be a point on the surface,

$$\frac{\cancel{p_0}^{\,0}}{\gamma} + H = \frac{\cancel{V^2}^{\,0}}{2g} + \frac{p}{\gamma}$$

where we have used gage pressure with $p_0 = 0$. This becomes, assuming $V^2/2g \cong 0$ at the tube's entrance,

$$p = \gamma H = 9800 \times 2 = 19\,600 \text{ Pa}$$

At the exit of the tube the pressure is zero; hence

$$\frac{\Delta p}{L} = \frac{19\,600}{1.2} = 16\,300 \text{ Pa/m (N/m}^3)$$

The average velocity is found to be

$$V = \frac{Q}{A} = \frac{6600 \times 10^{-9}/10}{\pi \times 0.001^2/4} = 0.840 \text{ m/s}$$

Check to make sure the velocity head is negligible: $V^2/2g = 0.036$ m compared with $p/\gamma = 2$ m, so the assumption of negligible velocity head is valid and our pressure calculation is acceptable. Using Eq. 7.3.14, we can find the viscosity of this assumed laminar flow to be

$$\mu = \frac{r_0^2}{8V}\frac{\Delta p}{L} = \frac{0.0005^2 \text{ m}^2}{8 \times 0.84 \text{ m/s}}(16\,300 \text{ N/m}^3) = 6.06 \times 10^{-4} \text{ N·s/m}^2$$

We should check the Reynolds number to determine if our assumption of a laminar flow is acceptable. It is

$$\text{Re} = \frac{\rho V D}{\mu} = \frac{1000 \text{ kg/m}^3 \times 0.84 \text{ m/s} \times 0.001\text{m}}{6.06 \times 10^{-4}\text{N·s/m}^2} = 1390$$

Where we use kg/m$^3$ = N·s$^2$/m$^4$. This is obviously a laminar flow since Re < 2000, so the calculations are valid providing the entrance length is not too long. It is

$$L_E = 0.065 \text{ Re} \times D = 0.065 \times 1390 \times 0.001 = 0.09 \text{ m}$$

This is approximately 8% of the total length, a sufficiently small quantity; hence the calculations are assumed reliable.

## Example 7.2

Derive an expression for the velocity distribution between horizontal, concentric pipes for a steady, incompressible developed flow (Fig. E7.2).

**Fig. E7.2**

### Solution

Let us use an elemental approach. The element is a hollow cylindrical shell as sketched in the figure. If we sum forces, we obtain

$$p2\pi r\,dr - (p + dp)2\pi r\,dr + \tau 2\pi r\,dx - (\tau + d\tau)2\pi(r + dr)\,dx = 0$$

Simplifying, there results, neglecting the term of differential magnitude,

$$\frac{dp}{dx} = -\frac{\tau}{r} - \frac{d\tau}{dr} - \underbrace{\frac{d\tau}{dr}}_{\text{neglect}}$$

Substituting $\tau = -\mu\,du/dr$ ($du/dr$ is negative near the outer wall where the element is sketched) we have

$$\frac{dp}{dx} = \mu\left(\frac{1}{r}\frac{du}{dr} + \frac{d^2u}{dr^2}\right)$$

$$= \frac{\mu}{r}\frac{d}{dr}\left(r\frac{du}{dr}\right)$$

Multiply both sides by $rdr$ and divide by $\mu$, then integrate:

$$r\frac{du}{dr} = \frac{1}{2\mu}\frac{dp}{dx}r^2 + A$$

Multiply both sides by $dr/r$ and integrate again:

$$u(r) = \frac{1}{4\mu}\frac{dp}{dx}r^2 + A\ln r + B$$

where $A$ and $B$ are arbitrary constants. They are found by setting $u = 0$ at $r = r_1$ and at $r = r_2$; that is,

$$0 = \frac{1}{4\mu}\frac{dp}{dx}r_1^2 + A\ln r_1 + B$$

$$0 = \frac{1}{4\mu}\frac{dp}{dx}r_2^2 + A\ln r_2 + B$$

Solve for $A$ and $B$:

$$A = \frac{1}{4\mu} \frac{dp}{dx} \frac{r_1^2 - r_2^2}{\ln(r_2/r_1)}$$

$$B = -A \ln r_2 - \frac{r_2^2}{4\mu} \frac{dp}{dx}$$

Thus

$$u(r) = \frac{1}{4\mu} \frac{dp}{dx} \left[ r^2 - r_2^2 + \frac{r_2^2 - r_1^2}{\ln(r_1/r_2)} \ln(r/r_2) \right]$$

This is integrated to give the flow rate:

$$Q = \int_{r_1}^{r_2} u(r) 2\pi r \, dr$$

$$= -\frac{\pi}{8\mu} \frac{dp}{dx} \left[ r_2^4 - r_1^4 - \frac{(r_2^2 - r_1^2)^2}{\ln(r_2/r_1)} \right]$$

As $r_1 \to 0$ the velocity distribution approaches the parabolic distribution of pipe flow. As $r_1 \to r_2$ this distribution approaches that of parallel-plate flow. These two conclusions are not obvious and are presented as a problem at the end of this chapter.

## 7.4 LAMINAR FLOW BETWEEN PARALLEL PLATES

Consider the incompressible, steady, developed flow of a fluid between parallel plates, with the upper plate moving with velocity $U$, as shown in Fig. 7.5. We will derive the velocity distribution by two methods; either can be used.

### 7.4.1 Elemental Approach

Consider an elemental volume of unit depth in the $z$-direction, as sketched in Fig. 7.5. If we sum forces in the $x$-direction, we can write

$$p \, dy - (p + dp) \, dy - \tau \, dx + (\tau + d\tau) \, dx + \gamma \, dx \, dy \sin \theta = 0 \quad (7.4.1)$$

The force sum is set equal to 0 since there is no acceleration. The pressure is assumed to depend only on $x$; variation with $y$ is assumed to be negligible since the dimension $a$ is quite small for most applications. After dividing by $dx \, dy$, the above is simplified to

$$\frac{d\tau}{dy} = \frac{dp}{dx} - \gamma \sin \theta \quad (7.4.2)$$

Since this is one-dimensional flow, the shear stress is

$$\tau = \mu \frac{du}{dy} \quad (7.4.3)$$

The element is sketched near the lower plate where $du/dy > 0$.

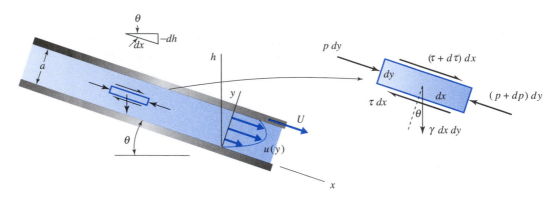

**Fig. 7.5** Developed flow between parallel plates.

Using this and $\sin\theta = -dh/dx$, there results

$$\mu \frac{d^2u}{dy^2} = \frac{dp}{dx} + \gamma\frac{dh}{dx} = \frac{d}{dx}(p + \gamma h) \qquad (7.4.4)$$

Since $u = u(y)$ for this developed flow, the left-hand side is only a function of $y$; because the right-hand side is a function of $x$ we conclude that it must be a constant. Hence, it can be integrated twice to yield (first, divide by $\mu$)

$$u(y) = \frac{y^2}{2\mu}\frac{d}{dx}(p + \gamma h) + Ay + B \qquad (7.4.5)$$

where $A$ and $B$ are integration constants. If we require $u = 0$ at $y = 0$ and $u = U$ at $y = a$, we have

$$A = \frac{U}{a} - \frac{a}{2\mu}\frac{d}{dx}(p + \gamma h) \quad \text{and} \quad B = 0 \qquad (7.4.6)$$

Thus the velocity distribution is the parabola

$$u(y) = \frac{1}{2\mu}\frac{d}{dx}(p + \gamma h)(y^2 - ay) + \frac{U}{a}y \qquad (7.4.7)$$

**Couette flow:** *A flow with a linear profile that results from the motion of the plate only.*

If the motion is due to the motion of the plate only (a linear profile), the flow is called a **Couette flow**; if the motion is due to the pressure gradient only, i.e., $U = 0$, it is Poiseuille flow.

Rather than summing forces on an element we can integrate the appropriate Navier–Stokes equation, as follows. If such an exercise is not of interest, go to Section 7.4.3.

## 7.4.2  Solving the Navier–Stokes Equations

For developed flow between parallel plates the streamlines are parallel to the plates so that $u = u(y)$ only and $v = w = 0$. The Navier–Stokes equation for the $x$-direction is (see Eq. 5.3.14)

$$\rho\left(\underbrace{\frac{\partial u}{\partial t}}_{0}^{\text{steady}} + u\underbrace{\frac{\partial u}{\partial x}}_{\text{flow}}^{\text{developed}} + \underbrace{v\frac{\partial u}{\partial y}}_{0} + \underbrace{w\frac{\partial u}{\partial z}}_{}\right)$$

$$= -\frac{\partial p}{\partial x} + \mu\left(\underbrace{\frac{\partial^2 u}{\partial x^2}}_{\text{developed}}^{} + \frac{\partial^2 u}{\partial y^2} + \underbrace{\frac{\partial^2 u}{\partial z^2}}_{\text{channel}}^{}\right) + \gamma \sin\theta \qquad (7.4.8)$$

If the parallel plates are the top and bottom of a channel, the channel must be wide; the analysis then applies to the midsection away from the side walls. Equation 7.4.8 reduces to, using $\sin\theta = -dh/dx$,

$$\frac{\partial^2 u}{\partial y^2} = \frac{1}{\mu}\frac{d}{dx}(p + \gamma h) \qquad (7.4.9)$$

The left-hand side is at most a function of $y$, and the right-hand side is at most a function of $x$. Hence we must have

$$\frac{\partial^2 u}{\partial y^2} = \lambda \qquad (7.4.10)$$

where $\lambda$ is a constant, since $x$ and $y$ are independent variables. This can be integrated twice to yield

$$u(y) = \frac{\lambda}{2}y^2 + Ay + B \qquad (7.4.11)$$

where $A$ and $B$ are arbitrary constants of integration. We demand that $u = 0$ at $y = 0$ and $u = U$ at $y = a$. This gives

$$A = \frac{U}{a} - \frac{\lambda a}{2} \quad \text{and} \quad B = 0 \qquad (7.4.12)$$

Using these constants, we can write Eq. 7.4.11 as

$$u(y) = \frac{\lambda}{2}(y^2 - ay) + \frac{U}{a}y$$

$$= \frac{1}{2\mu}\frac{d}{dx}(p + \gamma h)(y^2 - ay) + \frac{U}{a}y \qquad (7.4.13)$$

where we have used $\lambda$ as the right-hand side of Eq. 7.4.9. If the flow is due to the motion of the plate only (a linear profile), the flow is called a *Couette flow*, if the motion is due to the pressure gradient only, i.e., $U = 0$, it is Poiseuille flow.

### 7.4.3  Simplified Flow Situation

Using the results derived above, we can write the expression for the velocity distribution between fixed plates (let $U = 0$) as

$$u(y) = \frac{1}{2\mu} \frac{d(p + \gamma h)}{dx} (y^2 - ay) \tag{7.4.14}$$

Using this distribution, we can find the flow rate per unit width to be

$$Q = \int u \, dA$$

$$= \int_0^a \frac{1}{2\mu} \frac{d(p + \gamma h)}{dx} (y^2 - ay) \, dy = -\frac{a^3}{12\mu} \frac{d(p + \gamma h)}{dx} \tag{7.4.15}$$

The average velocity $V$ is found to be

$$V = \frac{Q}{a \times 1}$$

$$= -\frac{a^2}{12\mu} \frac{d(p + \gamma h)}{dx} \tag{7.4.16}$$

This can be expressed as the pressure drop in terms of the average velocity; for a horizontal[5] channel we have

$$\Delta p = \frac{12\mu V L}{a^2} \tag{7.4.17}$$

where we have used $\Delta p/L = -dp/dx$ since $dp/dx$ is constant for developed flow. Observe that the maximum velocity occurs at $y = a/2$ and from Eq. 7.4.14 is

$$u_{\max} = -\frac{a^2}{8\mu} \frac{dp}{dx} \tag{7.4.18}$$

Thus the average velocity is related to the maximum velocity by

$$V = \frac{2}{3} u_{\max} \tag{7.4.19}$$

---

[5]For plates on an incline, simply replace $p$ with $(p + \gamma h)$.

The shearing stress is found to be

$$\tau = \mu \frac{du}{dy} = \frac{1}{2} \frac{dp}{dx} (2y - a) \tag{7.4.20}$$

At the wall where $y = 0$ there results

$$\tau_0 = -\frac{a}{2} \frac{dp}{dx} \tag{7.4.21}$$

The pressure drop $\Delta p$ over a length $L$ of horizontal channel is found to be

$$\Delta p = \frac{2\tau_0}{a} L \tag{7.4.22}$$

recognizing that $dp/dx = -\Delta p/L$. This can be expressed in a more convenient form as

$$\frac{\Delta p}{\gamma} = f \frac{L}{2a} \frac{V^2}{2g} \tag{7.4.23}$$

if we introduce the friction factor $f$, defined by

$$f = \frac{\tau_0}{\frac{1}{8}\rho V^2} \tag{7.4.24}$$

**Friction factor:** *A dimentionless wall shear valid for both laminar and turbulent flow.*

In terms of the head loss (see Eq. 4.5.17), Eq. 7.4.23 becomes

$$h_L = f \frac{L}{2a} \frac{V^2}{2g} \tag{7.4.25}$$

If we combine Eqs. 7.4.17, 7.4.21, and 7.4.24, we find that

$$f = \frac{8}{\rho V^2} \left( -\frac{a}{2} \frac{dp}{dx} \right) = \frac{8}{\rho V^2} \left( -\frac{a}{2} \right) \left( -\frac{12\mu V}{a^2} \right) = \frac{48\mu}{\rho a V} = \frac{48}{\text{Re}} \tag{7.4.26}$$

where the Reynolds number $\text{Re} = \rho a V/\mu$ has been introduced. If this is substituted into Eq. 7.4.25, we see that

$$h_L = \frac{12\mu L V}{ra^2} \tag{7.4.27}$$

The head loss is directly proportional to the average velocity, a conclusion that is true for laminar flows, in general.

We have calculated most of the quantities of interest for laminar flow of an incompressible, steady flow between parallel plates. This would, of course, be an acceptable approximation for a wide channel, one for which the width exceeds the height by at least a factor of 8. For channels with smaller aspect ratios, the edge effects become significant and must be accounted for by adding an additional shear stress to the sides of the element in Fig. 7.5, or by maintaining $\partial^2 u/\partial z^2$ in Eq. 7.4.8. The solution is beyond the scope of this book.

**KEY CONCEPT** *A wide channel is one for which the width exceeds the height by at least a factor of 8.*

## Example 7.3

Water at 16°C flows with a Reynolds number of 1500 between the 50-cm-wide, horizontal plates shown in Fig. P7.3. Calculate (a) the flow rate, (b) the wall shear stress, (c) the pressure drop over 3 m, and (d) the velocity at $y = 0.5$ cm.

**Fig. E7.3**

### Solution

Since the Reynolds number is 1500, the laminar flow equations are assumed applicable.

(a) Using the definition of the Reynolds number, the average velocity is found as follows:

$$1500 = \frac{Va}{\nu}$$

$$\therefore V = \frac{1500\nu}{a} = \frac{1500 \times 1.11 \times 10^{-6}}{1.2 \times 10^{-2}} = 0.14 \text{ m/s}$$

Thus

$$Q = VA = 0.14 \times \frac{1.2 \times 50}{10^4} = 0.84 \times 10^{-3} \text{ m}^3/\text{s}$$

$$= 0.84 \text{ L/s}$$

(b) Using Eq. 7.4.17, the pressure gradient is

$$\frac{\Delta p}{L} = \frac{12\mu V}{a^2} = \frac{12 \times 1.11 \times 10^{-3} \times 0.14}{(0.012)^2} = 12.95 \text{ Pa/m}$$

The shearing stress at the wall is found, using Eq. 7.4.22, to be

$$\tau_0 = \frac{a}{2} \frac{\Delta p}{L} = \frac{0.012}{2} \times 12.95 = 0.078 \text{ Pa}$$

(c) The pressure drop over 3 m is found to be

$$\Delta p = 12.95 L = 12.95 \times 3 = 38.85 \text{ Pa}$$

(d) The velocity distribution of Eq. 7.4.14 is

$$u(y) = \frac{1}{2\mu} \frac{dp}{dx} (y^2 - ay)$$

$$= \frac{1}{2 \times 1.11 \times 10^{-3}} (-12.95) \left( y^2 - 0.012\, y \right) = -5833(y^2 - 0.012y)$$

where we have used $dp/dx = -\Delta p/L$. At $y = 0.5$ cm, the velocity is

$$u = -5833[(0.005)^2 - 0.012 \times 0.005] = 0.204 \text{ m/s}$$

We have used three significant digits since the fluid properties are assumed known to three significant digits.

# Example 7.4

Find an expression for the pressure gradient between two parallel plates that results in a zero shear stress at the lower wall, where $y = 0$; also, sketch the velocity profiles for a top plate speed of $U$ with various pressure gradients. Assume horizontal plates.

$$\frac{dp}{dx} = \frac{2\mu U}{a^2}$$

$$\frac{dp}{dx} > \frac{2\mu U}{a^2}$$

$$\frac{dp}{dx} = 0$$

$$\frac{dp}{dx} < 0$$

**Fig. E7.4**

## Solution

The velocity distribution for plates with the top plate moving with velocity $U$ is given by Eq. 7.4.17. Letting $dh/dx = 0$, we have

$$u(y) = \frac{1}{2\mu}\frac{dp}{dx}(y^2 - ay) + \frac{U}{a}y$$

The shear stress is

$$\tau = \mu\frac{du}{dy}$$

$$= \frac{1}{2}\frac{dp}{dx}(2y - a) + \mu\frac{U}{a}$$

If $\tau = 0$ at $y = 0$, then $du/dy = 0$ at $y = 0$ and the pressure gradient is

$$\frac{dp}{dx} = \frac{2\mu U}{a^2}$$

If $dp/dx$ is greater than this value, the slope $du/dy$ at $y = 0$ is negative and thus the velocity $u$ will be negative near $y = 0$. If $dp/dx = 0$, we observe that a linear velocity distribution results, namely,

$$u(y) = \frac{U}{a}y$$

If $dp/dx$ is negative, $u(y)$ is greater at each $y$-location than the linear distribution since $(y^2 - ay)$ is a negative quantity for all $y$s of interest.

All of the results above can be qualitatively displayed on a sketch of $u(y)$ for several $dp/dx$ as shown in Fig. E7.4.

## 7.5 LAMINAR FLOW BETWEEN ROTATING CYLINDERS

Flow between cylinders, 735

**KEY CONCEPT** *The laminar flow solution will be valid up to a Reynolds number of 1700.*

Fully developed, steady flow between concentric, rotating cylinders, as shown in Fig. 7.6, is another flow that yields a rather simple solution. It has particular application in the field of lubrication, where the fluid may be oil and the inner cylinder a rotating shaft. We will again use two approaches to find the velocity distribution. The laminar flow solution we will find will be valid up to a Reynolds number[6] of 1700, providing the angular velocity of the outer cylinder $\omega_2 = 0$, as is often the case. Above Re = 1700 a secondary laminar flow (a flow with a different velocity distribution) may develop, and eventually a turbulent flow develops. In fact, numerous laminar flows (all different) have been observed for Re > 1700.

### 7.5.1 Elemental Approach

Body forces will be neglected in this derivation, or the cylinders will be assumed to be vertical. Since pressure does not vary with $\theta$, an element in the form of a thin cylindrical shell will be used, as shown in Fig. 7.6b. The resultant torque acting on this element is zero because it has no angular acceleration; this is expressed as

$$\tau 2\pi r L \times r - (\tau + d\tau)2\pi(r + dr)L \times (r + dr) = 0 \qquad (7.5.1)$$

where $L$, the length of the cylinders, should be large with respect to the gap width $(r_2 - r_1)$ to avoid three-dimensional end effects. Equation 7.5.1 reduces to (neglecting the three higher-order terms which approach zero as $dr \to 0$)

$$2\tau + r\frac{d\tau}{dr} = 0 \qquad (7.5.2)$$

The one-dimensional constitutive equation (see Table 5.1), recognizing that $\tau = -\tau_{r\theta}$, provides the shear stress:

$$\tau = -\mu r\frac{d}{dr}\left(\frac{v_\theta}{r}\right) \qquad (7.5.3)$$

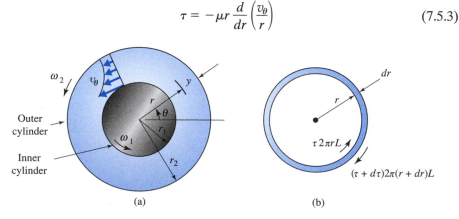

(a)　　　　　　(b)

**Fig. 7.6** Flow between concentric cylinders: (a) basic flow variables; (b) element from between the cylinders.

[6]The Reynolds number is defined as Re = $\omega_1 r_1 \delta/\nu$, where $\delta = r_2 - r_1$.

There results

$$-2\mu r \frac{d}{dr}\left(\frac{v_\theta}{r}\right) - r\mu \frac{d}{dr}\left[r\frac{d}{dr}\left(\frac{v_\theta}{r}\right)\right] = 0 \qquad (7.5.4)$$

Divide by $\mu r$, multiply by $dr$, and integrate to find

$$2\frac{v_\theta}{r} + r\frac{d}{dr}\left(\frac{v_\theta}{r}\right) = A \qquad (7.5.5)$$

This can be rearranged as $\left[\text{perform the differentiation: } \dfrac{d}{dr}\left(\dfrac{v_\theta}{r}\right) = \dfrac{1}{r}\dfrac{dv_\theta}{dr} - \dfrac{v_\theta}{r^2}\right]$

$$\frac{dv_\theta}{dr} + \frac{v_\theta}{r} = A \qquad (7.5.6)$$

or

$$\frac{1}{r}\frac{d}{dr}(rv_\theta) = A \qquad (7.5.7)$$

Multiply by $r\,dr$ and integrate again to find that

$$v_\theta(r) = \frac{A}{2}r + \frac{B}{r} \qquad (7.5.8)$$

The boundary conditions are $v_\theta = r_1\omega_1$ at $r = r_1$, and $v_\theta = r_2\omega_2$ at $r = r_2$. These conditions allow the constants to be evaluated as

$$A = 2\frac{\omega_2 r_2^2 - \omega_1 r_1^2}{r_2^2 - r_1^2} \qquad B = \frac{r_1^2 r_2^2(\omega_1 - \omega_2)}{r_2^2 - r_1^2} \qquad (7.5.9)$$

We can obtain the same result by integrating the appropriate Navier–Stokes equation, or if that is not of interest, go directly to Section 7.5.3.

## 7.5.2   Solving the Navier–Stokes Equations

For steady, laminar flow between concentric cylinders we assume the streamlines to be circular so that $v_r = v_z = 0$, $v_\theta = v_\theta(r)$ only, and $\partial p/\partial \theta = 0$. The $\theta$-component Navier–Stokes equation from Table 5.1 is

$$\left(\underset{\substack{\text{steady} \\ 0}}{\cancel{\frac{\partial v_\theta}{\partial t}}} + \underset{0}{\cancel{v_r}}\frac{\partial v_\theta}{\partial r} + \underset{\substack{\text{symmetric} \\ \text{flow}}}{\cancel{\frac{v_\theta}{r}\frac{\partial v_\theta}{\partial \theta}}} + \underset{0}{\cancel{v_z \frac{\partial v_\theta}{\partial z}}} + \underset{0}{\cancel{\frac{v_r v_\theta}{r}}}\right)$$

$$= -\underset{0}{\cancel{\frac{1}{r}\frac{\partial p}{\partial \theta}}} + \mu\left[\frac{\partial^2 v_\theta}{\partial r^2} + \frac{1}{r}\frac{\partial v_\theta}{\partial r} + \underset{\substack{\text{symmetric} \\ \text{flow}}}{\cancel{\frac{1}{r^2}\frac{\partial^2 v_\theta}{\partial \theta^2}}} + \underset{\substack{\text{long} \\ \text{cylinders}}}{\cancel{\frac{\partial^2 v_\theta}{\partial z^2}}} + \underset{\substack{\text{symmetric} \\ \text{flow}}}{\cancel{\frac{2}{r^2}\frac{\partial v_\theta}{\partial \theta}}} - \frac{v_\theta}{r^2}\right] \qquad (7.5.10)$$

This reduces to

$$0 = \frac{\partial^2 v_\theta}{\partial r^2} + \frac{1}{r}\frac{\partial v_\theta}{\partial r} - \frac{v_\theta}{r^2} \tag{7.5.11}$$

which can be written as

$$\frac{d^2 v_\theta}{dr^2} + \frac{d}{dr}\left(\frac{v_\theta}{r}\right) = 0 \tag{7.5.12}$$

Integrating once, there results

$$\frac{dv_\theta}{dr} + \frac{v_\theta}{r} = A \tag{7.5.13}$$

or

$$\frac{1}{r}\frac{d}{dr}(rv_\theta) = A \tag{7.5.14}$$

A second integration yields

$$v_\theta = \frac{A}{2}r + \frac{B}{r} \tag{7.5.15}$$

Applying the boundary conditions $v_\theta = r_1\omega_1$ at $r = r_1$ and $v_\theta = r_2\omega_2$ at $r = r_2$, we find that

$$A = 2\frac{r_2^2\omega_2 - r_1^2\omega_1}{r_2^2 - r_1^2} \qquad B = \frac{r_1^2 r_2^2(\omega_1 - \omega_2)}{r_2^2 - r_1^2} \tag{7.5.16}$$

### 7.5.3 Flow with the Outer Cylinder Fixed

For a number of situations, such as a shaft rotating in a bearing, the outer cylinder is fixed. Setting $\omega_2 = 0$, the velocity distribution is

$$v_\theta = \frac{r_1^2\omega_1}{r_2^2 - r_1^2}\left(\frac{r_2^2}{r} - r\right) \tag{7.5.17}$$

The shearing stress $\tau_1$ on the inner cylinder (see Table 5.1 and let $\tau = -\tau_{r\theta}$) is found to be

$$\tau_1 = -\left[\mu r \frac{d}{dr}\left(\frac{v_\theta}{r}\right)\right]_{r=r_1}$$

$$= \mu \frac{2}{r_1^2} \frac{r_1^2 r_2^2 \omega_1}{r_2^2 - r_1^2} = \frac{2\mu r_2^2 \omega_1}{r_2^2 - r_1^2} \qquad (7.5.18)$$

The torque $T$ necessary to rotate the inner cylinder of length $L$ is

$$T = \tau_1 A_1 r_1$$

$$= \frac{2\mu r_2^2 \omega_1}{r_2^2 - r_1^2} 2\pi r_1 L r_1 = \frac{4\pi \mu r_1^2 r_2^2 L \omega_1}{r_2^2 - r_1^2} \qquad (7.5.19)$$

The power $\dot{W}$ necessary to rotate the shaft is found by multiplying the torque by the rotational speed $\omega_1$; it is

$$\dot{W} = T\omega_1$$

$$= \frac{4\pi \mu r_1^2 r_2^2 L \omega_1^2}{r_2^2 - r_1^2} \qquad (7.5.20)$$

**KEY CONCEPT** *The power is found by multiplying the torque by the rotational speed.*

This power is necessary to overcome the resistance of viscosity and results in an increase in the internal energy and thus the temperature of the fluid. The removal of this energy from the fluid often requires special heat exchangers.

## Example 7.5

Estimate the viscosity of an oil contained in the annulus between two 25-cm-long cylinders, as shown in Fig. 7.6. The outer stationary cylinder is 8 cm in diameter. The 7.8-cm-diameter inner cylinder rotates at 3800 rpm when a torque of 0.12 N·m is applied. The specific gravity of the oil is 0.85. Neglect any torque due to the cylinder ends.

### Solution

Assuming that the Reynolds number is less than 1700, Eq. 7.5.19 provides

$$\mu = \frac{T(r_2^2 - r_1^2)}{4\pi r_1^2 r_2^2 L \omega_1}$$

$$= \frac{0.12 \text{ N·m}(0.04^2 - 0.039^2) \text{ m}^2}{4\pi \times 0.04^2 \text{ m}^2 \times 0.039^2 \text{ m}^2 \times 0.25 \text{ m} \times (3800 \times 2\pi/60) \text{ rad/s}} = 0.00312 \text{ N·s/m}^2$$

Check the Reynolds number using $\nu = \mu/\rho$:

$$\text{Re} = \frac{\omega_1 r_1 \delta}{\nu} = \frac{(3800 \times 2\pi/60) \text{ rad/s} \times 0.039 \text{ m} \times 0.002/2 \text{ m}}{0.0013/(1000 \times 0.85)\text{m}^2/\text{s}} = 845$$

This is less than 1700 so the calculation is acceptable.

## Example 7.5a  Taylor Cells, 24

## Example 7.6

Show that as the inner cylinder radius of Fig. P7.6 approaches the outer cylinder radius the velocity distribution approaches the linear distribution between parallel plates with one plate moving and a zero pressure gradient. This is Couette flow.

**Fig. E7.6**

### Solution

For this problem we will let $\omega_2 = 0$; the velocity distribution (7.5.17) is

$$v_\theta(r) = \frac{r_1^2\omega_1}{r_2^2 - r_1^2}\left(\frac{r_2^2}{r} - r\right)$$

$$= \frac{r_1^2\omega_1}{r_2^2 - r_1^2}\frac{r_2^2 - r^2}{r} = \frac{r_1^2\omega_1}{r_2^2 - r_1^2}(r_2 - r)\frac{r_2 + r}{r}$$

Introduce the independent variable $y$, measured from the outer cylinder defined by $r + y = r_2$ (see Fig. 7.6); let $\delta = r_2 - r_1$. Then the above can be written as

$$v_\theta(r) = \frac{r_1^2\omega_1(r_2 - r)}{(r_2 - r_1)(r_2 + r_1)}\frac{r_2 + r}{r}$$

$$= \frac{r_1^2\omega_1 y}{\delta(r_2 + r_1)}\frac{2r_2 - y}{r_2 - y}$$

As the inner radius approaches the outer radius we can write $r_1 \simeq r_2$. Letting $r_2 \simeq r_1 \simeq R$ we have $r_2 + r_1 \simeq 2R$ and

$$\frac{2R - y}{R - y} \simeq 2$$

since $y \ll R$. The velocity distribution then simplifies to

$$v_\theta(y) = \frac{R^2\omega_1 y}{\delta\, 2R} \times 2 = \frac{R\omega_1}{\delta}y$$

This is a linear distribution and is a good approximation to the flow whenever $\delta \ll R$.

## 7.6 TURBULENT FLOW IN A PIPE

The study of developed turbulent flow in a circular pipe is of substantial interest in actual flows since most flows encountered in practical applications are turbulent flows in pipes. Even though for carefully controlled laboratory conditions, laminar flows have been observed up to Reynolds numbers of 40 000 in a pipe flow, turbulent flow is assumed to occur in a pipe under standard operating conditions whenever the Reynolds number

$$\text{Re} = \frac{VD}{\nu}$$

exceeds 2000; between 2000 and 4000 it is possible for the flow to randomly oscillate between being laminar and turbulent. Consider, for example, water at 20°C flowing in a rather small 5-mm-diameter pipe; the average velocity need only be 0.8 m/s for turbulent flow to occur. This is the situation whenever we drink water from a drinking fountain. For larger-diameter pipes the average velocity is sufficiently large so that a turbulent flow is produced in most situations.

In a turbulent flow all three velocity components are nonzero. If we measure the components as a function of time, graphs similar to those shown in Fig. 7.7 would result for a flow in a pipe where $u, v, w$ are in the $x$-, $r$-, and $\theta$-directions, respectively. There is seldom any interest (to the engineer) in the details of the randomly fluctuating velocity components; hence we introduce the notion of a time-average quantity. The velocity components $(u, v, w)$ are written as

$$u = \bar{u} + u' \qquad v = \bar{v} + v' \qquad w = \bar{w} + w' \qquad (7.6.1)$$

where a bar over a quantity denotes a time average and a prime denotes the fluctuating part. Using the component $u$ as an example, the *time average* is defined as,

$$\bar{u} = \frac{1}{T} \int_0^T u(t)\, dt \qquad (7.6.2)$$

where $T$ is a time increment large enough[7] to eliminate all time dependence from $\bar{u}$. In a developed pipe flow $\bar{u}$ would be nonzero and $\bar{v} = \bar{w} = 0$, as is observed in Fig. 7.7.

**KEY CONCEPT** *Turbulent flow occurs whenever we drink water from a drinking fountain.*

**KEY CONCEPT** *A bar over a quantity denotes a time average.*

 Reynolds decomposition, 689

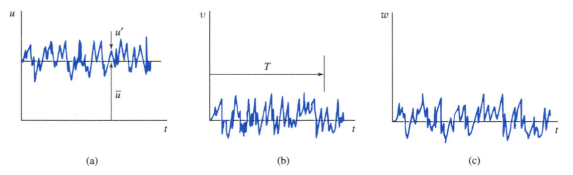

**Fig. 7.7**  Velocity components in a turbulent pipe flow: (a) $x$-component velocity; (b) $r$-component velocity; (c) $\theta$-component velocity.

[7]At a given location in a pipe this could be checked experimentally as follows: Start with a relatively large value of $T$ and then decrease $T$ for subsequent runs. If $\bar{u}$ remains unchanged as $T$ decreases, $T$ is sufficiently large. If $T$ is too small, $\bar{u}$ will be different for each run.

## Example 7.7

Show that $\overline{u'} = 0$ and $\dfrac{\overline{\partial u}}{\partial y} = \dfrac{\partial \overline{u}}{\partial y}$ for a turbulent flow.

**Solution**

To show that $\overline{u'} = 0$ we simply substitute the expression (7.6.1) for $u(t)$ into Eq. 7.6.2 and obtain

$$\overline{u} = \frac{1}{T} \int_0^T (\overline{u} + u')\,dt$$

$$= \frac{1}{T} \int_0^T \overline{u}\,dt + \frac{1}{T} \int_0^T u'\,dt$$

$$= \overline{u}\,\frac{1}{T} \int_0^T dt + \overline{u'}$$

$$= \overline{u} + \overline{u'}$$

Subtracting $\overline{u}$ from both sides results in

$$\overline{u'} = 0$$

Now, let us time average the derivative $\partial u / \partial y$. We have

$$\frac{\overline{\partial u}}{\partial y} = \frac{1}{T} \int_0^T \frac{\partial u}{\partial y}\,dt$$

$$= \frac{\partial}{\partial y}\left(\frac{1}{T} \int_0^T u\,dt\right) = \frac{\partial}{\partial y}\,\overline{u}$$

since $T$ is a constant. Thus

$$\frac{\overline{\partial u}}{\partial y} = \frac{\partial \overline{u}}{\partial y}$$

## 7.6.1 Differential Equation

The differential equation that must be solved to provide us with the time-average velocity distribution is derived by using a particle approach. The Navier–Stokes equations could be time averaged, arriving at the same differential equation; that approach is included in the problems.

Consider the situation for a turbulent flow in a horizontal pipe, as shown in Fig. 7.8. We use velocity components $u$ and $v$ in the $x$- and $y$-directions, respectively. Fluid particles[8] move randomly throughout the flow. At an instant in time a fluid particle moves through the incremental area $dA$ due to the velocity fluctuation $v'$; it enters a neighboring layer of fluid which is moving with a higher $x$-component velocity, thereby providing a retarding effect on the neighboring layer. A fluid particle that moves to a neighboring layer that is traveling with a lower $x$-component velocity would tend to accelerate the slower moving fluid. The $x$-component force that results due to the random motion of a fluid particle passing through the incremental area $dA$ would be (see Eq. 4.6.6)

Turbulent flow in a pipe, 687

$$dF = -\rho v' dA\, u' \tag{7.6.3a}$$

where $u'$ is the negative change in $x$-component velocity due to the momentum exchange and $\rho v' dA$ is the mass flux through the area; the negative sign provides a positive $dF$. If we divide both sides by the area $dA$, we obtain a "stress" that we call the *turbulent shear stress*. It is

$$\tau_{\text{turb}} = \frac{dF}{dA} = -\rho u' v' \tag{7.6.3b}$$

where we know that $(u'v')$ is, on the average, a negative quantity since a positive $v'$ produces a negative $u'$. This "shear stress" is actually a momentum exchange but since it has the same effect as a stress, we call it a shear stress.

The time-average turbulent shear stress, often called the *apparent shear stress*, which is of primary interest, is

**KEY CONCEPT** *A positive $v'$ produces a negative $u'$.*

$$\overline{\tau}_{\text{turb}} = -\rho \overline{u'v'} \tag{7.6.4}$$

where Eq. 7.6.2 would be used for the time averages. Note that $\overline{u'w'}$ would be zero since a $w'$-component (in the $\theta$-direction) would not move a fluid particle

**Fig. 7.8**   Turbulent flow in a horizontal pipe.

---

[8]A particle is a relatively small mass of the fluid, consisting of a large number of molecules, that tends to move as a unit.

into a layer of higher or lower $x$-component velocity. Also, $\overline{v'w'} = 0$ using the same logic.

The total shear stress at a particular location would be due to both the viscosity and the momentum exchange described above; that is,

$$\overline{\tau} = \overline{\tau}_{\text{lam}} + \overline{\tau}_{\text{turb}}$$

$$= \mu \frac{\partial \overline{u}}{\partial y} - \rho \overline{u'v'} \tag{7.6.5}$$

The shear stress can be related to the pressure gradient by summing forces on the horizontal cylindrical element shown in Fig. 7.8. There results

$$\overline{\tau} = -\frac{r}{2} \frac{d\overline{p}}{dx} = \frac{r\Delta\overline{p}}{2L} \tag{7.6.6}$$

which shows that the shear stress distribution is linear for a turbulent flow as well as a laminar flow (see Eq. 7.3.17). The turbulent shear obviously goes to zero at the wall since the velocity perturbations are zero at the wall, and the total shear is zero at the centerline where $r = 0$ or $y = r_0$, as shown in Fig. 7.9. The viscous shear is nonzero only in a very thin viscous wall layer, of thickness $\delta_\nu$, near the wall, as shown in part (b). Note that the turbulent shear reaches a maximum near the wall in the viscous wall layer.

The differential equation that must be solved if the time-average velocity distribution is to be determined is found by combining the two preceding equations; it is

$$\frac{r}{2} \frac{d\overline{p}}{dx} = \rho \overline{u'v'} + \mu \frac{d\overline{u}}{dr} \tag{7.6.7}$$

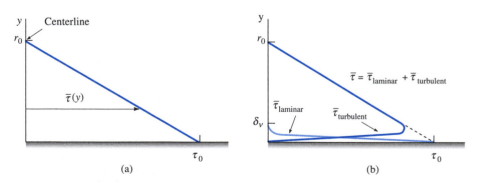

**Fig. 7.9** Shear stress distributions in a developed pipe flow.

where we have used $r + y = r_0$ so that $dy = -dr$. For developed flow we know that $d\bar{p}/dx = $ const.; hence if we know how $\overline{u'v'}$ varied with $r$, the differential equation could be solved. The quantity $\overline{u'v'}$ cannot be determined analytically, however, so the solution to Eq. 7.6.7 cannot be attempted until an empirical expression is found for $\overline{u'v'}$. Rather than finding an empirical expression for $\overline{u'v'}$ and then solving the differential equation for $\bar{u}$, we will simply present the empirical results obtained for the velocity profile $\bar{u}(y)$. However, before we do this in the next section we should introduce the eddy viscosity, the mixing length, and the correlation coefficient.

Instead of using the quantity $\overline{u'v'}$ as the unknown in Eq. 7.6.7, we often introduce the *eddy viscosity* $\eta$, defined by the relationship

$$\overline{u'v'} = \eta \frac{d\bar{u}}{dy} \qquad (7.6.8)$$

Note that it has the same dimensions as the kinematic viscosity. In terms of the eddy viscosity the differential equation becomes

$$\frac{r}{2}\frac{d\bar{p}}{dx} = \rho(\nu + \eta)\frac{d\bar{u}}{dr} \qquad (7.6.9)$$

If we view the turbulent process as the random and chaotic mixing of particles of fluid, we may choose to introduce the *mixing length* $l_m$, the distance a particle travels before interacting with another particle. Based on reasoning related to momentum interchange we relate the eddy viscosity to the mixing length with

$$\eta = l_m^2 \left|\frac{d\bar{u}}{dy}\right| \qquad (7.6.10)$$

The *correlation coefficient* $K_{uv}$, a normalized turbulent shear stress, often used when describing turbulent motions, has limits of $\pm 1$ and is defined by

$$K_{uv} = \frac{\overline{u'v'}}{\sqrt{\overline{u'^2}}\,\sqrt{\overline{v'^2}}} \qquad (7.6.11)$$

where the time-average quantities would be defined by Eq. 7.6.2

The quantities $\eta$, $l_m$, and $K_{uv}$ are functions of $r$ (or $y$) that simply replace the variable $\overline{u'v'}$. They do not simplify the differential equation 7.6.7; they allow the equation to take slightly different forms. Since we cannot derive an expression for $\eta$, $l_m$, or $K_{uv}$ we cannot find $\bar{u}(r)$ using analytical methods. We must rely on experimental data. It follows in the next section.

## Example 7.8

Note that in Fig. 7.9b there is a region near the wall where the turbulent shear is near its maximum and is relatively constant, as shown in Fig. P7.8, and the viscous shear is quite small. Assume that the mixing length is directly proportional to the distance from the wall. With this assumption show that $\bar{u}(y)$ is logarithmic in this region near the wall.

**Fig. E7.8**

### Solution

If the viscous shear is negligible, we have, combining Eqs. 7.6.5 and 7.6.8, and 7.6.10,

$$\bar{\tau}_{\text{turb}} = \rho \eta \frac{d\bar{u}}{dy} = \rho l_m^2 \left(\frac{d\bar{u}}{dy}\right)^2$$

Now, if $\bar{\tau}_{\text{turb}} = \text{const.} = c_1$ and we assume, as given in the problem statement, that

$$l_m = c_2 y$$

there results

$$c_1 = \rho c_2^2 y^2 \left(\frac{du}{dy}\right)^2$$

or

$$y \frac{d\bar{u}}{dy} = c_3$$

where $c_3 = \sqrt{c_1/\rho c_2^2}$. This is integrated to yield

$$\bar{u}(y) = c_3 \ln y + c_4$$

Hence, with the foregoing assumptions we see that a logarithmic profile is predicted for the region of constant turbulent shear near the wall. This is, in fact, observed from experimental data; so we conclude that the above assumptions are reasonable for a turbulent flow in a pipe.

 **Example 7.8a**  Turbulent Boundary Layer Lab, 858–860

### 7.6.2  Velocity Profile

The time-average velocity profile in a pipe is quite sensitive to the magnitude of the average wall roughness height $e$, as sketched in Fig. 7.10. Most materials are "rough" when viewed with sufficient magnification, with glass and plastic

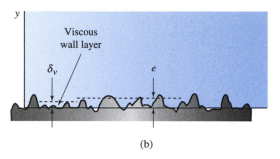

(a)                                         (b)

**Fig. 7.10**    (a) A smooth wall and (b) a rough wall.

assumed to be smooth with $e = 0$. (Values for $e$ are listed in Fig. 7.13 on page 308.) As noted in the preceding section, the laminar shear is significant only near the wall in the viscous wall layer with thickness $\delta_\nu$. If the thickness $\delta_\nu$ is sufficiently large, it submerges the wall roughness elements, as in Fig. 7.10a, so that they have negligible effect on the flow; it is as if the wall were smooth. Such a condition is often referred to as being *hydraulically smooth*. If the viscous wall layer is relatively thin, the roughness elements protrude out of this layer, as in Fig. 7.10b, and the wall is rough. The *relative roughness e/D* and the Reynolds number can be used to determine if a pipe is smooth or rough. This will be observed from the friction factor data presented in the next section.

> **KEY CONCEPT**    *If the thickness $\delta_\nu$ is sufficiently large, it is as if the wall were smooth. The pipe is hydraulically smooth.*

We do not solve for the velocity distribution in a developed turbulent flow since $\overline{u'v'}$ cannot be determined analytically. Thus we will present the empirical data for $\overline{u}(y)$ directly and not solve Eq. 7.6.7. The two most common expressions for turbulent flow are now presented.

The first method of empirically expressing the velocity distribution involves flows with smooth walls and flows with rough walls. If the flow has a smooth wall, as in Fig. 7.10a, we identify two regions of the flow, the *wall region* and the *outer region*. In the wall region the characteristic velocity and length are the **shear velocity** $u_\tau$ defined by[9] $u_\tau = \sqrt{\tau_0/\rho}$, and the **viscous length** $\nu/u_\tau$. The dimensionless velocity distribution in the wall region, for a smooth pipe, is

> **Shear velocity:**    *The quantity $\sqrt{\tau_0/\rho}$ is the shear velocity $u_\tau$.*

> **Viscous length:**    *The length $\nu/u_\tau$ is a characteristic length in a turbulent flow.*

$$\frac{\overline{u}}{u_\tau} = \frac{u_\tau y}{\nu} \text{ (viscous wall layer)} \qquad 0 \leq \frac{u_\tau y}{\nu} \lesssim 5 \qquad (7.6.12)$$

$$\frac{\overline{u}}{u_\tau} = 2.44 \ln \frac{u_\tau y}{\nu} + 4.9 \text{ (turbulent region)} \qquad 30 < \frac{u_\tau y}{\nu}, \frac{y}{r_0} < 0.15 \quad (7.6.13)$$

In the interval $5 < u_\tau y/\nu < 30$, the *buffer zone*, the experimental data do not fit either of the curves above but merge the two curves as shown in Fig. 7.11a. The viscous wall layer has thickness $\delta_\nu$ and it is in the viscous layer that turbulence is

> **KEY CONCEPT**    *It is in the viscous wall layer that turbulence is thought to be initiated.*

[9]The shear velocity $u_\tau$ is a fictitious velocity and has no relationship to an actual velocity in the flow. It is a quantity with dimensions of velocity that allows the experimental data to be presented in dimensionless form as *universal profiles* (profiles that are valid for all turbulent pipe flows). The viscous length $\nu/u_\tau$ is also a fictitious length.

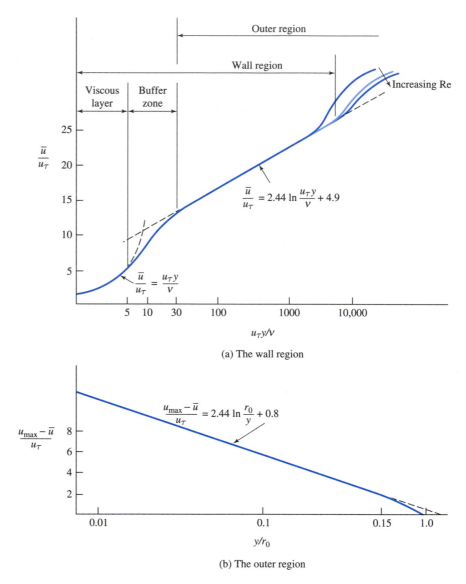

**Fig. 7.11** Empirical relations for turbulent flow in a smooth pipe: (a) wall region; (b) outer region. (Based on data from J. Laufer, The Structure of Turbulence in Fully Developed Pipe Flow, NACA Report 1174, 1954.)

thought to be initiated; this layer possesses a time-average, linear velocity distribution, but instantaneously the layer is very time dependent. The outer edge of the wall region is quite dependent on the Reynolds number, as shown; for low Re it may be located near $u_\tau y/\nu = 3000$.

For rough pipes the viscous wall layer does not play an important role since turbulence initiates from the protruding wall elements, so only a logarithmic profile is necessary in the wall region. The characteristic length is the average roughness height $e$; the dimensionless velocity profile for the rough pipe is

$$\frac{\bar{u}}{u_\tau} = 2.44 \ln \frac{y}{e} + 8.5 \qquad \frac{y}{r_0} < 0.15 \qquad \text{(wall region)} \qquad (7.6.14)$$

where the constants 8.5 and 2.44 allow a good fit with experimental data.

In the outer region, sketched in Fig. 7.11b, the characteristic length is $r_0$; the velocity defect $(u_{max} - \bar{u})$ is normalized with $u_\tau$, and the empirical relation, for both smooth and rough pipes, is

$$\frac{u_{max} - \bar{u}}{u_\tau} = 2.44 \ln \frac{r_0}{y} + 0.8 \qquad \frac{y}{r_0} \le 0.15 \qquad \text{(outer region)} \qquad (7.6.15)$$

An additional empirical equation is needed to complete the profile for $0.15 < y/r_0 \le 1$.

The wall region and the outer region overlap as shown in Fig. 7.11a. In this overlap region we can combine the equations above to obtain an expression for the maximum velocity; for a smooth pipe it is

$$\frac{u_{max}}{u_\tau} = 2.44 \ln \frac{u_\tau r_0}{\nu} + 5.7 \qquad \text{(smooth pipes)} \qquad (7.6.16)$$

and for a rough pipe we find that

$$\frac{u_{max}}{u_\tau} = 2.44 \ln \frac{r_0}{e} + 9.3 \qquad \text{(rough pipes)} \qquad (7.6.17)$$

Although we do not often desire the actual time-average velocity at a specified radial position in a pipe, the distributions above are of occasional use and are presented for completeness. Note, however, that before $u_{max}$ can be found we must know $u_\tau$; before $u_\tau$ can be found we must know $\tau_0$. To find $\tau_0$ we use the pressure gradient,

$$\tau_0 = -\frac{r_0}{2} \frac{dp}{dx} \qquad (7.6.18)$$

or the friction factor with Eq. 7.3.19. If neither $dp/dx$ nor $f$ is known, we may use the power-law form of the profile, described in the following paragraph, to approximate $f$.

An alternative, and simpler form that adequately describes the turbulent flow velocity distribution in a pipe is the *power-law profile*, namely,

**Fig. 7.12** Turbulent velocity profile.

$$\frac{\bar{u}}{u_{\max}} = \left(\frac{y}{r_0}\right)^{1/n} \tag{7.6.19}$$

where $y$ is measured from the pipe wall and $n$ is an integer between 5 and 10. Using this distribution the average velocity is found to be

$$V = \frac{1}{\pi r_0^2} \int_0^{r_0} \bar{u}(r)2\pi r\, dr = \frac{2n^2}{(n+1)(2n+1)}u_{\max} \tag{7.6.20}$$

This distribution is compared with a laminar profile in Fig. 7.12.

The value of $n$ in the exponent is related to the friction factor $f$ by the empirical expression

$$n = \frac{1}{\sqrt{f}} \tag{7.6.21}$$

The constant $n$ varies from 5 to 10 depending on the Reynolds number and the pipe wall roughness $e/D$. For smooth pipes, the exponent $n$ is related to the Reynolds number as shown in Table 7.1.

The power-law profile cannot be used to obtain the slope at the wall since it will always yield $(d\bar{u}/dy)_{\text{wall}} = \infty$ for all $n$. Thus it cannot be used to predict the wall shear stress. The wall shear is found by combining Eqs. 7.6.21 and 7.3.19. In addition, it gives a positive slope $d\bar{u}/dy$ at the centerline of the pipe, where the slope should be zero, so it is not valid near the centerline.

It should be noted that the kinetic-energy-correction factor $\alpha$ (see Eq. 4.5.29) in pipes is 1.11, 1.06, and 1.03 for $n = 5, 7$, and 10, respectively. Because it is close to unity for $n > 7$, it is often set equal to unity in the energy equation when working problems involving turbulent flow.

**KEY CONCEPT** *The power-law profile cannot be used to obtain the slope at the wall or the slope at the centerline.*

**TABLE 7.1** Exponent $n$ for Smooth Pipes

| Re = $VD/\nu$ | $4 \times 10^3$ | $10^5$ | $10^6$ | $>2 \times 10^6$ |
|---|---|---|---|---|
| $n$ | 6 | 7 | 9 | 10 |

## Example 7.9

Water at 20°C flows in a 10-cm-diameter pipe at an average velocity of 1.6 m/s. If the roughness elements are 0.046 mm high, would the wall be rough or smooth? Refer to Fig. 7.10.

### Solution

To determine if the wall is rough or smooth, we must compare the viscous wall layer thickness with the height of the roughness elements. So, let's find the viscous wall layer thickness. From Fig. 7.11 the viscous layer thickness is determined by letting $u_\tau y/\nu = 5$, where $y = \delta_\nu$. First, we must find $u_\tau$. The Reynolds number is

$$\text{Re} = \frac{VD}{\nu}$$

$$= \frac{1.6 \times 0.1}{10^{-6}} = 1.6 \times 10^5$$

From Table 7.1 $n \simeq 7.5$, so that, from Eq. 7.6.21,

$$f = \frac{1}{n^2}$$

$$= \frac{1}{7.5^2} = 0.018$$

The wall shear is calculated from Eq. 7.3.19:

$$\tau_0 = \frac{1}{8}\rho V^2 f$$

$$= \frac{1}{8} \times 1000 \times 1.6^2 \times 0.018 = 5.8 \text{ Pa}$$

The friction velocity is found from the definition of the shear velocity:

$$u_\tau = \sqrt{\frac{\tau_0}{\rho}}$$

$$= \sqrt{\frac{5.8}{1000}} = 0.076 \text{ m/s}$$

This allows us to calculate the viscous wall layer thickness using $y = \delta_\nu$:

$$\delta_\nu = \frac{5\nu}{u_\tau}$$

$$= \frac{5 \times 10^{-6}}{0.076} = 6.6 \times 10^{-5} \text{ m} \quad \text{or} \quad 0.066 \text{ mm}$$

Since the roughness elements are only 0.046 mm high, they are submerged in the viscous wall layer. Consequently, the wall is smooth (see Fig. 7.10a). If the pipe were made of cast iron with $e = 0.26$ mm, the wall would be rough.

   Note that the viscous wall layer, even at this relatively low velocity, is about 0.1% of the radius. The viscous wall layer is usually extremely thin.

## Example 7.10

The 4-cm-diameter smooth, horizontal pipe of Fig. E7.10 transports 0.004 m$^3$/s of water at 20°C. Using the power-law profile, approximate (a) the friction factor, (b) the maximum velocity, (c) the radial position where $u = V$, (d) the wall shear, (e) the pressure drop over a 10-m length, and (f) the maximum velocity using Eq. 7.6.16.

**Fig. E7.10**

### Solution

(a) The average velocity is calculated to be

$$V = \frac{Q}{A} = \frac{0.004}{\pi \times 0.02^2} = 3.18$$

The Reynolds number is

$$\text{Re} = \frac{VD}{\nu} = \frac{3.18 \times 0.04}{10^{-6}} = 1.27 \times 10^5$$

From Table 7.1 we see that $n \cong 7.5$ and from Eq. 7.6.21,

$$f = \frac{1}{n^2}$$

$$= \frac{1}{7.5^2} = 0.018$$

(b) The maximum velocity is found using Eq. 7.6.20 to be

$$u_{\max} = \frac{(n + 1)(2n + 1)}{2n^2} V$$

$$= \frac{8.5 \times 16}{2 \times 7.5^2} \times 3.18 = 3.84 \text{ m/s}$$

(c) The distance from the wall where $u = V = 3.18$ m/s is found using Eq. 7.6.19 as follows:

$$\frac{\bar{u}}{u_{max}} = \left(\frac{y}{r_0}\right)^{1/7.5}$$

$$\therefore y = r_0\left(\frac{u}{u_{max}}\right)^{7.5}$$

$$= 2\left(\frac{3.18}{3.84}\right)^{7.5} = 0.49 \text{ cm}$$

The radial position is thus

$$r = r_0 - y$$

$$= 2 - 0.49 = 1.51 \text{ cm}$$

(d) The wall shear is found using Eq. 7.3.19 and is

$$\tau_0 = \frac{1}{8}\rho V^2 f$$

$$= \frac{1}{8} \times 1000 \times 3.18^2 \times 0.018 = 23 \text{ Pa}$$

(e) The pressure drop is calculated using Eq. 7.6.18 with $\Delta p/L = -dp/dx$ to be

$$\Delta p = \frac{2\tau_0 L}{r_0}$$

$$= \frac{2 \times 23 \times 10}{0.02} = 23\,000 \text{ Pa} \quad \text{or} \quad 23 \text{ kPa}$$

(f) To use Eq. 7.6.16 we must know the shear velocity. It is

$$u_\tau = \sqrt{\frac{\tau_0}{\rho}}$$

$$= \sqrt{\frac{23}{1000}} = 0.152 \text{ m/s}$$

Then we find $u_{max}$ to be

$$u_{max} = 0.152\left(2.44 \ln \frac{0.152 \times 0.02}{10^{-6}} + 5.7\right) = 3.84 \text{ m/s}$$

This is the same as that given by the power-law formula in part (b). This answer is considered to be more accurate if it differs from that of Eq. 7.6.20. Note that the experimental data do not allow for accuracy in excess of three significant digits, and often to only two significant digits.

### 7.6.3 Losses in Developed Pipe Flow

Perhaps the most calculated quantity in pipe flow is the head loss. If the head loss is known in a developed flow, the pressure change can be calculated which allows pumps to be selected; for developed flow in a pipe the energy equation (4.5.17) provides us with

$$h_L = \frac{\Delta(p + \gamma h)}{\gamma} \qquad (7.6.22)$$

The head loss that results from the wall shear in a developed flow is related to the friction factor (see Eq. 7.3.20) by the Darcy–Weisbach equation, namely,

$$h_L = f\frac{L}{D}\frac{V^2}{2g} \qquad (7.6.23)$$

**KEY CONCEPT** *If the friction factor is known, we can find the head loss and the pressure drop.*

Consequently, if the friction factor is known, we can find the head loss and then the pressure drop.

The friction factor $f$ depends on the various quantities that affect the flow, written as

$$f = f(\rho, \mu, V, D, e) \qquad (7.6.24)$$

where the average wall roughness height $e$ accounts for the influence of the wall roughness elements. A dimensional analysis, following the steps of Section 6.2, provides us with

$$f = f\left(\frac{\rho VD}{\mu}, \frac{e}{D}\right) \qquad (7.6.25)$$

where $e/D$ is the relative roughness.

Experimental data that relate the friction factor to the Reynolds number have been obtained for fully developed pipe flow over a wide range of wall roughnesses. The results of these data are presented in Fig. 7.13, which is commonly referred to as the *Moody diagram*, named after Lewis F. Moody (1880–1953). There are several features of the Moody diagram that should be noted.

- For a given wall roughness, measured by the relative roughness $e/D$, there is a sufficiently large value of Re above which the friction factor is constant, thereby defining the *completely turbulent regime*. The average roughness element size $e$ is substantially greater than the viscous wall layer thickness $\delta_\nu$, so that viscous effects are not significant; the resistance to the flow is produced primarily by the drag of the roughness elements that protrude into the flow.

- For the smaller relative roughness $e/D$ values it is observed that, as Re decreases, the friction factor increases in the *transition zone* and eventually becomes the same as that of a smooth pipe. The roughness elements become submerged in the viscous wall layer so that they produce little effect on the main flow.

- For Reynolds numbers less than 2000, the friction factor of laminar flow is shown. The *critical zone* couples the turbulent flow to the laminar flow and may represent an oscillatory flow that alternately exists between turbulent and laminar flow.

- The $e$ values in this diagram are for new pipes. With age a pipe will corrode and become fouled, changing both the roughness and the pipe diameter, with a resulting increase in the friction factor. Such factors should be included in design considerations; they will not be reviewed here.

The following empirical equations represent the Moody diagram for Re > 4000:

Smooth pipe flow:
$$\frac{1}{\sqrt{f}} = 0.86 \ln \mathrm{Re}\, \sqrt{f} - 0.8 \qquad (7.6.26)$$

Completely turbulent zone:
$$\frac{1}{\sqrt{f}} = -0.86 \ln \frac{e}{3.7D} \qquad (7.6.27)$$

Transition zone:
$$\frac{1}{\sqrt{f}} = -0.86 \ln \left( \frac{e}{3.7D} + \frac{2.51}{\mathrm{Re}\,\sqrt{f}} \right) \qquad (7.6.28)$$

The transition zone equation (7.6.28) that couples the smooth pipe equation to the completely turbulent regime equation is known as the **Colebrook equation**. Note that Eq. 7.6.26 is the Colebrook equation with $e = 0$, and Eq. 7.6.27 is the Colebrook equation with Re = $\infty$.

**Colebrook equation:**   *The equation that couples the smooth pipe equation to the completely turbulent regime equation.*

**Fig. 7.13** Moody diagram. (From L. F. Moody, Trans. ASME, Vol. 66, 1944. Reproduced with permission of ASME.) (Note: If $e/D = 0.01$ and $\text{Re} = 10^4$, the dot locates $f = 0.043$.)

Three categories of problems can be identified for developed turbulent flow in a pipe of length $L$:

| Category | Known | Unknown |
|----------|-------|---------|
| 1 | $Q, D, e, \nu$ | $h_L$ |
| 2 | $D, e, \nu, h_L$ | $Q$ |
| 3 | $Q, e, \nu, h_L$ | $D$ |

A category 1 problem is straightforward and requires no iteration procedure when using the Moody diagram. Category 2 and 3 problems are more like problems encountered in engineering design situations and require an iterative trial-and-error process when using the Moody diagram. Each of these types will be illustrated with an example.

An alternative to using the Moody diagram that avoids any trial-and-error process is made possible by empirically derived formulas. Perhaps the best of such formulas were presented by Swamee and Jain (1976) for pipe flow; an explicit expression that provides an approximate value for the unknown in each category above is as follows:

$$h_L = 1.07 \frac{Q^2 L}{g D^5} \left\{ \ln\left[ \frac{e}{3.7D} + 4.62 \left( \frac{\nu D}{Q} \right)^{0.9} \right] \right\}^{-2} \qquad \begin{array}{l} 10^{-6} < e/D < 10^{-2} \\ 3000 < \text{Re} < 3 \times 10^8 \end{array} \qquad (7.6.29)$$

$$Q = -0.965 \left( \frac{g D^5 h_L}{L} \right)^{0.5} \ln\left[ \frac{e}{3.7D} + \left( \frac{3.17 \nu^2 L}{g D^3 h_L} \right)^{0.5} \right] \qquad \text{Re} > 2000 \qquad (7.6.30)$$

$$D = 0.66 \left[ e^{1.25} \left( \frac{L Q^2}{g h_L} \right)^{4.75} + \nu Q^{9.4} \left( \frac{L}{g h_L} \right)^{5.2} \right]^{0.04} \qquad \begin{array}{l} 10^{-6} < e/D < 10^{-2} \\ 5000 < \text{Re} < 3 \times 10^8 \end{array} \qquad (7.6.31)$$

Equation 7.6.30 is as accurate as the Moody diagram, and Eqs. 7.6.29 and 7.6.31 are accurate to within approximately 2% of the Moody diagram. These tolerances are acceptable for engineering calculations. It is important to realize that the Moody diagram is based on experimental data that likely is accurate to within no more than 5%. Hence the foregoing three formulas of Swamee and Jain, which can easily be input on a programmable hand-held calculator, are often used by design engineers. The following examples also illustrate the use of these approximate formulas.

## Example 7.11

Water at 24°C is transported for 450 m in a 3.8-cm-diameter wrought iron horizontal pipe with a flow rate of 2.75 L/s. Calculate the pressure drop over the 450-m length of pipe, using (a) the Moody diagram and (b) the alternate method.

### Solution

(a) The average velocity is

$$V = \frac{Q}{A} = \frac{2.75 \times 10^{-3}}{\pi \times (0.019)^2} = 2.43 \text{ m/s}$$

The Reynolds number is

$$\text{Re} = \frac{VD}{\nu} = \frac{2.43 \times 0.038}{9.1 \times 10^{-7}} = 1.01 \times 10^5$$

Obtaining $e$ from Fig. 7.13, we have, using $D = 0.038$ m,

$$\frac{e}{D} = \frac{0.000046}{0.038} = 0.0012$$

The friction factor is read from the Moody diagram to be

$$f = 0.023$$

The head loss is calculated as

$$h_L = f \frac{L}{D} \frac{V^2}{2g}$$

$$= 0.023 \frac{450}{0.038} \frac{(2.43)^2}{2 \times 9.8} = 82 \text{ m}$$

This answer is given to two significant numbers since the friction factor is known to at most two significant numbers. The pressure drop is found by Eq. 7.6.22 to be

$$\Delta p = \gamma h_L$$
$$= 9810 \text{ N/m}^3 \times 82 \text{ m} = 804420 \text{ N/m}^2 \quad \text{or} \quad 804.4 \text{ kPa}$$

(b) The alternate method for this Category 1 problem uses Eq. 7.6.29, with $D = 0.038$ m:

$$h_L = 1.07 \frac{(2.75 \times 10^{-3})^2 \times 450}{9.81 \times (0.038)^5} \left\{ \ln \left[ \frac{0.0012}{3.7} + 4.62 \left( \frac{9.1 \times 10^{-7} \times 0.038}{0.00275} \right)^{0.9} \right] \right\}^{-2}$$

$$= 82 \text{ m}$$

This much simpler method provides the same value as that found using the Moody diagram.

## Example 7.12

A pressure drop of 700 kPa is measured over a 300-m length of horizontal, 10-cm-diameter wrought iron pipe that transports oil ($S = 0.9$, $\nu = 10^{-5}$ m²/s). Calculate the flow rate using (a) the Moody diagram, and (b) the alternate method.

### Solution

(a) The relative roughness is

$$\frac{e}{D} = \frac{0.046}{100} = 0.00046$$

Assuming that the flow is completely turbulent (Re is not needed), the Moody diagram gives

$$f = 0.0165$$

The head loss is found to be

$$h_L = \frac{\Delta p}{\gamma_{\text{oil}}} = \frac{700\,000 \text{ N/m}^2}{9800 \text{ N/m}^3 \times 0.9} = 79.4 \text{ m}$$

The velocity is calculated from Eq. 7.6.23 to be

$$V = \left(\frac{2gDh_L}{fL}\right)^{1/2} = \left(\frac{2 \times 9.8 \text{ m/s}^2 \times 0.1 \text{ m} \times 79.4 \text{ m}}{0.0165 \times 300 \text{ m}}\right)^{1/2} = 5.61 \text{ m/s}$$

This provides us with a Reynolds number of

$$\text{Re} = \frac{VD}{\nu} = \frac{5.61 \text{ m/s} \times 0.1 \text{ m}}{10^{-5} \text{ m}^2/\text{s}} = 5.61 \times 10^4$$

Using this Reynolds number and $e/D = 0.00046$, the Moody diagram gives the friction factor as

$$f = 0.023$$

This corrects the original value for $f$. The velocity is recalculated to be

$$V = \left(\frac{2 \times 9.8 \times 0.1 \times 79.4}{0.023 \times 300}\right)^{1/2} = 4.75 \text{ m/s}$$

The Reynolds number is then

$$\text{Re} = \frac{4.75 \times 0.1}{10^{-5}} = 4.75 \times 10^4$$

From the Moody diagram $f = 0.023$ appears to be satisfactory. Thus the flow rate is

$$Q = VA = 4.75 \times \pi \times 0.05^2 = 0.037 \text{ m}^3/\text{s}$$

Only two significant numbers are given since $f$ is known to at most two significant numbers.

(b) The alternative method for this Category 2 problem uses the explicit relationship (7.6.30). We can directly calculate $Q$ to be

$$Q = -0.965\left(\frac{9.8 \times 0.1^5 \times 79.4}{300}\right)^{0.5} \ln\left[\frac{0.00046}{3.7} + \left(\frac{3.17 \times 10^{-10} \times 300}{9.8 \times 0.1^3 \times 79.4}\right)^{0.5}\right]$$

$$= -0.965 \times 5.096 \times 10^{-3} \times (-7.655) = 0.038 \text{ m}^3/\text{s}$$

This much simpler method produces a value essentially the same as that obtained using the Moody diagram.

### Example 7.13

Drawn tubing of what diameter should be selected to transport 0.002 m³/s of 20°C water over a 400-m length so that the head loss does not exceed 30 m? (a) Use the Moody diagram and (b) the alternative method.

### Solution

(a) In this problem we do not know $D$. Thus, a trial-and-error solution is anticipated. The average velocity is related to $D$ by

$$V = \frac{Q}{A} = \frac{0.002}{\pi D^2/4} = \frac{0.00255}{D^2}$$

The friction factor and $D$ are related as follows:

$$h_L = f\frac{L}{D}\frac{V^2}{2g}$$

$$30 = f\frac{400}{D}\frac{(0.00255/D^2)^2}{2 \times 9.8}$$

$$\therefore D^5 = 4.42 \times 10^{-6}f$$

The Reynolds number is

$$\mathrm{Re} = \frac{VD}{\nu} = \frac{0.00255D}{D^2 \times 10^{-6}} = \frac{2550}{D}$$

Now, let us simply guess a value for $f$ and check with the relations above and the Moody diagram. The first guess is $f = 0.03$, and the correction is listed in the following table. Note: the second guess is the value for $f$ found from the calculations of the first guess.

| $f$ | $D$(m) | Re | $e/D$ | $f$ (Fig. 7.13) |
|------|--------|------------------|----------|-----------------|
| 0.03 | 0.0421 | $6.06 \times 10^4$ | 0.000036 | 0.02 |
| 0.02 | 0.0388 | $6.57 \times 10^4$ | 0.000039 | 0.02 |

The value of $f = 0.02$ is acceptable, yielding a diameter of 3.88 cm. Since this diameter would undoubtedly not be standard, a diameter of

$$D = 4 \text{ cm}$$

would be the tube size selected. This tube would have a head loss less than the limit of $h_L = 30$ m imposed in the problem statement. Any larger-diameter tube would also satisfy this criterion but would be more costly, so it should not be selected.

(b) The alternative method for this Category 3 problem uses the explicit relationship (7.6.31). We can directly calculate $D$ to be

$$D = 0.66\left[(1.5 \times 10^{-6})^{1.25}\left(\frac{400 \times 0.002^2}{9.81 \times 30}\right)^{4.75} + 10^{-6} \times 0.002^{9.4}\left(\frac{400}{9.81 \times 30}\right)^{5.2}\right]^{0.04}$$

$$= 0.66[5.163 \times 10^{-33} + 2.102 \times 10^{-31}]^{0.04} = 0.039 \text{ m}$$

Hence $D = 4$ cm would be the tube size selected. This is the same tube size as that selected using the Moody diagram.

## 7.6.4   Losses in Noncircular Conduits

Good approximations can be made for the head loss in conduits with noncircular cross sections by using the *hydraulic radius R*, defined by

$$R = \frac{A}{P} \qquad (7.6.32)$$

where $A$ is the cross-sectional area and $P$ is the **wetted perimeter**, that perimeter where the fluid is in contact with the solid boundary. For a circular pipe flowing full the hydraulic radius is $R = r_0/2$. Hence we simply replace the radius $r_0$ with $2R$ and use the Moody diagram with

$$\mathrm{Re} = \frac{4VR}{\nu} \qquad \begin{array}{c}\text{relative}\\\text{roughness}\end{array} = \frac{e}{4R} \qquad (7.6.33)$$

The head loss becomes

$$h_L = f\frac{L}{4R}\frac{V^2}{2g} \qquad (7.6.34)$$

To use this hydraulic radius technique the cross section should be fairly "open," such as a rectangle with aspect ratio less than 4:1; an equilateral triangle, or an oval. For other shapes, such as an annulus, the error would be significant.

### Example 7.14

Air at standard conditions is to be transported through 500 m of a smooth, horizontal, 30 cm × 20 cm rectangular duct at a flow rate of 0.24 $m^3$/s. Calculate the pressure drop.

#### Solution

The hydraulic radius is

$$R = \frac{A}{P} = \frac{0.3 \times 0.2}{(0.3 + 0.2) \times 2} = 0.06 \text{ m}$$

The average velocity is

$$V = \frac{Q}{A} = \frac{0.24}{0.3 \times 0.2} = 4.0 \text{ m/s}$$

This gives a Reynolds number of

$$\mathrm{Re} = \frac{4VR}{\nu} = \frac{4 \times 4 \times 0.06}{1.5 \times 10^{-5}} = 6.4 \times 10^4$$

Using the smooth pipe curve of the Moody diagram, there results

$$f = 0.0196$$

Hence,

$$h_L = f\frac{L}{4R}\frac{V^2}{2g} = 0.0196 \frac{500 \text{ m}}{4 \times 0.06 \text{ m}} \frac{4^2 \text{ m}^2/\text{s}^2}{2 \times 9.8 \text{ m/s}^2} = 33.3 \text{ m}$$

The pressure drop is

$$\Delta p = \rho g h_L = 1.23 \times 9.8 \times 33.3 = 402 \text{ Pa}$$

### 7.6.5 Minor Losses in Pipe Flow

We now know how to calculate the losses due to a developed flow in a pipe. Pipe systems do, however, include valves, elbows, enlargements, contractions, inlets, outlets, bends, and other fittings that cause additional losses, referred to as *minor losses*, even though such losses can exceed the frictional losses of Eq. 7.6.23. Each of these elements causes a change in the magnitude and/or the direction of the velocity vectors and hence results in a loss. In general, if the flow is gradually accelerated by an element, the losses are very small; relatively large losses are associated with sudden enlargements or contractions because of the separated regions that result (a separated flow occurs when the primary flow separates from the wall).

A minor loss is expressed in terms of a loss coefficient $K$, defined by

$$h_L = K \frac{V^2}{2g} \qquad (7.6.35)$$

Values of $K$ have been determined experimentally for the various fittings and geometry changes of interest in piping systems. One exception is the sudden expansion from area $A_1$ to area $A_2$, for which the loss can be calculated; this was done in Example 4.14, where we found that

$$h_L = \left(1 - \frac{A_1}{A_2}\right)^2 \frac{V_1^2}{2g} \qquad (7.6.36)$$

Thus, for the sudden expansion

$$K = \left(1 - \frac{A_1}{A_2}\right)^2 \qquad (7.6.37)$$

If $A_2$ is extremely large (e.g., a pipe exiting into a reservoir), $K = 1.0$, since the entire kinetic energy is lost.

A pipe fitting that has a relatively large loss coefficient with no change in cross-sectional area is the pipe bend, or the elbow. This results primarily from the secondary flow caused by the fluid flowing from the high-pressure region to the low-pressure region (see Eq. 3.4.15), as shown in Fig. 7.14; this secondary flow is eventually dissipated after the fluid leaves the long sweep bend or elbow. In addition, a separated region occurs at the sharp corner in a standard elbow. Energy is needed to maintain a secondary flow and the flow in the separated region. This wasted energy is measured in terms of the loss coefficient.

The loss coefficients for various geometries are presented in Table 7.2 and Fig. 7.15. A globe valve may be used to control the flow rate by introducing large losses caused by partially closing the valve. The other types of valves should not be used to control the flow; damage could result.

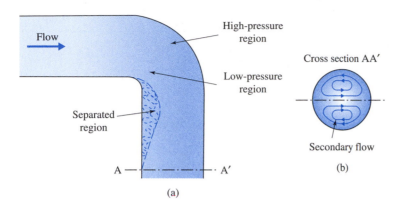

Flow

High-pressure region

Low-pressure region

Separated region

A ——— A′

(a)

Cross section AA′

Secondary flow

(b)

**Fig. 7.14**   Flow in an elbow.

$$\frac{D_2}{D_1} = 1.5$$

$$\frac{D_2}{D_1} = 3$$

$V_1$   $\theta$   $V_2$

$$h_L = K \frac{(V_1 - V_2)^2}{2g}$$

$K$

$\theta$

**Fig. 7.15**   Loss coefficients in a conical expansion.
(From A. H. Gibson, Vol. 93, 1912)

**TABLE 7.2** Nominal Loss Coefficients $K$ (Turbulent Flow)[a]

| Type of fitting | | Screwed | | Flanged | | |
|---|---|---|---|---|---|---|
| Diameter | | 2.5 cm | 10 cm | 5 cm | 10 cm | 20 cm |
| Globe valve | (fully open) | 8.2 | 5.7 | 8.5 | 6.0 | 5.8 |
| | (half open) | 20 | 14 | 21 | 15 | 14 |
| | (one-quarter open) | 57 | 40 | 60 | 42 | 41 |
| Angle valve (fully open) | | 4.7 | 1.0 | 2.4 | 2.0 | 2.0 |
| Swing check valve (fully open) | | 2.9 | 2.0 | 2.0 | 2.0 | 2.0 |
| Gate valve (fully open) | | 0.24 | 0.11 | 0.35 | 0.16 | 0.07 |
| Return bend | | 1.5 | 0.64 | 0.35 | 0.30 | 0.25 |
| Tee (branch) | | 1.8 | 1.1 | 0.80 | 0.64 | 0.58 |
| Tee (line) | | 0.9 | 0.9 | 0.19 | 0.14 | 0.10 |
| Standard elbow | | 1.5 | 0.64 | 0.39 | 0.30 | 0.26 |
| Long sweep elbow | | 0.72 | 0.23 | 0.30 | 0.19 | 0.15 |
| 45° elbow | | 0.32 | 0.29 | | | |

| | | |
|---|---|---|
| Square-edged entrance | | 0.5 |
| Reentrant entrance | | 0.8 |
| Well-rounded entrance | | 0.03 |
| Pipe exit | | 1.0 |

| | Area ratio | |
|---|---|---|
| Sudden contraction[b] | 2:1 | 0.25 |
| | 5:1 | 0.41 |
| | 10:1 | 0.46 |

| | Area ratio $A/A_0$ | |
|---|---|---|
| Orifice plate | 1.5:1 | 0.85 |
| | 2:1 | 3.4 |
| | 4:1 | 29 |
| | $\geq$ 6:1 | $2.78\left(\dfrac{A}{A_0} - 0.6\right)^2$ |

| | | |
|---|---|---|
| Sudden enlargement[c] | | $\left(1 - \dfrac{A_1}{A_2}\right)^2$ |
| 90° miter bend (without vanes) | | 1.1 |
| (with vanes) | | 0.2 |
| General contraction | (30° included angle) | 0.02 |
| | (70° included angle) | 0.07 |

[a]Values for other geometries can be found in *Technical Paper 410*, The Crane Company, 1957.
[b]Based on exit velocity $V_2$.
[c]Based on entrance velocity $V_1$.

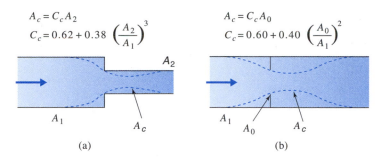

**Fig. 7.16**  Vena contractas in contractions and orifices: (a) sudden contraction; (b) concentric orifice.

Loss coefficients for sudden contractions and orifice plates can be approximated by neglecting the losses in the converging flow up to the vena contracta and calculating the losses in the diverging flow using the loss coefficient for a sudden expansion. Figure 7.16 provides the information necessary to establish the area $A_c$ of the **vena contracta**, the minimum area; this minimum area occurs where the converging streamlines begin to expand to fill the downstream area.

It is often the practice to express a loss coefficient as an *equivalent length* $L_e$ of pipe. This is done by equating Eq. 7.6.35 to Eq. 7.6.23:

$$K \frac{V^2}{2g} = f \frac{L_e}{D} \frac{V^2}{2g} \tag{7.6.38}$$

giving the relationship

$$L_e = K \frac{D}{f} \tag{7.6.39}$$

Hence the square-edged entrance of a 20-cm-diameter pipe with a friction factor of $f = 0.02$ could be replaced by an equivalent pipe length of $L_e = 5$ m.

Finally, a comment should be made concerning the magnitude of the minor losses. In piping systems involving intermediate lengths (i.e., 100 diameters) of pipe between minor losses, the minor losses may be of the same order of magnitude as the frictional losses; for relatively short lengths the minor losses may be substantially greater than the frictional losses; and for long lengths (e.g., 1000 diameters) of pipe, the minor losses are usually neglected.

**Vena contracta:**   *The minimum area in a sudden contraction.*

**KEY CONCEPT**   *For long lengths of pipe, the minor losses are usually neglected.*

## Example 7.15

If the flow rate through a 10-cm-diameter wrought iron pipe (Fig. E7.15) is 0.04 m³/s, find the difference in elevation $H$ of the two reservoirs.

**Fig. E7.15**

### Solution

The energy equation written for a control volume that contains the two reservoir surfaces (see Eq. 4.5.17), where $V_1 = V_2 = 0$ and $p_1 = p_2 = 0$, is

$$0 = z_2 - z_1 + h_L$$

Thus, letting $z_1 - z_2 = H$, we have

$$H = (K_{\text{entrance}} + K_{\text{valve}} + 2K_{\text{elbow}} + K_{\text{exit}}) \frac{V^2}{2g} + f \frac{L}{D} \frac{V^2}{2g}$$

The average velocity, Reynolds number, and relative roughness are

$$V = \frac{Q}{A} = \frac{0.04}{\pi \times 0.05^2} = 5.09 \text{ m/s}$$

$$\text{Re} = \frac{VD}{\nu} = \frac{5.09 \times 0.1}{10^{-6}} = 5.09 \times 10^5$$

$$\frac{e}{D} = \frac{0.046}{100} = 0.00046$$

From the Moody diagram we find that

$$f = 0.0173$$

Using the loss coefficients from Table 7.2 for an entrance, a globe valve, screwed 10-cm-diameter standard elbows, and an exit there results

$$H = (0.5 + 5.7 + 2 \times 0.64 + 1.0) \frac{5.09^2}{2 \times 9.8} + 0.0173 \frac{50}{0.1} \frac{5.09^2}{2 \times 9.8}$$

$$= 11.2 + 11.4 = 22.6 \text{ m}$$

Note: The minor losses are about equal to the frictional losses as expected, since there are five minor loss elements in 500 diameters of pipe length.

## Example 7.16

Approximate the loss coefficient for the sudden contraction $A_1/A_2 = 2$ by neglecting the losses in the contracting portion up to the vena contracta and assuming that all the losses occur in the expansion from the vena contracta to $A_2$ (see Fig. 7.16). Compare with that given in Table 7.2.

**Solution**

The head loss from the vena contracta to area $A_2$ is (see Table 7.2, sudden enlargement)

$$h_L = \left(1 - \frac{A_c}{A_2}\right)^2 \frac{V_c^2}{2g}$$

Continuity allows us to write

$$V_c = \frac{A_2}{A_c} V_2$$

Thus, the head loss based on $V_2$ is

$$h_L = \left(1 - \frac{A_c}{A_2}\right)^2 \left(\frac{A_2}{A_c}\right)^2 \frac{V_2^2}{2g}$$

so the loss coefficient of Eq. 7.6.35 is

$$K = \left(1 - \frac{A_c}{A_2}\right)^2 \left(\frac{A_2}{A_c}\right)^2$$

Using the expression of $C_c$ given in Fig. 7.16, we have

$$\frac{A_c}{A_2} = C_c = 0.62 + 0.38 \left(\frac{1}{2}\right)^3 = 0.67$$

Finally,

$$K = (1 - 0.67)^2 \frac{1}{0.67^2} = 0.24$$

This compares favorably with the value of 0.25 given in Table 7.2.

### 7.6.6   Hydraulic and Energy Grade Lines

When the energy equation is written in the form of Eq. 4.5.17, that is,

$$-\frac{\dot{W}_s}{\dot{m}g} = \frac{V_2^2 - V_1^2}{2g} + \frac{p_2 - p_1}{\gamma} + z_2 - z_1 + h_L \qquad (7.6.40)$$

the terms have dimensions of length. This has led to the conventional use of the hydraulic grade line and the energy grade line. The **hydraulic grade line** (HGL), the dashed line in Fig. 7.17, in a piping system is formed by the locus of points located a distance $p/\gamma$ above the center of the pipe, or $p/\gamma + z$ above a preselected datum; the liquid in a piezometer tube would rise to the HGL. The **energy grade line** (EGL), the solid line in Fig. 7.17, is formed by the locus of points a distance

**Hydraulic grade line (HGL):** *In a piping system, the HGL is located a distance p/γ above the center of the pipe.*

**Energy grade line (EGL):** *In a piping system, the EGL is located a distance V²/2g above the HGL.*

$V^2/2g$ above the HGL, or the distance $V^2/2g + p/\gamma + z$ above the datum; the liquid in a pitot tube would rise to the EGL. The following points are noted relating to the HGL and the EGL:

- As the velocity goes to zero, the HGL and the EGL approach each other. Thus, in a reservoir, they are identical and lie on the surface (see Fig. 7.17).
- The EGL and, consequently, the HGL slope downward in the direction of the flow due to the head loss in the pipe. The greater the loss per unit length, the greater the slope. As the average velocity in the pipe increases, the loss per unit length increases.
- A sudden change occurs in the HGL and the EGL whenever a loss occurs due to a sudden geometry change, as represented by the valve or the sudden enlargement of Fig. 7.17.
- A jump occurs in the HGL and the EGL whenever useful energy is added to the fluid, as occurs with a pump, and a drop occurs if useful energy is extracted from the flow, as occurs with a turbine.
- At points where the HGL passes through the centerline of the pipe, the pressure is zero. If the pipe lies above the HGL, there is a vacuum in the pipe, a condition that is often avoided, if possible, in the design of piping systems; an exception would be in the design of a siphon.

**KEY CONCEPT**  *If the pipe lies above the HGL, there is a vacuum in the pipe.*

The concepts of energy grade line and hydraulic grade line may also be applied to open-channel flows. The HGL coincides with the free surface and the EGL is a distance $V^2/2g$ above the free surface. Uniform flows in open channels will be discussed in the following section, and Chapter 10 is devoted to nonuniform open-channel flow.

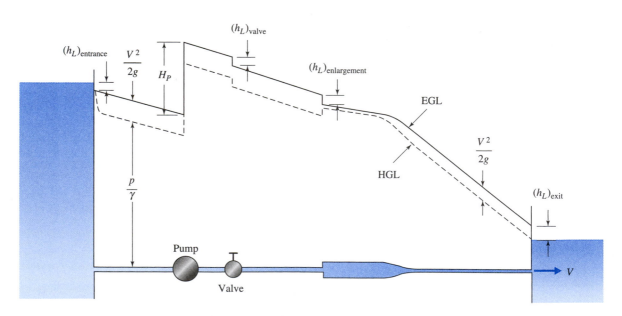

**Fig. 7.17**  Hydraulic grade line (HGL) and energy grade line (EGL) for a piping system.

# Example 7.17

Water at 20°C flows between two reservoirs at the rate of 0.06 m³/s as shown in Fig. E7.17. Sketch the HGL and the EGL. What is the minimum diameter $D_B$ allowed to avoid the occurrence of cavitation?

**Fig. E7.17**

## Solution

The EGL and the HGL are sketched on the figure, including sudden changes at the entrance, contraction, enlargement, and the exit. Note the large velocity head (the difference between the EGL and the HGL) in the smaller pipe because of the high velocity. The velocity, Reynolds number, and relative roughness in the 20-cm-diameter pipe are calculated to be

$$V = \frac{Q}{A} = \frac{0.06}{\pi \times 0.20^2/4} = 1.91 \text{ m/s}$$

$$\text{Re} = \frac{VD}{\nu} = \frac{1.91 \times 0.2}{10^{-6}} = 3.8 \times 10^5$$

$$\frac{e}{D} = \frac{0.26}{200} = 0.0013$$

Thus $f = 0.022$ from Fig. 7.13. The velocity, Reynolds number, and relative roughness in the smaller pipe are

$$V_B = \frac{0.06}{\pi D_B^2/4} = \frac{0.0764}{D_B^2}$$

$$\text{Re} = \frac{0.0764 \times D_B}{D_B^2 \times 10^{-6}} = \frac{76\,400}{D_B}$$

$$\frac{e}{D_B} = \frac{0.00026}{D_B}$$

The minimum possible diameter is established by recognizing that the water vapor pressure (2450 Pa absolute) at 20°C is the minimum allowable pressure. Since the

*(continued)*

distance between the pipe and the HGL is an indication of the pressure in the pipe, we can conclude that the minimum pressure will occur at section 2. Hence the energy equation applied between section 1, the reservoir surface, and section 2 gives

$$\frac{\cancel{V_1^2}^{\,0}}{2g} + \frac{p_1}{\gamma} + z_1 = \frac{V_B^2}{2g} + \frac{p_2}{\gamma} + \cancel{z_2}^{\,0} + K_{\text{ent}}\frac{V_A^2}{2g} + K_{\text{cont}}\frac{V_B^2}{2g} + f_A\frac{L_A}{D_A}\frac{V^2}{2g} + f_B\frac{L_B}{D_B}\frac{V_B^2}{2g}$$

where the subscript $A$ refers to the 20-cm-diameter pipe. This simplifies, using absolute pressure, to

$$\frac{101\,000}{9810} + 20 = \frac{(0.0764/D_B^2)^2}{2 \times 9.81}\left(1 + 0.25 + f_B\frac{20}{D_B}\right) + \frac{2450}{9810}$$

$$+ \left(0.5 + 0.022\frac{30}{0.2}\right)\frac{1.91^2}{2 \times 9.81}$$

$$98\,600 = \frac{1.25}{D_B^4} + f_B\frac{20}{D_B^5}$$

where we have used $K_{\text{ent}} = 0.5$ and assumed that $K_{\text{cont}} = 0.25$. This requires a trial-and-error solution. The following illustrates the procedure.

Let $D_B = 0.1$ m. Then $e/D = 0.0026$ and Re $= 7.6 \times 10^5$. Therefore, $f = 0.026$:

$$98\,600 \overset{?}{=} 12\,500 + 52\,000$$

Let $D_B = 0.09$ m. Then $e/D = 0.0029$ and Re $= 8.4 \times 10^5$. Therefore, $f = 0.027$:

$$98\,600 \overset{?}{=} 19\,000 + 91\,000$$

We see that 0.1 m is too large and 0.09 m is too small. In fact, the value of 0.09 m is only slightly too small. Consequently, to be safe we must select the next larger pipe size of 0.1 m diameter. If there were a pipe size of 9.5 cm diameter, that could be selected. Assuming that that size is not available, we select

$$D_B = 10 \text{ cm}$$

Note that the assumption of a 2:1 area ratio for the contraction is too small. It is actually 4:1. This would give $K_{\text{cont}} \simeq 0.4$. After a quick check we conclude that this value does not significantly influence the result.

## 7.6.7    Simple Pipe System with a Pump

The problems we have considered thus far in this section have not involved a pump. If a centrifugal pump is included in the piping system and the flow rate is specified, the solution is straightforward using the techniques we have already developed. If, on the other hand, the flow rate is not specified, as is often the case, a trial-and-error solution involving the centrifugal pump results since the head produced by a centrifugal pump and its efficiency $\eta_P$ (see Eq. 4.5.26) depend on the discharge, as shown by the pump characteristic curves, the solid curves in Fig. 7.18. Companies furnish such characteristic curves for each centrifugal pump manufactured. Such a curve provides one equation relating the flow rate $Q$ and pump head $H_P$. The other equation is provided by the energy equation, which can typically be written as (see the energy equation in the following example)

**KEY CONCEPT**    *If the flow rate is specified, the solution is straightforward.*

$$H_P = c_1 + c_2 Q^2 \tag{7.6.41}$$

This is the **system demand curve**. This along with the characteristic curve must be solved simultaneously to yield the desired flow rate. To determine the power requirement of the pump, the efficiency $\eta_P$ must be used.

**System demand curve:**  *The energy equation relating the pump head to the unknown flow rate.*

Note that for the piping system, the required pump energy head $H_P$, demanded by the energy equation, increases with $Q$ and from the pump characteristic curve we see that $H_P$ decreases with $Q$; hence the two curves will intersect at a point, called the *operating point* of the system. An example will illustrate the solution technique.

**Fig. 7.18**    Pump characteristic curves and the system demand curve.

## Example 7.18

Estimate the flow rate in the simple piping system of Fig. E7.18a if the pump characteristic curves are as shown in Fig. E7.18b. Also, find the pump power requirement.

**Fig. E7.18**

### Solution

We will assume that the Reynolds number is sufficiently large that the flow is completely turbulent. So, using $e/D = 0.046/200 = 0.00023$, the friction factor from the Moody diagram is

$$f = 0.014$$

The energy equation (see Eq. 7.6.40), with $H_P = -\dot{W}_S/\dot{m}g$, applied between the two surfaces, yields

$$H_P = \frac{\cancelto{0}{V_2^2} - \cancelto{0}{V_1^2}}{2g} + z_2 - z_1 + \frac{\cancelto{0}{p_2} - \cancelto{0}{p_1}}{\gamma} + h_L$$

or

$$H_P = 90 - 60 + \left( K_{\text{entrance}} + K_{\text{exit}} + f\frac{L}{D} \right) \frac{V^2}{2g}$$

$$= 30 + \left( 0.5 + 1.0 + 0.014\frac{400}{0.2} \right) \frac{Q^2}{2 \times 9.8 \times [\pi \times 0.1^2]^2}$$

$$= 30 + 1520Q^2$$

This equation, the system demand curve, and the characteristic curve $H_P(Q)$ of the pump are now solved simultaneously by trial and error. Actually, the curve could be plotted on the same graph as the characteristic curve, and the point of intersection, the operating point, would provide $Q$. Try $Q = 0.2$ m³/s: $(H_P)_{\text{energy}} = 91$ m, $(H_P)_{\text{char}} \cong 75$ m. Try $Q = 0.15$ m³/s: $(H_P)_{\text{energy}} = 64$ m, $(H_P)_{\text{char}} \cong 75$ m. Try $Q = 0.17$ m³/s: $(H_P)_{\text{energy}} = 74$ m, $(H_P)_{\text{char}} \cong 76$ m. This is our solution. We have

$$Q = 0.17 \text{ m}^3/\text{s}$$

Check the Reynolds number: $\text{Re} = DQ/A\nu = 0.2 \times 0.17/(\pi \times 0.1^2 \times 10^{-6}) = 1.08 \times 10^6$. This is sufficiently large, but marginally so.

The power requirement of the pump is given by Eq. 4.5.26:

$$\dot{W}_P = \frac{Q\gamma H_P}{\eta_P}$$

$$= \frac{0.17 \times 9800 \times 75}{0.65} = 198\,000 \text{ W} \quad \text{or} \quad 198 \text{ kW}$$

where the efficiency $\eta_P = 0.63$ is found from the characteristic curve at $Q = 0.17$ m³/s.
*Note:* Since $L/D > 1000$, minor losses due to the entrance and exit could have been neglected.

## 7.7  UNIFORM TURBULENT FLOW IN OPEN CHANNELS

The last internal flow situation we consider in this chapter is that of steady uniform (constant depth) flow in an open channel, shown in Fig. 7.19. We could already treat this flow by using the Darcy–Weisbach relation presented in Section 7.6.3. In fact, that technique predicts better results than does the more common method we present here. Both methods will be compared in two examples. However, unless otherwise stated, uniform flow in open, rough channels is commonly analyzed using the following less-complicated method.

If the energy equation is applied between two sections of the channel sketched in Fig. 7.19, we obtain

$$0 = \frac{V_2^2 - \cancelto{0}{V_1^2}}{2g} + \frac{p_2 - \cancelto{0}{p_1}}{\gamma} + z_2 - z_1 + h_L \tag{7.7.1}$$

which shows that the head loss is

$$h_L = z_1 - z_2$$
$$= L \sin \theta = LS \tag{7.7.2}$$

where $L$ is the length of the channel between the two sections and $S$ is the slope of the channel, which is assumed small, so that $\sin \theta \approx S$. (Do not confuse $S$ with specific gravity.)

The Darcy–Weisbach equation (7.6.31), with $h_L = LS$ from Eq. 7.7.2, takes the form

$$LS = f\frac{L}{4R}\frac{V^2}{2g} \qquad \therefore RS = \frac{f}{8g}V^2 \tag{7.7.3}$$

where $R$ is the hydraulic radius. Since open channels are usually quite large with large Reynolds numbers, the friction factor is invariably in the completely turbulent regime. Thus the equation above is written as

$$V = C\sqrt{RS} \tag{7.7.4}$$

> **KEY CONCEPT**  *Since open channels are quite large, with large Reynolds numbers, the friction factor is in the completely turbulent regime.*

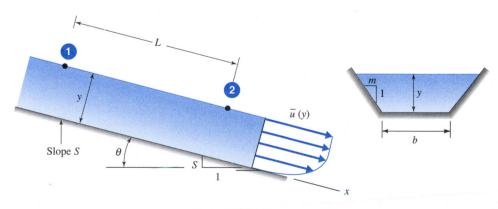

**Fig. 7.19**  Uniform flow in an open channel.

where the *Chezy coefficient, C* is a dimensional constant; the equation above is referred to as the *Chezy equation,* named after Antoine Chezy (1718–1798). The Chezy coefficient is related to the channel roughness and the hydraulic radius (much like *f* is in a pipe) by

$$C = \frac{c_1}{n} R^{1/6} \tag{7.7.5}$$

where the dimensional constant $c_1$ has a value of 1.0 if SI units are used. The dimensionless constant $n$ is directly related to the wall roughness; it is called the *Manning n*, named after Robert Manning (1816–1897). Values for various wall materials are listed in Table 7.3.

The flow rate, which is of primary interest in open-channel flow problems, is found to be

$$Q = \frac{c_1}{n} A R^{2/3} S^{1/2} \qquad c_1 = 1.0 \text{ for SI units} \tag{7.7.6}$$

This is the *Chezy–Manning equation.*

For smooth-surfaced channels, the use of the Chezy–Manning equation is discouraged since it implicitly assumes a rough wall. Calculations for smooth-surfaced channels such as glass or plastic should be based on the Darcy–Weisbach relation using *f* from Fig. 7.13; see Section 7.6.3.

**TABLE 7.3** Average Values[a] of the Manning $n$

| Wall material | Manning n |
|---|---|
| Planed wood | 0.012 |
| Unplaned wood | 0.013 |
| Finished concrete | 0.012 |
| Unfinished concrete | 0.014 |
| Sewer pipe | 0.013 |
| Brick | 0.016 |
| Cast iron, wrought iron | 0.015 |
| Concrete pipe | 0.015 |
| Riveted steel | 0.017 |
| Earth, straight | 0.022 |
| Corrugated metal flumes | 0.025 |
| Rubble | 0.03 |
| Earth with stones and weeds | 0.035 |
| Mountain streams | 0.05 |

[a]The values in this table result in flow rates too large for hydraulic radii greater than about 3 m. The Manning $n$ should be increased by 10 to 15% for such large conduits.

## Example 7.19

The depth of water at 16°C flowing in a 3.6-m-wide rectangular, finished concrete channel is measured to be 1.2 m. The slope is measured to be 0.0016. Estimate the flow rate using (a) the Chezy–Manning equation and (b) the Darcy–Weisbach equation.

### Solution

The hydraulic radius is calculated to be

$$R = \frac{A}{P} = \frac{yb}{2y + b} = \frac{1.2 \times 3.6}{2 \times 1.2 \times 3.6} = 0.72 \text{ m}$$

(a) Using the Chezy–Manning equation, with $n = 0.012$ from Table 7.3 and $c = 1$, we have

$$Q = \frac{1}{n} AR^{2/3}S^{1/2}$$

$$= \frac{1}{0.012} \times (1.2 \times 3.6) \text{ m}^2 \times (0.72)^{2/3} \text{ m}^{2/3} \times 0.0016^{1/2} = 11.57 \text{ m}^3/\text{s}$$

(b) The relative roughness is, using a low value $e = 0.00045$ m (it is finished concrete) shown on the Moody diagram:

$$\frac{e}{4R} = \frac{0.00045}{4 \times 0.72} = 0.00016$$

Assuming a completely turbulent flow, the Moody diagram gives the friction factor as

$$f = 0.013$$

The Darcy–Weisbach equation (7.7.3) then yields the velocity as follows:

$$\therefore V = \left(\frac{8RgS}{f}\right)^{1/2}$$

$$= \left(\frac{8 \times 0.72 \text{ m} \times 9.81 \text{ m/s}^2 \times 0.0016}{0.013}\right)^{1/2} = 2.64 \text{ m/s}$$

The flow rate is calculated as

$$Q = VA = 2.64 \times 1.2 \times 3.6 = 11.4 \text{ m}^3/\text{s}$$

These two values are within 2%, an acceptable engineering tolerance for this type of problem. That found using the Moody diagram is considered to be more accurate, however.

## Example 7.20

A 1.0-m-diameter concrete pipe transports 20°C water at a depth of 0.4 m. If the slope is 0.001, find the flow rate using (a) the Chezy–Manning equation and (b) the Darcy–Weisbach equation.

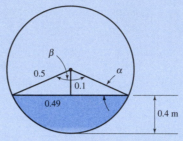

**Fig. E7.20**

### Solution

From the sketch of the pipe in Fig. E7.20 the following are calculated:

$$\alpha = \sin^{-1}\frac{0.1}{0.5} = 11.54°$$

$$\therefore \beta = 180 - 2 \times 11.54 = 156.9°$$

$$\therefore A = \pi \times 0.5^2 \times \frac{156.9}{360} - 0.49 \times 0.1 = 0.2933 \text{ m}^2$$

$$P = 2\pi \times 0.5 \times \frac{156.9}{360} = 1.369 \text{ m}$$

The hydraulic radius is found, using the above calculations, to be

$$R = \frac{A}{P} = \frac{0.2933}{1.369} = 0.2142 \text{ m}$$

(a) The Chezy–Manning equation yields, with $n$ from Table 7.3 and $c_1 = 1.0$,

$$Q = \frac{1.0}{n} AR^{2/3}S^{1/2} = \frac{1.0}{0.015} \times 0.2933 \times 0.2142^{2/3} \times 0.001^{1/2} = 0.22 \text{ m}^3/\text{s}$$

(b) The relative roughness is, using a relatively rough value for concrete pipe from Fig. 7.13, as suggested by Table 7.3, of $e = 2.0$ mm,

$$\frac{e}{4R} = \frac{2}{4 \times 214.2} = 0.0023$$

Assuming completely turbulent flow, the Moody diagram yields

$$f = 0.025$$

The Darcy–Weisbach equation (7.7.3) then gives the following:

$$\therefore V = \left(\frac{8RgS}{f}\right)^{1/2} = \left(\frac{8 \times 0.2142 \times 9.81 \times 0.001}{0.025}\right)^{1/2} = 0.820 \text{ m/s}$$

The flow rate is

$$Q = VA = 0.820 \times 0.2933 = 0.24 \text{ m}^3/\text{s}$$

This is within 8% of the result above, an acceptable tolerance for this type of problem. The second method, which is more difficult to apply, is considered to be more accurate, however.

## 7.8 SUMMARY

The laminar entrance lengths for a pipe and wide channel are, respectively,

$$\frac{L_E}{D} = 0.065 \, \text{Re} \qquad \frac{L_E}{h} = 0.04 \, \text{Re} \qquad\qquad (7.8.1)$$

For a high-Reynolds-number turbulent pipe flow the entrance length is

$$\frac{L_E}{D} = 120 \qquad\qquad (7.8.2)$$

For laminar flow in a pipe and a wide channel the pressure and friction factor are, respectively,

$$\Delta p = \frac{8\mu V L}{r_0^2} \qquad f = \frac{64}{\text{Re}} \qquad\qquad \text{pipe} \qquad (7.8.3)$$

$$\Delta p = \frac{12\mu V L}{a^2} \qquad f = \frac{48}{\text{Re}} \qquad\qquad \text{channel} \quad (7.8.4)$$

where $a$ is the channel height.

The torque required to rotate an inner cylinder with the outer cylinder fixed is

$$T = \frac{4\pi\mu r_1^2 r_2^2 L w_1}{r_2^2 - r_1^2} \qquad\qquad (7.8.5)$$

In a turbulent flow the head loss is calculated using

$$h_L = f \frac{L}{D} \frac{V^2}{2g} \qquad\qquad (7.8.6)$$

where $f$ is found using the Moody diagram of Fig. 7.13. Minor losses are included using

$$h_L = K \frac{V^2}{2g} \qquad\qquad (7.8.7)$$

where many loss coefficients $K$ are listed in Table 7.2.

To include a pump in a piping system when the flow rate is not known, it is necessary to have pump characteristic curves, such as those of Fig. E7.18. Example 7.18 illustrated the procedure.

The flow rate in an open channel is most often estimated using the equation

$$Q = \frac{c_1}{n} A R^{2/3} S^{1/2} \qquad c_1 = 1.0 \text{ for SI units} \qquad (7.8.8)$$

where $n$ is found in Table 7.3.

## Fundamentals of Engineering Exam Review Problems

**7.1** A parabolic profile is measured in a pipe flow. The flow is:

I. Laminar      **(A)** I and III
II. Developed     **(B)** I, II, and III
III. Steady        **(C)** I, II, and IV
IV. Symmetric    **(D)** I, II, III, and IV

**7.2** The velocity profile between parallel plates is calculated to be $Vy/h$, where $y$ is measured from the lower plate and $h$ is the distance between the plates. We know that

I. The flow is laminar
II. The lower plate is moving with velocity $V$ and the other is stationary
III. Lower plate is stationary and the other is moving with velocity $V$
IV. The flow is steady

**(A)** I, III, and IV      **(B)** II and III
**(C)** I and III          **(D)** I and IV

**7.3** A liquid flows in a pipe at a Reynolds number of 4000.

**(A)** The flow is laminar.
**(B)** The flow is turbulent.
**(C)** The flow is transitory, osscilating between laminar and turbulent.
**(D)** The flow could be any of the above.

**7.4** In a turbulent flow in a pipe, the head loss:

**(A)** Varies with the velocity squared
**(B)** Is directly proportional to the flow rate
**(C)** Decreases with increase with Reynolds number
**(D)** Is directly proportional to the length of the pipe

**7.5** The curves for the friction factor $f$ on the Moody diagram become horizontal for sufficiently large Reynolds numbers because:

**(A)** Wall roughness elements project through the Wall viscous layer.
**(B)** The wall viscous layer completely covers the wall roughness elements.
**(C)** The viscous effects become dominant in the flow.
**(D)** Inertial effects cease to become significant in the flow.

**7.6** The pressure drop over 15 m of 2-cm-diameter galvanized iron pipe is measured to be 60 Pa. If the pipe is horizontal, estimate the flow rate of water. Use $\nu = 10^{-6}$ m²/s.

**(A)** 6.82 L/s       **(B)** 2.18 L/s
**(C)** 0.682 L/s     **(D)** 0.218 L/s

**7.7** Water flows in an 8-cm-diameter pipe down a 30° incline and the pressure remains constant. Estimate the average velocity in the cast iron pipe. Use $\nu = 10^{-6}$ m²/s.

**(A)** 0.055 m/s
**(B)** 0.174 m/s
**(C)** 1.75 m/s
**(D)** 5.5 m/s

**7.8** A liquid with a density of 900 kg/m³ flows straight down in a 6-cm-diameter cast iron pipe. Estimate the pressure rise over 20 m of pipe if the average velocity is 4 m/s. Assume $\nu = 8 \times 10^{-6}$ m²/s.

**(A)** 250 kPa
**(B)** 100 kPa
**(C)** 77 kPa
**(D)** 10.2 kPa

**7.9** The flow of water occurs in a square cast iron conduit, 4 cm on a side. If the horizontal conduit transports 0.02 m³/s, estimate the pressure drop over 40 m. Use $\nu = 10^{-6}$ m²/s.

**(A)** 162 Pa        **(B)** 703 Pa
**(C)** 1390 Pa      **(D)** 1590 Pa

**7.10** A new design of a valve is to be tested. Which of the following parameters is the most important if liquid benzene flows through the valve?

**(A)** Froude number
**(B)** Reynolds number
**(C)** Mach number
**(D)** Euler number

**7.11** A water system is to be installed in a community that is quite hilly. The design engineer must be certain that:

**(A)** The hydraulic grade line must always be above the pipe line.
**(B)** The energy grade line must always be above the pipe line.
**(C)** The stagmation pressure must remain positive in the pipe line
**(D)** The elevation of the pipe must not be allowed to be negative.

**7.12** Water flows in a 2.4-m-wide, rectanglular, finished concrete channel at a depth of 80 cm. If the slope is 0.002, the flow rate is nearest

**(A)** 2.2 m³/s
**(B)** 3.4 m³/s
**(C)** 4.6 m³/s
**(D)** 6.2 m³/s

## PROBLEMS

### Laminar or Turbulent Flow

**7.13** Calculate the maximum average velocity $V$ with which 20°C water can flow in a pipe in the laminar state if the critical Reynolds number (Re = $VD/\nu$) at which transition occurs is 2000; the pipe diameter is:

(a) 2 m    (b) 2 cm    (c) 2 mm

**7.14** Water at 20°C flows in a wide river. Using a critical Reynolds number (Re = $Vh/\nu$) at which transition occurs of 1500, calculate the average velocity $V$ that will result in a laminar flow if the depth $h$ of the river is:

(a) 4 m    (b) 1 m    (c) 0.3 m

**7.15** Water at 10°C is flowing in a thin sheet in a parking lot at a depth of 0.5 cm with an average velocity of 0.45 m/s. Is the flow laminar or turbulent?

**7.16** Water is flowing, apparently quite placidly, in a river that is 20 m wide and 1.4 m deep. A leaf floating in the river is observed to move 1 m in 2 s. Is the flow laminar or turbulent? See Problem 7.14 for the definition of the Reynolds number.

**7.17** Flow occurs in a 2-cm-diameter pipe. What is the maximum velocity that can occur for water at 20°C for a laminar flow if:

(a) Re = 2000?    (b) Re = 40 000?

### Entrance and Developed Flow

**7.18** Calculate the laminar entrance length in a 4-cm-diameter pipe if $2 \times 10^{-4}$ m³/s of water is flowing at:

(a) 10°C        (b) 20°C
(c) 40°C        (d) 80°C

**7.19** A laminar flow is to occur in an experimental facility with 20°C air flowing through a 4-cm-diameter pipe. Calculate the average velocity, the inviscid core length, and the entrance length if the Reynolds number is:

(a) 1000        (b) 80 000

**7.20** A 6-cm-diameter pipe originates in a tank and delivers 0.025 m³/s of water at 20°C to a receiver 50 m away. Is the assumption of developed flow acceptable?

**7.21** A laboratory experiment is designed to create a laminar flow in a 2-mm-diameter tube shown in Fig P.7.21. Water flows from a reservoir through the tube. If 18 L is collected in 2 h, can the entrance length be neglected?

**7.22** Air at 23°C is used as the working fluid in a parallel-plate research project. If the plates are 1.2 cm apart, how long is the longest possible entrance length for laminar flow? What is the shortest entrance length?

**7.23** Air at 25°C can exist in either the laminar state or the turbulent state (a trip wire near the entrance is used to make it turbulent) for flow in a 6-cm-diameter pipe in a research lab. If the average velocity is 5 m/s, compare the length of the entrance region for the laminar flow with that of the turbulent flow.

**7.24** Water at 20°C flows with an average velocity of 0.2 m/s from a reservoir through a 4-cm-diameter pipe. Estimate the inviscid core length and the entrance length if the flow is:

(a) Laminar
(b) Turbulent

**7.25** Draw an incremental control volume with length $\Delta x$ and radius $r_0$ and show that for a laminar flow $(\Delta p/\Delta x)_{\text{entrance}} > (\Delta p/\Delta x)_{\text{developed}}$.

**7.26** Explain the pressure variations observed for a turbulent flow in Fig. 7.3 for:

(a) A high-Re flow (Re > 300 000)
(b) A low-Re flow (Re $\simeq$ 10 000)
(c) An intermediate-Re flow

**Fig. P7.21**

## Laminar Flow in a Pipe

**7.27** Define $p_k = p + \gamma h$ as the kinetic pressure and write Eq. 7.3.5 or 7.3.11 in terms of $p_k$. Can we let $dp_x/dx = \Delta p_k/L$, where $L$ is the length over which $\Delta p_k$ is measured? If so, express $u(r)$ in terms of $\Delta p_k/L$.

**7.28** Verify that Eq. 7.3.13 is, in fact, correct.

**7.29** A pressure drop of 0.5 kPa occurs over a section of 2-cm-diameter pipe transporting water at 20°C. Determine the length of the horizontal section if the Reynolds number is 1600. Also, find the shear stress at the wall and the friction factor.

**7.30** Find the angle $\theta$ of the 10-mm-diameter pipe of Fig. P7.30 in which water at 40°C is flowing with Re = 1500 such that no pressure drop occurs. Also, find the flow rate.

**Fig. P7.30**

**7.31** A liquid is pumped through a 2-cm-diameter pipe at a flow rate of 12 L/min. Calculate the pressure drop in a 10-m horizontal section if the liquid is:
**(a)** SAE-10W oil at 20°C
**(b)** Water at 20°C
**(c)** Glycerine at 40°C
Is the assumption of laminar flow acceptable?

**7.32** A liquid flows with no pressure drop in a vertical 2-cm-diameter pipe. Find the flow rate if, assuming a laminar flow, the liquid is:
**(a)** Water at 5°C
**(b)** SAE-30W oil at 25°C
**(c)** Glycerine at 20°C
Is the assumption of laminar flow acceptable?

**7.33** A laminar flow is to exist in a pipe transporting 3.3 L/s of SAE-10W oil at 20°C. What is the maximum allowable diameter? What is the pressure drop over 10 m of horizontal pipe for this diameter?

**7.34** Estimate the flow rate through the smooth pipe shown in Fig. P7.34. How long is the entrance region? Assume a laminar flow.

**Fig. P7.34**

**7.35** A manufacturer of small-diameter tubes wishes to know if the diameters are, in fact, accurate. An experimental setup, like that in Fig. P7.34, is used with a 4-m-long horizontal tube transporting 20°C water with a head of 4 m. If 3.4 L of water is collected in 60 min, what is the inside diameter of the tube, neglecting the effect of the entrance region? Is the effect of entrance region actually negligible?

**7.36** Air at 20°C flows in a horizontal 2-cm-diameter pipe. Calculate the maximum pressure drop in a 10-m section for a laminar flow. Assume that $\rho = 1.27$ kg/m³.

**7.37** Water at 20°C flows in the 4-mm-diameter pipe of Fig. P7.37. The pressure rise over the 10-m section is 6 kPa. Find the Reynolds number of the flow and the wall shear stress. Assume laminar flow.

**Fig. P7.37**

**7.38** A research experiment requires a laminar flow of air at 20°C through a 10-cm-diameter pipe at a Reynolds number of 40 000. What maximum velocity is to be expected? What would be the pressure drop over a 10-m horizontal length of developed flow? How long would the entrance region be? Use $\rho = 1.2$ kg/m³.

**7.39** Calculate the radius where a pitot probe must be placed in the laminar liquid flow of Fig. P7.39 so that the flow rate is given by $\pi R^2 \sqrt{2gH}$.

**Fig. P7.39**

**7.40** A laminar flow of water at 20°C occurs in a vertical 2-mm-diameter pipe. Calculate the flow rate if the pressure is constant. Is it reasonable to assume a laminar flow?

**7.41** Water flows downward at a rate of 4.0 L/min in a 40-mm-diameter vertical pipe. (a) Determine the pressure drop over a distance of 10 meters. (b) Calculate the friction head loss per unit length. (c) What is the shear stress at the pipe wall? Use $\mu = 1.14 \times 10^{-6} \, \text{N} \cdot \text{s/m}^2$.

**7.42** Find the radius in a developed laminar flow in a pipe where:
**(a)** The velocity is equal to the average velocity.
**(b)** The shear stress is equal to one-half the wall shear stress.

**7.43** Find the ratio of the total flow rate through a pipe of radius $r_0$ to the flow rate through an annulus with inner and outer radii of $r_0/2$ and $r_0$. Assume developed laminar flow with the same pressure gradient.

**7.44** A laminar flow of water at 16°C is obtained in a research lab at Re = 20 000 in a horizontal 5-cm-diameter pipe. Calculate the head loss in a 10-m section of developed flow, the wall shear stress, and the length of the entrance region.

**7.45** Water at 20°C flows between the two concentric horizontal pipes of Fig. P7.45 with diameters of 2 cm and 3 cm. A pressure drop of 100 Pa is measured over a 10-m section of developed laminar flow. Find the flow rate and the shear stress on the inner pipe.

**Fig. P7.45**

**7.46** Air at 20°C is to flow in the annulus between two concentric, horizontal pipes, with respective diameters of 2 cm and 3 cm, such that a pressure drop of 10 Pa occurs over a 10-m length. Find the average velocity and the shear stress at the inner pipe. Assume a developed laminar flow.

**7.47** Fluid flows in the annulus between two concentric horizontal pipes. The inner pipe is maintained at a higher temperature than the outer pipe so that the viscosity in the annulus cannot be assumed to be constant but $\mu = \mu(T)$. What differential equation would be solved to yield $u(r)$ assuming a developed laminar flow?

**7.48** Show that the velocity distribution of Example 7.2 approaches that of pipe flow as $r_1 \to 0$ and approaches that of parallel-plate flow as $r_1 \to r_2$.

---

### Laminar Flow Between Parallel Plates

**7.49** A flow occurs in a horizontal channel 1 cm × 40 cm with Re = 2000. Calculate the flow rate if the fluid is:
**(a)** Water at 15°C
**(b)** Atmospheric air at 15°C

**7.50** A board 1 m × 1 m that weighs 40 N moves down the incline shown in Fig. P7.50 with a velocity of $V = 0.2$ m/s. Estimate the viscosity of the fluid if $\theta$ is:
**(a)** 20°
**(b)** 30°

**Fig. P7.50**

**7.51** Water at 20°C exists between the plate and the surface of Problem 7.50. Calculate the velocity of the plate for an angle $\theta$ of:
 (a)  20°
 (b)  30°

**7.52** Water at 20°C flows down a 20° incline at a depth of 6 mm and a width of 50 m. Calculate the flow rate and the Reynolds number assuming laminar flow. Also, find the maximum velocity and the wall shear.

**7.53** Water at 20°C flows down a 100-m-wide parking lot on a slope of 0.00015 at a depth of 10 mm. Determine the flow rate and the Reynolds number assuming laminar flow. Also, calculate the friction factor and the wall shear.

**7.54** A pressure drop of 50 Pa is measured over a 60-m length of a 90 cm × 2 cm rectangular horizontal channel transporting 20°C air. Calculate the maximum flow rate and the associated Reynolds number. Use $\rho = 1.2$ kg/m$^3$.

**7.55** In Fig. P7.55, a pressure difference $p_A - p_B$ is measured to be 96 kPa. Find the friction factor for the wide channel assuming laminar flow. The flow direction is unknown.

**Fig. P7.55**

**7.56** A slit with dimensions 0.5 mm × 100 mm exists in the 5-cm-thick side of a pressure vessel containing SAE-10W oil at 27°C and 4.2 MPa. What is the maximum flow rate that can exist from the slit? Assume developed laminar flow.

**7.57** Air flows between the parallel plates as shown in Fig. P7.57. Find the pressure gradient such that:
 (a)  The shear stress at the upper surface is zero.
 (b)  The shear stress at the lower surface is zero.
 (c)  The flow rate is zero.
 (d)  The velocity at $y = 2$ mm is 4 m/s.

**Fig. P7.57**

**7.58** A pressure gradient of $-20$ Pa/m exists in 50°C air flowing between horizontal parallel plates spaced 6 mm apart. Find the velocity of the upper plate so that:
 (a)  The shear stress at the upper plate is zero.
 (b)  The shear stress at the lower plate is zero.
 (c)  The flow rate is zero.
 (d)  The velocity at $y = 2$ mm is 2 m/s.

**7.59** Using the Navier–Stokes equations (a) determine an expression for the velocity profile of a pressure driven flow between two parallel horizontal plates placed 10-mm apart. The lower plate is fixed, and the upper plate is moving with a constant velocity of 2 m/s. Assume steady, laminar incompressible flow of oil with viscosity of 0.4 N·s/m$^2$. (b) Based on your answer in part (a), determine the pressure gradient needed to cause a zero velocity midway between the plates.

**7.60** Consider steady, laminar, fully developed, incompressible flow between two inclined parallel plates separated by a distance $h$. The upper plate of Fig. P7.60 is moving upward with a constant velocity $U$, and the lower plate is fixed. Starting with the Navier–Stokes equations, obtain an expression for the velocity in the fluid between the two plates. Consider that $dp/dx =$ constant with the fluid flowing downward.

**Fig. P7.60**

**7.61** Oil with $\mu = 5 \times 10^{-3}$ kg·s/m$^2$ fills the concentric space between the rod and the surface shown in Fig. P7.61. Find the force $F$ if $V = 15$ m/s. Assume that $dp/dx = 0$.

**Fig. P7.61**

**7.62** Calculate the torque $T$ necessary to rotate the rod shown in Fig. P7.62 at 30 rad/s if the fluid filling the gap is SAE-10W oil at 20°C. Assume a linear velocity profile.

**Fig. P7.62**

**7.63** Oil with $\mu = 0.01$ N·s/m$^2$ fills the gap shown in Fig. P7.63. Estimate the torque necessary to rotate the disk shown, assuming a linear velocity profile. Is the assumption of laminar flow valid? Use $S = 0.86$.

**Fig. P7.63**

**7.64** Approximate the torque necessary to rotate the inner 20-cm-diameter cylinder shown in Fig. P7.64. SAE-30W oil at 20°C fills the gap. Assume a linear velocity profile.

**Fig. P7.64**

**7.65** Find the torque needed to rotate the cone shown in Fig. P7.65 if oil with $\mu = 0.01$ N·s/m$^2$ fills the gap as shown. Assume a linear velocity profile.

**Fig. P7.65**

**7.66** To create a high-Reynolds-number flow, the channel setup shown in Fig. P7.66 was proposed by Prof. John Foss of Michigan State University. It is a pressurized channel, thereby avoiding fatal leaks that always are present in a suction channel. (An upstream fan would produce blade vortices that make a high Re impossible to attain.) Estimate the power requirement of the 70%-efficient fan if the channel is 1.2 m wide and Re = 7000.

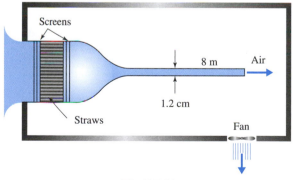

**Fig. P7.66**

## Laminar Flow Between Rotating Cylinders

**7.67** A long cylinder of radius $R$ rotates in a large container of liquid. What is the velocity distribution in the laminar flow? Calculate the torque needed to rotate a 1-m-long, 5-cm-diameter cylinder at 1000 rpm if the liquid is water at 10°C. Assume a laminar flow.

**7.68** SAE-10W oil at 40°C fills the gap between two concentric, 40-cm-long cylinders with respective radii of 2 cm and 3 cm. What torque is necessary to rotate the inner cylinder at 3000 rpm if the outer cylinder is fixed? What power is required? Verify that Eq. 7.5.17 is, in fact, the velocity profile.

**7.69** A torque of 0.015 N·m is required to rotate a 4-cm-radius cylinder inside a fixed 5-cm-radius cylinder at 40 rad/s. The concentric cylinders are 50 cm long. Calculate the viscosity of the fluid. Use $S = 0.9$. Verify that Eq. 7.5.17 is, in fact, the velocity profile.

**7.70** Find an expression for the torque necessary to rotate the outer cylinder if the inner cylinder of Fig. 7.6 is fixed.

**7.71** Rework Problem 7.62 using the velocity distribution of Eq. 7.5.17 and calculate the percent error assuming the linear velocity profile.

### Turbulent Flow

**7.72** Time average the differential continuity equation for an incompressible flow and show that two continuity equations result: the instantaneous continuity equation

$$\frac{\partial u'}{\partial x} + \frac{\partial v'}{\partial y} + \frac{\partial w'}{\partial z} = 0$$

and the time-average continuity equation

$$\frac{\partial \bar{u}}{\partial x} + \frac{\partial \bar{v}}{\partial y} + \frac{\partial \bar{w}}{\partial z} = 0$$

**7.73** Find an expression for the difference between the time-average acceleration $\overline{Du}/Dt$ and the quantity $D\bar{u}/Dt$ using the fact that

$$\overline{u'\frac{\partial u'}{\partial x}} + \overline{v'\frac{\partial u'}{\partial y}} + \overline{w'\frac{\partial u'}{\partial z}} = \frac{\partial}{\partial x}\overline{u'^2} + \frac{\partial}{\partial y}\overline{u'v'} + \frac{\partial}{\partial z}\overline{u'w'}$$

**7.74** Prove that the equation written in Problem 7.73 is indeed a fact. (*Hint:* Use the instantaneous continuity equation.)

**7.75** Show that the time-averaged x-component Navier–Stokes equation results in

$$\rho\frac{\partial}{\partial y}\overline{u'v'} = -\frac{\partial \bar{p}}{\partial x} + \mu\frac{\partial^2 \bar{u}}{\partial y^2}$$

for developed flow in a wide horizontal channel. Using $\bar{\tau}_{lam} = \mu(\partial\bar{u}/\partial y)$ and $\bar{\tau}_{turb} = -\rho\overline{u'v'}$, write the time-averaged Navier–Stokes equation in terms of stresses.

**7.76** The velocity components at a point in a turbulent flow are given in Table 1. Find $\bar{u}, \bar{v}$, $\overline{u'^2}, \overline{v'^2}$, and $\overline{u'v'}$ at that point.

**Table 1**

| $t$ (s) | $u$ (m/s) | $v$ (m/s) | $t$ (s) | $u$ (m/s) | $v$ (m/s) |
|---|---|---|---|---|---|
| 0.00 | 16.1 | 1.6 | 0.06 | 17.1 | −1.4 |
| 0.01 | 25.7 | −5.4 | 0.07 | 28.6 | 6.7 |
| 0.02 | 10.6 | −8.6 | 0.08 | 6.7 | −5.2 |
| 0.03 | 17.3 | 3.5 | 0.09 | 19.2 | −8.2 |
| 0.04 | 5.2 | 4.1 | 0.10 | 21.6 | 1.5 |
| 0.05 | 10.2 | − 6.0 | | | |

**7.77** Over a small radial distance in a developed turbulent flow, the time-average velocity is as given in the following table. The pressure drop in a 9-m horizontal section is measured to be 400 Pa. Find $\overline{u'v'}$ at $r = 0.21$ m. Air with $\rho = 1.85$ kg/m³ and $\nu = 1.46 \times 10^{-5}$ m²/s is flowing.

| $r$ (m) | 0.18 | 0.19 | 0.21 | 0.22 |
|---|---|---|---|---|
| $\bar{u}$ (m/s) | 25 | 23 | 21 | 18.2 |

**7.78** If at $r = 0.21$ m for the data of Problem 7.77, we measure $\overline{u'^2} = 29$ m²/s² and $\overline{v'^2} = 14$ m²/s², what are the magnitudes of the eddy viscosity, the correlation coefficient, and the mixing length?

**7.79** The velocity components are measured at a point in a laminar flow to be as shown in Fig. P7.79. Find $\overline{u'v'}$, $\eta$, $l_m$, and $K_{uv}$ if $d\bar{u}/dy = -10$ s⁻¹ at the point.

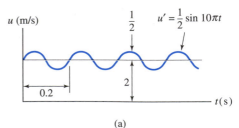

(a)

(*continued on next page*)

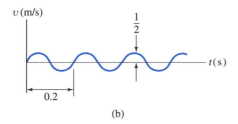

**Fig. P7.79**

**7.80**   A 20-cm-diameter pipe with $e = 0.26$ mm transports water at 20°C. Determine if the pipe is
smooth or rough if the average velocity is:

**(a)**   0.02 m/s     **(b)**   0.2 m/s     **(c)**   2 m/s

**7.81**   SAE-30 oil at 40°C is transported in a 10-cm-
diameter pipe at an average velocity of 6 m/s.
What is the largest size allowed for the roughness
element if the pipe is hydraulically smooth?

**7.82**   Calculate the maximum velocity in the pipe of:

**(a)**   Problem 7.80a
**(b)**   Problem 7.80c

**7.83**   The velocity profile for water at 20°C in a turbulent flow in a 10-cm-diameter smooth pipe is
approximated by $\bar{u} = 9.2y^{1/7}$ m/s. Find:

**(a)**   The wall shear
**(b)**   The velocity gradient $d\bar{u}/dy$ at the wall
**(c)**   The pressure gradient
**(d)**   The value of $\eta$ at $r = 2.5$ cm

**7.84**   Water at 20°C flows in a 12-cm-diameter horizontal pipe at the rate of 70 L/s. Find the constant $n$ in
the exponent of Eq. 7.6.19. What is the maximum
velocity?

**7.85**   Show that the kinetic-energy correction factor
is 1.10 for $n = 5$ and 1.03 for $n = 10$ using
$u = u_{max}(y/r_0)^{1/n}$ in a circular pipe.

**7.86**   Water at 20°C flows in a 10-cm-diameter pipe
with an average velocity of 10 m/s. Using
$u = u_{max}(y/r_0)^{1/n}$ with $n = 7$, plot the viscous shear
and the turbulent shear as a function of $r$. Also,
find $d\bar{p}/dx$.

**7.87**   SAE-10W oil at 10°C is transported in a smooth
80-cm-diameter pipe at the rate of 1.2 m³/s.
**(a)**   Find the Reynolds number.
**(b)**   Find the friction factor.
**(c)**   Find the maximum velocity using Eq. 7.6.20.
**(d)**   Find the viscous wall layer thickness.
**(e)**   Compare part (c) with the solution using the
logarithmic velocity profile.

**7.88**   Let the pipe in Problem 7.87 be a cast iron pipe
(Fig. 7.13 provides a value for $e$). Estimate the
maximum velocity using the logarithmic velocity
profile.

**7.89**   A pressure drop of 10 kPa is measured with gages
placed 5 m apart on a smooth horizontal 10-cm-
diameter pipe transporting water at 40°C.
Estimate:
**(a)**   The wall shear
**(b)**   The maximum velocity
**(c)**   The average velocity
**(d)**   The Reynolds number
**(e)**   The flow rate

**7.90**   A 12-cm-diameter horizontal pipe transports
SAE-10 oil at 10°C. Calculate the wall shear, the
average velocity, and the flow rate if the pressure
drop over a 10-m section of pipe is measured
to be:
**(a)**   5 kPa     **(b)**   20 kPa     **(c)**   200 kPa

**7.91**   Make a linear plot (not a semilog plot) of the
velocity profile of the flow in Problem 7.89 using:
**(a)**   The log profile
**(b)**   The power-law profile

## Turbulent Flow in Pipes and Conduits

**7.92**   Water at 20°C flows in an 8-cm-diameter plastic pipe
with a flow rate of 20 L/s. Determine the friction factor using (a) the Moody diagram, and (b) Eq. 7.6.26.

**7.93**   A flow rate of 0.03 m³/s of 15°C water occurs in a
10-cm-diameter cast iron pipe. Determine the friction factor using (a) the Moody diagram, and
(b) one of the equations (7.6.26) through (7.6.28).

**7.94**   Water at 20°C flows in a 4-cm-diameter cast iron
pipe. Determine the friction factor, using the
Moody diagram, if the average velocity is:
**(a)**   0.025 m/s          **(b)**   0.25 m/s
**(c)**   2.5 m/s            **(d)**   25 m/s

**7.95**    A flow rate of 0.02 m³/s occurs in a 10-cm-diameter wrought iron pipe. Using the Moody diagram, calculate the pressure drop over a 100-m horizontal section if the pipe transports:

(a)   Water at 20°C
(b)   Glycerine at 60°C
(c)   SAE-30W at 30°C
(d)   Kerosene at 10°C

Compare each answer with that using Eq. 7.6.29.

**7.96**    Water at 15°C flows in a 3.8-cm-diameter pipe with a flow rate of 1.65 L/s. Using the Moody diagram, determine the head loss in a 180-m section if the pipe is:

(a)   Cast iron          (b)   Galvanized iron
(c)   Wrought iron       (d)   Plastic

**7.97**    A mass flux of 1.2 kg/s occurs in a 10-cm-diameter plastic pipe at 20°C and 500 kPa absolute. Assume an incompressible flow and, using the Moody diagram, find the pressure drop in a 100-m section of pipe if the fluid flowing is:

(a)   Air
(b)   Carbon dioxide
(c)   Hydrogen

Compare each answer with that using Eq. 7.6.29.

**7.98**    SAE-30W oil flows at the rate of 0.08 m³/s in a 15-cm-diameter horizontal galvanized iron pipe. Find the pressure drop in 100 m if the temperature of the oil is:

(a)   0°C          (b)   30°C
(c)   60°C         (d)   90°C

Compare each answer with that using Eq. 7.6.29.

**7.99**    Select the material, listed in Fig. 7.13, from which each of the following pipes is probably constructed. Each 5-cm-diameter pipe is tested with water at 20°C using a flow rate of 400 L/min. The following pressure drops are measured over a 10-m length of horizontal pipe:

(a)   Pipe 1: 36 kPa
(b)   Pipe 2: 24 kPa
(c)   Pipe 3: 19 kPa

**7.100**   Water at 10°C flows up a 30° incline in a 6-cm-diameter plastic pipe with a flow rate of 8.3 L/s. Find the pressure change over a 90-m length of pipe.

**7.101**   Water at 40°C flows in a horizontal section of 5-cm-diameter wrought iron pipe with a flow rate of 0.02 m³/s. Does the pipe behave as a smooth pipe, or is the roughness significant?

**7.102**   An 80-cm-diameter concrete pipe is to transport storm water at 20°C at a rate of 5 m³/s. What pressure drop is to be expected over a 100-m section of horizontal pipe?

**7.103**   A pressure drop of 500 kPa is not to be exceeded over a 200-m length of horizontal 10-cm-diameter cast iron pipe. Calculate the maximum flow rate if the fluid is:

(a)   Water at 20°C
(b)   Glycerine at 20°C
(c)   SAE-10W oil at 20°C
(d)   Kerosene at 20°C

**7.104**   A pressure drop of 200 kPa is not to be exceeded over a 100-m length of horizontal 4-cm-diameter pipe. Estimate the maximum flow rate if water at 20°C is transported and the pipe is:

(a)   Cast iron
(b)   Wrought iron
(c)   Plastic

**7.105**   Neglecting all losses except that due to wall friction, estimate the flow rate through the pipe shown in Fig. P7.105 if the diameter is:

(a)   4 cm
(b)   8 cm
(c)   12 cm
(d)   16 cm

**Fig. P7.105**

**7.106**   A pressure drop of 400 Pa is allowable in gas flow in a 400-m horizontal section of 12-cm-diameter wrought iron pipe. If the temperature and pressure are 40°C and 200 kPa absolute, find the maximum mass flux if the gas is:

(a)   Air          (b)   Carbon dioxide
(c)   Hydrogen

**7.107**   A pressure drop of 210 kPa is not to be exceeded over a 180-m length of horizontal 1.2-m-diameter concrete pipe transporting water at 15°C. What flow rate can be accommodated? Use:

(a)   The Moody diagram
(b)   Equation 7.6.30

**7.108** Estimate the size of plastic tubing that should be selected if 0.002 m³/s of fluid is to be transported such that the pressure drop does not exceed 200 kPa in a 100-m horizontal section. The fluid is:
   **(a)** Water at 20°C
   **(b)** Glycerine at 60°C
   **(c)** Kerosene at 20°C
   **(d)** SAE-10W oil at 40°C

**7.109** Select a concrete pipe size that will transport 5 m³/s of 20°C water so that the head loss does not exceed 20 m in a 300-m horizontal pipe section. Use:
   **(a)** The Moody diagram
   **(b)** Equation 7.6.31

**7.110** A farmer wishes to siphon 10°C water from a lake 1200 m from a field positioned a distance of 3 m below the surface of the lake. What size of drawn tubing should be selected if 400 L of water is desired each minute? Use:
   **(a)** The Moody diagram
   **(b)** Equation 7.6.31
Neglect all losses except that due to wall friction. Is the exiting kinetic energy also negligible?

**7.111** Atmospheric air at 30°C is to be transported through a square sheet metal (smooth) conduit at the rate of 4 m³/s. What should be the conduit dimensions so that the head loss does not exceed 10 m over a horizontal length of 200 m?

**7.112** Water at 20°C is transported through a 2 cm × 4 cm smooth conduit and experiences a pressure drop of 80 Pa over a 2-m horizontal length. What is the flow rate?

**7.113** A plastic conduit 4 cm × 10 cm transports water at 20°C. If a pressure drop of 100 Pa is measured by gages spaced 5 m apart on a horizontal section, find the flow rate.

**7.114** An open rectangular 1.2-m-wide concrete (use $e = 1.5$ mm) channel transports 20°C water from a reservoir to a location 10 000 m away. Using the Moody diagram, estimate the flow rate if the channel is on a slope of 0.0015 and the water depth is:
   **(a)** 0.3 m          **(b)** 0.6 m
   **(c)** 0.9 m

## Minor Losses

**7.115** If the loss coefficient for a sudden expansion is based on the exit velocity $V_2$, determined the loss coefficient in terms of $A_1$ and $A_2$.

**7.116** Explain, with reference to Eq. 3.4.15, why the high- and low-pressure regions of Fig. 7.14 exist. Also, sketch a velocity profile from the inside corner of the bend to the outside of the bend along a line from $B$ to $C$ as sketched in Fig. P7.116. Explain why a secondary flow results after the bend.

**Fig. P7.116**

**7.117** For each system shown in Fig. P7.117, find $p_2$ if $Q = 0.02$ m³/s of air at 20°C and $p_1 = 50$ kPa.

(a)

(b)

**Fig. P7.117**

**7.118** Replace the sudden enlargement of Problem 7.117 with a 20° expansion angle and rework the problem.

**7.119** For each system shown in Fig. P7.119, estimate the loss coefficient based on $V_2$ using the data of Fig. 7.16.

(a)             (b)

(c)

**Fig. P7.119**

**7.120**   The flow rate is measured to be 3.3 L/s from the pipe shown in Fig. P7.120. Find the loss coefficient of the valve. Neglect wall friction.

**7.121**   The flow rate is measured to be 6 L/s in the pipe shown in Fig. P7.121. Find the loss coefficient of the valve if $H$ is:
(a)   4 cm          (b)   8 cm

**Fig. P7.120**

**Fig. P7.121**

## Simple Piping Systems

**7.122**   Find the flow rate from the pipe shown in Fig. P7.122. Sketch the EGL and the HGL.

**Fig. P7.122**

**7.123**   Water at 21°C flows from a 10-cm-diameter 360-m length of cast iron horizontal pipe that is attached to a reservoir with a square-edged entrance. A screwed globe valve that controls the flow is half open. Find the flow rate if the reservoir elevation above the pipe exit is:
(a)   4.5 m          (b)   9 m          (c)   18 m

**7.124**   Estimate the flow rate to be expected through the plastic siphon shown in Fig. P7.124 if the diameter is:
(a)   4 cm          (b)   8 cm          (c)   12 cm

**Fig. P7.124**

**7.125**   For the cast-iron piping shown in Fig. P7.125, calculate the flow rate and minimum pressure and sketch the HGL and the EGL if:
(a)   $H = 10$ m
(b)   $H = 20$ m
(c)   $H = 30$ m

**Fig. P7.125**

**7.126**   Water at 35°C flows from a bathtub through a 2.5-cm-diameter plastic pipe to a large air-filled drain pipe. There are two screwed elbows in the 10 m of pipe. If the drain pipe is 0.8 m below the water level in the tub, estimate how long it will take to drain 10 L of water.

**7.127**   A 3-cm-diameter plastic tube with screwed elbows is used to siphon the water as shown in Fig. P7.127. Estimate the maximum height $H$ for which the siphon will operate.

**Fig. P7.127**

**7.128** A 12-L lawn sprayer is filled with 8 L of 20°C water. It is 1.2 m high and has an 8-mm-diameter copper tube ($e \approx 0$) that reaches the bottom (it is a little short). A 1.2-m-long 5-mm-diameter smooth hose is attached to the copper tube. The hose ends with a 2-mm-diameter nozzle. If the sprayer is pressurized to 100 kPa, estimate the initial velocity exiting the nozzle.

**7.129** What is the maximum flow rate through the pipe shown in Fig. P7.129 if the elevation difference of the reservoir surfaces is:
(a) 80 m
(b) 150 m
(c) 200 m

**7.130** A 9.4-mm-diameter 60-m-long plastic pipe transports 20°C water from a spring to a pond 3 m below, similar to the situation shown in Problem 7.129. It is observed that the water alternates between a relatively fast moving stream to a relatively slow moving stream. Explain this phenomenon with supporting calculations.

**7.131** Water at 15°C is to be pumped through 300 m of cast iron pipe from a reservoir to a device that is 10 m above the reservoir surface. It is to enter the device at 200 kPa. Screwed components include two elbows, a square-edged entrance, and an angle valve. If the flow rate is to be 16 L/s, what pump power is needed (assume 80% efficiency) if the pipe diameter is:
(a) 4 cm  (b) 8 cm  (c) 12 cm

Valve (K = 1.0)

Water 15°C

2000 m

80-cm-dia. concrete pipe

**Fig. P7.129**

**7.132** What pump power (85% efficient) is needed for a flow rate of 0.01 m³/s in the pipe shown in Fig. P7.132? What is the greatest distance from the left reservoir that the pump can be located?

Water 15°C

400 m

T

90-cm-dia. cast iron pipe

**Fig. P7.133**

**7.134** What pump power (75% efficient) is needed in the piping system shown in Fig. P7.134? What is the greatest distance from the reservoir that the pump can be located?

el 80 m

el 10 m

4-cm-dia. drawn tubing

Water 15°C

P

800 m

**Fig. P7.132**

**7.133** A flow rate of 2 m³/s exists in the pipe shown in Fig. P7.133. What is the expected power output of the turbine (85% efficient) if the elevation difference of the reservoir surfaces is:
(a) 20 m  (b) 60 m  (c) 100 m

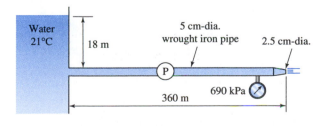

Water 21°C

18 m

5 cm-dia. wrought iron pipe

2.5 cm-dia.

P

360 m

690 kPa

**Fig. P7.134**

**7.135** The pump shown in Fig. P7.135 has the characteristic curves shown in Example 7.18.
  **(a)** Estimate the flow rate and the power required by the pump.
  **(b)** Sketch the EGL and the HGL.
  **(c)** If cavitation is possible, determine the maximum distance from the reservoir to locate the pump.

**Fig. P7.135**

**7.136** Reverse the flow direction in Problem 7.135 and redo the problem.

**7.137** A turbine replaces the pump of Problem 7.135. Estimate the power output if $\eta_T = 0.88$. The turbine's characteristic curve is $H_T = 0.8Q$, where $H_T$ is measured in meters and $Q$ in L/s.

**7.138** The pump shown in Fig. P7.138 has characteristic curves shown in Example 7.18. Estimate the flow rate and
  **(a)** Calculate the pump power requirement.
  **(b)** Calculate the pressure at the pump inlet.
  **(c)** Calculate the pressure at the pump outlet.
  **(d)** Sketch the EGL and the HGL.

**FIG. P7.138**

**7.139** Reverse the flow direction in Problem 7.138 and redo the problem.

**7.140** In the filtration system shown in Fig. P7. 140, water is circulated steadily through a filter using a pump. The pump characteristic curve is given by $H_p = 10 + 12Q - 150Q^2$, where $H_p$ is in meters and $Q$ is in m³/s. A 10-cm-diameter pipe is used in the system. The total length of pipe used is 60 m, and the friction factor of the pipe is $f = 0.04$. Determine the flow rate and power input to the pump. All bends are 90 miter bend without vanes.

**Fig. P7.140**

**7.141** A turbine with a characteristic curve shown in Fig. P7.141 is inserted in the pipeline. Calculate the turbine power output. Assume that $\eta_T = 0.90$.

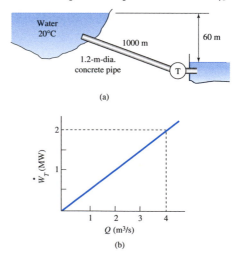

**Fig. P7.141**

**Open-Channel Flow**

**7.142** Using a control volume surrounding a finite length of the liquid in a channel flowing at constant depth, find the average shear stress on the walls if water flows in a 3-m-wide rectangular channel at a depth of 1.8 m on a 0.001 slope.

**7.143** Using the approach mentioned in Problem 7.142, determine the average shear stress on the portion of the wall of a 40-cm-diameter circular conduit in contact with the water, which is flowing at a constant depth of 10 cm. The slope is 0.0016.

**7.144** Calculate the flow rate in a 2-m-wide rectangular, planed wood channel on a slope of 0.001 if the depth is 60 cm. Use:
**(a)** The Chezy–Manning equation
**(b)** The Darcy–Weisbach equation

**7.145** Water flows in a 1.8-m-diameter finished concrete conduit on a 0.0012 slope. Predict the flow rate if the depth of flow is:
**(a)** 1.8 m minus a little
**(b)** 1.7 m
**(c)** 0.9 m
**(d)** 0.45 m
**(e)** 0.15 m

**7.146** For the channel shown in Fig. P7.146, find the flow rate and average velocity if $S = 0.001$ using:
**(a)** The Chezy–Manning equation
**(b)** The Darcy–Weisbach equation

**Fig. P7.146**

**7.147** At what depth will 5 m³/s of water flow in a 2-m-wide rectangular brick channel with $S = 0.001$? Use:
**(a)** The Chezy–Manning equation
**(b)** The Darcy–Weisbach equation

**7.148** The cross section of a straight river is approximated as shown in Fig. P7.148. At what depth will 100 m³/s of water flow? The slope is 0.001.

**Fig. P7.148**

**7.149** For the channel shown in Fig. P7.149, if $S = 0.0016$, find the flow depth if $Q = 10$ m³/s and the channel is constructed with:
**(a)** Planed wood
**(b)** Brick

**Fig. P7.149**

**7.150** Water flows in a 1.2-m-diameter sewer pipe at a flow rate of 660 L/s. Estimate the depth if the slope is 0.001.

**7.151** Water flows in a 80-cm-diameter sewer pipe at a flow rate of 0.2 m³/s. Determine the depth if the slope is 0.001.

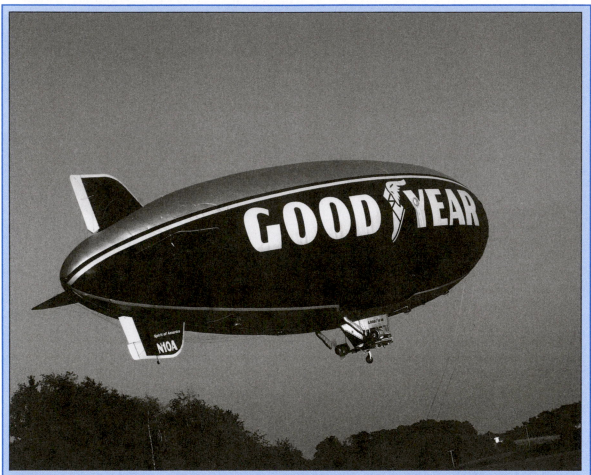

A blimp is a lighter-than-air-vessel that is streamlined to reduce the drag forces encountered during its forward motion. Large blimps, perhaps 1000 m long, could be used as cruise ships to tour the globe; sea sickness would be avoided! (Courtesy of The Goodyear Tire & Rubber Company)

# External Flows

## Outline

## Chapter Objectives

The objectives of this chapter are to:

▲ Discuss separated and attached flows.

▲ Introduce the lift and drag coefficients.

▲ Determine the drag on a variety of bodies.

▲ Study the influence of vortex shedding and streamlining.

▲ Determine when cavitation is present.

▲ Calculate the lift and drag on airfoils.

▲ Superimpose simple potential flows to construct flows of interest.

▲ Analyze both laminar and turbulent boundary layers on a flat plate.

▲ Provide numerous examples and problems demonstrating how the various quantities of interest are determined for the many exernal flows discussed in this chapter.

## 8.1 INTRODUCTION

The study of external flows is of particular importance to the aeronautical engineer in the analysis of airflow around the various components of an aircraft. In fact, much of the present knowledge of external flows has been obtained from studies motivated by such aerodynamic problems. There is, however, substantial interest by other engineers in external flows; the flow of fluid around turbine blades, automobiles, buildings, athletic stadiums, smokestacks, spray droplets, bridge abutments, submarine pipelines, river sediment, and red blood cells suggests a variety of phenomena that can be understood only from the perspective of external flows.

It is a difficult task to determine the flow field external to a body and the pressure distribution on the surface of a body, even for the simplest geometry. To discuss this subject, consider low-Reynolds-number flows (Re < 5, or so) and high-Reynolds-number flows (Re > 1000). Low-Reynolds-number flows, called *creeping flows* or *Stokes flows*, seldom occur in engineering applications (flow around spray droplets and red blood cells, lubrication in small gaps, and flow in porous media would be exceptions) and are not presented in this book; they are left to the specialist. We will direct our attention to high-Reynolds-number flows only. We do, however, show a Stokes flow in Fig. 8.1.

High-Reynolds-number flows can be subdivided into three major categories: (1) incompressible immersed flows involving such objects as automobiles, helicopters, submarines, low-speed aircraft, take-off and landing of commercial aircraft, buildings, and turbine blades; (2) flows of liquids that involve a free surface as experienced by a ship or a bridge abutment; and (3) compressible flows involving high-speed objects ($V > 100$ m/s) such as aircraft, missiles, and bullets. We will focus our attention on the first category of flows in this chapter and

**KEY CONCEPT** *Low-Reynolds-number flows seldom occur in engineering applications.*

**Fig. 8.1**   Flow past a circular cylinder at Re = 0.16. The flow is from left to right. It resembles superficially the pattern of potential flow. The flow of water is shown by aluminum dust. (Photograph by Sadatoshi Taneda. From Album of Fluid Motion, 1982, The Parabolic Press, Stanford, California.)

consider cases in which the object is far from a solid boundary or other objects. The flow becomes significantly influenced by the presence of a boundary or another object, as shown in Fig. 8.2; in part (d) the slender object must be at least five body lengths below the free surface before free surface effects can be neglected. Flows such as those shown in Fig. 8.2 are not included in an introductory presentation.

High-Reynolds-number incompressible immersed flows are divided into two categories: flows around blunt bodies and flows around streamlined bodies, as displayed in Fig. 8.3. The boundary layer (see Section 3.3.2) near the stagnation point is a laminar boundary layer and, for a sufficiently large Reynolds number, undergoes transition downstream to a turbulent boundary layer, as shown; the flow may separate from the body and form a **separated region**, a region of recirculating flow, as shown for the blunt body, or it simply leaves the streamlined body at the trailing edge (there may be a small separated region here). The **wake**, which is characterized by a *velocity defect* is a growing (diffusion) region and trails the body, as shown. The boundaries of the wake, the separated region, and the turbulent boundary layer are quite time dependent; in the sketch the time-average location of the wake is shown by the dashed lines. Shear stresses due to viscosity are concentrated in the thin boundary layer, the separated region, and the wake; outside these regions the flow is approximated by an inviscid flow.

From the figure it could be assumed that the separated region does not exchange mass with the free stream since mass does not cross a streamline. When viewed instantaneously, however, the separation streamline is highly time dependent, and due to this unsteady character, the separated region is able to exchange mass slowly with the free stream.

Flow past a cylinder, 116, 131, 190

**Separated region:** *A region of recirculating flow.*

**Wake:** *A region of velocity defect that grows due to diffusion.*

Flow over an airfoil, 649

(a)

(b)

(c)

(d)

**Fig. 8.2**  Examples of complicated immersed flows: (a) flow near a solid boundary; (b) flow between two turbine blades; (c) flow around an automobile; (d) flow near a free surface.

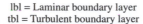

lbl = Laminar boundary layer
tbl = Turbulent boundary layer

(a)

(b)

**Fig. 8.3** Flow around a blunt body and a streamlined body.

**Drag:** *The force the flow exerts in the direction of the flow.*

**Lift:** *The force the flow exerts normal to the direction of flow.*

Several comments should be made regarding the separated region and the wake. The separated region eventually closes; the wake keeps diffusing into the main flow and eventually disappears as its area becomes exceedingly large (the fluid regains the free-stream velocity). Time-average streamlines do not enter a separated region; they do enter a wake. The separated region is always submerged within the wake.

Flow around a blunt object is usually treated empirically, as was done for a turbulent flow in a conduit. We follow this procedure here. We are interested primarily in the **drag**, the force the flow exerts on the body in the direction of the flow. The **lift**, which acts normal to the direction of flow, will be of interest for airfoil shapes, as presented in Section 8.4. The actual details of the flow field are seldom of interest and are not presented in this introductory text. We present the drag $F_D$ and lift $F_L$ in terms of dimensionless coefficients: the *drag coefficient* and *lift coefficient*, defined as

$$C_D = \frac{F_D}{\frac{1}{2}\rho V^2 A} \qquad C_L = \frac{F_L}{\frac{1}{2}\rho V^2 A} \qquad\qquad (8.1.1)$$

where $A$ is most often the projected area (projected on a plane normal to the direction of the flow); for airfoil shapes, the area is based upon the chord (see Fig. 8.4). Drag coefficients for several common shapes are presented in Section 8.3.1. Since the drag on a blunt object is dominated by the flow in the separated region, there is little interest in studying the boundary layer growth on the front part of a blunt body and the associated viscous shear at the wall. Hence interest is focused on the empirical data that provide the drag coefficient.

Flow around a streamlined body—the separated region is insignificantly small or nonexistent—provides the motivation for detailed study of the laminar and turbulent boundary layers. A boundary layer that develops on a plane streamlined surface, such as an airfoil, is usually sufficiently thin that the curvature of the surface can be ignored and the problem can be treated as a boundary layer developing on a flat plate with a nonzero pressure gradient. We will provide a detailed study of flow on a flat plate with zero pressure gradient; once that problem is understood, the influence of a pressure gradient can be discussed. If the flow in the boundary layer on a streamlined body can be determined, the drag can be calculated, since the drag is a result of the shear stress and pressure force acting on the body surface.

Outside the boundary layer there exists an inviscid, *free-stream flow*, as shown in Figs. 8.3 and 8.4. Initially, we will assume that the free-stream flow is known. Before the velocity profile in the boundary layer can be determined,

> **KEY CONCEPT**   *The drag on a blunt object is dominated by the flow in the separated region.*

 Separated flow over an airfoil, 167

> **KEY CONCEPT**   *The inviscid, free-stream flow exists outside the boundary layer.*

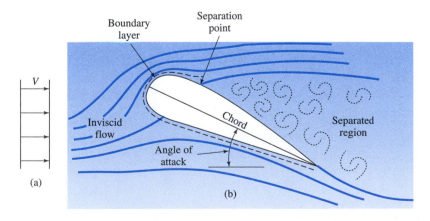

**Fig. 8.4**   Streamlined body that is stalled.

it is necessary that the inviscid flow solution be known. It is found by completely ignoring the boundary layer, since it is so thin, and solving the appropriate inviscid equations. The inviscid flow solution is then used to provide the lift on the body, and the two quantities used in the boundary layer flow solution: the pressure gradient and the velocity at the boundary. With the inviscid flow known and the boundary layer flow determined, quantities of interest can be obtained in the flow about a streamlined body.

## 8.2 SEPARATION

**Chord:** *A line connecting the trailing edge with the nose.*

**Stall:** *Flow condition where separation occurs on a streamlined body near the forward portion.*

> **KEY CONCEPT** *For blunt bodies, separation is unavoidable at high Reynolds numbers.*

Separated flow past sharp edges, 662, 664, 666

Before we present the empirical data associated with the flow around blunt bodies, the general nature of separation will be discussed. Separation occurs when the main stream flow leaves the body, resulting in a separated region of flow, as sketched in Fig. 8.3a. When separation occurs on a streamlined body near the forward portion of an airfoil, as it will with a sufficiently large angle of attack (the angle the oncoming flow makes with the **chord**, a line connecting the trailing edge with the nose), the flow situation is referred to as **stall**, as shown in Fig. 8.4. Stall is highly undesirable on aircraft at cruise conditions and leads to inefficiencies when it occurs on turbine blades. It is used, however, to provide the high drag needed when landing an aircraft, or in certain maneuvers by stunt planes. For blunt bodies, however, separation is unavoidable at high Reynolds numbers and its effect must be understood.

The location of the separation point is dependent primarily on the geometry of the body; if the body has an abrupt change in geometry, such as that shown in Fig. 8.5, separation will occur at, or near, the abrupt change; however, it will also occur upstream on the flat surface as shown. In addition, reattachment will occur at some location, as shown. Let us establish the criterion used in predicting the location of the separation point on a surface with no abrupt geometry change.

Consider the flow on the flat surface just before the step of Fig. 8.5. The region near the forward separation point is enlarged and shown in Fig. 8.6; the $y$-coordinate is normal to the wall and the $x$-coordinate is measured along the wall. Downstream of the separation point the $x$-component velocity near the wall is in the negative $x$-direction and thus at the wall $\partial u/\partial y$ must be negative.

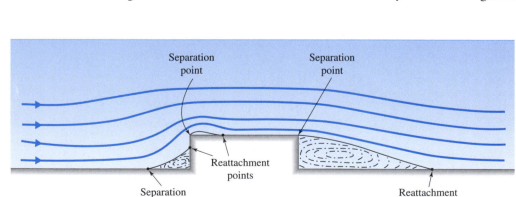

**Fig. 8.5**    Separation due to abrupt geometry changes.

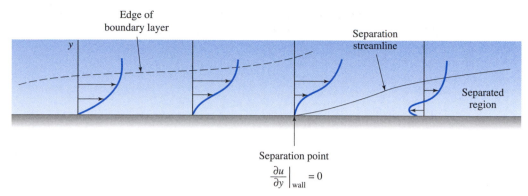

**Fig. 8.6**   Flow separation on a flat surface due to an adverse pressure gradient.

Upstream of the separation point the $x$-component velocity near the wall is in the positive $x$-direction, demanding that $\partial u/\partial y$ at the wall be positive. Hence we conclude that the separation point is defined as that point where $(\partial u/\partial y)_{\text{wall}} = 0$.

Observe that separation on the flat surface occurs as the flow is approaching a stagnation region where the velocity is low and the pressure is high. As the flow approaches the stagnation region the pressure increases, that is, $\partial p/\partial x > 0$; the pressure gradient is positive. Since separation is often undesirable, a positive pressure gradient is called an *adverse pressure gradient*; a negative gradient is a *favorable pressure gradient*. In general, the effect of an adverse pressure gradient results in decreasing velocities in the streamwise direction; if an adverse pressure gradient acts on a surface over a sufficient distance, separation may result. This is true even if the surface is a flat plate, such as the wall of a diffuser. See Section 8.6.7 for further discussion.

In addition to geometry and pressure gradient, several other parameters influence separation. These include the Reynolds number as a very important parameter, with the wall roughness, the *free-stream fluctuation intensity*[1] (the intensity of the disturbances that exist away from the boundary), and the wall temperature having less but occasionally significant influence.[2] Visualize, for example, flow around a sphere; at sufficiently low Reynolds numbers no separation will occur. As the Reynolds number increases to a particular value, separation will occur over a small area at the rear; this area will become larger and larger as the Reynolds number increases until at a sufficiently large Reynolds number no additional increase in separation area will be observed. The boundary layer before separation will still be laminar. An interesting phenomenon takes place as the boundary layer before separation becomes turbulent; there is a sudden movement of the separation point to the rear of the sphere, resulting in a substantial reduction in the separation area and thus the drag. This phenomenon is explained by comparing the velocity profile of a laminar boundary layer with that of a turbulent boundary layer, as displayed in Fig. 8.7. As was true in a pipe flow, the turbulent profile has a much larger gradient near the wall (much larger

**KEY CONCEPT**  *The separation point is defined as that point where $(\partial u/\partial y)_{\text{wall}} = 0$.*

**KEY CONCEPT**  *As the boundary layer before separation becomes turbulent, the separation point moves to the rear.*

---

[1] Free-stream fluctuation intensity is defined as $\sqrt{\overline{u'^2}}/V$, where $u'$ is the fluctuation. A value of 0.001 is quite low, and 0.1 is quite high.

[2] If a fluid is flowing past a body that is rigidly supported, the vibration level of the support system will also influence the separation phenomenon. External sound waves can also be significant.

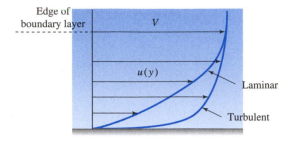

**Fig. 8.7** Comparison of laminar and turbulent velocity profiles.

wall shear stress) and thus the momentum of the fluid near the wall is substantially larger in the turbulent boundary layer. For a given geometry a greater distance is required to reduce the velocity near the wall to zero, resulting in the movement of the separation point to the rear, as can be observed in Fig. 8.8, where both spheres are moving with the same velocity (the sphere in (b) has sandpaper attached in the nose region). In Fig. 8.8a it is observed that separation occurs on the front half of the sphere, in a region of favorable pressure gradient. This separation is due to the centrifugal effects as the fluid moves around the sphere. This phenomenon of drag reduction is observed in the drop in the drag coefficient curves for a sphere and a cylinder, presented in the following section.

## 8.3 FLOW AROUND IMMERSED BODIES

### 8.3.1 Drag Coefficients

From our study of dimensional analysis, recall that for a steady, incompressible flow in which gravity, thermal, and surface tension effects are neglected, the primary flow parameter that influences the flow is the Reynolds number; other occasionally important parameters include the relative wall roughness and the free-stream velocity fluctuation intensity.

The drag coefficient curves for two bodies that do not exhibit sudden geometric changes will be presented; the drag coefficients for the smooth sphere and the long smooth cylinder are shown in Fig. 8.9 (on page 354) over a large range of Reynolds numbers. At $\text{Re} < 1$ creeping flow with no separation results. For the sphere, this creeping flow problem has been solved, with the result that

$$C_D = \frac{24}{\text{Re}} \qquad \text{Re} < 1 \tag{8.3.1}$$

Separation is observed at $\text{Re} \simeq 10$ over a very small area on the rear of the body. The separated area increases as the Reynolds number increases until $\text{Re} \simeq 1000$, where the separated region ceases to enlarge; during this growth of the separated region the drag coefficient decreases. At $\text{Re} = 1000$, 95% of the drag is due to

(a)

(b)

Flow around immersed bodies, 652, 657, 694

**Fig. 8.8**  Effect of boundary layer transition on separation: (a) laminar boundary layer before separation; (b) turbulent boundary layer before separation. (U.S. Navy photograph.)

**Fig. 8.9** Drag coefficients for flow around a long cylinder and a sphere. (See E. Achenbach, *J. Fluid Mech.*, Vol. 46, 1971, and Vol. 54, 1972.)

form drag (the drag force due to the pressure acting on the body), and 5% is due to frictional drag (the drag force due to the shear stresses acting on the body).

The drag coefficient curve is relatively flat for smooth bodies over the range $10^3 < \text{Re} < 2 \times 10^5$. The boundary layer before the point of separation is laminar and the separated region is as shown in Fig. 8.8a. At $\text{Re} \simeq 2 \times 10^5$, for a smooth surface and with low free-stream fluctuation intensity, the boundary layer before separation undergoes transition to a turbulent state and the increased momentum in the boundary layer "pushes" the separation back, as shown in Fig. 8.8b, with a substantial decrease (a 60 to 80% drop) in the drag. If the surface is rough (dimples on a golf ball) or the free stream has high free-stream fluctuation intensity, the drop in the $C_D$ curve may occur at $\text{Re} \simeq 8 \times 10^4$. Since a lower drag is usually desirable, surface roughness is often added; the dimples on the golf ball may increase the flight distance by 50 to 100%.

After the sudden drop in drag, the $C_D$ curve is observed to again increase with increased Reynolds number. Experimental data are not readily available for $\text{Re} > 10^6$ for a sphere and $\text{Re} > 6 \times 10^7$ for a cylinder; however, a value of $C_D = 0.2$ for a sphere at large Reynolds number appears acceptable. Some engineers use $C_D = 0.4$ for cylinders at large Reynolds numbers; however, the data presented here suggests that that is too low. Additional experimental data is needed.

For cylinders of finite length and for elliptic cylinders, the drag coefficients are presented in Table 8.1. The finite-length cylinders are assumed to have two free ends. If one end is fixed to a solid surface, its length must be doubled when using Table 8.1. Blunt objects with sudden geometry changes have separated regions that are relatively insensitive to the Reynolds number; the drag coefficients for some common shapes are given in Table 8.2.

> **KEY CONCEPT** *Dimples on the golf ball may increase the flight distance by 50 to 100%.*

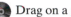 Drag on a golf ball, 265

**TABLE 8.1**   Drag Coefficients of Finite-Length Circular Cylinders[a] with Free Ends[b] and of Infinite-Length Elliptic Cylinders

| Circular cylinder | | Elliptic cylinder[c] | | |
|---|---|---|---|---|
| $\dfrac{Length}{Diameter}$ | $\dfrac{C_D}{C_{D\infty}}$ | $\dfrac{Major\ axis}{Minor\ axis}$ | Re | $C_D$ |
| $\infty$ | 1 | 2 | $4 \times 10^4$ | 0.6 |
| 40 | 0.82 | 4 | $10^5$ | 0.46 |
| 20 | 0.76 | 4 | $2.5 \times 10^4$ to $10^5$ | 0.32 |
| 10 | 0.68 | 8 | $2.5 \times 10^4$ | 0.29 |
| 5 | 0.62 | 8 | $2 \times 10^5$ | 0.20 |
| 3 | 0.62 | | | |
| 2 | 0.57 | | | |
| 1 | 0.53 | | | |

[a] $C_{D\infty}$ is the drag coefficient for the infinite-length circular cyulinder obtained in Fig. 8.9.
[b] If one end is fixed to a solid surface, double the length of the cylinder.
[c] Flow is in the direction of the major axis.

**TABLE 8.2** Drag Coefficients for Various Blunt Objects

| Object | | Re | $C_D$ |
|---|---|---|---|
| Square cylinder → □ $w$ | $L/w = \begin{cases} \infty \\ 1 \text{ (cube)} \end{cases}$ | $> 10^4$ <br> $> 10^4$ | 2.0 <br> 1.1 |
| rounded corners ($r = 0.2w$) | $L/w = \infty$ | $10^5$ | 1.2 |
| Rectangular plates →\| $w$ | $L/w = \begin{cases} \infty \\ 20 \\ 5 \\ 1 \end{cases}$ | $> 10^3$ <br> $> 10^3$ <br> $> 10^3$ <br> $> 10^3$ | 2.0 <br> 1.5 <br> 1.2 <br> 1.1 |
| Circular cylinder →□ $D$ | $L/D = \begin{cases} 0.1 \text{ (disk)} \\ 4 \\ 7 \end{cases}$ | $> 10^3$ <br> $> 10^3$ <br> $> 10^3$ | 1.1 <br> 0.9 <br> 1.0 |
| Semicircular cylinder → D <br> → C | | $> 10^4$ <br> $> 10^4$ | 2.2 <br> 1.2 |
| Semicircular shell → ) <br> → ( | | $2 \times 10^4$ <br> $2 \times 10^4$ | 2.3 <br> 1.1 |
| Equilateral cylinders → ◁ <br> → ▷ | 2.0 <br> 1 | $> 10^4$ <br> $> 10^4$ | 2.0 <br> 1.4 |
| Cone → ◁ $\alpha$ | $\alpha = \begin{cases} 30° \\ 60° \\ 90° \end{cases}$ | $> 10^4$ <br> $> 10^4$ <br> $> 10^4$ | 0.6 <br> 0.8 <br> 1.2 |
| Solid hemisphere → ▶ <br> → ◀ | | $> 10^4$ <br> $> 10^4$ | 1.2 <br> 0.4 |
| Hollow hemisphere → D <br> → C | | $> 10^4$ <br> $> 10^4$ | 1.4 <br> 0.4 |
| Parachute | | $> 10^7$ | 1.4 |
| Automobile <br> 1920 <br> Modern, with square corners <br> Modern, with rounded corners | <br> — <br> — <br> — | <br> $> 10^5$ <br> $> 10^5$ <br> $> 10^5$ | <br> 0.80 <br> 0.30 <br> 0.29 |
| Van | | $> 10^5$ | 0.42 |
| Bicycle, upright rider <br> racing, bent over <br> racing, drafting | | | 1.1 <br> 0.9 <br> 0.5 |
| Semitruck, standard <br> with streamlined deflector <br> with deflector and gap seal | | | 0.96 <br> 0.76 <br> 0.70 |

# Example 8.1

A square sign, 3 m × 3 m, is attached to the top of a 18-m-high pole which is 30 cm in diameter (Fig. E8.1). Approximate the maximum moment that must be resisted by the base for a wind speed of 30 m/s.

3 m

3 m

18 m

30 cm

**Fig. E8.1**

## Solution

The maximum force $F_1$ acting on the sign occurs when the wind is normal to the sign; it is

$$F_1 = C_D \times \tfrac{1}{2}\, \rho V^2 A$$
$$= 1.1 \times \tfrac{1}{2} \times 1.2 \text{ kg/m}^3 \times 30^2 \text{ m}^2/\text{s}^2 \times 3^2 \text{ m}^2 = 5346 \text{ N}$$

where $C_D$ is found in Table 8.2 and we use the standard value $\rho = 1.2$ kg/m$^3$ since it was not given. The force $F_2$ acting on the cylindrical pole is (using the projected area as $A = 18 \times 0.3$ m$^2$)

$$F_2 = C_D \times \tfrac{1}{2}\, \rho V^2 A$$
$$= 0.8 \times \tfrac{1}{2} \times 1.2 \times 30^2 \times 5.4 = 2332.8 \text{ N}$$

where $C_D$ is found from Fig. 8.9 with Re $= 30 \times 0.3/1.46 \times 10^{-5} = 6.2 \times 10^5$, assuming a high-intensity fluctuation level (i.e., a rough cylinder); since neither end is free, we do not use the multiplication factor of Table 8.1.

The resisting moment that must be supplied by the supporting base is

$$M = d_1 F_1 + d_2 F_2$$
$$= 19.5 \times 5346 + 9 \times 2332.8 = 125.24 \text{ kN} \cdot \text{m}$$

assuming that the forces act at the centers of their respective areas.

# Example 8.1a    Forces on an Airfoil, 924–931

## Example 8.2

Determine the terminal velocity of a 30-cm-diameter smooth sphere ($S = 1.02$) if it is released from rest in (a) air at 20°C and (b) water at 20°C.

### Solution

(a) When terminal velocity is reached by a falling object, the weight of the object is balanced by the drag force acting on the object. Using $\Sigma F = 0$ and Eq. 8.1.1, we have

$$W = F_D$$

$$\therefore \gamma_{\text{sphere}} \times \frac{4}{3}\pi R^3 = C_D \times \frac{1}{2}\rho V^2 A$$

Using $\gamma_{\text{sphere}} = S\gamma_{\text{water}}$ and projected area $A = \pi R^2$, this becomes

$$S\gamma_{\text{water}} \times \frac{4}{3}\pi R^3 = C_D \times \frac{1}{2}\rho V^2 \pi R^2$$

The velocity can now be expressed as

$$V = \left(\frac{8RS\gamma_{\text{water}}}{3\rho C_D}\right)^{1/2} = \left(\frac{8 \times 0.15\ \text{m} \times 1.02 \times 9800\ \text{N/m}^3}{3 \times 1.20\ \text{kg/m}^3 \times C_D}\right)^{1/2} = \frac{57.7}{\sqrt{C_D}}$$

The Reynolds number should be quite large so $C_D = 0.2$ from Fig. 8.9. Then

$$V = \frac{57.7}{\sqrt{0.2}} = 129\ \text{m/s}$$

We must check the Reynolds number to verify the $C_D$ value assumed. It is

$$\text{Re} = \frac{VD}{\nu} = \frac{129 \times 0.3}{1.6 \times 10^{-5}} = 2.42 \times 10^6$$

This is beyond the end of the curve where data are unavailable; we will assume that the drag coefficient is unchanged at 0.2, so the terminal velocity is 129 m/s.
(b) For the sphere falling in water, we must include the buoyancy force $B$ acting in the same direction as the drag force $F_D$. Hence the summation of forces yields

$$W = F_D + B$$

$$\therefore \gamma_{\text{sphere}} \times \frac{4}{3}\pi R^3 = C_D \times \frac{1}{2}\rho V^2 A + \gamma_{\text{water}} \times \frac{4}{3}\pi R^3$$

This gives

$$(S-1)\gamma_{\text{water}} \times \frac{4}{3}\pi R^3 = C_D \times \frac{1}{2}\rho V^2 \pi R^2$$

Using $\rho = 1000\ \text{kg/m}^3$, there results

$$V = \left(\frac{8R(S-1)\gamma_{\text{water}}}{3\rho C_D}\right)^{1/2} = \left(\frac{8 \times 0.15 \times 0.02 \times 9800}{3 \times 1000 \times C_D}\right)^{1/2} = \frac{0.28}{\sqrt{C_D}}$$

We anticipate the Reynolds number being lower than in part (a), so let's assume that it is in the range $2 \times 10^4 < \text{Re} < 2 \times 10^5$. Then $C_D = 0.5$ and there results

$$V = 0.40\ \text{m/s}$$

This gives a Reynolds number of

$$\text{Re} = \frac{VD}{\nu} = \frac{0.40 \times 0.3}{10^{-6}} = 1.2 \times 10^5$$

This is in the required range, so the terminal velocity is expected to be 0.40 m/s. Of course, if the sphere were roughened (sand glued to the surface), the $C_D$ value would be less and the velocity would be greater.

**Example 8.2a**   Cylinder-wake Virtual Lab, 936–938

### 8.3.2   Vortex Shedding

Long blunt objects, such as circular cylinders, exhibit a particularly interesting phenomenon when placed normal to a fluid flow; vortices or eddies (regions of circulating fluid) are shed from the object, regularly and alternately from opposite sides, as shown in Fig. 8.10. The resulting flow downstream is often referred to as a *Kármán vortex street*, named after Theodor von Kármán (1881–1963). The vortices are shed in the Reynolds number range $40 < \text{Re} < 10\,000$, and are accompanied by turbulence above $\text{Re} = 300$. Photographs of high- and low-Reynolds-number vortex shedding are presented in Fig. 8.11.

> **KEY CONCEPT**   *Vortices are shed regularly and alternately from opposite sides from circular cylinders.*

**Fig. 8.10**   Vortex shedding from a cylinder: (a) vortex shedding; (b) Strouhal number versus Reynolds number. (Roshko, A. (1952-01-01) On the development of turbulent wakes from vortex streets.)

(a)

(b)

**Fig. 8.11**    Vortex shedding at high and low Reynolds numbers: (a) Re = 10 000 (Photograph by Thomas Corke and Hassan Nagib.); (b) Re = 140 (Photograph by Sadatoshi Taneda. From Album of Fluid Motion, 1982, The Parabolic Press, Stanford, California.)

Vortex shedding, 81, 216

Dimensional analysis may be applied to find an expression for the shedding frequency. For high-Reynolds-number flows, that is, flows with insignificant viscous forces, the shedding frequency $f$, in hertz, depends only on the velocity and diameter. Thus $f = f(V, D)$. Using dimensional analysis we can show that $fD/V =$ const. The shedding frequency, expressed as a dimensionless quantity, is expressed as the Strouhal number,

$$\text{St} = \frac{fD}{V} \tag{8.3.2}$$

From the experimental results of Fig. 8.10, we observe that the Strouhal number is essentially constant (0.21) over the range $300 < \text{Re} < 10\,000$; hence, the frequency is directly proportional to the velocity over this relatively large Reynolds number range.

The engineer or architect must be very careful when designing structures, such as towers and bridges, that shed vortices. When a vortex is shed, a small force is applied to the structure; if the frequency of shedding is close to the natural frequency[3] (or one of the harmonics) of the structure, the phenomenon of resonance may occur in which the response to the applied force is multiplied by a large factor. For example, when resonance occurs on a television tower, the deflection of the tower due to the applied force may become so large that the supporting cables fail, leading to collapse of the structure. This has occurred many times, leading to severe damage and numerous deaths and injuries. The collapse of the Tacoma Narrows suspension bridge is undoubtedly the most spectacular failure due to vortex shedding. "Galloping" power lines, in which a power line alternates between the usual catenary and an inverted catenary, is another example that may lead to significant damage; this may occur when a power line has iced up, providing a much larger cross-sectional area for the wind.

> **KEY CONCEPT** *If the phenomenon of resonance occurs, the response to an applied force is multiplied by a large factor.*

## Example 8.3

The velocity of a slow-moving, 30°C air stream is to be measured using a cylinder and a pressure tap located between points $A$ and $B$ on the cylinder in Fig. 8.10. The velocity range is expected to be $0.1 < V < 1$ m/s. What size cylinder should be selected and what frequency would be observed by the pressure-measuring device for $V = 1$ m/s?

### Solution

The Reynolds number should be in the vortex shedding range, say 4000. For the maximum velocity the diameter would be found as follows:

$$4000 = \frac{VD}{\nu}$$

$$= \frac{1.0 \times D}{1.6 \times 10^{-5}}$$

$$\therefore D = 0.064 \text{ m} \qquad \text{Select } D = 6 \text{ cm}$$

At $V = 0.1$ m/s the Reynolds number is $0.1 \times 0.06/1.6 \times 10^{-5} = 375$. Vortex shedding would occur, so this is acceptable. The expected vortex shedding frequency at $V = 1.0$ m/s is found using a Strouhal number from Fig. 8.10 of 0.21. Hence

$$0.21 = \frac{fD}{V}$$

$$= \frac{f \times 0.06}{1.0}$$

$$\therefore f = 3.5 \text{ hertz}$$

 **Example 8.3a**   Example of Vortex Shedding, 195–197

---

[3] The natural frequency is that frequency with which a structure vibrates when given a "knock."

### 8.3.3 Streamlining

**Streamlining:** *Reducing the high pressure at the rear of the object allowing slow-moving surface flow to move rearward.*

> **KEY CONCEPT** *The angle at the trailing edge must not be greater than about 20° for streamlining to be effective.*

If the flow is to remain attached to the surface of a blunt object, such as a cylinder or a sphere, it must move into regions of higher and higher pressure as it progresses to the rear stagnation point. At sufficiently high Reynolds numbers (Re > 10) the slow-moving boundary layer flow near the surface is unable to make its way into the high-pressure region near the rear stagnation point, so it separates from the object. **Streamlining** reduces the high pressure at the rear of the object so that the slow-moving flow near the surface is able to negotiate its way into a slightly higher-pressure region. The fluid may not be able to make it all the way to the trailing edge of the streamlined object, but the separation region will be reduced to only a small percentage of the initial separated region on the blunt object. The included angle at the trailing edge must not be greater than about 20° or the separation region will be too large and the effect of streamlining will be negated. Drag coefficients for streamlined cylinders and spheres are displayed in Fig. 8.9.

When a body is streamlined, the surface area is increased substantially. This eliminates the majority of the pressure drag but increases the shear drag on the surface. To minimize drag, the idea is to minimize the sum of the pressure drag and the shear drag. Consequently, the streamlined body cannot be so long that the shear drag is larger than the pressure drag plus the shear drag for a shorter body. An optimization procedure is required. Such a procedure would lead to a thickness-to-chord length ratio of about 0.25 for a strut.

Obviously, for a low-Reynolds-number flow (Re < 10) the drag is due primarily to shear drag and thus streamlining is unnecessary; it would undoubtedly lead to increased drag since the surface area would be increased.

Finally, it should be pointed out that another advantage of streamlining is that the periodic shedding of vortices is usually eliminated. The vibrations produced by vortex shedding are often undesirable, so streamlining not only decreases drag but can eliminate the vibrations.

 Streamlining, 651

---

## Example 8.4

A strut on a stunt plane traveling at 60 m/s is 4 cm in diameter and 24 cm long. Calculate the drag force acting on the strut as a circular cylinder, and as a streamlined strut, as shown in Fig. E8.4. Would you expect vortex shedding from the circular cylinder?

Fig. E8.4(a)    Fig. E8.4(b)

**Solution**

The Reynolds number associated with the cylinder and the streamlined strut is, assuming that $T = 20°C$,

$$\text{Re} = \frac{VD}{\nu}$$

$$= \frac{60 \times 0.04}{1.5 \times 10^{-5}} = 1.6 \times 10^5$$

Assuming a smooth surface as in (a), the drag coefficient is $C_D = 1.2$ from Fig. 8.9. The drag force is then

$$F_D = C_D \times \frac{1}{2}\rho V^2 A$$

$$= 1.2 \times \frac{1}{2} \times 1.20 \text{ kg/m}^3 \times 60^2 \text{ m}^2/\text{s}^2 \times (0.24 \times 0.04) \text{ m}^2 = 24.9 \text{ N}$$

For the streamlined strut of (b), Fig. 8.9 yields $C_D = 0.04$. The drag force is

$$F_D = C_D \times \frac{1}{2}\rho V^2 A$$

$$= 0.04 \times \frac{1}{2} \times 1.20 \times 60^2 \times (0.24 \times 0.04) = 0.82 \text{ N}$$

This is a reduction of 97% in the drag, a rather substantial reduction.

Vortex shedding is not to be expected on the circular cylinder; the Reynolds number is too high. (See Fig. 8.10.)

## 8.3.4  Cavitation

**Cavitation** is a very rapid change of phase from liquid to vapor which occurs in a liquid whenever the local pressure is equal to or less than the vapor pressure. The first appearance of cavitation is at the position of lowest pressure in a field of flow. Four types of cavitation have been identified:

**Cavitation:**  *A change of phase from liquid to vapor which occurs whenever the local pressure is less than the vapor pressure.*

1. *Traveling cavitation*, which exists when vapor bubbles or cavities are formed, are swept downstream, and collapse.
2. *Fixed cavitation*, which exists when a fixed cavity of vapor exists as a separated region. The separated region may reattach to the body, or the separated region may enclose the rear of the body and be closed by the main flow, in which case it is referred to as *supercavitation.*
3. *Vortex cavitation*, which is found in the high-velocity, and thus low-pressure, core of a vortex, often observed in the tip vortex leaving a propeller.
4. *Vibratory cavitation*, which may exist when a pressure wave moves in a liquid. A pressure wave consists of a pressure pulse, which has a high pressure followed by a low pressure. The low-pressure part of the wave (or vibration) can result in cavitation.

The first type of cavitation, in which vapor bubbles are formed and collapse, is associated with potential damage. The instantaneous pressures resulting from the collapse can be extremely high (perhaps 1400 MPa) and may cause damage to stainless steel components, as happens on the propellers of ships.

Cavitation occurs whenever the *cavitation number* $\sigma$, defined by

$$\sigma = \frac{p_\infty - p_v}{\frac{1}{2}\rho V^2} \tag{8.3.3}$$

is less than the critical cavitation number $\sigma_{crit}$, which depends on the geometry of the body and the Reynolds number. Here $p_\infty$ is the absolute pressure in the undisturbed free stream and $p_v$ is the vapor pressure. As $\sigma$ decreases below $\sigma_{crit}$, the cavitation increases in intensity, moving from traveling cavitation to fixed cavitation to supercavitation.

**TABLE 8.3**   Drag Coefficients for Zero Cavitation Number for Blunt Objects

| Two-dimensional body | | | Axisymmetric body | | |
|---|---|---|---|---|---|
| *Geometry* | $\theta$ | $C_D(0)$ | *Geometry* | $\theta$ | $C_D(0)$ |
| Flat plate | — | 0.88 | Disk | — | 0.8 |
| Circular cylinder | — | 0.50 | Sphere | — | 0.30 |
| | 120 | 0.74 | Cone | 120 | 0.64 |
| Wedge | 90 | 0.64 | | 90 | 0.52 |
| | 60 | 0.49 | | 60 | 0.38 |
| | 30 | 0.28 | | 30 | 0.20 |

The drag coefficient of a body is dependent on the cavitation number and for small cavitation numbers is given by

$$C_D(\sigma) = C_D(0)(1 + \sigma) \tag{8.3.4}$$

where some values of $C_D(0)$ for common shapes are listed in Table 8.3 for Re $\simeq 10^5$.

The hydrofoil, an airfoil-type body that is used to lift a vessel out of the water, is a shape that is invariably associated with cavitation. Drag and lift coefficients and critical cavitation numbers are given in Table 8.4 for a typical hydrofoil with $10^5 <$ Re $< 10^6$, where the Reynolds number is based on the chord length and the area used with $C_D$ and $C_L$ is the chord times the length.

**TABLE 8.4**  Drag and Lift Coefficients and Critical Cavitation Number for a Typical Hydrofoil

| Angle (°) | Lift coefficient $C_L$ | Drag coefficient $C_D$ | Critical cavitation number $\sigma_{crit}$ |
|---|---|---|---|
| −2 | 0.2 | 0.014 | 0.5 |
| 0 | 0.4 | 0.014 | 0.6 |
| 2 | 0.6 | 0.015 | 0.7 |
| 4 | 0.8 | 0.018 | 0.8 |
| 6 | 0.95 | 0.022 | 1.2 |
| 8 | 1.10 | 0.03 | 1.8 |
| 10 | 1.22 | 0.04 | 2.5 |

## Example 8.5

A hydrofoil is to operate 50 cm below the surface of 15°C water at an angle of attack of 8° and travel at 14 m/s. If its chord length is 60 cm and it is 1.8 m long, calculate its lift and drag. Is cavitation present?

**Solution**

The absolute pressure $p_\infty$ is

$$p_\infty = \gamma h + p_{atm}$$

$$= 9810 \times 0.5 + 1.01 \times 10^5 = 105.9 \text{ kPa absolute}$$

The vapor pressure is $p_v = 1.765$ kPa, so

$$\sigma = \frac{p_\infty - p_v}{\frac{1}{2}\rho V^2}$$

$$= \frac{(105.9 - 1.765) \times 10^3}{\frac{1}{2} \times 1000 \times 14^2} = 1.06$$

Answering the last question first, we see that this is less than 1.8; hence cavitation exists.
The lift force is, finding $C_L$ in Table 8.4,

$$F_L = C_L \times \frac{1}{2}\rho V^2 A$$

$$= 1.1 \times \frac{1}{2} \times 1000 \text{ kg/m}^3 \times 14^2 \text{ m}^2/\text{s}^2 \times (0.6 \times 1.8) \text{ ft}^2 = 116.4 \text{ kN}$$

The drag force is, finding $C_D$ in Table 8.4,

$$F_D = C_D \times \frac{1}{2}\rho V^2 A$$

$$= 0.03 \times \frac{1}{2} \times 1000 \times 14^2 \times (0.6 \times 1.8) = 3.17 \text{ kN}$$

### 8.3.5 Added Mass

The previous sections in this chapter have dealt with bodies moving at constant velocity. In this section we consider bodies accelerating from rest in a fluid. When a body accelerates, we say that an unbalanced force acts on the body; not only does the body accelerate but so does some of the fluid surrounding the body. The acceleration of the surrounding fluid requires an added force over and above the force required to accelerate only the body. A relatively simple way of accounting for the fluid mass being accelerated is to add a mass, called the *added mass $m_a$*, to the mass of the body. Summing forces in the direction of motion for a symmetrical body moving in the direction of its axis of symmetry we have, for horizontal motion,

$$F - F_D = (m + m_a)\frac{dV_B}{dt} \qquad (8.3.5)$$

where $V_B$ is the velocity of the body and $F_D$ is the drag force. For an initial acceleration from rest $F_D$ would be zero.

The added mass is related to the mass of the fluid $m_f$ displaced by the body by the relationship

$$m_a = km_f \qquad (8.3.6)$$

where $k$ is an added mass coefficient. For a sphere $k = 0.5$; for an ellipsoid with major axis twice the minor axis and moving in the direction of the major axis, $k = 0.2$; for a long cylinder moving normal to its axis, $k = 1.0$. These values were calculated for inviscid flows and thus are applicable for motions starting from rest so that viscous forces are negligible.

For dense bodies accelerating in the atmosphere the added mass is negligibly small and is typically ignored. Masses accelerating from rest in a liquid are more influenced by the added mass and it must usually be accounted for. Offshore structures that are subjected to oscillatory wave motions experience periodic forces the determination of which must include the added mass effect.

### Example 8.6

A sphere with specific gravity 2.5 is released from rest in water. Calculate its initial acceleration. What is the percentage error if the added mass is ignored?

**Solution**

The summation of forces in the vertical direction, with zero drag, is

$$W - B = (m + m_a)\frac{dV_B}{dt}$$

where $B$ is the buoyant force. Substituting in the appropriate quantities gives, letting $V$ = sphere volume,

$$S\gamma_{\text{water}}V - \gamma_{\text{water}}V = (\rho_{\text{water}}SV + 0.5\rho_{\text{water}}V)\frac{dV_B}{dt}$$

This gives

$$g(S - 1) = (S + 0.5)\frac{dV_B}{dt}$$

Hence

$$\frac{dV_B}{dt} = \frac{g(S-1)}{S+0.5} = \frac{9.8(2.5-1)}{2.5+0.5} = 4.90 \text{ m/s}^2$$

If we ignored the added mass, the acceleration would be

$$\frac{dV_B}{dt} = \frac{g(S-1)}{S} = \frac{9.8(2.5-1)}{2.5} = 5.88 \text{ m/s}^2$$

This is an error of 20%.

## 8.4 LIFT AND DRAG ON AIRFOILS

Separation occurs on a blunt body, such as a cylinder, due to the strong adverse pressure gradient in the boundary layer on the rear of the body. An airfoil is a streamlined body designed to reduce the adverse pressure gradient so that separation will not occur, usually with a small angle of attack, as shown in Fig. 8.12. Without separation the drag is due primarily to the wall shear stress, which results from viscous effects in the boundary layer.

The boundary layer on an airfoil is very thin, and thus it can be ignored when solving for the flow field (the streamline pattern and the pressure distribution) surrounding the airfoil. Since the boundary layer is so thin, the pressure on the wall is not significantly influenced by the boundary layer's existence. Hence the lift on an airfoil can be approximated by integrating the pressure distribution as given by the inviscid flow solution on the wall. In the next section we will demonstrate how this is done; in this section we simply give empirical results.

**KEY CONCEPT**  *The lift on an airfoil can be approximated by integrating the inviscid flow solution.*

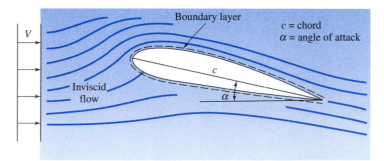

**Fig. 8.12**    Flow around an airfoil at an angle of attack.

The drag on an airfoil can be predicted by solving the boundary layer equations (simplified Navier–Stokes equations) for the shear stress on the wall and performing the appropriate integration. The inviscid flow field must be known before the boundary layer equations can be solved since the pressure gradient on the wall and the inviscid flow velocity at the wall[4] are needed as inputs in solving for the boundary layer flow. Boundary layer calculations will be presented in Section 8.6; in this section we present the empirical results for the drag.

The drag coefficient presented may seem quite low compared with the coefficients of the preceding section. For airfoils a much larger projected area is used, namely, the plan area, which is the chord $c$ (see Fig. 8.12) times the length $L$ of the airfoil. Thus the drag and lift coefficients are defined as

$$C_D = \frac{F_D}{\frac{1}{2}\rho V^2 cL} \qquad C_L = \frac{F_L}{\frac{1}{2}\rho V^2 cL} \qquad\qquad (8.4.1)$$

**KEY CONCEPT** *The design lift coefficient is near the minimum drag coefficient condition.*

For a typical airfoil the lift and drag coefficients are given in Fig. 8.13. For a specially designed airfoil the drag coefficient may be as low as 0.0035, but the maximum lift coefficient is about 1.5. The design lift coefficient (cruise condition) is about 0.3, which is near the minimum drag coefficient condition. This corresponds to an angle of attack of about 2°, far from the stall condition of about 16°.

Conventional airfoils are not symmetric; hence there is a positive lift coefficient at zero angle of attack. The lift is directly proportional to the angle of attack but deviates from the straight-line function just before stall. The drag coefficient also increases linearly up to an angle of attack of about five degrees for a conventional airfoil; then it increases in a nonlinear relation with angle of attack.

**KEY CONCEPT** *Slots allow high-pressure air to energize the slow-moving air, thereby preventing separation from the flap.*

To take off and land at relatively low speeds, it is necessary to attain significantly higher lift coefficients than the maximum of 1.7 of Fig. 8.13. Or if a relatively low lift coefficient is to be accepted, the area $c \times L$ must be enlarged. Both are actually accomplished. Flaps are moved out from a section of each airfoil, resulting in an increased chord, and the angle of attack of the flap is also increased. Slots are used to move high-pressure air from the underside into the relatively low momentum boundary layer flow on the top side, as shown in Fig. 8.14; this prevents separation from the flap, thereby maintaining high lift. The lift coefficient can reach 2.5 with a single-slotted flap and 3.2 with a double-slotted flap. On some modern aircraft there may be three flaps in series with three slots along with a nose flap, to ensure that the boundary layer does not separate from the upper surface of the airfoil.

---

[4] Because the boundary layer in very thin, the velocity at its outer edge is taken to be the velocity at the wall from the inviscid flow solution.

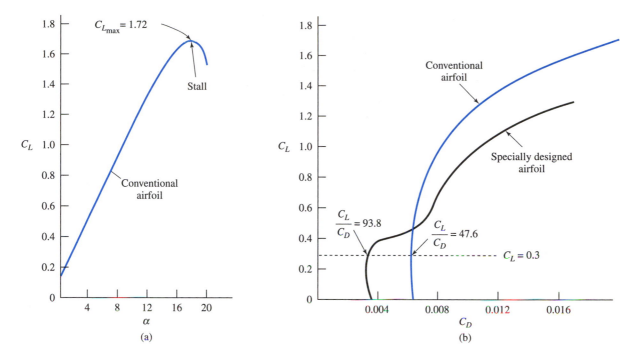

Fig. 8.13   Lift and drag coefficients for airfoils with Re $= Vc/\nu \simeq 9 \times 10^6$.

The total lift on an aircraft is supplied primarily by the airfoil. The effective length of the airfoil when calculating the lift is taken to be the tip-to-tip distance, the **wingspan**, since the fuselage acts to produce lift on the midsection of the aircraft. The drag calculation must include the shear acting on the airfoil, the fuselage, and the tail section.

The drag coefficient is essentially constant on airfoils up to a Mach number of about 0.75. Then a sudden rise occurs until the Mach number reaches unity; see Fig. 8.15. The drag coefficient then slowly falls. Obviously, the condition of M = 1 is to be avoided. Thus aircraft fly at either M < 0.75 or M > 1.5 or so, to avoid the high drag coefficients near M = 1. Near M = 1 there are also regions of flow that oscillate from subsonic to supersonic. Such oscillations create forces that are best avoided.

**Wingspan:**  *The effective length of an airfoil is the tip-to-tip distance.*

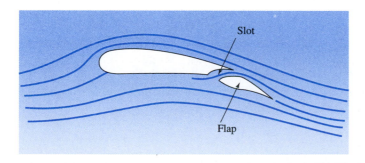

Fig. 8.14   Flapped airfoil with slot for separation control.

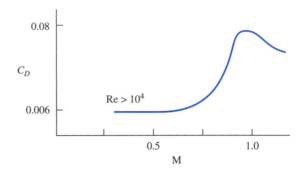

**Fig. 8.15** Drag coefficient as a function of Mach number (speed) for a typical unswept airfoil.

It is useful to use swept-back airfoils since it is the component of velocity normal to the leading edge of the airfoil that must be used in calculating the Mach number in Fig. 8.15. Cruise speeds at M = 0.8 with swept-back wings are not uncommon. It should be pointed out, though, that fuel consumption depends on power required, and power is drag force times velocity; hence fuel consumption depends on the velocity cubed since the drag force depends on the velocity squared, assuming all other parameters are constant. A lower velocity results in a fuel savings even though the engines must operate longer when traveling a fixed distance.

A final comment on airfoils regards the influence of a finite airfoil. To understand flow around a finite airfoil we make reference to a vortex. Fluid particles rotate about the center of a vortex as they travel along in the flow field. There is a high pressure on the bottom and a low pressure on the top side of the airfoil sketched in Fig. 8.16 with a model airfoil shown in Fig. 8.17. This results in a movement of air from the bottom side to the top side around the ends of the airfoil, as shown, resulting in a strong tip vortex. Distributed vortices are also shed all along the airfoil, and they all collect into two large trailing vortices. On a clear day, the two trailing vortices may show up as visible white streaks of water vapor behind a high-flying aircraft. The trailing vortices persist for a considerable distance (perhaps 15 km) behind a large aircraft, and their 90-m/s velocities can

**KEY CONCEPT** *The component of velocity normal to the leading edge is used in calculating the Mach number.*

Trailing vortex, 725, 577

**Fig. 8.16** Trailing vortex.

**Fig. 8.17**   Trailing vortices from a rectangular wing. The flow remains attached over the entire wing surface. The centers of the vortex cores leave the trailing edge at the tips. The model is tested in a smoke tunnel at Reynolds number 100 000. (Photograph by M. R. Head. From Album of Fluid Motion, 1982, The Parabolic Press, Stanford, California.)

cause a small trailing aircraft to flip over. Also, the trailing vortices induce a downwash, a downward velocity component, that must be accounted for in the design of the aircraft. The tail section is located up high to minimize the effect of this downwash.

> **KEY CONCEPT**   *Trailing vortices behind a large aircraft can cause a small aircraft to lose control.*

## Example 8.7

A light airplane weighs 10 000 N, its wingspan measures 12 m, its chord measures 1.8 m, and a payload of 2000 N is anticipated. Predict (a) the takeoff speed if an angle of attack of 8° is desired, (b) the stall speed of the conventional airfoil, and (c) the power required by the airfoil during cruise at 50 m/s.

### Solution

(a) The lift on an airplane is equal to its weight. With the payload the total weight is 12 000 N; hence the lift-coefficient equation (8.4.1) gives the following:

$$V = \left( \frac{F_L}{\frac{1}{2}\rho C_L c L} \right)^{1/2}$$

$$= \left( \frac{12\,000 \text{ N}}{\frac{1}{2} \times 1.20 \text{ kg/m}^3 \times 1.0 \times 1.8 \text{ m} \times 12 \text{ m}} \right)^{1/2} = 30.4 \text{ m/s}$$

where we have used $C_L = 1.0$ at $\alpha = 8°$ from Fig. 8.13, and $\rho = 1.2$ kg/m³ since light aircraft take off at ground level, and the elevation is not given.

*(continued)*

(b) The stall speed is found using a maximum lift coefficient of 1.72 from Fig. 8.13:

$$V_{\text{stall}} = \left( \frac{F_L}{\frac{1}{2}\rho C_L cL} \right)^{1/2}$$

$$= \left( \frac{12\,000}{\frac{1}{2} \times 1.20 \times 1.72 \times 1.8 \times 12} \right)^{1/2} = 23.2 \text{ m/s}$$

(c) The power demanded by the airfoil during cruise is equal to the drag force times the velocity. The design lift coefficient is assumed equal to 0.3, and thus from Fig. 8.13, assuming a conventional airfoil, $C_D = 0.0063$. This gives

$$F_D = \frac{1}{2}\rho V^2 cL C_D$$

$$= \frac{1}{2} \times 1.20 \text{ kg/m}^3 \times 50^2 \text{ m}^2/\text{s}^2 \times 1.8 \text{ m} \times 12 \text{ m} \times 0.0063 = 204 \text{ N}$$

The power is then

$$\text{power} = F_D \times V$$

$$= 204 \text{ N} \times 50 \text{ m/s} = 10\,200 \text{ W} \qquad \text{or} \quad 13.7 \text{ hp}$$

The total power would be significantly larger since the drag on the fuselage and tail section must be included.

## 8.5 POTENTIAL-FLOW THEORY

### 8.5.1 Basic Flow Equations

Inviscid flow exists outside the boundary layer and the wake in high-Reynolds-number flow around bodies. For an airfoil the boundary layer is quite thin, and the inviscid flow provides a good approximation to the actual flow; it is used to predict the pressure distribution on the surface, thereby giving a good estimate of the lift. It will also provide us with the velocity to be used as a boundary condition in the boundary layer solution of Section 8.6; from such a solution we can estimate the drag and predict possible points of separation. Consequently, the inviscid flow solution is very important in our study of external flows. Obviously, if we use the empirical results of previous sections, the details of the inviscid flow solution are unnecessary. If, on the other hand, we wish to predict quantities such as the lift and drag and locate possible points of separation by using the required differential equations, the inviscid flow solution is essential.

**KEY CONCEPT** *The inviscid flow solution is very important in our study of external flows.*

Consider a velocity field that is given by the gradient of a scalar function $\phi$, that is,

$$\mathbf{V} = \nabla\phi \tag{8.5.1}$$

in which $\phi$ is called a *velocity potential function*. Such a velocity field is called a **potential flow** (or an *irrotational flow*) and possesses the property that the vorticity $\boldsymbol{\omega}$, which is the curl of the velocity vector, is zero; this is expressed by

**Potential flow:** *A flow with zero vorticity.*

$$w = \nabla \times \mathbf{V} = 0 \tag{8.5.2}$$

The fact that the vorticity is zero for a potential flow can be shown by letting $\mathbf{V} = \nabla\phi$ and expanding Eq. 8.5.2 in rectangular coordinates. A fluid particle that does not possess vorticity (i.e., is not rotating) cannot obtain vorticity without the action of viscosity; normal pressure forces and body forces that act through the mass center cannot impart rotation to a fluid particle. This result is also observed from the *vorticity equation*, which is obtained by taking the curl of the Navier–Stokes equation (5.3.17) (see Section 5.3.4 if the details are of interest); the vorticity equation is

**KEY CONCEPT** *The vorticity is zero for a potential flow.*

Potential flows, 123, 271

$$\frac{D\boldsymbol{\omega}}{Dt} = (\boldsymbol{\omega} \cdot \nabla)\, \mathbf{V} + \nu\nabla^2\boldsymbol{\omega} \tag{8.5.3}$$

Note that if $\boldsymbol{\omega} = 0$, the only way that $D\boldsymbol{\omega}/Dt$ can be nonzero is for viscous effects to act through the last term. If viscous effects are absent, as in an inviscid flow, then $D\boldsymbol{\omega}/Dt = 0$ and vorticity must remain zero.

With the velocity given by the gradient of a scalar function, the differential continuity equation (5.2.10), for an incompressible flow, gives

$$\nabla \cdot \nabla\phi = \nabla^2\phi = 0 \tag{8.5.4}$$

which is known as *Laplace's equation*, named after Pierre S. Laplace (1749–1827). In rectangular coordinates this is

$$\frac{\partial^2\phi}{\partial x^2} + \frac{\partial^2\phi}{\partial y^2} + \frac{\partial^2\phi}{\partial z^2} = 0 \tag{8.5.5}$$

With the appropriate boundary conditions this equation could be solved. However, three-dimensional problems are quite difficult, so we focus our attention on plane flows in which the velocity components $u$ and $v$ depend on $x$ and $y$. This is acceptable for two-dimensional airfoils and other plane flows, such as flow around cylinders.

Before we attempt a solution to Eq. 8.5.5, let us define another scalar function that will aid us in our study of plane fluid flows. The continuity equation (5.2.9)

$$\frac{\partial u}{\partial x} + \frac{\partial v}{\partial y} = 0 \tag{8.5.6}$$

motivates the definition. If we let

$$u = \frac{\partial\psi}{\partial y} \quad \text{and} \quad v = -\frac{\partial\psi}{\partial x} \tag{8.5.7}$$

**Stream function:** *The stream function is constant along a streamline.*

we observe that the continuity equation is automatically satisfied; the scalar function $\psi(x, y)$ is called a **stream function**. By using the mathematical description of a streamline, $\mathbf{V} \times d\mathbf{r} = 0$, we see that, for a plane flow, $u\,dy - v\,dx = 0$. Substituting from Eqs. 8.5.7, this becomes

$$\frac{\partial \psi}{\partial y}\,dy + \frac{\partial \psi}{\partial x}\,dx = 0 \tag{8.5.8}$$

This is, by definition, $d\psi = 0$. Thus $\psi$ is constant along a streamline. Example 8.9 will show that the difference $(\psi_2 - \psi_1)$ between any two streamlines is equal to the flow rate per unit depth between the two streamlines.

The vorticity vector for a plane flow has only a $z$-component since $w = 0$ and there is no variation with $z$. The vorticity is

$$\omega_z = (\nabla \times \mathbf{V})_z = \frac{\partial v}{\partial x} - \frac{\partial u}{\partial y} \tag{8.5.9}$$

For our potential flow we demand that the vorticity be zero, so Eq. 8.5.9 gives, using Eqs. 8.5.7,

$$\frac{\partial^2 \psi}{\partial x^2} + \frac{\partial^2 \psi}{\partial y^2} = 0 \tag{8.5.10}$$

So we see that both the stream function $\psi$ and the potential function $\phi$ satisfy Laplace's equation for this plane flow.

Rather than attempt a solution of Laplace's equation for a particular flow of interest, we will use a different technique; we will identify some relatively simple functions that satisfy Laplace's equation and then superimpose these simple functions to create flows of interest. It is possible to generate any plane flow desired using this technique. Hence we will not actually solve Laplace's equation.

**KEY CONCEPT** *We will superimpose simple functions to create flows of interest.*

Before we present some simple functions, some additional observations concerning $\phi$ and $\psi$ will be made. Using Eqs. 8.5.1 and 8.5.7, we see that

$$u = \frac{\partial \phi}{\partial x} = \frac{\partial \psi}{\partial y} \qquad v = \frac{\partial \phi}{\partial y} = -\frac{\partial \psi}{\partial x} \tag{8.5.11}$$

These relationships between the derivatives of $\phi$ and $\psi$ are the famous *Cauchy–Riemann equations*, named after Augustin L. Cauchy (1789–1857) and Georg F. Riemann (1826–1866), from the theory of complex variables. The functions $\phi$ and $\psi$ are harmonic functions since they satisfy Laplace's equation and form an *analytic function* $(\phi + i\psi)$ called the *complex velocity potential*. The theory of complex variables with all its powerful theorems is thus applicable to this restricted class of problems: namely, plane, incompressible, potential flows. An example will show that the streamlines and constant-potential lines intersect each other at right angles.

**KEY CONCEPT** Streamlines and constant-potential lines intersect at right angles.

## Example 8.8

A scalar potential function is given by $\phi = A \tan^{-1}(y/x)$. Find the stream function $\psi(x, y)$.

### Solution

The relationship between $\phi$ and $\psi$ is given by Eq. 8.5.11. We have

$$\frac{\partial \psi}{\partial y} = \frac{\partial \phi}{\partial x} = \frac{\partial}{\partial x}\left(A \tan^{-1}\frac{y}{x}\right) = -\frac{Ay}{x^2 + y^2}$$

This can be integrated as follows:

$$\int \frac{\partial \psi}{\partial y}\, dy = -A \int \frac{y}{x^2 + y^2}\, dy$$

$$\therefore \psi = -\frac{A}{2}\ln(x^2 + y^2) + f(x)$$

A function of $x$ rather than a constant must be added since partial derivatives are being used. Now, let us differentiate this expression with respect to $x$. There results

$$\frac{\partial \psi}{\partial x} = -A\frac{x}{x^2 + y^2} + \frac{df}{dx}$$

This must equal $(-\partial\phi/\partial y)$ as demanded by Eq. 8.5.11; that is,

$$-\frac{Ax}{x^2 + y^2} + \frac{df}{dx} = -\frac{Ax}{x^2 + y^2}$$

Thus

$$\frac{df}{dx} = 0 \quad \text{or} \quad f = C$$

Since $\phi$ and $\psi$ are used to find the velocity components by differentiation, the constant $C$ is of no concern; it is usually set equal to zero. Hence

$$\psi = -\frac{A}{2}\ln(x^2 + y^2)$$

## Example 8.9

Show that the difference in the stream function between any two streamlines is equal to the flow rate per unit depth between the two streamlines. The flow rate per unit depth is denoted by $q$.

### Solution

Consider the flow between two streamlines infinitesimally close, as shown in Fig. E8.9a. The flow rate per unit depth through the elemental area is, referring to Fig. E8.9b,

*(continued)*

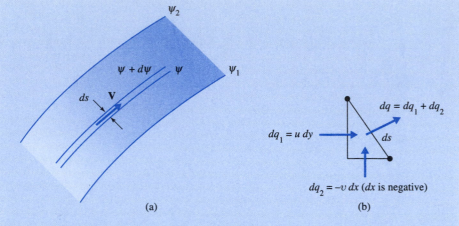

(a)                                   (b)

**Fig. E8.9**

$$dq = dq_1 + dq_2$$

$$= u\,dy - v\,dx$$

$$= \frac{\partial \psi}{\partial y}\,dy + \frac{\partial \psi}{\partial x}\,dx = d\psi$$

If this is integrated between two streamlines with $\psi = \psi_1$ and $\psi = \psi_2$, there results

$$q = \psi_2 - \psi_1$$

thereby proving the statement of the example.

---

## Example 8.10

Show that the streamlines and equipotential lines of a plane, incompressible, potential flow intersect one another at right angles.

### Solution

If, at a point, the slope of a streamline is the negative reciprocal of the slope of an equipotential line, the two lines are perpendicular to each other. The slope of a stream-line (see Fig. E8.10a) is given by

**Fig. E8.10a**

$$\frac{dy}{dx}\bigg|_{\psi \,=\, const} = \frac{v}{u}$$

The slope of an equipotential line is found from

$$d\phi = \frac{\partial \phi}{\partial x} dx + \frac{\partial \phi}{\partial y} dy = 0$$

since $\phi = $ constant along an equipotential line. This gives

$$\frac{dy}{dx}\bigg|_{\phi \,=\, const} = -\frac{\partial \phi/\partial x}{\partial \phi/\partial y} = -\frac{u}{v}$$

Hence we see that the slope of the streamline is the negative reciprocal of the slope of the equipotential line; that is,

$$\frac{dy}{dx}\bigg|_{\phi \,=\, const} = -\left(\frac{dy}{dx}\bigg|_{\psi \,=\, const}\right)^{-1}$$

Thus, whenever the streamlines intersect the equipotential lines, they must do so at right angles. A sketch of the streamlines and equipotential lines (equally spaced at large distances from the body), known as a *flow net*, is shown in Fig. E8.10b for flow over a weir. Such a carefully constructed sketch can be used to approximate the velocities at points of interest in an inviscid flow. Pressures can then be estimated using Bernoulli's equation.

$\psi = $ const
$\phi = $ const

**Fig. E8.10b**

## 8.5.2   Simple Solutions

Now, let us identify some relatively simple functions that satisfy Laplace's equation. However, before we do, it is often more convenient to use polar coordinates. Laplace's equation, the continuity equation, and the velocity components take the following forms:

$$\nabla^2 \psi = \frac{1}{r}\frac{\partial}{\partial r}\left(r\frac{\partial \psi}{\partial r}\right) + \frac{1}{r^2}\frac{\partial^2 \psi}{\partial \theta^2} = 0 \tag{8.5.12}$$

$$\nabla \cdot \mathbf{V} = \frac{1}{r}\frac{\partial}{\partial r}(rv_r) + \frac{1}{r}\frac{\partial v_\theta}{\partial \theta} = 0 \tag{8.5.13}$$

$$v_r = \frac{1}{r}\frac{\partial\psi}{\partial\theta} = \frac{\partial\phi}{\partial r} \qquad v_\theta = -\frac{\partial\psi}{\partial r} = \frac{1}{r}\frac{\partial\phi}{\partial\theta} \qquad (8.5.14)$$

Simple potential flows, 277

We will introduce the names of four simple flows, sketched in Fig. 8.18 and their corresponding functions, each of which satisfies Laplace's equation. The names and functions are:

**Uniform flow:** $\quad\quad\quad \psi = U_\infty y \quad\quad\quad \phi = U_\infty x$ $\quad\quad\quad$ (8.5.15)

**Line source:** $\quad\quad\quad \psi = \dfrac{q}{2\pi}\theta \quad\quad\quad \phi = \dfrac{q}{2\pi}\ln r$ $\quad\quad\quad$ (8.5.16)

**Irrotational vortex:** $\quad \psi = \dfrac{\Gamma}{2\pi}\ln r \quad\quad \phi = \dfrac{\Gamma}{2\pi}\theta$ $\quad\quad\quad$ (8.5.17)

**Doublet:** $\quad\quad\quad\quad \psi = -\dfrac{\mu\sin\theta}{r} \quad\quad \phi = -\dfrac{\mu\cos\theta}{r}$ $\quad\quad\quad$ (8.5.18)

The uniform flow velocity $U_\infty$ is assumed to be in the $x$-direction; if a $y$-component is desired, an appropriate term is added. The *source strength q* is the volume rate of flow per unit depth issuing from the source; a negative value would create a *sink*. The *vortex strength* $\Gamma$ is the *circulation* about the origin, defined by

$$\Gamma = \oint_L \mathbf{V}\cdot d\mathbf{s} \qquad (8.5.19)$$

where $L$ must be a closed curve (a circle is usually used) around the origin and clockwise is positive. The *doublet strength* $\mu$ is for a doublet oriented in the negative $x$-direction; note the large arrow (in Fig. 8.18d) showing the direction of the doublet. Doublets oriented in other directions are seldom of interest and are not considered here.

Of the four flows presented above, the doublet is rather mysterious; it can be visualized as a source and a sink of equal strength separated by a very small distance. Its usefulness is in the creation of certain other flows of interest. The irrotational vortex is encountered when water swirls down a drain or into the turbine of a hydropower dam, or more spectacularly, in a tornado.

**KEY CONCEPT** *The doublet can be visualized as a source and sink of equal strength separated by a very small distance.*

The velocity components for the four simple flows are shown, using Eqs. 8.5.11 and 8.5.14 for rectangular and polar coordinates, to be as follows:

**Uniform flow:** $\quad\quad u = U_\infty \quad\quad\quad\quad v = 0$

$\quad\quad\quad\quad\quad\quad\quad\quad v_r = U_\infty\cos\theta \quad\quad v_\theta = -U_\infty\sin\theta$ $\quad\quad$ (8.5.20)

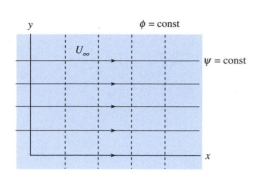

(a) Uniform flow in $x$-direction

(b) Line source

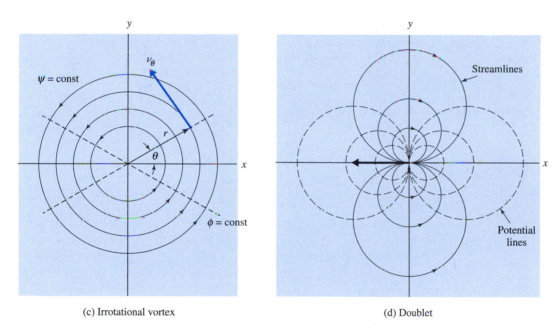

(c) Irrotational vortex

(d) Doublet

**Fig. 8.18**   Four simple potential flows.

$$\text{Line source:} \qquad v_r = \frac{q}{2\pi r} \qquad v_\theta = 0 \qquad\qquad (8.5.21)$$

$$u = \frac{q}{2\pi}\frac{x}{x^2 + y^2} \qquad v = \frac{q}{2\pi}\frac{y}{x^2 + y^2}$$

Irrotational vortex: $v_r = 0$     $v_\theta = -\dfrac{\Gamma}{2\pi r}$

$$u = -\frac{\Gamma}{2\pi}\frac{y}{x^2 + y^2} \qquad v = \frac{\Gamma}{2\pi}\frac{x}{x^2 + y^2}$$

(8.5.22)

Doublet:     $v_r = -\dfrac{\mu\cos\theta}{r^2} \qquad v_\theta = -\dfrac{\mu\sin\theta}{r^2}$

$$u = -\mu\frac{x^2 - y^2}{(x^2 + y^2)^2} \qquad v = -\mu\frac{2xy}{(x^2 + y^2)^2}$$

(8.5.23)

A doublet, 281

## Example 8.11

The pressure far from an irrotational vortex (a simplified tornado) in the atmosphere is zero gage. If the velocity at $r = 20$ m is 20 m/s, estimate the velocity and the pressure at $r = 2$ m. (The irrotational vortex ceases to be a good model for a tornado when $r$ is small. In the "eye" of the tornado the motion is approximated by rigid-body motion.)

### Solution

For an irrotational vortex, we know that

$$v_\theta = -\frac{\Gamma}{2\pi r}$$

Hence

$$\Gamma = -2\pi r v_\theta$$
$$= -2\pi \times 20 \times 20 = -800\pi \text{ m}^2/\text{s}$$

The velocity at $r = 2$ m is then

$$v_\theta = -\frac{-800\pi}{2\pi \times 2} = 200 \text{ m/s}$$

Bernoulli's equation for this incompressible, inviscid, steady flow gives the pressure as follows assuming a stagnant atmosphere away from the tornado:

$$\cancel{p_\infty}^{0} + \cancel{\frac{U_\infty^2}{2}}^{0}\rho = p + \frac{v_\theta^2}{2}\rho$$
$$\therefore p = -\frac{1}{2}\rho v_\theta^2$$
$$= -\frac{1}{2} \times 1.20 \times 200^2 = -24\,000 \text{ Pa}$$

The negative sign denotes a vacuum. It is this vacuum that causes roofs of buildings to blow off during a tornado.

### 8.5.3  Superposition

The simple flows presented in Section 8.5.2 are of particular interest because they can be superimposed with each other to form more complicated flows of engineering importance. In fact, the most complicated, incompressible plane flow can be constructed using these simple flows. For example, suppose that the flow around an airfoil with a slotted flap is desired. We could divide the surface of the airfoil into a relatively large number (say, 200) of panels, locate a source or a sink (or alternatively, a doublet) at the center of each panel, add a uniform flow and an irrotational vortex, and then by adjusting[5] the panel source strengths, the desired inviscid flow could be created. The development of the model and the computer routine necessary to perform the calculations are considered beyond the scope of this book.

> **KEY CONCEPT**  *The most complicated incompressible, plane flow can be constructed using these simple flows.*

In this section we demonstrate superposition by creating flow around a circular cylinder with and without circulation. First, superimpose a uniform flow and a doublet; there results

$$\psi = U_\infty y - \frac{\mu \sin \theta}{r} \tag{8.5.24}$$

The velocity component $v_r$ is (let $y = r \sin \theta$)

$$v_r = \frac{1}{r} \frac{\partial \psi}{\partial \theta}$$

$$= U_\infty \cos \theta - \frac{\mu}{r^2} \cos \theta \tag{8.5.25}$$

Let us ask the question: Is there a radius $r_c$ for which $v_r = 0$? If we set $v_r = 0$ we find that

$$r_c = \sqrt{\frac{\mu}{U_\infty}} \tag{8.5.26}$$

At this radius, $v_r$ is identically zero for all $\theta$ and thus the circle $r = r_c$ must be a streamline.

The stagnation points are found by setting $v_\theta = 0$ on the circle $r = r_c$. There results

$$v_\theta = -\frac{\partial \psi}{\partial r}$$

$$= -U_\infty \sin \theta - \frac{\mu \sin \theta}{r_c^2} = -2U_\infty \sin \theta = 0 \tag{8.5.27}$$

---

[5] The source strengths are adjusted so that the normal component of velocity at the center of each panel is zero. The vortex strength is adjusted so that the rear stagnation point occurs at the trailing edge.

Thus we see that $v_\theta = 0$ at $\theta = 0°$ and $180°$. The flow is as sketched in Fig. 8.19a. We only have an interest in the flow external to the circular streamline $r = r_c$.

If the pressure distribution were desired on the cylinder, Bernoulli's equation could be used between the stagnation point where $V = 0$ and $p = p_0$ and some arbitrary point on the cylinder to yield

$$p_c = p_0 - \rho \frac{v_\theta^2}{2}$$

$$= p_0 - 2\rho U_\infty^2 \sin^2\theta \tag{8.5.28}$$

This provides a symmetric pressure distribution that yields zero drag and zero lift. The zero-lift prediction is acceptable for a real flow, but the zero-drag result is unacceptable. This may suggest that we ignore the inviscid flow solution; how-ever, compared with the actual flow situation of Fig. 8.19b, the measured pressure distribution on the cylinder up to the point of separation is very nearly the same

**KEY CONCEPT**   *The pressure distribution up to the point of separation is very nearly the same as that predicted by potential flow.*

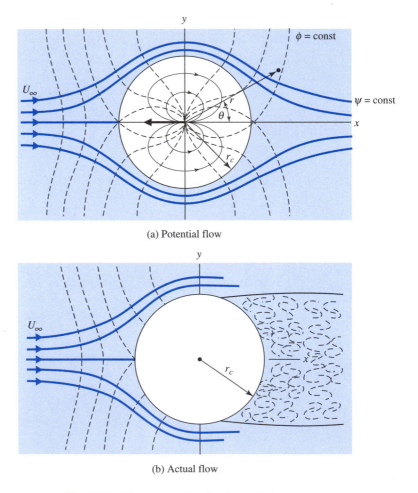

(a) Potential flow

(b) Actual flow

**Fig. 8.19**   Flow around a circular cylinder.

as that predicted by the potential flow solution. Hence the potential flow solution is quite useful to us, even for blunt bodies that experience a separated flow. At low Reynolds numbers, viscous effects are not confined to a thin boundary layer, so potential-flow theory is not useful.

Let us now consider flow around a rotating cylinder. This is accomplished by adding an irrotational vortex to the doublet and uniform flow so that

$$\psi = U_\infty y - \frac{\mu \sin \theta}{r} + \frac{\Gamma}{2\pi} \ln r \qquad (8.5.29)$$

Since the vortex flow, consisting of circular streamlines, does not influence the velocity component $v_r$, the cylinder $r = r_c$ remains unchanged. The stagnation points, however, are changed and are located by setting $v_\theta = 0$ on $r = r_c$; that is, letting $\mu = U_\infty r_c^2$,

$$v_\theta = -\frac{\partial \psi}{\partial r}$$

$$= -2U_\infty \sin \theta - \frac{\Gamma}{2\pi r_c} = 0 \qquad (8.5.30)$$

This gives the location of the stagnation points as shown in Fig. 8.20. In (a) the stagnation points are on the cylinder where $r = r_c$, but in (b) the circulation is sufficiently large that a single stagnation point is off the cylinder with $\theta = 270°$.

Bernoulli's equation gives the pressure distribution as

$$p_c = p_0 - \rho \frac{U_\infty^2}{2} \left( 2 \sin \theta + \frac{\Gamma}{2\pi r_c U_\infty} \right)^2 \qquad (8.5.31)$$

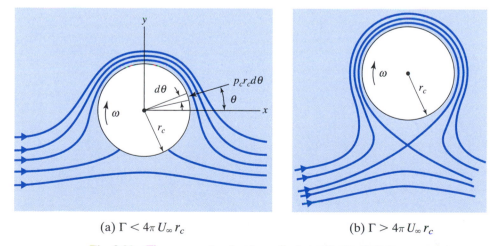

(a) $\Gamma < 4\pi U_\infty r_c$          (b) $\Gamma > 4\pi U_\infty r_c$

**Fig. 8.20**  Flow around a circular cylinder with circulation.

This can be integrated to give drag = 0 and the lift per unit length as

$$F_L = -\int_0^{2\pi} p_c \sin\theta \, r_c d\theta$$

$$= \rho U_\infty \Gamma \tag{8.5.32}$$

This expression for the lift provides an excellent approximation to the lift for all cylinders, including the airfoil. Along with the zero-drag conclusion, it forms the *Kutta-Joukowsky theorem*.

Other superpositions of the simple flows are included in the problems.

## Example 8.12

An 20-cm-diameter cylinder rotates clockwise at 1000 rpm in a 15°C-atmospheric airstream flowing at 4.5 m/s. Locate any stagnation points and find the minimum pressure on the cylinder.

### Solution

The circulation is calculated (see Eq. 8.5.19) to be

$$\Gamma = \oint_L \mathbf{V} \cdot d\mathbf{s}$$

$$= 2\pi r_c^2 \omega = 2\pi \times (0.1)^2 \times \frac{1000 \times 2\pi}{60} = 6.57 \ \mathrm{m^2/s}$$

This is greater than $4\pi U_\infty r_c = 4\pi \times 4.5 \times 0.1 = 5.65 \ \mathrm{m^2/s}$; hence the stagnation point is off the cylinder (see Fig. 8.20b) at $\theta = 270°$. The cylinder's radius is

$$r_0 = -\frac{\Gamma}{4\pi U_\infty \sin 270°} = \frac{6.57}{4\pi \times 4.5 \times (-1)} = 0.116 \ \mathrm{m}$$

Only one stagnation point exists.

The minimum pressure is located on the top of the cylinder where $\theta = 90°$. Using Bernoulli's equation from the free stream to that point, we have, letting $p_\infty = 0$,

$$\cancel{p_\infty} + \rho \frac{U_\infty^2}{2} = p_{\min} + \frac{\rho}{2}(v_\theta)_{\max}^2$$

$$\therefore p_{\min} = \frac{\rho}{2}[U_\infty^2 - (v_\theta)_{\max}^2] = \frac{\rho}{2}\left[U_\infty^2 - \left(-2U_\infty \sin 90° - \frac{\Gamma}{2\pi r_c}\right)^2\right]$$

$$= \frac{1.2 \ \mathrm{kg/m^3}}{2}\left[(4.5)^2 - \left(2 \times 4.5 + \frac{6.57}{2\pi \times 0.1}\right)^2\right] \ \mathrm{m^2/s^2} = -215.1 \ \mathrm{Pa}$$

 **Example 8.12a** Potential flow virtual lab, 295

## 8.6 BOUNDARY-LAYER THEORY

### 8.6.1  General Background

In our study of high-Reynolds-number external flows we have observed that the viscous effects are confined to a thin layer of fluid, a boundary layer, next to the body and to the wake downstream of the body. For a streamlined body such as an airfoil, a good approximation to the drag can be obtained by integrating the viscous shear stress on the wall. To predict the wall shear, the velocity gradient at the wall must be known. This requires a complete solution of the flow field (i.e., a solution of the Navier–Stokes equations) inside the boundary layer. Such a solution also allows us to predict locations of possible separation. In this section we derive the integral and differential equations and provide solution techniques for a boundary-layer flow on a flat plate with a zero pressure gradient; this simplified flow has many applications. Nonzero-pressure-gradient flows on flat plates and flows on curved surfaces will not be considered in this introductory presentation.

Let us now discuss some of the characteristics of a boundary layer. The edge of the boundary layer, with thickness designated by $\delta(x)$, cannot be observed in an actual flow; we arbitrarily define it to be that locus of points where the velocity is equal to 99% of the free-stream velocity [the free-stream velocity is the inviscid flow wall-velocity $U(x)$], as shown in Fig. 8.21. Since the boundary layer is thin, the pressure in the boundary layer is assumed to be the pressure $p(x)$ at the wall, as predicted by the inviscid flow solution.

The boundary layer begins as a laminar flow with zero thickness at the leading edge of a flat plate, as shown in Fig. 8.22, or with some finite thickness at the stagnation point of a blunt object or an airfoil (see Fig. 8.3 on the next page). After a distance $x_T$, which depends on the free-stream velocity, the viscosity, the pressure gradient, the wall roughness, the free-stream fluctuation level, and wall rigidity, the laminar flow undergoes a transition process that results, after a short distance, in a turbulent flow, as sketched. For flow over a flat plate with zero pressure gradient, this transition process occurs when $U_\infty x_T/\nu = 3 \times 10^5$ for flow on rough plates or with high free-stream fluctuation intensity ($\sqrt{\overline{u'^2}}/U_\infty \simeq 0.1$), or $U_\infty x_T/\nu = 5 \times 10^5$

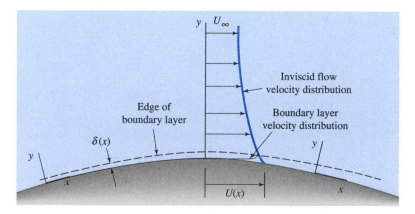

**Fig. 8.21**   Boundary layer on a curved surface.

**Fig. 8.22** Boundary layer with transition.

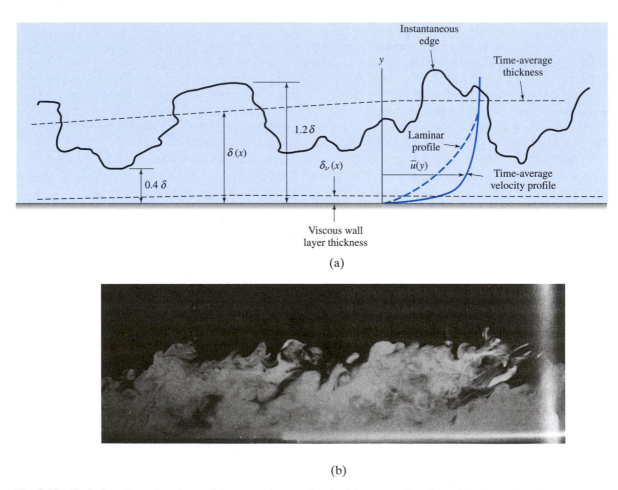

(a)

(b)

**Fig. 8.23** Turbulent boundary layer: (a) nomenclature sketch; (b) streamwise slice of the boundary layer. (Photograph by R. E. Falco) for flow on smooth rigid plates with low free-stream fluctuation intensity. For extremely low fluctuation levels in research labs, laminar flows have been observed on smooth rigid plates with carefully designed leading edges up to $U_\infty x_T / \nu = 10^6$.

The quantity $U_\infty x / \nu$ is the *local Reynolds number* and $U_\infty x_T / \nu$ is the *critical Reynolds number*. For a smooth rigid flat plate and a very low free-stream fluctuation level, a zero-pressure-gradient flow becomes unstable (i.e., small disturbances will grow) at a local Reynolds number of about $6 \times 10^4$. The small disturbances grow initially as a two-dimensional wave, then a three-dimensional wave, and finally burst into a turbulent spot; the initial burst forms the beginning of the transition region. The transition region is relatively short and is usually ignored in calculations. The flow up to $x_T$ is assumed to be laminar, and the flow after $x_T$ is considered to be turbulent.

The turbulent boundary layer thickens much more rapidly than the laminar layer. It also has a substantially greater wall shear. A sketch of a turbulent boundary layer with its submerged viscous wall layer is shown in Fig. 8.23a, and an actual photograph in Fig. 8.23b. The time-average thickness is $\delta(x)$, and the time-average viscous wall–layer thickness is $\delta_\nu(x)$. Both layers are actually quite time dependent. The instantaneous boundary-layer thickness varies between $0.4\delta$ and $1.2\delta$, as shown. The turbulent profile has a greater slope at the wall than does a laminar profile with the same boundary–layer thickness, as is sketched in Fig. 8.23a.

Finally, we should emphasize that the boundary layer is quite thin. A thick boundary layer is drawn to scale in Fig. 8.24. We have assumed a laminar flow up to $x_T$ and a turbulent flow thereafter. For higher velocities the boundary-layer thickness decreases. If we assumed that $U_\infty = 100$ m/s, the boundary layer would hardly be noticed drawn to the same scale, yet all the viscous effects are confined in that thin layer; the velocity is brought to rest with very large gradients. Viscous dissipative effects in this thin layer are large enough to cause sufficiently high temperatures that satellites burn up on reentry.

> Boundary layers, 163, 260, 602, 671

> **KEY CONCEPT**   *The transition region is relatively short and is usually ignored in calculations.*

> **KEY CONCEPT**   *A turbulent profile has a greater slope at the wall than does a laminar profile.*

> **KEY CONCEPT**   *Viscous effects in the boundary layer cause satellites to burn up.*

### 8.6.2   Von Kármán Integral Equation

From the velocity profile in the boundary layer of Fig. 8.24, it is observed that the velocity goes from $u = 0.99U_\infty$ at $y = \delta$ to $u = 0$ at $y = 0$ over a very short distance

**Fig. 8.24**   Boundary layer in air with $\text{Re}_{\text{crit}} = 3 \times 10^5$.

(the thickness of the boundary layer). Hence it is not surprising that we can approximate the velocity profile, for both laminar and turbulent flow with a good deal of accuracy. If the velocity profile can be assumed known, the continuity and momentum integral equations will enable us to predict the boundary-layer thickness and the wall shear and thus the drag. So let us develop the integral equations for the boundary layer.

Consider an infinitesimal control volume, shown in Fig. 8.25a. The integral continuity equation allows us to find $\dot{m}_{top}$ (see Fig. 8.25b). It is, assuming unit depth,

$$\dot{m}_{top} = \dot{m}_{out} - \dot{m}_{in}$$

$$= \frac{\partial}{\partial x} \int_0^{\delta} \rho u \, dy \, dx \tag{8.6.1}$$

The integral momentum equation takes the form

$$\Sigma F_x = \dot{mom}_{out} - \dot{mom}_{in} - \dot{mom}_{top} \tag{8.6.2}$$

where $\dot{mom}$ represents the momentum flux in the $x$-direction. Referring to Figs. 8.25c and 8.25d this becomes, neglecting high-order terms,

$$-\delta \, dp - \tau_0 dx = \frac{\partial}{\partial x} \int_0^{\delta} \rho u^2 \, dy \, dx - \left( \frac{\partial}{\partial x} \int_0^{\delta} \rho u \, dy \, dx \right) U(x) \tag{8.6.3}$$

(a) Control volume

(b) Mass flux

(c) Forces

(d) Momentum flux

**Fig. 8.25**   Control volume for a boundary layer with variable $U(x)$.

where $\dot{m}_{\text{top}}$ is given in Eq. 8.6.1. We have not assumed that $U(x)$ is constant. Divide through by $(-dx)$ and obtain

$$\tau_0 + \delta \frac{dp}{dx} = U(x) \frac{d}{dx} \int_0^\delta \rho u \, dy - \frac{d}{dx} \int_0^\delta \rho u^2 \, dy \qquad (8.6.4)$$

where we have used ordinary derivatives since the integrals are only functions of $x$. This equation is often referred to as the *von Kármán integral equation*.

For flow over a flat plate with zero pressure gradient, so that $dp/dx = 0$ and $U(x) = U_\infty$, von Kármán's integral equation takes the simplified form

$$\tau_0 = \frac{d}{dx} \int_0^\delta \rho u U_\infty dy - \frac{d}{dx} \int_0^\delta \rho u^2 \, dy$$

$$= \frac{d}{dx} \int_0^\delta \rho u (U_\infty - u) \, dy \qquad (8.6.5)$$

If the velocity profile can be assumed, this equation along with $\tau_0(x) = \mu \partial u/\partial y|_{y=0}$ allows us to solve for both $\delta(x)$ and $\tau_0(x)$. This will be demonstrated in the following sections for both a laminar and a turbulent boundary layer.

Before we do this, however, there are two additional lengths often used in boundary-layer theory. They are the *displacement thickness* $\delta_d$ and the *momentum thickness* $\theta$, defined by

$$\delta_d = \frac{1}{U} \int_0^\delta (U - u) \, dy \qquad (8.6.6)$$

$$\theta = \frac{1}{U^2} \int_0^\delta u(U - u) \, dy \qquad (8.6.7)$$

The displacement thickness is the displacement of the streamlines in the free stream as a result of velocity deficits in the boundary layer, as can be shown by continuity considerations. The momentum-layer thickness is the equivalent thickness of a fluid layer with velocity $U$ with momentum equal to the momentum lost due to friction; the momentum thickness is often used as a characteristic length in turbulent boundary-layer studies. The end-of-chapter problems will demonstrate the use of $\delta_d$ and $\theta$. It should be noted, though, that von Kármán's integral equation (8.6.5) takes the form, assuming that $\rho = \text{const.}$,

$$\tau_0 = \rho U_\infty^2 \frac{d\theta}{dx} \qquad (8.6.8)$$

### 8.6.3   Approximate Solution to the Laminar Boundary Layer

It is possible to use the von Kármán integral equation and obtain a fairly accurate approximation to the laminar boundary layer on a flat plate with zero pressure gradient. We have four conditions that a proposed velocity profile should satisfy:

$$
\begin{aligned}
u &= 0 & &\text{at} \quad y = 0 \\
u &= U_\infty & &\text{at} \quad y = \delta \\
\frac{\partial u}{\partial y} &= 0 & &\text{at} \quad y = \delta \\
\frac{\partial^2 u}{\partial y^2} &= 0 & &\text{at} \quad y = 0
\end{aligned}
\tag{8.6.9}
$$

The first three of these conditions are obvious from a sketch of a velocity profile; the fourth condition comes from the $x$-component Navier–Stokes equation (5.3.14) since $u = v = 0$ at the wall, $\partial^2 u / \partial x^2 = 0$ on the wall, and $dp/dx = 0$ for the steady flow on the flat plate under consideration.

A cubic polynomial can satisfy the four conditions above; let us assume that

$$
\frac{u}{U_\infty} = A + By + Cy^2 + Dy^3
\tag{8.6.10}
$$

where $A$, $B$, $C$, and $D$ can be functions of $x$. Using the four conditions, we see that

$$
A = 0 \qquad B = \frac{3}{2\delta} \qquad C = 0 \qquad D = -\frac{1}{2\delta^3}
\tag{8.6.11}
$$

Hence a good approximation for the velocity profile in a laminar flow is

$$
\frac{u}{U_\infty} = \frac{3}{2}\frac{y}{\delta} + \frac{1}{2}\left(\frac{y}{\delta}\right)^3
\tag{8.6.12}
$$

Let us now use this velocity profile to find $\delta(x)$ and $\tau_0(x)$. Von Kármán's integral equation (8.6.5) gives

$$
\begin{aligned}
\tau_0 &= \frac{d}{dx}\int_0^\delta \rho\left(\frac{3y}{2\delta} - \frac{y^3}{2\delta^3}\right)\left(1 - \frac{3y}{2\delta} - \frac{y^3}{2\delta^3}\right)U_\infty^2\,dy \\
&= 0.139\rho U_\infty^2 \frac{d\delta}{dx}
\end{aligned}
\tag{8.6.13}
$$

At the wall we know that $\tau_0 = \mu\,\partial u/\partial y\,|_{y=0}$, or using the cubic profile (8.6.12),

$$
\tau_0 = \mu\left(U_\infty \frac{3}{2\delta}\right)
\tag{8.6.14}
$$

Equating the foregoing expressions for $\tau_0(x)$, we find that

$$\delta \, d\delta = \frac{\frac{3}{2}\mu U_\infty}{0.139\rho U_\infty^2} \, dx = 10.8 \frac{\nu}{U_\infty} \, dx \qquad (8.6.15)$$

Using $\delta = 0$ at $x = 0$ (the leading edge), Eq. 8.6.15 is integrated to give

$$\delta = 4.65 \sqrt{\frac{\nu x}{U_\infty}} = 4.65 \frac{x}{\sqrt{\mathrm{Re}_x}} \qquad (8.6.16)$$

where $\mathrm{Re}_x$ is the local Reynolds number. This is substituted back into Eq. 8.6.14, giving the wall shear as

$$\tau_0 = 0.323\rho U_\infty^2 \sqrt{\frac{\nu}{xU_\infty}}$$

$$= \frac{0.323\rho U_\infty^2}{\sqrt{\mathrm{Re}_x}} \qquad (8.6.17)$$

The shearing stress is made dimensionless by dividing by $\frac{1}{2}\rho U_\infty^2$. The **local skin friction coefficient** $c_f$ results; it is

**Local skin friction coefficient:** *A dimensionless wall shearing stress.*

$$c_f = \frac{\tau_0}{\frac{1}{2}\rho U_\infty^2}$$

$$= \frac{0.646}{\sqrt{U_\infty x/\nu}} = \frac{0.646}{\sqrt{\mathrm{Re}_x}} \qquad (8.6.18)$$

If the wall shear is integrated over the length $L$, there results, per unit width,

$$F_D = \int_0^L \tau_0 \, dx = 0.646\rho U_\infty \sqrt{U_\infty L \nu}$$

$$= \frac{0.646\rho U_\infty^2 L}{\sqrt{\mathrm{Re}_L}} \qquad (8.6.19)$$

or in terms of the **skin friction coefficient** $C_f$,

**Skin friction coefficient:** *A dimensionless drag force.*

$$C_f = \frac{F_D}{\frac{1}{2}\rho U_\infty^2 L}$$

$$= \frac{1.29}{\sqrt{U_\infty L/\nu}} = \frac{1.29}{\sqrt{\mathrm{Re}_L}} \qquad (8.6.20)$$

where $Re_L$ is the Reynolds number at the end of the flat plate.

Note that the shearing stress $\tau_0$ becomes unbounded as $x \to 0$. Hence we would not expect $\tau_0(x)$ to be a very good approximation to the wall shear near the leading edge. The expression for the drag is acceptable, however.

The results above are quite good when compared with results from a solution of the differential equations (refer to Section 8.6.6). The boundary layer thickness is 7% too low; the constant in Eq. 8.6.16 should be 5 if an exact solution were obtained. The wall shear is less than 3% too low; the constant in Eq. 8.6.17 should be 0.332.

---

## Example 8.13

Assume that the velocity profile in a boundary-layer flow can be approximated by a parabolic velocity profile. Calculate the boundary-layer thickness of Eq. 8.6.16 and the wall shear of Eq. 8.6.17. Compare with those calculated above for the cubic profile.

### Solution

The parabolic velocity profile is assumed to be

$$\frac{u}{U_\infty} = A + By + Cy^2$$

The fourth condition, which would be impossible to satisfy, of (8.6.9) is omitted; this leaves

$$0 = A$$

$$1 = A + B\delta + C\delta^2$$

$$0 = B + 2C\delta$$

A simultaneous solution provides

$$A = 0 \qquad B = \frac{2}{\delta} \qquad C = -\frac{1}{\delta^2}$$

The velocity profile is then

$$\frac{u}{U_\infty} = 2\frac{y}{\delta} - \frac{y^2}{\delta^2}$$

This is substituted into von Kármán's integral equation (8.6.5) to obtain

$$\tau_0 = \frac{d}{dx} \int_0^\delta \rho U_\infty^2 \left(2\frac{y}{\delta} - \frac{y^2}{\delta^2}\right)\left(1 - \frac{2y}{\delta} + \frac{y^2}{\delta^2}\right) dy$$

$$= \frac{2}{15} \rho U_\infty^2 \frac{d\delta}{dx}$$

We also use $\tau_0 = \mu\, \partial u/\partial y\,|_{y=0}$; that is,

$$\tau_0 = \mu U_\infty \frac{2}{\delta}$$

Equating the two expressions above, we obtain

$$\delta \, d\delta = 15 \frac{\nu}{U_\infty} dx$$

Using $\delta = 0$ at $x = 0$, this is integrated to

$$\delta = 5.48 \sqrt{\frac{\nu x}{U_\infty}}$$

This is 18% higher than the value using the cubic but only 10% higher than the more accurate result of $5\sqrt{\nu x / U_\infty}$.

The wall shear is found to be

$$\tau_0 = \frac{2\mu U_\infty}{\delta}$$

$$= 0.365 \rho U_\infty^2 \sqrt{\frac{\nu}{x U_\infty}}$$

This is 13% higher than the value using the cubic and 10% higher than the more accurate value of $0.332 \, \rho U_\infty^2 \sqrt{\nu / x U_\infty}$. Because the boundary layer is so thin, there is little difference between a cubic and a parabola or the actual profile; refer to the profile in Fig. 8.24.

**Example 8.13a**    Viscous Layer Growth, 619–621

## 8.6.4    Turbulent Boundary Layer: Power-Law Form

For turbulent boundary-layer flow we have two methods for obtaining the information desired. Both methods utilize experimental data, but the one that we present in this section is the simpler of the two. The second method, to be presented in the next section, provides us with more information than we usually desire for most applications; however, it is more accurate.

In the method to be presented first we fit the data for the velocity profile with a power-law equation. The power-law form is

$$\frac{\overline{u}}{U_\infty} = \left(\frac{y}{\delta}\right)^{1/n} \qquad n = \begin{cases} 7 & \mathrm{Re}_x < 10^7 \\ 8 & 10^7 < \mathrm{Re}_x < 10^8 \\ 9 & 10^8 < \mathrm{Re}_x < 10^9 \end{cases} \qquad (8.6.21)$$

where

$$\mathrm{Re}_x = \frac{U_\infty x}{\nu} \qquad (8.6.22)$$

Von Kármán's integral equation can now be applied following the steps used for laminar flow, except when the shear stress at the wall is evaluated. The power-law form (8.6.21) yields $(\partial \overline{u}/\partial y)_{y=0} = \infty$; hence the profile gives poor results near the

> **KEY CONCEPT**    *The power-law form gives poor results near the wall.*

wall, especially for the shear stress. So rather than using $\tau_0 = (\mu \; \partial\bar{u}/\partial y)_{y = 0}$, we use an empirical relation; the *Blasius formula*, named after Paul R. H. Blasius (1883–1970), which relates the local skin friction coefficient to the boundary-layer thickness, is

$$c_f = 0.046 \left(\frac{\nu}{U_\infty \delta}\right)^{1/4} \qquad (8.6.23)$$

or, relating $\tau_0$ to $c_f$ using the definition of $c_f$ in Eq. 8.6.18, the shear stress relation is

$$\tau_0 = 0.023 \; \rho U_\infty^2 \left(\frac{\nu}{U_\infty \delta}\right)^{1/4} \qquad (8.6.24)$$

Von Kármán's integral equation provides us with a second expression for $\tau_0$. Substitute the velocity profile of (8.6.21) with $\mathrm{Re}_x < 10^7$ into Eq. 8.6.5 and obtain

$$\begin{aligned}
\tau_0 &= \frac{d}{dx} \int_0^\delta \rho U_\infty^2 \left(\frac{y}{\delta}\right)^{1/7} \left[1 - \left(\frac{y}{\delta}\right)^{1/7}\right] dy \\
&= \frac{7}{72} \; \rho U_\infty^2 \frac{d\delta}{dx}
\end{aligned} \qquad (8.6.25)$$

Combining the two expressions above for $\tau_0$, we find that

$$\delta^{1/4} \; d\delta = 0.237 \left(\frac{\nu}{U_\infty}\right)^{1/4} dx \qquad (8.6.26)$$

Assuming a turbulent flow from the leading edge (the laminar portion is often quite short, i.e., $L \gg x_T$), there results

$$\begin{aligned}
\delta &= 0.38x \left(\frac{\nu}{U_\infty x}\right)^{1/5} \\
&= 0.38x \mathrm{Re}_x^{-1/5} \qquad \mathrm{Re}_x < 10^7
\end{aligned} \qquad (8.6.27)$$

Substituting this expression for $\delta$ back into Eq. 8.6.23, we find that

$$c_f = 0.059 \; \mathrm{Re}_x^{1/5} \qquad \mathrm{Re}_x < 10^7 \qquad (8.6.28)$$

and, performing the required integration, there results, with $n = 7$,

$$C_f = 0.073 \; \mathrm{Re}_L^{-1/5} \qquad \mathrm{Re}_L < 10^7 \qquad (8.6.29)$$

where $\mathrm{Re}_L = U_\infty L/\nu$. The relations above can be stretched to $\mathrm{Re}_x \approx 10^8$ without substantial error.

If $L$ is not much larger than $x_T$, say $L = 3x_T$, then there is a significant laminar portion on the leading part of a flat plate, and the skin friction coefficient can be modified as

$$C_f = 0.073\mathrm{Re}_L^{-1/5} - 1700\mathrm{Re}_L^{-1} \qquad \mathrm{Re}_L < 10^7 \qquad (8.6.30)$$

This relationship is based on transition occurring at $\mathrm{Re}_{\mathrm{crit}} = 5 \times 10^5$. If $\mathrm{Re}_{\mathrm{crit}} = 3 \times 10^5$, the constant of 1700 is replaced by 1060; if $\mathrm{Re}_{\mathrm{crit}} = 6 \times 10^5$, it is replaced by 2080.

Finally, the displacement thickness and momentum thickness can be evaluated, using $n = 7$, to be

$$\delta_d = 0.048x\mathrm{Re}_x^{-1/5} \qquad (8.6.31)$$
$$\theta = 0.037x\mathrm{Re}_x^{-1/5}$$

## Example 8.14

Estimate the boundary-layer thickness at the end of a 4-m-long flat surface if the free-stream velocity is $U_\infty = 5$ m/s. Use atmospheric air at 30°C. Also, predict the drag force if the surface is 5 m wide. (a) Neglect the laminar portion of the flow and (b) account for the laminar portion using $\mathrm{Re}_{\mathrm{crit}} = 5 \times 10^5$.

**Fig. E8.14**

### Solution

(a) Let us first assume turbulent flow from the leading edge. The boundary-layer thickness is given by Eq. 8.6.27. It is

$$\delta = 0.38x\mathrm{Re}_x^{-1/5}$$
$$= 0.38 \times 4 \times \left(\frac{5 \times 4}{1.6 \times 10^{-5}}\right)^{-1/5} = 0.0917 \text{ m}$$

The drag force is, using Eq. 8.6.29,

$$F_D = C_f \times \frac{1}{2}\rho U_\infty^2 Lw$$
$$= 0.073\left(\frac{5 \times 4}{1.6 \times 10^{-5}}\right)^{-1/5} \times \frac{1}{2} \times 1.16 \text{ kg/m}^3 \times 5^2 \text{ m}^2/\text{s}^2 \times 4 \text{ m} \times 5 \text{ m} = 1.28 \text{ N}$$

*(continued)*

The predictions above assume that $Re_L < 10^7$. The Reynolds number is

$$Re_L = \frac{5 \times 4}{1.6 \times 10^{-5}} = 1.25 \times 10^6$$

Hence the calculations are acceptable.

(b) Now let us account for the laminar portion of the boundary layer. Referring to Fig. E8.14, the distance $x_T$ is found as follows:

$$Re_{crit} = 5 \times 10^5 = \frac{U_\infty x_T}{\nu}$$

$$\therefore x_T = 5 \times 10^5 \times 1.6 \times \frac{10^{-5}}{5} = 1.6 \text{ m}$$

The boundary-layer thickness at $x_T$ is, replacing the constant of 4.65 in Eq. 8.6.16 with the more accurate value of 5,

$$\delta = 5\sqrt{\frac{x\nu}{U_\infty}}$$

$$= 5\sqrt{\frac{1.6 \text{ m} \times 1.6 \times 10^{-5} \text{ m}^2/\text{s}}{5 \text{ m/s}}} = 0.0113 \text{ m}$$

The location of the fictitious origin of the turbulent flow (see Fig. E8.14) is found using Eq. 8.6.27 to be

$$x'^{4/5} = \frac{\delta}{0.38}\left(\frac{U_\infty}{\nu}\right)^{1/5}$$

$$\therefore x' = \left(\frac{0.0113}{0.38}\right)^{5/4}\left(\frac{5}{1.6 \times 10^{-5}}\right)^{1/4} = 0.292 \text{ m}$$

The distance $x_{turb}$ is then $x_{turb} = 4 - 1.6 + 0.292 = 2.69$ m. Using Eq. 8.6.27, the thickness at the end of the surface is

$$\delta = 0.38x\left(\frac{\nu}{U_\infty x}\right)^{1/5}$$

$$= 0.38 \times 2.69 \times \left(\frac{1.6 \times 10^{-5}}{5 \times 2.69}\right)^{1/5} = 0.067 \text{ m}$$

The value of part (a) is 37% too high when compared with this more accurate value.

The more accurate drag force is found using Eq. 8.6.30 to be

$$F_D = C_f \times \frac{1}{2}\rho U_\infty^2 Lw$$

$$= [0.073 \, Re_L^{-1/5} - 1700 Re_L^{-1}] \times \frac{1}{2}\rho U_\infty^2 Lw$$

$$= \left[0.073\left(\frac{5 \times 4}{1.6 \times 10^{-5}}\right)^{-1/5} - 1700\left(\frac{5 \times 4}{1.6 \times 10^{-5}}\right)^{-1}\right] \times \frac{1}{2} \times 1.16 \times 5^2 \times 4 \times 5$$

$$= 0.88 \text{ N}$$

The prediction of part (a) is 45% too high. For relatively short surfaces it is obvious that significant errors result if the thinner laminar portion with its smaller shear stress is neglected.

**Example 8.14a**   Profiles in a Turbulent Plume, 861–864

## 8.6.5 Turbulent Boundary Layer: Empirical Form

The second method of predicting turbulent flow quantities on a flat plate with zero pressure gradient is based entirely on data. It is more accurate than the power-law form but also more complicated. The time-average turbulent velocity profile can be divided into two regions, the *inner region* and the *outer region*, as shown in Fig. 8.26. The inner region is characterized by the *self-similar* (the dimensionless dependent variable depends on only one dimensionless independent variable) relation,

$$\frac{\overline{u}}{u_\tau} = f\left(\frac{u_\tau y}{\nu}\right) \tag{8.6.32}$$

in which $u_\tau$ is the *shear velocity*, given by[6]

$$u_\tau = \sqrt{\frac{\tau_0}{\rho}} \tag{8.6.33}$$

**KEY CONCEPT** $u_\tau$ *is a ficticious velocity called the shear velocity.*

The velocity profile in the outer region is given by the self-similar relation

$$\frac{U_\infty - \overline{u}}{u_\tau} = f\left(\frac{y}{\delta}\right) \tag{8.6.34}$$

where $(U_\infty - \overline{u})$ is called the *velocity defect*.

The inner region has three distinct zones: the viscous wall layer, the buffer zone, and the turbulent zone, as shown in Fig. 8.26b. The highly fluctuating *viscous wall layer* has a linear time-average profile given by

**KEY CONCEPT** *The viscous wall layer is highly fluctuating.*

$$\frac{\overline{u}}{u_\tau} = \frac{u_\tau y}{\nu} \tag{8.6.35}$$

The quantity $\nu/u_\tau$ is the characteristic length in the turbulent inner region; hence the dimensionless distance from the wall is denoted by

$$y^* = \frac{u_\tau y}{\nu} \tag{8.6.36}$$

The viscous wall layer is very thin, extending up to $y^* \simeq 5$. A logarithmic profile exists from $y^* \simeq 50$ to $y/\delta \simeq 0.15$. In this self-similar *turbulent zone,*

---

[6] The shear velocity is a fictitious velocity and is defined because the quantity $\sqrt{\tau_0/\rho}$ occurs often in empirical relations in turbulent boundary-layer flows.

(a) Standard profile

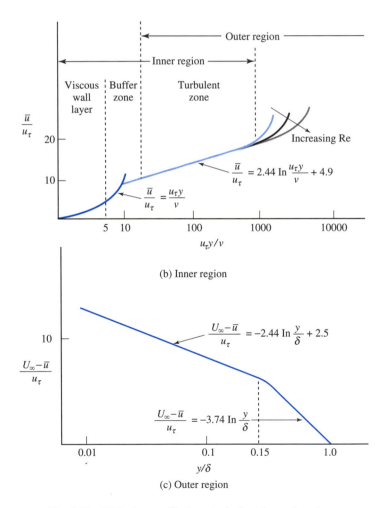

(b) Inner region

(c) Outer region

**Fig. 8.26**   Velocity profile in a turbulent boundary layer.

$$\frac{\bar{u}}{u_\tau} = 2.44 \ln \frac{u_\tau y}{\nu} + 4.9 \qquad 50 < \frac{u_\tau y}{\nu} \qquad \frac{y}{\delta} < 0.15 \qquad (8.6.37)$$

The location of the outer edge of the turbulent zone is strongly dependent on the Reynolds number. The value of $u_\tau y/\nu$ that locates the outer edge increases as the Reynolds number increases, as shown. A *buffer zone*, with no specified velocity profile, connects the two self-similar zones.

The outer region relates the velocity defect to $y/\delta$. In the turbulent zone the velocity-defect profile is sketched in Fig. 8.26c and is

$$\frac{U_\infty - \bar{u}}{u_\tau} = -2.44 \ln \frac{y}{\delta} + 2.5 \qquad 50 < \frac{u_\tau y}{\nu} \qquad \frac{y}{\delta} < 0.15 \qquad (8.6.38)$$

Between $y/\delta = 0.15$ and $y/\delta = 1$ researchers fit the data with several relations; the one selected here is

$$\frac{U_\infty - \bar{u}}{u_\tau} = -3.74 \ln \frac{y}{\delta} \qquad \frac{y}{\delta} > 0.15 \qquad (8.6.39)$$

The equations above involve the shear velocity $u_\tau$, which depends on the wall shear $\tau_0$. The equation that is valid at the wall is Eq. 8.6.35, which, using $\tau_0 = \mu \, \partial u/\partial y|_{y=0}$, simply provides us with an identity. It does not allow us to calculate $\tau_0$. Hence a relationship is necessary to provide us with $\tau_0$ (or equally $c_f$). There are several such relationships used; one that gives excellent results is

$$c_f = \frac{0.455}{(\ln 0.06 \mathrm{Re}_x)^2} \qquad (8.6.40)$$

This local skin friction coefficient allows us to determine $\tau_0$ and thus $u_\tau$ at any location of interest. The velocity profiles can be used to calculate quantities of interest, but $u_\tau$ must be known.

Assuming turbulent flow from the leading edge, the shear stress can be integrated to yield the drag. Then the skin friction coefficient becomes

**KEY CONCEPT**  *This local skin friction coefficient allows us to determine $\tau_0$ and thus $u_\tau$.*

$$C_f = \frac{0.523}{(\ln 0.06 \mathrm{Re}_L)^2} \qquad (8.6.41)$$

This relation is very good and can be used up to $\mathrm{Re}_L = 10^9$ with an error of 2% or less. Even at $\mathrm{Re}_L = 10^{10}$ the error is about 4%. To account for a laminar portion, the same term included in Eq. 8.6.30 can be subtracted from Eq. 8.6.41.

In concluding this section, a very useful relationship can be obtained by combining the two logarithmic profiles for the common turbulent zone. Substitute Eq. 8.6.37 into Eq. 8.6.38 and obtain

$$\frac{U_\infty}{u_\tau} = 2.44 \ln \frac{u_\tau \delta}{\nu} + 7.4 \qquad (8.6.42)$$

This equation allows an easy calculation of $\delta$ knowing $u_\tau$.

---

## Example 8.15

Estimate the thickness $\delta_\nu$ of the viscous wall layer and the boundary-layer thickness at the end of a 4.5-m-long flat plate if $U_\infty = 30$ m/s in 15°C–atmospheric air. Also, calculate the drag force on one side if the plate is 3 m wide. Use the empirical data.

### Solution

To find the viscous wall layer thickness we must know the shear velocity and hence the wall shear. The wall shear, using Eq. 8.6.40, and the shear velocity at $x = 4.5$ m are

$$\tau_0 = \frac{1}{2} \rho U_\infty^2 c_f$$

$$= \frac{1}{2} \rho U_\infty^2 \frac{0.455}{(\ln 0.06 \mathrm{Re}_x)^2}$$

$$= \frac{1}{2} \times 1.2 \ \mathrm{kg/m^3} \times 30^2 \ \mathrm{m^2/s^2} \frac{0.455}{\left(\mathrm{In}\ 0.06 \dfrac{30 \times 4.5}{1.46 \times 10^{-5}}\right)^2} = 1.404 \ \mathrm{Pa}$$

$$u_\tau = \sqrt{\frac{\tau_0}{\rho}} = \sqrt{\frac{1.404 \ \mathrm{N/m^2}}{1.2 \ \mathrm{kg/m^3}}} = 1.082 \ \mathrm{m/s}$$

The viscous wall-layer thickness is determined using Eq. 8.6.36 with $y^* = 5$ as follows:

$$\frac{u_\tau \delta_\nu}{\nu} = 5$$

$$\delta_\nu = \frac{5\nu}{u_\tau} = \frac{5 \times 1.46 \times 10^{-5}}{1.082} = 6.75 \times 10^{-5} \ \mathrm{m}$$

The boundary-layer thickness is found using Eq. 8.6.42:

$$\frac{U_\infty}{u_\tau} = 2.44 \ln \frac{u_\tau \delta}{\nu} + 7.4$$

$$\frac{30}{1.082} = 2.44 \ln \frac{1.082 \times \delta}{1.46 \times 10^{-5}} + 7.4 \qquad \therefore \delta = 0.056 \ \mathrm{m} = 5.6 \ \mathrm{cm}$$

The drag force is calculated using Eq. 8.6.41 to be

$$F_D = C_f \times \frac{1}{2}\rho U_\infty^2 Lw$$

$$= \frac{0.523}{(\ln 0.06 \mathrm{Re}_L)^2} \times \frac{1}{2}\rho U_\infty^2 Lw$$

$$= \frac{0.523}{\left(\ln 0.06 \dfrac{30 \times 4.5}{1.46 \times 10^{-5}}\right)^2} \times \frac{1}{2} \times 1.2\ \mathrm{kg/m^3} \times 30^2\ \mathrm{m^2/s^2}\,(4.5 \times 3)\ \mathrm{m^2} = 21.8\ \mathrm{N}$$

The laminar portion of the boundary layer has been neglected.

---

## Example 8.16

Estimate the maximum boundary-layer thickness and the drag due to friction on the side of a ship that measures 40 m long with a submerged depth of 8 m assuming the side of the ship is approximated as a flat plate. The ship travels at 10 m/s. (a) Use the empirical methods and (b) compare with the results using the power law model.

### Solution

(a) The boundary-layer thickness is found from Eq. 8.6.42. First we must find $\tau_0$ from Eq. 8.6.40 and then $u_\tau$ as follows:

$$\tau_0 = \frac{1}{2}\rho U_\infty^2 \frac{0.455}{(\ln 0.06 \mathrm{Re}_L)^2}$$

$$= \frac{1}{2} \times 1000\ \mathrm{kg/m^3} \times 10^2\ \mathrm{m^2/s^2}\ \frac{0.455}{\left(\ln 0.06 \dfrac{10 \times 40}{10^{-6}}\right)^2} = 78.8\ \mathrm{Pa}$$

$$\therefore u_\tau = \sqrt{\frac{\tau_0}{\rho}}$$

$$= \sqrt{\frac{78.8\ \mathrm{N/m^2}}{1000\ \mathrm{kg/m^3}}} = 0.28\ \mathrm{m/s}$$

The maximum boundary-layer thickness is found using Eq. 8.6.42:

$$\frac{U_\infty}{u_\tau} = 2.44 \ln \frac{u_\tau \delta}{\nu} + 7.4$$

$$\frac{10}{0.28} = 2.44 \ln \frac{0.28\delta}{10^{-6}} + 7.4 \qquad \therefore \delta = 0.39\ \mathrm{m}$$

The drag is

$$F_D = C_f \times \frac{1}{2}\rho U_\infty^2 Lw$$

$$= \frac{0.523}{\left(\ln 0.06 \dfrac{10 \times 40}{10^{-6}}\right)^2} \times \frac{1}{2} \times 1000 \times 10^2 \times 40 \times 8 = 29\,000\ \mathrm{N}$$

*(continued)*

(b) First, calculate the Reynolds number: $\text{Re} = 10 \times 40/10^{-6} = 4 \times 10^8$. We select $n = 9$. Equation (8.6.25) becomes

$$\tau_0 = \frac{d}{dx} \int_0^{\delta} \rho U_\infty^2 \left(\frac{y}{\delta}\right)^{1/9} \left[1 - \left(\frac{y}{\delta}\right)^{1/9}\right] dy$$

$$= \frac{9}{110} \rho U_\infty^2 \frac{d\delta}{dx}$$

Equating this to the $\tau_0$ of Eq. 8.6.24, we find that

$$\delta^{1/4} \, d\delta = 0.281 \, (\nu/U_\infty)^{1/4} \, dx$$

Assume $\delta = 0$ at $x = 0$ and integrate. This provides

$$\delta = 0.433 x \, \text{Re}_x^{-1/5}$$

$$= 0.433(40) \left(\frac{10 \times 40}{10^{-6}}\right)^{-1/5} = 0.33 \text{ m}$$

This value is 15% too low.
   The drag force is found to be

$$F_D = 0.071 \text{Re}_L^{-1/5} \times \tfrac{1}{2} \rho U_\infty^2 \, Lw$$

$$= 0.071 \left(\frac{10 \times 40}{10^{-6}}\right)^{-1/5} \times \frac{1}{2} \times 1000 \times 10^2 \times 40 \times 8 = 21\ 600 \text{ N}$$

This value is 25% too low. Obviously, the power-law equations are in significant error.

### 8.6.6  Laminar Boundary-Layer Equations

The solution presented in Section 8.6.3 for the laminar boundary layer was an approximate solution using a cubic polynomial to approximate the velocity profile. In this section we simplify the Navier–Stokes equations and present a more accurate solution for the laminar boundary layer on a flat plate with a zero pressure gradient.

   The $x$-component Navier–Stokes equation for a steady, incompressible, plane flow is (see Eq. 5.3.14 and ignore the gravity term)

$$u \frac{\partial u}{\partial x} + v \frac{\partial u}{\partial y} = -\frac{1}{\rho} \frac{\partial p}{\partial x} + \nu \left(\frac{\partial^2 u}{\partial x^2} + \frac{\partial^2 u}{\partial y^2}\right) \tag{8.6.43}$$

**KEY CONCEPT**   *There is no pressure variation in the y-direction in the boundary layer.*

In boundary-layer theory the boundary layer is assumed to be very thin (see Fig. 8.22), so there is no pressure variation in the $y$-direction in the boundary layer; that is, $p = p(x)$. In addition (this is a very important point), the pressure $p(x)$ is given by the inviscid flow solution as the wall pressure; hence the pressure is not an unknown. This leaves only two unknowns, $u$ and $v$. Equation 8.6.43 provides us with one equation and the continuity equation

$$\frac{\partial u}{\partial x} + \frac{\partial v}{\partial y} = 0 \tag{8.6.44}$$

provides us with the other. The $y$-component Navier–Stokes equation is not of use in boundary layer theory since all the terms are negligibly small ($v \ll u$ as inferred from Fig. 8.24).

In addition to the simplification that a known pressure gradient provides, $\partial^2 u/\partial x^2$ is much, much less than the large gradients that exist in the $y$-direction (refer to the sketch of Fig. 8.24); consequently, neglecting $\partial^2 u/\partial x^2$, the boundary-layer equation that must be solved is

$$u \frac{\partial u}{\partial x} + v \frac{\partial u}{\partial y} = -\frac{1}{\rho} \frac{dp}{dx} + v \frac{\partial^2 u}{\partial y^2} \qquad (8.6.45)$$

where the pressure gradient $dp/dx$ is assumed to be known from the inviscid flow solution. This is often referred to as the *Prandtl boundary-layer equation*, named after Ludwig Prandtl (1875–1953). Neither term on the left can be neglected; the $y$-component $v$ may be small, but the velocity gradient $\partial u/\partial y$ is quite large; hence the product must be retained.

Let us focus on flow over a flat plate with zero pressure gradient. In addition, let us introduce the stream function:

$$u = \frac{\partial \psi}{\partial y} \qquad v = -\frac{\partial \psi}{\partial x} \qquad (8.6.46)$$

The boundary-layer equation in terms of the stream function, becomes

$$\frac{\partial \psi}{\partial y} \frac{\partial^2 \psi}{\partial x \partial y} - \frac{\partial \psi}{\partial x} \frac{\partial^2 \psi}{\partial y^2} = v \frac{\partial^3 \psi}{\partial y^3} \qquad (8.6.47)$$

In this form the $x$ and $y$ dependence cannot be separated. If we transform this equation (such transformations are selected by trial and error and experience) by letting

$$\xi = x \qquad \eta = y \sqrt{\frac{U_\infty}{vx}} \qquad (8.6.48)$$

there results

$$-\frac{1}{2\xi} \left( \frac{\partial \psi}{\partial \eta} \right)^2 + \frac{\partial \psi}{\partial \eta} \frac{\partial^2 \psi}{\partial \xi \partial \eta} - \frac{\partial \psi}{\partial \xi} \frac{\partial^2 \psi}{\partial \eta^2} = v \frac{\partial^3 \psi}{\partial \eta^3} \sqrt{\frac{U_\infty}{v\xi}} \qquad (8.6.49)$$

This equation may appear to be a more difficult equation to solve than Eq. 8.6.47, but by observing the position of $\xi$ in this equation, we separate variables by letting

$$\psi(\xi, \eta) = \sqrt{U_\infty v \xi}\, F(\eta) \qquad (8.6.50)$$

The velocity components can then be shown to be, using Eqs. 8.6.48 and 8.6.50,

$$
\begin{aligned}
u &= \frac{\partial \psi}{\partial y} = U_\infty F'(\eta) \\
v &= -\frac{\partial \psi}{\partial x} = \frac{1}{2}\sqrt{\frac{\nu U_\infty}{x}}\,(\eta F' - F)
\end{aligned}
\tag{8.6.51}
$$

Substitute Eq. 8.6.50 into Eq. 8.6.49 and an ordinary nonlinear, differential equation results; it is

$$
F\frac{d^2 F}{d\eta^2} + 2\frac{d^3 F}{d\eta^3} = 0
\tag{8.6.52}
$$

This equation replaces the partial differential equation (8.6.47). Now let us state the boundary conditions.

The boundary conditions $[u(x,0) = 0, v(x,0) = 0$ and $u(x, y > \delta) = U_\infty]$ take the form

$$
F = F' = 0 \text{ at } \eta = 0 \quad \text{and} \quad F' = 1 \text{ at large } \eta
\tag{8.6.53}
$$

The boundary-value problem, consisting of the ordinary differential equation (8.6.52) and boundary conditions (8.6.53), can now be solved numerically. The results are tabulated in Table 8.5. The last two columns are used to provide $v$ and $\tau_0$, respectively.

Defining the boundary-layer thickness to be the point where $u = 0.99U_\infty$, we see from Table 8.5 that this occurs where $\eta = 5$. Hence, letting $\eta = 5$ and $y = \delta$ in Eq. 8.6.48, we have

$$
\delta = 5\sqrt{\frac{\nu x}{U_\infty}}
\tag{8.6.54}
$$

Using

$$
\frac{\partial u}{\partial y} = \frac{\partial u}{\partial \eta}\frac{\partial \eta}{\partial y} = U_\infty F'' \sqrt{\frac{U_\infty}{\nu x}}
\tag{8.6.55}
$$

the wall shear in a laminar boundary layer with $dp/dx = 0$ is

$$
\tau_0 = \mu\left.\frac{\partial u}{\partial y}\right|_{y=0} = 0.332\,\rho U_\infty^2 \sqrt{\frac{\nu}{x U_\infty}}
\tag{8.6.56}
$$

**TABLE 8.5**    Solution for the Laminar Boundary Layer with $dp/dx = 0$

| $\eta = y \sqrt{\dfrac{U_\infty}{\nu x}}$ | $F$ | $F' = u/U_\infty$ | $\frac{1}{2}(\eta F' - F)$ | $F''$ |
|:---:|:---:|:---:|:---:|:---:|
| 0 | 0 | 0 | 0 | 0.3321 |
| 1 | 0.1656 | 0.3298 | 0.0821 | 0.3230 |
| 2 | 0.6500 | 0.6298 | 0.3005 | 0.2668 |
| 3 | 1.397 | 0.8461 | 0.5708 | 0.1614 |
| 4 | 2.306 | 0.9555 | 0.7581 | 0.0642 |
| 5 | 3.283 | 0.9916 | 0.8379 | 0.0159 |
| 6 | 4.280 | 0.9990 | 0.8572 | 0.0024 |
| 7 | 5.279 | 0.9999 | 0.8604 | 0.0002 |
| 8 | 6.279 | 1.0000 | 0.8605 | 0.0000 |

The local skin friction coefficient is

$$c_f = \frac{0.664}{\sqrt{\mathrm{Re}_x}} \qquad (8.6.57)$$

and the skin friction coefficient is

$$C_f = \frac{1.33}{\sqrt{\mathrm{Re}_L}} \qquad (8.6.58)$$

Numerically integrating Eqs. 8.6.6 and 8.6.7, the displacement and momentum thicknesses are found to be

$$\delta_d = 1.72 \sqrt{\frac{\nu x}{U_\infty}} \qquad \theta = 0.644 \sqrt{\frac{\nu x}{U_\infty}} \qquad (8.6.59)$$

## Example 8.17

Atmospheric air at 30°C flows over a 8-m-long, 2-m-wide flat plate at 2 m/s. Assume that laminar flow exists in the boundary layer over the entire length. At $x = 8$ m, calculate (a) the maximum value of $v$, (b) the wall shear, and (c) the flow rate through the layer. (d) Also, calculate the drag force on the plate.

*(continued)*

**Solution**

(a) The $y$-component of velocity has been assumed to be small in boundary-layer theory. Its maximum value at $x = 8$ m is found, using 8.6.51, to be

$$v = \sqrt{\frac{\nu U_\infty}{x}} \times \frac{1}{2}(\eta F' - F)$$

$$= \sqrt{\frac{1.6 \times 10^{-5} \times 2}{8}} \times 0.86 = 0.00172 \text{ m/s}$$

where 0.86 comes from Table 8.5. Compare $v$ with $U_\infty = 2$ m/s.

(b) The wall shear at $x = 8$ m is found using Eq. 8.6.56 to be

$$\tau_0 = 0.332 \rho U_\infty^2 \sqrt{\frac{\nu}{U_\infty x}}$$

$$= 0.332 \times 1.16 \text{ kg/m}^3 \times 2^2 \text{ m}^2/\text{s}^2 \sqrt{\frac{1.6 \times 10^{-5} \text{ m}^2/\text{s}}{2 \text{ m}^2/\text{s} \times 8 \text{ m}}}$$

$$= 0.00154 \text{ Pa}$$

(c) The flow rate through the boundary layer at $x = 8$ m is given by

$$Q = \int_0^\delta u \times w\, dy = w \sqrt{\frac{\nu x}{U_\infty}} \int_0^5 U_\infty F'\, d\eta$$

where we have substituted for $u$ and $y$ from Eqs. 8.6.51 and 8.6.48. Recognizing that $\int F'\, d\eta = F$, the flow rate is

$$Q = wU_\infty \sqrt{\frac{\nu x}{U_\infty}} [F(5) - \overset{0}{F(\emptyset)}]$$

$$= 2 \text{ m} \times 2 \text{ m/s} \sqrt{\frac{1.6 \times 10^{-5} \text{ m}^2/\text{s} \times 8 \text{ m}}{2 \text{ m/s}}} \times 3.28 = 0.105 \text{ m}^3 \cdot \text{s}$$

(d) The drag force is determined to be

$$F_D = \frac{1}{2} \rho U_\infty^2 L w C_f$$

$$= \frac{1}{2} \times 1.16 \text{ kg/m}^3 \times 2^2 \text{ m}^2/\text{s}^2 \times 8 \text{ m} \times 2 \text{ m} \times \frac{1.33}{\sqrt{2 \times 8/1.6 \times 10^{-5}}}$$

$$= 0.049 \text{ N}$$

 **Example 8.17a**    Laminar Boundary Layer Growth, 625–627

### 8.6.7  Pressure-Gradient Effects

In the foregoing sections we have concentrated our study of boundary layers on the flat plate with a zero pressure gradient. This is the simplest boundary-layer flow and allows us to model many flows of engineering interest. The inclusion of a pressure gradient, even though relatively low, markedly alters the boundary-layer flow. In fact, a strong negative pressure gradient (such as flow in a contraction) may relaminarize a turbulent boundary layer; that is, the turbulence production in the viscous wall layer which sustains the turbulence ceases and a laminar boundary layer is reestablished. A positive pressure gradient quickly causes the boundary layer to thicken and eventually to separate. These two effects are shown in the photographs of Fig. 8.27.

(a)

(b)

**Fig. 8.27**   Influence of a strong pressure gradient on a turbulent flow: (a) a strong negative pressure gradient may relaminarize a flow; (b) a strong positive pressure gradient causes a strong boundary layer to thicken. (Photograph by R. E. Falco)

The flow about any plane body with curvature, such as an airfoil, can be modeled as flow over a flat plate with a nonzero pressure gradient. The boundary-layer thickness is so much smaller than the radius of curvature that the additional curvature terms drop out of the differential equations. The inviscid flow solution at the wall provides the pressure gradient $dp/dx$ and the velocity $U(x)$ at the edge of the boundary-layer. For axisymmetric flows, such as flow over the nose of an aircraft, boundary layer equations in cylindrical coordinates must be utilized.

The pressure gradient determines the value of the second derivative $\partial^2 u/\partial y^2$ at the wall. From the boundary-layer equation (8.6.45) at the wall, $u = v = 0$, so that

$$\frac{dp}{dx} = \mu \left.\frac{\partial^2 u}{\partial y^2}\right|_{y=0} \tag{8.6.60}$$

for either a laminar or a turbulent boundary-layer flow. For a zero pressure gradient, the second derivative is zero at the wall; then since the first derivative is greatest at the wall and decreases as $y$ increases, the second derivative must be negative for positive $y$. The profiles are sketched in Fig. 8.28a.

For a negative (favorable) pressure gradient, the slope of the velocity profile near the wall is relatively large with a negative second derivative at the wall and throughout the layer. The momentum near the wall is larger than that of the zero pressure gradient flow, as shown in Fig. 8.28b, and thus there is a reduced tendency for the flow to separate. Turbulence production is discouraged, and relaminarization may occur for a sufficiently large negative pressure gradient over a sufficient distance.

**KEY CONCEPT**   *For a negative pressure gradient, there is a reduced tendency for the flow to separate.*

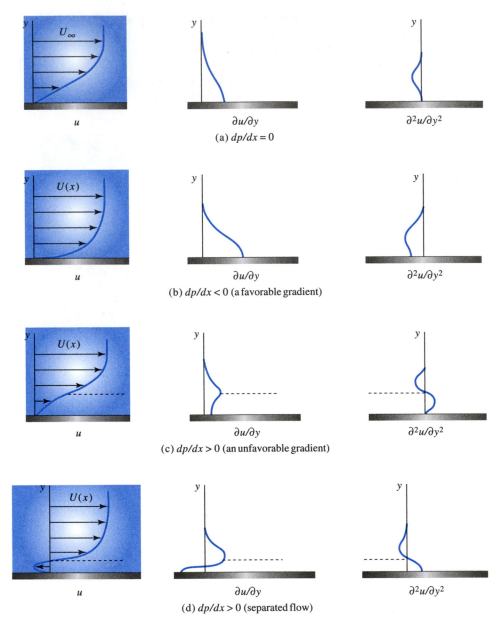

**Fig. 8.28** Influence of the pressure gradient.

If a positive (unfavorable) pressure gradient is imposed on the flow, the second derivative at the wall will be positive and the flow will be as sketched in part (c) or (d). If the unfavorable pressure gradient acts over a sufficient distance, part (d) will probably represent the flow situation with the flow separated from the surface. Near the wall the higher pressure downstream will drive the low momentum flow near the wall in the upstream direction, resulting in flow reversal, as shown. The point at which $\partial u/\partial y = 0$ at the wall locates the point of separation.

The problem of a laminar boundary layer with a pressure gradient can be solved using conventional numerical techniques. The procedure is relatively simple using the simplified boundary-layer equation (8.6.45) with a known pressure gradient. For a turbulent flow the Reynolds stress term must be included; research continues to develop models of turbulent quantities that will result in acceptable numerical solutions. Experimental results are often necessary for turbulent-flow problems, as was the situation for internal flows.

## 8.7 SUMMARY

The drag and lift coefficients are defined as

$$C_D = \frac{\text{Drag}}{\frac{1}{2}\rho V^2 A} \qquad C_L = \frac{\text{Lift}}{\frac{1}{2}\rho V^2 A} \tag{8.7.1}$$

where the area is the projected area for blunt objects, and the chord times the length for an airfoil.

Vortex shedding occurs from a cylinder whenever the Reynolds number is in the range $300 < \text{Re} < 10\,000$. The frequency of shedding is found from the Strouhal number

$$\text{St} = \frac{fD}{V} \tag{8.7.2}$$

where $f$ is the frequency, in hertz.

Plane potential flows are constructed by superimposing the following simple flows:

Uniform flow: $\psi = U_\infty y$

Line source: $\psi = \dfrac{q}{2\pi\theta}$

Irrotational vortex: $\psi = \dfrac{\Gamma}{2\pi}\ln r$     (8.7.3)

Doublet: $\psi = -\dfrac{\mu}{r}\sin\theta$

The stream function for the rotating cylinder is given by

$$\psi_{\text{cylinder}} = U_\infty y - \frac{\mu}{r}\sin\theta + \frac{\Gamma}{2\pi}\ln r \tag{8.7.4}$$

where the cylinder radius is

$$r_c = \sqrt{\frac{\mu}{U_\infty}} \tag{8.7.5}$$

The velocity components are

$$u = \frac{\partial \psi}{\partial y} \qquad v = -\frac{\partial \psi}{\partial x}$$

$$v_r = \frac{1}{r}\frac{\partial \psi}{\partial \theta} \qquad v_\theta = -\frac{\partial \psi}{\partial r} \qquad (8.7.6)$$

For a laminar boundary layer on a flat plate with zero pressure gradient the exact solution provides

$$\delta = 5\sqrt{\frac{\nu x}{U_\infty}} \qquad c_f = 0.664\sqrt{\frac{\nu}{xU_\infty}} \qquad C_f = 1.33\sqrt{\frac{\nu}{LU_\infty}} \qquad (8.7.7)$$

For a turbulent flow from the leading edge, the power-law profile with $\eta = 7$ gives

$$\delta = 0.38x\left(\frac{\nu}{xU_\infty}\right)^{1/5} \qquad c_f = 0.059\left(\frac{xU_\infty}{\nu}\right)^{1/5} \qquad C_f = 0.073\left(\frac{\nu}{LU_\infty}\right)^{1/5} \qquad (8.7.8)$$

where the wall shear and drag force per unit width are, respectively,

$$\tau_0 = \frac{1}{2}c_f\rho U_\infty^2 \qquad F_D = \frac{1}{2}C_f\rho U_\infty^2 L \qquad (8.7.9)$$

---

## FUNDAMENTALS OF ENGINEERING EXAM REVIEW PROBLEMS

**8.1** The drag force on a streamlined shape is due primarily to:
  (A) The wake
  (B) The component of the pressure force acting in the flow direction
  (C) The shear stress
  (D) The separated region near the trailing edge

**8.2** A golf ball has dimples to increase its flight distance. Select the best reason that accounts for the longer flight distance of a dimpled ball compared to that of a smooth ball
  (A) The shearing stress is smaller on the dimpled ball
  (B) The dimpled ball has an effective smaller diameter
  (C) The wake on the dimpled ball is smaller
  (D) The pressure on the front of the smooth ball is larger

**8.3** Flood waters at 10°C flow over an 8-mm diameter fence wire with a velocity of 0.8 m/s. Which of the following is true?
  (A) It is a Stokes flow with no separation
  (B) The separated region covers most of the rear of the wire
  (C) The drag is due to the relatively low pressure in the separated region
  (D) The separated region covers only a small area at the rear of the wire

**8.4** The drag on a 10-m-diameter spherical water storage tank in an 80 km/h wind is approximately:
  (A) 6300 N       (C) 3200 N
  (B) 4700 N       (D) 2300 N

**8.5** A 4-m-long smooth cylinder experiences a drag of 60 N when subjected to an atmospheric air speed of 40 m/s. Estimate the diameter of the cylinder.
  (A) 127 mm       (C) 26 mm
  (B) 63 mm        (D) 4.1 mm

**8.6** Vortices are shed from a 2-cm-diameter cylinder due to a 4 m/s air stream. How far apart would you expect the vortices to be downstream of the cylinder?
(A) 44 cm    (C) 9 cm
(B) 23 cm    (D) 4 cm

**8.7** Streamlining reduces drag primarily by:
(A) Reducing the wall shear
(B) Reducing the pressure in the stagnation region
(C) Reducing the separated flow area
(D) Eliminating the wake

**8.8** Estimate the takeoff speed needed for a 1200-kg aircraft (including payload) if the angle of attack at takeoff is to be 10°. The effective wing area (chord times length) is 16 m².
(A) 22 m/s    (B) 33 m/s
(C) 44 m/s    (D) 55 m/s

---

## PROBLEMS

### Separated Flows

**8.9** Sketch the flow over an airfoil at a large angle of attack for both attached flow and separated flow. Also, sketch the expected pressure distributions on the top and bottom surfaces for both flows. Identify regions of favorable and unfavorable pressure gradients.

**8.10** A spherical particle is moving in 20°C atmospheric air at a velocity of 20 m/s. What must its diameter be for Re = 5 and Re = $10^5$? Sketch the expected flow field at these Reynolds numbers. Identify all regions of the flow.

**8.11** Sketch the flow that is expected over a semitruck (a tractor and trailer) where the trailer is substantially higher than the tractor with and without an air deflector attached to the roof of the tractor. Sketch a side view indicating any separated regions, boundary layers, and the wake.

**8.12** Air is blowing by a long, rectangular building with the wind blowing parallel to the long sides. Sketch the top view showing the separated flow regions, the inviscid flow region, the boundary layers, and the wake.

**8.13** A 2-cm-diameter sphere is to move with Re = 5. At what velocity does it travel if it is submerged in:
(a) Water at 15°C?
(b) Water at 80°C?
(c) Standard air at 15°C?

**8.14** Air at 20°C flows around a cylindrical body at a velocity of 20 m/s. Calculate the Reynolds number if the body is:
(a) A 6-m-diameter smokestack
(b) A 6-cm-diameter flagpole
(c) A 6-mm-diameter wire

Use Re = $VD/\nu$. Would a separated flow be expected?

**8.15** The pressure distribution on the front of a 2-m-diameter disk (Fig. P8.15) is approximated by $p(r) = p_0(1 - r^2)$. If $V$ = 20 m/s in this 20°C–atmospheric airflow, estimate the drag force and the drag coefficient for this disk. Assume that the pressure on the back side is zero.

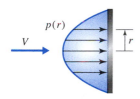

**Fig. P8.15**

**8.16** A flat plate 30 cm × 30 cm acts as a hydrofoil. If it is oriented at an angle of attack of 10°, estimate the lift and the drag if the pressure on the bottom side is 20 kPa and on the top side there is a vacuum of 10 kPa; neglect the effect of shear stress. Also, estimate the lift and drag coefficients if the hydrofoil velocity is 5 m/s. Use the surface area of the plate in the definition of the lift and drag coefficients.

**8.17** The symmetric airfoil shown in Fig. P8.17 flies at an altitude of 12 000 m with an angle of attack of 5°. If $p_l$ = 26 kPa and $p_u$ = 8 kPa, estimate the lift and drag coefficients neglecting shear stresses.

**Fig. P8.17**

**8.18** If the drag coefficient for a 10-cm-diameter sphere is given by $C_D = 1.0$, calculate the drag if the sphere is falling in the atmosphere:
  **(a)** At sea level  **(c)** In 10°C water
  **(b)** At 30 000 m

**8.19** Calculate the drag on a smooth 50-cm-diameter sphere when subjected to a 20°C–atmospheric airflow of:
  **(a)** 6 m/s
  **(b)** 15 m/s
  **(c)** Flow past a sphere, page 5

**8.20** A 4.45-cm-diameter golf ball is roughened to reduce the drag during its flight. If the Reynolds number at which the sudden drop occurs is reduced from $3 \times 10^5$ to $6 \times 10^4$ by the roughening (dimples), would you expect this to lengthen the flight of a golf ball significantly? Justify your reasoning with appropriate calculations.

**8.21** A 10-cm-diameter smooth sphere experiences a drag of 2.2 N when placed in standard 15°C air.
  **(a)** What is the velocity of the airstream?
  **(b)** At what increased speed will the sphere experience the same drag?

**8.22** A smooth 20-cm-diameter sphere experiences a drag of 4.2 N when placed in a 20°C water channel. Calculate the drag coefficient and the Reynolds number.

**8.23** A 2-m-diameter smokestack stands 60 m high. It is designed to resist a 40-m/s wind. At this speed, what total force would be expected, and what moment would the base be required to resist? Assume atmospheric–20°C air.

**8.24** A flagpole is composed of three sections: a 5-cm-diameter top section that is 10 m long, a 7.5-cm-diameter middle section 15 m long, and a 10-cm-diameter bottom section 20 m long. Calculate the total force acting on the flagpole and the resisting moment provided by the base when subjected to a 25-m/s wind speed. Make the calculations for:
  **(a)** A winter day at −30°C
  **(b)** A summer day at 35°C

**8.25** A drag force of 45 N is desired at Re $= 10^5$ on a 1.8-m-long cylinder in a 15°C–atmospheric airflow. What velocity should be selected, and what should be the cylinder diameter?

**8.26** A 20-m-high structure is 2 m in diameter at the top and 8 m in diameter at the bottom as shown in Fig. P8.26. If the diameter varies linearly with height, estimate the total drag force due to a 30-m/s wind. Use atmospheric air at 20°C.

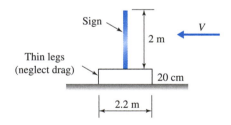

Fig. P8.26

**8.27** A steel sphere ($S = 7.82$) is dropped in water at 20°C. Calculate the terminal velocity if the diameter of the sphere is:
  **(a)** 10 cm  **(c)** 1 cm
  **(b)** 5 cm  **(d)** 2 mm

**8.28** Estimate the terminal velocity of a 50-cm-diameter sphere as it falls in a 15°C atmosphere near the Earth if it has a specific gravity of:
  **(a)** 0.005  **(b)** 0.02  **(c)** 1.0

**8.29** Estimate the terminal velocity of a skydiver by making reasonable approximations of the arms, legs, head, and body. Assume air at 20°C.

**8.30** Assuming that the drag on a modern automobile at high speeds is due primarily to form drag, estimate the power (horsepower) needed by an automobile with a 3.2-m$^2$ cross-sectional area to travel at:
  **(a)** 80 km/h  **(b)** 90 km/h
  **(c)** 100 km/h

**8.31** The 2 m × 3 m sign shown in Fig. P8.31 weighs 400 N. What wind speed is required to blow over the sign?

Fig. P8.31

**8.32** Calculate the drag force on a 60-cm-diameter, 6-m-long cylinder if 20°C-air is blowing normal to its axis at 40 km/h and the cylinder:
  **(a)** Is a section of a very long cylinder
  **(b)** Has both ends free
  **(c)** Is fixed to the ground with the top free.

**8.33** An 80-kg paratrooper leaps from an elevation of 3000 m. Estimate the paratrooper's landing speed if she:
  **(a)** Curls up as tightly as possible
  **(b)** Uses an 8-m-diameter lightweight parachute.
  **(c)** Uses a 2-m-diameter safety parachute

**8.34** A semitruck travels 200 000 km each year at an average speed of 90 km/h. Estimate the savings if a streamlined deflector is added to reduce the drag coefficient. Fuel costs $0.40 per liter and the truck without the deflector averages 1.2 km per liter of fuel.

**8.35** A rectangular car carrier has a cross section 1.8 m × 0.6 m. Estimate the minimum added power (horsepower) required to travel at 90 km/h because of the car carrier.

**8.36** Assume that the velocity at the corners of an automobile where the rear-view mirrors are located is 1.6 times the automobile's velocity. How much horsepower is required by the two 10-cm-diameter rearview mirrors for an automobile speed of 100 km/h?

**8.37** Atmospheric air at 25°C blows normal to a 4-m-long section of a cone, 30 cm in diameter at one end and 2 m in diameter at the other end, with a velocity of 20 m/s. Predict the drag on the object. Assume that $C_D = 0.4$ for each cylindrical element of the object.

**8.38** An 80-cm-diameter balloon (Fig. P8.38) that weighs 0.5 N is filled with 20°C helium at 20 kPa. Neglecting the weight of the string, estimate V if α equals:
  **(a)** 80°
  **(b)** 70°
  **(c)** 60°
  **(d)** 50°

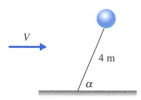

**Fig. P8.38**

**8.39** For the newly planted tree shown in Fig. P8.39, the soil–ball interface is capable of resisting a moment of 5000 N·m. Predict the minimum wind speed that might possibly tip the tree over. Assume that $C_D = 0.4$ for a cylinder in this air flow.

**Fig. P8.39**

**8.40** A sign 1.2 m × 0.6 m is attached to the top of a pizza delivery car. The car travels 10 hours a day, 6 days a week. Estimate the added cost in a year that the sign adds to the fuel used by the car. The average speed of the car is 40 km/h, fuel costs $0.60 per liter, the engine/drive train is 30% efficient, and the fuel contains 12 000 kJ/kg.

**8.41** A bike rider is able to travel at an average speed of 25 mph while riding upright. It is determined that the rider's projected area is 0.56 m². If the rider assumes a racing position so that his projected area is 0.40 m², estimate his increased speed if his drag coefficient decreases 20%, assuming the same energy expenditure.

**8.42** An automobile with a cross-sectional area of 3 m² is powered by a 40-hp engine. Estimate the maximum possible speed if the drive train is 90% efficient. (The engine is rated by the power produced before the transmission.)

## Vortex Shedding

**8.43** Over what range of velocities would you expect vortex shedding from a telephone wire 3 mm in diameter? Could you hear any of the vorticies being shed? (Human beings can hear frequencies between 20 and 20 000 Hz.)

**8.44** A wire is being towed through 15°C water normal to its axis at a speed of 1.8 m/s. What diameter (both large and small) could the wire have so that vortex shedding would not occur?

**8.45** It is quite difficult to measure low velocities. To determine the velocity in a low-speed airflow, the vorticies being shed from a 10-cm-diameter cylinder are observed to occur at 0.2 Hz. Estimate the airspeed if the temperature is 20°C.

**8.46** Motion pictures show that vortices are shed from a 2-m-diameter cylinder at 0.002 Hz while it is moving in 20°C water. What is the cylinder velocity?

**8.47** The cables supporting a suspended bridge (Fig. P8.47) have a natural frequency of $T/(\pi \rho L^2 d^2)$ hertz, where $T$ is the tension, $\rho$ the cable density, $d$ its diameter, and $L$ its length. The vortices shed from the cables may lead to resonance and possible failure. A certain 1.6-cm-diam- eter steel cable is subjected to a force of 30 000 N. What length cable would result in resonance in a 10-m/s wind? (*Note:* The third and fifth harmonics may also lead to resonance. Calculate all three lengths.)

**Fig. P8.47**

## Streamlining

**8.48** A 15-cm-diameter smokestack on a semitruck extends 1.8 m straight up into the free stream. Estimate the horsepower needed because of the stack for a speed of 90 km/h. If the stack were streamlined, estimate the reduced horsepower. Also calculate the power values in kW.

**8.49** A wind speed of 3 m/s blows normal to an 8-cm-diameter smooth cylinder that is 2 m long. Calculate the drag force. The cylinder is now streamlined. What is the percentage reduction in the drag? Assume that $T = 20°C$.

**8.50** Water flows by an 80-cm-diameter cylinder that protrudes 2 m up from the bottom of a river. For an average water speed of 2 m/s, estimate the drag on the cylinder. If the cylinder were streamlined, what would be the percentage reduction in drag?

**8.51** Circular 2-cm-diameter tubes are used as supports on an ultralight airplane, designed to fly at 50 km/h. If there are 20 linear meters of the tubes, estimate the horsepower needed by the tubes. If the tubes were streamlined, estimate the reduced horsepower required of the tubes.

**8.52** A bike racer is able to ride 50 km/h at top speed. Estimate the drag force just due to his head. If he wore a streamlined, tight-fitting helmet, estimate the reduced drag force.

## Cavitation

**8.53** The critical cavitation number for a streamlined strut is 0.7. Find the maximum velocity of the body to which the strut is attached if cavitation is to be avoided. The body is traveling 5 m beneath a water surface.

**8.54** A lift force of 200 kN is desired at a speed of 12 m/s on the hydrofoil of Fig. P8.54, designed to operate at a depth of 40 cm. It has a 40-cm chord and is 10 m long. Calculate the angle of attack and the drag force. Is cavitation present under these conditions?

Angle of attack

**Fig. P8.54**

**8.55** A hydrofoil, designed to operate at a depth of 40 cm, has a 40-cm chord and is 9 m long. A lift force of 200 kN is desired at a speed of 10 m/s. Calculate the angle of attack and the drag force. Is cavitation present under these conditions?

**8.56**  A body resembling a sphere has a diameter of approximately 0.8 m. It is towed at a speed of 20 m/s, 5 m below the water surface. Estimate the drag acting on the body.

**8.57**  A 2200-kg mine sweeper is designed to ride above the water with hydrofoils at the four corners providing the lift. If the hydrofoils have a 40-cm chord length, what total hydrofoil length is required if they are to operate 60 cm below the surface at a 6° angle of attack? The vehicle is to travel at 50 m/s.

## Added Mass

**8.58**  A 40-cm-diameter sphere, which weighs 400 N, is released from rest while submerged in water. Calculate its initial acceleration:
 **(a)**  Neglecting the added mass
 **(b)**  Including the added mass

**8.59**  A submersible, whose length is twice its maximum diameter, resembles an ellipsoid. If its added mass is ignored, what is the percentage error in a calculation of its initial acceleration if its specific gravity is 1.2?

## Lift and Drag on Airfoils

**8.60**  A rectangular wing on a small airplane has a 1.3 m chord and a 10 m span. When flying in air at 220 km/h, the wing experiences a total aerodynamic force of 18 kN. If the lift to drag ratio is 3, determine the lift coefficient of the wing.

**8.61**  In a lift-and-drag experiment performed in a wind tunnel, an NACA0012 airfoil was used. The airfoil has a 15cm chord and a span length of 45cm. Using a force measuring device, a lift force of 60 N was measured on the airfoil for a Reynolds number of $4.586 \times 10^5$. The lift coefficient for this airfoil is given by $C_L = 2\pi \sin\alpha$, where $\alpha$ is the angle of attack. Under the given conditions, what is the angle of attack in degrees?

**8.62**  An aircraft with a mass, including payload, of 1000 kg is designed to cruise at a speed of 80 m/s at 10 km. The effective wing area is approximately 15 m$^2$. Determine the lift coefficient and the angle of attack. What power is required by the airfoil during cruise? Assume a conventional airfoil.

**8.63**  A 1500 kg aircraft is designed to carry a payload of 3000 N when cruising at 80 m/s at a height 10 km. The effective wing area is 20 m$^2$. Assuming a conventional airfoil, calculate:
 **(a)**  The takeoff speed if an angle of attack of 10° is desired
 **(b)**  The stall speed when landing
 **(c)**  The power required at cruise if 45% of the power is needed to move the airfoil

**8.64**  In Problem 8.63 it was necessary to assume that takeoff was at sea level. What would be the takeoff speed in Wyoming, where the elevation is 2000 m?

**8.65**  The aircraft of Problem 8.63 is flown at 2 km rather than 10 km. Estimate the percentage increase or decrease in power required at cruise.

**8.66**  An additional load of 6000 N is added to the aircraft of Problem 8.63. Estimate the takeoff speed if the angle of attack remains at 10°.

**8.67**  Estimate the minimum landing speed for a 250 000-kg aircraft in an emergency situation if the angle of attack is selected to be near stall and:
 **(a)**  No slotted flaps are used
 **(b)**  One slotted flap is used
 **(c)**  Two slotted flaps are used

Its wingspan is 60 m and the airfoil chord averages 8 m.

**8.68**  In Problem 8.67 it is assumed that the aircraft lands at standard sea-level conditions since neither elevation nor temperature is given. Calculate the percentage increase or decrease in the emergency landing speed if the aircraft is to land:
 **(a)**  In Denver where the elevation is 1600 m
 **(b)**  At sea level when the temperature is very cold at −40°C
 **(c)**  At sea level when the temperature is hot at 50°C

**8.69** A proposed aircraft is to resemble a huge airfoil, a flying wing (Fig. P8.69). Its wingspan will be 200 m and its chord will average 30 m. Estimate, assuming a conventional airfoil, the total mass of the aircraft, including payload, for a design speed of 800 km/h at an elevation of 8 km. Also, calculate the power requirement.

**Fig. P8.69**

## Vorticity, Velocity Potential, and Stream Function

**8.70** Take the curl of the Navier–Stokes equation and show that the vorticity equation (8.5.3) results. (See Eq. 5.3.20.)

**8.71** Write the three component vorticity equations contained in Eq. 8.5.3 using rectangular coordinates. Use $\boldsymbol{\omega} = \omega_x\hat{\mathbf{i}} + \omega_y\hat{\mathbf{j}} + \omega_z\hat{\mathbf{k}}$.

**8.72** Simplify the vorticity equation 8.5.3 for a plane flow ($w = 0$ and $\partial/\partial z = 0$). Use $\boldsymbol{\omega} = (\omega_x, \omega_y, \omega_z)$. What conclusion can be made about the magnitude of $\omega_z$ in an inviscid, plane flow (such as flow through a short contraction) that contains vorticity?

**8.73** Consider the flow in the contracting channel shown below. It is assumed that a vortex tube ($y$-vorticity) exists in the boundary layer on the upper and lower plates. Using the vorticity transport equation, explain the existence of $x$-vorticity downstream of the obstruction.

**Fig. P8.73**

**8.74** Determine which of the following flows are irrotational and incompressible and find the velocity potential function, should one exist for each incompressible flow.
(a) $\mathbf{V} = 10x\hat{\mathbf{i}} + 20y\hat{\mathbf{j}}$
(b) $\mathbf{V} = 8y\hat{\mathbf{i}} + 8x\hat{\mathbf{j}} - 6z\hat{\mathbf{k}}$
(c) $\mathbf{V} = (x\hat{\mathbf{i}} + y\hat{\mathbf{j}})/\sqrt{x^2 + y^2}$
(d) $\mathbf{V} = (x\hat{\mathbf{i}} + y\hat{\mathbf{j}})/(x^2 + y^2)$

**8.75** An attempt is to be made to solve Laplace's equation for flow around a circular cylinder of radius $r_c$ oriented in the center of a channel of height $2h$. The velocity profile far from the cylinder is uniform. State the necessary boundary conditions. Assume that $\psi = 0$ at $y = -h$. The origin of the coordinate system is located at the center of the cylinder.

**8.76** State the stream function and velocity potential corresponding to a uniform velocity of $100\hat{\mathbf{i}} + 50\hat{\mathbf{j}}$ using rectangular coordinates.

**8.77** A flow is represented by the stream function $\psi = 40 \tan^{-1}(y/x)$.
(a) Express the stream function in polar form
(b) Is this an incompressible flow? Show why
(c) Determine the velocity potential
(d) Find the radius where the acceleration is $-10$ m/s$^2$

**8.78** The stream function for a flow is
$\psi = 20 \ln(x^2 + y^2)$ m$^2$/s.

Determine the complex velocity potential for this incompressible flow. If the pressure at a large distance from the origin is 20 kPa, what is the pressure at the point (0, 20 cm) if water is flowing?

**8.79** A stream function is given by

$$\psi = 10y - \frac{10y}{x^2 + y^2}$$

(a) Show that this satisfies $\nabla^2\psi = 0$
(b) Find the velocity potential $\phi(x, y)$
(c) Assuming water to be flowing, find the pressure along the $x$-axis if $p = 50$ kPa at $x = -\infty$
(d) Locate any stagnation points.

**8.80** The velocity potential for a flow is

$$\phi = 10x + 5 \ln (x^2 + y^2)$$

 (a) Show that this function satisfies Laplace's equation
 (b) Find the stream function $\psi (x, y)$
 (c) Assume that water is flowing and find the pressure along the $x$-axis if $p = 100$ kPa at $x = -\infty$

 (d) Locate any stagnation points
 (e) Find the acceleration at $x = -2$ m, $y = 0$

**8.81** The velocity profile in a wide 0.2-m-high channel is given by $u(y) = y - y^2/0.2$. Determine the stream function for this flow. Calculate the flow rate by integrating the velocity profile and by using $\Delta \psi$. Explain why a velocity potential does not exist by referring to Eq. 8.5.2.

## Superposition of Simple Flows

**8.82** The body formed by superimposing a source at the origin of strength $0.45\pi$ m²/s and a uniform flow of 9 m/s is shown in Fig. P8.82.
 (a) Locate any stagnation points
 (b) Find the $y$-intercept $y_B$ of the body
 (c) Find the thickness of the body at $x = \infty$
 (d) Find $u$ at $x = -0.3$ m, $y = 0$

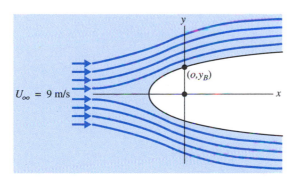

**Fig. P8.82**

**8.83** A source with strength $\pi$ m²/s and a sink of equal strength are located at $(-1 \text{ m}, 0)$ and $(1 \text{ m}, 0)$, respectively. They are combined with a uniform flow $U_\infty = 10$ m/s to form a *Rankine oval*. Calculate the length and maximum thickness of the oval. If $p = 10$ kPa at $x = -\infty$, find the minimum pressure if water is flowing.

**8.84** An oval is formed from a source and sink of strengths $2\pi$ m²/s located at $(-1, 0)$ and $(1, 0)$, respectively, combined with a uniform flow of 2 m/s. Locate any stagnation points, and find the velocity at $(-4, 0)$ and $(0, 4)$. Distances are in meters.

**8.85** Two sources of strength $2\pi$ m²/s are located at $(0, 1)$ and $(0, -1)$, respectively. Sketch the resulting flow and locate any stagnation points. Find the velocity at $(1, 1)$. Distances are in meters.

**8.86** The two sources of Problem 8.85 are superimposed with a uniform flow. Sketch the flow, locate any stagnation points, and find the $y$-intercept of the body formed if:
 (a) $U_\infty = 10$ m/s
 (b) $U_\infty = 1$ m/s
 (c) $U_\infty = 0.2$ m/s

**8.87** A doublet of strength 60 m³/s is superimposed with a uniform flow of 8 m/s of water. Calculate:
 (a) The radius of the resulting cylinder
 (b) The pressure increase from $x = -\infty$ to the stagnation point
 (c) The velocity $v_\theta (\theta)$ on the cylinder.
 (d) The pressure decrease from the stagnation point to the point of minimum pressure on the cylinder

**8.88** A sink of strength $4\pi$ m²/s is superimposed with a vortex of strength $20\pi$ m²/s.
 (a) Sketch a pathline of a particle initially occupying the point $(x = 0, y = 1 \text{ m})$. Use tangents at every 45° with straight-line extensions
 (b) Calculate the acceleration at $(0, 1)$
 (c) If $p(10, 10) = 20$ kPa, what is $p (0, 0.1 \text{ m})$ if atmospheric air is flowing?

**8.89** The cylinder shown in Fig. P8.89 is formed by combining a doublet of strength 40 m³/s with a uniform flow of 10 m/s.
 (a) Sketch the velocity along the $y$-axis from the cylinder to $y = \infty$
 (b) Calculate the velocity at $(x = -4$ m, $y = 3$ m)
 (c) Calculate the drag coefficient for the cylinder assuming potential flow over the front half and constant pressure over the back half

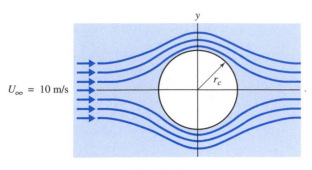

**Fig. P8.89**

**8.90** A 2-m-diameter cylinder is placed in a uniform water flow of 4 m/s.
   **(a)** Sketch the velocity along the x-axis from the cylinder to $x = -\infty$
   **(b)** Find $v_\theta$ on the front half of the cylinder
   **(c)** Find $p(\theta)$ on the front half of the cylinder if $p = 50$ kPa at $x = -\infty$
   **(d)** Estimate the drag force on a 1-m length of the cylinder if the pressure over the rear half is constant and equal to the value at $\theta = 90°$

**8.91** Superimpose a free stream $U_\infty = 9$ m/s, a doublet $\mu = 36$ m$^2$/s, and a vortex $\Gamma = 90$ m$^2$/s. Locate any stagnation points and predict the minimum and maximum pressure on the surface of the cylinder if $p = 0$ at $x = -\infty$ and atmospheric air is flowing.

**8.92** A 0.8-m-diameter cylinder is placed in a 20-m/s atmospheric airflow. At what rotational speed should the cylinder rotate so that only one stagnation point exists on its surface? Calculate the minimum pressure acting on the cylinder if $p = 0$ at $x = -\infty$.

**8.93** A 1.2-m-diameter cylinder rotates at 120 rpm in a 3-m/s atmospheric airstream. Locate any stagnation points and calculate the minimum and maximum pressures on the cylinder if $p = 0$ at $x = -\infty$.

**8.94** The circulation around a 18-m-long airfoil (measured tip to tip) is calculated to have a value of 1400 m$^2$/s. Estimate the lift generated by the airfoil if the aircraft is flying at 9000 m with a speed of 105 m/s. Assume the flow to be incompressible.

**8.95** The velocity field due to a source of strength $2\pi$ m$^2$/s per meter, located at (2 m, 2 m) in a 90° corner, is desired. Use the *method of images*, that is, add one or more sources at the appropriate locations, and determine the velocity field by finding $u(x, y)$ and $v(x, y)$.

**8.96** Flow through a porous medium is modeled with Laplace's equation and an associated velocity potential function. Natural gas can be stored in certain underground rock structures for use at a later time. A well is placed next to an impervious rock formation, as shown in Fig. P8.96. If the well is to extract 0.2 m$^3$/s per meter, predict the velocity to be expected at the point (4 m, 3 m). See Problem 8.95 for the method of images.

(6 m, 2 m)

**Fig. P8.96**

## Boundary Layers

**8.97** How far from the leading edge can turbulence be expected on an airfoil traveling at 90 m/s if the elevation is:
   **(a)** 0 m?    **(b)** 3600 m?    **(c)** 9000 m?

Use $Re_{crit} = 6 \times 10^5$ and assume a flat plate with zero pressure gradient.

**8.98** An airfoil on a commercial jetliner can be crudely approximated by a flat plate. Determine the distance from the leading edge at which transition to turbulence can be expected. The jet is traveling at 800 km/h at an altitude of 9000 m, where the pressure and temperature are 31 kPa and $-43°$C, respectively.

**8.99** The boundary layer on a flat plate with zero pressure gradient is to be studied in a tunnel transporting 20°C air. How far from the leading edge would you expect to find turbulent flow if $U_\infty = 10$ m/s and:
   **(a)** The plate is held rigid with high free-stream disturbance level?
   **(b)** The plate is held rigid with low free-stream disturbance level?

**(c)** The plate is vibrated with low free-stream disturbance level?

**(d)** The plate is vibrated with high free-stream disturbance level?

**(e)** At what distance would you expect a small disturbance to grow for the flow of part (b)?

**8.100** Repeat Problem 8.99 but place the flat plate in a channel transporting 20°C water.

**8.101** A laminar region is desired to be at least 2 m long on a smooth rigid flat plate. A wind tunnel and a water channel are available. What maximum speed can be selected for each? Assume low free-stream fluctuation intensity and 20°C for each.

**8.102** Determine the pressure $p(x)$ in the boundary layer and the velocity $U(x)$ at the edge of the boundary layer that would be present on the front of the cylinder of Problem 8.89. Let the pressure at the stagnation point be 20 kPa with water flowing. Let $x$ be measured from the stagnation point; see Figure 8.19.

**8.103** A boundary layer would develop as shown in Fig. P8.103 from the front stagnation point of Problem 8.90. Determine $U(x)$ and $p(x)$ that would be needed to solve for the boundary-layer growth on the cylinder front. Measure $x$ from the stagnation point.

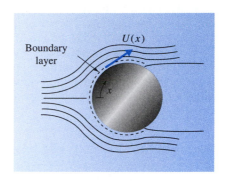

**Fig. P8.103**

**8.104** Assuming inviscid uniform flow of air through the contraction shown in Figure P8.104, estimate $U(x)$ and $dp/dx$, which are necessary to solve for the boundary-layer growth on the flat plate. Assume a one-dimensional flow with $\rho = 1.0$ kg/m³.

**Fig. P8.104**

## Von Kármán Integral Equation

**8.105** Provide the detailed steps to the final form of Eq. 8.6.1 and Eq. 8.6.3. Refer to Figure 8.23.

**8.106** Show that the von Kármán integral equation 8.6.4 can be put in the form

$$\tau_0 = -\delta\frac{dp}{dx} + \rho\frac{d}{dx}\int_0^\delta u(U-u)\,dy - \rho\frac{dU}{dx}\int_0^\delta u\,dy$$

Note that the quantity $\int_0^\delta u\,dy$ is only a function of $x$.

**8.107** Show that the von Kármán integral equation of Problem 8.106 can be written as

$$\tau_0 = \rho\frac{d}{dx}(\theta U^2) + \rho\delta_d U\frac{dU}{dx}$$

To accomplish this, we must show that Bernoulli's equation $p + \rho U^2/2 = $ const. can be differentiated to yield

$$\frac{dp}{dx} = -\rho U\frac{dU}{dx} = -\frac{\rho}{\delta}\frac{dU}{dx}\int_0^\delta U\,dy$$

**8.108** Assume that $u = U_\infty \sin(\pi y/2\delta)$ in a zero pressure gradient boundary layer. Calculate:
**(a)** $\delta(x)$
**(b)** $\tau_0(x)$
**(c)** $v$ at $y = \delta$ and $x = 3$ m

**8.109** Assume a linear velocity profile and find $\delta(x)$ and $\tau_0(x)$. Compute the percentage error when compared with the exact expressions for a laminar flow. Use $dp/dx = 0$.

**8.110** A boundary-layer profile is approximated with:

$$u = 3\,U_\infty\frac{y}{\delta} \qquad\qquad 0 < y \le \delta/6$$

$$u = U_\infty\left(\frac{y}{\delta} + \frac{1}{3}\right) \qquad \delta/6 < y \le \delta/2$$

$$u = U_\infty\left(\frac{y}{3\delta} + \frac{2}{3}\right) \qquad \delta/2 < y \le \delta$$

Determine $\delta(x)$ and $\tau_0(x)$; compute the percentage error when compared with the exact expressions for a laminar flow.

**8.111** If the walls in a wind-tunnel test section are parallel, the velocity in the center portion of the tunnel will accelerate as shown in Fig. P8.111. To maintain a constant tunnel velocity so that $dp/dx = 0$, show that the walls should be displaced outward a distance $\delta_d(x)$. If a wind tunnel were square, how far should one wall be displaced outward for $dp/dx = 0$?

**Fig. P8.111**

**8.112** The velocity profile at a given $x$-location in the boundary layer (Fig. P8.112) is assumed to be

$$u(y) = 10 \left( 2\frac{y}{\delta} - \frac{y^2}{\delta^2} \right)$$

A streamline is 2 cm from the flat plate at the leading edge. How far is it from the plate when $x = 3$ m (i.e., what is $h$)? Also, calculate the displacement thickness at $x = 3$ m. Compare the displacement thickness to $(h - 2)$ cm.

**Fig. P8.112**

**8.113** It is desired that the test section in a wind tunnel experience a zero pressure gradient. If the test section has a square cross section, what should be the equation of the displacement of one of the walls (three walls will be straight and parallel or perpendicular) if the 30°C–air is pressurized to 160 kPa absolute? Assume a boundary layer profile of $u/U_\infty = 2y/\delta - y^2/\delta^2$. Assume that $y = 0$ at $x = 0$ of the displacement equation $y(x)$.

**8.114** Find $\delta_d$ and $\theta$ for a laminar boundary layer assuming:
**(a)** A cubic profile
**(b)** A parabolic profile
**(c)** That $u = U_\infty \sin(\pi y/2\delta)$
Compute percentage errors when compared with the exact values of $\delta_d = 1.72\sqrt{vx/U_\infty}$ and $\theta = 0.644\sqrt{vx/U_\infty}$.

**8.115** A laminar flow is maintained in a boundary layer on a 6-m-long, 4.5-m-wide flat plate with 15°C–atmospheric air flowing at 3.6 m/s. Assuming a parabolic profile, calculate:
**(a)** $\delta$ at $x = 6$ m
**(b)** $\tau_0$ at $x = 6$ m
**(c)** The drag force on one side
**(d)** $v$ at $y = \delta$ and $x = 3$ m

**8.116** Work Problem 8.115, but assume a cubic profile.

## Laminar and Turbulent Boundary Layers

**8.117** Atmospheric air at 20°C flows at 10 m/s over a 2-m-long 4-m-wide flat plate. Calculate the maximum boundary-layer thickness and the drag force on one side assuming:
**(a)** Laminar flow over the entire length
**(b)** Turbulent flow over the entire length

**8.118** Fluid flows over a flat plate at 20 m/s. Determine $\delta$ and $\tau_0$ at $x = 6$ m if the fluid is:
**(a)** Atmospheric air at 20°C
**(b)** Water at 20°C
Neglect the laminar portion

**8.119** Assume a turbulent velocity profile $\bar{u} = U_\infty(y/\delta)^{1/7}$. Does this profile satisfy the conditions at $y = \delta$? Can it give the shear stress at the wall? Plot both a cubic laminar profile and the one-seventh power-law profile on the same graph assuming the same boundary-layer thickness.

**8.120** Estimate the drag on one side of a 3.6-m-long, 15-m-wide flat plate if atmospheric air at 15°C is flowing with a velocity of 6 m/s. Assume that:
  (a) $Re_{crit} = 3 \times 10^5$
  (b) $Re_{crit} = 5 \times 10^5$
  (c) $Re_{crit} = 6 \times 10^5$

**8.121** A 1-m-long flat plate with a sharp leading edge that is 2 m wide is towed parallel to itself in 20°C water at 1.2 m/s. Estimate the total drag if:
  (a) $Re_{crit} = 3 \times 10^5$
  (b) $Re_{crit} = 6 \times 10^5$
  (c) $Re_{crit} = 9 \times 10^5$

**8.122** Air moving at 60 km/h is considered to have a zero boundary-layer thickness at a distance of 100 km from shore. At the beach, estimate the boundary-layer thickness and the wall shear using:
  (a) A one-seventh power law
  (b) Empirical data
  Use $T = 20°C$

**8.123** For the conditions of Problem 8.122, calculate:
  (a) The thickness of the viscous wall layer
  (b) The displacement thickness at the beach

**8.124** Atmospheric air at 15°C flows over a flat plate at 90 m/s. At $x = 6$ m, estimate:
  (a) The local skin friction coefficient
  (b) The wall shear
  (c) The viscous wall-layer thickness
  (d) The boundary-layer thickness

**8.125** Water at 20°C flows over a flat plate at 10 m/s. At $x = 3$ m, estimate:
  (a) The viscous wall layer thickness
  (b) The velocity at the edge of the viscous wall layer
  (c) The value of $y$ at the outer edge of the turbulent zone
  (d) The boundary-layer thickness

**8.126** Estimate the total shear drag on a ship traveling at 10 m/s if the sides are flat plates 10 m × 100 m with zero pressure gradients. What is the maximum boundary-layer thickness?

**8.127** The large 600-m-long, 100-m-diameter, cigar-shaped dirigible of Fig. P8.127 is planned for cruises for retired engineers. As a first estimate the drag is calculated assuming it to be a flat plate with zero pressure gradient, with the drag on the nose and rear areas neglected.
  (a) Estimate the power required of each of four engines if it is to cruise at 15 m/s
  (b) Estimate the payload if one half of its volume is filled with helium, and its engines, equipment, and structure have a mass of $1.2 \times 10^6$ kg

**Fig. P8.127**

---

## Laminar Boundary-Layer Equations

**8.128** Assuming that $dp/dx = 0$, show that Eq. 8.6.47 follows from Eq. 8.6.45.

**8.129** Recalling from calculus that

$$\frac{\partial \psi}{\partial y} = \frac{\partial \psi}{\partial \xi}\frac{\partial \xi}{\partial y} + \frac{\partial \psi}{\partial \eta}\frac{\partial \eta}{\partial y}$$

so that

$$\frac{\partial^2 \psi}{\partial y^2} = \frac{\partial}{\partial y}\left(\frac{\partial \psi}{\partial y}\right) = \frac{\partial(\partial\psi/\partial y)}{\partial \xi}\frac{\partial \xi}{\partial y} + \frac{\partial(\partial\psi/\partial y)}{\partial \eta}\frac{\partial \eta}{\partial y}$$

show that Eq. 8.6.49 follows from Eq. 8.6.47.

**8.130** Show that Eqs. 8.6.51 follow from the preceding equations.

**8.131** Solve Eq. 8.6.52 with the appropriate boundary conditions, using a third-order Runge–Kutta scheme (or any other appropriate numerical algorithm) and verify the results of Table 8.5. (This was a Ph.D. thesis project for Blasius before the advent of the computer!)

**8.132** A laminar boundary layer exists on a flat plate with atmospheric air at 20°C moving at 5 m/s. At $x = 2$ m, find:
  (a) The wall shear
  (b) The boundary-layer thickness
  (c) The maximum value of $v$
  (d) The flow rate through the boundary layer

**8.133** A laminar boundary layer exists on a flat plate with atmospheric air at 15°C moving at 4.5 m/s. At $x = 1.8$ m, find:
  **(a)** The wall shear
  **(b)** The boundary-layer thickness
  **(c)** The maximum value of $v$
  **(d)** The flow rate through the boundary layer

**8.134** Water at 20°C flows over a flat plate with zero pressure gradient at 5 m/s. At $x = 2$ m, find:
  **(a)** The wall shear
  **(b)** The boundary-layer thickness
  **(c)** The flow rate through the boundary-layer

**8.135** If, when we defined the boundary layer thickness, we defined $\delta$ to be that $y$-location where

$u = 0.999U_\infty$, estimate the thickness of the boundary layer of:
  **(a)** Problem 8.132     **(b)** Problem 8.133

**8.136** Find the $y$-location where $u = 0.5U_\infty$ for the boundary layer of Problem 8.132. What is the value of $v$ at that $y$-location? What is the shear stress there?

**8.137** Assume that $v$ remains unchanged between $y = \delta$ and $y = 10\delta$. Sketch $u(y)$ for $0 \leq y < 10\delta$ for a flat plate with zero pressure gradient. Now, assume that $v = 0$ at $y = 10\delta$. Again sketch $u(y)$. Explain by referring to appropriate equations.

**8.138** Sketch the boundary-layer velocity profile near the end of the flat plate of Problem 8.115 and show a Blasius profile of the same thickness on the same sketch.

---

## Pressure-Gradient Effects

**8.139** Sketch the velocity profiles near and normal to the cylinder's surface at each of the points indicated in Fig. P8.139. The flow separates at $C$.

**Fig. P8.139**

**8.140** Sketch the expected boundary-layer profiles at each of the points indicated in Fig. P8.140, showing relative thicknesses. The flow undergoes transition to turbulence just after point $A$. It separates at $D$. Show all profiles on the same plot. Indicate the sign of the pressure gradient at each point.

**Fig. P8.140**

A new-generation space vehicle is rendered by an artist as it reenters the earth's atmosphere. A shock wave is shown to be generated at the leading edge of the vessel, signifying that it is traveling faster than the speed of sound. (U.S. National Aeronautics and Space Administration)

# Compressible Flow

## Outline

## Chapter Objectives

The objectives of this chapter are to:

▲ Present the various equations needed to solve the uniform-flow problems of a compressible gas flow.

▲ Apply the basic equations to isentropic flow through a nozzle.

▲ Introduce the normal shock wave.

▲ Solve for velocities and pressures when a shock wave exists in a converging-diverging nozzle.

▲ Analyze the supersonic flow of steam through a nozzle.

▲ Study the oblique shock wave needed to turn a supersonic flow through an angle.

▲ Calculate the angle through which expansion waves can turn a flow around a convex corner.

▲ Provide numerous examples and problems that illustrate isentropic flow, flow through normal and oblique shock waves, and the turning of a supersonic flow around a convex corner.

## 9.1 INTRODUCTION

**Compressible flow:** *The flow of gas in which the density changes significantly between points on a streamline.*

> **KEY CONCEPT** *Not all gas flows are compressible flows.*

In this chapter we consider flows of gases in which the density changes significantly between points on a streamline; such flows are called **compressible flows**. We will consider those problems that can be solved using the integral equations: the continuity equation, the energy equation, and for some problems, the momentum equation.

Not all gas flows are compressible flows, neither are all compressible flows gas flows. At low speeds, less than a Mach number ($M = V/\sqrt{kRT}$) of about 0.3, gas flows may be treated as incompressible flows. This is justified because the density variations caused by the flow are insignificant (less than about 3%). Incompressible gas flows occur in a large number of situations of engineering interest; many of these have been considered in earlier chapters. There are many flows, however, in which the density variations must be accounted for. Included among these are airflows around commercial and military aircraft, airflow through jet engines, and the flow of a gas in compressors and turbines.

There are examples of compressible effects important in liquid flows; water hammer and compression waves due to underwater blasts are examples of compressible liquid flows. Compressibility of rock accounts for the propagation through the earth's surface of longitudinal waves due to an earthquake. In this chapter we are concerned with compressibility effects in gas flows.

Let us introduce the effects of compressibility into the simplest of flow situations. The velocity at a given streamwise location in a conduit is assumed to be uniform and hence does not vary normal to the flow direction. For this simple uniform flow we recall that the continuity equation takes the form (see Eq. 4.4.5)

$$\dot{m} = \rho_1 A_1 V_1 = \rho_2 A_2 V_2 \tag{9.1.1}$$

The momentum equation for the uniform, compressible flow takes the form (see Eq. 4.6.5)

$$\Sigma \mathbf{F} = \dot{m}(\mathbf{V}_2 - \mathbf{V}_1) \tag{9.1.2}$$

The energy equation, neglecting potential energy changes, is written (see Eqs. 4.5.17 and 4.5.18)

$$\frac{\dot{Q} - \dot{W}_S}{\dot{m}} = \frac{V_2^2 - V_1^2}{2} + h_2 - h_1 \tag{9.1.3}$$

where we have used $h = \tilde{u} + p/\rho$. Assuming an ideal gas with constant specific heats, the energy equation takes the form

$$\frac{\dot{Q} - \dot{W}_S}{\dot{m}} = \frac{V_2^2 - V_1^2}{2} + c_p(T_2 - T_1) \tag{9.1.4}$$

or

$$\frac{\dot{Q} - \dot{W}_S}{\dot{m}} = \frac{V_2^2 - V_1^2}{2} + \frac{k}{k-1}\left(\frac{p_2}{\rho_2} - \frac{p_1}{\rho_1}\right) \qquad (9.1.5)$$

where we have used the thermodynamic relations

$$h_2 - h_1 = c_p(T_2 - T_1) \qquad (9.1.6a)$$

$$c_p = R + c_v \qquad (9.1.6b)$$

$$k = \frac{c_p}{c_v} \qquad (9.1.6c)$$

and the ideal-gas law

$$p = \rho RT \qquad (9.1.7)$$

If we have an interest in calculating the entropy change between two sections, we will use the definition of entropy as

$$\Delta S = \int \frac{\delta Q}{T}\bigg|_{\text{reversible}} \qquad (9.1.8)$$

where $\delta Q$ represents the differential heat transfer. Using the first law, this becomes, for an ideal gas with constant specific heat,

$$\Delta s = c_p \ln \frac{T_2}{T_1} - R \ln \frac{p_2}{p_1} \qquad (9.1.9)$$

If a process is adiabatic ($Q = 0$) and reversible (no losses), Eq. 9.1.8 shows that the entropy change is zero (i.e., the flow is *isentropic*). If the flow is isentropic, the relationship above can be used, along with the ideal-gas law and Eqs. 9.1.6b and c, to show that

$$\frac{T_2}{T_1} = \left(\frac{p_2}{p_1}\right)^{(k-1)/k} \qquad \frac{p_2}{p_1} = \left(\frac{\rho_2}{\rho_1}\right)^{k} \qquad (9.1.10)$$

*Note*: Temperatures and pressures must be measured on absolute scales.

Let us now introduce a parameter, the Mach number, that will be of special interest throughout our study of compressible flow.

## 9.2 SPEED OF SOUND AND THE MACH NUMBER

The speed of sound is the speed at which a pressure disturbance of small amplitude travels through a fluid. It is analogous to the small ripple, a gravity wave, that travels radially outward when a pebble is dropped into a pond. To determine

**Sound wave:** *It travels with a velocity c relative to a stationary observer.*

Speed of sound, 57

the speed of sound consider a small pressure disturbance, called a **sound wave**, to be passing through a pipe, as shown in Fig. 9.1. It travels with a velocity $c$ relative to a stationary observer as shown in part (a); the pressure, density, and temperature will change by the small amounts $\Delta p$, $\Delta \rho$, and $\Delta T$, respectively. There will also be an induced velocity $\Delta V$ in the fluid immediately behind the sound wave. To simplify the problem we will create a steady flow by having the observer travel at the speed of the wave so that the sound wave appears to be stationary. The flow will then approach the wave from the right, with the speed of sound $c$, as shown in Fig. 9.1b. The flow properties will all change across the wave and the velocity in the downstream flow will be expressed as $c + \Delta V$, where $\Delta V$ is the small change in velocity. If the flow speed is reduced, the wave would move to the right, and if the flow speed is increased, it would move to the left. For the flow speed equal to the speed of sound, $V = c$, the sound wave would be stationary, as shown.

Let us apply the continuity equation and the energy equation to a small control volume enclosing the sound wave, shown in Fig. 9.1c. The continuity equation (9.1.1) takes the form

$$\rho A c = (\rho + \Delta \rho)A(c + \Delta V) \tag{9.2.1}$$

This can be rewritten as

$$\rho \Delta V = -c \Delta \rho \tag{9.2.2}$$

where we have neglected the higher-order term $\Delta \rho \, \Delta V$; that is, $\Delta \rho$ represents a small percentage change in $\rho$ so that $\Delta \rho \ll \rho$; also, $\Delta V \ll$ c.

The streamwise-component momentum equation yields

$$pA - (p + \Delta p)A = \rho A c(c + \Delta V - c) \tag{9.2.3}$$

which simplifies to,

$$-\Delta p = \rho c \, \Delta V \tag{9.2.4}$$

**Fig. 9.1** Sound wave: (a) stationary observer; (b) observer moving with the wave; (c) control volume enclosing the wave.

Combining this with Eq. 9.2.2 results in

$$c = \sqrt{\frac{\Delta p}{\Delta \rho}} \qquad (9.2.5)$$

Since the changes $\Delta p$ and $\Delta \rho$ are quite small, we can write

$$\frac{\Delta p}{\Delta \rho} \simeq \frac{dp}{d\rho} \qquad (9.2.6)$$

Small-amplitude, moderate-frequency waves (up to about 18 000 Hz) travel with no change in entropy (isentropically), so that

$$\frac{p}{\rho^k} = \text{const.} \qquad (9.2.7)$$

This can be differentiated to give

$$\frac{dp}{d\rho} = k \frac{p}{\rho} \qquad (9.2.8)$$

Using this in Eq. 9.2.5, the speed of sound $c$ is given by

$$c = \sqrt{\frac{kp}{\rho}} \qquad (9.2.9)$$

or, using the ideal-gas law,

$$c = \sqrt{kRT} \qquad (9.2.10)$$

At high frequency, sound waves generate friction and the process ceases to be isentropic. It is better approximated by an isothermal process. For an ideal gas an isothermal approximation would lead to

$$c = \sqrt{RT} \qquad (9.2.11)$$

For small waves traveling through liquids or solids the *bulk modulus* is used; it has dimensions of pressure and is equal to $\rho \, dp/d\rho$. For water it has a nominal value of 2110 MPa. It does vary slightly with temperature and pressure. Using Eq. 9.2.5, this leads to a speed of propagation of 1450 m/s for a small-amplitude pressure wave in water.

**KEY CONCEPT** *A small-amplitude pressure wave in water travels 1450 m/s.*

An important quantity used in the study of compressible flow is the dimensionless velocity called the *Mach number*, introduced in Eq. 3.3.3 as

$$M = \frac{V}{c} \qquad (9.2.12)$$

If M $< 1$, the flow is a *subsonic* flow, and if M $> 1$, it is a *supersonic* flow. If M $> 5$ it may be referred to as *hypersonic* flow; such flows will not be considered in this book.

If a source of sound waves is at a fixed location, the waves travel radially away from the source with the speed of sound $c$. Figure 9.2a shows the position of the sound waves after a time increment $\Delta t$ and after multiples of $\Delta t$. Part (b) shows a source moving with a velocity $V$ which is less than the speed of sound. Note that the sound waves always propagate ahead of the source so that an airplane traveling at a speed less than the speed of sound will always "announce" its approach. This is not true, however, for an object traveling at a speed greater than the speed of sound, as shown in Fig. 9.2c. The region outside the cone is a *zone of silence*, so that an approaching object moving at a supersonic speed could not be heard until it passed overhead and the *Mach cone*, the cone shown, intercepted the observer. From the figure the angle $\alpha$ of the Mach cone is given by

> **KEY CONCEPT** *The approach of an object moving at* M $> 1$ *cannot be sensed until it has passed by.*

$$\alpha = \sin^{-1}\frac{c}{V} = \sin^{-1}\frac{1}{M} \tag{9.2.13}$$

The discussion above is limited to small-amplitude sound waves, often called *Mach waves*. They are formed on the needle nose of an aircraft, or on the leading edge of an airfoil if the leading edge is sufficiently sharp. If the nose is blunt or if the leading edge is not sufficiently sharp, a supersonic aircraft will produce a large-amplitude wave called a **shock wave**. The shock wave will also form a zone of silence, but the initial angle at the source created by the shock wave will be larger than that of a Mach wave. Shock waves will be considered in subsequent sections.

> **Shock wave:** *A large-amplitude wave which may be created by a blunt object.*

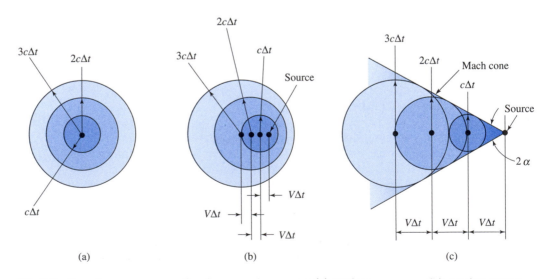

**Fig. 9.2** Sound waves propagating from a noise source: (a) stationary source; (b) moving source, $V < c$; (c) moving source, $V > c$.

## Example 9.1

A needle-nose projectile traveling at a speed with M = 3 passes 200 m above the observer of Fig. E9.1. Calculate the projectile's velocity and determine how far beyond the observer the projectile will first be heard.

**Fig. E9.1**

### Solution

At a Mach number of 3 the velocity is

$$V = \text{M}c = \text{M}\sqrt{kRT}$$

$$= 3\sqrt{1.4 \times 287 \text{ J/kg} \cdot \text{K} \times 288 \text{ K}} = 1021 \text{ m/s}$$

where a standard temperature of 15°C has been assumed since no temperature is given. (Remember, kg = N $\cdot$ s$^2$/m and J = N $\cdot$ m.) Using $h$ as the height and $L$ as the distance beyond the observer (refer to Fig. 9.2c), we have

$$\sin \alpha = \frac{h}{\sqrt{L^2 + h^2}} = \frac{1}{\text{M}}$$

With the information given,

$$\frac{200}{\sqrt{L^2 + 200^2}} = \frac{1}{3}$$

giving

$$L = 566 \text{ m}$$

*Note:* The units on $kRT$ are $\dfrac{\text{N} \cdot \text{m}}{\text{kg} \cdot \text{K}} \times \text{K} = \dfrac{\text{N} \cdot \text{m}}{\text{N} \cdot \text{s}^2/\text{m}} = \dfrac{\text{m}^2}{\text{s}^2}.$ The quantity $k$ is

dimensionless.

## 9.3 ISENTROPIC NOZZLE FLOW

There are many applications where gas flows through a section of tube or conduit that has a changing area in which a steady, uniform, isentropic flow is a good approximation to the actual flow situation. The diffuser near the front of a jet engine, exhaust gases passing through the blades of a turbine, the nozzles on a rocket engine, a broken natural gas line, and gas flow measuring devices are all examples of situations that can be modeled with a steady, uniform, isentropic

**KEY CONCEPT**  *Steady, uniform, isentropic flow is a good approximation to many flow situations.*

flow. Consider the flow through the infinitesimal control volume shown in Fig. 9.3. With a changing area the continuity equation

$$\rho A V = \text{const.} \tag{9.3.1}$$

applied between two sections a distance $dx$ apart takes the form

$$\rho A V = (\rho + d\rho)(A + dA)(V + dV) \tag{9.3.2}$$

Keeping only first-order terms in the differential quantities, Eq. 9.3.2 can be put in the form

$$\frac{dV}{V} + \frac{dA}{A} + \frac{d\rho}{\rho} = 0 \tag{9.3.3}$$

The energy equation can be written (see Eq. 9.1.5)

$$\frac{V^2}{2} + \frac{k}{k-1}\frac{p}{\rho} = \text{const.} \tag{9.3.4}$$

For the present application we have

$$\frac{V^2}{2} + \frac{k}{k-1}\frac{p}{\rho} = \frac{(V + dV)^2}{2} + \frac{k}{k-1}\frac{p + dp}{\rho + d\rho} \tag{9.3.5}$$

or again, retaining only first-order terms,

$$V\,dV + \frac{k}{k-1}\frac{\rho\,dp - p\,d\rho}{\rho^2} = 0 \tag{9.3.6}$$

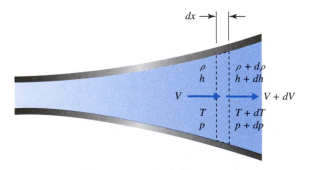

**Fig. 9.3** Uniform, isentropic flow.

For an isentropic process we use Eq. 9.2.8 and there results

$$V \, dV + k \frac{p}{\rho^2} d\rho = 0 \tag{9.3.7}$$

Substituting for $d\rho/\rho$ from Eq. 9.3.3, the above becomes

$$\frac{dV}{V}\left(\frac{\rho V^2}{kp} - 1\right) = \frac{dA}{A} \tag{9.3.8}$$

In terms of the speed of sound this is written as

$$\frac{dV}{V}\left(\frac{V^2}{c^2} - 1\right) = \frac{dA}{A} \tag{9.3.9}$$

Introduce the Mach number and we have the very important relationship

$$\frac{dV}{V}(M^2 - 1) = \frac{dA}{A} \tag{9.3.10}$$

for an isentropic uniform flow in a changing area.

From Eq. 9.3.10 we can make the following observations:

1. If the area is increasing, $dA > 0$, and $M < 1$, we see that $dV$ must be negative, that is, $dV < 0$. The flow is decelerating for this subsonic flow.
2. If the area is increasing and $M > 1$, we see that $dV > 0$; hence the flow is accelerating in the diverging section for this supersonic flow.
3. If the area is decreasing and $M < 1$, then $dV > 0$, resulting in an accelerating flow.
4. If the area is decreasing and $M > 1$, then $dV < 0$, indicating a decelerating flow.
5. At a throat where $dA = 0$, either $dV = 0$ or $M = 1$, or possibly both.

If we define a **nozzle** as a device that accelerates the flow, we see that observations 2 and 3 describe a nozzle and observations 1 and 4 describe a **diffuser**, a device that decelerates the flow. The supersonic flow leads to rather surprising results: an accelerating flow in an enlarging area and a decelerating flow in a decreasing area. This is, in fact, the situation encountered in a traffic flow on a freeway; hence a supersonic flow might be used to model a traffic flow.

Note that the observations above prohibit a supersonic flow in a converging section attached to a reservoir. If a supersonic flow is to be generated by releasing a gas from a reservoir, there must be a converging section in which a subsonic flow accelerates to the throat where $M = 1$ followed by a diverging section in which the flow continues to accelerate with $M > 1$, as shown in Fig. 9.4. This is the type of nozzle observed on rockets used to place satellites into orbit.

**Nozzle:**   *A device that accelerates the flow.*

**Diffuser:**   *A device that decelerates the flow.*

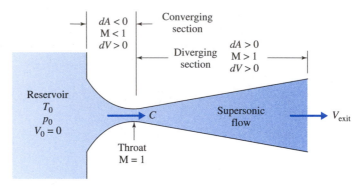

**Fig. 9.4** Supersonic nozzle.

The isentropic flow in the nozzle will now be considered in more detail. The energy equation between the reservoir where $V_0 = 0$ and any section can be written in the form

$$c_p T_0 = \frac{V^2}{2} + c_p T \tag{9.3.11}$$

Quantities with a zero subscript are often called **stagnation quantities** since they occur at a location where $V = 0$. Recognizing that $V = \mathrm{M}c$, $c_p/c_v = k$, $c_p = c_v + R$, and $c = \sqrt{kRT}$, we can write the energy equation as

$$\frac{T_0}{T} = 1 + \frac{k-1}{2}\mathrm{M}^2 \tag{9.3.12}$$

For our isentropic flow, the pressure and density ratios are expressed as

$$\frac{p_0}{p} = \left(1 + \frac{k-1}{2}\mathrm{M}^2\right)^{k/(k-1)}$$
$$\frac{\rho_0}{\rho} = \left(1 + \frac{k-1}{2}\mathrm{M}^2\right)^{1/(k-1)} \tag{9.3.13}$$

If a supersonic flow occurs downstream of the throat, then $\mathrm{M} = 1$ at the throat, and denoting this *critical area* with an asterisk (*) superscript, we have the following critical ratios:

$$\frac{T^*}{T_0} = \frac{2}{k+1}$$
$$\frac{p^*}{p_0} = \left(\frac{2}{k+1}\right)^{k/(k-1)}$$
$$\frac{\rho^*}{\rho_0} = \left(\frac{2}{k+1}\right)^{1/(k-1)} \tag{9.3.14}$$

We often make reference to the critical area even though an actual throat does not occur; we can imagine a throat occurring and call it the critical area. In fact, the isentropic flow table, Table D.1 in Appendix D, includes just such an area. For air with $k = 1.4$ the critical values are

$$p^* = 0.5283\, p_0 \qquad T^* = 0.8333\, T_0 \qquad \rho^* = 0.6340\, \rho_0 \qquad (9.3.15)$$

We can determine an expression for the mass flux through the nozzle from the equation

$$\dot{m} = \rho A V$$
$$= \frac{p}{RT} A M \sqrt{kRT} = \frac{p}{\sqrt{T}} \sqrt{\frac{k}{R}} A M \qquad (9.3.16)$$

Using Eqs. 9.3.12 and 9.3.13, this can be expressed as

$$\dot{m} = \frac{p_0 \left(1 + \dfrac{k-1}{2} M^2\right)^{k/(1-k)}}{\sqrt{T_0}\left(1 + \dfrac{k-1}{2} M^2\right)^{-1/2}} \sqrt{\frac{k}{R}}\, A M$$

$$= p_0 \sqrt{\frac{k}{RT_0}}\, MA \left(1 + \frac{k-1}{2} M^2\right)^{(k+1)/2(1-k)} \qquad (9.3.17)$$

If we choose the critical area where $M^* = 1$, we see that

$$\dot{m} = p_0 A^* \sqrt{\frac{k}{RT_0}}\left(\frac{k+1}{2}\right)^{(k+1)/2(1-k)} \qquad (9.3.18)$$

This shows that the mass flux in the nozzle is only dependent on the reservoir conditions and the critical area $A^*$. By combining Eqs. 9.3.17 and 9.3.18, the area ratio $A/A^*$ can be written in terms of the Mach number as

$$\frac{A}{A^*} = \frac{1}{M}\left[\frac{2 + (k-1)M^2}{k+1}\right]^{(k+1)/2(k-1)} \qquad (9.3.19)$$

This ratio is included in Table D.1 for airflow.

As a final consideration in our study of isentropic nozzle flow, we will present the influence of reservoir pressure and receiver pressure on the mass flux. First, the converging nozzle will be presented; then the converging–diverging nozzle will follow. The converging nozzle is assumed to be attached to a reservoir, as shown in Fig. 9.5a, with fixed conditions; the pressure in the receiver can be lowered to provide an increasing mass flux through the nozzle, as shown by the left curve in Fig. 9.5b. When the receiver pressure $p_r$ reaches the critical pressure

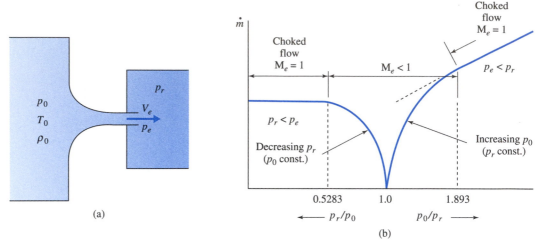

**Fig. 9.5** Converging nozzle.

(for air $p_r = 0.5283\, p_0$), the Mach number $M_e$ at the throat (the exit) is unity. As $p_r$ is reduced below this critical value the mass flux will not increase, and the condition of *choked flow* occurs, as shown by the left curve in Fig. 9.5b. If the throat is a critical area, that is, $M = 1$ at the throat, then $\dot{m}$ is only dependent on the throat area $A^*$ and the reservoir conditions, as indicated by Eq. 9.3.18. Hence, reducing $p_r$ below the critical pressure has no effect on the upstream flow. This is reasonable since disturbances travel at the speed of sound; if $p_r$ is reduced, the disturbances that would travel upstream, thereby changing conditions, cannot do so since the velocity of the stream at the exit is equal to the speed of sound thereby preventing any disturbances from propagating upstream.

If, in the converging nozzle, we keep $p_r$ constant and increase the reservoir pressure (keep $T_0$ constant also), a choked flow again occurs when $M_e = 1$; however, when $p_0$ is increased still further, we see from Eq. 9.3.18 that the mass flux will increase, as shown by the right curve of Fig. 9.5b. The pressure $p_e$ is equal to $p_r$ until the Mach number $M_e$ is just equal to unity. This begins the condition of choked flow. The exit pressure $p_e$ for the choked flow condition will be greater than the receiver pressure $p_r$, a condition that occurs when a gas line ruptures.

A note may be in order explaining how it is possible for the flow exit pressure $p_e$ to exceed the receiver pressure $p_r$. If $p_e > p_r$, the flow exiting the nozzle is able to turn rather sharply, causing a flow pattern sketched in Fig. 9.6. This possible flow situation will be studied in Section 9.8.

Now, consider the converging–diverging nozzle, shown in Fig. 9.7, with reservoir and receiver as indicated. We will only present the condition of constant reservoir pressure and reduced receiver pressure. For this nozzle we sketch the pressure ratio $p/p_0$ as a function of location in the nozzle for various receiver pressure ratios $p_r/p_0$. If $p_r/p_0 = 1$, no flow occurs, corresponding to curve $A$. If $p_r$ is reduced a small amount, curve $B$ results and a subsonic flow exists throughout the nozzle with a minimum pressure occurring at the throat. As the pressure is reduced still further, a pressure is reached that will result in the Mach number at

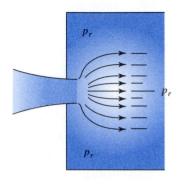

**Fig. 9.6**  Nozzle exit flow for $p_e > p_r$.

the throat just being unity, as sketched by curve $C$; the flow remains subsonic throughout, however. Another particular receiver pressure exists, considerably below that of curve $C$, that will also produce an isentropic flow throughout; it results in curve $D$. Any receiver pressure in between these two particular pressures will result in a nonisentropic flow in the nozzle; a shock wave, to be studied later, occurs which renders our assumption of isentropic flow invalid. If the receiver pressure is below that associated with curve $D$, we again find that the nozzle exit pressure $p_e$ is greater than the receiver pressure $p_r$. The mass flux in the nozzle increases from curve $A$ to curve $C$; but as the receiver pressure is reduced below that of curve $C$, no increase in mass flux can occur since the throat condition will remain unchanged. The mass flux is given by Eq. 9.3.17.

Final notes regard nozzle and diffuser effectiveness. The purpose of a nozzle is to convert enthalpy (which can be thought of as stored energy) into kinetic

> **KEY CONCEPT**  *The purpose of a nozzle is to convert stored energy into kinetic energy.*

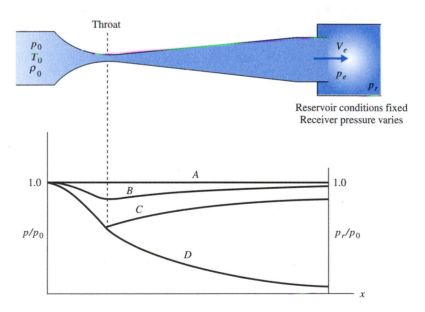

**Fig. 9.7**  Converging–diverging nozzle.

energy as described by Eq. 9.1.3 with $\dot{Q} = \dot{W}_S = 0$. The efficiency $\eta_N$ of a nozzle is defined as

$$\eta_N = \frac{(\Delta KE)_{\text{actual}}}{(\Delta KE)_{\text{isentropic}}}$$

$$= \frac{h_0 - h_e}{h_0 - h_{es}} \tag{9.3.20}$$

where $h_e$ is the actual exit enthalpy and $h_{es}$ is the isentropic exit enthalpy. Efficiencies are between 90 and 99% with larger nozzles having the higher percentages because the viscous wall effects that account for most of the losses are relatively small with the larger nozzles.

**KEY CONCEPT** *The purpose of a diffuser is to recover the pressure.*

The purpose of a diffuser is to slow down the fluid and recover the pressure. For a diffuser we define the *pressure recovery factor* $C_p$ to be

$$C_p = \frac{\Delta p_{\text{actual}}}{\Delta p_{\text{isentropic}}} \tag{9.3.21}$$

Such factors vary between 40% when the flow actually separates from the wall (the included angle should be less than about 10° for a subsonic section to avoid this separation) to 85%. Vanes can be installed in wide-angled subsonic diffusers such that each passage between vanes expands at an angle of 10° or less. Viscous effects are significantly greater in a diffuser than in a nozzle because of the thicker viscous wall layers.

**KEY CONCEPT** *Supersonic nozzles are constructed with large included angles.*

The reader may think that the flow in the supersonic diverging nozzle may also tend to separate from the wall; this is not the case. Expansion fans, a phenomenon provided by nature, to be studied in Section 9.8, allow the supersonic flow to turn rather sharp angles so that supersonic nozzles are constructed with large included angles, such as those attached to the rockets that propel satellites into orbit.

## Example 9.2

Air exits from a reservoir maintained at 20°C and 500 kPa absolute into a receiver maintained at (a) 300 kPa absolute and (b) 200 kPa absolute. Estimate the mass flux if the exit area is 10 cm$^2$. Use the equations first and then the isentropic flow table, Table D.1. Refer to Fig. 9.5.

### Solution

To estimate the mass flux we will assume isentropic flow. For air the receiver pressure that would result in $M_e = 1$ is

$$p_r = 0.5283\, p_0 = 0.5283 \times 500 = 264.2 \text{ kPa}$$

For part (a) $M_e < 1$ since $p_r > 264.2$ kPa, and for part (b) choked flow occurs and $M_e = 1$ since $p_r < 264.2$ kPa.

(a) To find the exit Mach number, Eq. 9.3.13 gives

$$1 + \frac{k-1}{2} M^2 = \left(\frac{p_0}{p}\right)^{(k-1)/k}$$

or

$$M_e^2 = \frac{2}{k-1} \left(\frac{p_0}{p}\right)^{(k-1)/k} - \frac{2}{k-1}$$

$$= \frac{2}{0.4} \left(\frac{500}{300}\right)^{0.2857} - \frac{2}{0.4} = 0.7857 \qquad \therefore M_e = 0.8864$$

The mass flux is given by Eq. 9.3.17 and is found to be

$$\dot{m} = p_0 \sqrt{\frac{k}{RT_0}} MA \left(1 + \frac{k-1}{2} M^2\right)^{(k+1)/2(1-k)}$$

$$= 500\,000 \sqrt{\frac{1.4}{287 \times 293}} \times 0.8864 \times 0.001 \left(1 + \frac{0.4}{2} \times 0.8864^2\right)^{-2.4/0.8}$$

$$= 1.167 \text{ kg/s}$$

(b) Choked flow occurs and thus $M_e = 1$; Eq. 9.3.18 yields

$$\dot{m} = p_0 A^* \sqrt{\frac{k}{RT_0}} \left(\frac{k+1}{2}\right)^{(k+1)/2(1-k)}$$

$$= 500\,000 \times 0.001 \sqrt{\frac{1.4}{287 \times 293}} \left(\frac{2.4}{2}\right)^{-2.4/0.8} = 1.181 \text{ kg/s}$$

Now, let us use the isentropic flow table (Table D.1) and solve parts (a) and (b).
(a) For a pressure ratio of $p/p_0 = 300/500 = 0.6$, we interpolate to find

$$M_e = \frac{0.6041 - 0.6}{0.6041 - 0.5913} \times 0.02 + 0.88 = 0.886$$

$$\frac{T_e}{T_0} = \frac{0.6041 - 0.6}{0.6041 - 0.5913} (0.8606 - 0.8659) + 0.8659 = 0.864$$

$$T_e = 0.864 \times 293 = 253 \text{ K}$$

The velocity and density are, respectively,

$$V = Mc = 0.886 \sqrt{1.4 \times 287 \times 253} = 282 \text{ m/s}$$

$$\rho = \frac{p}{RT} = \frac{300}{0.287 \times 253} = 4.13 \text{ kg/m}^3$$

The mass flux is then

$$\dot{m} = \rho A V$$

$$= 4.13 \times 0.001 \times 282 = 1.165 \text{ kg/s}$$

(b) For choked flow we know that $M_e = 1$. The table gives

(continued)

$$\frac{T_e}{T_0} = 0.8333 \quad \text{and} \quad \frac{p_e}{p_0} = 0.5283$$

Thus the temperature, velocity, and density are, respectively,

$$T = 0.8333 \times 293 = 244.2 \text{ K}$$

$$V = Mc = 1 \times \sqrt{1.4 \times 287 \times 244.2} = 313.2 \text{ m/s}$$

$$\rho = \frac{p}{RT} = \frac{0.5283 \times 500}{0.287 \times 244.2} = 3.769 \text{ kg/m}^3$$

The mass flux is calculated to be

$$\dot{m} = \rho AV$$
$$= 3.769 \times 0.001 \times 313.2 = 1.180 \text{ kg/s}$$

The results using the equations are essentially the same as those using the tables.

## Example 9.3

A converging–diverging nozzle, with an exit area of 40 cm$^2$ and a throat area of 10 cm$^2$, is attached to a reservoir with $T = 20°C$ and $p = 500$ kPa absolute. Determine the two exit pressures that result in M = 1 at the throat for an isentropic flow. Also, determine the associated exit temperatures and velocities. See Fig. 9.7.

### Solution

The exit pressures we seek are associated with curves $C$ and $D$ of Fig. 9.7. The area ratio is

$$\frac{A}{A^*} = \frac{40}{10} = 4$$

We could solve Eq. 9.3.19 for M using a trial-and-error technique; however, let us use the isentropic flow table, Table D.1. There are two entries for $A/A^* = 4$. Interpolation gives

$$\left(\frac{p}{p_0}\right)_C = \frac{4.182 - 4.0}{4.182 - 3.673}(0.9823 - 0.9864) + 0.9864 = 0.9849$$

$$\left(\frac{p}{p_0}\right)_D = \frac{4.0 - 3.999}{4.076 - 3.999}(0.02891 - 0.02980) + 0.02980 = 0.02979$$

Hence the two exit pressures that will result in isentropic flow are

$$p_C = 492.4 \text{ kPa} \quad \text{and} \quad p_D = 14.9 \text{ kPa}$$

Note the very small pressure difference (7.6 kPa) between receiver and reservoir necessary to create the flow condition of curve $C$ of Fig. 9.7.

The exit temperature ratios and Mach numbers are interpolated to be

$$\left(\frac{T}{T_0}\right)_C = 0.3576(0.9949 - 0.9961) + 0.9961 = 0.9957$$

$$\left(\frac{T}{T_0}\right)_D = 0.01299(0.3633 - 0.3665) + 0.3665 = 0.3665$$

$$M_C = 0.3576 \times 0.02 + 0.14 = 0.147$$

$$M_D = 0.01299 \times 0.02 + 2.94 = 2.94$$

The exit temperatures associated with curves $C$ and $D$ are thus

$$T_C = 0.9957 \times 293 = 291.7 \text{ K}$$

$$T_D = 0.3665 \times 293 = 107.4 \text{ K}$$

The exit velocities are found from $V = Mc$ to be

$$V_C = 0.147 \sqrt{1.4 \times 287 \times 291.7} = 50.3 \text{ m/s}$$

$$V_D = 2.94 \sqrt{1.4 \times 287 \times 107.4} = 611 \text{ m/s}$$

## Example 9.4

Gas flows are assumed to be incompressible flows at Mach numbers less than about 0.3. Determine the error involved in calculating the stagnation pressure for an airflow with $M = 0.3$.

### Solution

For an incompressible airflow the energy equation (4.5.20) with no losses would give

$$p_0 = p_1 + \rho \frac{V_1^2}{2}$$

where $V_0 = 0$ at the stagnation point.

The isentropic flow equation (9.3.13), with $k = 1.4$, gives

$$p_0 = p_1(1 + 0.2M_1^2)^{3.5}$$

Use the binomial theorem $(1 + x)^n = 1 + nx + n(n - 1)x^2/2! + \cdots$ and express this as, letting $x = 0.2M_1^2$,

$$p_0 = p_1(1 + 0.7M_1^2 + 0.175M_1^4 + \cdots)$$

This can be written as (see Eqs. 9.2.9 and 9.2.12)

$$p_0 - p_1 = p_1 M_1^2(0.7 + 0.175M_1^2 + \cdots)$$

$$= \frac{\rho_1 V_1^2}{1.4}(0.7 + 0.175\mathrm{M}_1^2 + \cdots)$$

$$= \rho_1 \frac{V_1^2}{2}(1 + 0.25\mathrm{M}_1^2 + \cdots)$$

Substituting $\mathrm{M}_1 = 0.3$, we see that

$$p_0 - p_1 = \rho_1 \frac{V_1^2}{2}(1 + 0.0225 + \cdots)$$

Comparing this with the incompressible flow equation, we see that the error is only slightly greater than 2%. Hence it is reasonable to approximate a gas flow below $\mathrm{M} = 0.3$ ($V \cong 100$ m/s for air at standard conditions) with an incompressible flow for many engineering applications.

## 9.4 NORMAL SHOCK WAVE

Small-amplitude disturbances travel at the speed of sound, as was established in Section 9.2. In this section a large-amplitude disturbance will be studied. Its speed of propagation and its effect on other flow properties, such as pressure and temperature, will be considered. Large-amplitude disturbances occur in a number of situations. Examples include the flow in a gun barrel ahead of the projectile, the exit flow from a rocket or jet engine nozzle, the airflow around a supersonic aircraft, and the expanding front due to an explosion. Such large disturbances propagating through a gas are called **shock waves**. They can be oriented normal to the flow or at oblique angles. In this section we consider only the normal shock wave that occurs in a tube or directly in front of a blunt object; the photograph in Fig. 9.8 shows the shock wave in front of a sphere.

The property changes that occur across a shock wave take place over an extremely short distance. For usual conditions the distance is only several mean free paths of the molecules, on the order of $10^{-4}$ mm. Phenomena such as viscous dissipation and heat conduction that occur within the shock wave will not be studied here. We will treat the shock wave as a zero-thickness discontinuity in the flow and allow our integral control volume equations to relate the global quantities of interest.

Consider a normal shock wave moving with velocity $V_1$. We can make it stationary by moving the flow in a tube at velocity $V_1$, as shown in Fig. 9.9. The continuity equation, recognizing that $A_1 = A_2$, is

$$\rho_1 V_1 = \rho_2 V_2 \tag{9.4.1}$$

The energy equation (9.1.5), with $\dot{Q} = \dot{W}_S = 0$, is

$$\frac{V_2^2 - V_1^2}{2} + \frac{k}{k-1}\left(\frac{p_2}{\rho_2} - \frac{p_1}{\rho_1}\right) = 0 \tag{9.4.2}$$

**Shock wave:** *A large disturbance that propagates through a gas.*

**KEY CONCEPT** *The changes that occur across a shock wave take place over an extremely short distance.*

**Fig. 9.8**  A shock wave is observed in front of a sphere at M = 1.53. (Photograph by A. C. Charters. From Album of Fluid Motion, 1982, The Parabolic Press, Stanford, California.)

The momentum equation (9.1.2), with only pressure forces, becomes

$$p_1 - p_2 = \rho_1 V_1 (V_2 - V_1) \tag{9.4.3}$$

where the areas have divided out. These three equations allow us to determine three unknown quantities; if $\rho_1$, $V_1$, and $p_1$ are known, we can find $\rho_2$, $V_2$, $p_2$, and subsequently $T_2$ and $M_2$, using the appropriate equations.

    It is convenient, however, to express the equations in terms of the Mach numbers $M_1$ and $M_2$. This results in a set of equations that are easier to solve than solving the three simultaneous equations listed above. To do this we write Eq. 9.4.3, using $\rho_1 V_1 = \rho_2 V_2$, in the form

**KEY CONCEPT**  *The three equations allow us to determine three unknowns.*

$$p_1 \left( 1 + \frac{\rho_1 V_1^2}{p_1} \right) = p_2 \left( 1 + \frac{\rho_2 V_2^2}{p_2} \right) \tag{9.4.4}$$

Control volume

$V_1$        $V_2$

$p_1$   $\rho_1$        $p_2$   $\rho_2$

**Fig. 9.9**  Stationary shock wave in a tube.

Introducing $M^2 = V^2 \rho / p k$ (combine Eq. 9.2.9 and 9.2.12), the momentum equation becomes

$$\frac{p_2}{p_1} = \frac{1 + k M_1^2}{1 + k M_2^2} \tag{9.4.5}$$

Similarly, the energy equation (9.4.2), with $p = \rho R T$, is written as

$$T_1\left(1 + \frac{k-1}{kRT_1}\frac{V_1^2}{2}\right) = T_2\left(1 + \frac{k-1}{kRT_2}\frac{V_2^2}{2}\right) \tag{9.4.6}$$

or, substituting $M^2 = V^2/kRT$:

$$\frac{T_2}{T_1} = \frac{1 + \dfrac{k-1}{2} M_1^2}{1 + \dfrac{k-1}{2} M_2^2} \tag{9.4.7}$$

If we substitute $\rho = p/RT$ into the continuity equation (9.4.1), we have

$$\frac{p_1 V_1}{RT_1} = \frac{p_2 V_2}{RT_2} \tag{9.4.8}$$

which becomes, using $V = M\sqrt{kRT}$,

$$\frac{p_2}{p_1}\frac{M_2}{M_1}\sqrt{\frac{T_1}{T_2}} = 1 \tag{9.4.9}$$

Substituting for the pressure and temperature ratios from Eqs. 9.4.5 and 9.4.7, the continuity equation takes the form

$$\frac{M_1\left(1 + \dfrac{k-1}{2} M_1^2\right)^{1/2}}{1 + k M_1^2} = \frac{M_2\left(1 + \dfrac{k-1}{2} M_2^2\right)^{1/2}}{1 + k M_2^2} \tag{9.4.10}$$

Hence the downstream Mach number is related to the upstream Mach number by

$$M_2^2 = \frac{M_1^2 + \dfrac{2}{k-1}}{\dfrac{2k}{k-1} M_1^2 - 1} \tag{9.4.11}$$

This allows us to express the pressure and temperature ratios in terms of $M_1$ only. The momentum equation (9.4.5) takes the form

$$\frac{p_2}{p_1} = \frac{2k}{k+1} M_1^2 - \frac{k-1}{k+1} \tag{9.4.12}$$

and the energy equation becomes

$$\frac{T_2}{T_1} = \frac{\left(1 + \dfrac{k-1}{2} M_1^2\right)\left(\dfrac{2k}{k-1} M_1^2 - 1\right)}{\dfrac{(k+1)^2}{2(k-1)} M_1^2} \tag{9.4.13}$$

For air, with $k = 1.4$, the preceding three equations reduce to

$$M_2^2 = \frac{M_1^2 + 5}{7M_1^2 - 1}$$

$$\frac{p_2}{p_1} = \frac{7M_1^2 - 1}{6} \tag{9.4.14}$$

$$\frac{T_2}{T_1} = \frac{(M_1^2 + 5)(7M_1^2 - 1)}{36M_1^2}$$

From the first of these three equations, we observe:

- If $M_1 = 1$, then $M_2 = 1$ and no shock wave exists.
- If $M_1 > 1$, then $M_2 < 1$ and the normal shock wave converts a supersonic flow into a subsonic flow.
- If $M_1 < 1$, then $M_2 > 1$ and a subsonic flow appears to be converted into a supersonic flow by the presence of a normal shock wave. This possibility is eliminated by the second law since it would demand a decrease in entropy by a process in an isolated system, an impossibility.

**KEY CONCEPT** *A finite wave that converts a subsonic flow into a supersonic flow is an impossibility.*

The impossibility stated above is observed by considering the entropy increase, given by

$$s_2 - s_1 = c_p \ln \frac{T_2}{T_1} - R \ln \frac{p_2}{p_1}$$

$$= c_p \ln \frac{2 + (k-1)M_1^2}{2 + (k-1)M_2^2} - R \ln \frac{1 + kM_1^2}{1 + kM_2^2} \tag{9.4.15}$$

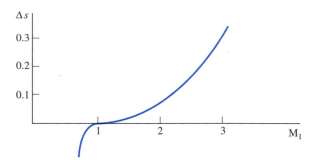

**Fig. 9.10**  Entropy change for a normal shock in air.

For air, with $k = 1.4$, this is plotted in Fig. 9.10, relating $M_2$ to $M_1$, with Eq. 9.4.11. Note the impossible negative entropy change whenever $M_1 < 1$.

The relationship among the thermodynamic properties, listed in the equations above, can be demonstrated graphically with reference to the $T$–$s$ diagram in Fig. 9.11. The conditions upstream of the normal shock are designated by state 1, and downstream by state 2. Note the dashed line from state 1 to state 2 for the irreversible process that occurs inside the shock wave. The energy equation, with $\dot{Q} = \dot{W}_S = 0$, can be written as

$$\frac{V_1^2}{2c_p} + T_1 = \frac{V_2^2}{2c_p} + T_2 \tag{9.4.16}$$

The stagnation temperature is defined as the temperature that would exist if the flow is brought to rest isentropically. Thus the energy equation gives

$$T_{01} = T_{02} \tag{9.4.17}$$

as shown in the figure. The substantial decrease in stagnation pressure, $p_{02} < p_{01}$, is also observed in Fig. 9.11. If the entropy increases from state 1 to state 2, as it must, the stagnation pressure $p_{02}$ must decrease, as shown, if we are to maintain $T_{01} = T_{02}$.

Gas tables are available that give the pressure ratio, the temperature ratio, the downstream Mach number, and the stagnation pressure ratio as a function of the upstream Mach number. Table D.2 is such a table for $k = 1.4$ and includes the ratios given by Eq. 9.4.14. Note that $M_2$ is always less than unity, $p_2$ is always greater than $p_1$, $T_2$ is always greater than $T_1$, and $p_{02}$ is always less than $p_{01}$.

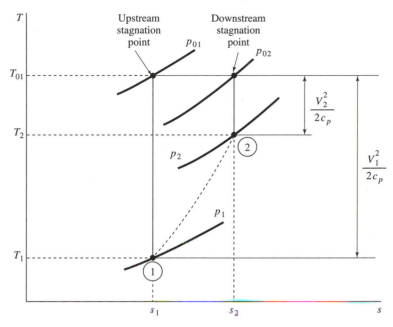

**Fig. 9.11** $T$–$s$ diagram for a normal shock wave.

## Example 9.5

A normal shock wave passes through stagnant air at 16°C and atmospheric pressure of 82 kPa with a speed of 453 m/s. Calculate the pressure and temperature downstream of the shock wave. Use (a) the equations and (b) the gas tables.

**Fig. E9.5**

### Solution

We consider the shock wave to be stationary with $V_1 = 453$ m/s and $p_1 = 82$ kPa (abs.)
(a) To use the simplified equations (9.4.14) we must know the upstream Mach number. It is

$$M_1 = \frac{V_1}{c_1} = \frac{V_1}{\sqrt{kRT_1}}$$

$$= \frac{453 \text{ m/s}}{\sqrt{1.4 \times 287 \text{ J/kg} \cdot \text{K} \times 289 \text{ K}}} = 1.33$$

(*continued*)

The pressure and temperature are then found to be

$$p_2 = \frac{p_1(7M_1^2 - 1)}{6}$$

$$= \frac{82(7 \times 1.33^2 - 1)}{6} = 155.5 \text{ kPa (abs.)}$$

$$T_2 = \frac{T_1(M_1^2 + 5)(7M_1^2 - 1)}{36M_1^2}$$

$$= \frac{289(1.33^2 + 5)(7 \times 1.33^2 - 1)}{36 \times 1.33^2} = 349.6 \text{ K}$$

(b)  From part (a), we use $M_1 = 1.33$. Interpolation in Table D.2 yields

$$\frac{p_2}{p_1} = \frac{1.33 - 1.32}{1.34 - 1.32}(1.928 - 1.8) + 1.866 = 1.897$$

$$\frac{T_2}{T_1} = \frac{1.33 - 1.32}{1.34 - 1.32}(1.216 - 1.204) + 1.204 = 1.21$$

Using the information given, we have

$$p_2 = 82 \times 1.897 = 155.5 \text{ kPa}$$

$$T_2 = 289 \times 1.21 = 349.7 \text{ K}$$

$$= 79.7°C$$

## Example 9.6

A normal shock wave propagates through otherwise stagnant air at standard conditions at a speed of 700 m/s. Determine the speed induced in the air immediately behind the shock wave as shown in Fig. E9.6. Use the equations.

**Fig. E9.6**

### Solution

For standard conditions the temperature is 15°C. The upstream Mach number is thus

$$M_1 = \frac{V_1}{c_1} = \frac{V_1}{\sqrt{kRT_1}} = \frac{700}{\sqrt{1.4 \times 287 \times 288}} = 2.06$$

Using Eq. 9.4.14, we find that

$$M_2 = \left(\frac{M_1^2 + 5}{7M_1^2 - 1}\right)^{1/2} = \left(\frac{2.06^2 + 5}{7 \times 2.06^2 - 1}\right)^{1/2} = 0.567$$

$$T_2 = \frac{T_1(M_1^2 + 5)(7M_1^2 - 1)}{36M_1^2}$$

$$= \frac{288(2.06^2 + 5)(7 \times 2.06^2 - 1)}{36 \times 2.06^2} = 500.2 \text{ K}$$

This allows us to calculate

$$V_2 = M_2c_2 = 0.567 \sqrt{1.4 \times 287 \times 500.2} = 254.4 \text{ m/s}$$

This velocity assumes a flow with the shock wave stationary and the air approaching the shock wave at 700 m/s. If we superimpose a velocity of 700 m/s moving opposite to $V_1$, we find the induced velocity to be

$$V_{\text{induced}} = V_2 - V_1$$
$$= 254.4 - 700 = -446 \text{ m/s}$$

where the negative sign means that the induced velocity would be moving to the left if $V_1$ is to the right. The induced velocity would be in the same direction as the propagation of the shock wave. Such large induced velocities are responsible for much of the damage away from the bomb center caused by explosions of high-power bombs.

## 9.5 SHOCK WAVES IN CONVERGING–DIVERGING NOZZLES

The converging–diverging nozzle has already been presented for an isentropic flow; for receiver-to-reservoir pressure ratios between those of curves $C$ and $D$ of Figs. 9.7 and 9.12 (on the next page), shock waves exist in the flow either inside or outside the nozzle. If $p_r/p_0 = a$ (locate $a$ on the vertical axis on the right of Fig. 9.12), a normal shock wave would exist at an internal location in the diverging portion of the nozzle. Usually, the location of the shock wave is prescribed in student problems since a trial-and-error solution is necessary to locate the shock. When $p_r/p_0 = b$ the normal shock wave is located in the exit plane of the nozzle. For a pressure ratio less than $b$ but greater than $e$, two types of oblique shock wave patterns are observed, one with a central normal shock wave, as sketched for $p_r/p_0 = c$, and one with only oblique waves, as sketched for $p_r/p_0 = d$. The pressure ratios that result in oblique shock waves will not be considered here. As we move from $d$ to $e$, the oblique shock waves become weaker and weaker until the isentropic flow is again realized at $p_r/p_0 = e$ with all shock waves absent. For pressure ratios below $e$ a very complicated flow exists. The flow turns the corner at the nozzle exit rather abruptly due to expansion waves (isentropic waves to be considered in Section 9.8), then turns back due to the same expansion waves, resulting in a billowing out of the exhaust flow, as is visible from high-altitude satellite rocket engines. Let us work some examples for the converging–diverging nozzle; no new equations are necessary.

**KEY CONCEPT** *The location of the shock wave is prescribed in student problems.*

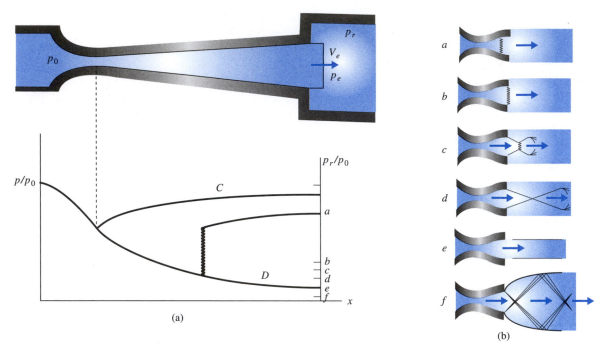

**Fig. 9.12** Converging–diverging nozzle.

## Example 9.7

A converging–diverging nozzle has a throat diameter of 5 cm and an exit diameter of 10 cm. The reservoir is the laboratory, maintained at atmospheric conditions of 20°C and 90 kPa absolute. Air is constantly pumped from a receiver so that a normal shock wave stands across the exit plane of the nozzle. Determine the receiver pressure and the mass flux.

### Solution

Isentropic flow occurs from the reservoir, to the throat, to the exit plane in front of the normal shock wave at state 1. Supersonic flow occurs downstream of the throat making the throat the critical area. Hence

$$\frac{A_1}{A^*} = \frac{10^2}{5^2} = 4$$

Interpolation in the isentropic flow table (Table D.1) gives

$$M_1 = 2.94 \qquad \frac{p_1}{p_0} = 0.0298$$

Hence the pressure in front of the normal shock is

$$p_1 = p_0 \times 0.0298$$
$$= 90 \times 0.0298 = 2.68 \text{ kPa}$$

From the normal shock table (Table D.2), using $M_1 = 2.94$, we find that

$$\frac{p_2}{p_1} = 9.918$$

$$\therefore p_2 = 9.918 \times 2.68 = 26.6 \text{ kPa}$$

This is the receiver pressure needed to orient the shock across the exit plane as shown for $p_r/p_0 = b$ in Fig. 9.12.

To find the mass flux through the nozzle, we need only consider the throat. Recognizing that $M_t = 1$, so that $V_t = c_t$, we can write

$$\dot{m} = \rho_t A_t V_t = \frac{p_t}{RT_t} A_t \sqrt{kRT_t} = p_t A_t \sqrt{\frac{k}{RT_t}}$$

The isentropic flow table yields

$$\frac{p_t}{p_0} = 0.5283 \qquad \frac{T_t}{T_0} = 0.8333$$

Thus the mass flux becomes

$$\dot{m} = (0.5283 \times 90\,000) \times \frac{\pi \times 0.05^2}{4} \sqrt{\frac{1.4}{287 \times (0.8333 \times 293)}}$$

$$= 0.417 \text{ kg/s}$$

Remember, the pressure must be measured in pascals in the equation above. Check the units to make sure that the units on $\dot{m}$ are kg/s.

## Example 9.8

Air flows from a reservoir at 20°C and 200 kPa absolute through a 5-cm-diameter throat and exits from a 10-cm-diameter nozzle. Calculate the exit pressure $p_e$ needed to locate a normal shock wave at a position where the diameter is 7.5 cm.

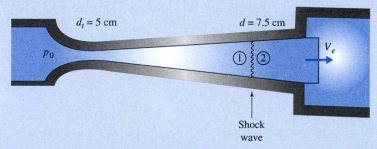

Fig. E9.8

### Solution

We will use the gas tables for this flow represented by the curve established by $p_r/p_0 = a$ in Fig. 9.12. The throat is a critical area since for a supersonic flow $M_t = 1$. The area ratio is

$$\frac{A_1}{A^*} = \frac{7.5^2}{5^2} = 2.25$$

(continued)

For this area ratio we find, from Table D.1, that

$$M_1 = 2.33$$

Then from Table D.2, for this Mach number, we obtain

$$M_2 = 0.531 \qquad \frac{p_{02}}{p_{01}} = 0.570$$

The reservoir pressure $p_0 = p_{01}$. Thus

$$p_{02} = 0.570 \times 200 = 114 \text{ kPa}$$

Isentropic flow occurs from state 2 immediately after the normal shock wave to the exit. Hence, for $M_2 = 0.531$, we find from Table D.1 that

$$\frac{A_2}{A^*} = 1.285$$

If $A_e$ is the exit area, then

$$\frac{A_e}{A^*} = \frac{A_2}{A^*} \times \frac{A_e}{A_2} = 1.285 \times \frac{10^2}{7.5^2} = 2.284$$

The Mach number and pressure ratio corresponding to this area ratio are

$$M_e = 0.265 \qquad \frac{p_e}{p_{0e}} = 0.952$$

For isentropic flow between the shock and the exit, we know that $p_{02} = p_{0e}$; thus

$$p_e = p_{02} \times 0.952 = 114 \times 0.952 = 109 \text{ kPa}$$

Note the usefulness of the critical area ratio in obtaining the desired results.

## Example 9.9

A pitot probe, the device used to measure the stagnation pressure in a flow, is inserted into an airstream and measures 300 kPa absolute, as shown in Fig. E9.9. The pressure in the flow is measured to be 75 kPa absolute. If the temperature at the stagnation point of the probe is measured as 150°C, determine the free-stream velocity $V$.

**Fig. E9.9**

**Solution**

When a blunt object is placed in a supersonic flow a detached shock wave forms around the object, as it does around the front of the pitot probe shown. The flow that meets the front of the pitot probe at the stagnation point passes through a normal shock wave from state 1 to state 2; the subsonic flow at state 2 then decelerates isentropically to state 3, the stagnation point.

For the isentropic flow from state 2 to state 3 we can use Eq. 9.3.13,

$$\frac{p_3}{p_2} = \left(1 + \frac{k-1}{2}\,\mathrm{M}_2^2\right)^{k/(k-1)}$$

Across the normal shock we know that (see Eq. 9.4.12)

$$\frac{p_2}{p_1} = \frac{2k}{k+1}\,\mathrm{M}_1^2 - \frac{k-1}{k+1}$$

Also, the Mach numbers are related by Eq. 9.4.11,

$$\mathrm{M}_2^2 = \frac{(k-1)\mathrm{M}_1^2 + 2}{2k\mathrm{M}_1^2 - k + 1}$$

The three equations above can be combined, with some algebraic manipulation, to yield the *Rayleigh-pitot-tube formula* for supersonic flows, namely,

$$\frac{p_3}{p_1} = \frac{\left(\dfrac{k+1}{2}\,\mathrm{M}_1^2\right)^{k/(k-1)}}{\left(\dfrac{2k\mathrm{M}_1^2}{k+1} - \dfrac{k-1}{k+1}\right)^{1/(k-1)}}$$

Substituting $k = 1.4$, $p_3 = 300$ kPa, and $p_1 = 75$ kPa, we have

$$\frac{300}{75} = \frac{(1.2\mathrm{M}_1^2)^{3.5}}{(1.167\mathrm{M}_1^2 - 0.1667)^{2.5}}$$

This can be solved by trial and error to give

$$\mathrm{M}_1 = 1.65$$

*(continued)*

The Mach number, after the normal shock is interpolated from the shock table to be

$$M_2 = 0.654$$

Using the isentropic flow table with this Mach number, we interpolate the temperature at state 2 to be

$$T_2 = T_3 \times 0.921$$

$$= 423 \times 0.921 = 389.6 \text{ K}$$

The temperature in front of the normal shock is found by using the normal shock table as follows:

$$\frac{T_2}{T_1} = 1.423$$

$$\therefore T_1 = \frac{T_2}{1.423} = 274 \text{ K}$$

Finally, the velocity before the normal shock is given by

$$V_1 = M_1 c_1 = M_1 \sqrt{kRT_1}$$

$$= 1.65 \sqrt{1.4 \times 287 \times 274} = 547 \text{ m/s}$$

## 9.6  VAPOR FLOW THROUGH A NOZZLE

The flow of vapor through a nozzle forms a very important engineering problem. High-pressure steam flows through the nozzles of turbines in electrical generating plants; in this section we present the technique for analyzing such a problem. We recall, however, that vapor that is not substantially superheated does not behave very well as an ideal gas; the vapor tables must be consulted since $c_p$ and $c_v$ cannot be assumed to be constant. Consider the problem of a superheated vapor entering the nozzle of Fig. 9.12. The flow would be isentropic unless a shock wave were encountered, and the expansion would be as sketched on the $T$–$s$ diagram of Fig. 9.13. Assume that the flow initiates in a reservoir with stagnation conditions, flows through a throat indicated by state $t$, and exits out a diverging section to the exit at state $e$.

Note that the exit state could very likely be in the quality region with the possibility of condensation of liquid droplets; however, for sufficiently high flow velocities there may not be sufficient time for the formation of the droplets and the associated heat transfer process. This produces a situation called *supersaturation* and a condition of *metastable equilibrium* exists; that is, the nonequilibrium state $e$ is reached rather than the equilibrium state $e'$. To estimate the temperature of the metastable state $e$, we assume that the steam behaves as an ideal gas,

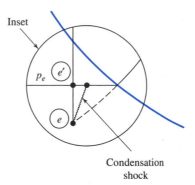

**Fig. 9.13** Isentropic expansion of a vapor.

so that $Tp^{k/(k-1)} = $ const. If the exit state is sufficiently far into the quality region, a *condensation shock* will be experienced, a phenomenon not considered in this book.

Because of supersaturation it is possible to model the isentropic flow of a vapor through a nozzle with acceptable accuracy by considering the ratio of specific heats to be constant. For steam $k = 1.3$ gives acceptable results over a considerable range of temperatures. The critical pressure ratio given by Eq. 9.3.14 for steam becomes

$$\frac{p^*}{p_0} = \left(\frac{2}{k+1}\right)^{k/(k-1)} = 0.546 \qquad (9.6.1)$$

If saturated steam enters the nozzle, some liquid droplets will be formed and entrained by the vapor; provided that a condensation shock does not exist, a good approximation to the critical pressure ratio is 0.577 corresponding to $k = 1.14$.

## Example 9.10

Steam is to be expanded isentropically from reservoir conditions of 300°C and 800 kPa absolute to an exit condition of 100 kPa absolute. If supersonic flow is desired, calculate the necessary throat and exit diameters if a mass flux of 2 kg/s is demanded.

### Solution

From the steam tables (found in any thermodynamics textbook) we find that

$$s_0 = s_e = 7.2336 \text{ kJ/kg} \cdot \text{K}$$
$$h_0 = 3056.4 \text{ kJ/kg}$$

To estimate the temperature of the metastable exit state, we use

$$T_e = T_0 \left(\frac{p_e}{p_0}\right)^{(k-1)/k} = 593 \left(\frac{100}{800}\right)^{0.3/1.3} = 367 \text{ K} \qquad \text{or} \qquad 94°C$$

*(continued)*

Using the steam tables at this temperature, we interpolate, using the exit quality $x_e$, to find that

$$7.2336 = 1.239 + 6.90x_e$$

$$\therefore x_e = 0.968$$

Thus we have the enthalpy and specific volume at the exit:

$$h_e = 394 + 0.968 \times 2273 = 2594 \text{ kJ/kg}$$

$$v_e = 0.001 + 0.968 \times (2.06 - 0.001) = 1.99 \text{ m}^3\text{/kg}$$

Using the energy equation, the exit velocity is estimated as follows:

$$\frac{V_0^{\cancel{2}\,0}}{2} + h_0 = \frac{V_e^2}{2} + h_e$$

$$\therefore V_e = \sqrt{2(h_0 - h_e)} = \sqrt{2(3056 - 2594) \times 1000} = 961 \text{ m/s}$$

where the 1000 converts kJ to J. From the definition of mass flux we have

$$m_e = \rho_e A_e V_e$$

$$2 = \frac{1}{1.99} \times \frac{\pi d_e^2}{4} \times 961$$

$$\therefore d_e = 0.0726 \text{ m} \quad \text{or} \quad 7.26 \text{ cm}$$

To determine the diameter of the throat, we recognize the throat to be the critical area; thus Eq. 9.6.1 gives

$$p^* = 0.546\, p_0 = 437 \text{ kPa}$$

Using this pressure and $s^* = s_0 = 7.2336$ kJ/kg·K we could use the steam tables to find $h^*$ and $v^*$; the energy equation would then allow us to find $V^*$ and thus $d_t$. However, a simpler, approximate technique, assuming constant specific heats, is to use Eq. 9.3.18 with $k = 1.3$ and obtain the following:

$$\dot{m} = p_0 A^* \sqrt{\frac{1.3}{RT_0}} \left(\frac{2.3}{2}\right)^{2.3/-0.6}$$

$$2 = 800\,000\, \frac{\pi d_t^2}{4} \sqrt{\frac{1.3}{462 \times 573}} \times 0.585$$

$$\therefore d_t = 0.049 \text{ m} \quad \text{or} \quad 4.9 \text{ cm}$$

This is reasonable since we have already assumed constant specific heats in predicting $T_e$. Obviously, the above is approximate; using $k = 1.3$ does, however, give reasonable predictions.

## 9.7 OBLIQUE SHOCK WAVE

In this section we investigate the oblique shock wave, a finite-amplitude wave that is not normal to the incoming flow. The flow approaching the oblique shock wave will be assumed to be in the $x$-direction. After the oblique shock the velocity vector

(a)    (b)

**Fig. 9.14**    Oblique shock waves in a supersonic flow: (a) flow over a symmetrical wedge; (b) flow in a corner.

will have a component normal to the flow direction. We will continue to assume that the flow before and after the oblique shock is uniform and steady.

Oblique shock waves form on the leading edge of a supersonic airfoil or in an abrupt corner, as sketched in Fig. 9.14. Oblique shock waves may also be found on axisymmetric bodies such as a nose cone or a bullet traveling at supersonic speeds. In this book, we consider only plane flows.

The function of the oblique shock wave is to turns the flow so that the velocity vector $\mathbf{V}_2$ is parallel to the plane wall. The angle between the two velocity vectors introduces another variable into our analysis. The problem remains solvable, however, with the additional tangential momentum equation.

To analyze the oblique shock wave, consider a control volume enclosing a portion of the shock as shown in Fig. 9.15. The velocity vector upstream is assumed to be in the $x$-direction only; the oblique shock wave makes an angle $\beta$ with the upstream velocity vector and turns the flow through the *deflection angle* or *wedge angle* $\theta$ so that $V_2$ is parallel to the wall. The components of the velocity vectors are shown normal and tangential to the oblique shock wave. The tangential components do not cause fluid to flow through the shock; hence the continuity equation, with $A_1 = A_2$, gives

> **KEY CONCEPT**  *The oblique shock wave turns the flow so that the velocity vector is parallel to the plane wall.*

$$\rho_1 V_{1n} = \rho_2 V_{2n} \qquad (9.7.1)$$

where the normal components $V_{1n}$ and $V_{2n}$ are displayed in Fig. 9.15. The pressure forces act normal to the oblique shock and produce no tangential components. Thus the momentum equation expressed in the tangential direction requires that the tangential momentum into the control volume equal the tangential momentum leaving the control volume; that is,

$$\dot{m}_1 V_{1t} = \dot{m}_2 V_{2t} \qquad (9.7.2)$$

or, using $\dot{m}_1 = \dot{m}_2$, we demand that

$$V_{1t} = V_{2t} \qquad (9.7.3)$$

**Fig. 9.15** Control volume enclosing a small portion of an oblique shock wave.

The normal momentum equation takes the form

$$p_1 - p_2 = \rho_2 V_{2n}^2 - \rho_1 V_{1n}^2 \tag{9.7.4}$$

The energy equation, using $V^2 = V_n^2 + V_t^2$, may be written as

$$\frac{V_{1n}^2}{2} + \frac{k}{k-1}\frac{p_1}{\rho_1} = \frac{V_{2n}^2}{2} + \frac{k}{k-1}\frac{p_2}{\rho_2} \tag{9.7.5}$$

**KEY CONCEPT** *The tangential components of the two velocity vectors do not enter the equations.*

where the tangential component terms have been canceled from both sides. Note that the tangential components of the two velocity vectors do not enter the continuity, normal momentum, or energy equations, the three equations used in the solution of the normal shock wave. Hence we may substitute $V_{1n}$ and $V_{2n}$ for $V_1$ and $V_2$, respectively, in the normal shock wave equations and obtain a solution. Either the normal shock wave equations or the normal shock wave table (Table D.2) may be used. We also replace $M_1$ and $M_2$ with $M_{1n}$ and $M_{2n}$, respectively.

It is useful to relate the oblique shock angle $\beta$ to the deflection angle $\theta$. Using the continuity equation (9.7.1), with reference to Fig. 9.15, there results

$$\frac{\rho_2}{\rho_1} = \frac{V_{1n}}{V_{2n}} = \frac{V_{1t}\tan\beta}{V_{2t}\tan(\beta-\theta)} = \frac{\tan\beta}{\tan(\beta-\theta)} \tag{9.7.6}$$

From the normal shock wave equations (9.4.12) and (9.4.13) we can find the density ratio to be

$$\frac{\rho_2}{\rho_1} = \frac{p_2 T_1}{p_1 T_2} = \frac{(k+1)M_{1n}^2}{(k-1)M_{1n}^2 + 2} \tag{9.7.7}$$

Substituting this into Eq. 9.7.6 gives

$$\tan(\beta - \theta) = \frac{\tan \beta}{k + 1}\left[k - 1 + \frac{2}{M_1^2 \sin^2 \beta}\right] \qquad (9.7.8)$$

For a flow with a given $M_1$, this equation relates the oblique shock angle $\beta$ to the wedge or corner angle $\theta$. The three variables $\beta$, $\theta$, and $M_1$ in the equation above are often plotted as in Fig. 9.16. We can observe several phenomena by studying the figure.

- For a specified upstream Mach number $M_1$ and a given wedge angle $\theta$ there are two possible oblique shock angles $\beta$, the larger one corresponding to a "strong" shock and the smaller corresponding to a "weak" shock.
- For a given wedge angle $\theta$ there is a minimum Mach number for which there is only one oblique shock angle $\beta$.
- For a given wedge angle $\theta$, if $M_1$ is less than the minimum for that particular curve, no oblique shock wave exists and the shock wave becomes detached, as shown in Fig. 9.17. Also, for a given $M_1$ there is a sufficiently large $\theta$ that will result in a detached shock wave.

The pressure rise across the oblique shock wave determines whether a weak shock or a strong shock occurs. For a relatively small pressure rise a weak shock will occur with $M_2 > 1$. If the pressure rise is relatively large a strong shock occurs with $M_2 < 1$. Note that for the detached shocks around bodies, a normal shock exists for the stagnation streamline; this is followed away from the stagnation point by a strong oblique shock, then a weak oblique shock, and eventually a Mach wave. For blunt bodies moving at supersonic speeds the shock wave is always detached.

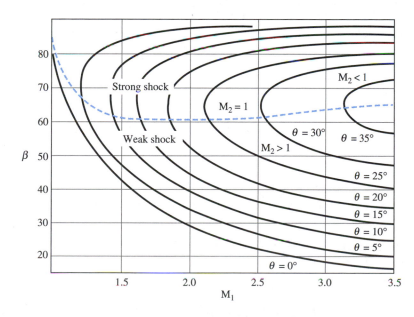

**Fig. 9.16**   Oblique shock wave relationships for $k = 1.4$.

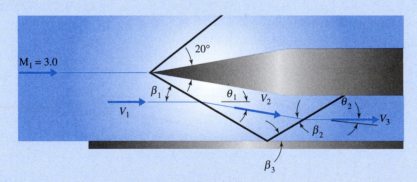

**Fig. 9.17** Detached shock waves: (a) flow around a wedge; (b) flow around a blunt object.

## Example 9.11

Air flows over a wedge with $M_1 = 3$, as shown in Fig. E9.11. A weak shock reflects from the wall. Determine the values of $M_3$ and $\beta_3$ for the reflected wave.

**Fig. E9.11**

### Solution

From Fig. 9.16 with $\theta_1 = 10°$ and $M_1 = 3.0$, we find for the weak shock that $\beta_1 = 27.5°$. This yields

$$M_{1n} = 3 \sin 27.5° = 1.39$$

From the shock table we interpolate to find

$$M_{2n} = 0.744 = M_2 \sin(27.5° - 10°)$$

$$\therefore M_2 = 2.48$$

The reflected shock must again turn the flow through an angle of 10°, that is, $\theta_2 = 10°$. For this wedge angle and $M_2 = 2.48$ from Fig. 9.16 for a weak shock, we see that $\beta_2 = 33°$. This results in

$$M_{2n} = 2.48 \sin 33° = 1.35$$

From the shock table

$$M_{3n} = 0.762 = M_3 \sin 23°$$

$$\therefore M_3 = 1.95$$

The desired angle is calculated to be

$$\beta_3 = \beta_2 - 10° = 23°$$

Note that Fig. 9.16 does not allow precise calculations. Equation 9.7.8 could be used, by trial and error, to improve the accuracy of the $\beta$'s and hence the quantities that follow.

## 9.8 ISENTROPIC EXPANSION WAVES

In this section we consider the supersonic flow around a convex corner, as shown in Fig. 9.18. Let us first attempt to accomplish such a flow with the finite ampli-tude wave of Fig. 9.18a. The flow must turn the angle $\theta$ so that $V_2$ is parallel to the wall. The tangential component must be conserved because of momentum con-servation. This would result in $V_2 > V_1$, as is obvious from the sketch. This would be the situation if a subsonic flow, $M_{1n} < 1$, could experience a finite increase to a supersonic flow, $M_{2n} > 1$. This, of course, is impossible because of the second law, as was noted in the discussion associated with Fig. 9.10. Consequently, we consider the turning of a flow around a convex corner using a finite wave as an impossibility.

Consider a second possible mechanism that would allow the flow to turn the corner, a fan composed of an infinite number of Mach waves, emanating from the

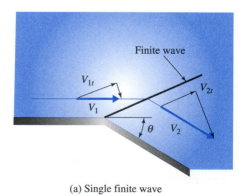

(a) Single finite wave    (b) Infinite number of Mach waves

**Fig. 9.18**   Supersonic flow around a convex corner.

**Fig. 9.19**   A single Mach wave.

corner, as shown in Fig. 9.18b. The second law would not be violated with such a mechanism since each Mach wave is an isentropic wave. We will determine the effect of a single Mach wave on the flow and then integrate to obtain the total effect. Figure 9.19 shows the infinitesimal velocity change due to a single Mach wave. For the control volume enclosing the Mach wave, we know that the tangential momentum is conserved; thus the tangential velocity component remains unchanged as shown, as in the oblique shock wave. From the triangles in the figure we can write

$$V_t = V \cos \mu = (V + dV) \cos(\mu + d\theta) \tag{9.8.1}$$

Since $d\theta$ is small, this becomes,[1] using $\cos(\mu + d\theta) = \cos \mu - d\theta \sin \mu$,

$$V \sin \mu \, d\theta = \cos \mu \, dV \tag{9.8.2}$$

Substituting $\sin \mu = 1/M$ (see Eq. 9.2.13) and $\cos \mu = (\sqrt{M^2 - 1})/M$, we have

$$d\theta = \sqrt{M^2 - 1} \, \frac{dV}{V} \tag{9.8.3}$$

The relationship $V = M \sqrt{kRT}$ can be differentiated and rearranged to give

$$\frac{dV}{V} = \frac{dM}{M} + \frac{1}{2} \frac{dT}{T} \tag{9.8.4}$$

The energy equation, in the form $\frac{1}{2} V^2 + kRT/(k - 1) = $ const., may also be differentiated to yield

$$\frac{dV}{V} + \frac{1}{(k - 1)M^2} \frac{dT}{T} = 0 \tag{9.8.5}$$

---

[1] Recall the trigonometric identity, $\cos(\alpha + \beta) = \cos \alpha \cos \beta - \sin \alpha \sin \beta$. Then using $\cos d\theta = 1$ and $\sin d\theta = d\theta$, we have $\cos(\mu + d\theta) = \cos \mu - d\theta \sin \mu$.

Eliminating $dT/T$ by combining the two preceding equations, there results

$$\frac{dV}{V} = \frac{2}{2 + (k-1)M^2} \frac{dM}{M} \qquad (9.8.6)$$

This can be substituted into Eq. 9.8.3, allowing a relationship between $\theta$ and M to be obtained. We have

$$d\theta = \frac{2\sqrt{M^2 - 1}}{2 + (k-1)M^2} \frac{dM}{M} \qquad (9.8.7)$$

This can be integrated, using $\theta = 0$ at M = 1, to provide a relationship between the resulting Mach number ($M_2$ in Fig. 9.18) and the angle, provided that the incoming Mach number is unity; the relationship is

$$\theta = \left(\frac{k+1}{k-1}\right)^{1/2} \tan^{-1}\left[\frac{k-1}{k+1}(M^2 - 1)\right]^{1/2} - \tan^{-1}(M^2 - 1)^{1/2} \qquad (9.8.8)$$

The angle $\theta$, which is a function of M, is the **Prandtl–Meyer function**. It is tabulated for $k = 1.4$ in Table D.3 so that trial-and-error solutions to Eq. 9.8.8 for M are not necessary. Other changes that may be desired, such as the pressure or temperature changes, can be found from the isentropic-flow equations. The collection of Mach waves that turn the flow are referred to as an *expansion fan*.

We will see, in working the examples and problems, that both the Mach number and the velocity increase as supersonic flow turns the convex corner. The flow remains attached to the wall as it turns the corner, even for large angles, a phenomenon not observed in a subsonic flow; a subsonic flow would separate from the abrupt corner, even for small angles.

If we substitute M = $\infty$ in Eq. 9.8.8, we find the angle of maximum turning to be $\theta = 130.5°$. This would mean that both the temperature and pressure would be absolute zero; obviously, the gas would become a liquid before this would be possible. The angle of 130.5° is, however, an upper limit. The point is, rather large turning angles are possible in supersonic flows, angles that may exceed 90°, a rather surprising result. This introduces a design constraint on the exhaust nozzle of rocket engines that exhaust into the vacuum of space; the exhaust gases may turn through an angle so that impingement on the spacecraft body would occur if not designed properly.

**Prandtl–Meyer function:** *The angle $\theta$ through which the supersonic flow turns.*

**KEY CONCEPT** *The flow remains attached to the wall as it turns the corner, even for large angles.*

**KEY CONCEPT** *Rather large turning angles are possible in supersonic flows.*

## Example 9.12

Air at a Mach number of 2.0 and a temperature and pressure of 500°C and 200 kPa absolute, respectively, flows around a corner with a convex angle of 20° (Fig. E9.12). Find $M_2, p_2, T_2$, and $V_2$.

**Fig. E9.12**

### Solution

Table D.3 uses M = 1 as a reference condition; thus we visualize the flow as originating from a flow with M = 1 and turning through the angle $\theta_1$ to $M_1 = 2$, as shown in the sketch. From the table, adding an additional 20° to the deflection angle, we find that $\theta_2 = 46.4°$. This would be equivalent to the flow at M = 1 turning a convex corner with $\theta = 46.4°$. Since the flow is isentropic, we can simply superimpose in this manner. Now, for an angle of $\theta = 46.4°$ from the table, we find that

$$M_2 = 2.83$$

From the isentropic flow table (Table D.1) we find that

$$p_2 = p_1 \frac{p_0}{p_1} \frac{p_2}{p_0}$$

$$= 200 \times \frac{1}{0.1278} \times 0.0352 = 55.1 \text{ kPa}$$

$$T_2 = T_1 \frac{T_0}{T_1} \frac{T_2}{T_0}$$

$$= 773 \times \frac{1}{0.5556} \times 0.3844 = 534.8 \text{ K} \quad \text{or} \quad 261.8°C$$

The velocity $V_2$ is found to be

$$V_2 = M_2 \sqrt{kRT_2}$$

$$= 2.83 \sqrt{1.4 \times 287 \times 534.8} = 1312 \text{ m/s}$$

## 9.9 SUMMARY

The zone of silence of a supersonic object that produces only Mach waves exists outside a cone with included angle $\alpha$ found from

$$\sin \alpha = \frac{1}{M} \tag{9.9.1}$$

where the Mach number is $M = V/c$ and $c = \sqrt{kRT}$.

The relationship

$$\frac{dV}{V}(M^2 - 1) = \frac{dA}{A} \tag{9.9.2}$$

allows us to predict how subsonic and supersonic flows behave in converging and diverging nozzles. The mass flux through a nozzle with throat area $A*$ where $M* = 1$ is given by

$$\dot{m} = p_0 A* \sqrt{\frac{k}{RT_0}} \left(\frac{k+1}{2}\right)^{(k+1)/2(1-k)} \tag{9.9.3}$$

where $p_0$ and $T_0$ are the reservoir conditions. Temperature, pressure, and velocity in an isentropic flow are found from

$$\frac{T_0}{T} = 1 + \frac{k-1}{2}M^2, \quad \frac{p_0}{p} = \left(1 + \frac{k-1}{2}M^2\right)^{k/(k-1)}, \quad \frac{V^2}{2} = c_p(T_0 - T) \tag{9.9.4}$$

The flow variables across a normal shock are found from continuity, momentum, and energy:

$$\rho_1 V_1 = \rho_2 V_2, \qquad p_1 - p_2 = \rho_1 V_1(V_2 - V_1),$$
$$\frac{V_2^2 - V_1^2}{2} + \frac{k}{k-1}\left(\frac{p_2}{\rho_2} - \frac{p_1}{\rho_1}\right) = 0 \tag{9.9.5}$$

Rather than solve the above equations, we often use the normal-shock flow Table D.2.

Across an oblique shock the tangential velocity component does not change. The normal velocity component $V_n$ simply replaces the velocity $V$ in the normal-shock wave equations above, and Table D.2 may be used with $M_n$ replacing M. The wedge angle $\theta$ through which the flow is turned is related to the oblique shock angle $\beta$ by

$$\tan(\beta - \theta) = \frac{\tan \beta}{k+1}\left[k - 1 + \frac{2}{M_1^2 \sin^2\beta}\right] \tag{9.9.6}$$

or Fig. 9.16 may be used to avoid a trial-and-error solution if $\beta$ is desired.

In an expansion wave, the angle $\theta$ through which a flow with M = 1 can be turned is the Prandtl–Meyer function:

$$\theta = \left(\frac{k+1}{k-1}\right)^{1/2} \tan^{-1}\left[\frac{k-1}{k+1}(M^2-1)\right]^{1/2} - \tan^{-1}(M^2-1)^{1/2} \quad (9.9.7)$$

If we know $\theta$ and desire M, a trial-and-error solution is required. This is avoided by using Table D.3.

## PROBLEMS

**9.1** In thermodynamics $c_p$ for air is often used as 0.24 kcal/kg·°C. From Table B.4 $c_p$ is 1.004 kJ/kg·K. Show these two values to be equal. Also, calculate $c_v$ in both sets of units.

**9.2** Show that $c_p = Rk/(k-1)$.

**9.3** Show that Eqs. 9.1.10 follow from Eqs. 9.1.9 and 9.1.7.

**9.4** Verify the various forms of the energy equation in Eqs. 9.1.3, 9.1.4, and 9.1.5.

### Speed of Sound

**9.5** Show that Eq. 9.2.8 follows from Eq. 9.2.7.

**9.6** Show that the velocity of sound in a high-frequency wave (i.e., a dog whistle) travels with speed $c = \sqrt{RT}$ if the process is assumed to be isothermal.

**9.7** Show that for a small adiabatic disturbance in a steady flow, the energy equation takes the form $\Delta h = -c\,\Delta V$.

**9.8** Verify that the speed of propagation of a small wave through water is about 1450 m/s.

**9.9** Two rocks are slammed together on one side of a lake. An observer on the other side with his head under water "hears" the disturbance 0.6 s later. How far is it across the lake?

**9.10** You and a friend are standing 10 m apart in waist-deep water. You slam two rocks together under water. How long after the rocks are slammed together does your friend hear the interaction if your friend's head is underwater?

**9.11** Calculate the Mach number for an aircraft if it is flying at:
(a) Sea level at 200 m/s
(b) 4500 m at 180 m/s
(c) 10 000 m at 200 m/s
(d) 18 000 m at 180 m/s
(e) 10 000 m at 200 m/s

**9.12** A wood chopper is chopping wood some distance away. You carefully observe, using your numerical wrist timer, that it takes 1.21 s for the sound of the axe to reach your ears. Calculate the distance you are from the chopper if the temperature is −10°C.

**9.13** You see a bolt of lightning and 2 s later you hear the sound of thunder. How far away did the lightning bolt strike?

**9.14** A needle-nosed projectile passes over you on a military testing field at a speed of 1000 m/s (Fig. P9.14). You know that it has an elevation of 1000 m where $T = -10°C$. How long after it passes overhead will you hear its sound? How far away will it be? Calculate its Mach number.

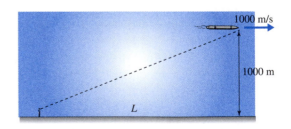

**Fig. P9.14**

**9.15** A special camera is capable of showing the Mach angle of a sharp-nosed bullet passing through the test section of a wind tunnel. If the Mach angle is measured to be 22°, calculate the speed of the bullet. Assume standard conditions.

**9.16** A small-amplitude wave passes through the standard atmosphere at sea level with a pressure rise of 14 Pa. Estimate the associated induced velocity and temperature rise.

## Isentropic Flow

**9.17** Provide all the steps needed to proceed from:
  **(a)** Eq. 9.3.2 to Eq. 9.3.3
  **(b)** Eq. 9.3.5 to Eq. 9.3.10
  **(c)** Eq. 9.3.11 to Eq. 9.3.12
  **(d)** Eq. 9.3.16 to Eq. 9.3.18
  **(e)** Eq. 9.3.17 to Eq. 9.3.19

**9.18** A pitot probe, an instrument that measures stagnation pressure, is used to determine the velocity of an airplane. Such a probe, attached to an aircraft, measures 10 kPa. Determine the speed of the aircraft if it is flying at an altitude of:
  **(a)** 3000 m           **(b)** 10 000 m

  Assume an isentropic process from the free stream to the stagnation point.

**9.19** A pitot probe, used to measure the stagnation pressure, indicates a pressure of 4 kPa at the nose of a surface vehicle traveling in 15°C–atmospheric air. Calculate its speed assuming:
  **(a)** Isentropic flow
  **(b)** Incompressible flow

  Compute the percent error for part (b).

**9.20** A converging nozzle with an exit diameter of 2 cm is attached to a reservoir maintained at 25°C and 200 kPa absolute. Using equations only, determine the mass flux of air if the receiver pressure is:
  **(a)** 100 kPa absolute
  **(b)** 130 kPa absolute

**9.21** The converging nozzle of Fig. P9.21 is attached to a reservoir maintained at 20°C and 200 kPa (abs.). Using equations only, determine the mass flux of air if the receiver pressure is:
  **(a)** 100 kPa           **(b)** 150 kPa

**Fig. P9.21**

**9.22** Rework Problem 9.20 using the isentropic flow table.

**9.23** Rework Problem 9.21 using the isentropic flow table.

**9.24** Air flows from a reservoir ($T_0 = 30°C, p_0 = 400$ kPa absolute) out a converging nozzle with a 10-cm-diameter exit. What exit pressure would just result in $M_e = 1$? Determine the mass flux for this condition. Use the isentropic flow table.

**9.25** Air flows from a converging nozzle attached to a reservoir with $T_0 = 10°C$. What reservoir pressure is necessary to just cause $M_e = 1$ if the 6-cm-diameter nozzle exits to atmospheric pressure? Calculate the mass flux for this condition.

**9.26** Air flows from a converging nozzle attached to a reservoir with $T_0 = 280$ K. If the 6-cm-diameter nozzle exits to atmospheric pressure, what reservoir pressure is necessary to just cause $M_e = 1$? Calculate the mass flux for this condition. Now double the reservoir pressure and determine the increased mass flux.

**9.27** A 25-cm-diameter air line is pressurized to 500 kPa absolute and suddenly bursts (Fig. P9.27). The exit area is later measured to be 30 cm². If 6 min elapsed before the 10°C air was turned off, estimate the cubic meters of air that was lost.

25 cm dia.

Air          10 °C

**Fig. P9.27**

**9.28** A converging nozzle is attached to a reservoir containing helium with $T_0 = 27°C$ and $p_0 = 200$ kPa absolute. Determine the receiver pressure that will just give $M_e = 1$. Now attach a diverging section with a 15-cm-diameter exit to the 6-cm-diameter throat. What is the highest receiver pressure that will give $M_t = 1$?

**9.29** A venturi tube is used to measure the mass flux of air in a pipe by reducing the diameter from 10 cm to 5 cm and then back to 10 cm. The pressure at the inlet is 300 kPa and at the minimum diameter section it is 240 kPa. If the upstream temperature is 20°C, determine the mass flux.

**9.30** Air enters a converting-diverging nozzle from a reservoir at 200 kPa (gage) and 22°C. the area at the throat is 9.7 cm², and at the exit it is 13 cm². Considering isentropic flow throughout with M = 1 at the throat, determine the velocity of air at the nozzle exit.

**9.31** Consider the nozzle flow of Problem 9.30, but in this case the Mach number at the throat is 0.72. Determine the velocity at the nozzle exit for the given condition.

**9.32** The mass flux of air flowing through the pipe of Fig. P9.32 is desired. The pipe is reduced in diameter from 10 cm to 5 cm and then back to 10 cm. The pressure at the inlet is 310 kPa gauge and at the minimum diameter section it is 250 kPa gauge. If the upstream temperature is 16°C, determine the mass flux.

310 kPa          250 kPa

5 cm

10 cm diameter

**Fig. P9.32**

**9.33** Air flows from a reservoir maintained at 30°C and 200 kPa absolute through a converging–diverging nozzle that has a 10-cm-diameter throat. Determine the diameter where M = 3. Use equations only.

**9.34** Air flows in a converging–diverging nozzle from a reservoir maintained at 20°C and 500 kPa absolute. The throat and exit diameters are 5 cm and 15 cm, respectively. What two receiver pressures will result in M = 1 at the throat if isentropic flow occurs throughout? Use equations only.

**9.35** Rework Problem 9.34 using the isentropic flow table.

**9.36** Air flows from a nozzle with a mass flux of 14.6 kg/s. If $T_0 = 16°C$, $p_0 = 830$ kPa (abs.), and $p_e = 103$ kPa (abs.). Calculate the throat and exit diameters for an isentropic flow. Also, determine the exit velocity.

**9.37** Air flows from a reservoir, maintained at 20°C and 2 MPa absolute, and exits from a nozzle with $M_e = 4$. The receiver pressure is then raised until the flow is just subsonic throughout the entire nozzle. Calculate this receiver pressure.

**9.38** A small wind tunnel uses compressed air from a reservoir at 3.5 MPa and 320 K. The air flows isentropically from the reservoir through a nozzle, as shown in Fig. P9.38. If the Mach number in the test section is 2.8, calculate the pressure and velocity in the test section

Reservoir
3.5 MPa
320 K

Test section
M = 2.8

**Fig. P9.38**

**9.39** Consider isentropic air flow from a reservoir (where $p_0 = 600$ kPa, $T_0 = 30°C$) through a converging-diverging nozzle. At a section in the converging part of the nozzle before the throat, the Mach number is 0.50, and the cross-sectional area is 12.4 cm$^2$.

(a) If the area at the throat is 10 cm$^2$, calculate the pressure, temperature, and velocity at the throat.

(b) If M = 1 at the throat, what should be the receiver pressure to just produce supersonic flow at the exit?

(c) If the Mach number is 2.0 at the exit, what should be the exit area and the mass flux?

**9.40** Air at 30°C flows in a 10-cm-diameter pipe at a velocity of 150 m/s. A venturi tube is used to measure the flow rate. What should be the minimum diameter of the tube so that supersonic flow does not occur?

**9.41** For a nozzle efficiency of 96%, rework Problem 9.23.

**9.42** Nitrogen enters a diffuser at 100 kPa absolute and 100°C with a Mach number of 3.0. The mass flux is 10 kg/s and the exit velocity is small. Sketch the diffuser, then determine the throat area and the exit pressure and temperature, assuming isentropic flow.

**9.43** Nitrogen enters a diffuser at 103 kPa (abs.) absolute and 95°C with a Mach number of 3.0. The mass flux is 3 kg/s and the exit velocity is small. Sketch the diffuser, then determine the throat area and the exit pressure and temperature, assuming isentropic flow.

**9.44** A rocket has a mass of 80 000 kg and is to be lifted vertically from a platform with six nozzles exiting

exhaust gases with $T_e = 1000°C$. What should the exit velocity be from each 50-cm-diameter nozzle if the exhaust gases are assumed to be carbon dioxide?

**9.45** A 100-kg man straps a small air-breathing jet engine on his back and just lifts off the ground vertically (Fig. P9.45). The engine has an exit area of 200 cm$^2$. With what velocity must the 600°C exhaust gases exit the engine?

**Fig. P9.45**

**9.46** A converging–diverging nozzle is bolted into a reservoir at a diameter of 40 cm. The throat and exit diameters are 5 cm and 10 cm, respectively. If $T_0 = 27°C$ and $p_e = 100$ kPa absolute and isentropic flow of air occurs throughout the supersonic nozzle, calculate the force necessary to hold the nozzle onto the reservoir.

**9.47** What is the maximum speed in (a) m/s and (b) km/h that an aircraft can have during take-off or landing if the airflow around the aircraft is to be modeled as an incompressible flow? Allow a 3% error in the pressure from free stream to the stagnation point. Assume standard conditions.

## Normal Shock

**9.48** The pressure, temperature, and velocity before a normal shock wave are 80 kPa absolute, 10°C, and 1000 m/s, respectively. Calculate $M_1$, $M_2$, $p_2$, $T_2$, and $\rho_2$ for air. Use:

(a) Basic equations

(b) The normal shock table

**9.49** The pressure, temperature, and velocity before a normal shock wave are 83 kPa, 278 K, and 900 m/s, respectively. Calculate $M_1$, $M_2$, $p_2$, $T_2$, and $\rho_2$ for air. Use:

(a) Basic equations

(b) The normal shock table

**9.50** Derive the *Rankine–Hugoniot relationship*,

$$\frac{\rho_2}{\rho_1} = \frac{(k+1)\, p_2/p_1 + k - 1}{(k-1)\, p_2/p_1 + k + 1}$$

which relates the density ratio to the pressure ratio across a normal shock wave. Find the limiting density ratio for air across a strong shock for which $p_2/p_1 \gg 1$.

**9.51** An explosion occurs just above the surface of the Earth, producing a shock wave that travels radially outward. At a given location it has a Mach number of 2.0. Determine the pressure just behind the shock and the induced velocity.

**9.52** Air at 200 kPa absolute and 20°C passes through a normal shock wave with strength so that $M_2 = 0.5$. Calculate $V_1$, $p_2$, and $\rho_2$.

**9.53** Air at 200 kPa and 16°C passes through a normal shock wave with strength so that $M_2 = 0.5$. Calculate $V_1$, $p_2$, and $\rho_2$.

**9.54** A blunt object travels at 1000 m/s at an elevation of 10 000 m. The flow approaching the stagnation point passes through a normal shock wave and then decelerates isentropically to the stagnation point. Calculate $p_0$ and $T_0$ at the stagnation point.

**9.55** A pitot probe is inserted in an airflow in a pipe in which $p = 800$ kPa absolute, $T = 40°C$, and $M = 3.0$ (Fig. P9.55). What pressure does it measure?

**Fig. P9.55**

**9.56** Air flows from a 25°C reservoir to the atmosphere through a nozzle with a 5-cm-diameter throat and a 10-cm-diameter exit. What reservoir pressure will just result in $M = 1$ at the throat? Also, calculate the mass flux. Maintaining this reservoir pressure, reduce the throat diameter to 4 cm and determine the resulting mass flux. Sketch the pressure distribution as in Fig. 9.12.

**9.57** Air flows from a 20°C reservoir to the atmosphere through a nozzle with a 5-cm-diameter throat and a 10-cm-diameter exit. What reservoir pressure is necessary to locate a normal shock wave at the exit? Also, calculate the velocity and pressure at the throat, before the shock, and after the shock.

**9.58** Air flows from a 17°C reservoir to the atmosphere through a nozzle with a 5-cm-diameter throat and a 10-cm-diameter exit. What reservoir pressure is necessary to locate a normal shock wave at the exit? Also, calculate the velocity and pressure at the throat, before the shock, and after the shock.

**9.59** Air flows from a reservoir maintained at 25°C and 500 kPa absolute. It flows through a nozzle with throat and exit diameters of 5 cm and 10 cm, respectively. What receiver pressure is necessary to locate a normal shock wave at a location where the diameter is 8 cm? Also, determine the velocity before the shock and at the exit.

## Vapor Flow

**9.60** Steam flows at a rate of 4 kg/s from reservoir conditions of 400°C and 1.2 MPa absolute through a converging–diverging nozzle to the atmosphere. Determine the throat and exit diameters if supersonic, isentropic flow exists throughout the diverging section.

**9.61** Steam flows from reservoir conditions of 350°C and 1000 kPa absolute to the atmosphere at the rate of 15 kg/s. Estimate the converging nozzle exit diameter.

**9.62** Steam flows from reservoir conditions of 370°C and 1 MPa (abs.) to the atmosphere at the rate of 3.65 kg/s. Estimate the converging nozzle exit diameter.

**9.63** A header supplies steam at 400°C and 1.2 MPa absolute to a set of nozzles with throat diameters of 1.5 cm. The nozzles exhaust to a pressure of 120 kPa absolute. If the flow is approximately isentropic, calculate the mass flux and the exit temperature.

## Oblique Shock Wave

**9.64** An airflow with velocity, temperature, and pressure of 800 m/s, 30°C, and 40 kPa absolute, respectively, is turned with an oblique shock wave emanating from the wall, which makes an abrupt 20° corner.

**(a)** Find the downstream Mach number, pressure, and velocity for a weak shock.

**(b)** Find the downstream Mach number, pressure, and velocity for a strong shock.

**(c)** If the concave corner angle were 35°, sketch the corner flow situation.

**9.65** Two oblique shocks intersect as shown in Fig. P9.65. Determine the angle of the reflected shocks if the airflow must leave parallel to its original direction. Also find $M_3$.

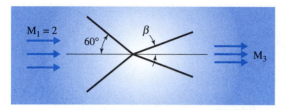

**Fig. P9.65**

**9.66** An oblique shock wave at a 35° angle is reflected from a plane wall. The upstream Mach number $M_1$ is 3.5 and $T_1 = 0°C$. Find $V_3$ after the reflected oblique shock wave for the airflow.

**9.67** A supersonic inlet can be designed to have a normal shock wave oriented at the inlet, or a wedge can be used to provide a weak oblique shock wave, as shown in Fig. P9.68. Compare the pressure $p_3$ from the flow shown to the pressure that would exist behind a normal shock wave with no oblique shock.

**Fig. P9.67**

## Expansion Waves

**9.68** A supersonic airflow with $M_1 = 3$, $T_1 = -20°C$, and $p_1 = 20$ kPa absolute turns a convex corner of 25°. Calculate $M_2, p_2, T_2$, and $V_2$ after the expansion fan. Also calculate the included angle of the fan.

**9.69** A supersonic airflow with $M_1 = 2$, $T_1 = 0°C$, and $p_1 = 20$ kPa absolute turns a convex corner. If $M_2 = 4$, what angle $\theta$ should the corner have? Also calculate $T_2$ and $V_2$.

**9.70** A supersonic airflow with $M_1 = 2$, $T_1 = -2°C$, and $p_1 = 35$ kPa (abs.) turns a convex corner. If $M_2 = 4$, what angle $\theta$ should the corner have? Also calculate $T_2$ and $V_2$.

**9.71** The flat plate shown in Fig. P9.72 is used as an airfoil at an angle of attack of 5°. Oblique shock waves and expansion fans allow the air to remain attached to the plate with the flow behind the airfoil parallel to the original direction. Calculate

(a) The pressures on the upper and lower sides of the plate

(b) The downstream Mach numbers $M_{2u}$ and $M_{2l}$

(c) The lift coefficient defined by $C_L =$ lift$/(\frac{1}{2}\rho_1 V_1^2 A)$. Note that $\rho_1 V_1^2 = kM_1^2 \rho_1$

**Fig. P9.71**

**9.72** The supersonic airfoil shown in Fig. P9.73 is to fly at zero angle of attack. Calculate the drag coefficient $C_D = $ drag$/(\frac{1}{2}\rho_1 V_1^2 A)$. Note that $\rho_1 V_1^2 = kM_1^2 \rho_1$.

**Fig. P9.72**

**9.73** The airfoil in Problem 9.73 flies at an angle of attack of 5°. Determine the lift and drag coefficients, $C_L$ and $C_D$. See Problems 9.72 and 9.73 for definitions of $C_L$ and $C_D$.

The Wahluke Branch Canal, part of the Columbia Basin Project, is an example of an engineered canal. Water is delivered over a distance of many kilometers. (U.S. Bureau of Reclamation)

# 10

# Flow in Open Channels

## Outline

## Chapter Objectives

The objectives of this chapter are to:

▲ Describe various types of open-channel, free-surface flows

▲ Apply the Chezy–Manning equation with various geometric cross-sections for uniform flow applications

▲ Derive energy and momentum principles for rapidly varied flow situations

▲ Derive the differential equation for nonuniform gradually varied flow

▲ Present the method of profile synthesis—the qualitative description of open-channel flow, including establishment of controls and classification of water surface profiles

▲ Develop and present methods to numerically compute gradually varied flow

▲ Present numerous examples of uniform, rapidly varied, and gradually varied flow

▲ Detail several examples of profile synthesis and numerical analysis of complex open-channel flow with emphasis on design application

## 10.1 INTRODUCTION

Free-surface flow is probably the most commonly occurring flow phenomenon that we encounter on the surface of the earth. Ocean waves, river currents, and overland flow of rainfall are examples that occur in nature. Man-made situations include flows in canals and culverts, drainage over impervious materials such as roofs and parking lots, and wave motion in harbors.

In all of these situations, the flow is characterized by an interface between the air and the upper layer of water, which is termed the **free surface**. At the free surface, the pressure is constant, and for nearly all situations, it is atmospheric. In such a case, the hydraulic grade line and the liquid free surface coincide. In engineering practice, most open channels convey water as the fluid. However, the principles developed and implemented in this chapter are also applicable for other liquids that flow with a free surface.

Generally, the elevation of the free surface does not remain constant; it can vary along with the fluid velocities. Another complexity is that the flow is often three-dimensional. Fortunately, there are many instances in which two-dimensional and even one-dimensional simplifications can be made. The flow patterns in estuaries[1] and lakes under certain circumstances can be treated as two-dimensional in the horizontal plane, averaged vertically over the depth. River and channel flows are usually treated as one-dimensional with respect to the position coordinate along the streambed. In this chapter we restrict our consideration to one-dimensional flows.

Figure 10.1a shows a representative centerline velocity distribution in a channel. The velocity profile is three-dimensional at a given cross section (Fig. 10.1b). The boundary shear stress is nonuniform; at the free surface the shear

**Free surface:** *The interface between the air and the upper layer of water.*

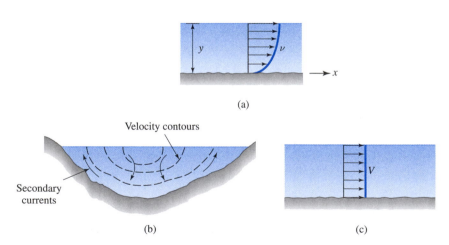

**Fig. 10.1**   Free-surface flow: (a) centerline velocity distribution; (b) cross section; (c) one-dimensional model.

---

[1]An estuary is the lower course of a river that is influenced by the ocean tides.

stress is negligible, yet the shear stress varies around the wetted perimeter. In some circumstances the presence of secondary currents will force the maximum velocity to occur slightly below the free surface. By convention, $y$ is defined as the depth from the deepest location to the free surface; note that $y$ is not a coordinate. The mean velocity is given by the relation

$$V = \frac{1}{A} \int_A v \, dA \qquad (10.1.1)$$

In the one-dimensional model, we assume the velocity to be equal to $V$ everywhere at a given cross section. This model provides excellent results and is used widely. Flows in open channels are most likely to be turbulent, and the velocity profile can be assumed to be approximately constant, as in Fig. 10.1c, without significant error. Hence the one-dimensional model is employed.

## 10.2 OPEN-CHANNEL FLOWS

### 10.2.1 Classification of Free-Surface Flows

Flow in a channel is characterized by the mean velocity, even though a velocity profile exists at a given section, as shown in Fig. 10.1. The flow is classified as a combination of steady or unsteady, and uniform or nonuniform. Steady flow signifies that the mean velocity $V$, as well as the depth $y$, is independent of time, whereas unsteady flow necessitates that time be considered as an independent variable. Uniform flow implies that $V$ and $y$ are independent of the position coordinate in the direction of flow; nonuniform flow signifies that $V$ and $y$ vary in magnitude along that coordinate. The possible combinations are shown in Table 10.1; the position coordinate is designated as $x$.

Steady, uniform flow is the situation where terminal velocity has been reached in a channel of constant cross section; not only is the mean velocity constant, but the depth is invariant as well. Steady, nonuniform flow is a common occurrence in rivers and in man-made channels. In those situations, it will be found to occur in two ways. In relatively short reaches, called *transitions*, there is a rapid change

**KEY CONCEPT** *In relatively short reaches, called transitions, there is a rapid change in depth and velocity.*

**Table 10.1** Combinations of One-Dimensional Free-Surface Flows

| Type of flow | Average velocity | Depth |
|---|---|---|
| Steady, uniform | $V = $ const. | $y = $ const. |
| Steady, nonuniform | $V = V(x)$ | $y = y(x)$ |
| Unsteady, uniform | $V = V(t)$ | $y = y(t)$ |
| Unsteady, nonuniform | $V = V(x,t)$ | $y = y(x,t)$ |

**Rapidly varied flow:** *A rapid change in depth and velocity.*

in depth and velocity; such flow is termed **rapidly varied flow**. Examples are the hydraulic jump (shown in Example 4.12), flow entering a steep channel from a lake or reservoir, flow close to a free outfall from a channel, and flow in the vicinity of an obstruction such as a bridge pier or a sluice gate.

Along more extensive reaches of channel the velocity and depth may not vary rapidly, but rather, change in a slow manner. Here the water surface can be considered continuous, and the regime is called **gradually varied flow**. Examples of gradually varied steady flow are the backwater created by a dam placed in a river and the drawdown of a water surface as flow approaches a falls. Figure 10.2 illustrates how both rapidly varied flow (RVF) and gradually varied flow (GVF) can occur simultaneously in a reach of channel. Note that the vertical scale is larger than the horizontal scale; such scale distortion is common when representing open-channel flow situations.

**Gradually varied flow:** *In extensive reaches of a channel, the velocity and depth change in a slow manner.*

Unsteady uniform flow rarely occurs, but unsteady nonuniform flow is common. Flood waves in rivers, hydraulic bores, and regulated flow in canals are all examples of the latter category. In many situations, these flows may be considered to behave sufficiently like steady uniform flow or steady nonuniform flow to justify treating them as such. Unsteady flows are beyond the scope of a fundamental treatment and will not be presented in this book; the one exception is the hydraulic bore—a moving hydraulic jump—which can be analyzed in a quasi-steady fashion.

### 10.2.2 Significance of Froude Number

The primary mechanism for sustaining flow in an open channel is gravitational force. For example, the elevation difference between two reservoirs will cause water to flow through a canal that connects them. The parameter that represents this gravitational effect is the *Froude number*,

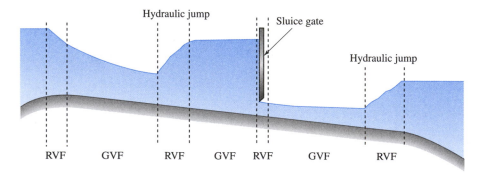

**Fig. 10.2**  Steady nonuniform flow in a channel.

$$\text{Fr} = \frac{V}{\sqrt{gL}} \tag{10.2.1}$$

which has been observed in Chapter 6 to be the ratio of inertial force to gravity force. In the context of open-channel flow, $V$ is the mean cross-sectional velocity, and $L$ is a representative length. For a channel of rectangular cross section, $L$ is the depth $y$ of the flow.

The Froude number plays the dominant role in open-channel flow analysis. It appears in a number of relations that will be developed later in this chapter. Furthermore, by knowing its magnitude, one can ascertain significant character-istics regarding the flow regime. For example, if $\text{Fr} > 1$, the flow possesses a rel-atively high velocity and shallow depth; on the other hand, when $\text{Fr} < 1$, the velocity is relatively low and the depth is relatively deep. Except in the vicinity of rapids, cascades, and waterfalls, most rivers experience a Froude number less than unity. Constructed channels may be designed for Froude numbers to be greater or less than unity, or to vary from greater than unity to less than unity along the length of the channel.

### 10.2.3  Hydrostatic Pressure Distribution

Consider a channel in which the flow is nearly horizontal, as shown in Fig. 10.3. In this case, there is little or no vertical acceleration of fluid within the reach, and the streamlines remain nearly parallel. Such a condition is common in many open-channel flows, and indeed, if slight variations exist, the streamlines are assumed to behave as if they are parallel. Since the vertical accelerations are nearly zero, one can conclude that in the vertical direction the pressure distribu-tion is hydrostatic. As a result, the sum $(p + \gamma z)$ remains a constant at any depth, and the hydraulic grade line coincides with the water surface. It is customary in open-channel flow to designate $z$ as the elevation of the channel bottom and $y$ as the depth of flow. Since at the channel bottom $p/\gamma = y$, the hydraulic grade line

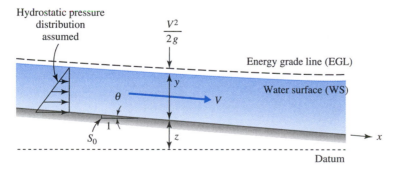

**Fig. 10.3**   Reach of open-channel flow.

is given by the sum $(y + z)$. The concepts in this chapter are developed assuming a hydrostatic pressure distribution.

## 10.3  UNIFORM FLOW

Before we study nonuniform flow in Section 10.6, let us focus our attention on the simpler condition of steady, uniform flow or simply uniform flow. Such a flow is rare, but if it were to occur in a channel, the depth and velocity would not vary along its length, or in other words, terminal conditions would have been reached. In addition to uniform flow, this section covers the cross-sectional geometry of open channels; the formulations will apply to both uniform and nonuniform flow. Some of that material was covered in Section 7.7; it is reviewed here for completeness.

> **KEY CONCEPT**  *Channel cross sections can be considered regular or irregular.*

### 10.3.1  Channel Geometry

Channel cross sections can be considered to be either regular or irregular. A **regular section** is one whose shape does not vary along the length of the channel, whereas an irregular section will have changes in its geometry. In this chapter we consider mostly regular channel shapes; three common geometries are shown in Fig. 10.4.

> **Regular section:**  *Section whose shape does not vary along the length of the channel.*

The simplest channel shape is a rectangular section. The cross-sectional area is given by

$$A = by \tag{10.3.1}$$

in which $b$ is the width of the channel bottom (see Fig. 10.4a). Additional parameters of importance for open-channel flow are the wetted perimeter, the hydraulic radius, and the width of the free surface. The **wetted perimeter** $P$ is the length of the line of contact between the liquid and the channel; for a rectangular channel it is

> **Wetted perimeter:**  *The length of the line of contact between the liquid and the channel.*

$$P = b + 2y \tag{10.3.2}$$

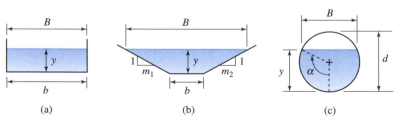

**Fig. 10.4**  Representative regular cross sections: (a) rectangular; (b) trapezoidal; (c) circular.

The **hydraulic radius** $R$ is the area divided by the wetted perimeter, that is,

*Hydraulic radius:    The area divided by the wetted perimeter.*

$$R = \frac{A}{P} = \frac{by}{b + 2y} \tag{10.3.3}$$

The free-surface width $B$ is equal to the bottom width $b$ for a rectangular section.

A trapezoidal section (Fig. 10.4b) has the added feature of sloped side walls. If $m_1$ is the ratio of the horizontal to vertical change of the wall on one side, and $m_2$ is the corresponding quantity on the other wall, the area, wetted perimeter, and free-surface width are given as

$$A = by + \frac{1}{2}y^2(m_1 + m_2) \tag{10.3.4}$$

$$P = b + y\left(\sqrt{1 + m_1^2} + \sqrt{1 + m_2^2}\right) \tag{10.3.5}$$

$$B = b + y(m_1 + m_2) \tag{10.3.6}$$

Note that the rectangular section is embodied in the trapezoidal definition, since for vertical side walls $m_1$ and $m_2$ are zero, and the relations for $A, P$, and $B$ become identical to those for the rectangular channel. In addition, if $b$ is set equal to zero, Eqs. 10.3.4 to 10.3.6 describe the geometry of a triangular-shaped channel.

The circular cross section is an important one to consider, since many free-surface flows in drainage and sewer systems are conveyed in circular conduits. If $d$ is the conduit diameter, the area, wetted perimeter, and free-surface width are given by

$$A = \frac{d^2}{4}(\alpha - \sin \alpha \cos \alpha) \tag{10.3.7}$$

$$P = \alpha d \tag{10.3.8}$$

$$B = d \sin \alpha \tag{10.3.9}$$

where

$$\alpha = \cos^{-1}\left(1 - 2\frac{y}{d}\right) \tag{10.3.10}$$

The angle $\alpha$ is defined in Fig. 10.4c.

A generalized cross-sectional geometry can be expressed in functional form as $A(y), P(y), R(y)$, and $B(y)$. The functional representations encompass all of the analytical forms given above, and they also may be used to describe an irregular channel. For example, at a river section, one can describe the area and wetted perimeter in tabular form and utilize techniques such as curve

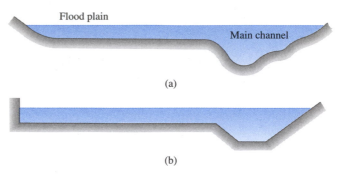

**Fig. 10.5**   Generalized section representation: (a) actual cross section; (b) composite cross section.

fitting or interpolation to extract the numerical information as functions of the depth. Such procedures are useful for computer-based analyses.

A composite section is one made up of several subsections; usually these subsections are of analytic form. The example shown in Fig. 10.5a consists of a main channel and a floodplain. The main channel is approximated by a trapezoid and the floodplain by a rectangle, Fig. 10.5b. One could derive analytical expressions for such a composite section; however, it may be more useful to consider the functional forms for the geometric parameters. Note that the functions will be discontinuous at depths where the two sections are matched.

Most of the theoretical developments in this chapter focus on cross sections that are rectangular. Such an assumption allows one to simplify the mathematics associated with open-channel flow analysis. Even though the equations will be simplified relative to more complicated geometries, the physical understanding of the phenomena and conclusions reached will apply to most generalized prismatic cross sections. A clear distinction will be made between rectangular and other types of geometry when various developments and concepts are presented.

### 10.3.2   Equation for Uniform Flow

Uniform flow occurs in a channel when the depth and velocity do not vary along its length, that is, when terminal conditions have been reached in the channel. Under such conditions, the energy grade line, water surface, and channel bottom are all parallel. Uniform flow can be predicted by an equation of the form

$$V = C \sqrt{RS_0} \qquad (10.3.11)$$

in which $S_0$ is the slope of the channel bottom and $C$ is the *Chezy coefficient*, which is independent of the Reynolds number since the flow is considered completely turbulent. It has become common engineering practice to relate $C$ to the channel roughness and the hydraulic radius by use of the *Manning relation*

$$C = \frac{c_1}{n} R^{1/6} \qquad (10.3.12)$$

where $c_1 = 1$ for SI units. Combining Eqs. 10.3.11 and 10.3.12 with the definition of discharge results in the *Chezy–Manning equation*

$$Q = \frac{c_1}{n} A R^{2/3} \sqrt{S_0} \qquad (10.3.13)$$

Values of the Manning coefficient $n$ are given in Table 7.3.

    The depth associated with uniform flow is designated $y_0$; it is called either *uniform depth* or *normal depth*. Uniform flow rarely occurs in rivers because of the irregularity of the geometry. In man-made channels it is not always present, since the presence of controls such as sluice gates, weirs, or outfalls will cause the flow to become gradually varied. It is, however, necessary to determine $y_0$ when analyzing gradually varied flow conditions, since it provides a basis for evaluating the type of water surface that may exist in the channel. The design of gravity flow sewer networks is often based on assuming uniform flow and the use of Eq. 10.3.13, even though much of the time the flow in such systems may be nonuniform.

    An examination of Eq. 10.3.13 reveals that it can be solved explicitly for $Q$, $n$, or $S_0$. Examples 7.19 and 7.20 provide illustrations. Use of a trial-and-error solution or equation solver is necessary when it is required to find $y_0$ with the remaining parameters given.

> **KEY CONCEPT** *The depth associated with uniform flow is called either uniform depth or normal depth.*

---

## Example 10.1

Water is flowing at a rate of 4.5 m³/s in a trapezoidal channel (Fig. 10.4b) whose bottom width is 2.4 m and side slopes are 1 vertical to 2 horizontal. Compute $y_0$ if $n = 0.012$ and $S_0 = 0.0001$.

**Solution**

Given geometrical data are $b = 2.4$ m and $m_1 = m_2 = 2$. Rearrange Eq. 10.3.13, noting that $R = A/P$ and $c_1 = 1$:

$$\frac{A^{5/3}}{P^{2/3}} = \frac{nQ}{\sqrt{S_0}}$$

Substituting in the known data and trapezoidal geometry, one has

$$\frac{\left[ 2.4y_0 + \frac{1}{2}y_0^2(2 + 2) \right]^{5/3}}{[2.4 + y_0(2\sqrt{1 + 2^2})]^{2/3}} = \frac{0.012 \times 4.5}{\sqrt{0.0001}}$$

Solving for $y_0$, either by trial-and-error or by use of computational software, yields $y_0 = 1.28$ m.

---

## Example 10.2

Uniform flow occasionally occurs in a 5-m-diameter circular concrete conduit (Fig 10.4c), but the depth of flow can vary. The Manning coefficient is $n = 0.013$, and the channel slope is $S_0 = 0.0005$. (a) Calculate the discharge for y = 3 m, (b) Plot the discharge-depth curve.

### Solution

(a) First, use Eq. 10.3.10 to find the angle $\alpha$:

$$\alpha = \cos^{-1}(1 - 2y/d)$$
$$= \cos^{-1}(1 - 2 \times 3/5) = 101.54° \text{ or } 101.54 \times \pi/180 = 1.772 \text{ rad}$$

Using Eqs. 10.3.7 and 10.3.8, the area and wetted perimeter are

$$A = \frac{d^2}{4}(\alpha - \sin\alpha \cos\alpha) = \frac{5^2}{4}(1.772 - \sin 101.54° \cos 101.54°) = 12.3 \text{ m}^2$$

$$p = \alpha d = 1.772 \times 5 = 8.86 \text{ m}$$

The hydraulic radius is then

$$R = \frac{A}{P} = \frac{12.3}{8.86} = 1.388 \text{ m}$$

Finally, the discharge, when $y = 3$ m, is

$$Q = \frac{1}{n}AR^{2/3}S^{1/2} = \frac{1}{0.013} \times 12.3 \times 1.388^{2/3} \times 0.0005^{0.5} = 26.3 \text{ m}^3/\text{s}$$

(b) Mathcad is employed to generate the curve. Note that the solution is generalized, so that any diameter, Manning coefficient, or channel slope can be entered into the algorithm. Equations 10.3.7 and 10.3.8 are used to define the area and wetted perimeter, respectively. Either SI or English units can be employed by properly defining the parameter $c_1$. In this problem a value of 1.0 is used. A MATLAB solution is shown in Appendix E, Fig E.1.

**Input diameter, Manning coefficient, and channel slope:**

$$d := 5 \qquad n := 0.013 \qquad S_0 := 0.0005 \qquad c_1 := 1.0$$

**Define geometric functions:**

$$\alpha(y) := \text{acos}\left(1 - 2 \cdot \frac{y}{d}\right)$$

$$A(y) := \frac{d^2}{4} \cdot (\alpha(y) - \sin(\alpha(y)) \cdot \cos(\alpha(y)))$$

$$P(y) := \alpha(y) \cdot d$$

$$R(y) := \frac{A(y)}{P(y)}$$

**Define discharge function, (i.e., Manning's equation):**

$$Q(y) := \frac{c_1}{n} \cdot A(y) \cdot R(y)^{\frac{2}{3}} \cdot \sqrt{S_0}$$

**Plot depth versus discharge:**

$y := 0, 0.01 .. d$

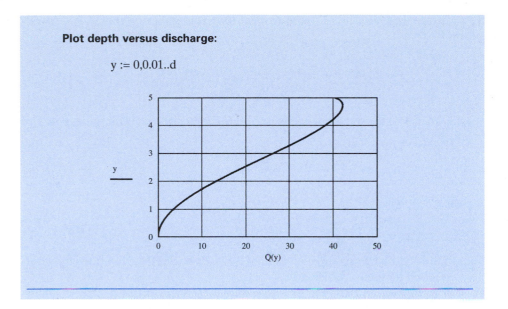

### 10.3.3  Most Efficient Section

Design of a channel to convey uniform flow typically consists of selecting or specifying the appropriate geometrical cross section provided that $Q$, $n$, and $S_0$ are known. Once chosen, an optimum sizing of the cross section can be based on the criteria of minimum resistance to flow. The flow resistance per unit length equals wall shear stress times wetted perimeter. Using a control volume for uniform flow, assume a small slope $S_0$ so that $\sin \theta = S_0$; then, by summing forces on

$$\tau_0 P = \gamma A S_0 \tag{10.3.14}$$

Hence, a least resistance criterion is equivalent to requiring a minimum cross-sectional area with respect to the parameters defining the area. Furthermore, we have to satisfy the Chezy–Manning equation. Since $Q$, $n$, and $S_0$ are given and $R = A/P$, Eq. 10.3.13 can be written as

$$P = cA^{5/2} \tag{10.3.15}$$

in which $c$ is a constant. As an example, consider a rectangular channel with width $b$ and depth $y$. The best hydraulic cross section is obtained by rewriting Eq. 10.3.15 such that $A$ becomes a function of $b$ only. Therefore, we express $P$ in terms of $A$ and $b$, using $A = by$ and $P = b + 2y$:

$$P = b + \frac{2A}{b} \tag{10.3.16}$$

Substitution of this equation into Eq. 10.3.15 yields

$$b + \frac{2A}{b} = cA^{5/2} \tag{10.3.17}$$

Now, differentiate Eq. 10.3.17 with respect to $b$, remembering that $A$ is dependent on $b$:

$$1 + \frac{2}{b}\frac{dA}{db} - \frac{2A}{b^2} = \frac{5}{2}cA^{3/2}\frac{dA}{db} \tag{10.3.18}$$

Set $dA/db = 0$, since the objective is to find the value of $b$ that minimizes $A$. The result is

$$\frac{2A}{b^2} = 1 \tag{10.3.19}$$

or, using $A = by$,

$$b = 2y \tag{10.3.20}$$

Thus, if the width of a rectangular channel is twice the depth of the flowing water, the water will flow most efficiently.

For a trapezoidal cross section, it is simpler to begin with Eq. 10.3.15, eliminate $b$, and express $P$ as a function of $A$, $y$, and $m$, where $m = m_1 = m_2$. Then by considering $A$ as a function of $m$ and $y$, the minimum of $A$ is found by setting the gradient vector of $A$ to zero. The result is $m = \sqrt{3}/3$ or a side slope angle of $60°$ with the horizontal. The resulting hexagonal shape is a trapezoid that best approximates a semicircle. The evaluation is left as an exercise for the reader.

The optimum criterion based on flow resistance cannot always be used, and in fact may be less significant than other design factors. Additional aspects to be considered are the type of excavation, and if the channel is unlined, the slope stability of the side walls and the possibility of erosion of the bed.

## 10.4 ENERGY CONCEPTS

The energy head at any position along the channel is the sum of the vertical distance measured from a horizontal datum $z$, the depth of flow $y$, and the kinetic energy head $V^2/2g$. That sum defines the energy grade line and is termed the **total energy** $H$:

**Total energy:** *The sum of the vertical distance to the channel bottom measured from a horizontal datum, the depth of flow, and the kinetic energy.*

$$H = z + y + \frac{V^2}{2g} \tag{10.4.1}$$

The kinetic-energy-correction factor associated with the $V^2/2g$ term is assumed to be unity (see Section 4.4.4); this is common practice for most prismatic channels of simple geometry since the velocity profiles are nearly uniform for the turbulent flows involved. The fundamental energy equation, developed in Section 4.4, states that losses will occur for a real fluid between any two sections of the channel, and hence the total energy will not remain constant. The energy balance is given simply by the relation

$$H_1 = H_2 + h_L \tag{10.4.2}$$

in which $h_L$ is the head loss. The only manner in which energy can be added to an open-channel flow system is for mechanical pumping or lifting of the liquid to take place. Equation 10.4.2 is applicable for rapidly varied as well as gradually varied flow situations; it will be used in conjunction with the momentum and continuity equations in a number of applications.

### 10.4.1  Specific Energy

It is convenient in open-channel flow to measure the energy relative to the bottom of the channel; it provides a useful means to analyze complex flow situations. Such a measure is termed **specific energy** and is designated as $E$:

**Specific energy:** *Measurement of energy relative to the bottom of the channel.*

$$E = y + \frac{V^2}{2g} \qquad (10.4.3)$$

Specific energy then is the sum of the flow depth $y$ and kinetic energy head $V^2/2g$.

**Rectangular Sections.**   For a rectangular section, the specific energy can be expressed as a function of the depth $y$. The **specific discharge** $q$ is defined as the total discharge divided by the channel width, that is,

**Specific discharge:** *The total discharge divided by the channel width (valid only for a rectangular channel).*

$$q = \frac{Q}{b} = Vy \qquad (10.4.4)$$

The specific energy for a rectangular channel can thus be put in the form

$$E = y + \frac{q^2}{2gy^2} \qquad (10.4.5)$$

This $E$–$y$ relation is shown in Fig. 10.6a. We may observe that a specific discharge requires at least a minimum energy; this minimum energy is referred to as *critical energy, $E_c$.* The corresponding depth $y_c$ is called the **critical depth**. If the specific energy is greater than $E_c$, two depths are possible; those depths are referred to as **alternate depths**. For constant $q$, Eq. 10.4.5 is a cubic equation in $y$ for a given value of $E$ which is greater than $E_c$. The two positive solutions of $y$ are the alternate depths.[2] Another way to express Eq. 10.4.5 is to consider $E$ constant and vary $q$. Equation 10.4.5 can be solved for $q$ as

**Critical depth:** *Critical depth is the depth for which specific energy is a minimum.*

**Alternate depths:** *The two depths of flow that are possible for a given specific energy and discharge.*

$$q = \sqrt{2gy^2 (E - y)} \qquad (10.4.6)$$

---

[2]The $E$-$y$ curve falls between the two asymptotes $E = y$ and $y = 0$. Another curve defined by the relation exists for negative $y$; it is not considered since it has no physical meaning for open-channel flow.

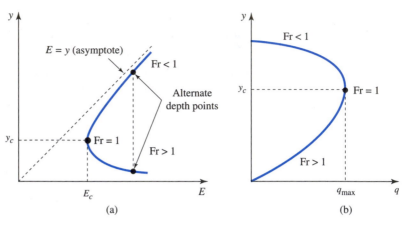

**Fig. 10.6** Variation of specific energy and specific discharge with depth: (a) $E$ versus $y$ for constant $q$; (b) $q$ versus $y$ for constant $E$.

This relation is shown in Fig. 10.6b. This form of the energy relation is useful for analyzing flows in which the specific energy remains nearly constant throughout the transition region; examples are a change in the width of a channel and the variation of depth with discharge at the entrance to a channel. Note that maximum unit discharge, $q_{max}$, occurs at critical depth.

The critical depth $y_c$ can be evaluated from Eq. 10.4.5 by setting the derivative of $E$ with respect to $y$ equal to zero:

$$\frac{dE}{dy} = 1 - \frac{q^2}{gy^3} = 0 \qquad (10.4.7)$$

Since $q = Vy$, the condition of minimum $E$ is

$$1 - \frac{V^2}{gy} = 1 - Fr^2 = 0 \qquad (10.4.8)$$

where the Froude number in a rectangular channel is

$$Fr = \frac{V}{\sqrt{gy}} = \frac{q}{\sqrt{gy^3}} \qquad (10.4.9)$$

Thus, from Eq. 10.4.8 the Froude number is equal to unity for minimum energy. Solving for the depth in Eq. 10.4.7 in terms of $q$, we have

$$y = y_c = \left(\frac{q^2}{g}\right)^{1/3} \qquad (10.4.10)$$

This relation gives the critical depth in terms of the specific discharge. Note that at critical flow conditions, $E_c$ can conveniently be expressed by combining Eqs. 10.4.5 and 10.4.10 to eliminate $q$, resulting in

$$E_c = \frac{3}{2} y_c \qquad (10.4.11)$$

On the $E$–$y$ curve, for any depth greater than $y_c$, the flow is relatively slow or tranquil, and $\text{Fr} < 1$; such a state is termed *subcritical flow*. Conversely, for a depth less than critical, the flow is relatively rapid or shooting, with $\text{Fr} > 1$, and the regime is one of *supercritical flow*. The $E$–$y$ diagram is a representation of the change in specific energy as the depth is varied, given a constant specific discharge. It is possible for $q$ to vary, as when the width of a rectangular section changes in a transition region. As $q$ increases, the $E$–$y$ curve shifts to the right in Fig. 10.6a. It is left as an exercise for the reader to show that for a given specific energy, maximization of $q$ of Eq. 10.4.6 will produce critical conditions at maximum discharge (see Fig. 10.6b).

**Generalized Cross Section.**   For a generalized section, the specific energy is written in terms of the total discharge $Q$ and the cross-sectional area $A$ as

$$E = y + \frac{Q^2}{2gA^2} \qquad (10.4.12)$$

The minimum-energy condition is obtained by differentiating Eq. 10.4.12 with respect to $y$ to obtain

$$\frac{dE}{dy} = 1 - \frac{Q^2}{gA^3} \frac{dA}{dy} \qquad (10.4.13)$$

For incremental changes in depth, the corresponding change in area is $dA = B\,dy$. Thus setting Eq. 10.4.13 to zero, the minimum-energy condition becomes

$$1 - \frac{Q^2 B}{gA^3} = 0 \qquad (10.4.14)$$

By analogy to Eq. 10.4.8, the second term in Eq. 10.4.14 is the square of the Froude number; hence

$$\text{Fr} = \sqrt{\frac{Q^2 B}{gA^3}} = \frac{Q/A}{\sqrt{gA/B}} = \frac{V}{\sqrt{gA/B}} \qquad (10.4.15)$$

The ratio $A/B$ is termed the *hydraulic depth*, and its use allows the definition of the Froude number to be generalized. The reader can verify that $A/B$ is equal to $y$ for a rectangular channel.

The specific energy diagram of Fig. 10.6 provides a useful means of visualizing the solution to a transition problem. Even though one probably would solve the problem numerically, an assessment of the graphical solution may provide useful physical insight as well as prevent the incorrect root from being selected.

## Example 10.3

Water is flowing in a triangular channel with $m_1 = m_2 = 1.0$ at a discharge of $Q = 3$ m³/s. If the water depth is 2.5 m, determine the specific energy, Froude number, hydraulic depth, and alternate depth.

### Solution

Recognizing that $b = 0$, the flow area and top width are computed from Eqs. 10.3.4 and 10.3.6 as follows:

$$A = \frac{1}{2}y^2(m_1 + m_2)$$

$$= \frac{1}{2} \times 2.5^2 \times (1 + 1) = 6.25 \text{ m}^2$$

$$B = (m_1 + m_2)y$$

$$= (1 + 1) \times 2.5 = 5.0 \text{ m}$$

Using Eqs. 10.4.12 and 10.4.15, $E$ and Fr are found to be

$$E = y + \frac{Q^2}{2gA^2}$$

$$= 2.5 + \frac{3^2}{2 \times 9.81 \times 6.25^2} = 2.51 \text{ m}$$

$$\text{Fr} = \sqrt{\frac{Q^2 B}{gA^3}}$$

$$= \sqrt{\frac{3^2 \times 5}{9.81 \times 6.25^3}} = 0.137$$

The hydraulic depth is

$$\frac{A}{B} = \frac{6.25}{5.0} = 1.25 \text{ m}$$

The alternate depth is calculated using the energy equation. Recognizing that $A = y^2$, we have

$$2.51 = y + \frac{3^2}{2 \times 9.81 \times (y^2)^2}$$

$$= y + \frac{0.459}{y^4}$$

A trial-and-error solution provides $y = 0.71$ m.

## 10.4.2   Use of the Energy Equation in Transitions

As mentioned in Section 10.2.1, a transition is a relatively short reach of channel where the depth and velocity change, creating a nonuniform, rapidly varied flow. The mechanism for such changes in the flow is usually an alteration of one

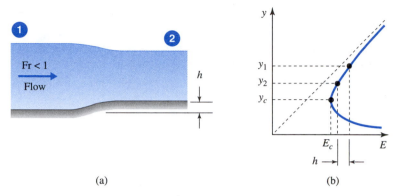

(a)  (b)

**Fig. 10.7**  Channel constriction: (a) raised channel bottom; (b) specific energy diagram.

or more geometric parameters of the channel.[3] Within such regions, the energy equation can be used effectively to analyze the flow in a transition or to aid in the design of a transition. Two applications are given in this section to demonstrate the methodology; other types of transitions can be treated similarly.

**Channel Constriction.**  Consider a rectangular channel whose bottom is raised by a distance $h$ over a short region (Fig. 10.7a). The change in depth in the transition can be analyzed by use of the energy equation, and as a first approximation, one can neglect losses. Assume that the specific energy upstream of the transition is known. Recognizing that $H = E + z$, Eq. 10.4.2 is applied from location 1 to the end of the transition region, location 2:

$$E_1 = E_2 + h \qquad (10.4.16)$$

The depth $y_2$ at the end of the transition can be visualized by inspection of the $E$–$y$ diagram (Fig. 10.7b). If the flow at location 1 is subcritical, $y_1$ is located on the upper leg as shown in the specific energy diagram. The magnitude of $h$ is selected to be relatively small such that $y_2 > y_c$, hence the flow at location 2 is similarly subcritical; however, as $h$ is increased still further, a state of minimum energy is ultimately reached in the transition. The condition of minimum energy is sometimes referred to as a *choking condition* or as *choked flow*. Once choked flow occurs, as $h$ increases, the variations in depth and velocity are no longer localized in the vicinity of the transition. Influences may be observed for significant distances both upstream and downstream of the transition.

The reader should verify that in a transition region, a narrowing of the channel width will create a situation similar to an elevation of the channel bottom. The most general transition region is one that possesses a change in both width and in bottom elevation.

**KEY CONCEPT**  *The condition of choked flow or a choking condition implies that minimum specific energy exists within the transition.*

---

[3]One notable exception to this is the hydraulic jump, which requires use of the momentum principle and will be considered in Section 10.5.2.

## Example 10.4

A rectangular channel 3 m wide is conveying water at a depth $y_1 = 1.55$ m, and velocity $V_1 = 1.83$ m/s. The flow enters a transition region as shown in Fig. E10.4, in which the bottom elevation is raised by $h = 0.20$ m. Determine the depth and velocity in the transition, and the value of $h$ for choking to occur.

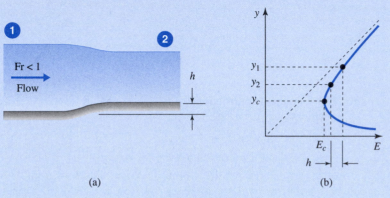

**Fig. E10.4**

### Solution

Use Eq. 10.4.4 to find the specific discharge to be

$$q = V_1 y_1$$
$$= 1.83 \times 1.55 = 2.84 \text{ m}^2\text{/s}$$

The Froude number at location 1 is

$$\text{Fr} = \frac{V_1}{\sqrt{gy_1}}$$

$$= \frac{1.83}{\sqrt{9.81 \times 1.55}} = 0.47$$

which is less than unity. Hence, the flow at location 1 is subcritical. The specific energy at location 1 is found, using Eq. 10.4.3, to be

$$E_1 = y_1 + \frac{V_1^2}{2g}$$

$$= 1.55 + \frac{1.83^2}{2 \times 9.81} = 1.72 \text{ m}$$

The specific energy at location 2 is found, using Eq. 10.4.16, to be

$$E_2 = E_1 - h$$
$$= 1.72 - 0.20 = 1.52 \text{ m}$$

If $E_2 > E_c$, it is possible to find the depth $y_2$. Therefore, $E_c$ is calculated first. From Eqs. 10.4.10 and 10.4.11 the critical conditions are

$$y_c = \left(\frac{q^2}{g}\right)^{1/3} = \left(\frac{2.84^2}{9.81}\right)^{1/3} = 0.94 \text{ m}$$

$$E_c = \frac{3y_c}{2} = 3 \times \frac{0.94}{2} = 1.41 \text{ m}$$

Hence, since $E_2 > E_c$, we can proceed with calculating $y_2$. The depth $y_2$ can be evaluated by substituting known values into Eq. 10.4.5:

$$1.52 = y_2 + \frac{2.84^2}{2 \times 9.81 \times y_2^2}$$

The solution is

$$y_2 = 1.26 \text{ m}$$

$$\therefore V_2 = \frac{q}{y_2}$$

$$= \frac{2.84}{1.26} = 2.25 \text{ m/s}$$

The flow at location 2 is subcritical since there is no way in which the flow can become supercritical in the transition with the given geometry.

The value of $h$ for critical flow to appear at location 2 is determined by setting $E_2 = E_c$ in Eq. 10.4.16:

$$h = E_1 - E_c = 1.72 - 1.40 = 0.31 \text{ m}$$

**Channel Entrance with Critical Flow.**   Consider flow entering a channel from a lake or reservoir over a short rounded crest, as sketched in Fig. 10.8. If the channel slope is steep, the flow will discharge freely into the channel and supercritical flow will occur downstream of the entrance region. Upstream of the crest the flow may be considered as subcritical. Since the flow regime at the crest changes from subcritical to supercritical, the flow at the crest must be critical. To determine the discharge, we assume that rapidly varied flow takes place over the crest in conjunction with the critical flow condition at the crest. An example illustrates the procedure.

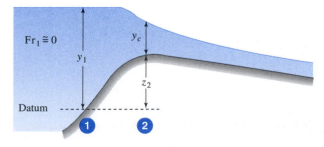

**Fig. 10.8**   Outflow from a reservoir with critical flow at the channel entrance.

## Example 10.5

Water flows freely from a reservoir into a trapezoidal channel with bottom width $b = 5.0$ m and side slope parameters $m_1 = m_2 = 2.0$. The elevation of the water surface in the reservoir is 2.3 m above the entrance crest. Assuming negligible losses in the transition and a negligible velocity in the reservoir upstream of the entrance, find the critical depth at the transition and the discharge into the channel.

**Fig. E10.5**

### Solution

The total energy at location 1 in Fig. E10.5 is $y_1$ since the kinetic energy in the reservoir is negligible ($V_1 \approx 0$). Equating the total energies at locations 1 and 2 gives

$$y_1 = E_2 + z_2$$

Since critical conditions occur at location 2, Eqs. 10.4.12 and 10.4.14 can be combined to eliminate the discharge, with the result

$$E_2 = y_c + \frac{A}{2B}$$

Elimination of $E_2$ in the two equations yields the expression

$$y_1 - z_2 = y_c + \frac{A}{2B} = y_c + \frac{by_c + \frac{1}{2}(m_1 + m_2)y_c^2}{2[b + (m_1 + m_2)y_c]}$$

or, with the given data, the expression becomes

$$2.3 = y_c + \frac{5y_c + \frac{1}{2}(2 + 2)y_c^2}{2[5 + (2 + 2)y_c]}$$

The relation above is a quadratic in $y_c$. The positive root is chosen, which gives

$$y_c = 1.70 \text{ m}$$

Subsequently, one can find that $A = 14.28$ m$^2$ and $B = 11.80$ m. Use Eq. 10.4.14 to find the discharge to be

$$Q = \sqrt{\frac{gA^3}{B}}$$

$$= \sqrt{\frac{9.8 \times 14.3^3}{11.8}} = 49.3 \text{ m}^3/\text{s}$$

**Energy Losses.**    Energy losses in expansions and contractions are known to be relatively small when the flow is subcritical; however, it may be necessary under certain circumstances that the losses be considered. Equation 10.4.2 includes a loss term that can account for transitional losses. The following experimentally derived formulas can be employed. For a channel expansion use

$$h_L = K_e \left( \frac{V_1^2}{2g} - \frac{V_2^2}{2g} \right) \tag{10.4.17}$$

and for a channel contraction use

$$h_L = K_c \left( \frac{V_2^2}{2g} - \frac{V_1^2}{2g} \right) \tag{10.4.18}$$

In Eq. 10.4.17, $K_e$ is an expansion coefficient; it has been suggested (King and Brater, 1963) to use $K_e = 1.0$ for sudden or abrupt expansions, and $K_e = 0.2$ for well-designed or rounded expansions. For the contraction coefficient $K_c$ in Eq. 10.4.18, use $K_c = 0.5$ for sudden contractions, and $K_c = 0.10$ for well-designed contractions. When the flow is supercritical, significant standing-wave patterns can be generated throughout and downstream of the transition region; for such flows, proper design requires that wave mechanics be considered (Chow, 1959; Henderson, 1966).

### 10.4.3   Flow Measurement

The most common means of metering discharge in an open channel is to use a weir. Basically, a **weir** is a device placed in the channel that forces the flow through an opening or aperture designed to measure the discharge. Specialized weirs have been designed for specific needs; in this section two fundamental types—broad-crested and sharp-crested—will be presented.

A properly designed weir will exhibit subcritical flow upstream of the structure, and the flow will converge and accelerate to a critical condition near the top or *crest* of the weir. As a result, a correlation between discharge and a depth upstream of the weir can be made. The downstream overflow is termed the *nappe*, which usually discharges freely into the atmosphere. There are a number of factors that affect the performance of a weir; significant among them are the three-dimensional flow pattern, the effects of turbulence, frictional resistance, surface tension, and the amount of ventilation beneath the nappe. The simplified derivations presented here are based on the Bernoulli equation; the other effects can be accounted for by modifying the ideal discharge with a *discharge coefficient, $C_d$*. The actual discharge is the ideal discharge multiplied by the discharge coefficient. When possible, it is advantageous to calibrate a particular weir in place in order to obtain the desired accuracy.

**Weir:**   *A device placed in a channel that forces the flow through an opening or aperture, often designed to measure the discharge.*

**KEY CONCEPT**   *Flow will converge and accelerate to a critical condition near the crest of the weir.*

**Broad-Crested Weir.** A broad-crested weir is shown in Fig. 10.9. It has sufficient elevation above the channel bottom to choke the flow, and it is long enough so that the overflow streamlines become parallel, resulting in a hydrostatic pressure distribution. Then, at some position, say location 2, a critical flow condition exists.

Assume a rectangular horizontal channel and let the height of the weir be $h$. Location 1 is a point upstream of the weir where the flow is relatively undisturbed, and $Y$ is the vertical distance from the top of the weir to the free surface at that point. Applying Bernoulli's equation from location 1 to location 2 at the free surface and neglecting the kinetic energy head at location 1 ($V_1 \approx 0$), one has the result

$$h + Y = h + y_c + \frac{V_c^2}{2g} \tag{10.4.19}$$

Solving for $V_c$,

$$V_c = \sqrt{2g(Y - y_c)} \tag{10.4.20}$$

For a weir whose breadth normal to the flow is $b$, the ideal discharge is

$$Q = by_cV_c = by_c \sqrt{2g(Y - y_c)} \tag{10.4.21}$$

Recognizing that $Y = E_c$, Eq. 10.4.11 is used to relate $y_c$ to $Y$, and when substituted into Eq. 10.4.21, the result is

$$Q = \frac{2}{3}\sqrt{\frac{2}{3}g}\, bY^{3/2} \tag{10.4.22}$$

For a properly rounded upstream edge on the weir, Eq. 10.4.22 is accurate to within several percent of the actual flow; hence, a discharge coefficient is not applied.

**Sharp-Crested Weir.** A sharp-crested weir is a vertical plate placed normal to the flow containing a sharp-edge crest so that the nappe will behave as a free jet. Figure 10.10 shows a rectangular weir with a horizontal crest extending across the entire width of the channel. Because of the presence of the side walls, lateral contractions are not present.

**Fig. 10.9** Broad-crested weir.

(a)          (b)

**Fig. 10.10**   Rectangular sharp-crested weir: (a) ideal flow; (b) actual flow.

Let us define an idealized flow situation: The flow in the vertical plane does not contract as it passes over the crest so that the streamlines are parallel, atmospheric pressure is present in the nappe, and uniform flow exists at location 1 with negligible kinetic energy ($v_1 \approx 0$). The Bernoulli equation is applied along a representative streamline (Fig. 10.10a) and solved for $v_2$, the local velocity in the nappe:

$$v_2 = \sqrt{2g\eta} \tag{10.4.23}$$

If $b$ is the width of the crest normal to the flow, the ideal discharge is given as

$$Q = b \int_0^Y v_2 \, d\eta = b \int_0^Y \sqrt{2g\eta} \, d\eta = b\frac{2}{3}\sqrt{2g}\, Y^{3/2} \tag{10.4.24}$$

Experiments have shown that the magnitude of the exponent is nearly correct but that a discharge coefficient $C_d$ must be applied to accurately predict the real flow, shown in Fig. 10.10b:

$$Q = C_d \frac{2}{3}\sqrt{2g}\, b\, Y^{3/2} \tag{10.4.25}$$

The discharge coefficient accounts for the effect of contraction, velocity of approach, viscosity, and surface tension. An experimentally (Chow, 1959) obtained formula for $C_d$ has been given as

$$C_d = 0.61 + 0.08 \frac{Y}{h} \tag{10.4.26}$$

Normally, for small $Y/h$ ratio, $C_d = 0.61$. If the weir crest does not extend to the side walls but allows for end contractions to appear, as in Fig. 10.11a, the effective width of the weir can be approximated by $(b - 0.2Y)$.

(a)          (b)

**Fig. 10.11**   Contracted rectangular and V-notch weirs: (a) rectangular; (b) V-notch.

The V-notch weir (Fig. 10.11b) is more accurate than the rectangular weir for measurement of low discharge. In a manner similar to the development of the rectangular weir relation, the idealized discharge is found by integrating the local velocity throughout the nappe above the crest. A discharge coefficient is applied to yield

$$Q = C_d \frac{8}{15} \sqrt{2g} \left( \tan \frac{\theta}{2} \right) Y^{5/2} \qquad\qquad (10.4.27)$$

For use with water, and for $\theta$ varying from 22.5° to 120°, experiments (King and Brater, 1963) have shown that a value of $C_d = 0.58$ is acceptable for engineering calculations. Derivation of Eq. 10.4.27 is left as an exercise.

The above equations can, as usual, be used with either set of units.

## Example 10.6

Determine the discharge of water over a rectangular sharp-crested weir, $b = 1.25$ m, $Y = 0.35$ m, $h = 1.47$ m, with side walls and with end contractions. If a 90° V-notch weir were to replace the rectangular weir, what would be the required $Y$ for a similar discharge?

### Solution
For the rectangular weir, using Eq. 10.4.26, the discharge coefficient is

$$C_d = 0.61 + 0.08 \frac{Y}{h} = 0.61 + 0.08 \times \frac{0.35}{1.47} = 0.63$$

Substitute into Eq. 10.4.25 and calculate

$$Q = C_d \frac{2}{3} \sqrt{2g}\, bY^{3/2}$$
$$= 0.63 \times \frac{2}{3} \times \sqrt{2 \times 9.81} \times 1.25 \times 0.35^{3/2}$$
$$= 0.48 \text{ m}^3/\text{s}$$

With end contractions the effective width of the weir is reduced by $0.2Y$, resulting in

$$Q = C_d \frac{2}{3} \sqrt{2g}\, (b - 0.2Y)\, Y^{3/2}$$
$$= 0.63 \times \frac{2}{3} \times \sqrt{2 \times 9.81} \times (1.25 - 0.2 \times 0.35) \times 0.35^{3/2}$$
$$= 0.45 \text{ m}^3/\text{s}$$

With a discharge of $Q = 0.48$ m³/s, use Eq. 10.4.27 to find $Y$ for the 90° V-notch weir:

$$Y = \left[ \frac{Q}{C_a \times \dfrac{8}{15} \times \sqrt{2g}\, \tan(\theta/2)} \right]^{2/5}$$

$$= \left[ \frac{0.482}{0.58 \times \dfrac{8}{15} \times \sqrt{2 \times 9.81} \times \tan 45°} \right]^{2/5} = 0.66 \text{ m}$$

**Additional Methods of Flow Measurement.**  Other types of weirs include those whose faces are inclined in the upstream and downstream direction (triangular, trapezoidal, irregular). In addition, the spillway section of a dam can be considered as a weir with a rounded crest. King and Brater (1963) give details on the selection and use of these types.

A special type of open flume is one in which the geometry of the throat is con-stricted in such a manner as to choke the flow, thereby creating critical flow followed by a hydraulic jump. When manufactured using a particular standardized section, the flume is called a **Parshall flume,** shown in Fig. 10.12. Extensive calibrations have established reliable empirical formulas to predict the discharge. For throat widths 0.3 to 2.4 m, the discharge is given by the formula

$$Q = 4BH^{1.522B^{0.026}} \qquad (10.4.28)$$

> **Parshall flume:**  *An open flume where the throat is constricted to choke the flow to create critical flow followed by a hydraulic jump.*

where $H = H'/0.3048$ and $B = B'/0.3048$ in which $H'$ is the depth measured at the upstream location shown in Fig. 10.12. Note that $H'$ and $B'$ are measured in meters, and $Q$ is in cubic meters per second. The other dimensions are used to develop Eq. 10.4.28.

In a natural section of a river it may be impractical to place a weir; in that case stream gaging can be used to measure the discharge. A control location is established upstream of the gaging site, and for a given depth, or *stage*, of the river, the two-dimensional velocity profile is measured using current meters.

**Fig. 10.12**  Parshall flume. (HENDERSON, OPEN CHANNEL FLOW, 1st,©1966. Printed and Electronically reproduced by permission of Pearson Education, Inc., Upper Saddle River, New Jersey.)

Subsequently, the profile is numerically integrated to yield the discharge. A series of such measurements will produce a stage-discharge curve, which can then be employed to estimate the discharge by measuring the river stage.

## 10.5 MOMENTUM CONCEPTS

In the preceding section we have seen how the energy equation is applied to rapidly varied flow situations, and in particular, how it is utilized to analyze flows in transition regions. The momentum equation is also employed to study certain phenomena in those situations. When used in conjunction with the energy and continuity relations, the momentum equation gives the user a concise means of analyzing nearly all significant transition problems, including problems that involve a hydraulic jump.

### 10.5.1   Momentum Equation

Consider the open-channel reach with supercritical flow upstream of a submerged obstacle, as depicted in Fig. 10.13a. It represents a rapidly varied flow with an abrupt change in depth but no change in width. In general, such a change may result from an obstacle in the flow or from a hydraulic jump. The upstream flow is supercritical and the downstream flow is subcritical. Other flow regimes may also be considered; for example, the conditions could be subcritical throughout the entire control volume. Each situation should be approached as a unique formulation.

The generalized flow situation may be used to develop the equation of motion for transition regions. The control volume corresponding to Fig. 10.13a is shown in Fig. 10.13b. The pressure distribution is assumed hydrostatic, and the resultant hydrostatic forces are given by $\gamma A \bar{y}$, where the distance $\bar{y}$ to the centroid of the cross-sectional area is measured from the free surface. The submerged obstacle (e.g., a rock or a two-dimensional obstacle) imparts a force $F$ on the control volume

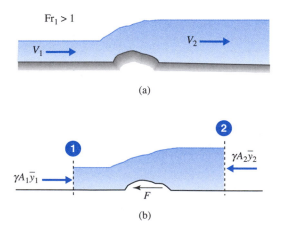

**Fig. 10.13**   Channel flow over an obstacle: (a) idealized flow; (b) control volume.

with a direction opposite to the direction of flow. The linear momentum equation, presented in Section 4.5, is applied to the control volume in the $x$-direction to give

$$\gamma A_1 \bar{y}_1 - \gamma A_2 \bar{y}_2 - F = \rho Q \, (V_2 - V_1) \qquad (10.5.1)$$

Note that frictional forces have not been included in Eq. 10.5.1; they are usually quite small relative to the other terms, so they can be ignored. Similarly, gravitational forces in the flow direction are negligible for the small channel slopes considered. Equation 10.5.1 can be rearranged in the form

$$M_1 - M_2 = \frac{F}{\gamma} \qquad (10.5.2)$$

in which $M_1$ and $M_2$ are terms that contain the hydrostatic force and the momentum flux at locations 1 and 2, respectively. The quantity $M$ is called the *momentum function*, and for a general prismatic section it is given by

$$M = A\bar{y} + \frac{Q^2}{gA} \qquad (10.5.3)$$

For a rectangular section, $A\bar{y} = by^2/2$, and the momentum function is

$$M = \frac{by^2}{2} + \frac{bq^2}{gy} = b\left(\frac{y^2}{2} + \frac{q^2}{gy}\right) \qquad (10.5.4)$$

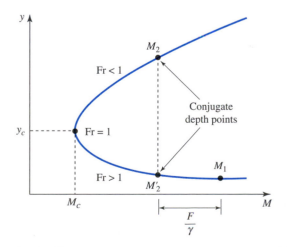

**Fig. 10.14**   Variation of the momentum function with depth.

**Conjugate or sequent depths:**
*The two depths of flow that are possible for a given value of the momentum function and discharge.*

Equation 10.5.4 is sketched in Fig. 10.14. Two positive roots $y$ exist for a given $M$ and $q$; they are termed either **conjugate** or **sequent** depths. The upper leg of the curve ($y > y_c$) applies to subcritical flow and the lower leg ($y < y_c$) to supercritical flow. The values of $M_1$ and $M_2$ for the example of Fig. 10.13a are shown, indicating supercritical flow upstream and subcritical flow downstream of the submerged obstacle. The horizontal distance between $M_1$ and $M_2$ is equal to $F/\gamma$. The downstream flow would be supercritical if no hydraulic jump occurred; the corresponding value of the momentum function is indicated by $M_2'$.

The depth associated with a minimum $M$ is found by differentiating $M$ with respect to $y$ in Eq. 10.5.3:

$$\frac{dM}{dy} = A - \frac{BQ^2}{gA^2} = 0 \qquad (10.5.5)$$

Note[4] that $d(A\bar{y})/dy = A$. The condition for minimum $M$ is thus

$$Q^2 B = gA^3 \qquad (10.5.6)$$

This result is identical to Eq. 10.4.14. Hence the condition of minimum $M$ is equivalent to that of minimum energy: critical flow with a Froude number equal to unity.

**Momentum Equation Applied to a Transition Region.**   Quite often the momentum equation is applied in situations where it is desired to determine the resultant force acting at a specific location, or to find the change in depth or velocity when there is a significant undefined loss across the transition region. It is important to remember that the energy and continuity equations are also at our disposal, and we must determine which relations are necessary. In some instances, along with the continuity equation, both energy and momentum equations must be applied. An example is given below to illustrate the technique.

---

[4]This can be observed from the definition $\bar{y} = (1/A) \int (y - \eta)\, dA$. Differentiation of $(\bar{y}A)$ using Leibnitz's rule from calculus yields

$$\frac{d}{dy}(\bar{y}A) = \frac{d}{dy}\int_0^y (y - \eta)w(\eta)d\eta$$

$$= (y - \eta)w(\eta)\Big|_{\eta=y} + \int_0^y w(\eta)d\eta$$

$$= 0 + A = A$$

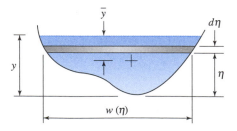

## Example 10.7

In a rectangular 5-m-wide channel, water is discharging at 14.0 m³/s (Fig. E10.7). Find the force exerted on the sluice gate when $y_1 = 2$ m and $y_2 = 0.5$ m.

$Fr_1 < 1$

$Q$

$Fr_2 > 1$

1          2

**Fig. E10.7**

**Solution**

Using Eq. 10.5.3, the momentum functions at 1 and 2 are

$$M_1 = A_1\bar{y}_1 + \frac{Q^2}{gA_1}$$

$$= 5 \times 2 \times 1 + \frac{(14)^2}{9.81 \times 5 \times 2} = 12.0 \text{ m}^3$$

$$M_2 = A_2\bar{y}_2 + \frac{Q^2}{gA_2}$$

$$= 5 \times 0.5 \times 0.25 + \frac{(14)^2}{9.81 \times 5 \times 0.5} = 8.62 \text{ m}^3$$

The resultant force acting on the fluid control volume is determined, using Eq. 10.5.2, to be

$$F = \gamma(M_1 - M_2)$$

$$= 9800 \times (12.0 - 8.62) = 33\,100 \text{ N}$$

Hence the force on the gate acts in the downstream sense with a magnitude of 33.1 kN.

## 10.5.2 Hydraulic Jump

A **hydraulic jump** is a phenomenon in which fluid flowing at a supercritical state will undergo a transition to a subcritical state. Boundary conditions upstream and downstream of the jump will dictate its strength as well as its location. An idealized hydraulic jump is shown in Fig. 10.15. The strength of the jump varies widely, as shown in Table 10.2, with relatively mild disturbances occurring at one extreme, and significant separation and eddy formation taking place at the other. As a result, the energy loss associated with the jump is considered to be unknown, so the energy equation is not used in the initial analysis. Assuming no

**Hydraulic jump:** *A phenomenon where fluid flowing at a supercritical state will undergo a transition to a subcritical state.*

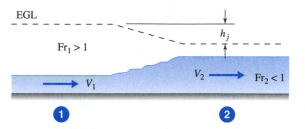

**Fig. 10.15**    Idealized hydraulic jump.

friction along the bottom and no submerged obstacle; that is, setting $F$ equal to zero, Eq. 10.5.2 shows that $M_1 = M_2$. For a rectangular section, Eq. 10.5.4 can be substituted into the relation, allowing one to obtain

$$\frac{q^2}{g}\left(\frac{1}{y_1} - \frac{1}{y_2}\right) = \frac{1}{2}\left(y_2^2 - y_1^2\right) \qquad (10.5.7)$$

Rearranging, factoring, and noting that $\mathrm{Fr}_1^2 = q^2/gy_1^3$ there results

$$\mathrm{Fr}_1^2 = \frac{1}{2}\frac{y_2}{y_1}\left(\frac{y_2}{y_1} + 1\right) \qquad (10.5.8)$$

This equation is dimensionless and it relates the Froude number upstream of the jump with the ratio of the downstream to upstream depths. It can be seen that Eq. 10.5.8 is quadratic with respect to $y_2/y_1$ provided that $\mathrm{Fr}_1$ is known. Solving for $y_2/y_1$, one obtains

$$\frac{y_2}{y_1} = \frac{1}{2}\left(\sqrt{1 + 8\mathrm{Fr}_1^2} - 1\right) \qquad (10.5.9)$$

The positive sign in front of the radical has been chosen to yield a physically meaningful solution.

It is worth noting that Eq. 10.5.9 is also valid if the subscripts on the depths and Froude number are reversed:

$$\frac{y_1}{y_2} = \frac{1}{2}\left(\sqrt{1 + 8\mathrm{Fr}_2^2} - 1\right) \qquad (10.5.10)$$

The theoretical energy loss associated with a hydraulic jump in a rectangular channel can be determined once the depths and flows at locations 1 and 2 are known. The energy equation is applied from 1 to 2, including the head loss $h_j$

across the jump, as displayed in Fig. 10.15. Combining the energy equation with Eq. 10.5.7 and the continuity equation, one can show after some algebra that

$$h_j = \frac{(y_2 - y_1)^3}{4y_1 y_2}$$    (10.5.11)

Equations 10.5.8 through 10.5.11 are useful forms for solving most hydraulic jump problems in a rectangular channel.

Table 10.2 shows the various forms that a hydraulic jump may assume relative to the upstream Froude number. A steady, well-established jump, with $4.5 < \text{Fr} < 9.0$, is often used as an energy dissipator downstream of a dam or spillway. It is characterized by the existence of breaking waves and rollers accompanied by a submerged jet with significant turbulence and dissipation of energy in the main body of the jump; downstream, the water surface is relatively smooth. For Froude numbers outside the range 4.5 to 9.0, less desirable jumps exist and may create undesirable downstream surface waves. The length of a jump is the distance from the front face to just downstream where smooth water exists; a steady jump has an approximate length of six times the upstream depth.

**Table 10.2**    Hydraulic Jumps in Horizontal Rectangular Channels

| Upstream Fr | Type | Description | |
|---|---|---|---|
| 1.0–1.7 | Undular | Ruffled or undular water surface; surface rollers form near Fr = 1.7 | |
| 1.7–2.5 | Weak | Prevailing smooth flow; low energy loss | |
| 2.5–4.5 | Oscillating | Intermittent jets from bottom to surface, causing persistent downstream waves | |
| 4.5–9.0 | Steady | Stable and well-balanced; energy dissipation contained in main body of jump | |
| >9.0 | Strong | Effective, but with rough, wavy surface downstream | |

*Source:* Adapted with permission from Chow, 1959. (Adapted from Chow, 1959)

## Example 10.8

A hydraulic jump is situated in a 4-m-wide rectangular channel. The discharge in the channel is 7.5 m$^3$/s, and the depth upstream of the jump is 0.20 m. Determine the depth downstream of the jump, the upstream and downstream Froude numbers, and the rate of energy dissipated by the jump.

### Solution

Find the unit discharge and upstream Froude number:

$$q = \frac{Q}{b}$$

$$= \frac{7.5}{4} = 1.88 \text{ m}^2/\text{s}$$

$$Fr_1 = \frac{q}{\sqrt{gy_1^3}}$$

$$= \frac{1.88}{\sqrt{9.81 \times 0.20^3}} = 6.71$$

The downstream depth is computed, using Eq. 10.5.9, to be

$$y_2 = \frac{y_1}{2}\left(\sqrt{1 + 8Fr_1^2} - 1\right)$$

$$= \frac{0.20}{2}\left(\sqrt{1 + 8 \times 6.71^2} - 1\right) = 1.80 \text{ m}$$

The downstream Froude number is

$$Fr_2 = \frac{q}{\sqrt{gy_2^3}}$$

$$= \frac{1.88}{\sqrt{9.81 \times 1.80^3}} = 0.25$$

The head loss in the jump is given by Eq. 10.5.11:

$$h_j = \frac{(y_2 - y_1)^3}{4y_1 y_2}$$

$$= \frac{(1.80 - 0.20)^3}{4 \times 0.20 \times 1.80} = 2.84 \text{ m}$$

Hence, the rate of energy dissipation in the jump is (see Eq. 4.5.25)

$$\gamma Q h_j = 9800 \times 7.5 \times 2.84 = 2.09 \times 10^5 \text{ W} \quad \text{or} \quad 209 \text{ kW}$$

**KEY CONCEPT**   *A translating hydraulic jump is a positive surge wave, maintaining a stable front as it propagates into an undisturbed region.*

**Translating Hydraulic Jump.**   A translating hydraulic jump, alternately termed a *surge* or a *hydraulic bore,* is shown in Fig. 10.16a; it is termed a positive surge wave in the sense that it maintains a stable front as it propagates at speed $w$ into an undisturbed region. Such a wave can be generated by abruptly

**Fig. 10.16**   Translating hydraulic jump: (a) front moving upstream; (b) front appears stationary by superposition.

lowering a downstream gate or by rapidly releasing water at an upstream location into a channel. This unsteady rapidly varied flow situation can be conveniently analyzed as a steady-state problem by superposing the bore speed $w$ in the opposite sense on the control volume (Fig. 10.16b). The front appears stationary, and the relative velocities at locations 1 and 2 are equal to $V_1 + w$ and $V_2 + w$, respectively.

Assume a bore translating in a rectangular, frictionless, horizontal channel. Equation 10.5.9 can be applied by substituting $V_1 + w$ for $V_1$ in the definition of $Fr_1$ to obtain

$$\frac{y_2}{y_1} = \frac{1}{2}\left[\sqrt{1 + 8\frac{(V_1 + w)^2}{gy_1}} - 1\right] \tag{10.5.12}$$

The continuity relation applied to the control volume of Fig. 10.16b is

$$y_1(V_1 + w) = y_2(V_2 + w) \tag{10.5.13}$$

Equations 10.5.12 and 10.5.13 contain five parameters: $y_1$, $y_2$, $V_1$, $V_2$, and $w$. Three of them must be known to solve for the remaining two. Depending on which variables are unknown, the solution of Eqs. 10.5.12 and 10.5.13 will be either explicit or based on a trial-and-error procedure.

## Example 10.9

Water flows in a rectangular channel with a velocity of 2.5 m/s and a depth of 1.5 m. A gate is suddenly completely closed, forming a surge that travels upstream. Find the speed of the surge and the depth behind the surge.

**Solution**

Since the gate is closed, the downstream velocity is $V_2 = 0$. Equations 10.5.12 and 10.5.13 contain the two unknowns $w$ and $y_2$. Combining them to eliminate $y_2$ and substituting $V_1 = 2.5$ and $y_1 = 1.5$ results in the relation

$$2.5 + w = \frac{w}{2}\left[\sqrt{1 + 8\frac{(2.5 + w)^2}{9.81 \times 1.5}} - 1\right]$$

*(continued)*

A trial-and-error solution, yields $w = 3.41$ m/s. Use Eq. 10.5.13 to compute $y_2$:

$$y_2 = y_1 \frac{V_1 + w}{V_2 + w}$$

$$= 1.5 \times \frac{2.5 + 3.41}{3.41} = 2.60 \text{ m}$$

**Drag on Submerged Objects.** If an object is submerged in the flow it is possible to describe the drag force in the manner

$$F = C_D A \rho \frac{V^2}{2} \tag{10.5.14}$$

in which $C_D$ is the drag coefficient and $A$ is the projected area normal to the flow. A discussion of drag coefficients for various immersed objects is given in Section 8.3.1. In open channel flow, since a free surface is present, the drag coefficient must be modified to account for wave drag as well as drag due to friction and separation. Examples of submerged objects in open channel flow include pipelines and bridge piers; in these situations, a hydraulic jump may not be present.

A design application is illustrated in Fig. 10.17. Baffle blocks are devices placed in a reach of channel known as a *stilling basin* to stabilize the location of a hydraulic jump and to aid in the dissipation of flow energy. Typically they are used for $\text{Fr}_1$ greater than 4.5. Example 10.10 demonstrates how baffle blocks reduce the magnitude of the momentum function downstream of a hydraulic jump, resulting in reduced downstream depth and increased energy dissipation. In practice, Eq. 10.5.14 is seldom employed to analyze or design a stilling basin, since other factors such as additional appurtenances, approach velocity, scour, and cavitation must also be considered. Instead, design standards have been established based on observations of existing basins and laboratory model studies (U.S. Department of Interior, 1974; Roberson et al., 1988).

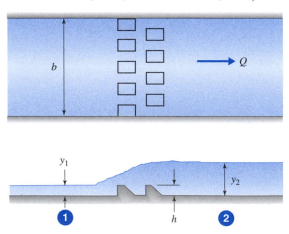

**Fig. 10.17** Stilling basin with baffle blocks.

## Example 10.10

In the flow situation presented in Example 10.8, a series of baffle blocks is placed in the channel as shown in Fig. E10.10. Laboratory experimentation has shown that the arrangement has an effective drag coefficient of 0.25, provided that the blocks are submerged in the flow. If the blocks are 0.15 m high, and if the discharge, upstream depth and width remain the same as in Example 10.8, determine the depth downstream of the jump and the rate of energy dissipated by the jump.

**Fig. E10.10**

### Solution

It is necessary to use Eq. 10.5.2 since obstacles (i.e., the baffle blocks) are placed within the control volume. The upstream velocity is

$$V_1 = \frac{Q}{A_1}$$

$$= \frac{7.5}{4 \times 0.2} = 9.38 \text{ m/s}$$

The force $F$ due to the presence of the baffle blocks is computed using Eq. 10.5.14:

$$F = C_D A \rho \frac{V_1^2}{2}$$

$$= 0.25 \times (4 \times 0.15) \times 1000 \times \frac{9.38^2}{2} = 6600 \text{ N}$$

Note that the frontal area is the width of the channel multiplied by the height of the blocks. Substituting known conditions into Eq. 10.5.2, making use of Eq. 10.5.4 which defines $M$ for a rectangular channel, and noting that $q = 7.5/4 = 1.88$ m²/s, we find

$$b\left(\frac{y_1^2}{2} + \frac{q^2}{gy_1}\right) - b\left(\frac{y_2^2}{2} + \frac{q^2}{gy_2}\right) = \frac{F}{\gamma}$$

$$4\left(\frac{0.2^2}{2} + \frac{1.88^2}{9.81 \times 0.2}\right) - 4\left(\frac{y_2^2}{2} + \frac{1.88^2}{9.81 y_2}\right) = \frac{6600}{9800}$$

*(continued)*

The relation reduces to

$$y_2^2 + \frac{0.721}{y_2} = 3.31$$

The solution for $y_2$ is 1.70 m. The change in specific energy between locations 1 and 2 is

$$E_1 - E_2 = y_1 + \frac{q^2}{2gy_1^2} - \left(y_2 + \frac{q^2}{2gy_2^2}\right)$$

$$= 0.2 + \frac{1.88^2}{2 \times 9.81 \times 0.2^2} - \left(1.70 + \frac{1.88^2}{2 \times 9.81 \times 1.70^2}\right)$$

$$= 2.94 \text{ m}$$

The rate of energy dissipation, therefore, is

$$\gamma Q(E_1 - E_2) = 9800 \times 7.5 \times 2.94$$

$$= 2.16 \times 10^5 \text{ W} \quad \text{or} \quad 216 \text{ kW}$$

### 10.5.3 Numerical Solution of the Momentum Equation

For nonrectangular channels, the momentum relation can be used directly to analyze the hydraulic jump or other problems requiring the momentum equation; the technique is demonstrated as follows. Consider a trapezoidal channel with conditions known at location 1 upstream of the jump. Consequently, $M_1$ and $F$ are evaluated as constants and Eq. 10.5.2 can be written in the form

$$M_2 - M_1 + \frac{F}{\gamma} = 0 \tag{10.5.15}$$

in which $M_2$ is a function of $y_2$. Introducing the trapezoidal geometry at location 2, with $m_1 = m_2 = m$, the relation above can be written as

$$\frac{y_2^2}{6}(2my_2 + 3b) + \frac{Q^2}{g(by_2 + my_2^2)} - M_1 + \frac{F}{\gamma} = 0 \tag{10.5.16}$$

This can be solved for $y_2$ by a suitable numerical technique such as interval halving, false position, or Newton's method, which are discussed in any book on numerical methods (see, e.g., Chapra and Canale, 1998). Note that by setting $F = 0$, it becomes the relation for finding the conditions downstream of a hydraulic jump, and that by additionally letting $m = 0$, a rectangular hydraulic jump problem can be solved.

# Example 10.11

A hydraulic jump occurs in a triangular channel with $m_1 = m_2 = 2.5$. The discharge is 20 m$^3$/s and $y_c = 1.67$ m. Upstream of the jump the following parameters are given: $y_1 = 0.75$ m, $Fr_1 = 7.42$, and $M_1 = 29.35$ m$^3$. Determine the conjugate depth $y_2$ downstream of the jump.

## Solution

Use Eq. 10.5.16 with $F = 0$:

$$\frac{y_2^2}{6} \times 2 \times 2.5 y_2 + \frac{20^2}{9.81 \times 2.5 y_2^2} - 29.35 = 0$$

The relation reduces to

$$f(y_2) = y_2^3 + \frac{19.58}{y_2^2} - 35.23 = 0$$

The false position method is chosen to find $y_2$. The first step is to set the upper and lower limits, termed $y_u$ and $y_l$. Since $Fr_1 > 1$, and $y_2 > y_c$, an appropriate lower limit is $y_l = y_c = 1.67$ m. The upper limit is assumed to be 5 m.

| Iteration | $y_u$ | $y_l$ | $f(y_u)$ | $f(y_l)$ | $y_r$ | $f(y_r)$ | Sign of $f(y_l) \times f(y_r)$ | $\varepsilon$ |
|-----------|-------|-------|----------|----------|-------|----------|-------------------------------|---------------|
| 1 | 5 | 1.67  | 90.55 | −23.55 | 2.357 | −18.61 | — | |
| 2 | 5 | 2.357 | 90.55 | −18.61 | 2.808 | −10.61 | — | |
| 3 | 5 | 2.808 | 90.55 | −10.61 | 3.038 | −5.076 | — | |
| 4 | 5 | 3.038 | 90.55 | −5.067 | 3.142 | −2.231 | — | |
| 5 | 5 | 3.142 | 90.55 | −2.231 | 3.187 | −0.944 | — | |
| 6 | 5 | 3.187 | 90.55 | −0.944 | 3.205 | −0.393 | — | |
| 7 | 5 | 3.205 | 90.55 | −0.393 | 3.213 | −0.162 | — | $2.5 \times 10^{-3}$ |
| 8 | 5 | 3.213 | 90.55 | −0.162 | 3.216 | −0.0672 | — | $9.4 \times 10^{-4}$ |
| 9 | 5 | 3.216 | 90.55 | −0.0672 | 3.218 | −0.0277 | — | $4.9 \times 10^{-4}$ |

The solution is tabulated above. In each iteration a new estimate $y_r$ of the root is made:

$$y_r = \frac{y_u f(y_l) - y_l f(y_u)}{f(y_l) - f(y_u)}$$

The product $f(y_l) \times f(y_r)$ is formed to determine on which subinterval the root will be found. If $f(y_l) \times f(y_r) < 0$, then $y_u = y_r$; otherwise, $y_l = y_r$. It is required that initially $f(y_u)$ and $f(y_l)$ be of opposite sign. Iterations continue until a relative error $\varepsilon$, defined by

$$\varepsilon = \left| \frac{y_r^{new} - y_r^{old}}{y_r^{old}} \right|$$

is less than a specified value, which in the example is 0.0005. The result after nine iterations is $y_2 = 3.22$ m, rounded off to three significant figures.

Using Mathcad or MATLAB, one finds the solution to be less time consuming; see Appendix E, Figs. E.2 and E.3.

## 10.6  NONUNIFORM GRADUALLY VARIED FLOW

The evaluation of many open-channel flow situations must include accurate analyses of relatively long reaches where the depth and velocity may vary but do not exhibit rapid or sudden changes. In the preceding two sections we emphasized nonuniform, rapidly varied flow phenomena that occur over relatively short reaches, or transitions, in open channels. Attention is now focused on nonuniform, gradually varied flow, where the water surface is continuously smooth. One significant difference between the two is that for rapidly varied flow, losses may often be neglected without severe consequences, whereas for gradually varied flow, it is necessary to include losses due to shear stress distributed along the channel length. The shear stress is the primary mechanism that resists the flow.

Gradually varied flow is a type of steady, nonuniform flow in which $y$ and $V$ do not exhibit sudden or rapid changes, but instead vary so gradually that the water surface can be considered continuous. As a result, it is possible to develop a differential equation that describes the incremental variation of $y$ with respect to $x$, the distance along the channel. An analysis of this relation will enable one to predict the various trends that the water surface profile may assume based on the channel geometry, magnitude of the discharge, and the known boundary conditions. Numerical evaluation of the same equation will provide engineering design criteria.

### 10.6.1  Differential Equation for Gradually Varied Flow

A representative nonuniform gradually varied flow is shown in Fig. 10.18. Over the incremental distance $\Delta x$, the depth and velocity are known to change slowly. The slope of the energy grade line is designated as $S$. In contrast to uniform flow, the slopes of the energy grade line, water surface, and channel bottom are no longer parallel. Since the changes in $y$ and $V$ are gradual, the energy loss over the incremental length $\Delta x$ can be represented by the Chezy–Manning equation. This means that Eq. 10.3.13, which is valid for uniform flow, can also be used to evaluate $S$ for a gradually varied flow situation, and that the roughness coefficients presented in Table 7.3 are applicable. Additional assumptions include a regular cross section, small channel slope, hydrostatic pressure distribution, and one-dimensional flow.

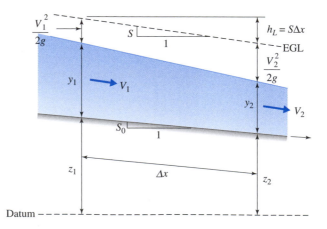

**Fig. 10.18**  Nonuniform gradually varied flow.

The energy equation is applied from location 1 to location 2, with the loss term $h_L$ given by $S\Delta x$. If the total energy at location 2 is expressed as the energy at location 1 plus the incremental change in energy over the distance $\Delta x$, Eq. 10.4.2 becomes

$$H_1 = H_2 + S\,\Delta x = H_1 + \frac{dH}{dx}\Delta x + S\,\Delta x \qquad (10.6.1)$$

Substitute $H = y + z + V^2/2g$ and $dz/dx = -S_0$ into this relation and rearrange to find

$$S - S_0 = -\frac{d}{dx}\left(y + \frac{V^2}{2g}\right) \qquad (10.6.2)$$

The right-hand term is $-dE/dx$, and it is transformed to

$$\frac{dE}{dx} = \frac{dE}{dy}\frac{dy}{dx} = (1 - \mathrm{Fr}^2)\frac{dy}{dx} \qquad (10.6.3)$$

(Recall from Eqs. 10.4.13 and 10.4.15 that $dE/dy = 1 - Q^2B/gA^3 = 1 - \mathrm{Fr}^2$.) Finally, upon substitution into the energy relation and solving for the slope $dy/dx$ of the water surface one finds that

$$\frac{dy}{dx} = \frac{S_0 - S}{1 - \mathrm{Fr}^2} \qquad (10.6.4)$$

where $S$ and $\mathrm{Fr}$ depend on $x$. This is the differential equation for gradually varied flow and is valid for any regular channel shape.

## Example 10.12

Using an appropriate control volume for gradually varied flow, show that the slope $S$ of the energy grade line is equivalent to $\tau_0/\gamma R$.

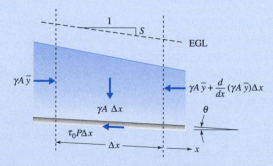

**Fig. E10.12**

*(continued)*

**Solution**

The control volume is shown in Fig. E10.12. The resultant force acting on the control volume is due to the incremental change in hydrostatic pressure $[\gamma d(A\bar{y})/dx]\Delta x$, the component of weight in the $x$-direction $\gamma A \sin\theta\Delta x$, and the resistance term $\tau_0 P\Delta x$. Using the momentum equation

$$\Sigma F_x = \dot{m}(V_{2x} - V_{1x})$$

with $V_{2x} - V_{1x} = (dV/dx)\,\Delta x$ results in

$$-\gamma\frac{d}{dx}(A\bar{y})\,\Delta x + \gamma A\sin\theta\,\Delta x - \tau_0 P\,\Delta x = \rho VA\frac{dV}{dx}\Delta x$$

This relation can be simplified by noting that

$$\frac{d(A\bar{y})}{dx} = \frac{d(A\bar{y})}{dy}\frac{dy}{dx} = A\frac{dy}{dx}$$

and $P = A/R$. Substitute and divide the equation by $\gamma A\Delta x$, the weight of the control volume, and find that

$$-\frac{dy}{dx} + \sin\theta - \frac{\tau_0}{\gamma R} = \frac{V}{g}\frac{dV}{dx}$$

Since $\sin\theta \simeq S_0$ for small $\theta$, the equation above can be rearranged in the form

$$\frac{\tau_0}{\gamma R} - S_0 = -\frac{dy}{dx} - \frac{V}{g}\frac{dV}{dx}$$

$$= -\frac{d}{dx}\left(y + \frac{V^2}{2g}\right)$$

Upon comparison with Eq. 10.6.2, it is seen that the right-hand side is equivalent to $S - S_0$, and consequently,

$$\frac{\tau_0}{\gamma R} - S_0 = S - S_0$$

or

$$S = \frac{\tau_0}{\gamma R}$$

## 10.6.2 Water Surface Profiles

It is possible to identify a series of water surface profiles based on an evaluation of Eq. 10.6.4 (Bakhmeteff, 1932). Essential to the development is the determination of normal and critical depths. Note that $y_0$ and $y_c$ are uniquely determined once the channel properties and discharge are established. Table 10.3 shows the classification of the water surface profiles. Associated with $y_c$ is a critical slope $S_c$, which is found by substituting $y_c$ into the Chezy–Manning equation and solving for the slope. The channel slope can be designated as mild, steep, or critical, depending on whether $S_0$ is less than, greater than, or equal to $S_c$, respectively. A horizontal slope exists when $S_0 = 0$ and an adverse slope when $S_0 < 0$. Inspection of Table 10.3 shows that there are 12 possible profiles. Each profile is classified by a letter/number combination. The letter refers to the channel slope: M for mild, S for steep, C for critical, H for horizontal, and A for adverse.

The numerical subscript designates the range of $y$ relative to $y_0$ and $y_c$. Flow can occur at depths above or below $y_c$ and at depths above or below $y_0$.

The variation of $y$ with respect to $x$ for each profile in Table 10.3 can now be found. For a given $Q$, $n$, $S_0$, and channel geometry, the analysis reduces to the determination of how $S$ and Fr vary with $y$. An inspection of the Chezy–Manning equation will reveal that $S$ decreases with increasing $y$; similarly, the Froude number decreases as $y$ increases. The numerator in Eq. 10.6.4 will assume the following inequalities: $(S_0 - S) > 0$ for $y > y_0$, and $(S_0 - S) < 0$ for $y < y_0$. In addition, the denominator varies in the manner $(1 - \text{Fr}^2) > 0$ for $y > y_c$ and $(1 - \text{Fr}^2) < 0$ for $y < y_c$. With these criteria, the sign of $dy/dx$ can be evaluated. In addition, with the use of Eq. 10.6.2, the sign of $dE/dx$ is revealed.

**Table 10.3**  Classification of Surface Profiles

| Channel slope | Profile type | Depth range | Fr | $\dfrac{dy}{dx}$ | $\dfrac{dE}{dx}$ | |
|---|---|---|---|---|---|---|
| Mild $S_0 < S_c$ $y_0 > y_c$ | $M_1$ | $y > y_0 > y_c$ | $<1$ | $>0$ | $>0$ | |
| | $M_2$ | $y_0 > y > y_c$ | $<1$ | $<0$ | $<0$ | |
| | $M_3$ | $y_0 > y_c > y$ | $>1$ | $>0$ | $<0$ | |
| Steep $S_0 > S_c$ $y_0 < y_c$ | $S_1$ | $y > y_c > y_0$ | $<1$ | $>0$ | $>0$ | |
| | $S_2$ | $y_c > y > y_0$ | $>1$ | $<0$ | $>0$ | |
| | $S_3$ | $y_c > y_0 > y$ | $>1$ | $>0$ | $<0$ | |
| Critical $S_0 = S_c$ $y_0 = y_c$ | $C_1$ | $y > y_c$ or $y_0$ | $<1$ | $>0$ | $>0$ | |
| | $C_3$ | $y_c$ or $y_0 > y$ | $>1$ | $>0$ | $<0$ | |
| Horizontal $S_0 = 0$ $y_0 \to \infty$ | $H_2$ | $y > y_c$ | $<1$ | $<0$ | $<0$ | |
| | $H_3$ | $y_c > y$ | $>1$ | $>0$ | $<0$ | |
| Adverse $S_0 < 0$ $y_0$ undefined | $A_2$ | $y > y_c$ | $<1$ | $<0$ | $<0$ | |
| | $A_3$ | $y_c > y$ | $>1$ | $>0$ | $<0$ | |

The boundaries of the profiles can be established as follows[5]:

1. As $y$ tends to $y_0$, $S$ tends to $S_0$. Hence $dy/dx$ approaches zero; in other words, the water surface approaches $y_0$ asymptotically. This applies to curves $M_1$, $M_2$, $S_2$, and $S_3$.

2. As $y$ becomes large, the velocity becomes small and $S$ and Fr tend to zero, so that $dy/dx$ approaches $S_0$. Hence the surface approaches a horizontal asymptote; curves $M_1$, $S_1$, and $C_1$ are of this type.

3. As $y$ approaches $y_c$, $dy/dx$ becomes infinite, a limit that is never reached. For supercritical flow, as the $M_3$, $H_3$, and $A_3$ curves approach $y_c$, a hydraulic jump will form and create a discontinuity in the water surface; at the beginning of the curve, rapid acceleration occurs with nonparallel streamlines. When the flow is subcritical ($M_2$, $H_2$, $A_2$), rapid drawdown takes place close to $y_c$, and the streamlines are no longer parallel. For all of these situations, Eq. 10.6.4 is invalid, since the flow is no longer one-dimensional.

For the profiles shown in Table 10.3, there is no physical significance to the theoretical limit of $y$ approaching zero, since a finite depth is necessary for the existence of flow. It is left as an exercise to show that for the critical slope condition, $dy/dx > S_c$ as $y$ approaches $y_c$ from either a $C_1$ or $C_2$ profile, and that $dy/dx$ approaches a horizontal asymptote as $y$ becomes very large.

## Example 10.13

By assuming a wide rectangular channel, develop the right-hand side of Eq. 10.6.4 to show how $dy/dx$ varies with $y$.

### Solution

For a wide rectangular channel, assume that $b \gg y$, so that the wetted perimeter is approximated by $P \simeq b$. The hydraulic radius then becomes $R = A/P \simeq (by)/b = y$. Noting that $Q = qb$, the Chezy–Manning equation, used to evaluate $S$, simplifies to

$$S = \frac{(qbn)^2}{(by^{5/3})^2} = \frac{(qn)^2}{y^{10/3}}$$

It is assumed that in the Chezy–Manning equation $c_1 = 1$. For a rectangular section the square of the Froude number (see Eq. 10.4.9) is

$$\text{Fr}^2 = \frac{q^2}{gy^3}$$

Substituting into Eq. 10.6.4 gives

$$\frac{dy}{dx} = \frac{S_0 - (qn)^2/y^{10/3}}{1 - q^2/(gy)^3}$$

---

[5]Note that $x$ is measured parallel to the channel bottom, and that $y$ is measured vertically from the channel bottom; hence $dy/dx$ and $dE/dx$ are evaluated relative to the channel bottom and not to a horizontal datum.

Since $(qn)^2 y_0^{-10/3} = S_0$ and $\mathrm{Fr}_c^2 = q^2/(g y_c^3) = 1$, the relation can be written as

$$\frac{dy}{dx} = S_0 \frac{1 - (y_0/y)^{10/3}}{1 - (y_c/y)^3}$$

This equation can be used as an alternative to Eq. 10.6.4 to evaluate the water surface profiles shown in Table 10.3.

### 10.6.3   Controls and Critical Flow

The existence of the various profiles shown in Table 10.3 depend on the boundary conditions that are specified at given locations in the channel. Quite often, a control will define the boundary condition. A *control* exists when a depth-discharge relationship can be established at a section. The manner in which the control section affects the water surface away from its specific location can be studied by examining flow near the critical state in a rectangular channel.

KEY CONCEPT   *A control is a channel feature that establishes a depth-discharge relationship in its vicinity.*

Equation 10.5.12 is the expression for a surge of finite magnitude translating in the upstream direction in a rectangular channel. As $y_2$ approaches $y_1$, the magnitude of the surge becomes infinitesimal; these waves may be generated by the presence of control structures and other transition sections that tend to "disturb" the flow. In Eq. 10.5.12, one can replace $y_1$ and $y_2$ by $y$, and $V_1$ by $V$. In addition, let the wave travel in the downstream direction, so that the equation becomes

$$1 = \frac{1}{2}\left[\sqrt{1 + 8\frac{(V - w)^2}{gy}} - 1\right] \tag{10.6.5}$$

Solving for $w$ provides

$$w = V \pm \sqrt{gy} = V \pm c \tag{10.6.6}$$

in which $c = \sqrt{gy}$, termed the *celerity*. The celerity is the speed at which an infinitesimal wave will travel into an undisturbed region, that is, a region with zero velocity. For a nonrectangular channel, $c = \sqrt{gA/B}$. Thus if a disturbance is created at a midstream location in a channel, two infinitesimal waves are generated. One wave front will tend to propagate upstream at the speed $V - c$, and the second will move downstream at the speed $V + c$. These observations are made relative to an observer in a fixed position, that is, to one who is standing on the shore observing the wave motion.

In a rectangular channel with critical flow conditions, $\mathrm{Fr} = V/\sqrt{gy} = 1$, or $V = \sqrt{gy} = c$. Hence, at critical flow, the first wave front generated by the disturbance would not move upstream but would appear stationary and become a so-called standing wave. The opposite front would be swept downstream at the speed $2c$. If $\mathrm{Fr} < 1$, the first wave would travel upstream, and the opposite wave would travel downstream at a speed less than $2c$. For $\mathrm{Fr} > 1$ in the channel, since $V > c$, both waves are swept downstream. Thus, since it contains a mechanism that disturbs the flow, a control will influence upstream conditions only when the

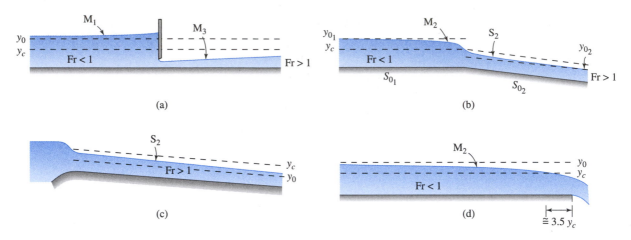

**Fig. 10.19** Representative controls: (a) sluice gate; (b) change in slope from mild ($S_{01}$) to steep ($S_{02}$); (c) entrance to a steep channel; (d) free outfall.

flow is subcritical (Fr < 1). Similarly, when the flow is supercritical (Fr > 1), the control can influence only the downstream flow conditions. This is illustrated in Figure 10.19a, where a sluice gate is positioned in a channel with subcritical flow upstream and supercritical flow downstream. An $M_1$ profile is generated above the gate, and an $M_3$ profile exists below it. Any movement of the gate would influence the nature of the two profiles; lowering the gate would lengthen their range, and raising it would produce the opposite effect.

The profiles shown in Table 10.3 are all influenced by the presence of controls. In Fig. 10.19, several controls are shown that create a variety of profiles. Critical depth often is associated with an effective control. Examples include sluice gates, weirs, dams, and flumes, all of which force critical flow to occur somewhere in the transition region. In addition, a control can be located at a break in channel slope from mild upstream to steep downstream (Fig. 10.19b). The entrance to a steep channel (Fig. 10.19c) is an example of an upstream control, with critical flow occurring at the crest. Critical flow will occur a short distance upstream from a free outfall on a mild slope (Fig. 10.19d). Other controls, not shown in Fig. 10.19, are a channel constriction acting as a choke, and for a mild slope, the existence of uniform flow at some location.

### 10.6.4 Profile Synthesis

Identification of controls and their interaction with the possible profiles is a requirement for successful understanding and correct design and analysis of open-channel flow. Since controls are essentially transition sections, the rapidly varied flow principles presented in Sections 10.4 and 10.5 can be used to determine the necessary depth–discharge relations. Once the controls have been identified, the profiles can be selected and the range of influence of the controls established.

As an example, consider the situation shown in Fig. 10.20. Flow enters the steep channel from a reservoir, so that critical flow exists at the channel entrance.

**Fig. 10.20**   Example of profile synthesis.

Both the magnitude of the discharge and the $S_2$ profile are influenced by the depth at the entrance; therefore, the depth acts as a control. At the downstream end of the channel, the lower reservoir acts as a control to establish an $S_1$ profile that projects upstream. At some interior location, a hydraulic jump occurs to allow the flow to pass from a supercritical to a subcritical state. The location of the jump can be found by plotting a curve of the depth conjugate to the $S_2$ profile and finding its point of intersection with the $S_1$ curve. A lowering of the lower reservoir elevation would cause the jump to move downstream and ultimately be swept out of the channel. Increasing the lower reservoir elevation would move the jump upstream; if it were to move into the entrance region, the upstream control would no longer exist. Additional examples of profile synthesis are given in Section 10.7.

## Example 10.14

In a rectangular channel, $b = 3$ m, $n = 0.015$, $S_0 = 0.0005$, and $Q = 5$ m³/s. At the entrance to the channel, flow issues from a sluice gate at a depth of 0.15 m. The channel is sufficiently long that uniform flow conditions are established away from the entrance region, Fig. E10.14a. Find the nature of the water surface profile in the vicinity of the entrance.

**Fig. E10.14**

### Solution

First find $y_0$ and $y_c$ to determine the type of channel. To find $y_0$, follow the method shown in Example 10.1. Substitute known data into the Chezy–Manning equation:

*(continued)*

$$\frac{(3y_0)^{5/3}}{(3 + 2y_0)^{2/3}} = \frac{0.015 \times 5}{\sqrt{0.0005}} = 3.354$$

Solving gives $y_0 = 1.39$ m. Next, the critical depth is computed to be

$$y_c = \left(\frac{q^2}{g}\right)^{1/3}$$

$$= \left[\frac{(5/3)^2}{9.81}\right]^{1/3} = 0.66 \text{ m}$$

Since $y_0 > y_c$, a mild slope condition exists. The gate is a control and there will be an $M_3$ profile beginning at the entrance, terminated by a hydraulic jump. Downstream of the jump, the condition of uniform flow acts as a control, so at that location the depth is $y_0$, and the Froude number is

$$\text{Fr}_0 = \frac{q}{\sqrt{gy_0^3}}$$

$$= \frac{5/3}{\sqrt{9.81 \times 1.39^3}} = 0.325$$

Using Eq. 10.5.10, the depth before the jump is

$$y_1 = \frac{y_0}{2}\left(\sqrt{1 + 8\text{Fr}_0^2} - 1\right)$$

$$= \frac{1.39}{2}\left(\sqrt{1 + 8 \times 0.325^2} - 1\right) = 0.25 \text{ m}$$

The depths $y_c$ and $y_1$ have been calculated to two significant figures, since the Manning coefficient is known to only two significant figures. Figure E10.14b shows the profile.

## 10.7 NUMERICAL ANALYSIS OF WATER SURFACE PROFILES

Section 10.6 dealt with the understanding and interpretation of various aspects of gradually varied flow. An important procedure was outlined: A water surface profile can be synthesized, or predicted, by making use of relevant information about the channel geometry, roughness, and flow, and by determining or assuming the appropriate controls. Once a satisfactory profile synthesis has been conducted, we are in a position to numerically calculate the desired water surface profiles and accompanying energy gradelines.

Regardless of the type of method chosen to numerically evaluate gradually varied flow, analysis of a water surface profile on a reach of channel with constant slope usually follows these steps:

1.  The channel geometry, channel slope $S_0$, roughness coefficient $n$, and discharge $Q$ are given or assumed.

2.  Determine normal depth $y_0$ and critical depth $y_c$.
3.  Establish the controls (i.e., the depth of flow) at the upstream and down-stream ends of the channel reach.
4.  Integrate Eq. 10.6.4 to find $y$ and subsequently $E$ as functions of $x$, allow-ing for the possibility of a hydraulic jump to occur within the reach.

For a prismatic channel, $y_0$ can be evaluated by applying the Chezy–Manning equation and finding the root of the function

$$\frac{Qn}{c_1 AR^{2/3}\sqrt{S_0}} - 1 = 0 \qquad (10.7.1)$$

Similarly, $y_c$ is found from application of the Froude number in the form

$$\frac{Q^2 B}{gA^3} - 1 = 0 \qquad (10.7.2)$$

Numerical methods, such as the false position or interval halving (Chapra and Canale, 1988), or computational software can be used to solve Eqs. 10.7.1 and 10.7.2.

The Chezy–Manning equation, given here in the form of Eq. 10.7.1, is used to represent the depth–discharge variation for nonuniform as well as for uniform flow. Thus one can replace $S_0$ by the slope $S$ of the energy grade line in Eq. 10.7.1 and solve for $S$ in the manner

$$S(y) = \frac{Q^2 n^2}{c_1^2 [A(y)]^2 [R(y)]^{4/3}} \qquad (10.7.3)$$

Since $Q$ and $n$ are either given or assumed, the right-hand side of Eq. 10.7.3 is a function of $y$ alone.

Two numerical methods are presented in this section to compute water sur-face profiles along with normal and critical depths. The first, termed the *standard step method*, is the most widely used, and the second employs an accurate numer-ical integration scheme. Solutions will be illustrated using Excel, Mathcad, and MATLAB software. The third method is analytical, and uses Eq. 10.6.4 in inte-grated form, assuming simplified geometric properties of the channel cross section.

## 10.7.1   Standard Step Method

To develop a numerical procedure for solving gradually varied flow problems, we use Eq. 10.6.2; it is applied over the reach of channel shown in Fig. 10.21, resulting in

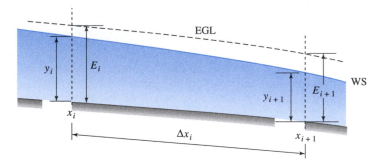

**Fig. 10.21**   Notation for computing gradually varied flow.

$$\frac{dE}{dx} = \frac{d}{dx}\left( y + \frac{V^2}{2g} \right)$$

$$= S_0 - S(y) \tag{10.7.4}$$

For small changes in $y$ and $V$ between $x_i$ and $x_{i+1}$, Eq. 10.7.4 can be approximated by

$$E_{i+1} - E_i = \int_{x_i}^{x_{i+1}} [S_0 - S(y)]\, dx$$

$$\simeq (x_{i+1} - x_i)[S_0 - S(y_m)] \tag{10.7.5}$$

in which $y_m = (y_{i+1} + y_i)/2$. Equation 10.7.5 is solved for $x_{i+1}$ to yield

$$x_{i+1} = x_i + \frac{E_{i+1} - E_i}{S_0 - S(y_m)} \tag{10.7.6}$$

The calculations take place stepwise, beginning at a control point or other location where the depth is known. Assume that it is desired to generate a water surface profile along the channel, and compute values of $x_i$, $y_i$, and $E_i$, for $i = 1, \ldots, k$, where $k$ is the location at the opposite end of the reach. Other than $x_k$, it is not necessary that location $x_i$ be fixed, so it can take on any value. Beginning at location $i$, the evaluation of conditions at location $i + 1$ proceeds as follows:

1.   Choose $y_{i+1}$.
2.   Compute $E_{i+1}$ from Eq. 10.4.12, $y_m$ knowing $y_i$ and $y_{i+1}$, and $S(y_m)$ from Eq. 10.7.3.
3.   Compute $x_{i+1}$ from Eq. 10.7.6.
4.   At location $k$, trial values of $y_k$ are assumed until Eq. 10.7.6 is satisfied for the known value of $x_k$.

The standard step method can be carried out using spreadsheet analysis.

# Example 10.15

Water is flowing at $Q = 22$ m³/s in a long trapezoidal channel, $b = 7.5$ m, $m_1 = m_2 = 2.5$. A free overfall is located at the downstream end of the channel, where $x = 2000$ m. For $n = 0.015$, $S_0 = 0.0006$, find the water surface profile and energy grade line for a distance of approximately 800 m upstream from the free outfall.

**Solution**

Equations 10.7.1 and 10.7.2 are used to evaluate $y_0$ and $y_c$ by substituting in known data:

$$\frac{22 \times 0.015 \times \left[ 7.5 + 2y_0 \sqrt{1 + (2.5)^2} \right]^{2/3}}{\left[ 7.5y_0 + \frac{1}{2}y_0^2(2.5 + 2.5) \right]^{5/3} \sqrt{0.0006}} - 1 = 0$$

$$\frac{(22)^2 \times \left[ 7.5 + y_c(2.5 + 2.5) \right]}{9.8 \times \left[ 7.5y_c + \frac{1}{2}y_c^2(2.5 + 2.5) \right]^3} - 1 = 0$$

The roots of these equations can be found using a routine such as Excel Solver®; the solutions are $y_0 = 1.29$ m and $y_c = 0.86$ m. Hence the channel is a mild type, and control will be close to the free overfall at the downstream end of the channel. Without any serious loss of accuracy, one can assume that critical conditions will exist at the free overfall. Referring to Table 10.3, the profile upstream of the overfall will be of type $M_2$. An Excel spreadsheet solution is shown in Table E10.15. The upper part shows the values of critical and normal depths found by using Solver. In the residual column are very small numbers that should be close to zero; see the two above equations. The lower part of the table shows the step method solution. Calculations proceed from station 1 to station 5 in a straightforward manner, with arbitrary values of depth selected and placed in the $y$ column. The beginning value of $x$ (2000 m) is placed in the first cell of the $x$ column, and the remaining distances are computed as explained on the previous page. At station 6, different values of $y$ are chosen until the distance is close to the desired value of 1200 m; a depth of 1.27 m results in a distance of 1230 m, which is acceptable. The spreadsheet equations for computing normal and critical depths, and for the step method, are provided in Appendix E. In addition, a MATLAB solution to this problem is shown in Appendix E, Fig. E.5.

**Table E10.15**

|          | Depth [m] | Residual  |
|----------|-----------|-----------|
| Critical | 0.865     | 1.087E-06 |
| Normal   | 1.292     | 1.812E-06 |

| Station | $y$ [m] | $A$ [m²] | $V$ [m/s] | $E$ [m] | $y_m$ [m] | $S(y_m)$  | $\Delta x$ [m] | $x$ [m] |
|---------|---------|----------|-----------|---------|-----------|-----------|----------------|---------|
| 1       | 0.865   | 8.358    | 2.632     | 1.218   |           |           |                | 2000    |
| 2       | 0.950   | 9.381    | 2.345     | 1.230   | 0.908     | 2.165E-03 | −8             | 1992    |
| 3       | 1.050   | 10.631   | 2.069     | 1.268   | 1.000     | 1.527E-03 | −41            | 1951    |
| 4       | 1.150   | 11.931   | 1.844     | 1.323   | 1.100     | 1.081E-03 | −114           | 1837    |
| 5       | 1.250   | 13.281   | 1.656     | 1.390   | 1.200     | 7.866E-04 | −357           | 1480    |
| 6       | 1.270   | 13.557   | 1.623     | 1.404   | 1.260     | 6.574E-04 | −250           | 1230    |

## Example 10.16

A trapezoidal channel 300 m long conveys water flowing at $Q = 25$ m³/s. The inverse side slopes of the cross section are $m_1 = m_2 = 2.5$, the bottom width is 5 m, the bottom slope is $S_0 = 0.005$, and the Manning coefficient is $n = 0.013$. Compute the water surface profile and energy grade line if the upstream depth $y_u = 0.55$ m and the downstream depth $y_d = 2.15$ m.

**Fig. E10.16**

## Solution

An Excel spreadsheet solution using the step method is shown in Table E10.16. First, input data is entered, including the gravitational constant and the corresponding constant $c_1$ in the Manning equation. Since the two inverse side slopes are equal, a single value $m$ is entered. Then critical and normal depths are evaluated using Excel Solver. Since $y_c > y_0$, the slope is steep. One control point is at the upstream end, and an $S_3$ profile exists in the upper reach of the channel. The profile is computed from station 1 beginning with $y_u$ to station 7 where the depth is 0.85 m and the distance is 288 m. Then an $S_1$ profile is calculated from station 9 at the end of the channel beginning with $y_d$, and terminating when $y$ reaches critical depth. A hydraulic jump is located approximately 190 m downstream from the channel entrance. It is located by plotting the curve of depth conjugate to the $S_3$ profile and finding its intersection with the $S_1$ profile as shown in Fig. E10.16. The conjugate depths $y_{cj}$ are computed using Excel Solver.

**Table E10.16**

| $Q =$ | 25 | $n =$ | 0.013 | $S_0 =$ | 0.005 | $L =$ | 300 | $g =$ | 9.81 |
| $b =$ | 5 | $m =$ | 2.5 | $y_n =$ | 0.55 | $y_d =$ | 2.15 | $c_1 =$ | 1.00 |

| | | Depth | Residual |
|---|---|---|---|
| $y_c =$ | | 1.124 | −2.26E-07 |
| $y_0 =$ | | 0.864 | 4.363E-07 |

| Station | $y$ | $A$ | $V$ | $E$ | $y_m$ | $S(y_m)$ | $\Delta x$ | $x$ | $y_{cj}$ | Residual |
|---|---|---|---|---|---|---|---|---|---|---|
| 1 | 0.550 | 3.506 | 7.130 | 3.141 | | | | | | |
| 2 | 0.600 | 3.900 | 6.410 | 2.694 | 0.575 | 2.189E-02 | 26 | 26 | 1.901 | 8.81E-07 |
| 3 | 0.650 | 4.306 | 5.806 | 2.368 | 0.625 | 1.624E-02 | 29 | 56 | 1.802 | 5.19E-07 |
| 4 | 0.700 | 4.725 | 5.291 | 2.127 | 0.675 | 1.231E-02 | 33 | 88 | 1.711 | 1.68E-07 |
| 5 | 0.750 | 5.156 | 4.848 | 1.948 | 0.725 | 9.500E-03 | 40 | 128 | 1.628 | 9.30E-07 |
| 6 | 0.800 | 5.600 | 4.464 | 1.816 | 0.775 | 7.449E-03 | 54 | 182 | 1.550 | 3.85E-07 |
| 7 | 0.850 | 6.056 | 4.128 | 1.719 | 0.825 | 5.923E-03 | 105 | 288 | 1.477 | 5.07E-07 |
| | | | | | | | | | | |
| 9 | 2.150 | 22.306 | 1.121 | 2.214 | | | | 300 | | |
| 10 | 2.050 | 20.756 | 1.204 | 2.124 | 2.100 | 1.575E-04 | −19 | 281 | | |
| 11 | 1.950 | 19.256 | 1.298 | 2.036 | 2.000 | 1.924E-04 | −18 | 263 | | |
| 12 | 1.850 | 17.806 | 1.404 | 1.950 | 1.900 | 2.371E-04 | −18 | 245 | | |
| 13 | 1.750 | 16.406 | 1.524 | 1.868 | 1.800 | 2.951E-04 | −17 | 228 | | |
| 14 | 1.650 | 15.056 | 1.660 | 1.791 | 1.700 | 3.713E-04 | −17 | 211 | | |
| 15 | 1.550 | 13.756 | 1.817 | 1.718 | 1.600 | 4.728E-04 | −16 | 195 | | |
| 16 | 1.450 | 12.506 | 1.999 | 1.654 | 1.500 | 6.102E-04 | −15 | 180 | | |
| 17 | 1.350 | 11.306 | 2.211 | 1.599 | 1.400 | 7.998E-04 | −13 | 167 | | |
| 18 | 1.250 | 10.156 | 2.462 | 1.559 | 1.300 | 1.067E-03 | −10 | 157 | | |
| 19 | 1.124 | 8.778 | 2.848 | 1.537 | 1.187 | 1.514E-03 | −6 | 151 | | |

## 10.7.2 Numerical Integration Method

There are a number of numerical methods available to solve Eq. 10.6.4 for channels with regular cross sections, any of which would provide a solution with sufficient accuracy for design and analysis purposes. Worthy of mention are integration techniques using trapezoidal or Simpson's rules, and solution of the differential form using a Runge–Kutta procedure. The underlying theory of these methods and algorithms for their implementation can be found in texts on numerical methods. McBean and Perkins (1975) discuss convergence problems associated with use of the trapezoidal rule applied to gradually varied flow.

A useful integration scheme is the two-point Gauss–Legendre quadrature (Chapra and Canale, 1998). It is particularly well suited for computer use and is usually more accurate than either the trapezoidal or Simpson's rule methods. Equation 10.6.4 is written in the integral form

$$x_{i+1} = x_i + \int_{y_i}^{y_{i+1}} \frac{1 - \mathrm{Fr}^2}{S_0 - S}\, dy$$

$$= x_i + \int_{y_i}^{y_{i+1}} G(y)\, dy \qquad (10.7.7)$$

The last integral is approximated by the Gauss–Legendre formula, giving the result

$$\int_{y_i}^{y_{i+1}} G(y)\, dy = \frac{y_{i+1} - y_i}{2}\left[ G\!\left( \frac{y_{i+1} + y_i - \sqrt{3}/3(y_{i+1} - y_i)}{2} \right)\right.$$

$$\left. + G\!\left( \frac{y_{i+1} + y_i + \sqrt{3}/3(y_{i+1} - y_i)}{2} \right) \right] \qquad (10.7.8)$$

The following example illustrates use of the algorithm.

## Example 10.17

Rework Example 10.15 using Gauss–Legendre quadrature.

### Solution

A Mathcad solution is shown below. After the known data is entered and functions are defined, normal and critical depths are computed. Subsequently, the depth is iterated to find corresponding values of distance using Eqs. 10.7.7 and 10.7.8.

### Input data:

$$Q := 22 \qquad M := 2.5 \qquad b := 7.5 \qquad L := 2000$$

$$S_0 := 0.0006 \qquad n := 0.015 \qquad g := 9.81$$

### Define functions:

$$A(y) := b \cdot y + M \cdot y^2 \qquad B(y) := b + 2 \cdot M \cdot y \qquad P(y) := b + 2 \cdot y \cdot \sqrt{1 + M^2}$$

$$S(y) := \frac{Q^2 \cdot n^2}{A(y)^{3.3333} \cdot P(y)^{-1.3333}} \qquad Fr(y) := \sqrt{\frac{Q^2 \cdot B(y)}{g \cdot A(y)^3}} \qquad G(y) := \frac{1 - Fr(y)^2}{S_0 - S(y)}$$

$$E(y) := y + \frac{Q^2}{2 \cdot g \cdot A(y)^2}$$

### Compute normal and critical depths:

$$y_n := \text{root}(S(y) - S_0, y, 0.01, 5) \qquad y_n = 1.292$$

$$y_c := \text{root}(Fr(y) - 1, y, 0.01, 5) \qquad y_c = 0.865$$

### Compute S2 profile from upstream location:

$$X_0 := 2000 \qquad y_0 := y_c \qquad \Delta y := 0.10 \qquad N := 4$$

$$i := 1, 2 .. N$$

$$y_i := y_{i-1} + \Delta y$$

$$x_i := x_{i-1} + \frac{\Delta y}{2} \cdot \left( G\left( \frac{y_i + y_{i-1} - \frac{\sqrt{3}}{3} \cdot \Delta y}{2} \right) + G\left( \frac{y_i + y_{i-1} + \frac{\sqrt{3}}{3} \cdot \Delta y}{2} \right) \right)$$

### Computed depth, distance and specific energy:

$$y = \begin{pmatrix} 0.865 \\ 0.965 \\ 1.065 \\ 1.165 \\ 1.265 \end{pmatrix} \qquad x = \begin{pmatrix} 2 \times 10^3 \\ 1.988 \times 10^3 \\ 1.938 \times 10^3 \\ 1.798 \times 10^3 \\ 1.257 \times 10^3 \end{pmatrix} \qquad E(y) = \begin{pmatrix} 1.218 \\ 1.235 \\ 1.275 \\ 1.333 \\ 1.4 \end{pmatrix}$$

### 10.7.3   Irregular Channels

The method demonstrated in Example 10.15 works well with regular channels. When calculating the profile for a reach of river channel in which the cross-sectional data are irregular and usually given at fixed locations, a trial solution is necessary at each location. Henderson (1966) outlines a method of computation that includes corrections for composite sections and eddy losses that occur at bends and expansions. For a composite section, the slope of the energy grade line is given by

$$S = \frac{Q^2}{(\Sigma\, K_i)^2} \tag{10.7.9}$$

in which $K_i$ is termed the *conveyance* of subsection $i$:

$$K_i = \left(\frac{c_1 A R^{2/3}}{n}\right)_i \tag{10.7.10}$$

A kinetic-energy coefficient must be applied to the kinetic-energy term; for a composite section use

$$\alpha = \frac{(\Sigma\, A_i)^2}{(\Sigma\, K_i)^3} \Sigma \left(\frac{K_i^3}{A_i^2}\right) \tag{10.7.11}$$

With such complexity, it is most convenient to use the step method procedure in computer-coded form. The United States Army Corps of Engineers has developed the algorithm "HEC-RAS River Analysis System" for use in natural channels. The solution is based on the standard step method and is available for use on a personal computer. The program and documentation can be downloaded from the Website *www.hec.usace.army.mil/software/hec-ras/*.

### 10.7.4   Direct Integration Methods

Equation 10.6.4 can be integrated in a straightforward manner if one assumes a wide rectangular, horizontal channel. A more general integration procedure is available for nonhorizontal prismatic channels of various shapes (Chow, 1959). Assume that in the Chezy–Manning equation the product $A^2 R^{4/3}$ is proportional to $y^N$, and in the Froude number the ratio $A^3/B$ is proportional to $y^M$, where $M$ and $N$ are constants. Then Eq. 10.6.4 can be written

$$dx = \frac{1}{S_0}\left[\frac{1 - (y_c/y)^M}{1 - (y_0/y)^N}\right] dy \tag{10.7.12}$$

The solution of Eq. 10.7.12 is

$$x = \frac{y_0}{S_0}\left[ u - F(u, N) + \left(\frac{y_c}{y_0}\right)^M \frac{J}{N} F(v, J) \right]$$  (10.7.13)

in which $u = y/y_0$, $v = u^{N/J}$, $J = N/(N - M + 1)$, and $F$ is the varied flow function

$$F(u, N) = \int_0^u \frac{d\eta}{1 - \eta^N}$$  (10.7.14)

Equation 10.7.14 can be evaluated numerically and the results are presented as the varied flow function in Appendix E, Table E.1. Values of $N = 2\frac{1}{2}, 3$, and $3\frac{1}{3}$ are commonly used since they are consistent with a wide rectangular channel assumption (see Example 10.13). Note that the function $F$ may have any desired constant of integration assigned to it. In the varied flow function table the constant is adjusted so that $F(0, N) = F(\infty, N) = 0$. Since integration from one location to another is direct, no intermediate calculations are necessary. This method of evaluation is in direct contrast to the numerical methods, where intermediate steps cannot be avoided without the loss of precision.

## Example 10.18

A wide rectangular channel conveys a discharge of $q = 3.72$ m³/s per meter width on a slope of $S_0 = 0.001$. At a given location the depth is 3 m. Determine the distance upstream where the depth is 2.5 m. The Manning coefficient is 0.025.

### Solution
From Example 10.13, for a wide rectangular channel we found that $N = 3.33$ and $M = 3$. Therefore, one can calculate $J$ to be

$$J = \frac{N}{N - M + 1} = \frac{3.33}{3.33 - 3 + 1} = 2.5$$

Also from Example 10.13, we can determine $y_0$ in the manner

$$y_0 = \left(\frac{q^2 n^2}{S_0}\right)^{3/10} = \left(\frac{3.72^2 \times 0.025^2}{0.001}\right)^{3/10} = 1.91 \text{ m}$$

Furthermore,

$$y_c = \left(\frac{q^2}{g}\right)^{1/3} = \left(\frac{3.72^2}{9.81}\right)^{1/3} = 1.12 \text{ m}$$

Substitute these values into Eq. 10.7.13 and simplify:

$$x = \frac{1.91}{0.001}\left[ u - F(u, 3.33) + \left(\frac{1.12}{1.91}\right)^3 \times \frac{2.5}{3.33} F(v, 2.5) \right]$$
$$= 1910[u - F(u, 3.33) + 0.151F(v, 2.5)]$$

Since $y_0 > y_c$, and at the downstream location $y > y_0$, the profile is an $M_1$ curve. At the downstream location where the depth $y = 3$ m,

*(continued)*

$$u = \frac{y}{y_0} = \frac{3}{1.91} = 1.57 \quad \text{and} \quad v = u^{N/J} = 1.57^{3.33/2.5} = 1.825$$

From Appendix E, Table E.1, making use of linear interpolation between recorded values we find that

$$F(1.57, 3.33) = 0.166 \quad \text{and} \quad F(1.825, 2.5) = 0.300$$

Considering $x$ as a distance measured from an arbitrary datum, we find that

$$x = 1910(1.57 - 0.166 + 0.151 \times 0.300) = 2763 \text{ m}$$

To determine the distance upstream of the weir where the depth is 2.5 m, we perform the following calculations in a manner analogous to those at the downstream location:

$$u = \frac{2.5}{1.91} = 1.31 \quad \text{and} \quad v = 1.31^{3.33/2.5} = 1.43$$

$$F(1.31, 3.33) = 0.578 \quad \text{and} \quad F(1.43, 2.5) = 0.474$$

$$x = 1910(1.31 - 0.578 + 0.151 \times 0.474) = 1536 \text{ m}$$

Hence, the distance between the two locations, from a depth of 3 m to where the depth is 2.5 m, is $2763 - 1536 = 1227$ m, or approximately 1230 m.

## 10.8 SUMMARY

As opposed to pipe flows, open channel flow is complicated by the presence of a free surface. In addition to velocity, the depth of flow is variable, so that even though it may be steady, the flow in a channel is generally nonuniform. In Section 10.3, as well in Chapter 7, we have introduced uniform flow as a limiting case—one that does not often occur but nonetheless is significant. Friction losses in open channel flow are described mathematically by the Chezy–Manning equation, which relates velocity or discharge to the geometric and hydraulic properties of the channel cross section. Energy and momentum principles were introduced in Sections 10.4 and 10.5 and applied to rapidly varied nonuniform flow that occurs at channel transitions and at a hydraulic jump. Critical and conjugate depths have been derived along with the definition of the Froude number. We have seen that specific energy is a useful concept for dealing with transition analysis. Much of the development in this chapter has used a rectangular cross section; this was done primarily for mathematical expediency. The guidelines that were developed and conclusions made using the rectangular section apply to nonrectangular sections as well. Table 10.4 summarizes several of the formulas developed in the chapter.

Gradually varied nonuniform flow occurs where the depth and velocity vary continuously with distance along a channel. In Section 10.6, application of the energy equation to gradually varied flow resulted in a mathematical description of the water surface profile. We have presented a classification of the various types of flow regimes that occur along a channel reach. In addition we have shown how the understanding of rapidly varied and gradually varied flow enable one to predict the manner in which water surfaces will behave, a task that is required prior to numerical computation of water surface profiles. Several numerical methods were introduced in Section 10.7 to integrate the gradually varied flow equation, and examples of solutions were provided using spreadsheets, computational software, and approximate integrals. Perhaps the most useful means of solution is the spreadsheet, in which a generalized algorithm can be readily adapted for use in different problems.

**Table 10.4**   Formulas for Rectangular and General Sections

| Section | Fr | $y_c$ | E | M |
|---|---|---|---|---|
| Rectangular | $\dfrac{q}{\sqrt{gy^3}}$ | $\left(\dfrac{q^2}{g}\right)^{1/3}$ | $y + \dfrac{q^2}{2gy^2}$ | $\dfrac{by^2}{2} + \dfrac{q^2}{gy}$ |
| General | $\dfrac{Q/A}{\sqrt{gA/B}}$ | $\dfrac{Q^2B}{gA^3} = 1$ | $y + \dfrac{Q^2}{2gA^2}$ | $A\bar{y} + \dfrac{Q^2}{gA}$ |

## PROBLEMS

**10.1**   Give an example of each of the following types of flow for both closed-conduit and free-surface conditions. Include a sketch with each representation:
   **(a)**   Steady, uniform
   **(b)**   Unsteady, nonuniform
   **(c)**   Steady, nonuniform
   **(d)**   Unsteady, uniform

**10.2**   Water is flowing with a free surface at 4 m³/s in a circular conduit. Find the diameter $d$ such that the critical depth $y_c = 0.3d$. Under critical flow conditions, $Q^2B/(gA^3) = 1$.

**10.3**   Classify the following flows as steady or unsteady, and uniform or nonuniform:
   **(a)**   Water flowing through a logjam in a river with the observer standing on the shore.
   **(b)**   Water flowing through a river rapids with the observer on a raft traveling with the current.
   **(c)**   A flood wave traveling in a river which is generated by a gentle rainfall/runoff event, with the observer standing on the bank.
   **(d)**   A moving hydraulic jump in a prismatic channel, with the observer running alongside the bank at the same speed as the jump.

### Uniform Flow

**10.4**   For each regular channel section, construct the curves $R$ versus $y$ and $AR^{2/3}$ versus $y$:

   **(a)**   Circular, $d = 2$ m

   **(b)**   Trapezoidal, $b = 3$ m, $m_1 = m_2 = 2.5$

   **(c)**   Composite (see Fig. P10.4)

**Fig. P10.4**

**10.5**   Determine the most efficient section based on flow resistance for a trapezoidal channel. Assume equal side slopes, that is, $m_1 = m_2$.

**10.6**   Determine the uniform depth $y_0$, area $A$, and wetted perimeter $P$ for the following conditions:
   **(a)**   Trapezoidal channel, $m_1 = 2.5$, $m_2 = 3.5$, $b = 4.5$ m, $Q = 35$ m$^3$/s, $n = 0.015$, $S_0 = 0.00035$
   **(b)**   Circular channel, $d = 2.1$ m, $Q = 4.25$ m$^3$/s, $n = 0.012$, $S_0 = 0.001$
   **(c)**   Circular channel, $d = 1.8$ m, $Q = 3.3$ m$^3$/s, $n = 0.012$, $S_0 = 0.001$
   **(d)**   Trapezoidal channel, $Q = 15$ m$^3$/s, $n = 0.011$, $S_0 = 0.0013$, "most efficient" cross section
   **(e)**   Rectangular channel, $b = 7.5$ m, $Q = 33$ m$^3$/s, $n = 0.020$, $S_0 = 0.0004$

**10.7**   Uniform flow occurs in a trapezoidal channel with $m_1 = m_2 = 1.75$. The channel is to convey a discharge of 4.0 m$^3$/s, with an average velocity of 1.4 m/s. If $n = 0.015$ and $S_0 = 0.001$, find the bottom width and the uniform depth.

**10.8**   A channel cross section, commonly called a *gutter*, forms at the side of a street next to a curb during rainfall conditions (Fig. P10.8). The slope along the roadway is $S_0 = 0.0005$, and the Manning roughness coefficient is $n = 0.015$. Assuming that uniform flow conditions occur:
   **(a)**   Determine the discharge if the depth of flow is $y_0 = 12$ cm
   **(b)**   If $Q = 80$ L/s, what is the flow depth $y_0$?

**Fig. P10.8**

**10.9**   A sewer pipe of circular cross section is made of concrete ($n = 0.014$). It is to convey water at uniform flow conditions such that the pipe flows half full, that is, $y = d/2$, where $d$ = diameter of the pipe. Evaluate the following conditions:

   **(a)**   Given $S_0 = 0.0003$ and $Q = 1.75$ m$^3$/s, find $d$
   **(b)**   Given $S_0 = 0.00005$ and $d = 1.3$ m, find $Q$
   **(c)**   Given $d = 0.75$ m and $Q = 0.45$ m$^3$/s, find $S_0$

---

## Energy Concepts

**10.10**   For a given specific energy in a rectangular channel, show that critical flow exists when the specific discharge is at its maximum value $q_{max}$.

**10.11**   The specific energy relation for a rectangular section, Eq. 10.4.5, can be made dimensionless by normalizing it with respect to critical depth $y_c$. Prepare such a plot with $y/y_c$ as the ordinate and $E/y_c$ as the abscissa. Locate the minimum point mathematically.

**10.12**   Water is flowing in a rectangular channel at a velocity of 3 m/s and a depth of 2.5 m. Determine the changes in water surface elevation for the following alterations in the channel bottom:
   **(a)**   An increase (upward step) of 20 cm, neglecting losses.
   **(b)**   A "well-designed" increase of 15 cm.
   **(c)**   The maximum increase allowable for the specified upstream flow conditions to remain unchanged, neglecting losses.

   **(d)**   A "well-designed" decrease (downward step) of 20 cm.

**10.13**   Water is flowing in a rectangular channel whose width is 5 m. The depth of flow is 2 m and the discharge is 25 m$^3$/s. Determine the changes in depth for the following alterations in the channel width:
   **(a)**   An increase of 50 cm, neglecting losses
   **(b)**   A decrease of 25 cm, assuming a "well-designed" transition

**10.14**   Flow in a rectangular channel of width $b$ takes place with a known depth $y$ and velocity $V$. Downstream of this location, there is an upward step of height $h$. Neglecting losses, determine the change in width that must take place simultaneously for critical flow to occur within the transition.

$$b = 3 \text{ m}, y = 3 \text{ m},$$
$$V = 3 \text{ m/s}, h = 70 \text{ cm}$$

**10.15**   Water is flowing in a rectangular channel 2.0 m wide. At a transition section the channel bottom is lowered by $h = 0.1$ m for a short distance, and then is raised back to the original elevation (Fig. P10.15). If $y = 1.22$ m and $Q = 4.8$ m³/s, then, with losses neglected:

   **(a)**   Find the change in channel width necessary to maintain a horizontal water surface through the transition.

   **(b)**   What change in width would cause critical flow to occur in the transition?

**Fig. P10.15**

**10.16**   Water flows at a depth of 2.15 m and a unit discharge of 5.5 m²/s in a rectangular channel. Energy losses can be neglected.

   **(a)**   What is the maximum height $h$ of a raised bottom that will permit the flow to pass over it without increasing the upstream depth?

   **(b)**   Show the solution on an $E$–$y$ diagram.

   **(c)**   Sketch the water surface and energy grade line.

   **(d)**   If the channel bottom is raised greater than $h$, discuss a type of change that may take place upstream of the transition.

**10.17**   A lake discharges into a steep channel. At the channel entrance the lake level is 2.5 m above the channel bottom. Neglecting losses, find the discharge for the following geometries:

   **(a)**   Rectangular section, $b = 4$ m

   **(b)**   Trapezoidal section, $b = 3$ m, $m_1 = m_2 = 2.5$

   **(c)**   Circular section, $d = 3.5$ m

**10.18**   Steep flow conditions exist at the outlet from a reservoir into an open canal. The canal width is not yet fixed, but the depth $y$ at the entrance and the discharge $Q$ are known. Determine the necessary widths for the following geometries:

   **(a)**   Rectangular section, $y = 1$ m, $Q = 18$ m³/s

   **(b)**   Trapezoidal section, $m_1 = m_2 = 3$, $y = 1$ m, $Q = 18$ m³/s

**10.19**   Rapidly varied flow occurs over a rectangular sill (Fig. P10.19), with free outfall conditions at location C. Throughout the reach losses can be neglected.

   **(a)**   Compute the discharge.

   **(b)**   Evaluate the depths at locations A, B, and C.

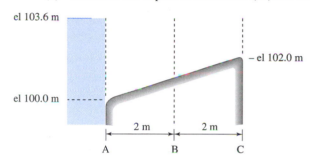

**Fig. P10.19**

**10.20**   Prepare a computer algorithm that evaluates the critical depth of a regular channel using the false position technique or a method of your choice. Verify the program with the following data:

   **(a)**   Circular section, $Q = 5$ m³/s, $d = 2.5$ m

   **(b)**   Rectangular section, $Q = 3.5$ m³/s, $b = 3.6$ m

   **(c)**   Trapezoidal section, $Q = 120$ m³/s, $b = 10$ m, $m_1 = m_2 = 5$

   **(d)**   Triangular section, $Q = 7$ m³/s, $m_1 = 2.5, m_2 = 3.0$

**10.21**   For the channel cross section shown in Fig. P10.21, determine the critical depth if the discharge is 16.5 m³/s.

**Fig. P10.21**

**10.22** For the composite channel section shown in Fig. P10.22, find the critical depth $y_c$ if:
  (a)  $Q = 55$ m³/s    (b)  $Q = 3.5$ m³/s

$b = 3\sqrt{y}$ m

1 m

$y$

10 m          10 m

**Fig. P10.22**

**10.23** A parabolic channel (Fig. P10.23) is described by the function $\eta = 0.1x^2$. If $Q = 25$ m³/s and $y = 2.00$ m, find the alternate depth.

$\eta$

$dA = 2x\,d\eta$

$d\eta$

$y$

$x$

**Fig. P10.23**

**10.24** Prepare a computer algorithm that determines the alternate depth, given the geometry of a prismatic channel, the discharge, and the depth of flow. Use the interval-halving technique or a solution of your choice.

**10.25** Derive the relation for a V-notch weir that expresses $Q$ as a function of $Y$, Eq. 10.4.27.

**10.26** Debris blocks the entire width of a river as shown in Fig. P10.26. The cross section of the river is approximately rectangular, where $b = 15$ m, $Q = 22.5$ m³/s, $y_0 = 1.2$ m, and $h = 0.2$ m.
  (a)  Neglecting losses, analyze the rapidly varied water surface in the vicinity of the debris.
  (b)  Sketch the energy grade line and water surface.

$y_0$

$Q$

$h$

① ② ③

**Fig. P10.26**

**10.27** Develop the expression for the theoretical discharge for the weir shown in Fig. P10.27.

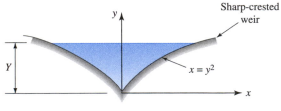

$y$

Sharp-crested weir

$Y$

$x = y^2$

$x$

**Fig. P10.27**

**10.28** A Parshall flume is placed in a small stream to measure the discharge. It has a width of 1 m at its throat.

  (a)  Compute the discharge if the measured upstream depth is 0.36 m.

  (b)  Prepare a plot of $Q$ versus $H$ where $H$ ranges from 0.15 m to 0.45 m.

**10.29** Water is flowing at a given depth and discharge in a 3.5-m-wide rectangular channel. A 60-cm-high submerged raised section is placed in the channel as shown in Fig. P10.29. The raised section spans the entire width of the channel, and is designed such that losses can be neglected. Calculate the water surface and energy grade line:

**(a)**   When $y = 1.5$ m and $Q = 4$ m³/s

**(b)**   When $y = 1.0$ m and $Q = 3$ m³/s
**(c)**   Under what given conditions, (a) or (b) above, would the raised section act similar to a broad crested weir (explain your answer in words only)?

**Fig. P10.29**

**10.30**  Design a broad-crested weir to convey a river discharge that varies between $Q_1$ and $Q_2$. The maximum water depth upstream of the weir is not to exceed $y_2$ and the minimum depth is not to be less than $y_1$.

$Q_1 = 0.15$ m³/s, $Q_2 = 30$ m³/s, $y_1 = 1.05$ m, $y_2 = 1.75$ m

**10.31**  Design a canal to deliver water between the two reservoirs shown in Fig. P10.31. The horizontal distance between the reservoirs is $L$. It is required that uniform flow exist throughout the reach. The water surface in reservoir A is to be no more than the distance $h$ above the bottom of the canal at the entrance. The canal is rectangular in cross section, and made of concrete. Neglect entrance and exit losses, and assume

uniform, subcritical flow conditions at the channel entrance.

$El_A = 501.8$ m, $El_B = 500.2$ m,
$L = 1500$ m, $h = 2$ m, $b = 2.5$ m

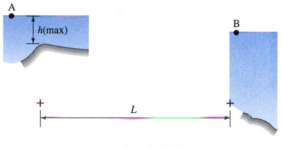

**Fig. P10.31**

---

## Momentum Concepts

**10.32**  The momentum function for a rectangular section, Eq. 10.5.4, can be made dimensionless by normalizing it with respect to $b(y_c)^2$. Prepare such a plot using $y/y_c$ as ordinate and $M/(by_c)^2$ as abscissa.

**10.33**  Rework Problem 10.24 to evaluate the conjugate depth in place of the alternate depth.

**10.34**  Flow occurs in a wide rectangular channel at a depth $y$. A construction project over this channel requires that cofferdams spaced at distance $w$ on centerlines be placed in the channel (Fig. P10.34). Assuming that the cofferdams create no energy losses between locations 1 and 2, determine the maximum permissible diameter of the cofferdams without creating upstream backwater effects (i.e., without increasing the upstream depth), and the

resultant drag force on each cofferdam (assume that $C_D = 0.15$).

$q = 1.5$ m²/s, $y = 1.8$ m, $w = 6$ m

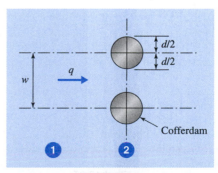

**Fig. P10.34**

**10.35** A temporary pipeline (diameter = 200 mm $C_D = 0.30$) is to be placed on a river bed normal to the direction of flow. The river is approximately rectangular in section, and is 100 m wide. The depth of flow downstream of the pipe is 2.5 m and the mean velocity is 3 m/s. Neglecting bed slope and resistance, determine the following:
   **(a)** Flow conditions (i.e., depth and velocity) upstream of the pipe once it is in place
   **(b)** Resultant drag force on the pipe

**10.36** A hydraulic jump occurs over a sill located in a triangular channel with water flowing as shown in Fig. P10.36. The inverse side slopes are $m_1 = m_2 = 3$, the drag coefficient $C_D = 0.40$, and the height of the sill $h = 0.3$ m. Determine the discharge $Q$ if $y_1 = 0.50$ m and $y_2 = 1.8$ m.

**Fig. P10.36**

**10.37** Water is flowing in a rectangular channel at a depth of 1.6 m and a velocity of 0.85 m/s. At a downstream location the discharge is suddenly reduced to zero, causing a surge to propagate upstream. Find the depth and velocity behind the surge, and the speed of the surge wave.

**10.38** In a rectangular channel, water is flowing at a depth $y$ and a velocity $V$. At a downstream location, the discharge is suddenly reduced by 60% (i.e., to 40% of the original value), causing a surge to propagate upstream. Determine the depth and velocity behind the surge, and the speed of the surge wave.
   $y = 1.5$ m, $V = 1$ m/s

**10.39** Water enters a reach of rectangular channel where $y_1 = 0.5$ m, $b = 7.5$ m, and $Q = 20$ m$^3$/s (Fig. P10.39). It is desired that a hydraulic jump occur upstream (location 2) of the sill and on the sill critical conditions exist (location 3). Other than across the jump, losses can be neglected. Determine the following:
   **(a)** Depths at locations 2 and 3
   **(b)** Required height of the sill, $h$
   **(c)** Resultant force acting on the sill
   **(d)** Sketch the water surface and energy grade line.

**(e)** Describe the nature and character of the jump.

**Fig. P10.39**

**10.40** Water is flowing as shown in Fig. P10.40 under the sluice gate in a horizontal rectangular channel that is 5 m wide. The depths $y_1$ and $y_2$ are 2.5 m and 10 cm, respectively. The horizontal distances between locations 1, 2, and 3 are sufficiently short that rapidly varied flow conditions can be assumed to occur. Determine the following:
   **(a)** The discharge
   **(b)** The depth downstream of the jump at location 3
   **(c)** The power lost in the hydraulic jump

**Fig. P10.40**

**10.41** A transition section is located at the entrance to a rectangular channel as shown in Fig. P10.41. At location 1, the depth is sufficiently large so that the velocity is negligible. If $b = 3$ m, Fr$_3 = 0.75$, and $Q = 5.55$ m$^3$/s, determine the following:
   **(a)** The water surface elevation relative to the datum at locations 1, 2, and 3. Sketch the water surface and energy grade lines between locations 1 and 3.
   **(b)** The resultant horizontal force acting on the channel bottom between locations 2 and 3

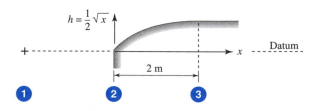

**Fig. P10.41**

**10.42** Water is flowing in a trapezoidal channel, $b = 5$ m, $m_1 = m_2 = 3$. A stationary hydraulic jump occurs, with the upstream depth $y_1 = 1.1$ m and the discharge $Q = 60$ m$^3$/s. Find the downstream depth $y_2$ and the power dissipated by the jump.

**10.43** A 4-m-wide rectangular channel contains an upward rectangular step whose width is the same as the channel. The given flow conditions are $y_1 = 0.5$ m, $V_1 = 8$ m/s, and $y_2 = 2$ m.
  **(a)** What is the height $h$ of the step if $C_D = 1.2$?
  **(b)** What would be $y_2$ if the step were not present?

**10.44** A rectangular channel (Fig. P10.44) has a sudden upward step of 0.17 m. A hydraulic jump occurs above the step. At location 2, downstream of the jump, the depth is $y_2 = 1.5$ m, and the Froude number there is Fr$_2 = 0.40$. If the width of the channel is 5 m and the drag coefficient on the step is $C_D = 0.35$, find the discharge $Q$ and the depth $y_1$ upstream of the jump.

**Fig. P10.44**

## Nonuniform Gradually Varied Flow

**10.45** The channel whose bottom profile is shown in Fig. P10.45 has a rectangular cross section, with $b = 8$ m, $n = 0.014$, and $S_0 = 0.004$. Determine the following:
  **(a)** The discharge in the channel
  **(b)** Sketch the water surface and energy grade line.
  **(c)** The possible existence of a hydraulic jump
  (*Hint:* Assume critical flow conditions at the channel entrance.)

**Fig. P10.45**

**10.46** Uniform flow occurs in a long rectangular channel of width $b$, $S_0 = 0.0163$ and $n = 0.012$. The channel is altered by raising the channel bottom by $\Delta z$ for a short distance. The alteration is intended to act as a "smooth transition region" so that losses can be considered negligible. Determine what changes take place due to the presence of the transition. In your analysis include all relevant rapidly varied flow calculations, identify any gradually varied flow profiles, and sketch the water surface and energy grade line.
$$Q = 0.35 \text{ m}^3/\text{s}, b = 1.80 \text{ m}, \Delta z = 100 \text{ mm}$$

**10.47** Rework Problem 10.46 if $S_0 = 0.0013$.

**10.48** Flow of water takes place in a rectangular channel with $b = 4$ m, $n = 0.012$, $L = 500$ m, $S_0 = 0.00087$. The discharge is 33 m$^3$/s. At the entrance to the channel, the depth is 0.68 m, and at the downstream end a free-outfall condition exists. Perform a profile synthesis to determine the nature of the water surface and energy grade line.

**10.49** A flow of 8.5 m$^3$/s occurs in a long rectangular channel 3 m wide with $y_0 = 1.54$ m. There is a smooth constriction in the channel to 1.8 m width.
  **(a)** Determine the depths to be expected in and just upstream of the constriction, neglecting losses. Show the solution on an $E-y$ diagram.
  **(b)** Classify the gradually varied flow profile upstream of the constriction.

**10.50** Water is discharging at 20 m$^3$/s in a triangular channel with $m_1 = 3.5$, $m_2 = 2.5$, $S_0 = 0.001$, $n = 0.014$. At the entrance to the channel the depth is 0.50 m, and at a distance 300 m downstream the depth is 2.5 m. Determine the nature of the water surface over the 300-m reach.

**10.51** A long channel has a change in bottom slope 500 m from its downstream end; upstream of the transition the slope is $S_{01} = 0.0003$, and downstream of it the slope is $S_{02} = 0.005$. The downstream reach of channel terminates at a reservoir whose elevation is 3 m above the channel bottom at that location. The upper reach is quite long, and very far upstream from the transition normal flow conditions exist. The channel section is trapezoidal, with a bottom width of 3 m, $m_1 = m_2 = 1.8$, $n = 0.012$, and a discharge of 17.5 m³/s. Determine the nature of the water surface and energy grade line from a location far upstream of the transition to the end of the channel.

**10.52**  A rectangular channel with $b = 4$ m has a change in slope such that the normal depths are $y_{01} = 0.93$ m and $y_{02} = 1.42$ m. At locations far upstream and far downstream of the transition, flow occurs at the respective uniform depths that serve as controls. The discharge is 15 m³/s. Find the variation of the water surface and energy grade line throughout the region.

**10.53**  The partial water profile shown in Fig. P10.53 is for a rectangular channel of width $b = 3$ m, in which water is flowing at a discharge of $Q = 5$ m³/s.
  **(a)**  Does a hydraulic jump occur in the channel? If so, is it located upstream or downstream of location A?
  **(b)**  Sketch the water surface and energy grade line, and identify any known water surface profiles.

$y_0 = 0.4$ m
(uniform depth)

A

1.6 m

(horizontal slope)

**Fig. P10.53**

**10.54**  A very long rectangular channel (Fig. P10.54) has normal flow conditions at locations A and C, and a change in the bottom slope occurs at location B.
  **(a)**  Somewhere between locations A and C a hydraulic jump will take place. Explain why this statement is true.
  **(b)**  Classify and sketch the two possible water surface profiles that exist between locations A and C.
  **(c)**  By appropriate reasoning and calculation, determine whether the jump will occur upstream of location B, or downstream of location B.

**10.55**  A long rectangular channel (Fig. P10.55) has normal flow conditions at locations A and D, respectively, far upstream and downstream of location C. At C a change in slope occurs as shown.
  **(a)**  Classify the water surface, and sketch the water surface and the energy grade line between A and D.
  **(b)**  Determine the height $h$ of a transition such that normal flow conditions occur between A and B, which is a short distance upstream of A.
  **(c)**  Explain what type of profile now exists downstream of B.

$y_{01} = 1.0$ m        $y_{02} = 1.8$ m        $q = 5.75$ m²/s

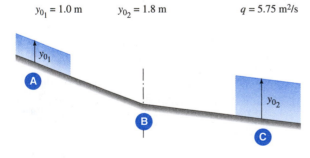

$y_{01}$

A

B

$y_{02}$

C

**Fig. P10.54**

$y_{01} = 1.6$ m        $y_{02} = 0.8$ m        $q = 4$ m²/s

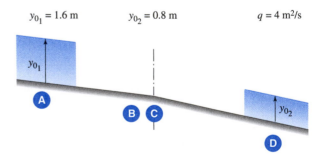

$y_{01}$

A

B  C

$y_{02}$

D

**Fig. P10.55**

**10.56**  Water is flowing at a discharge $Q$ in a circular conduit of diameter $d$ and length $L$, with $S_0 = 0.001$, and $n = 0.015$. At the inlet the water depth is $y_0$ and at the outlet a free overfall exists. Determine the nature of the water surface in the conduit.

$$Q = 2.5 \text{ m}^3/\text{s}, L = 500 \text{ m}, d = 2.5 \text{ m},$$
$$y_0 = 0.4 \text{ m}$$

**10.57**  With the results from Problem 10.51, compute the water surface profile and energy grade line using a numerical method of your choice.

**10.58**  A rectangular channel ($b = 5$ m, $S_0 = -0.001$) conveys flow at a rate of $Q = 3$ m$^3$/s. At a given location the depth is $y_1 = 2$ m and downstream of that location the depth is $y_2 = 1.8$ m. The Manning coefficient is $n = 0.02$.

   **(a)**  Determine the distance between locations 1 and 2.

   **(b)**  What type of water surface profile exists in the reach?

**10.59**  A rectangular channel (Fig. P10.59) has the following properties: $b = 1$ m, $S_0 = 0.0002$, $n = 0.017$. A 120° V-notch weir is placed in the channel to measure the discharge, with the bottom notch of the weir located 0.5 m above the channel floor. Upstream of the weir the depth $Y$ is measured to be 0.37 m.

   **(a)**  What is the discharge in the channel?

   **(b)**  If uniform flow conditions exist very far upstream of the weir, classify the nature of the water surface.

$Y$

0.5 m

$b = 1$ m

**Fig. P10.59**

**10.60**  Gradually varied flow occurs over a reach of rectangular channel with $n = 0.013$, $S_0 = 0.005$, and $b = 2.5$ m. At location 1 the depth is 1.05 m and at location 2 the depth is 1.2 m. If the two locations are 50 m apart, find the discharge and identify the profile type.

**10.61**  Over a portion of a very long rectangular channel of width $b$ (Fig. P10.61), uniform flow occurs with discharge $Q$ and depth $y_0$. At the end of the channel the flow is terminated by a free outfall. In addition, at the outlet the channel width is reduced to $b_1$ m for a very short distance. Determine and describe as completely as possible the changes that take place in the water surface between locations A and B. Use an $E$–$y$ diagram for illustration.

$$Q = 5.5 \text{ m}^3/\text{s}, y_0 = 0.5 \text{ m}, b = 3 \text{ m},$$
$$b_1 = 1.5 \text{ m}.$$

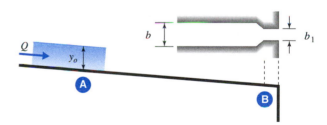

$Q$

$y_o$

A

$b$

$b_1$

B

**Fig. P10.61**

**10.62**  A sluice gate is placed in a long rectangular channel, $b = 4$ m, $n = 0.014$, and $S_0 = 0.0008$. The depth upstream of the gate is $y_1 = 1.85$ m and downstream of the gate the depth is $y_2 = 0.35$ m. Neglecting losses, identify and compute the water surface profiles on either side of the gate.

**10.63** With the following data,[7] identify the profiles and plot the water surface and energy grade line to scale.

| $Q = 3.5$ | $n = 0.011$ | $S_0 = 0.001$ | $L = 200$ | $g = 9.81$ |
|---|---|---|---|---|
| $b = 5$ | $m = 2.5$ | $y_u = 0.80$ | $y_d = 2.00$ | $c_1 = 1.00$ |

|  | Depth | Residual |
|---|---|---|
| $y_c =$ | 1.356 | $-3.44$E-07 |
| $y_0 =$ | 1.442 | $-4.23$E-07 |

| Station | $y$ | $A$ | $V$ | $E$ | $y_m$ | $S(y_m)$ | $\Delta x$ | $x$ | $y_{cj}$ | Residual |
|---|---|---|---|---|---|---|---|---|---|---|
| 1 | 0.800 | 5.600 | 6.250 | 2.791 | | | | | | |
| 2 | 0.850 | 6.056 | 5.779 | 2.552 | 0.825 | 8.311E-03 | 33 | 33 | 2.044 | $-5.98$E-07 |
| 3 | 0.900 | 6.525 | 5.364 | 2.366 | 0.875 | 6.689E-03 | 33 | 65 | 1.961 | $-3.09$E-07 |
| 4 | 0.950 | 7.006 | 4.996 | 2.222 | 0.925 | 5.443E-03 | 33 | 98 | 1.883 | 8.24E-07 |
| 5 | 1.000 | 7.500 | 4.667 | 2.110 | 0.975 | 4.472E-03 | 32 | 130 | 1.809 | $-7.46$E-07 |
| 6 | 1.050 | 8.006 | 4.372 | 2.024 | 1.025 | 3.706E-03 | 32 | 162 | 1.740 | $-9.44$E-08 |
| 7 | 1.100 | 8.525 | 4.106 | 1.959 | 1.075 | 3.097E-03 | 31 | 193 | 1.674 | 2.14E-07 |
| 8 | 1.150 | 9.056 | 3.865 | 1.911 | 1.125 | 2.606E-03 | 30 | 223 | 1.611 | 4.94E-07 |
| | | | | | | | | | | |
| 7 | 2.000 | 20.000 | 1.750 | 2.156 | | | | 200 | | |
| 8 | 1.975 | 19.627 | 1.783 | 2.137 | 1.988 | 2.770E-04 | $-26$ | 174 | | |
| 9 | 1.950 | 19.256 | 1.818 | 2.118 | 1.963 | 2.916E-04 | $-26$ | 147 | | |
| 10 | 1.925 | 18.889 | 1.853 | 2.100 | 1.938 | 3.073E-04 | $-27$ | 121 | | |
| 11 | 1.900 | 18.525 | 1.889 | 2.082 | 1.913 | 3.240E-04 | $-27$ | 94 | | |
| 12 | 1.875 | 18.164 | 1.927 | 2.064 | 1.888 | 3.417E-04 | $-27$ | 67 | | |
| 13 | 1.850 | 17.806 | 1.966 | 2.047 | 1.863 | 3.607E-04 | $-27$ | 40 | | |
| 14 | 1.825 | 17.452 | 2.006 | 2.030 | 1.838 | 3.810E-04 | $-27$ | 13 | | |

---

[7]The data was generated by Excel spreadsheet.

**10.64** With the following data,[8] identify the profiles and plot the water surface and energy grade line to scale.

**(a)**

| $Q =$ 3.5 | $n =$ 0.013 | $S_0 =$ 0.005 | $L =$ 100 | $g =$ 9.81 |
|---|---|---|---|---|
| $b =$ 2.5 | $m =$ 0 | $y_u =$ 0.30 | $y_d =$ 0.95 | $c_1 =$ 1.00 |

|  | Depth | Residual |
|---|---|---|
| $y_c =$ | 0.585 | −8.58E-07 |
| $y_0 =$ | 0.508 | −1.42E-07 |

| Station | $y$ | $A$ | $V$ | $E$ | $y_m$ | $S(y_m)$ | $\Delta x$ | $x$ | $y_{cj}$ | Residual |
|---|---|---|---|---|---|---|---|---|---|---|
| 1 | 0.300 | 0.750 | 4.667 | 1.410 |  |  |  |  |  |  |
| 2 | 0.325 | 0.813 | 4.308 | 1.271 | 0.313 | 2.154E-02 | 8 | 8 | 0.985 | 6.45E-07 |
| 3 | 0.350 | 0.875 | 4.000 | 1.165 | 0.338 | 1.702E-02 | 9 | 17 | 0.932 | −4.60E-07 |
| 4 | 0.375 | 0.938 | 3.733 | 1.085 | 0.363 | 1.369E-02 | 9 | 26 | 0.884 | 6.34E-07 |
| 5 | 0.400 | 1.000 | 3.500 | 1.024 | 0.388 | 1.119E-02 | 10 | 36 | 0.840 | 3.55E-07 |
| 6 | 0.425 | 1.063 | 3.294 | 0.978 | 0.413 | 9.272E-03 | 11 | 47 | 0.799 | 2.00E-07 |
| 7 | 0.450 | 1.125 | 3.111 | 0.943 | 0.438 | 7.774E-03 | 13 | 60 | 0.762 | 4.40E-07 |
| 8 | 0.475 | 1.188 | 2.947 | 0.918 | 0.463 | 6.587E-03 | 16 | 76 | 0.727 | 9.05E-07 |
| 9 | 0.500 | 1.250 | 2.800 | 0.900 | 0.488 | 5.635E-03 | 29 | 104 | 0.694 | 2.22E-07 |
| 10 | 0.950 | 7.006 | 0.500 | 0.963 |  |  |  | 100 |  |  |
| 11 | 0.850 | 6.056 | 0.578 | 0.867 | 0.900 | 9.698E-04 | −24 | 76 |  |  |
| 12 | 0.800 | 5.600 | 0.625 | 0.820 | 0.825 | 1.236E-03 | −13 | 64 |  |  |
| 13 | 0.750 | 5.156 | 0.679 | 0.773 | 0.775 | 1.474E-03 | −13 | 51 |  |  |
| 14 | 0.700 | 4.725 | 0.741 | 0.728 | 0.725 | 1.781E-03 | −14 | 36 |  |  |
| 15 | 0.650 | 4.306 | 0.813 | 0.684 | 0.675 | 2.183E-03 | −16 | 21 |  |  |
| 16 | 0.600 | 3.900 | 0.897 | 0.641 | 0.625 | 2.725E-03 | −19 | 2 |  |  |

**(b)**

| $Q =$ 125 | $n =$ 0.013 | $S_0 =$ 0.005 | $L =$ 300 | $g =$ 32.2 |
|---|---|---|---|---|
| $b =$ 8 | $m =$ 0 | $y_u =$ 1.00 | $y_d =$ 3.00 | $c_1 =$ 1.49 |

|  | Depth | Residual |
|---|---|---|
| $y_c =$ | 1.965 | −1.04E-07 |
| $y_0 =$ | 1.710 | −4.30E-04 |

| Station | $y$ | $A$ | $V$ | $E$ | $y_m$ | $S(y_m)$ | $\Delta x$ | $x$ | $y_{cj}$ | Residual |
|---|---|---|---|---|---|---|---|---|---|---|
| 1 | 1.000 | 8.000 | 15.625 | 4.791 |  |  |  |  |  |  |
| 2 | 1.100 | 8.800 | 14.205 | 4.233 | 1.050 | 2.155E-02 | 34 | 34 | 3.311 | −3.50E-07 |
| 3 | 1.200 | 9.600 | 13.021 | 3.833 | 1.150 | 1.634E-02 | 35 | 69 | 3.102 | 8.20E-07 |
| 4 | 1.300 | 10.400 | 12.019 | 3.543 | 1.250 | 1.269E-02 | 38 | 107 | 2.914 | 5.79E-07 |
| 5 | 1.400 | 11.200 | 11.161 | 3.334 | 1.350 | 1.007E-02 | 41 | 148 | 2.744 | 3.35E-07 |
| 6 | 1.500 | 12.000 | 10.417 | 3.185 | 1.450 | 8.135E-03 | 48 | 195 | 2.589 | 2.31E-07 |
| 7 | 1.600 | 12.800 | 9.766 | 3.081 | 1.550 | 6.674E-03 | 62 | 258 | 2.447 | 3.00E-07 |
| 8 | 1.700 | 13.600 | 9.191 | 3.012 | 1.650 | 5.549E-03 | 126 | 384 | 2.317 | 2.58E-07 |
| 9 | 3.000 | 37.500 | 3.333 | 3.173 |  |  |  | 300 |  |  |
| 10 | 2.800 | 33.600 | 3.720 | 3.015 | 2.900 | 1.105E-03 | −40 | 260 |  |  |
| 11 | 2.600 | 29.900 | 4.181 | 2.871 | 2.700 | 1.349E-03 | −39 | 220 |  |  |
| 12 | 2.400 | 26.400 | 4.735 | 2.748 | 2.500 | 1.674E-03 | −37 | 183 |  |  |
| 13 | 2.200 | 23.100 | 5.411 | 2.655 | 2.300 | 2.121E-03 | −32 | 151 |  |  |
| 14 | 2.000 | 20.000 | 6.250 | 2.607 | 2.100 | 2.751E-03 | −21 | 129 |  |  |
| 15 | 1.965 | 19.478 | 6.417 | 2.605 | 1.983 | 3.247E-03 | −1 | 128 |  |  |

---

[8]The data was generated by Excel spreadsheet.

**10.65** Consider a backwater curve situated upstream of a dam in a rectangular reservoir (Fig. P10.65). Given data are: normal depth $y_0 = 3.92$ m, $Q = 125$ m$^3$/s, $S_0 = 0.0004$, $n = 0.025$, $y_A = 11$ m, $y_B = 12$ m

**(a)** Identify the type of profile that exists in the reservoir.
**(b)** Compute the distance between A and B.
**(c)** Sketch the water surface and energy grade line.

**Fig. P10.65**

**10.66** A rectangular channel has a change in slope as shown in Fig. P10.66. The channel is 3.66 m wide, with $n = 0.017$, $S_{02} = 0.00228$, and $Q = 15.38$ m$^3$/s.

**(a)** Determine the depth that must exist in the downstream channel for a hydraulic jump to terminate at uniform flow conditions.

**(b)** If $y_{01} = 0.6$ m, calculate the length to the jump, $L_j$, using several increments of depth in a step calculation.

**(c)** Sketch the water surface and energy grade line, and identify all gradually varied flow profiles.

**10.67** In the three channels shown in Fig. P10.67, normal depths and critical depths are indicated by dashed (- - -) and dotted ($\cdot \cdot$) lines, respectively. Sketch one possible composite profile for each system, and identify all gradually varied surfaces.

(a)

(b)

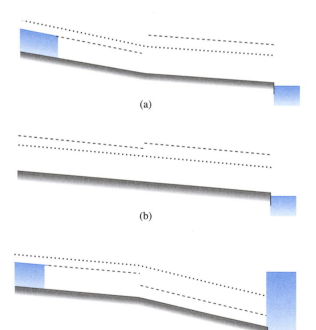

(c)

**Fig. P10.67**

**Fig. P10.66**

**10.68** A design flow of $Q = 19$ m³/s occurs in a river that can be approximated as a wide rectangular channel with the following properties: $b = 20$ m, $S_0 = 0.0005$, $n = 0.016$. A small dam is to be placed in the channel as part of a flood control plan (Fig. P10.68). The height of the dam is $h = 6$ m, and the depth of flow at the toe of the dam is $y_{toe} = 0.10$ m. A concrete spillway apron ($n = 0.014$) is to be situated downstream of the dam. The purpose of the apron is to keep any hydraulic jump that may occur downstream of the dam off the river bed for the given design conditions to prevent erosion of the river bed. The depth behind the jump at the end of the apron is $y_2 = 0.75$ m. Determine the following:

(a) Power in kilowatts dissipated by the jump
(b) Design length of the apron
(c) The gradually-varied-flow profiles upstream of the dam, and between the toe and the hydraulic jump

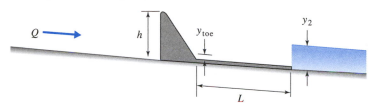

**Fig. P10.68**

**10.69** A very wide rectangular channel has the following properties: $S_0 = 0.0005$, $n = 0.017$, $q = 1.2$ m²/s. At some location, say location A, the depth is 0.65 m, and at location B, the depth is 0.90 m.

(a) Is location B upstream or downstream of location A?
(b) Using the varied flow function determine the distance between locations A and B.

**10.70** An estuary to a river empties to the ocean. The estuary can be approximated as a very wide channel with a depth of 7 m at its outlet, and with $S_0 = 0.0001$, $n = 0.015$. If the discharge into the ocean from the estuary is $q = 1.5$ m³/s per meter width, determine the distances upstream from the ocean where the depth is equal to 6, 5, 4, and 2 m.

**10.71** A river cross section can be approximated by two rectangles as shown in Fig. P10.71. The channel properties and dimensions are: $b_1 = 150$ m, $z_1 = 4$ m, $n_1 = 0.03$, $b_2 = 5$ m, $z_2 = 0$ m, and $n_2 = 0.02$. If the slope of the energy grade line is $S = 0.0005$ and the depth $y = 5$ m, find:

(a) The energy coefficient for the composite section
(b) The discharge

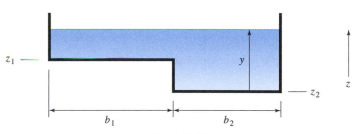

**Fig. P10.71**

**10.72** Consider the same river cross section of Problem 10.71. In the accompanying table, data are provided for two locations $x$ along the river:

At the downstream location $x = 400$ m, the depth $y = 3.0$ m when the discharge $Q = 280$ m³/s. Using a trial-and-error procedure, determine the depth $y$ at the upstream location $x = 0$.

| $x$ (m) | $b_1$ (m) | $b_2$ (m) | $z_1$ (m) | $z_2$ (m) | $n_1$ | $n_2$ |
|---|---|---|---|---|---|---|
| 0 | 115 | 107 | 16.7 | 15.0 | 0.05 | 0.03 |
| 400 | 149 | 91 | 17.7 | 15.1 | 0.05 | 0.03 |

In a complex of industrial piping such as this one, engineering design requires that steady flows be analyzed for correct pipe sizing and placement of pumps. In addition, analysis is occasionally undertaken to mitigate unsteady, or transient, excitations to the system. (Marafona/Shutterstock)

# 11

# Flows in Piping Systems

## Outline

## Chapter Objectives

The objectives of this chapter are to:

▲ Compare empirical frictional loss equations in piping

▲ Introduce the concept of hydraulic grade line as a variable for piping analysis

▲ Describe ad hoc methods of computing discharges and pressures in simple pipe systems

▲ Present linearized solutions (i.e., Hardy Cross method) for pipe networks using spreadsheet analysis

▲ Show use of computational software for pipe network analysis

▲ Introduce simplified analysis for unsteady flows in simple pipe systems

## 11.1 INTRODUCTION

Internal flows in pipelines and ducts are commonly encountered in all parts of our industrialized society. From delivering potable water to transporting chemicals and other industrial liquids, engineers have designed and constructed untold kilometers of relatively large-scale piping systems. Smaller piping units are also in abundance: in hydraulic controls, in heating and air conditioning systems, and in cardiovascular and pulmonary flow systems, to name only a few. These flows can be either steady or unsteady, uniform or nonuniform. The fluid can be either incompressible or compressible, and the piping material can be elastic, inelastic, or perhaps, viscoelastic. In this chapter we concentrate primarily on incompressible, steady flows in rigid piping. The piping system may be relatively simple, such that the variables can be solved rather easily using a calculator, or it may be sufficiently complicated that it is more convenient to use computational software.

Piping systems are considered to be composed of elements and components. Basically, pipe **elements** are reaches of constant-diameter piping, and the **components** consist of valves, tees, bends, reducers, or any other device that may create a loss to the system. In addition to components and elements, pumps add energy to the system and turbines extract energy. The elements and components are linked at junctions. Fig. 11.1 illustrates several types of piping systems.

Following a discussion of piping losses, several pipe systems, including series, branch, and parallel configurations, are analyzed. Attention is then directed to comprehensive network systems, where several methods of solution are presented. Most of the piping problems analyzed are those where the discharge is the unknown variable; this type of problem is classified as category 2 in Section 7.6.3.

Finally, a brief introduction to unsteady flow in pipelines is presented. This topic is one that has been important for many years. It is taking on increasing importance as the demand for the construction of more cost-effective piping continues and the manner in which piping, valves, pumps, and other components interact and become more sophisticated. We address only the fundamental aspects of unsteady flows, focusing on two assumptions in a single pipe of constant diameter: an incompressible flow in an inelastic pipe, and a compressible liquid flow in an elastic pipe.

**Elements:** *Reaches of constant-diameter piping.*

**Components:** *Valves, tees, bends, reducers, or any other device that creates a loss to the system.*

## 11.2 LOSSES IN PIPING SYSTEMS

Losses can be divided into two categories: (a) those due to wall shear in pipe elements and (b) those due to piping components. The former are distributed along the length of pipe elements. The latter are treated as discrete discontinuities in the hydraulic and energy grade lines and are commonly referred to as minor losses; they are due primarily to separated or secondary flows.

The fundamental mechanics of wall shear and development of the empirical relations relating to pipe losses are treated in Chapter 7. Minor losses are covered in detail in Section 7.6.4 and are not discussed further here. The following material focuses on the treatment of losses in the analysis of piping systems.

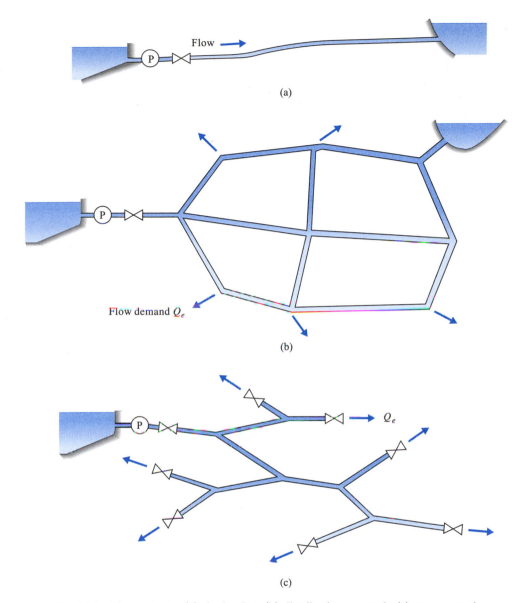

**Fig. 11.1**   Pipe systems: (a) single pipe; (b) distribution network; (c) tree network.

## 11.2.1   Frictional Losses in Pipe Elements

It is convenient to express the pipe element frictional loss in the exponential form

$$h_L = RQ^\beta \qquad (11.2.1)$$

in which $h_L$ is the head loss over length $L$ of pipe, $R$ is the *resistance coefficient*, $Q$ is the discharge in the pipe, and $\beta$ is an exponent. Depending on the formulation chosen, the resistance coefficient may be a function of pipe roughness, Reynolds number, or length and diameter of the pipe element. In particular, the Darcy–Weisbach relation, Eq. 7.6.23, can be substituted into Eq. 11.2.1. Then $\beta = 2$, and the resulting expression for $R$ is

$$R = \frac{fL}{2gDA^2}$$

$$= \frac{8fL}{g\pi^2 D^5} \tag{11.2.2}$$

**KEY CONCEPT** *Frictional losses in piping are commonly evaluated using the Darcy–Weisbach or Hazen–Williams equation. The Darcy–Weisbach formulation provides a more accurate estimation.*

where $f$ is the friction factor. The characteristics of $f$ for commercial pipe flows are developed in Section 7.6.3. Specifically, the Moody diagram, Fig. 7.13, presents a comprehensive picture of just how the friction factor varies over a wide range of Reynolds numbers and for a range of relative roughnesses. For pipe network analysis, it is convenient to express the behavior of $f$ using approximate, equivalent empirical formulas in which the friction factor can be obtained directly in terms of the Reynolds number and relative roughness. A number of relations have been developed and shown to be reasonably accurate for engineering calculations (Benedict, 1980). In particular, the formulas of Swamee and Jain (1976) are presented in Section 7.6.3 and are shown to accurately represent the Colebrook relation, Eq. 7.6.28. The friction factor formula developed by Swamee and Jain is

$$f = 1.325 \left\{ \ln\left[ 0.27 \left(\frac{e}{D}\right) + 5.74 \left(\frac{1}{\mathrm{Re}}\right)^{0.9} \right] \right\}^{-2} \tag{11.2.3}$$

Combining Eqs. 11.2.2 and 11.2.3, one finds that

$$R = 1.07 \left(\frac{L}{gD^5}\right) \left\{ \ln\left[ 0.27 \left(\frac{e}{D}\right) + 5.74 \left(\frac{1}{\mathrm{Re}}\right)^{0.9} \right] \right\}^{-2} \tag{11.2.4}$$

Equations 11.2.3 and 11.2.4 are valid over the ranges $0.01 > e/D > 10^{-8}$, and $10^8 > \mathrm{Re} > 5000$. The fully rough regime, where Re has a negligible effect on $f$, begins at a Reynolds number given by

$$\mathrm{Re} = \frac{200D}{e\sqrt{f}} \tag{11.2.5}$$

For values of Re greater than this, the friction factor is a function only of $e/D$, and is given by

$$f = 1.325 \left\{ \ln\left[ 0.27 \left(\frac{e}{D}\right)\right] \right\}^{-2} \tag{11.2.6}$$

Two additional expressions for pipe frictional losses, which have found wide use, are the Hazen–Williams and Chezy–Manning formulas. For water flow, the value of $R$ in Eq. 11.2.1 for the Hazen–Williams relation is

$$R = \frac{K_1 L}{C^{\beta} D^m} \qquad K_1 = 10.59 \tag{11.2.7}$$

in which the exponents are $\beta = 1.85$ and $m = 4.87$, and $C$ is the Hazen–Williams coefficient dependent only on the roughness. Values of the Hazen–Williams roughness coefficient $C$ are given in Table 11.1.

The Chezy–Manning equation is more commonly associated with open-channel flow. However, in sewage and drainage systems in particular, it has been applied to conduits flowing under *surcharge*, that is, under pressurized conditions. The Chezy–Manning equation was introduced in Section 7.7. For a circular pipe flowing full, Eq. 7.7.6 can be substituted into Eq. 11.2.1 and solved for $R$:

$$R = \frac{10.29 n^2 L}{K_2 D^{5.33}} \qquad K_2 = 1 \tag{11.2.8}$$

in which $n$ is the Manning roughness coefficient. In Eq. 11.2.1, the exponent $\beta = 2$.

An advantage to using Eq. 11.2.7 or Eq. 11.2.8 as opposed to Eq. 11.2.4 is that in the first two, $C$ and $n$ are dependent on roughness only, while in the last, $f$ depends on the Reynolds number as well as the relative roughness. However, Eq. 11.2.4 is recommended since it provides a more precise representation of pipe frictional losses. Note that the Hazen–Williams and Chezy–Manning relations are dimensionally inhomogeneous,[1] whereas the Swamee–Jain equation is dimensionally homogeneous and contains the two parameters $e/D$ and Re, which appropriately influence the losses.

**Table 11.1**    Nominal Values of the Hazen–Williams Coefficient $C$

| Type of pipe | C |
|---|---|
| Extremely smooth; asbestos-cement | 140 |
| New or smooth cast iron; concrete | 130 |
| Wood stave; newly welded steel | 120 |
| Average cast iron; newly riveted steel; vitrified clay | 110 |
| Cast iron or riveted steel after some years of use | 95–100 |
| Deteriorated old pipes | 60–80 |

---

[1] In a dimensionally inhomogeneous equation, the constants in the equation have units assigned to them. In a dimensionally homogeneous equation the constants are dimensionless.

The limitations of the Hazen–Williams and Chezy–Manning formulas are demonstrated as follows. Beginning with Eq. 11.2.1, the head loss $h_L$ based on the Darcy–Weisbach resistance coefficient, Eq. 11.2.2, with the exponent $\beta = 2$ can be equated with $h_L$ based on the Hazen–Williams coefficient, Eq. 11.2.7, with $\beta = 1.85$. Introducing the Reynolds number to eliminate $Q$ and solving for the friction factor $f$ results in

$$f = \frac{1.28 g K_1}{C^{1.85} D^{0.02} (\mathrm{Re}\ \nu)^{0.15}} \tag{11.2.9}$$

With SI units and for water at 20°C, Eq. 11.2.9 reduces to

$$f = \frac{1056}{C^{1.85} D^{0.02} \mathrm{Re}^{0.15}} \tag{11.2.10}$$

Note that $f$ is weakly dependent upon $D$ in this equation. In a similar fashion, Eq. 11.2.8 can be substituted into Eq. 11.2.1 with $\beta = 2$ and equated to $h_L$ based on the Darcy–Weisbach formulation to yield

$$f = \frac{124.5 n^2}{D^{0.33}} \tag{11.2.11}$$

Comparisons between the original Colebrook formula (Eq. 7.6.28), the formula of Swamee and Jain (Eq. 11.2.3), and the equivalent expressions for the Hazen–Williams and Manning coefficients (Eqs. 11.2.10 and 11.2.11) are shown in Fig. 11.2 for a 1-m-diameter concrete pipe flowing full with water ($C = 130$,

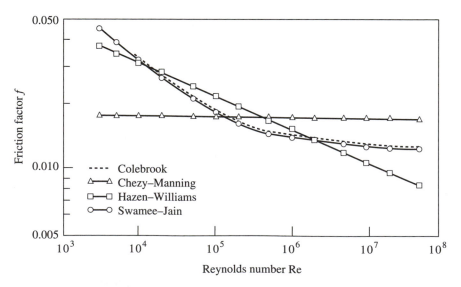

**Fig. 11.2** Comparison of several approximate formulas with the original Colebrook formula.

$n = 0.012$, $e/D = 0.00015$). It is evident that the Hazen–Williams and Chezy–Manning relations are valid over a limited range of Re, and that Eq. 11.2.3 provides a more versatile and accurate estimate of pipe losses. In the remainder of this chapter most of the frictional losses will be treated using the Darcy–Weisbach formulation, Eqs. 11.2.2 to 11.2.6. Any exceptions will be noted.

## Example 11.1

A pipeline is conveying 0.05 $m^3$/s of water, at 30°C. The length of the line is 300 m, and the diameter is 0.25 m. Estimate the head loss due to friction using the three formulas (a) Darcy–Weisbach ($e = 0.5$ mm), (b) Hazen–Williams ($C = 110$), and (c) Chezy–Manning ($n = 0.012$).

### Solution

The kinematic viscosity of water at 30°C is $\nu = 0.804 \times 10^{-6}$ $m^2$/s. The relative roughness, velocity, and Reynolds number are computed to be

$$\frac{e}{D} = 0.002 \qquad V = 1.02 \text{ m/s} \qquad \text{Re} = 3.17 \times 10^5$$

(a)  Substitute values into Eq. 11.2.4:

$$R = 1.07 \frac{300}{9.81 \times (0.25)^5}$$

$$\times \{\ln[0.27 \times 0.002 + 5.74 (3.17 \times 10^5)^{-0.9}]\}^{-2} = 610$$

Then, with Eq. 11.2.1, the friction loss based on the Darcy–Weisbach formulation is

$$h_L = RQ^2$$

$$= 610 (0.05)^2 = 1.52 \text{ m}$$

(b)  For the Hazen–Williams formulation, use Eq. 11.2.7:

$$R = \frac{10.59 \times 300}{(110)^{1.85}(0.25)^{4.87}} = 454$$

$$h_L = 454 (0.05)^{1.85} = 1.78 \text{ m}$$

(c)  Use Eq. 11.2.8 for the Chezy–Manning formulation:

$$R = \frac{10.29(0.012)^2(300)}{1 \times (0.25)^{5.33}} = 719$$

$$h_L = 719 (0.05)^2 = 1.80 \text{ m}$$

Part (a) with $h_L = 1.52$ m is the most accurate. The Hazen–Williams result is 17% too high, and the Chezy–Manning result is 18% too high.

## 11.3 SIMPLE PIPE SYSTEMS

The analysis of a single pipeline is presented in Section 7.6.4; it would be appropriate here for the reader to review the three categories of pipe problems in that section. For more complex piping systems, the methodology is similar. With relatively simple networks, such as series, parallel, and branching systems, ad hoc solutions can be developed that are suitable for use with calculators, spreadsheet algorithms, or computational software. Such approaches are relevant since they make use of the solver's ingenuity and require an understanding of the nature of the flow and piezometric head distributions for the particular piping arrangement. For systems of greater complexity, an alternative means of solving such problems is the Hardy Cross method, presented in Section 11.4.

The fundamental principle in the ad hoc approach is to identify all of the unknowns and write an equivalent number of independent equations to be solved. Subsequently, the system is simplified by eliminating as many unknowns as possible and reducing the problem to a series of single pipe problems—either category 1 or category 2, Section 7.6.3.

### 11.3.1  Series Piping

Consider the series system shown in Fig. 11.3. It consists of $N$ pipe elements with a specified number of minor-loss components $\Sigma K$ associated with each $i$th pipe element. A single minor loss is described in Section 7.6.5 to be equal to $h_L = KV^2/2g$. It is convenient here to express the minor loss in terms of the discharge rather than the velocity, so that $h_L = KQ^2/2gA^2$. For many flow situations, it is common practice to neglect the kinetic-energy terms at the inlet and outlet; they would be significant only if the velocities were relatively high. Assuming that the exponent in Eq. 11.2.1 is $\beta = 2$, the energy equation applied from location A to location B of Fig. 11.3 is

$$\left(\frac{p}{\gamma}+z\right)_A - \left(\frac{p}{\gamma}+z\right)_B = \left(R_1 + \frac{\Sigma K}{2gA_1^2}\right)Q_1^2 + \left(R_2 + \frac{\Sigma K}{2gA_2^2}\right)Q_2^2$$

$$+ \cdots + \left(R_N + \frac{\Sigma K}{2gA_N^2}\right)Q_N^2$$

$$= \sum_{i=1}^{N}\left(R_i + \frac{\Sigma K}{2gA_i^2}\right)Q_i^2 \qquad (11.3.1)$$

in which $R_i$ is the resistance coefficient for pipe $i$.

**Fig. 11.3**  Series piping system.

The statement of continuity for the series system is that the discharge in every element is identical, or

$$Q_1 = Q_2 = \cdots = Q_i = \cdots = Q_N = Q \qquad (11.3.2)$$

Replacing $Q_i$ with $Q$, Eq. 11.3.1 becomes

$$\left(\frac{p}{\gamma} + z\right)_A - \left(\frac{p}{\gamma} + z\right)_B = \left[\sum_{i=1}^{N}\left(R_i + \frac{\Sigma K}{2gA_i^2}\right)\right]Q^2 \qquad (11.3.3)$$

Equation 11.2.4 can be substituted for $R_i$, or alternatively, one may use Eq. 11.2.3 or the Moody diagram, Fig. 7.13, to obtain values of the friction factor. The reader should recognize that in a series system, the discharge remains constant from one pipe element to another, and the losses are accumulative; that is, they are the sum of the minor loss components and the pipe frictional losses.

For a category 1 problem, the right-hand side of Eq. 11.3.3 is known and the solution is straightforward. For a category 2 problem, in which $Q$ is unknown, a trial-and-error solution is required, since the Reynolds number, in terms of the unknown discharge (Re $= 4Q/\pi v D$), is present in the friction factor relation. Note that if flow in the fully rough zone is assumed, $f$ is independent of $Q$ and Eq. 11.3.3 reduces to a quadratic in $Q$. A category 3 problem is not encountered in this type of analysis. A series piping system with minor losses and constant-diameter piping is applied to a category 1 problem in Example 7.13. The solution for a somewhat more complex series system is illustrated in Example 11.2.

## Example 11.2

For the system shown in Fig. E11.2, find the required power to pump 100 L/s of liquid ($S = 0.85$, $v = 10^{-5}$ m²/s). The pump is operating at an efficiency $\eta = 0.75$. Pertinent data are given in the figure.

Line 1:   $L = 10$ m, $D = 0.20$ m, $e = 0.05$ mm, $K_1 = 0.5$, $K_v = 2$
Line 2:   $L = 500$ m, $D = 0.25$ m, $e = 0.05$ mm, $K_e = 0.25$, $K_2 = 1$

**Fig. E11.2**

*(continued)*

**Solution**

This is a category 1 problem. The energy relation, Eq. 11.3.3, for the system is

$$\left(\frac{p}{\gamma} + z\right)_A + H_P = \left(\frac{p}{\gamma} + z\right)_B + \left[R_1 + \frac{K_1 + K_v}{2gA_1^2} + R_2 + \frac{2K_e + K_2}{2gA_2^2}\right]Q^2$$

The resistance coefficients, $R_1$ and $R_2$, are calculated with Eq. 11.2.4 after first evaluating Re and $e/D$:

$$\text{Re}_1 = \frac{4Q}{\pi D_1 \nu} = \frac{4 \times 0.10}{\pi \times 0.20 \times 10^{-5}} = 6.37 \times 10^4 \qquad \left(\frac{e}{D}\right)_1 = \frac{0.05}{200} = 0.00025$$

$$\text{Re}_2 = \frac{4Q}{\pi D_2 \nu} = \frac{4 \times 0.10}{\pi \times 0.25 \times 10^{-5}} = 5.09 \times 10^4 \qquad \left(\frac{e}{D}\right)_2 = \frac{0.05}{250} = 0.0002$$

$$R_1 = 1.07\left(\frac{10}{9.81 \times 0.20^5}\right)$$

$$\times \{\ln[0.27 \times 0.00025 + 5.74 \times (6.37 \times 10^4)^{-0.9}]\}^{-2} = 53.4$$

$$R_2 = 1.07\left(\frac{500}{9.81 \times 0.25^5}\right)$$

$$\times \{\ln[0.27 \times 0.0002 + 5.74 \times (5.09 \times 10^4)^{-0.9}]\}^{-2} = 904$$

The minor loss coefficient terms are calculated to be

$$\frac{K_1 + K_v}{2gA_1^2} = \frac{0.5 + 2}{2 \times 9.81 \times [\pi/4 \times 0.20^2]^2} = 129.1$$

$$\frac{2K_e + K_2}{2gA_2^2} = \frac{2 \times 0.25 + 1}{2 \times 9.81 \times [\pi/4 \times 0.25^2]^2} = 31.7$$

Substitute these values into the energy equation and obtain

$$0 + 10 + H_P = \frac{200 \times 10^3}{0.85 \times 9800} + 20 + (53.4 + 129.1 + 904 + 31.7)\,0.1^2$$

This relation reduces to

$$10 + H_P = 24 + 20 + 11.2$$

Solving for the head across the pump gives $H_P = 45.2$ m. The required input power is

$$\dot{W}_P = \frac{\gamma Q H_P}{\eta}$$

$$= \frac{(9800 \times 0.85) \times 0.10 \times 45.2}{0.75}$$

$$= 5.0 \times 10^4 \text{ W} \qquad \text{or} \qquad 50 \text{ kW}$$

## 11.3.2    Parallel Piping

A parallel piping arrangement is shown in Fig. 11.4; it is essentially an arrangement of $N$ pipe elements joined at A and B with $\Sigma K$ minor loss components associated with each pipe element $i$. The continuity equation applied at either location A or B is given by

$$Q = \sum_{i=1}^{N} Q_i \tag{11.3.4}$$

The algebraic sum of the energy grade line around any defined loop must be zero. As in the case of series piping, it is customary to assume that $V^2/2g \ll (p/\gamma + z)$. Hence, for any pipe element $i$, the energy equation from location A to B is

$$\left(\frac{p}{\gamma} + z\right)_{\!A} - \left(\frac{p}{\gamma} + z\right)_{\!B} = \left(R_i + \frac{\Sigma K}{2gA_i^2}\right) Q_i^2 \qquad i = 1, \ldots, N \tag{11.3.5}$$

The unknowns in Eqs. 11.3.4 and 11.3.5 are the discharges $Q_i$ and the difference in piezometric head between A and B; the discharge $Q$ into the system is assumed known. It is possible to convert the minor loss terms using an equivalent length as defined in Section 7.6.4. For each pipe element $i$ the equivalent length $L_e$ for $\Sigma K$ minor loss components is

$$(L_e)_i = \frac{D_i}{f_i} \Sigma K \tag{11.3.6}$$

Thus Eq. 11.3.5 simplifies to the form

$$\left(\frac{P}{\gamma} + z\right)_{\!A} - \left(\frac{P}{\gamma} + z\right)_{\!B} = \overline{R}_i Q_i^2 \tag{11.3.7}$$

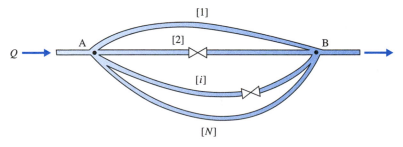

**Fig. 11.4**    Parallel piping system.

in which the modified pipe resistance coefficient $\overline{R}_i$ is given by

$$\overline{R}_i = \frac{8f_i[L_i + (L_e)_i]}{g\pi^2 D_i^5} \qquad (11.3.8)$$

Note that the right-hand side of Eq. 11.3.8 is equivalent to the term $R_i + \Sigma K/(2gA_i^2)$, where Eq.11.2.2 expresses $R_i$.

A solution employing the method of successive substitution is developed in the following manner. Define the variable $W$ to be the change in hydraulic grade line between A and B; that is, $W = (p/\gamma + z)_A - (p/\gamma + z)_B$. Then Eq. 11.3.7 can be solved for $Q_i$ in terms of $W$ as

$$Q_i = \sqrt{\frac{W}{\overline{R}_i}} \qquad (11.3.9)$$

Equations 11.3.4 and 11.3.9 are combined to eliminate the unknown discharges $Q_i$, resulting in

$$Q = \sum_{i=1}^{N} \sqrt{\frac{W}{\overline{R}_i}} = \sqrt{W} \sum_{i=1}^{N} \frac{1}{\sqrt{\overline{R}_i}} \qquad (11.3.10)$$

The remaining unknown $W$ is taken out of the summation sign since it is the same in all pipes. Solving for $W$ in Eq. 11.3.10 results in

$$W = \left( \frac{Q}{\sum\limits_{i=1}^{N} \dfrac{1}{\sqrt{\overline{R}_i}}} \right)^2 \qquad (11.3.11)$$

An iterative procedure can be formulated to solve for $W$ and the discharges $Q_i$ as follows:

1.  Assume flows in each line to be in the completely rough zone, and compute an initial estimate of the friction factors in each line using Eq. 11.2.6.
2.  Compute $\overline{R}_i$ for each pipe and evaluate $W$ with Eq. 11.3.11.
3.  Compute $Q_i$ in each pipe with Eq. 11.3.9.
4.  Update the estimates of the friction factors in each line using the current values of $Q_i$ and Eq. 11.2.3.
5.  Repeat steps 2 to 4 until the unknowns $W$ and $Q_i$ do not vary within a desired tolerance.

Note that if friction factors are in the completely rough zone so that they are independent of the discharge and therefore constant, steps 4 and 5 are unnecessary and a solution results on the first iteration. The technique is illustrated in Example 11.3.

# Example 11.3

Find the distribution of flow and the drop in hydraulic grade line for the three-parallel-pipe arrangement shown in Fig. E11.3. Use variable friction factors with $\nu = 10^{-6}$ m²/s. The total water discharge is $Q = 0.020$ m³/s.

| Pipe | L (m) | D (m) | e (mm) | ΣK |
|------|-------|-------|--------|-----|
| 1 | 100 | 0.05 | 0.1 | 10 |
| 2 | 150 | 0.075 | 0.2 | 3 |
| 3 | 200 | 0.085 | 0.1 | 2 |

Fig. E11.3

## Solution

The initial estimates of $f$ are based on Eq. 11.2.6. Preliminary calculations provide the following:

| Pipe | e/D | $f$ (Eq. 11.2.6) | $L_e$ (Eq. 11.3.6) | $\overline{R}$ (Eq. 11.3.8) |
|------|-----|------------------|---------------------|------------------------------|
| 1 | 0.002 | 0.023 | 21.7 | $7.40 \times 10^5$ |
| 2 | 0.0027 | 0.025 | 9.0 | $1.38 \times 10^5$ |
| 3 | 0.0012 | 0.021 | 8.1 | $8.14 \times 10^4$ |

Apply Eq. 11.3.11, and the first estimate of $W$ is

$$W = \left[ \frac{0.020}{(7.40 \times 10^5)^{-1/2} + (1.38 \times 10^5)^{-1/2} + (8.14 \times 10^4)^{-1/2}} \right]^2$$

$$= 7.39 \text{ m}$$

Then with Eq. 11.3.9, estimates of $Q_i$ are made:

$$Q_1 = \left( \frac{7.39}{7.40 \times 10^5} \right)^{1/2} = 0.00316 \text{ m}^3/\text{s}$$

$$Q_2 = \left( \frac{7.39}{1.38 \times 10^5} \right)^{1/2} = 0.00732 \text{ m}^3/\text{s}$$

$$Q_3 = \left( \frac{7.39}{8.14 \times 10^4} \right)^{1/2} = 0.00953 \text{ m}^3/\text{s}$$

A continuity check is made using Eq. 11.3.4:

$$\sum_{i=1}^{3} Q_i = 0.00316 + 0.00732 + 0.00953 = 0.0200 \text{ m}^3/\text{s}$$

Even though the sum satisfies the solution, another iteration will be carried out to study the convergence of the solution technique. First, the $\overline{R}$-values are updated using Eq. 11.2.3 to evaluate the friction factors.

*(continued)*

| Pipe | $\mathrm{Re} = \dfrac{4Q}{\pi D \nu}$ | $f$ (Eq. 11.2.3) | $\overline{R}$ (Eq. 11.3.8) |
|---|---|---|---|
| 1 | $8.05 \times 10^4$ | 0.026 | $8.20 \times 10^5$ |
| 2 | $1.24 \times 10^5$ | 0.027 | $1.49 \times 10^5$ |
| 3 | $1.43 \times 10^5$ | 0.022 | $8.51 \times 10^4$ |

Evaluating $W$ using Eq. 11.3.11 yields $W = 7.88$ m, and application of Eq. 11.3.9 gives the new estimates of $Q_i$:

$$Q_1 = 0.00310 \text{ m}^3/\text{s} \qquad Q_2 = 0.00727 \text{ m}^3/\text{s} \qquad Q_3 = 0.00962 \text{ m}^3/\text{s}$$

A continuity check shows that

$$\sum_{i=1}^{3} Q_i = 0.0200 \text{ m}^3/\text{s}$$

which is the same as in the first iteration.

A solution of Example 11.3 using Mathcad is shown in Appendix F, Fig. F1.

### 11.3.3 Branch Piping

The branching network, illustrated in Fig. 11.5a, is made up of three elements connected to a single junction. In contrast to the parallel system shown in Fig. 11.4, no closed loops exist. In the analysis, one assumes the direction of flow in each element; then the energy equation for each element is written using an equivalent length to account for minor losses:

$$\left(\frac{p}{\gamma} + z\right)_A - \left(\frac{p}{\gamma} + z\right)_B = \overline{R}_1 Q_1^2 \tag{11.3.12}$$

$$\left(\frac{p}{\gamma} + z\right)_B - \left(\frac{p}{\gamma} + z\right)_C = \overline{R}_2 Q_2^2 \tag{11.3.13}$$

$$\left(\frac{p}{\gamma} + z\right)_B - \left(\frac{p}{\gamma} + z\right)_D = \overline{R}_3 Q_3^2 \tag{11.3.14}$$

The piezometric heads at locations A, C, and D are considered known. The unknowns are the piezometric head at B and the discharges $Q_1$, $Q_2$, and $Q_3$. The additional relation is the continuity balance at location B, which is

$$Q_1 - Q_2 - Q_3 = 0 \tag{11.3.15}$$

Thus there are four equations with four unknowns. One convenient ad hoc method of solution is outlined below and illustrated in Example 11.4:

1.  Assume a discharge $Q_1$ in element 1 (with or without a pump). Establish the piezometric head $H$ at the junction by solving Eq. 11.3.12 (a category 1 problem).

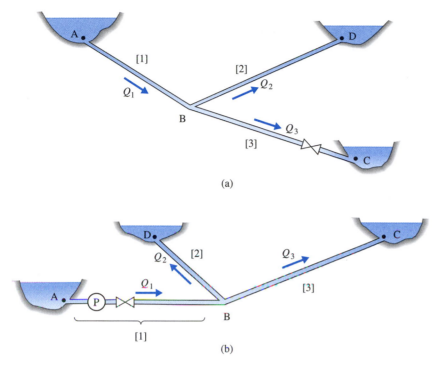

**Fig. 11.5**   Branch piping systems: (a) gravity flow; (b) pump-driven flow.

2.   Compute the discharge $Q_i$ in the remaining branches using Eqs. 11.3.13 and 11.3.14 (a category 2 problem).
3.   Substitute the $Q_i$ into Eq. 11.3.15 to check for continuity balance. Generally, the flow imbalance at the junction $\Delta Q$ will be nonzero. In Eq. 11.3.15, $\Delta Q = Q_1 - Q_2 - Q_3$.
4.   Adjust the flow $Q_1$ in element 1 and repeat steps 2 and 3 until $\Delta Q$ is within desired limits.

If a pump exists in pipe 1 (see Fig. 11.5b), Eq. 11.3.12 is altered in the manner

$$\left(\frac{p}{\gamma} + z\right)_A - \left(\frac{p}{\gamma} + z\right)_B + H_P = \overline{R}_1 Q_1^2 \qquad (11.3.16)$$

An additional unknown, namely, the pump head $H_P$, is introduced. The additional necessary relationship is the head-discharge curve for the pump; see, for example, Fig. 7.18. The solution can proceed in a manner similar to that described in Example 11.5. It may be convenient to follow the solution graphically by plotting the assumed discharge in line 1 versus either the piezometric head at B or the imbalance of flow at B. Such a step is helpful to determine in what manner the discharge in line 1 should be altered for the next iteration.

An alternative method of solution for a single branching system is to eliminate all of the variables except the piezometric head $H$ (= $p/\gamma + z$) at the junction, location B in Fig. 11.5. Then a numerical solving technique can be employed. An additional requirement is to assume the direction of flow in each pipe. It may become necessary to correct the sign in one or more of the equations if during the solution, $H$ moves from below one of the reservoir elevations to above it, or vice versa. Example 11.4 illustrates the procedure.

Examples 11.4 and 11.5 represent a level of complexity that one may not want to exceed employing a calculator-based solution. To include additional pumps, reservoirs or pipes would make either ad hoc or simplified numerical approaches too cumbersome. For such systems, the more generalized network analysis described in Section 11.4 is recommended.

## Example 11.4

For the three-branch piping system shown in Fig. E11.4 we have the following data:

| Pipe | L (m) | D (m) | f | ΣK |
|------|-------|-------|-------|-----|
| 1 | 500 | 0.10 | 0.025 | 3 |
| 2 | 750 | 0.15 | 0.020 | 2 |
| 3 | 1000 | 0.13 | 0.018 | 7 |

**Fig. E11.4**

Determine the flow rates $Q_i$ and the piezometric head $H$ at the junction. Assume constant friction factors.

### Solution
The equivalent lengths and resistance coefficients are

$$(L_e)_1 = \frac{0.10}{0.025} \times 3 = 12 \text{ m} \qquad \overline{R}_1 = \frac{8 \times 0.025 \times 512}{9.81 \times \pi^2 \times (0.10)^5} = 1.06 \times 10^5 \text{ s}^2/\text{m}^5$$

$$(L_e)_2 = \frac{0.15}{0.020} \times 2 = 15 \text{ m} \qquad \overline{R}_2 = \frac{8 \times 0.020 \times 765}{9.81 \times \pi^2 \times (0.15)^5} = 1.66 \times 10^4 \text{ s}^2/\text{m}^5$$

$$(L_e)_3 = \frac{0.13}{0.018} \times 7 = 51 \text{ m} \qquad \overline{R}_3 = \frac{8 \times 0.018 \times 1051}{9.81 \times \pi^2 \times (0.13)^5} = 4.21 \times 10^4 \text{ s}^2/\text{m}^5$$

With the flow directions assumed as shown, the energy equation is written for each pipe and solved for the unknown discharge:

$$Q_1 = \left(\frac{H - 5}{\overline{R}_1}\right)^{1/2} \qquad Q_2 = \left(\frac{20 - H}{\overline{R}_2}\right)^{1/2} \qquad Q_3 = \left(\frac{H - 13}{\overline{R}_3}\right)^{1/2}$$

The continuity equation is $-Q_1 + Q_2 - Q_3 = 0$. Eliminating $Q_1, Q_2,$ and $Q_3$ with the energy relations results in an algebraic equation in terms of $H$:

$$w(H) = -\left(\frac{H - 5}{1.06 \times 10^5}\right)^{1/2} + \left(\frac{20 - H}{1.66 \times 10^4}\right)^{1/2} - \left(\frac{H - 13}{4.21 \times 10^4}\right)^{1/2} = 0$$

Even though this can be solved as a quadratic, the method of false position is chosen to compute $H$, which would be required if the friction factors varied. The procedure is presented in Example 10.11. In the present example the recurrence formula is

$$H_r = \frac{H_l\, w(H_u) - H_u\, w(H_l)}{w(H_u) - w(H_l)}$$

The solution is shown in the table below. Note that with the initial guesses of $H_l$ and $H_u$, the sign conventions in $w$ require that $20 > H > 13$. Iteration continues until the convergence criterion shown in the last column becomes less than the arbitrary value of 0.005.

| Iteration | $H_u$ | $H_l$ | $w(H_u)$ | $w(H_l)$ | $H_r$ | $w(H_r)$ | Sign of $w(H_l) \times w(H_r)$ | $\varepsilon = \left\lvert \dfrac{H_r^{new} - H_r^{old}}{H_r^{old}} \right\rvert$ |
|---|---|---|---|---|---|---|---|---|
| 1 | 18 | 13 | −0.01100 | 0.01185 | 15.59 | −0.00154 | + | |
| 2 | 15.59 | 13 | −0.00154 | 0.01185 | 15.29 | −0.000384 | + | 0.019 |
| 3 | 15.29 | 13 | −0.000384 | 0.01185 | 15.22 | −0.000112 | + | 0.0046 |

Hence $H \simeq 15.2$ m. The discharges are now computed:

$$Q_1 = \left(\frac{15.2 - 5}{1.06 \times 10^5}\right)^{1/2} = 0.0098 \text{ m}^3/\text{s}$$

$$Q_2 = \left(\frac{20 - 15.2}{1.66 \times 10^4}\right)^{1/2} = 0.0170 \text{ m}^3/\text{s}$$

$$Q_3 = \left(\frac{15.2 - 13}{4.21 \times 10^4}\right)^{1/2} = 0.0072 \text{ m}^3/\text{s}$$

Note that continuity is satisfied.

## Example 11.5

For the system shown in Fig. E11.5, determine the flow distribution $Q_i$ of water and the piezometric head $H$ at the junction. The fluid power input by the pump is constant, equal to $\gamma Q H_P = 20$ kW. Assume constant friction factors.

| Pipe | L (m) | D (m) | f | ΣK |
|---|---|---|---|---|
| 1 | 50 | 0.15 | 0.02 | 2 |
| 2 | 100 | 0.10 | 0.015 | 1 |
| 3 | 300 | 0.10 | 0.025 | 1 |

Fig. E11.5

### Solution
The equivalent lengths and resistance coefficients are computed from Eqs. 11.3.6 and 11.3.8 to be

*(continued)*

$$(L_e)_1 = \frac{0.15}{0.02} \times 2 = 15 \text{ m} \qquad \overline{R}_1 = \frac{8 \times 0.02 \times 65}{9.81 \times \pi^2 \times (0.15)^5} = 1.42 \times 10^3 \text{ s}^2/\text{m}^5$$

$$(L_e)_2 = \frac{0.10}{0.015} \times 1 = 6.7 \text{ m} \qquad \overline{R}_2 = \frac{8 \times 0.015 \times 106.7}{9.81 \times \pi^2 \times (0.10)^5} = 1.32 \times 10^4 \text{ s}^2/\text{m}^5$$

$$(L_e)_3 = \frac{0.10}{0.025} \times 1 = 4 \text{ m} \qquad \overline{R}_3 = \frac{8 \times 0.025 \times 304}{9.81 \times \pi^2 \times (0.10)^5} = 6.28 \times 10^4 \text{ s}^2/\text{m}^5$$

Assume flow directions as shown. The energy equation for pipe 1 from the reservoir to the junction B is

$$z_1 + H_P = H + \overline{R}_1 Q_1^2$$

in which $H$ is the piezometric head at B. Substituting in known parameters and solving for $H$ results in

$$H = 10 + \frac{20 \times 10^3}{9800 Q_1} - 1.42 \times 10^3 Q_1^2$$

$$= 10 + \frac{2.04}{Q_1} - 1420 Q_1^2$$

An iterative solution is shown in the accompanying table. For each iteration, a value of $Q_1$ is estimated. Then $H$ is calculated and $Q_2$ and $Q_3$ are evaluated from the relations

$$Q_2 = \left(\frac{H - z_2}{\overline{R}_2}\right)^{1/2} = \left(\frac{H - 30}{1.32 \times 10^4}\right)^{1/2}$$

$$Q_3 = \left(\frac{H - z_3}{\overline{R}_3}\right)^{1/2} = \left(\frac{H - 15}{6.28 \times 10^4}\right)^{1/2}$$

In the last column of the table, a continuity balance is employed to check the accuracy of the estimate of $Q_1$. The third estimate of $Q_1$ is based on a linear interpolation by setting $\Sigma Q = 0$ and using values of $Q_1$ and $\Delta Q$ from the first two iterations.

| Iteration | $Q_1$ | $H$ | $Q_2$ | $Q_3$ | $\Delta Q = Q_1 - Q_2 - Q_3$ |
|-----------|-------|-------|--------|--------|------------------------------|
| 1 | 0.050 | 47.25 | 0.0362 | 0.0227 | −0.0089 |
| 2 | 0.055 | 42.80 | 0.0311 | 0.0210 | +0.0029 |
| 3 | 0.054 | 43.64 | 0.0322 | 0.0214 | +0.0004 |

The approximate solution is $H = 43.6$ m, $Q_1 = 54$ L/s, $Q_2 = 32$ L/s, and $Q_3 = 21$ L/s. If greater precision is desired, a solution should be employed similar to that shown in Example 11.4.

Solutions for Examples 11.4 and 11.5 using Mathcad and MATLAB are provided in Appendix F, Figs. F2 to F5. In those solutions note that the term $\overline{R}Q^2$ is replaced by $\overline{R}Q|Q|$, which automatically accounts for changes in flow direction. In addition to the computational software solutions in Appendix F, the more general pipe network solutions described in Section 11.4 can be applied to the problems described in this section.

## 11.4  ANALYSIS OF PIPE NETWORKS

Piping systems more complicated than those considered in Section 11.3 are best analyzed by formulating the solution for a network. Before considering a generalized set of network equations, it is worthwhile to examine a specific example of a piping network to observe the degree of complexity involved.

Fig. 11.6a shows a relatively simple network consisting of seven pipes, two reservoirs, and one pump. The hydraulic grade lines at A and F are assumed known; these locations are termed *fixed-grade nodes*. Outflow demands are present at nodes C and D. Nodes C and D, along with nodes B and E are called

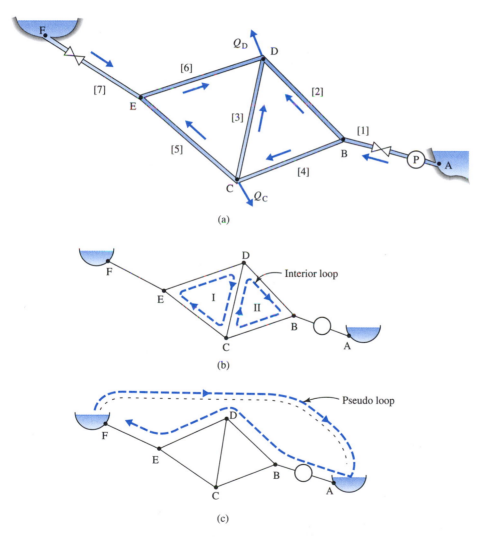

(a)

(b)

(c)

**Fig. 11.6**   Representative piping network: (a) assumed flow directions and numbering scheme; (b) designated interior loops; (c) path between two fixed-grade nodes.

*interior nodes* or *junctions*. Flow directions, even though not initially known, are assumed to be in the directions shown. The system equations are given as follows:

1.  Energy balance for each pipe (seven equations):

$$H_A - H_B + H_P(Q_1) = \overline{R}_1 Q_1^2 \qquad H_C - H_E = \overline{R}_5 Q_5^2$$
$$H_B - H_D = \overline{R}_2 Q_2^2 \qquad H_E - H_D = \overline{R}_6 Q_6^2$$
$$H_C - H_D = \overline{R}_3 Q_3^2 \qquad H_F - H_E = \overline{R}_7 Q_7^2 \qquad (11.4.1)$$
$$H_B - H_C = \overline{R}_4 Q_4^2$$

2.  Continuity balance for each interior node (four equations):

$$Q_1 - Q_2 - Q_4 = 0$$
$$Q_2 + Q_3 + Q_6 = Q_D$$
$$Q_4 - Q_3 - Q_5 = Q_C \qquad (11.4.2)$$
$$Q_5 - Q_6 + Q_7 = 0$$

3.  Approximation of pump curve (one equation):

$$H_P(Q_1) = a_0 + a_1 Q_1 + a_2 Q_1^2 \qquad (11.4.3)$$

in which $a_0$, $a_1$, and $a_2$ are known constants.

The unknowns are $Q_1, \ldots, Q_7, H_B, H_C, H_D, H_E$, and $H_P$. Thus there are 12 unknowns and 12 equations to solve simultaneously. Since the energy equations and the pump equation are nonlinear, it is necessary to resort to some type of approximate numerical solution. The 12 equations can be reduced in number by combining the energy equations along special paths. Let the drop in hydraulic grade line for any pipe element $i$ be designated as $W_i$. Then

$$W_i = \overline{R}_i Q_1^2 \qquad (11.4.4)$$

For the system under consideration, two closed paths, or *interior loops*, can be identified (see Fig. 11.6b). Flow is considered positive in a clockwise sense around each loop. Energy balances, written around loops I and II, are

$$W_6 - W_3 + W_5 = 0$$
$$W_3 - W_2 + W_4 = 0 \qquad (11.4.5)$$

To account for the flow in pipes 1 and 7, a path can be defined along nodes A, B, D, E, and F in Fig. 11.6c. Then with the addition of the pump head, the energy balance from A to F is

$$H_A + H_P - W_1 - W_2 + W_6 + W_7 = H_F \qquad (11.4.6)$$

Note that the path energy equation connects two fixed-grade nodes. Such a path is sometimes termed a *pseudoloop*, since an imaginary pipe with infinite resistance, or no flow, can be considered to connect the two reservoirs. The imaginary pipe is shown by a dotted line in Fig. 11.6c. Substituting the pump equation and the friction equation into the energy relations above results in the following reduced set of equations:

$$-\overline{R}_3 Q_3^2 + \overline{R}_5 Q_5^2 + \overline{R}_6 Q_6^2 = 0$$

$$-\overline{R}_2 Q_2^2 + \overline{R}_3 Q_3^2 + \overline{R}_4 Q_4^2 = 0$$

$$-\overline{R}_1 Q_1^2 + (a_0 + a_1 Q_1 + a_2 Q_1^2) - \overline{R}_2 Q_2^2 + \overline{R}_6 Q_6^2 + \overline{R}_7 Q_7^2 + H_A - H_F = 0$$

$$Q_1 - Q_2 - Q_4 = 0$$

$$Q_2 + Q_3 + Q_6 = Q_D$$

$$Q_4 - Q_3 - Q_5 = Q_C$$

$$Q_5 - Q_6 + Q_7 = 0$$

$$(11.4.7)$$

There are now seven unknowns $(Q_1, \ldots, Q_7)$ and seven equations to solve. The energy relations are nonlinear since the loss terms and the pump head are represented as polynomials with respect to the discharges.

## 11.4.1   Generalized Network Equations

Networks of piping, such as those shown in Fig. 11.6, can be represented by the following equations.

1.  Continuity at the $j$th interior node:

$$\Sigma(\pm)_j Q_j - Q_e = 0 \qquad (11.4.8)$$

    in which the subscript $j$ refers to the pipes connected to a node, and $Q_e$ is the external demand. The algebraic plus or minus sign convention pertains to the assumed flow direction: Use the positive sign for flow into the junction, and the negative sign for flow out of the junction.

2.  Energy balance around an interior loop:

$$\Sigma(\pm)_i W_i = 0 \qquad (11.4.9)$$

    in which the subscript $i$ pertains to the pipes that make up the loop. There will be a relation for each of the loops. Here it is assumed that there are no pumps located in the interior of the network. The plus sign is used if the flow in the element is positive in the clockwise sense; otherwise, the minus sign is employed.

3. Energy balance along a unique path or pseudoloop connecting two fixed-grade nodes:

$$\Sigma(\pm)_i[W_i - (H_P)_i] + \Delta H = 0 \qquad (11.4.10)$$

where $\Delta H$ is the difference in magnitude of the two fixed-grade nodes in the path ordered in a clockwise fashion across the imaginary pipe in the pseudoloop. The term $(H_P)_i$ is the head across a pump that could exist in the $i$th pipe element. If $F$ is the number of fixed-grade nodes, there will be $(F - 1)$ unique path equations. The plus and minus signs in Eq. 11.4.10 follow the same argument given for Eq. 11.4.9.

Let $P$ be the number of pipe elements in the network, $J$ the number of interior nodes, and $L$ the number of interior loops. Then the following relation will hold if the network is properly defined:

$$P = J + L + F - 1 \qquad (11.4.11)$$

In Fig. 11.6 of the introductory example, $J = 4$, $L = 2$, $F = 2$, so that $P = 4 + 2 + 2 - 1 = 7$.

An additional necessary formulation is the relation between discharge and loss in each pipe; it is

$$W = RQ^\beta + \frac{\Sigma K}{2gA^2} Q^2 \qquad (11.4.12)$$

If the minor losses can be defined in terms of an equivalent length, Eq. 11.4.12 can be replaced by[2]

$$W = \overline{R}Q^\beta \qquad (11.4.13)$$

An approximate pump head-discharge representation is given by the polynomial

$$H_P(Q) = a_0 + a_1Q + a_2Q^2 \qquad (11.4.14)$$

---

[2] In Eq. 11.4.13, the exponent $\beta$ is used, since the Hazen–Williams formula is commonly employed in pipe network analysis. The equivalent length for use in the Hazen–Williams formulation is given by

$$L_e = 0.8106 \frac{\Sigma K}{K_1} D^{0.87} C^{1.85} Q^{0.15}$$

Note that it is necessary to estimate the discharge when using this equation. The Darcy–Weisbach equation with $\beta = 2$ is recommended. An alternative to using the equivalent-length concept is to account for pipe friction and minor losses separately in each line, making use of Eq. 11.4.12.

The coefficients $a_0$, $a_1$, and $a_2$ are assumed known; typically, they can be found by substituting three known data points from a specified pump curve and solving the three resulting equations simultaneously. In place of defining the pump head-discharge curve, an alternative means of including a pump in a line is to specify the useful power the pump puts into the system. The useful, or actual, power $\dot{W}_f$ is assumed to be constant and allows $H_P$ to be represented in the manner

$$H_P(Q) = \frac{\dot{W}_f}{\gamma Q} \qquad (11.4.15)$$

This equation is particularly useful when the specific operating characteristics of a pump are unknown.

## 11.4.2  Linearization of System Energy Equations

Equation 11.4.10 is a general relation that can be applied to any path or closed loop in a network. If it is applied to a closed loop, $\Delta H$ is set equal to zero, and if no pump exists in the path or loop, $(H_P)_i$ is equal to zero. Note that Eq. 11.4.9 can be considered to be a subset of Eq. 11.4.10. In the following development, Eq. 11.4.10 will be used to represent any loop or path in the network.

Define the function $\phi(Q)$ to contain the nonlinear terms $W(Q)$ and $H_P(Q)$ in the form

$$\begin{aligned} \phi(Q) &= W(Q) - H_P(Q) \\ &= \overline{R}Q^\beta - H_P(Q) \end{aligned} \qquad (11.4.16)$$

Equation 11.4.16 can be expanded in a Taylor series as

$$\phi(Q) = \phi(Q_0) + \left.\frac{d\phi}{dQ}\right|_{Q_0} (Q - Q_0) + \left.\frac{d^2\phi}{dQ^2}\right|_{Q_0} \frac{(Q - Q_0)^2}{2} + \cdots \qquad (11.4.17)$$

in which $Q_0$ is an estimate of $Q$. To approximate $\phi(Q)$ accurately, $Q_0$ should be chosen so that the difference $(Q - Q_0)$ is numerically small. Retaining the first two terms on the right-hand side of Eq. 11.4.17, and using Eq. 11.4.16, yields

$$\phi(Q) \simeq \overline{R}Q_0^\beta - H_P(Q_0) + \left[\beta\overline{R}Q_0^{\beta-1} - \left.\frac{dH_P}{dQ}\right|_{Q_0}\right](Q - Q_0) \qquad (11.4.18)$$

Note that the approximation to $\phi(Q)$ is now linear with respect to $Q$.

The parameter $G$ is introduced as

$$G = \beta\overline{R}Q_0^{\beta-1} - \left.\frac{dH_P}{dQ}\right|_{Q_0} \qquad (11.4.19)$$

Using Eq. 11.4.14 to represent the pump head, Eq. 11.4.19 becomes

$$G = \beta \overline{R} Q_0^{\beta-1} - (a_1 + 2a_2 Q_0) \qquad (11.4.20)$$

Alternatively, with Eq. 11.4.15 substituted into Eq. 11.4.19, one has

$$G = \beta \overline{R} Q_0^{\beta-1} + \frac{\dot{W}_f}{\gamma Q_0^2} \qquad (11.4.21)$$

Substituting Eq. 11.4.19 into Eq. 11.4.18 gives

$$\begin{aligned}\phi(Q) &= \overline{R} Q_0^{\beta} - H_P(Q_0) + (Q - Q_0)G \\ &= W_0 - H_{P0} + (Q - Q_0)G\end{aligned} \qquad (11.4.22)$$

in which $W_0 = W(Q_0)$ and $H_{P0} = H_P(Q_0)$. Finally, Eq. 11.4.22 is substituted into Eq. 11.4.10 to produce the linearized loop or path energy equation

$$\Sigma(\pm)_i[(W_0)_i - (H_{P0})_i] + \Sigma[Q_i - (Q_0)_i]G_i + \Delta H = 0 \qquad (11.4.23)$$

The second term does not contain the plus or minus sign since $G_i$ is a monotonically increasing function of the flow correction. Since $Q$ in the relations above can assume positive or negative values, $Q_0^{\beta}$ and $Q_0^{\beta-1}$ are often replaced by $Q_0|Q_0|^{\beta-1}$ and $|Q_0|^{\beta-1}$, respectively, in solution algorithms. This is done, for example, in Eq. 11.4.23, which forms the basis for the Hardy Cross solution outlined below.

### 11.4.3   Hardy Cross Method

In Eq. 11.4.23, let $(Q_0)_i$ be the estimates of discharge from the previous iteration, and let $Q_i$ be the new estimates of the discharge. Define a flow adjustment $\Delta Q$ for each loop to be

$$\Delta Q = Q_i - (Q_0)_i \qquad (11.4.24)$$

The adjustment is applied independently to all pipes in a given loop. Hence Eq. 11.4.23 can be written as

$$\Sigma(\pm)_i[(W_0)_i - (H_{P0})_i] + \Delta Q \Sigma G_i + \Delta H = 0 \qquad (11.4.25)$$

Solving for $\Delta Q$, one has

$$\Delta Q = \frac{-\Sigma(\pm)_i[(W_0)_i - (H_{P0})_i] - \Delta H}{\Sigma G_i} \qquad (11.4.26)$$

It is necessary for the algebraic sign of $Q$ to be positive in the direction of normal pump operation; otherwise, the pump curve will not be represented properly and

Eq. 11.4.26 will be invalid. Furthermore, it is important that the discharge $Q$ through the pump remain within the limits of the data used to generate the curve.

For a closed loop in which no pumps or fixed-grade nodes are present, Eq. 11.4.26 reduces simply to the form

$$\Delta Q = \frac{-\Sigma(\pm)_i(W_0)_i}{\Sigma \, G_i} \qquad (11.4.27)$$

The Hardy Cross iterative solution is outlined in the following steps:

1.  Assume an initial estimate of the flow distribution in the network that satisfies continuity, Eq. 11.4.8. The closer the initial estimates are to the correct values, the fewer will be the iterations required for convergence. One guideline to use is that in a pipe element, as $\overline{R}$ increases, $Q$ decreases.

2.  For each loop or path, evaluate $\Delta Q$ with Eq. 11.4.26 or 11.4.27. The numerators should approach zero as the loops or paths become balanced.

3.  Update the flows in each pipe in all loops and paths, that is, from Eq. 11.4.24.

$$Q_i = (Q_0)_i + \Sigma \Delta Q \qquad (11.4.28)$$

The term $\Sigma \Delta Q$ is used, since a given pipe may belong to more than one loop; hence the correction will be the sum of corrections from all loops to which the pipe is common.

4.  Repeat steps 2 and 3 until a desired accuracy is attained. One possible criterion to use is

$$\frac{\Sigma|Q_i - (Q_0)_i|}{\Sigma|Q_i|} \le \varepsilon \qquad (11.4.29)$$

in which $\varepsilon$ is an arbitrarily small number. Typically, $0.001 < \varepsilon < 0.005$.

The Hardy Cross method of analysis is a simplified version of the method of successive approximations applied to a set of linearized equations. It does not require the inversion of a matrix; hence it can be used to solve relatively small networks using either a calculator or a spreadsheet algorithm on a personal computer. The continuity relations, Eq. 11.4.8, are in a sense "decoupled" from the solution of the energy relations, Eq. 11.4.26. Continuity is satisfied initially with assumed flows and remains satisfied throughout the solution process. Essentially, one computes a correction $\Delta Q$ to the flows $Q_i$ in each loop separately, and then applies $\Delta Q$ to the entire network to bring flow through the loops into closer balance. In effect this is a superposition type of solution. Since the flow corrections $\Delta Q$ are applied to each loop independently, convergence may not be rapid.

## Example 11.6

For the piping system shown in Fig. E11.6a, determine the flow distribution and piezo-metric heads at the junctions using the Hardy Cross method of solution. Assume that losses are proportional to $Q^2$.

(a)

(b)

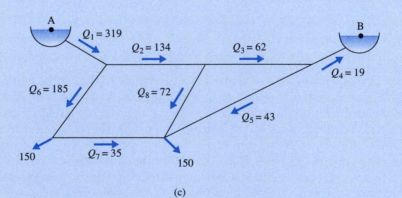

(c)

**Fig. E11.6**

## Solution

There are five junctions ($J = 5$), eight pipes ($P = 8$), and two fixed-grade nodes ($F = 2$). Hence the number of closed loops is $L = 8 - 5 - 2 + 1 = 2$, plus one pseudoloop. The three loops and assumed flow directions are shown in Fig. E11.6b. Equation 11.4.26 is applied to loop I:

$$\Delta Q_I = \frac{-(\pm W_4 \pm W_3 \pm W_2 \pm W_1) - (z_A - z_B)}{G_4 + G_3 + G_2 + G_1}$$

Apply Eq. 11.4.27 to loops II and III:

$$\Delta Q_{II} = \frac{-(\pm W_2 \pm W_8 \pm W_7 \pm W_6)}{G_2 + G_8 + G_7 + G_6}$$

$$\Delta Q_{III} = \frac{-(\pm W_3 \pm W_5 \pm W_8)}{G_3 + G_5 + G_8}$$

The problem is solved using an Excel spreadsheet, and the solution is shown in Table E11.6. The spreadsheet layout showing the equations used in the solution is provided in Appendix F, Fig. F.6. In the loop equations note that $W$ and $G$ have the correct sign attributed automatically using the relations $\pm W = \overline{R}Q|Q|$ and $G = 2\overline{R}|Q|$. In addition, the initial values of each $Q$ take on a positive or negative sign depending on the assumed flow direction relative to the positive clockwise direction for each loop. The initial flow estimates, shown under the heading "Iteration 1," are chosen to satisfy continuity. Note that flow adjustments to each pipe element are made after all of the $\Delta Q$'s have been computed; for example, relative to loop I, $Q_2 = (Q_0)_2 + \Delta Q_I - \Delta Q_{II}$. After four iterations the absolute magnitude of the $\Delta Q$'s are all less than 0.001, and the left-hand side of Eq. 11.4.29 is 0.0051.

The values of $Q_i$ after four iterations are shown in Fig. E11.6c, along with the final flow directions. Flow rates are in liters per second. The piezometric heads are evaluated by computing the energy drop along designated paths, beginning at a known fixed-grade node, in this case, node A:

$$H_C = H_A - R_1 Q_1^2 = 50 - 100(0.319)^2 = 39.8 \text{ m}$$

$$H_D = H_C - R_2 Q_2^2 = 39.8 - 500(0.134)^2 = 30.8 \text{ m}$$

$$H_E = H_D - R_3 Q_3^2 = 30.8 - 200(0.062)^2 = 30.0 \text{ m}$$

$$H_F = H_C - R_6 Q_6^2 = 39.8 - 300(0.185)^2 = 29.5 \text{ m}$$

$$H_G = H_D - R_8 Q_8^2 = 30.8 - 300(0.072)^2 = 29.2 \text{ m}$$

Note that there is negligible loss in element four.

*(continued)*

| | | | | Iteration 1 | | Iteration 2 | | | Iteration 3 | | | Iteration 4 | | | |
|---|---|---|---|---|---|---|---|---|---|---|---|---|---|---|---|
| | | R | Q | RQ\|Q\| | 2R\|Q\| | Q | RQ\|Q\| | 2R\|Q\| | Q | RQ\|Q\| | 2R\|Q\| | Q | RQ\|Q\| | 2R\|Q\| | Q |
| ΔH | | | | 20.000 | | | 20.000 | | | 20.000 | | | 20.000 | | |
| Loop 1 | Pipe4 | 100 | −0.020 | −0.040 | 4.000 | −0.022 | −0.051 | 4.495 | −0.019 | −0.038 | 3.899 | −0.020 | −0.039 | 3.968 | −0.019 |
| | Pipe3 | 200 | −0.060 | −0.720 | 24.000 | −0.064 | −0.809 | 25.440 | −0.062 | −0.770 | 24.817 | −0.063 | −0.787 | 25.098 | −0.062 |
| | Pipe2 | 500 | −0.130 | −8.450 | 130.000 | −0.137 | −9.433 | 137.352 | −0.133 | −8.907 | 133.466 | −0.135 | −9.150 | 135.276 | −0.134 |
| | Pipe1 | 100 | −0.320 | −10.240 | 64.000 | −0.322 | −10.399 | 64.495 | −0.319 | −10.208 | 63.899 | −0.320 | −10.230 | 63.968 | −0.319 |
| | | | | 0.550 | 222.000 | | −0.691 | 231.783 | | 0.078 | 226.081 | | −0.206 | 228.309 | |
| | | | | ΔQ = −2.48E-03 | | | ΔQ = 2.98E-03 | | | ΔQ = −3.44E-04 | | | ΔQ = 9.03E-04 | | |
| Loop 2 | Pipe2 | 500 | 0.130 | 8.450 | 130.000 | 0.137 | 9.433 | 137.352 | 0.113 | 8.907 | 133.466 | 0.135 | 9.150 | 135.276 | 0.134 |
| | Pipe8 | 300 | 0.070 | 1.470 | 42.000 | 0.074 | 1.632 | 44.251 | 0.071 | 1.530 | 42.853 | 0.073 | 1.578 | 43.519 | 0.072 |
| | Pipe7 | 400 | −0.040 | −0.640 | 32.000 | −0.035 | −0.494 | 28.101 | −0.036 | −0.519 | 28.823 | −0.035 | −0.478 | 27.650 | −0.035 |
| | Pipe6 | 300 | −0.190 | −10.830 | 114.000 | −0.185 | −10.281 | 111.075 | −0.186 | −10.382 | 111.617 | −0.185 | −10.219 | 110.738 | −0.185 |
| | | | | −1.550 | 318.000 | | 0.290 | 320.779 | | −0.464 | 316.759 | | 0.031 | 317.183 | |
| | | | | ΔQ = 4.87E-03 | | | ΔQ = −9.03E-04 | | | ΔQ = 1.47E-03 | | | ΔQ = −9.83E-05 | | |
| Loop 3 | Pipe3 | 200 | 0.060 | 0.720 | 24.000 | 0.064 | 0.809 | 25.440 | 0.062 | 0.770 | 24.817 | 0.063 | 0.787 | 25.098 | 0.062 |
| | Pipe5 | 400 | 0.040 | 0.640 | 32.000 | 0.041 | 0.676 | 32.898 | 0.043 | 0.724 | 34.039 | 0.043 | 0.736 | 34.325 | 0.043 |
| | Pipe8 | 300 | −0.070 | −1.470 | 42.000 | −0.074 | −1.632 | 44.251 | −0.071 | −1.530 | 42.853 | −0.073 | −1.578 | 43.519 | −0.072 |
| | | | | −0.110 | 98.000 | | −0.146 | 102.589 | | −0.036 | 101.710 | | −0.054 | 102.941 | |
| | | | | ΔQ = 1.12E-03 | | | ΔQ = 1.43E-03 | | | ΔQ = 3.57E-04 | | | ΔQ = 52.29E-04 | | |

A Mathcad solution for Example 11.6 is given in Appendix F, Fig. F.7.

## 11.4.4  Network Analysis Using Generalized Computer Software

Spreadsheets and computational software such as Mathcad or MATLAB can be used to solve for the flow distribution in small networks, as illustrated in Example 11.6. However, for larger networks that contain more than several loops, the task of programming becomes cumbersome, especially since a unique description is required for each system accompanied by a large number of equations to solve. An advantage of using these methods of solution is that the user is applying the fundamental relations directly and can easily discern convergence of the solution. On the other hand, generalized computer codes have robust solution algorithms that allow a wide variety of network systems to be analyzed, and provide user-friendly input/output schemes. They have become an indispensable tool for many engineers and practitioners in the water-supply industry as well as finding use in other industrial applications. However, the user must become familiar with the code so that the correctness of the solution is assured.

With the advancement of software development and the increase of personal computer memory and computational speed, it is now possible to analyze large, highly complex networks with relative ease. In addition to the Hardy Cross method outlined in Section 11.4.3, other linearized methods have been used in computer codes. Herein is described briefly the use of the program EPANET developed by the United States Environmental Protection Agency.

EPANET is a comprehensive program that simulates both hydraulic flow and water quality in pressurized piping networks. For the hydraulic analysis, it uses a hybrid node-loop algorithm termed the *gradient method* to solve for the unknown piezometric heads and discharges. The network can contain pipes, pipe junctions, pumps, and various types of valves, reservoirs, and storage water tanks. Friction head loss is computed using the Darcy–Weisbach, Hazen–Williams, or Chezy–Manning formulas. Either pump curve data or useful power can be accommodated. Further capabilities of the program, a comprehensive description, and details of running the code are provided in the EPANET Users Manual (Rossman, 2000). The source code and users manual is available from the EPA Web site *www.epa.gov*.

## Example 11.7

Rework Example 11.6 using EPANET.

### Solution
The user-defined network map is shown in Fig. E11.7, and a partial listing of the computed results are provided in Table E11.7. For coding details refer to the EPANET Users Manual.

*(continued)*

**Fig. E11.7**

**Table E11.7**

Link – Node Table:

| Link ID | Start Node | End Node | Length m | Diameter mm |
|---------|-----------|----------|----------|-------------|
| 1 | 1 | 2 | 66 | 250 |
| 2 | 2 | 3 | 330 | 250 |
| 3 | 3 | 4 | 130 | 250 |
| 4 | 4 | 7 | 66 | 250 |
| 5 | 4 | 5 | 260 | 250 |
| 6 | 3 | 5 | 200 | 250 |
| 7 | 2 | 6 | 200 | 250 |
| 8 | 6 | 5 | 260 | 250 |

Node Results:

| Node ID | Demand LPS | Head m | Pressure m | Quality | |
|---------|-----------|--------|------------|---------|---|
| 2 | 0.00 | 40.45 | 40.45 | 0.00 | |
| 3 | 0.00 | 30.96 | 30.96 | 0.00 | |
| 4 | 0.00 | 30.05 | 30.05 | 0.00 | |
| 5 | 150.00 | 29.16 | 29.16 | 0.00 | |
| 6 | 150.00 | 29.82 | 29.82 | 0.00 | |
| 1 | −319.72 | 50.00 | 0.00 | 0.00 | Reservoir |
| 7 | 19.72 | 30.00 | 0.00 | 0.00 | Reservoir |

Link Results:

| Link ID | Flow LPS | Velocity m/s | Headloss m/km | Status |
|---------|----------|--------------|---------------|--------|
| 1 | 319.72 | 6.52 | 144.71 | Open |
| 2 | 133.59 | 2.72 | 28.75 | Open |
| 3 | 62.19 | 1.27 | 6.98 | Open |
| 4 | 19.72 | 0.40 | 0.83 | Open |
| 5 | 42.47 | 0.87 | 3.44 | Open |
| 6 | 71.40 | 1.46 | 9.01 | Open |
| 7 | 186.12 | 3.79 | 53.13 | Open |
| 8 | 36.12 | 0.74 | 2.55 | Open |

In addition to Example 11.7, solutions for the two networks shown in Fig. 11.7 are provided in Appendix F, Figs. F.8 and F.9. For the branching or tree-like system shown in Fig. 11.7a, the Hazen–Williams friction formula is used with

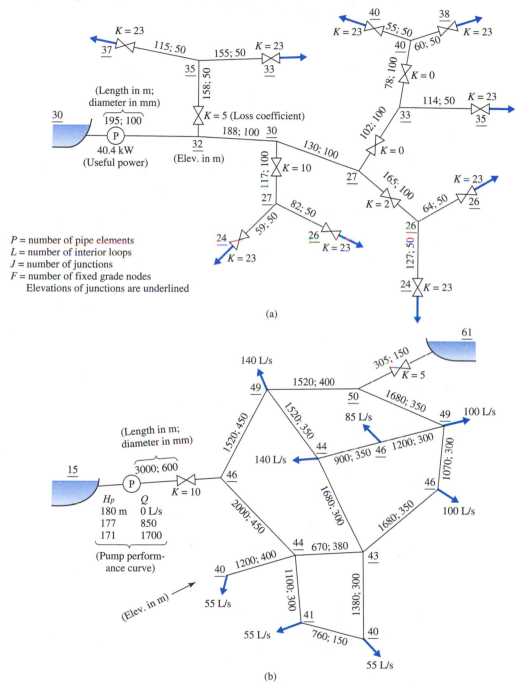

(a)

**Fig. 11.7** Two piping networks: (a) tree watering system: $P = 17, L = 0, J = 8, F = 10$; (b) water distribution system: $P = 17, L = 4, J = 12, F = 2$. (Wood, D. J. Algorithms for Pipe Network Analysis and Their Reliability, Research Report 127, Water Resources Research Institute, University of Kentucky, Lexington, KY., 1981) the coefficient $C = 130$. The useful input power of the pump is 40.4 kW. For the looping network shown in Fig. 11.7b, friction losses are based on the Darcy–Weisbach relation with an absolute roughness of 0.15 mm for all pipes and a kinematic viscosity of $10^{-6}$ m$^2$/s. The pump performance curve is defined by three settings of pump head and discharge.

## 11.5 UNSTEADY FLOW IN PIPELINES

**KEY CONCEPT**   *For unsteady flow to occur in piping, an excitation to the system is required (e.g., sudden closure of a valve).*

Unsteady, or transient flows in pipelines traditionally have been associated with hydropower piping and with long water and oil pipeline delivery systems. However, the application has broadened in recent years to include hydraulic control system operation, events that take place in power plant piping networks, fluid–structure interaction in liquid-filled piping, and pulsatile blood flow. For the unsteadiness to occur, some type of excitation to the system is necessary. Representative excitations are valve opening or closure, pump or turbine operations, pipe rupture or break, and cavitation events. In this section we examine only the most fundamental aspects of unsteady flow in piping. First we study unsteady flow in a single constant-diameter pipe assuming inelastic and incompressible conditions; that is followed by analysis of a system in which elasticity and compressibility play a significant role in the response of the pressure and velocity to an excitation. The first assumption results in the phenomenon called *surging*, while the second one leads to the phenomenon called *water hammer*.

### 11.5.1    Incompressible Flow in an Inelastic Pipe

Consider a single horizontal pipe of length $L$ and diameter $D$ (Fig. 11.8). The upstream end of the pipe is connected to a reservoir and a valve is situated at the downstream end. The upstream piezometric head is $H_1$, and downstream of the valve the piezometric head is $H_3$. Note that both $H_1$ and $H_3$ are constant time-independent reservoir elevations. The friction factor $f$ is assumed constant, and the loss coefficient for the valve is $K$. There are a number of possible excitations that could be considered, but we will only look at the situation where initially there is a steady velocity $V_0$, the valve is then instantaneously opened to a new position, and

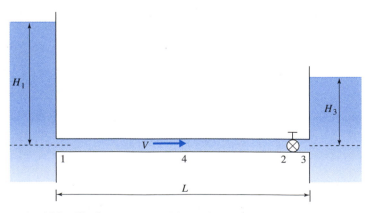

**Fig. 11.8**   Horizontal pipe with a valve at the downstream end.

subsequently the flow accelerates, increasing to a new steady-state velocity $V_{ss}$. For situations in which the valve is closed, either partially or completely, one should consider the possibility of water hammer taking place; this is treated in the next section. Here, we assume that the velocity does not vary with position, only with time.

In Fig. 11.8 define a control volume for the liquid in the pipe between locations 1 and 2, whose mass is $\rho AL$. Note that location 2 is upstream of the valve. At the two ends of the control volume the pressures are $p_1$ and $p_2$, respectively, and on the surface the wall shear stress is $\tau_0$. The conservation of momentum for that liquid volume is given by

$$A(p_1 - p_2) - \tau_0 \pi DL = \rho AL \frac{dV}{dt} \qquad (11.5.1)$$

Without serious consequences we can assume steady-flow conditions across the valve from location 2 to 3, and utilize the energy equation to provide

$$p_2 = p_3 + K \frac{\rho V^2}{2} \qquad (11.5.2)$$

It is reasonable to assume that the Darcy–Weisbach friction factor based on steady-state flow can be employed without undue error. Then, from Section 7.3.3, Eq. 7.3.19, we have the relation

$$\tau_0 = \frac{\rho f V^2}{8} \qquad (11.5.3)$$

Substituting Eqs. 11.5.2 and 11.5.3 into Eq. 11.5.1, dividing by the mass of the liquid column, and recognizing that $p_1 - p_3 = \rho g(H_1 - H_3)$, since the velocity heads are assumed to be negligible, there results

$$\frac{dV}{dt} + \left(\frac{f}{D} + \frac{K}{L}\right)\frac{V^2}{2} - g\frac{H_1 - H_3}{L} = 0 \qquad (11.5.4)$$

Equation 11.5.4 is the relation that represents incompressible unsteady flow in the pipe. The initial condition at $t = 0$ is a given velocity $V = V_0$. When the final steady-state condition is attained, $dV/dt = 0$, and that steady-state velocity, designated as $V_{ss}$, can be obtained by setting the derivative to zero in Eq. 11.5.4:

$$V_{ss} = \sqrt{\frac{2g(H_1 - H_3)}{fL/D + K}} \qquad (11.5.5)$$

Substituting Eq. 11.5.5 into Eq. 11.5.4, separating variables, and expressing the result in integral form, one has

$$\int_0^t dt = \frac{V_{ss}^2 L}{g(H_1 - H_3)} \int_{V_0}^V \frac{dV}{V_{ss}^2 - V^2} \qquad (11.5.6)$$

KEY CONCEPT   *The velocity does not vary with position, only with time.*

Upon integration, the resulting relation defines the velocity $V$ relative to time $t$ after the valve excitation:

$$t = \frac{V_{ss}L}{2g(H_1 - H_3)} \ln \frac{(V_{ss} + V)(V_{ss} - V_0)}{(V_{ss} - V)(V_{ss} + V_0)} \qquad (11.5.7)$$

There are some significant features to the solution. First, by studying Eq. 11.5.7 one concludes that infinite time is required to reach the steady-state velocity $V_{ss}$. In reality, $V_{ss}$ will be reached some time sooner, because the losses have not been completely accounted for. However, it is possible to determine the time when a percentage, for example 99%, of $V_{ss}$ has been reached that would be adequate for engineering purposes. In addition, it is possible for the fluid to be initially at rest, that is, $V_0 = 0$. Finally, remember that the solution is based on the assumptions that both liquid compressibility and pipe elasticity are ignored; the next section addresses the situation when those assumptions are invalid.

## Example 11.10

A horizontal pipe 1000 m in length, with a diameter of 500 mm, and a steady velocity of 0.5 m/s, is suddenly subjected to a new piezometric head differential of 20 m when the downstream valve suddenly opens and its coefficient changes to $K = 0.2$. Assuming a friction factor of $f = 0.02$, determine the final steady-state velocity, and the time when the actual velocity is 75% of the final value.

### Solution
Given are $L = 1000$ m, $D = 0.5$ m, $V_0 = 0.5$ m/s, $f = 0.02$, and $K = 0.2$. Setting $H_1 - H_3 = 20$ m, the final steady-state velocity $V_{ss}$ is found by direct substitution into Eq. 11.5.5:

$$V_{ss} = \sqrt{\frac{2 \times 9.81 \times 20}{0.02 \times 1000/0.5 + 0.2}} = 3.12 \text{ m/s}$$

The velocity that is 75% of $V_{ss}$ is $V = 0.75 \times 3.12 = 2.34$ m/s. Then, using Eq. 11.5.7, the time corresponding to that velocity is

$$t = \frac{3.12 \times 1000}{2 \times 9.81 \times 20} \ln \frac{(3.12 + 2.34) \times (3.12 - 0.5)}{(3.12 - 2.34) \times (3.12 + 0.5)} = 12.9 \text{ s}$$

Hence the final steady-state velocity is 3.12 m/s, and the time when 75% of that velocity is attained is approximately 13 s.

## 11.5.2   Compressible Flow in an Elastic Pipe

In contrast to the developments of Section 11.5.1, there are situations where the liquid is not incompressible and the piping is not rigid. Rather, the interaction between changes in momentum and applied forces causes the liquid to slightly compress and the pipe material to experience very small deformations. When this occurs, significant pressure changes can take place, and the phenomenon is termed water hammer. Water hammer is accompanied by pressure and velocity perturbations traveling at very high velocities, close to the speed of sound in the liquid. The resulting wave action occurs at relatively high frequencies. We will consider the situation in which a valve at the downstream end of a pipe closes or opens suddenly, either partially or completely, to initiate a water hammer response.

> **KEY CONCEPT** *Water hammer is accompanied by pressure and velocity perturbations that travel at high velocities.*

First let us develop the fundamental equations. Consider again the horizontal pipe shown in Fig. 11.8, where now the valve will close so rapidly that elastic effects cause water hammer to occur. The movement of the valve will cause an acoustic, or pressure, wave with speed $a$ to propagate upstream. A control volume of an incremental section of liquid contained in the pipe is shown in Fig. 11.9a, where the pressure wave at a given instant is located. The presence of the wave implies that an unsteady flow is taking place within the control volume; at the entrance the velocity is $V$, and at the exit it is $V + \Delta V$. Steady-state conservation laws can be applied if the wavefront is made to appear stationary by an observer moving with the wavespeed (Fig. 11.9b). (Refer to Section 4.5.3 for the development of inertial reference frames that move with a constant velocity.) In contrast to Fig. 11.9a, the entrance velocity is now $V + a$, and at the exit it is $V + \Delta V + a$. The pressure, pipe area, and density at the entrance are $p$, $A$, and $\rho$, respectively. Due to the passage of the pressure wave, at the exit the pressure, pipe area, and

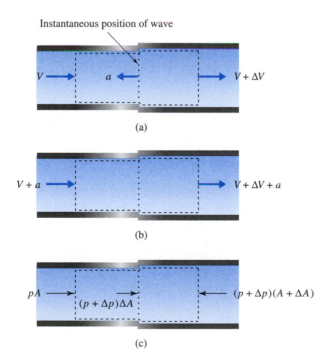

**Fig. 11.9**  Pressure wave moving through a segment of horizontal pipe: (a) pressure wave moving to the left at speed $a$; (b) pressure wave appears stationary using the principle of superposition; (c) pressure forces acting on the control volume.

liquid density are altered to $p + \Delta p$, $A + \Delta A$, and $\rho + \Delta \rho$, where $\Delta p$, $\Delta A$, and $\Delta \rho$ are the respective changes in pressure, area, and density.

Applying the conservation of mass across the control volume in Fig. 11.9b, we find that

$$0 = (\rho + \Delta\rho)(V + \Delta V + a)(A + \Delta A) - \rho(V + a)A \tag{11.5.8}$$

Neglecting frictional and gravitational forces, only pressure forces act on the control volume in the direction of flow, as shown in Fig. 11.9c. The conservation of momentum across the control volume is

$$pA + (p + \Delta p)\,\Delta A - (p + \Delta p)(A + \Delta A)$$
$$= \rho A(V + a)[V + \Delta V + a - (V + a)] \tag{11.5.9}$$

Equations 11.5.8 and 11.5.9 are expanded, and the terms containing factors of $\Delta^2$ and $\Delta^3$ are dropped, since they are much smaller in magnitude than the remaining ones. Then Eqs. 11.5.8 and 11.5.9 become

$$\rho A \Delta V + (V + a)(A \Delta \rho + \rho \Delta A) = 0 \tag{11.5.10}$$

and

$$- A \Delta p = \rho A(V + a)\,\Delta V \tag{11.5.11}$$

In nearly all situations $V \ll a$, so that Eq. 11.5.11 becomes

$$\Delta p = -\rho a \Delta V \tag{11.5.12}$$

**Joukowsky equation:** *Relates the pressure change to the pressure wave speed and the change in velocity.*

Equation 11.5.12, called the **Joukowsky equation**, relates the pressure change to the pressure wave speed and the change in velocity. Note that a velocity reduction (a negative $\Delta V$) produces a pressure rise (a positive $\Delta p$), and a positive $\Delta V$ yields a negative $\Delta p$.

Once the wave has passed through the control volume, the altered conditions $p + \Delta p$, $V + \Delta V$, $A + \Delta A$, and $\rho + \Delta \rho$ will persist until the wave reflects from the upstream boundary; this wave action will be discussed later. First it is useful to examine the nature of the pressure pulse wave speed $a$. Equations 11.5.10 and 11.5.11 are combined, again recognizing that $V \ll a$, to eliminate $\Delta V$, with the result that

$$\frac{\Delta p}{\rho a^2} = \frac{\Delta \rho}{\rho} + \frac{\Delta A}{A} \tag{11.5.13}$$

From the definition of the bulk modulus of elasticity $B$ for a fluid, Eq. 1.5.11, we can relate the change in density to the change in pressure as $\Delta\rho/\rho = \Delta p/B$. The change in pipe area can be related to the change in pressure by considering an instantaneous, elastic response of the pipe wall to pressure changes. Assuming a circular pipe cross section of radius $r$, we have $\Delta A/A = 2\Delta r/r$, and the change in circumferential strain $\varepsilon$ in the pipe wall is $\Delta\varepsilon = \Delta r/r$. For a thin-walled pipe whose thickness $e$ is much smaller than the radius, that is, $e \ll r$, the circumferential stress is given by $\sigma = pr/e$. For small changes in $r$ and $e$, $\Delta\sigma \approx (r/e)\Delta p$. The elastic modulus for the pipe wall material is the change in stress divided by the change in strain, or

$$E = \frac{\Delta\sigma}{\Delta\varepsilon} \approx \frac{(r/e)\Delta p}{\Delta r/r} = \frac{(2r/e)\Delta p}{\Delta A/A} \tag{11.5.14}$$

Solving for $\Delta A/A$, and substituting the result into Eq. 11.5.13, along with the relative change in density related to the change in pressure and bulk modulus, there results

$$\frac{\Delta\rho}{\rho a^2} = \frac{\Delta p}{B} + \frac{2r\Delta p}{E} \tag{11.5.15}$$

The parameter $\Delta p$ can be eliminated in Eq. 11.5.15, the diameter substituted for the radius, and the relation solved for $a$:

$$a = \sqrt{\frac{B/\rho}{1 + (D/e)(B/E)}} \tag{11.5.16}$$

Hence the pressure pulse wave speed is shown to be related to the properties of the liquid ($\rho$ and $B$) and those of the pipe wall ($D$, $e$, and $E$). If the pipe is very rigid, or stiff, then the term $DB/eE \ll 1$, and Eq. 11.5.16 becomes $a = \sqrt{B/\rho}$, which is the speed of sound in an unbounded liquid (see Eq. 1.5.12). Note that the effect of pipe elasticity is to reduce the speed of the pressure wave.

The use of Eqs. 11.5.12 and 11.5.16 will provide only the magnitude of the water hammer excitation. In addition, it is necessary to understand the nature of the pressure pulse wave as it travels throughout the pipe and is reflected from the boundaries. Consistent with the development of Eqs. 11.5.12 and 11.5.16, we neglect friction to simplify the analysis. Consider the situation in which the valve is suddenly closed and the pipe is rigid. Let us study in detail the sequence of events throughout one cycle of motion, as illustrated in Fig. 11.10.

In Fig. 11.10a an initial steady condition exists, the velocity is $V_0$, and the valve is suddenly closed at $t = 0$. Note that when friction is neglected, the hydraulic grade line appears horizontal. Subsequent to closure of the valve, the wave travels upstream (Fig. 11.10b). Behind the wave, the velocity is reduced to zero, the pressure rises by the amount $\Delta p$, the liquid has been compressed, and the pipe slightly expanded. At time $L/a$ the wave reaches the reservoir, and an

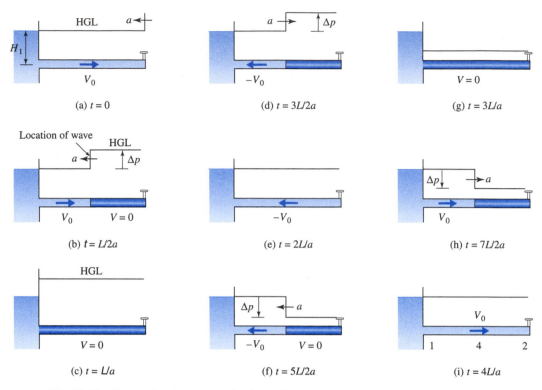

**Fig. 11.10**    One cycle of wave motion in a pipe due to a sudden valve closure.

unbalanced force occurs at the pipe inlet (Fig. 11.10c). In the pipe the pressure reduces to the reservoir pressure, and the velocity suddenly reverses direction; this process begins at the upstream end and propagates downstream at the wave speed $a$ (Fig. 11.10d).

When the wave reaches the valve at time $2L/a$, the velocity has magnitude $-V_0$ throughout the pipe (Fig. 11.10e). Adjacent to the valve, which is now closed, the velocity reduces to zero, and the pressure reduces by the amount $\Delta p$ (Fig. 11.10f). The low-pressure wave travels upstream at speed $a$, and behind the wave the liquid is expanded and the pipe wall is contracted. Note that if the pressure behind the wave reduces to vapor pressure, cavitation will occur, and some of the liquid will vaporize. When the pressure wave reaches the reservoir at time $3L/a$, an unbalanced condition occurs again, opposite in magnitude to that at time $L/a$ (Fig. 11.10g). Equilibrium of forces will now cause a wave to travel downstream, with the pressure increased by the amount $\Delta p$ and the velocity equal to $+V_0$ behind the front (Fig. 11.10h). When the wave reaches the valve at time $4L/a$, initial steady-state conditions prevail once again throughout the pipe (Fig. 11.10i).

The process repeats itself every $4L/a$ seconds. For the ideal, frictionless situation shown here, the motion will perpetuate. The pressure waveform at the valve and at the midpoint of the pipe and the velocity at the pipe entrance are shown in Fig. 11.11. For a real pipe system, friction, pipe motion, and inelastic behavior of the pipe material will eventually cause the oscillation to die down and cease (see Fig. 11.12).

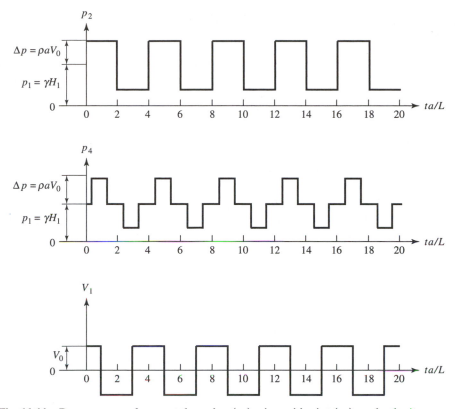

**Fig. 11.11** Pressure waveforms at the valve ($p_2$), pipe midpoint ($p_4$), and velocity waveform at the entrance to the pipe ($V_1$).

The pressure rise $\Delta p$ predicted by Eq. 11.5.13 is based on the assumption that the valve closes instantaneously. Actually, it can be used to predict the maximum pressure rise when the valve closes with any time less than $2L/a$, the time it takes for the pressure wave to travel from the valve to the reservoir and back again. For valve closure times greater than $2L/a$, a more comprehensive analysis is required (Wylie and Streeter, 1993). Keep in mind that the previous discussion

**Fig. 11.12** Pressure waveform at the valve for an actual pipe system following rapid valve closure. (After Martin, 1983.)(Martin, C. D., Experimental Investigation of Column Separation with Rapid Valve Closure, Proceedings, 4th International Conference on Pressure Surges, BHRA Fluid Engineering, Cranfield, England, 1983, pp. 77–88. Reproduced with permission.)

pertains only to a single horizontal pipe with a reservoir at the upstream end and a valve closing instantaneously at the downstream end and one that contains a frictionless liquid. Most of the water hammer analysis performed by engineers today makes use of computer-based numerical methods for complex piping systems, incorporating a variety of excitation mechanisms such as pumps, surge suppressers, and various types of valves (Wylie and Streeter, 1993).

## Example 11.11

A steel pipe ($E = 207$ MPa, $L = 1500$ m, $D = 300$ mm, $e = 10$ mm) conveys water at $20°C$. The initial velocity is $V_0 = 1$ m/s. A valve at the downstream end is closed so rapidly that the motion is considered to be instantaneous, reducing the velocity to zero. Determine the pressure pulse wave speed in the pipe, the speed of sound in an unbounded water medium, the pressure rise at the valve, the time it takes for the wave to travel from the valve to the reservoir at the upstream end, and the period of oscillation.

### Solution

The density and bulk modulus of water at $20°C$ are found in Table B.1: $\rho = 998$ kg/m$^3$ and $B = 220 \times 10^7$ Pa. The pressure wave speed, $a$, is given by Eq. 11.15.16:

$$a = \sqrt{\frac{220 \times 10^7/998}{1 + \frac{0.3}{0.01} \times \frac{220 \times 10^7}{207 \times 10^9}}} = 1290 \text{ m/s}$$

The speed of sound in an unbounded water medium is found by use of Eq. 1.5.12 to be

$$a = \sqrt{\frac{220 \times 10^7}{998}} = 1485 \text{ m/s}$$

Note that the sound speed is about 15% larger than the pressure wave speed. To compute the pressure rise at the valve, we recognize that the reduction in velocity is $-V_0 = -1$ m/s. Using Eq. 11.15.12, the increase in pressure upstream of the valve is

$$\Delta p = -998 \times 1290 \times (-1) = 1.29 \times 10^6 \text{ Pa} \quad \text{or} \quad 1290 \text{ kPa}$$

The wave travel time from the valve to the reservoir is $L/a = 1500/1290 = 1.16$ s, and the period of oscillation is $4L/a = 4 \times 1.16 = 4.65$ s.

## 11.6 SUMMARY

A methodology for dealing with losses in piping has been provided in Section 11.2 using the resistance coefficient as a generalized term that can include minor losses along with a chosen empirical friction loss. For accurate representation of losses, the Darcy–Weisbach friction loss is recommended, with losses proportional to the square of the velocity (or discharge). Equation 11.2.3 is useful for estimating the Darcy–Weisbach friction factor in an iterative, or trial, solution. In Section 11.3 simple pipe systems have been defined as those that contain one to several pipes, arranged singly, in series, parallel, or branched. Typically, a problem consists of finding the flow distribution in the piping; along with discharges the piezometric head often is an unknown. Calculator-based solutions or computational software are useful tools to analyze these systems.

For more complex piping networks, a systematic solution approach has been presented in Section 11.4. First the network equations were linearized, and then the Hardy Cross method was applied to solve for the network flows and piezometric heads. A spreadsheet solution was applied to a network using the Hardy Cross solution; it allows one to correctly apply the energy and continuity equations as well as track the convergence of the solution. Note that even the simple pipe systems presented in Section 11.3 can be solved using the Hardy Cross method. The computer program EPANET was also introduced in Section 11.4. This software is useful for large systems in which spreadsheet programming becomes cumbersome.

Finally, analysis of unsteady flows in pipes has been introduced in Section 11.5. We have focused on flow that behaves either in an incompressible or compressible manner; the latter behavior is commonly called "water hammer." Water hammer is a result of acoustic waves propagating in a pipe. The waves travel at the speed of sound in the contained liquid and produce a periodic response of both pressures and velocities throughout the pipe. The Joukowsky relation, Eq. 11.5.12, gives the magnitude of the pressure wave.

## PROBLEMS

### Steady Flows

**11.1** Using the results from Example 11.2, tabulate and plot to correct vertical scale the hydraulic and energy gradelines. Assume a 10-m length between the two elbows.

**11.2** A pump is situated between two sections in a horizontal pipeline. The diameter $D_1$ and pressure $p_1$ are given at the upstream section, and $D_2$ and $p_2$ are given at the downstream section. Determine the required fluid power of the pump for the following conditions:

$D_1 = 50$ mm, $p_1 = 350$ kPa, $D_2 = 80$ mm, $p_2 = 760$ kPa, $Q = 95$ L/min, $h_L = 6.6$ m, water is flowing at 20°C.

**11.3** An oil ($S = 0.82$) is pumped between two storage tanks in a pipe with the following characteristics: $L = 2440$ m, $D = 200$ mm, $f = 0.02$, $\Sigma K = 12.5$. The upper tank is 32 m higher than the lower one. Using the pump data provided, determine:
(a) The oil discharge in the pipe.
(b) The power requirement for the pump.

| $Q$ (L/s) | 0 | 15 | 30 | 45 | 60 | 75 | 100 |
|---|---|---|---|---|---|---|---|
| $H_P$ (m) | 55 | 54 | 53 | 52 | 49 | 44 | 35 |
| $\eta$ | 0 | 0.4 | 0.6 | 0.7 | 0.75 | 0.7 | 0.5 |

**11.4** Consider the simple pumping system shown in Fig. P11.4. Assume the following parameters to be known:

Reservoir elevations, $z_1$, $z_2$
Pipe length, $L$
Pipe roughness, $e$
Sum of minor loss coefficients, $\Sigma K$
Kinematic viscosity, $\nu$
Discharge, $Q$

Develop a numerical solution to find the required diameter. Use an equivalent length to represent the minor losses. Determine the diameter for the following data:

**(a)** $z_2 - z_1 = 36$ m, $L = 450$ m, $e = 0.0009$ m, $\Sigma K = 2.5$, water at 10°C, $Q = 950$ L/s, pump characteristic curve:

| $H_P$ (m) | 147 | 146 | 142 | 136 | 127 | 122 |
|---|---|---|---|---|---|---|
| $Q$ (L/s) | 0 | 252 | 505 | 757 | 1010 | 1136 |

**(b)** $z_2 - z_1 = 40$ m, $L = 500$ m, $e = 1$ mm, $\Sigma K = 2.5$, water at 20°C, $Q = 1.1$ m³/s, pump characteristic curve:

| $H_P$ (m) | 160 | 158 | 154 | 148 | 138 |
|---|---|---|---|---|---|
| $Q$ (L/s) | 0 | 400 | 800 | 1200 | 1600 |

**Fig. P11.4**

**11.5** Gasoline is being pumped at 400 L/s in a pipeline from location A to B, as shown in Fig. P11.5. The pipe follows the topography as shown, with the highest elevation shown at location C. The only contributions to minor losses are the two valves located at the ends of the pipe. If $S = 0.81$, $\nu = 4.26 \times 10^{-7}$ m²/s, $p_v = 55.2$ kPa absolute, and $p_{atm} = 100$ kPa,
**(a)** Determine the necessary power to be delivered to the system to meet the flow requirement.

**(b)** What is the maximum elevation possible at location C without causing vapor pressure conditions to exist?
**(c)** Sketch the hydraulic grade line.

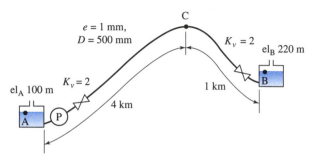

**Fig. P11.5**

**11.6** For the three pipes in series shown in Fig. P11.6, minor losses are proportional to the discharge squared, and the Hazen–Williams formula is used to account for friction losses. With the data given, use Newton's method to determine the discharge. Note that minor losses can be neglected in order to initially estimate $Q$.

$$(p/\gamma + z)_A = 250 \text{ m and } (p/\gamma + z)_B = 107 \text{ m}.$$

| Pipe | $L$ (m) | $D$ (mm) | $\Sigma K$ | $C$ (Hazen–Williams) |
|---|---|---|---|---|
| 1 | 200 | 200 | 2 | 100 |
| 2 | 150 | 250 | 3 | 120 |
| 3 | 300 | 300 | 0 | 90 |

**Fig. P11.6**

**11.7** A long oil pipeline consists of the three segments shown in Fig. P11.7. Each segment has a booster pump that is used primarily to overcome pipe friction. The two reservoirs are at the same elevation.

(a) Derive a single equation to determine the discharge in the system if the resistance factors $\bar{R}_i$ for each pipe and the useful power $\dot{W}_{fi}$ for each pump are known.

(b) Determine the discharge for the data given in the figure. The unit weight of oil is $\gamma = 8830$ N/m$^3$.

$Q \longrightarrow$

$\bar{R}_1 = 4 \times 10^4$ s$^2$/m$^5$    $\bar{R}_3 = 2 \times 10^5$ s$^2$/m$^5$

$\bar{R}_2 = 3 \times 10^4$ s$^2$/m$^5$

$\dot{W}_{f_1} = 200$ kW    $\dot{W}_{f_2} = 200$ kW    $\dot{W}_{f_3} = 200$ kW

**Fig. P11.7**

**11.8** A liquid with a specific gravity of 0.68 is pumped from a storage tank to a free jet discharge through a pipe of length $L$ and diameter $D$ (Fig. P11.8). The pump provides a known amount of fluid power $\dot{W}_f$ to the liquid. Assuming a constant friction factor of 0.015, determine the discharge for the following conditions:

(a) $z_1 = 24$ m, $p_1 = 110$ kPa, $z_2 = 18$ m, $L = 450$ m, $D = 300$ mm, $\dot{W}_f = 10$ kW.

(b) $z_1 = 22.5$ m, $p_1 = 100$ kPa, $z_2 = 18$ m, $L = 450$ m, $D = 200$ mm, $\dot{W}_f = 12$ kW.

**Fig. P11.8**

**11.9** Water at 20°C is being pumped through the three pipes in series as shown in Fig. P11.9. The power delivered to the pump is 1920 kW, and the pump efficiency is 0.82. Compute the discharge.

| Pipe | L (m) | D (mm) | e (mm) | Σ K |
|------|-------|--------|--------|-----|
| 1 | 200 | 1500 | 1 | 2 |
| 2 | 300 | 1000 | 1 | 0 |
| 3 | 120 | 1200 | 1 | 10 |

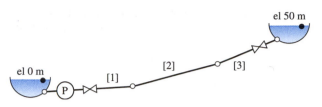

**Fig. P11.9**

**11.10** Find the water flow distribution in the parallel system shown in Fig. P11.10, and the required pumping power if the discharge through the pump is $Q_1 = 3$ m$^3$/s. The pump efficiency is 0.75. Assume constant friction factors.

| Pipe | L (m) | D (mm) | f | Σ K |
|------|-------|--------|-------|-----|
| 1 | 100 | 1200 | 0.015 | 2 |
| 2 | 1000 | 1000 | 0.020 | 3 |
| 3 | 1500 | 500 | 0.018 | 2 |
| 4 | 800 | 750 | 0.021 | 4 |

**Fig. P11.10**

**11.11** For the system shown in Fig. P11.11, determine the water flow distribution and the piezometric head at the junction using an ad hoc approach. Assume constant friction factors. The pump characteristic curve is $H_P = a - bQ^2$.

(a) $a = 20$ m, $b = 30$ s$^2$/m$^5$, $z_1 = 10$ m, $z_2 = 20$ m, $z_3 = 18$ m.

(b) $a = 16.5$ m, $b = 40$ s$^2$/m$^2$, $z_1 = 6$ m, $z_2 = 15$ m, $z_3 = 13.5$ m.

| Pipe | L (m) | D (cm) | f | K |
|------|-------|--------|-------|---|
| 1 | 30 | 24 | 0.020 | 2 |
| 2 | 60 | 20 | 0.015 | 0 |
| 3 | 90 | 16 | 0.025 | 0 |

**Fig. P11.11**

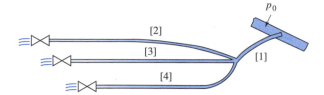

**Fig. P11.14**

**11.12** Work Problem 11.11 using a numerical approach:
(a) Newton's method.
(b) False position method.

**11.13** Find the flow distribution in the parallel network shown in Fig. P11.13. Assume constant friction factors. The change in hydraulic grade line between A and B is $(p/\gamma + z)_A - (p/\gamma + z)_B = 50$ m.

| Pipe | L (m) | D (mm) | e (mm) | $\Sigma K$ |
|------|-------|--------|--------|------------|
| 1 | 600 | 1000 | 0.1 | 2 |
| 2 | 1000 | 1200 | 0.15 | 0 |
| 3 | 550 | 850 | 0.2 | 4 |
| 4 | 800 | 1000 | 0.1 | 1 |

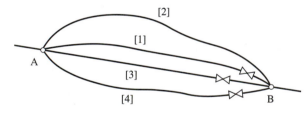

**Fig. P11.13**

**11.14** The water sprinkling system shown in Fig. P11.14 is applied from a large diameter pipe with constant internal pressure $p_0 = 300$ kPa. The system is positioned in a horizontal plane. Determine the flow distribution $Q_1, Q_2, Q_3, Q_4$ for the given data. Valve losses are included in the $\bar{R}$, values. (*Hint:* If you are clever, no trial-and-error solution is necessary!)

| Pipe | $\bar{R}$ (s²/m⁵) |
|------|-------------------|
| 1 | $1.6 \times 10^4$ |
| 2 | $5.3 \times 10^5$ |
| 3 | $1.0 \times 10^6$ |
| 4 | $1.8 \times 10^6$ |

**11.15** For the system shown in Fig. P11.15, a fluid flows from A to B. Determine in which pipe there is the greatest velocity.

| Pipe | L (m) | D (mm) | f |
|------|-------|--------|-------|
| 1 | 2000 | 450 | 0.012 |
| 2 | 650 | 150 | 0.020 |
| 3 | 1650 | 300 | 0.015 |

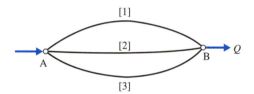

**Fig. P11.15**

**11.16** A proposed water irrigation system consists of one main pipe with a pump and three pipe branches (Fig. P11.16). Each branch is terminated by an orifice, and each orifice has the same elevation. It is apparent that the flow distribution can be solved by treating the piping arrangement as a branched system. However, the piping can also be treated as a parallel system in order to determine the flows.
(a) Identify the equations and unknowns to satisfy the solution for a parallel system. Why is it possible to treat the irrigation system as a parallel piping problem?
(b) Why would the solution to a parallel system be preferred to that of a branching system?
(c) Determine the flow distribution and sketch the hydraulic grade line.
(d) What part of the piping would one change to approximately double the discharge, assuming that the individual lengths and pump curve could not be altered?

$R_1 = 34\ 650$
$R_2 = 82\ 500$
$R_3 = 127\ 900$
$R_4 = 115\ 500$
el 0 m
el 0 m
P
$H_p = 45 - 1 \times 10^4 Q^2$

($H_p$ in m; $Q$ in m³/s; $\bar{R}$ in s²/m⁵, orifice loss included in $\bar{R}$)

**Fig. P11.16**

**11.17** Determine the required head ($H_P$) and discharge ($Q_1$) of water to be handled by the pump for the system shown in Fig. P11.17. The discharge in pipe 2 is $Q_2 = 35$ L/s in the direction shown.

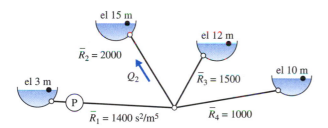

el 15 m
el 12 m
$\bar{R}_2 = 2000$
$Q_2$
el 3 m
$\bar{R}_3 = 1500$
el 10 m
P
$\bar{R}_1 = 1400$ s²/m⁵
$\bar{R}_4 = 1000$

**Fig. P11.17**

**11.18** Determine the flow distribution of water in the parallel piping system shown in Fig. P11.18.
   (a)  $Q_{in} = 600$ L/min
   (b)  $Q_{in} = 480$ L/min

| Pipe | L (m) | D (mm) | f | Σ K |
|------|-------|--------|-------|-----|
| 1 | 30 | 50 | 0.020 | 3 |
| 2 | 40 | 75 | 0.025 | 5 |
| 3 | 60 | 60 | 0.022 | 1 |

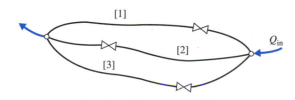

[1]
[2]
[3]
$Q_{in}$

**Fig. P11.18**

**11.19** Water is being pumped in the piping system shown in Fig. P11.19. The pump curve is approximated by the relation $H_P = 150 - 5Q_1^2$, with $H_P$ in meters and $Q_1$ in m³/s. The pump efficiency is $\eta = 0.75$. Compute the flow distribution and find the required pump power.

| Pipe | $\bar{R}$ (s²/m⁵) |
|------|-------------------|
| 1 | 400 |
| 2 | 1000 |
| 3 | 1500 |

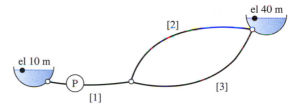

el 40 m
[2]
el 10 m
P
[1]
[3]

**Fig. P11.19**

**11.20** A pipeline consists of two pipe segments in series (Fig. P11.20). The specific gravity of the fluid is 0.81. If pump A has a constant power input of 1 MW, find the discharge, the pressure head in pumps A and B, and the required power for pump B. The minimum allowable pressure on the suction side of pump B is 150 kPa, and both pumps have an efficiency of 0.76.

| Pipe | L (m) | D (mm) | Σ K | f |
|------|-------|--------|-----|-------|
| 1 | 5000 | 750 | 2 | 0.023 |
| 2 | 7500 | 750 | 10 | 0.023 |

el 27 m
el 50 m
[2]
el 0 m
[1]
P
P
A
B

**Fig. P11.20**

**11.21** Determine the flow distribution of water in the system shown in Fig. P11.21. The equivalent roughness for all elements is 0.1 mm.

| Pipe | L (m) | D (mm) | Σ K |
|------|-------|--------|-----|
| 1 | 1000 | 200 | 3 |
| 2 | 200 | 25 | 0 |
| 3 | 250 | 25 | 0 |
| 4 | 340 | 30 | 2 |
| 5 | 420 | 40 | 0 |
| 6 | 500 | 175 | 5 |

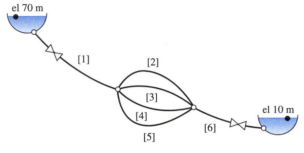

**Fig. P11.21**

**11.22** Determine the flow distribution of water in the branching pipe system shown in Fig. P11.22:
**(a)** Without a pump in line 1.
**(b)** Include a pump located in line 1 adjacent to the lower reservoir. The pump characteristic curve is given by $H_P = 250 - 0.4Q - 0.1Q^2$, with $H_P$ in meters and $Q$ in m$^3$/s.

The Hazen–Williams coefficient $C = 130$ for all pipes.

| Pipe | L (m) | D (mm) |
|------|-------|--------|
| 1 | 200 | 500 |
| 2 | 600 | 300 |
| 3 | 1500 | 300 |
| 4 | 1500 | 400 |

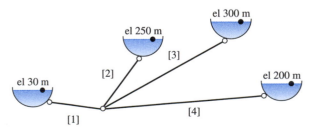

**Fig. P11.22**

**11.23** Determine the flow distribution of water in the system shown in Fig. P11.23. Assume constant friction factors, with $f = 0.02$. The head-discharge relation for the pump is $H_P = 60 - 10Q^2$, where $H_P$ is in meters and the discharge is in cubic meters per second.

| Pipe | L (m) | D (mm) | Σ K |
|------|-------|--------|-----|
| 1 | 100 | 350 | 2 |
| 2 | 750 | 200 | 0 |
| 3 | 850 | 200 | 0 |
| 4 | 500 | 200 | 2 |
| 5 | 350 | 250 | 2 |

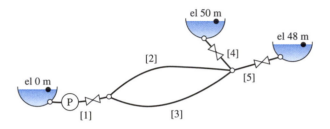

**Fig. P11.23**

**11.24** A pump, whose performance and efficiency curves are given in Fig. P11.24a has been selected to deliver water in a piping system. The piping consists of four pipes arranged as shown in Fig. P11.24b. Water at 15°C is being pumped from reservoir A and exits at either reservoir B or at location D, depending on whether the valves at those locations are open or closed. The pipe characteristics are shown in the accompanying table. All of the pipe diameters are 10 cm, and the friction factor in each pipe is assumed to be $f = 0.04$.
**(a)** If the discharge through the pump is 5.2 L/s, what is the head loss across pipe 2?
**(b)** Compute the discharge in the system, assuming that the valve at location D is closed.
**(c)** If the valve at location D is open and the discharge through the pump is 11.6 L/s, determine the discharge in pipe 4.

| Pipe | L (m) | Σ K |
|------|-------|-----|
| 1 | 3 | 1 |
| 2 | 150 | 2 |
| 3 | 600 | 2 |
| 4 | 225 | 4 |

(a)

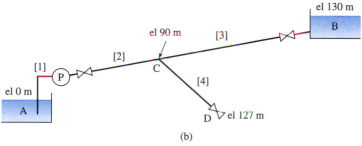

(b)

**Fig. P11.24**

**11.25**  Refer to the system and pump curves associated with Problem 11.24. Assume that the discharge through the pump is 12.6 L/s.

  **(a)**   What is the required power for the pump?
  **(b)**   Determine the pressure at location C.
  **(c)**   If the valve at location D is closed, what is the gage pressure at that location?

**11.26**  Refer to the system and pump curves associated with Problem 11.24.

  **(a)**   Assuming that the valve at location D is open and the head rise across the pump is 140 m, determine the discharge in pipe 3.
  **(b)**   If the valve at location B was closed and the valve at location D was open, what is the head rise across the pump?

**11.27** For the network shown in Fig. P11.27, perform the following:
  **(a)** Identify all of the unknown parameters and write the required system equations. Subsequently, reduce them to one equation in one unknown. Assume $\beta = 2$.
  **(b)** Prepare the network for a Hardy Cross analysis by making a sketch, identifying necessary loops and paths, and writing the appropriate equations.

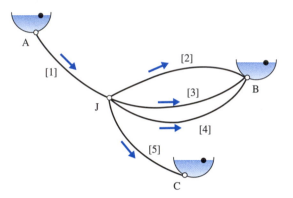

**Fig. P11.27**

**11.28** The branching system shown in Fig. P11.28a is to be solved by using the Hardy Cross method.
  **(a)** Write the equations for $\Delta Q_I$, $\Delta Q_{II}$, $W$, and $G$.
  **(b)** Perform two iterations by filling in the blanks in the table shown in Fig. P11.28b.

After two iterations, sketch the hydraulic grade line.

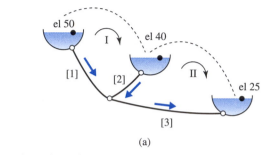

(a)

| Loop | Pipe | $\bar{R}$ | $2\bar{R}$ | $Q$ | $W$ | $G$ | $Q$ | $W$ | $G$ | $Q$ |
|------|------|-----------|------------|------|-----|-----|-----|-----|-----|-----|
| I | 1 | 2 | 4 | −3.5 | | | | | | |
|   | 2 | 2 | 4 | 0 | | | | | | |

$$\Delta Q_I = \qquad\qquad \Delta Q_I =$$

| Loop | Pipe | $\bar{R}$ | $2\bar{R}$ | $Q$ | $W$ | $G$ | $Q$ | $W$ | $G$ | $Q$ |
|------|------|-----------|------------|------|-----|-----|-----|-----|-----|-----|
| II | 2 | 2 | 4 | 0 | | | | | | |
|    | 3 | 2 | 4 | −3.5 | | | | | | |

$$\Delta Q_{II} = \qquad\qquad \Delta Q_{II} =$$

(b)

**Fig. P11.28**

**11.29** Determine the flow and the hydraulic grade line in the system shown in Fig. P11.29 using the Hardy Cross analysis method. Use $\gamma = 9800 \ \text{N/m}^3$.

**Fig. P11.29**

**11.30** The solution of flow of water in a pipe network is shown in Fig. P11.30. Compute:
  **(a)** The hydraulic grade line throughout the system.
  **(b)** The pressure at each node.

**Fig. P11.30**

**11.31** For the configuration shown in Problem 11.27, determine the water flow distribution and the hydraulic grade line at J. The system data are tabulated below.

| Reservoir | Elevation (m) | Line | L (m) | D (cm) | f | Σ K |
|-----------|---------------|------|-------|--------|-------|-----|
| A | 195 | 1 | 240 | 20 | 0.015 | 0 |
| B | 173 | 2 | 180 | 7.5 | 0.020 | 2 |
| C | 54 | 3 | 195 | 7.5 | 0.020 | 2 |
| | | 4 | 127 | 7.5 | 0.025 | 3 |
| | | 5 | 300 | 10 | 0.015 | 4 |

**11.32** Water is flowing in the piping system shown in Fig. P11.32. Determine the flow distribution using the Hardy Cross method:

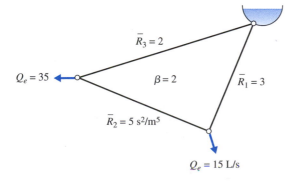

**Fig. P11.32**

**11.33** Determine the water flow distribution in the piping system shown in Fig. P11.33 and the pressures at the internal nodes employing the Hardy Cross method of solution.

| Pipe | L (m) | D (mm) | e (mm) | Σ K |
|------|-------|--------|--------|-----|
| 1 | 500 | 300 | 0.15 | 0 |
| 2 | 600 | 250 | 0.15 | 0 |
| 3 | 50 | 150 | 0.15 | 10 |
| 4 | 200 | 250 | 0.15 | 2 |
| 5 | 200 | 300 | 0.15 | 2 |

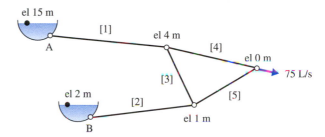

**Fig. P11.33**

**11.34** Solve Problem 11.9 by employing the Hardy Cross method. Use Eq. 11.2.6 to compute the friction factors.

**11.35** Determine the discharge in the piping system shown in Fig. P11.35:
  **(a)** With an exact solution.
  **(b)** Using the Hardy Cross method.

Water is the flowing liquid and the losses are proportional to the velocity squared ($\beta = 2$). The pump characteristic curve is $H_P = 100 - 826Q^2$ ($H_P$ in meters and $Q$ in m³/s).

**Fig. P11.35**

**11.36** The piping system shown in Fig. P11.36 delivers water to two side-branch sprinklers (C and E) and to a lower reservoir (F). The water is supplied by an upper reservoir (A). The sprinklers are represented hydraulically as orifices with relatively high $K$ values. Determine the flow distribution using the Hardy Cross method.

| Line | L (m) | D (mm) | e (mm) | ΣK |
|------|-------|--------|--------|----|
| 1 | 200 | 100 | 0.1 | 2 |
| 2 | 150 | 50 | 0.1 | 30 |
| 3 | 500 | 100 | 0.1 | 0 |
| 4 | 35 | 50 | 0.1 | 35 |
| 5 | 120 | 100 | 0.1 | 2 |

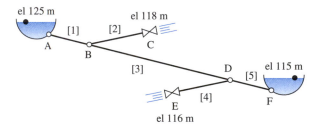

**Fig. P11.36**

**11.37** In Problem 11.36, if a pump supplying 10 kW of useful power was inserted near A, what changes in flow and pressure would occur?

**11.38** A cooling water condenser flow system for a thermal power plant is shown in Fig. P11.38. Water is pumped from a reservoir at location A through a large-diameter pipe [1], passing through the condenser [2], and discharging through another large-diameter pipe [3] into the receiving pond, location B. The condenser consists of a large number of elevated, small-diameter tubes placed in parallel; at either end of the condenser is a large water-filled box called a header, C and C'. Water surface elevations at A and B are known, as are the lengths and diameters of pipes 1 and 3. The condenser has $N$ identical tubes, each with the same known diameter $D_2$ and length $L_2$. The friction factor for all pipes is the same constant value, and the pump curve is approximated by the relation $H_P = a_0 + a_1 Q + a_2 Q^2 + a_3 Q^3$. Minor losses can be assumed negligible.

**(a)** Derive an equivalent single-pipe resistance coefficient $R_2$ for the condenser tubes.

**(b)** Write the energy equation for the entire system, using the required given variables and expressing flow resistance in terms of the resistance coefficients for pipes 1 and 3, and the equivalent resistance coefficient for pipe 2.

**(c)** Set up a Hardy Cross algorithm to determine the discharge through the system.

**(d)** If $z_A = 2$ m, $z_B = 0$ m, $L_1 = 100$ m, $D_1 = 2$ m, $L_2 = 15$ m, $D_2 = 0.025$ m, $N = 1000$, $L_3 = 200$ m, $D_3 = 2$ m, and $f = 0.02$, compute the discharge and hydraulic grade line. The coefficients for the pump curve are $a_0 = 30.4$, $a_1 = -31.8$, $a_2 = 18.6$, and $a_3 = -4.0$, where $Q$ is in cubic meters per second.

**(e)** If the top of the condenser is situated at an elevation of 6 m, what are the pressures at the top of the upstream and downstream headers?

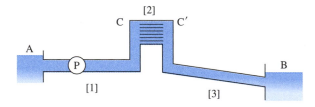

**Fig. P11.38**

**11.39** Determine the flow distribution in the branching or "tree-like" network shown in Fig. P11.39. Water is the flowing liquid: $\gamma = 9810$ N/m³. The source of flow is a large pipeline (location A) maintained at a constant pressure of 400 kPa. Each branch terminates at a location where the hydraulic grade line is known. (Courtesy of D. Wood)

**11.40** The 12-pipe system shown in Fig. P11.40 represents a low-pressure region that is connected to a high-pressure system by pressure regulators set at 350 kPa. Determine the pressure and water flow distribution for the demands shown. Assume Hazen–Williams $C = 120$ for all pipes. (Courtesy of D. Wood.)

105 kPa

9 m

$K = 10$

Hazen–Williams $C = 110$

140 kPa

12 m

30; 50

(Length in m; diameter in mm)

$K = 1$

24; 100  6 m

A    60; 100  45; 50    60; 100    210 kPa

15    15    12 m    $K = 2$

(Elev. in m)

400 kPa    60; 50    90; 50

$K = 5$    30    3 m    $K = 3$

70 kPa

**Fig. P11.39**

el 48

450; 25

55 L/S

27.5 L/S

Regulator
($p = 350$ kPa)    46.5    120; 25   45    180; 25    45

240; 25    240; 25    27.5 L/S    240; 25

45    135; 25    45    165; 25    27.5 L/S

225; 25    (Length in m; diameter in cm)    43.5

Regulator
($p = 350$ kPa)    435; 25    43.5    300; 25    210; 20    42    27.5 L/S

27.5 L/S

45    (Elevation in m)

**Fig. P11.40**

**11.41** Determine the flow distribution for the 14-pipe water supply system shown in Fig. P11.41. The characteristic curve for the pump is represented by the following data (courtesy of D. Wood):

| $H_P$(m) | 166 | 132 | 18 |
|---|---|---|---|
| $Q$ (L/s) | 0 | 600 | 1000 |

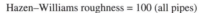

Hazen–Williams roughness = 100 (all pipes)

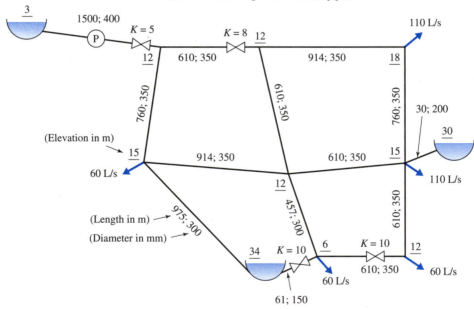

**Fig. P11.41**

**11.42** The layout of the required water demands for a proposed industrial complex are shown in Fig. P11.42. Design an appropriate network and determine the appropriate useful power for a pump to meet the demand needs. The design criteria are as follows:

1. The lower reservoir supplies water to the system. The upper reservoir supplies water for emergency use only, such as fire, line breaks, and so on; under normal operation, there is no flow into or out of it.

2. Under normal operation, pressures in the network can range between 560 kPa and 840 kPa.

3. Two gate valves are to be placed in each line (one at each end) to isolate it in case of a break or for needed maintenance.

4. All demands must be met in case of a single line break with minimum allowable pressures of 140 kPa.

5. Pipes are cast iron, with the following standard diameters available: 10, 15, 20, 30, 40, 50, 60, 75, 90, 105, and 120 cm. Pipes can be placed anywhere within the region.

**Fig. P11.42**

**11.43** Design an irrigation system for part or all of the 18-hole golf course shown in Fig. P11.43. The requirements are as follows:

1. Place the sprinkler couplings at approximately 30-m intervals along the fairway centerline. The flow rate from each sprinkler is to be 2.5 L/s. A maximum of two sprinklers on each fairway are to be in operation at any time.
2. Place four sprinklers around the periphery of each green. The flow rate from each sprinkler is to be 2.5 L/s. A maximum of three sprinklers on each green are to be in operation at any time. As an alternative to specifying the sprinkler flow rate, specify the loss coefficient across each sprinkler (e.g., try $K_v = 35$), and with successive trials, adjust it until the desired flow rate is achieved.
3. Water is to be pumped from the well shown. The maximum possible flow from the well is 125 L/s. Work with useful power in order to satisfy the system demand.
4. System pressure is to be 550 ± 70 kPa at all locations, and the design velocity in each pipe is not to exceed 1.5 m/s. Use smooth plastic piping and fittings throughout.

**Fig. P11.43**

## Unsteady Flow

**11.44**  An irrigation system has a nearly horizontal supply pipe that is connected to a tank at one end and a quick-opening valve at the other. The length of the pipe is 2500 m, and its diameter is 100 mm. If the elevation of water in the tank is 3 m, how long will it take for the flow to reach the 99% steady-state condition if the valve is suddenly opened from a closed position? Assume that the water is incompressible, the pipe is inelastic, $f = 0.025$, and $K = 0.15$ once the valve is opened.

**11.45**  In Problem 11.44 plot the velocity in the pipe as a function of time.

**11.46**  Gasoline is supplied by gravity without pumping from a storage tank through a 800-m-long 50-mm-diameter nearly horizontal pipe into a tanker truck. There is a quick-acting valve at the end of the pipe. The difference in elevations of gasoline between the reservoir and the truck tank is 8 m. Initially, the valve is partially closed so that $K = 275$. Then the operator decides to increase the discharge by opening the valve quickly to the position where $K = 5$. Assuming an incompressible fluid and an inelastic piping system, determine the new steady-state discharge and the time it takes to reach 95% of that value. Assume that $f = 0.015$.

**11.47**  Water is flowing through a pipe whose diameter is 200 mm and whose length is 800 m. The pipe is made of cast iron ($E = 150$ GPa) with a thickness of 12 mm. There is a reservoir at the upstream end of the pipe and a valve at the downstream end. Under steady-state conditions the discharge is $Q = 0.05$ m³/s, when a valve at the end of the pipe is actuated very rapidly so that water hammer occurs.
**(a)**  How long does it take for an acoustic wave to travel from the valve to the reservoir and back to the valve?

**(b)**  Determine the change in pressure at the valve if the valve is opened such that the discharge is doubled.
**(c)**  Determine the change in pressure at the valve if the valve is closed such that the discharge is halved.

**11.48**  When operating the system described in Problem 11.46, the operator decides to suddenly close the valve once steady-state flow has been reached. Assuming that the gasoline is slightly compressible and the piping is elastic, determine:
**(a)**  The acoustic wave speed.
**(b)**  The pressure rise at the valve once the valve is rapidly closed in a manner to cause water hammer.
**(c)**  If the maximum allowable pressure in the pipe is 250 kPa, what can you conclude about the outcome of the water (gasoline) hammer activity? The pipe is made of aluminum, $E = 70$ GPa, with 2.5-mm-thick walls, and the bulk modulus of elasticity of gasoline is $B = 1.05$ GPa.

**11.49**  Oil with a specific gravity of $S = 0.90$ is flowing at 550 L/s through a 50-cm-diameter 3900-m-long pipe. The elastic modulus of the steel pipe is $E = 200 \times 10^9$ N/m², its thickness is 10 mm and the bulk modulus of elasticity of the oil is $B = 1.5 \times 10^9$ N/m². A valve at the downstream end of the pipe is partially closed very rapidly so that a water hammer event is initiated and a pressure wave propagates upstream. If the magnitude of the wave is not to exceed 640 kPa, determine:
**(a)**  The percent decrease of flow rate tolerable during the valve closure.
**(b)**  The time it takes the pressure wave to reach the upstream end of the pipe.

Grand Coulee Dam is the largest hydropower producer in the United States with a total generating capacity of 6 809 MW. It is also part of the Columbia Basin Project, irrigating more than 242 800 hectares, and is the cornerstone for water control on the Columbia River in the United States. The third phase, shown on the left, was completed in 1978; it has Francis turbines with 9.7 m runner diameters, operating at a head of 87 m and power output of 820 MW. (Courtesy of Voith Hydro)

# 12

# Turbomachinery

## Outline

## Chapter Objectives

The objectives of this chapter are to:

▲ Develop fundamental pump theory using the moment-of-momentum principle

▲ Describe radial-flow, mixed-flow, and axial-flow pumps, including presentation of prototype data and deviation from ideal conditions

▲ Introduce dimensionless pump coefficients and show how they are used in conjunction with similarity rules to develop data for turbomachinery design and analysis

▲ Demonstrate how pumps are incorporated into piping design and analysis

▲ Present introductory material on turbines: basic theory, types of turbines, and their selection and operation in conjunction with piping systems

## 12.1 INTRODUCTION

Turbomachines are the commonly employed devices that either supply or extract energy from a flowing fluid by means of rotating propellers or vanes. A turbopump, more commonly called a pump, adds energy to a system, with the result that the pressure is increased; it also causes flow to occur or it increases the rate of flow. A turbine extracts energy from a system and converts it to some other useful form, typically, to electric power. Pumps are essential components of piping systems which are designed to convey liquids. Similarly, turbomachines are called blowers, fans, or compressors when performing work on air or other gases in ducts. A hydroturbine, or simply a turbine, is a machine that generates power from high-pressure water; relatively large conduits or tunnels deliver fluid to closed turbines in order to generate power. Steam and air turbines are also of substantial engineering importance, but such devices are considered in other courses, such as thermodynamics.

The examples above are all those of turbomachines designed to facilitate or utilize internal flows. A wind turbine, on the other hand, makes use of the surrounding external flow to convert the energy contained in the natural movement of atmospheric air into useful electrical power. As shown in Section 4.5.4, a propeller performs work on the surrounding fluid to provide thrust and propel an object along a desired path, while a stationary fan performs work to circulate air. All turbomachines are characterized as being capable of either adding or subtracting energy from the fluid by means of rotating propellers or vanes. There is no contained volume of fluid being transported, as in a positive-displacement piston pump.

The bulk of this chapter is devoted to pumps, in particular, those that convey liquids such as water or gasoline. The centrifugal, or radial-flow pump is presented in detail, and to a lesser extent, mixed-flow and axial-flow pumps are studied. Dimensional analysis, an important tool for the selection and design of turbomachines, is presented next, followed by a discussion of the proper selection and implementation of pumps in piping systems. Finally, turbines are introduced by considering their fundamental characteristics.

## 12.2 TURBOPUMPS

**Turbopump:** *Pump consisting of two major parts—an impeller and the pump housing.*

A **turbopump** consists of two principal parts: an *impeller*, which imparts a rotary motion to the liquid, and the pump *housing*, or *casing*, which directs the liquid into the impeller region and transports it away under a higher pressure. Fig. 12.1 shows a typical single-suction *radial-flow pump*. The impeller is mounted on a shaft and is often driven by an electric motor. The casing includes the suction and discharge nozzles and houses the impeller assembly. The portion of the casing surrounding the impeller is termed the *volute*. Liquid enters through the suction nozzle to the impeller eye and travels along the *shroud*, developing a rotary motion due to the impeller vanes. It leaves the volute casing peripherally at a higher pressure through the discharge nozzle. Some single-suction impellers are open, with the front shroud removed. Double-suction impellers have liquid entering from both sides.

**KEY CONCEPT** Common types of pumps are radial flow, mixed flow, and axial flow.

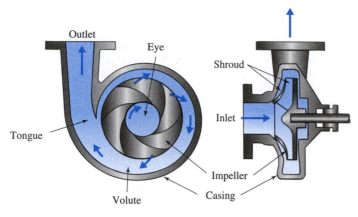

**Fig. 12.1**   Single-suction pump.

In a radial-flow pump, the impeller vanes are commonly curved backward and the impeller is relatively narrow. As the impeller becomes wider, the vanes have a double curvature, becoming twisted at the suction end. Such pumps convey liquids with a lower-pressure rise than radial-flow pumps and are called *mixed-flow pumps*. At the opposite extreme from the radial-flow pump is the *axial-flow pump*; it is characterized by the flow entering and leaving the impeller region axially, parallel to the shaft axis. Typically, an axial-flow pump delivers liquid with a relatively low pressure rise. For axial-flow pumps and some mixed-flow pumps, the impellers are open; that is, there is no shroud surrounding them. Various types of impellers are shown in Fig. 12.2.

## 12.2.1   Radial-Flow Pumps

A representative radial-flow pump is shown in Fig. 12.1; this type of pump is commonly called a *centrifugal pump*, and it is the most commonly used pump in existence today. An elementary analysis will be performed on a radial-flow pump, which will provide a theoretical relationship between the discharge and the developed head rise. In addition, a better understanding of the manner in which momentum exchange takes place in such a turbomachine will be provided. Actual pumps operate at efficiencies less than unity; that is, they do not perform at the theoretical, idealized conditions. It is necessary, therefore, to conduct experiments so that the true operating characteristics of a turbopump can be determined.

**KEY CONCEPT** *A radial-flow, or centrifugal, pump is designed to deliver relatively low discharge at a high head.*

**Fig.12.2**   Various types of pump impellers.

**Elementary Theory.** The actual flow patterns in a turbopump are highly three-dimensional with significant viscous effects and separation patterns taking place. To construct a simplified theory for the radial-flow pump, it is necessary to neglect viscosity and to assume idealized two-dimensional flow throughout the impeller region. Fig. 12.3a defines a control volume that encompasses the impeller region. Flow enters through the inlet control surface and exits through the outlet surface. Note that a series of vanes exists within the control volume, and that they are rotating about the axis with an angular speed $\omega$.

A portion of the control volume is shown at an instant in time in Fig. 12.3b. The idealized velocity vectors are diagrammed at the inlet, location 1, and the outlet, location 2. In the velocity diagrams, $V$ is the absolute fluid velocity, $V_t$ is the tangential component of $V$, and $V_n$ is the radial, or normal, component of $V$. The peripheral or circumferential speed of the blade is $u = \omega r$, where $r$ is the radius of the control surface. The angle between $V$ and $u$ is $\alpha$. The fluid velocity measured relative to the vane is $v$.

The relative velocity is assumed to be always tangent to the vane; that is, perfect guidance of the fluid throughout the control volume takes place. The angle between $v$ and $u$ is designated as $\beta$; since perfect guidance along the vane is assumed, $\beta$ designates the blade angle as well.

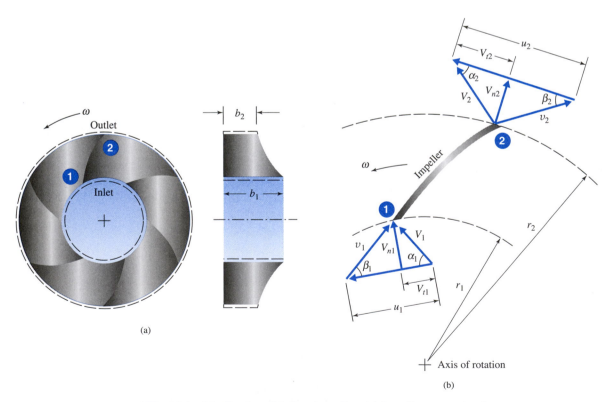

**Fig. 12.3** Idealized radial-flow impeller: (a) impeller control volume; (b) velocity diagrams at control surfaces.

The moment-of-momentum relation, Eq. 4.6.3, can be written for steady flow in the form

$$\Sigma \, \mathbf{M} = \int_{c.s.} \rho \mathbf{r} \times \mathbf{V}(\mathbf{V} \cdot \hat{n}) \, dA \qquad (12.2.1)$$

Applied to the control volume of Fig. 12.3, this becomes

$$T = \rho Q(r_2 V_{t2} - r_1 V_{t1}) \qquad (12.2.2)$$

in which $T$ is the torque acting on the fluid in the control volume, and the right-hand side represents the flux of angular momentum through the control volume.

The power delivered to the fluid is the product of $\omega$ and $T$:

$$\omega T = \rho Q(u_2 V_{t2} - u_1 V_{t1}) \qquad (12.2.3)$$

From the velocity vector diagrams in Fig. 12.3b, $V_{t1} = V_1 \cos \alpha_1$ and $V_{t2} = V_2 \cos \alpha_2$, so that Eq. 12.2.3 can be written as

$$\omega T = \rho Q(u_2 V_2 \cos \alpha_2 - u_1 V_1 \cos \alpha_1) \qquad (12.2.4)$$

For the idealized situation in which there are no losses, the delivered power must be equal to $\gamma Q H_t$, in which $H_t$ is the theoretical pressure head rise across the pump (see Eq. 4.5.26). Then there results Euler's turbomachine relation,

$$H_t = \frac{\omega T}{\gamma Q}$$

$$= \frac{u_2 V_2 \cos \alpha_2 - u_1 V_1 \cos \alpha_1}{g} \qquad (12.2.5)$$

Insight on the nature of flow through an impeller region can be obtained using Eq. 12.2.5. From the law of cosines we can write

$$v_1^2 = u_1^2 + V_1^2 - 2u_1 V_1 \cos \alpha_1$$

$$v_2^2 = u_2^2 + V_2^2 - 2u_2 V_2 \cos \alpha_2$$

These can be substituted into Eq. 12.2.5 to provide the relation

$$H_t = \frac{V_2^2 - V_1^2}{2g} + \frac{(u_2^2 - u_1^2) - (v_2^2 - v_1^2)}{2g} \qquad (12.2.6)$$

The first term on the right-hand side represents the gain in kinetic energy as the fluid passes through the impeller; the second term accounts for the increase in pressure across the impeller. This can be seen by applying the energy equation across the impeller and solving for $H_t$:

$$H_t = \frac{p_2 - p_1}{\gamma} + z_2 - z_1 + \frac{V_2^2 - V_1^2}{2g} \tag{12.2.7}$$

Eliminating $H_t$ between Eqs. 12.2.6 and 12.2.7 yields the expression

$$\frac{p_1}{\gamma} + z_1 + \frac{v_1^2 - u_1^2}{2g} = \frac{p_2}{\gamma} + z_2 + \frac{v_2^2 - u_2^2}{2g} = \text{const} \tag{12.2.8}$$

This relation has historically been called the Bernoulli equation in rotating coordinates. Since $z_2 - z_1$ is usually much smaller than $(p_2 - p_1)/\gamma$, it can be eliminated, and thus the pressure difference is

$$p_2 - p_1 = \frac{\rho}{2}\left[(v_1^2 - u_1^2) - (v_2^2 - u_2^2)\right] \tag{12.2.9}$$

Returning to Eq. 12.2.5, we see that a "best design" for a pump would be one in which the angular momentum entering the impeller is zero, so that maximum pressure rise can take place. Then in Fig. 12.3b, $\alpha_1 = 90°$, $V_{n1} = V_1$, and Eq. 12.2.5 becomes

$$H_t = \frac{u_2 V_2 \cos \alpha_2}{g} \tag{12.2.10}$$

From the triangle geometry of Fig. 12.3, $V_2 \cos \alpha_2 = u_2 - V_{n2} \cot \beta_2$, so that Eq. 12.2.10 takes the form

$$H_t = \frac{u_2^2}{g} - \frac{u_2 V_{n2} \cot \beta_2}{g} \tag{12.2.11}$$

Applying the continuity principle at the outlet region to the control volume provides the relation

$$V_{n2} = \frac{Q}{2\pi r_2 b_2} \qquad (12.2.12)$$

in which $b_2$ is the width of the impeller at location 2. Introducing Eq. 12.2.12 into Eq. 12.2.11, and recalling that $u_2 = \omega r_2$, one has the relation

$$H_t = \frac{\omega^2 r_2^2}{g} - \frac{\omega \cot \beta_2}{2\pi b_2 g} Q \qquad (12.2.13)$$

For a pump running at constant speed, Eq. 12.2.13 takes the form

$$H_t = a_0 - a_1 Q \qquad (12.2.14)$$

in which $a_0$ and $a_1$ are constants. Equation 12.2.13 is the *theoretical head curve* and is seen to be a straight line with a slope of $-a_1$, as shown in Fig. 12.4a. The effect of the blade angle $\beta_2$ is shown in Fig. 12.4b. A forward curving blade ($\beta_2 > 90°$) can be unstable and cause pump surge, where the pump oscillates in an attempt to establish an operating point. Backward-curving vanes ($\beta_2 < 90°$) are generally preferred.

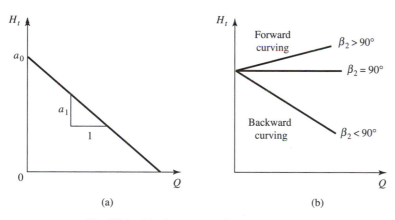

**Fig. 12.4**    Ideal pump performance curves.

## Example 12.1

A radial-flow pump has the following dimensions:

$$\beta_1 = 44° \qquad r_1 = 21 \text{ mm} \qquad b_1 = 11 \text{ mm}$$
$$\beta_2 = 30° \qquad r_2 = 66 \text{ mm} \qquad b_2 = 5 \text{ mm}$$

For a rotational speed of 2500 rpm, assuming ideal conditions (frictionless flow, negligible vane thickness, perfect guidance), with $\alpha_1 = 90°$ (no prerotation), determine (a) the discharge, theoretical head, required power, and pressure rise across the impeller, and (b) the theoretical head-discharge curve. Use water as the fluid.

### Solution

(a) Construct the velocity diagram at location 1, as shown in Fig. E12.1a. The rotational speed is converted to the appropriate units as

$$\omega = 2500 \frac{2\pi}{60} = 261.8 \text{ rad/s}$$

The impeller speed at $r_1$ is then

$$u_1 = \omega r_1 = 261.8 \times 0.021 = 5.50 \text{ m/s}$$

Fig. E12.1

From the velocity diagram we see that

$$V_1 = u_1 \tan \beta_1 = 5.50 \tan 44° = 5.31 \text{ m/s}$$

and since $\alpha_1 = 90°$, $V_1 = V_{n1}$, or $V_{n1} = 5.31$ m/s. The discharge is computed to be

$$Q = 2\pi r_1 b_1 V_{n1}$$
$$= 2\pi \times 0.021 \times 0.011 \times 5.31 = 7.71 \times 10^{-3} \text{ m}^3/\text{s}$$

The normal component of velocity at location 2 is

$$V_{n2} = \frac{Q}{2\pi r_2 b_2} = \frac{7.71 \times 10^{-3}}{2\pi \times 0.066 \times 0.005} = 3.72 \text{ m/s}$$

and the impeller speed at the outlet is

$$u_2 = \omega r_2 = 261.8 \times 0.066 = 17.28 \text{ m/s}$$

The velocity diagram at location 2 is now sketched as shown in Fig. E12.1b. From the velocity diagram we see that

$$u_2 - V_{t2} = \frac{V_{n2}}{\tan \beta_2} = \frac{3.72}{\tan 30°} = 6.44 \text{ m/s}$$

Therefore,

$$V_{t2} = u_2 - 6.44 = 17.28 - 6.44 = 10.84 \text{ m/s}$$

$$\alpha_2 = \tan^{-1} \frac{V_{n2}}{V_{t2}} = \tan^{-1} \frac{3.72}{10.84} = 18.9°$$

$$V_2 = \frac{V_{t2}}{\cos \alpha_2} = \frac{10.84}{\cos 18.9°} = 11.46 \text{ m/s}$$

The theoretical head is computed with Eq. 12.2.10:

$$H_t = \frac{u_2 V_2 \cos \alpha_2}{g} = \frac{17.28 \times 11.46 \times \cos 18.9°}{9.81} = 19.1 \text{ m}$$

Hence the theoretical required power is

$$\dot{W}_p = \gamma Q H_t = 9810 \times (7.71 \times 10^{-3}) \times 19.1 = 1440 \text{ W}$$

The pressure rise is determined from the energy equation as follows:

$$p_2 - p_1 = \left( H_t + \frac{V_1^2 - V_2^2}{2g} \right) \gamma$$

$$= \left[ 19.1 + \frac{(5.31)^2 - (11.46)^2}{2 \times 9.81} \right] \times 9810 = 1.36 \times 10^5 \text{ Pa}$$

(b) The theoretical head-discharge curve is Eq. 12.2.13. For the present example we have

$$H_t = \frac{(\omega r_2)^2}{g} - \frac{\omega \cot \beta_2}{2\pi b_2 g} Q$$

$$= \frac{(261.8 \times 0.066)^2}{9.81} - \frac{261.8 \cot 30°}{2\pi \times 0.005 \times 9.81} Q$$

$$= 30.4 - 1471 Q$$

The curve is shown in Fig. E12.1c.

**Head-Discharge Relations: Performance Curves.** For real fluid flow, the theoretical head curve cannot be achieved in practice, and it is necessary to resort to experimentation to determine the actual head-discharge curve. The energy equation written across a pump from the suction side (location 1, Fig. 12.3) to the discharge side (location 2) is

$$H_P = \left( \frac{p}{\gamma} + \frac{V^2}{2g} + z \right)_2 - \left( \frac{p}{\gamma} + \frac{V^2}{2g} + z \right)_1$$
$$= H_t - h_L \qquad (12.2.15)$$

in which $H_P$ is the actual head across the pump, and $h_L$ represents the losses through the pump. The actual power delivered to the fluid, designated as $\dot{W}_f$, is

$$\dot{W}_f = \gamma Q H_P \qquad (12.2.16)$$

**Brake power:** *Power delivered to the impeller.*

while the power delivered to the impeller is $\dot{W}_P$, often termed the **brake power**, and is given by

$$\dot{W}_P = \omega T \qquad (12.2.17)$$

If there were no losses, $\dot{W}_f$ would be equal to $\dot{W}_P$. Since in actuality $\dot{W}_f < \dot{W}_P$, the *pump efficiency*[1] $\eta_P$ is defined as

$$\eta_P = \frac{\dot{W}_f}{\dot{W}_P} = \frac{\gamma Q H_P}{\omega T} \qquad (12.2.18)$$

One objective of pump design is to make the efficiency as high as possible. The theoretical head curve is compared with the actual head curve in Fig. 12.5. The difference between the two can be attributed to the following effects: (1) prerotation of the fluid before entering the impeller region, (2) separation due to imperfect guidance of fluid as it enters the impeller region, and (3) separation due to expansion in the flow passages. Leakage and high shear rates created by the differences in velocity between fluid particles in the volute and impeller contribute further to the losses. Finally, impeller blades are designed to be most efficient at the so-called design discharge; at any other discharge—that is, "off-design"—the performance will deteriorate.

---

[1] More precisely, $\eta_P$ defined by Eq. 12.2.18 is termed the *overall efficiency*. It is possible to define three additional efficiencies for pumps as follows:

$$\text{Hydraulic efficiency:} \quad \eta_H = \frac{H_P}{H_t}$$

$$\text{Volumetric efficiency:} \quad \eta_V = \frac{Q}{Q + Q_L}, \qquad Q_L = \text{leakage flow rate}$$

$$\text{Mechanical efficiency:} \quad \eta_M = \frac{\gamma H_t (Q + Q_L)}{T \omega}$$

These are related to the overall efficiency by $\eta_P = \eta_H \, \eta_V \, \eta_M$.

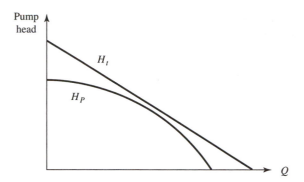

**Fig. 12.5** Comparison between theoretical and actual radial-flow pump performance curves.

A representative pump and its performance curve are shown in Fig. 12.6; this is the curve supplied by the pump manufacturer. Typically, more than one impeller diameter will be available; in the diagram, four different impellers are represented. In addition to the head-discharge performance curves, isoefficiency, power, and *net positive suction head* (*NPSH*) curves are usually provided. Net positive suction head is associated with concern for cavitation; it is discussed in Section 12.2.3.

### 12.2.2  Axial- and Mixed-Flow Pumps

The velocity diagram for an axial-flow pump is shown in Fig. 12.7. In the axial-flow pump, there is no radial flow and the liquid particles leave the impeller at the same radius at which they enter, so that $u_1 = u_2 = u$. Furthermore, assuming a uniform flow, continuity considerations give $V_{n1} = V_{n2} = V_n$. Equation 12.2.5, which is valid for an axial-flow pump as well as a radial-flow pump, can be combined with the identities $V_2 \cos \alpha_2 = u - V_n \cot \beta_2$ and $V_1 \cos \alpha_1 = V_n \cot \alpha_1$ to produce

> **KEY CONCEPT** *An axial-flow pump produces relatively large discharges at low heads.*

$$H_t = \frac{u^2}{g} - \frac{uV_n}{g}(\cot \alpha_1 + \cot \beta_2)$$  (12.2.19)

This form of the turbomachine relation is useful when the ideal absolute velocity entrance angle $\alpha_1$ is established by a fixed vane, or stator. If there is no prerotation, $\alpha_1 = 90°$ and the theoretical head relation, Eq. 12.2.19, becomes

$$H_t = \frac{u^2}{g} - \frac{uV_n \cot \beta_2}{g}$$  (12.2.20)

This relation is identical to Eq. 12.2.11, which relates the theoretical head to the impeller outlet parameters for a radial flow pump. Note, however, that Eq. 12.2.11 is valid at the periphery of the impeller for the radial-flow pump, and that all fluid particles reach the maximum head at that location. By contrast,

**Fig. 12.6** Radial-flow pump and performance curves for four different impellers with $N = 2900$ rpm ($\omega = 304$ rad/s). Water at 20°C is the pumped liquid. (Courtesy of Sulzer Pumps Ltd.)

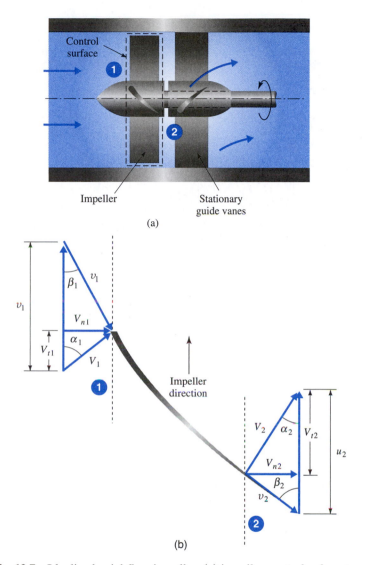

**Fig. 12.7** Idealized axial-flow impeller: (a) impeller control volume;
(b) velocity diagrams at control surface.

Eq. 12.2.20 pertains only at a specified radius for the axial-flow pump, since fluid
particles enter and leave the control volume at their respective radii. In this case
the head varies from a minimum at the axis to a maximum at the periphery, and
the total pump head is an integrated average. Axial-flow impellers are designed
to maintain a constant axial velocity throughout the impeller region; this requires
that the vane angles increase gradually from the periphery to the axis, as well as
from the inlet to the outlet region. Fig. 12.8 on the next page shows a representative axial-flow pump, and its performance curves.

**Fig. 12.8** Axial-flow pump and performance curves for four different vane angles with $N = 880$ rpm ($\omega = 92.2$ rad/s). Propeller diameter is 500 mm. Water at 20°C is the pumped liquid. (Courtesy of Sulzer Pumps Ltd.)

Generally, flow in an impeller region consists of two components: the circular, vortex-like motion due to the action of the vanes, and the net flow through the impeller. The vortex motion is superposed on the radial outward flow in a radial-flow pump, and on the axial flow in an axial-flow pump. The action of the mixed-flow pump lies between these two types. The previous discussions relating to radial- and axial-flow pumps pertain to mixed-flow pumps as well. A mixed-flow pump and its characteristic curves are shown in Fig. 12.9 on the next page.

As in the case for a radial-flow pump, the performance characteristics of axial- and mixed-flow pumps deviate from the idealized considerations. Generally, radial-flow pumps are designed to deliver relatively low discharges at high heads, and axial-flow pumps produce relatively large discharges at low heads. Mixed-flow pumps deliver heads and discharges between these two extremes. A method of selecting a pump appropriate for required design considerations is presented in Section 12.4.

**KEY CONCEPT** *A mixed-flow pump delivers heads and discharges between those of radial-flow and axial-flow pumps.*

## Example 12.2

An axial-flow pump is designed with a fixed guide vane, or stator blade, located upstream of the impeller. The stator imparts an angle $\alpha_1 = 75°$ to the fluid as it enters the impeller region. The impeller has a rotational speed of 500 rpm with a blade exit angle of $\beta_2 = 70°$. The control volume has an outer diameter of $D_o = 300$ mm and an inner diameter of $D_i = 150$ mm. Determine the theoretical head rise and power required if 150 L/s of liquid ($S = 0.85$) is to be pumped.

### Solution
First, the normal velocity component $V_n$ is

$$V_n = \frac{Q}{A} = \frac{Q}{(\pi/4)(D_o^2 - D_i^2)} = \frac{0.15}{(\pi/4)(0.3^2 - 0.15^2)} = 2.83 \text{ m/s}$$

The peripheral speed $u$ of the impeller is based on an average radius:

$$u \simeq \omega \frac{D_o + D_i}{4} = 500\left(\frac{2\pi}{60}\right)\frac{0.3 + 0.15}{4} = 5.89 \text{ m/s}$$

The theoretical head $H_t$ is computed with Eq. 12.2.19 to be

$$H_t = \frac{u}{g}[u - V_n(\cot \alpha_1 + \cot \beta_2)] = \frac{5.89}{9.81}[5.89 - 2.83(\cot 75° + \cot 70°)] = 2.46 \text{ m}$$

Finally, for the assumed ideal conditions, the required power is

$$\dot{W}_P = \gamma Q H_t = (9810 \times 0.85) \times 0.15 \times 2.46 = 3080 \text{ W} \quad \text{or} \quad 3.1 \text{ kW}$$

### 12.2.3 Cavitation in Turbomachines

**Cavitation** refers to conditions at certain locations within the turbomachine where the local pressure drops to the vapor pressure of the liquid, and as a result, vapor-filled cavities are formed. As the cavities are transported through the turbomachine into regions of greater pressure, they will collapse rapidly, generating extremely high localized pressures. Those bubbles that collapse close to solid boundaries can weaken the solid surface, and after repeated collapsing, pitting,

**Cavitation:** *Condition where local pressure drops to the vapor pressure of the liquid, forming vapor-filled cavities.*

**Fig. 12.9** Mixed-flow pump and performance curves for four different impellers and vane angles with $N = 970$ rpm ($\omega = 102$ rad/s). Water at 20°C is the pumped liquid. (Courtesy of Sulzer Pumps Ltd.)

**Fig. 12.10**   Cavitation bubble distribution in an impeller region. (Courtesy of Sulzer Pumps Ltd.)

erosion, and fatigue of the surface can occur. Signs of cavitation in turbopumps include noise, vibration, and lowering of the head-discharge and efficiency curves. Regions most susceptible to damage in a turbomachine are those slightly beyond the low-pressure zones on the back side of impellers (see Fig. 12.10). In general, sudden changes in direction, sudden increases in area, and lack of streamlining are the culprits causing cavitation damage in turbopumps (Karassik et al., 1986).

The proper design of turbopumps will have minimized the possibility for cavitation to occur. Under adverse operating conditions, however, the pressures may decrease and cavitation may occur. Two parameters are used to designate the potential for cavitation: the cavitation number and the net positive suction head. Their interrelationship and use are now discussed.

Consider a pump operating in the manner shown in Fig. 12.11. Location 1 is on the liquid surface on the suction side, and location 2 is the point of minimum pressure within the pump. Writing the energy equation from location 1 to location 2 and using an absolute pressure datum results in

$$\frac{V_2^2}{2g} = \frac{p_{\text{atm}} - p_2}{\gamma} - \Delta z - h_L \tag{12.2.21}$$

in which $h_L$ is the loss between location 1 and location 2, $\Delta z = z_2 - z_1$, and the kinetic energy at location 1 is assumed negligible. The minimum allowable pressure at location 2 is the vapor pressure $p_v$. Substituting this into the expression, one can say that the left-hand side of Eq. 12.2.21 represents the maximum kinetic energy head possible at location 2 when cavitation is imminent. Thus the *net positive suction head* (*NPSH*) is defined as

$$NPSH = \frac{p_{\text{atm}} - p_v}{\gamma} - \Delta z - h_L \tag{12.2.22}$$

**Fig. 12.11**   Cavitation setting for a pump.

The *NPSH* is also used for turbines; however, the sign of the $h_L$ term in Eq. 12.2.22 changes, and location 1 refers to the liquid surface on the discharge side of the machine. The design requirement for a pump is thus established as follows:

$$ NPSH \leq \frac{p_{\text{atm}} - p_v}{\gamma} - \Delta z - h_L \qquad (12.2.23) $$

The performance data supplied by turbomachinery manufacturers usually includes *NPSH* curves; these are developed by testing a given family in a laboratory environment. Figures 12.6, 12.8, and 12.9 show *NPSH* curves. The *NPSH* curve enables one to specify the required maximum value of $\Delta z$ to be used for a given turbomachine; note that it is necessary to estimate $h_L$ to obtain this.

The right-hand side of Eq. 12.2.22 can be divided by $H_P$, the total head across the pump to yield

$$ \sigma = \frac{1}{H_P} \frac{(p_{\text{atm}} - p_v)}{\gamma} - \Delta z - h_L \qquad (12.2.24) $$

in which $\sigma$ is the *Thoma cavitation number*. This parameter is used as an alternative to the *NPSH* to establish design criteria for cavitation. A *critical cavitation number*, which signals that cavitation is imminent, is determined experimentally. Thus, for no cavitation to occur, $\sigma$ must be greater than the critical cavitation number. The cavitation number is in dimensionless form, which is preferred to the dimensional form of the *NPSH*.

## Example 12.3

Determine the elevation at which the 240-mm-diameter pump of Fig. 12.6 can be situated above the water surface of the suction reservoir without experiencing cavitation. Water at 15°C is being pumped at 250 m³/h. Neglect losses in the system. Use $p_{atm} = 101$ kPa.

### Solution

From Fig. 12.6, at a discharge of 250 m³/h, the *NPSH* for the 240-mm-diameter impeller is approximately 7.4 m. For a water temperature of 15°C, $p_v = 1666$ Pa absolute, and $\gamma = 9800$ N/m³. Equation 12.2.22 with $h_L = 0$ is employed to compute $\Delta z$ to be

$$\Delta z = \frac{p_{atm} - p_v}{\gamma} - NPSH - h_L = \frac{101\,000 - 1666}{9800} - 7.4 - 0 = 2.74 \text{ m}$$

Thus the pump can be placed approximately 2.7 m above the suction reservoir water surface.

## 12.3 DIMENSIONAL ANALYSIS AND SIMILITUDE FOR TURBOMACHINERY

The development and utilization of turbomachinery in engineering practice has benefited greatly from the application of dimensional analysis, probably to an extent more than any other area of applied fluid mechanics. It has enabled pump and turbine manufacturers to test and develop a relatively small number of turbomachines and subsequently produce a series of commercial units that cover a broad range of head and flow demands. The material developed in Chapter 6, which covers similitude and dimensional analysis, is applied to turbomachines in this section.

### 12.3.1 Dimensionless Coefficients

The following parameters can be considered significant ones for a turbomachine: power, rotational speed, outer diameter of the impeller, discharge, pressure change across the impeller, density of the fluid, and viscosity of the fluid. These are given in Table 12.1 along with the relevant dimensions. In functional form, the relation between these variables is given by

$$f(\dot{W}, \omega, D, Q, \Delta p, \rho, \mu) = 0 \qquad (12.3.1)$$

Using $\omega, \rho,$ and $D$ as repeating variables, a set of convenient dimensionless groupings can be derived. They are given as follows:

$$C_{\dot{W}} = \frac{\dot{W}}{\rho\omega^3 D^5} \qquad \textbf{power coefficient} \qquad (12.3.2)$$

$$C_P = \frac{\Delta p}{\rho\omega^2 D^2} \qquad \textbf{pressure coefficient} \qquad (12.3.3)$$

$$C_Q = \frac{Q}{\omega D^3} \qquad \textbf{flow rate coefficient} \qquad (12.3.4)$$

$$\text{Re} = \frac{\omega D^2 \rho}{\mu} \qquad \textbf{Reynolds number} \qquad (12.3.5)$$

It is customary to equate $\Delta p$ to $\gamma H$, where $H$ represents either the pressure head rise in the case of a pump or the pressure head drop for a turbine. Then Eq. 12.3.3 is replaced by

$$C_H = \frac{gH}{\omega^2 D^2} \qquad \textbf{head coefficient} \qquad (12.3.6)$$

**TABLE 12.1**  Turbomachinery Parameters

| Parameter | Symbol | Dimensions |
|---|---|---|
| Power | $\dot{W}$ | $ML^2/T^3$ |
| Rotational speed | $\omega$ | $T^{-1}$ |
| Outer diameter of impeller | $D$ | $L$ |
| Discharge | $Q$ | $L^3/T$ |
| Pressure change | $\Delta p$ | $M/LT^2$ |
| Fluid density | $\rho$ | $M/L^3$ |
| Fluid viscosity | $\mu$ | $M/LT$ |

Another dimensionless parameter for a pump can be found by grouping $C_Q$, $C_{\dot{W}}$, and $C_H$ to form the ratio

$$\frac{C_Q C_H}{C_{\dot{W}}} = \frac{\gamma Q H}{\dot{W}} = \eta_P \qquad \text{(pump)} \qquad (12.3.7)$$

Hence the efficiency is also a similitude parameter. For a turbine the efficiency grouping is given by

$$\frac{C_{\dot{W}}}{C_Q C_H} = \frac{\dot{W}}{\gamma Q H} = \eta_T \qquad \text{(turbine)} \qquad (12.3.8)$$

in which $\dot{W}$ is the output power.[2]

The interrelationships between the dimensionless parameters are expressed in the form of dimensionless performance curves, as shown in Fig. 12.12[3]; these are for the same turbomachine as shown in Fig. 12.6. Dimensionless performance curves for the axial-flow pump of Fig. 12.8 and the mixed-flow pump of Fig. 12.9 are shown on the next page in Figs. 12.13 and 12.14, respectively. The vertical dashed lines designate the operating conditions at maximum or so-called "best" efficiency.

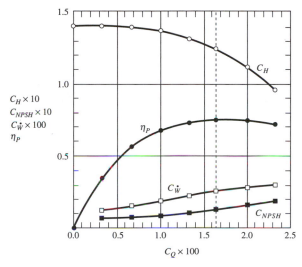

**Fig. 12.12**   Dimensionless radial-flow pump performance curves for the pump presented in Fig. 12.6; $D = 240$ mm; $\Omega_P = 0.61$.

Note that the Reynolds number is not present as a parameter in Figs. 12.12 to 12.14, even though viscous effects are significant in turbomachinery flow. In addition to the Reynolds number, one could include a relative roughness ratio. It is known that viscous and roughness effects will alter pump behavior; however, because of the difficulties involved, it is usual practice not to make a correlation between the Reynolds number, roughness, and pump losses. The flow regime in pump passages is usually very turbulent, and the roughness varies widely for different pumps. Furthermore, the friction losses may not be as significant as the losses due to eddy formation and separation caused by diffused flow occurring in the impeller and casing. As a result, the Re effect is usually not directly represented in similitude studies. Instead, the two following performance characteristics

---

[2] It is common industrial practice to use so-called *homologous units* in place of the dimensionless coefficients. The relationships between the two groupings are as follows:

| Dimensionless coefficient | $C_Q$ | $C_H$ | $C_{\dot{W}}$ |
|---|---|---|---|
| Homologous unit | $\dfrac{Q}{ND^3}$ | $\dfrac{H}{N^2D^2}$ | $\dfrac{\dot{W}}{N^3D^5}$ |

In the homologous units, density and gravitational acceleration have been eliminated, and the rotational speed is given in revolutions per minute, designated by $N$. Note that the homologous units are dimensional.

[3] In the caption, $\Omega_P$ is the specific speed, which will be defined in Section 12.3.3.

are recognized: (1) larger pumps exhibit relatively smaller losses and higher efficiencies than smaller pumps of the same family since they have smaller roughness and clearance ratios; and (2) higher viscosity liquids will cause reductions in pump head and efficiency and an increase in power requirement for a given discharge.

A dimensionless net positive suction head coefficient, $C_{NPSH}$, is also presented in Figs. 12.12 to 12.14. It is formed by replacing $H$ in Eq. 12.3.6 with the *NPSH* defined by Eq. 12.2.22.

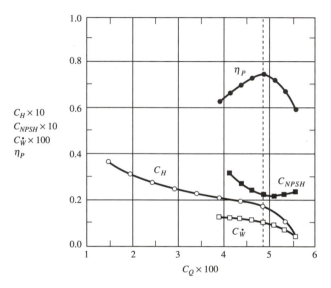

**Fig. 12.13** Dimensionless axial-flow pump performance curve for the pump presented in Fig. 12.8; $D = 500$ mm; vane angle $= 14°$; $\Omega_P = 4.5$.

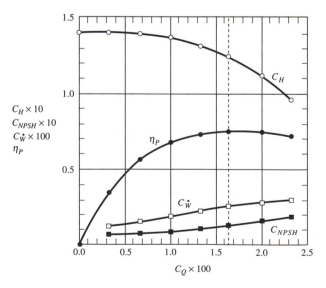

**Fig. 12.14** Dimensionless mixed-flow pump performance curve for the pump presented in Fig. 12.9; $D = 371$ mm; $\Omega_P = 3.1$.

## Example 12.4

Determine the speed, size, and required power for the axial-flow pump of Fig. 12.13 to deliver 0.65 m³/s of water at a head of 2.5 m.

**Solution**

The pump data are obtained from Fig. 12.13. Reading from the figure we find that at the design, or maximum, efficiency of $\eta_P = 0.75$

$$C_Q = 0.048 \qquad C_H = 0.018 \qquad C_{\dot{W}} = 0.0012$$

Equations 12.3.4 and 12.3.6 are employed to determine the two unknowns $\omega$ and $D$. Rearrange Eq. 12.3.6 to solve for $\omega D$:

$$\omega D = \sqrt{\frac{gH_P}{C_H}} = \sqrt{\frac{9.81 \times 2.5}{0.018}} = 36.9 \text{ m/s}$$

Equation 12.3.4 is rearranged to include $\omega D$ as a known quantity to provide $D$:

$$D = \sqrt{\frac{Q}{C_Q \omega D}} = \sqrt{\frac{0.65}{0.048 \times 36.9}} = 0.61 \text{ m}$$

Substituting $D = 0.61$ m into the relation $\omega D = 36.9$, one finds that

$$\omega = \frac{36.9}{D} = \frac{36.9}{0.61} = 60.5 \text{ rad/s} \qquad \text{or} \qquad 578 \text{ rpm}$$

The density for water is $\rho = 1000$ kg/m³. Then the pump power is determined by using Eq. 12.3.2:

$$\dot{W}_P = C_{\dot{W}} \rho \omega^3 D^5 = 0.0012 \times 1000 \times 60.5^3 \times 0.61^5 = 2.2 \times 10^4 \text{ W}$$

Thus the speed, size, and required power are approximately 580 rpm, 610 mm, and 22 kW, respectively.

### 12.3.2 Similarity Rules

Following the principles outlined in Chapter 6 for similitude, similarity relationships between any two pumps from the same geometric family can be developed. They are:

$$(C_{\dot{W}})_1 = (C_{\dot{W}})_2 \quad \text{or} \quad \frac{\dot{W}_2}{\dot{W}_1} = \frac{\rho_2}{\rho_1}\left(\frac{\omega_2}{\omega_1}\right)^3\left(\frac{D_2}{D_1}\right)^5 \tag{12.3.9}$$

$$(C_H)_1 = (C_H)_2 \quad \text{or} \quad \frac{H_2}{H_1} = \frac{g_1}{g_2}\left(\frac{\omega_2}{\omega_1}\right)^2\left(\frac{D_2}{D_1}\right)^2 \tag{12.3.10}$$

$$(C_Q)_1 = (C_Q)_2 \quad \text{or} \quad \frac{Q_2}{Q_1} = \frac{\omega_2}{\omega_1}\left(\frac{D_2}{D_1}\right)^3 \tag{12.3.11}$$

**KEY CONCEPT** *Similarity rules relate variables from a family of geometrically similar turbomachines, or are used to examine changes in variables for a given machine.*

These equations, called the *turbomachinery similarity rules*, are used to design or select a turbomachine from a family of geometrically similar units. Another use of them is to examine the effects of changing speed, fluid, or size on a given unit. They could also be used to design a pump to deliver flow on the moon or on a space station.

In light of Eq. 12.3.7, one should also expect that $\eta_1 = \eta_2$, but as stated previously, larger pumps are more efficient than smaller ones of the same geometric family. An acceptable empirical correlation (Stepanoff, 1957) relating efficiencies to size is

$$\frac{1 - (\eta_P)_2}{1 - (\eta_P)_1} = \left(\frac{D_1}{D_2}\right)^{1/4} \tag{12.3.12}$$

This relation can also be used for turbines by replacing $\eta_P$ by $\eta_T$.

### Example 12.5

Determine the performance curve for the mixed-flow pump whose characteristic curves are shown in Fig. 12.14. The required discharge is 2.5 m³/s with the pump operating at a speed of 600 rpm. At design efficiency, what are the power and the *NPSH* requirements? Water is being pumped.

#### Solution

From Fig. 12.14 the design efficiency is $\eta_P = 0.85$ and the corresponding design values for the coefficients are

$$C_Q = 0.15 \qquad C_H = 0.067 \qquad C_{\dot{W}} = 0.0115 \qquad C_{NPSH} = 0.035$$

The rotational speed in radians per second is

$$\omega = 600 \times \frac{\pi}{30} = 62.8 \text{ rad/s}$$

The pump diameter is computed with Eq. 12.3.4:

$$D = \left(\frac{Q}{\omega C_Q}\right)^{1/3}$$

$$= \left(\frac{2.5}{62.8 \times 0.15}\right)^{1/3} = 0.64 \text{ m}$$

With Eqs. 12.3.2 and 12.3.6, the required power and *NPSH* are computed:

$$\dot{W}_P = \rho\omega^3 D^5 C_{\dot{W}}$$

$$= 1000 \times 62.8^3 \times 0.64^5 \times 0.0115 = 3.1 \times 10^5 \text{ W}$$

$$NPSH = \frac{\omega^2 D^2}{g} C_{NPSH}$$

$$= \frac{62.8^2 \times 0.64^2}{9.81} \times 0.035 = 5.8 \text{ m}$$

Hence, the required power is 310 kW and the required *NPSH* is 5.8 m.
The performance curve is constructed using Eqs. 12.3.10 and 12.3.11:

$$H_2 = \frac{\omega_2^2 D_2^2}{g} (C_H)_1$$

$$= \frac{62.8^2 \times 0.64^2}{9.81} (C_H)_1 = 165(C_H)_1$$

$$Q_2 = \omega_2 D_2^3 (C_Q)_1$$

$$= 62.8 \times 0.64^3 (C_Q)_1 = 16.5(C_Q)_1$$

Values of $(C_H)_1$ and $(C_Q)_1$ are read from Fig. 12.14, along with $\eta_P$ and placed in the table shown below. Columns 4 and 5 are $Q_2$ and $H_2$ computed from the similarity equations. At best efficiency, $H_2 = 10.4$ m and $Q_2 = 2.54$ m³/s. The characteristic curve and efficiency curve are plotted in Fig. E12.5.

| $(C_Q)_1$ | $(C_H)_1$ | $\eta_P$ | $Q_2(m^3/s)$ | $H_2(m)$ |
|---|---|---|---|---|
| 0 | 0.138 | | 0 | 22.8 |
| 0.019 | 0.123 | | 0.31 | 20.3 |
| 0.039 | 0.110 | | 0.64 | 18.2 |
| 0.058 | 0.099 | | 0.96 | 16.3 |
| 0.077 | 0.091 | 0.59 | 1.27 | 15.0 |
| 0.096 | 0.088 | 0.70 | 1.58 | 14.5 |
| 0.116 | 0.085 | 0.78 | 1.91 | 14.0 |
| 0.135 | 0.076 | 0.84 | 2.23 | 12.5 |
| 0.154 | 0.063 | 0.85 | 2.54 | 10.4 |
| 0.174 | 0.046 | 0.79 | 2.87 | 7.6 |

**Fig. E12.5**

### 12.3.3 Specific Speed

**Specific speed:** *Dimensionless number that characterizes a turbomachine at maximum efficiency.*

It is possible to correlate a turbomachine of a given family to a dimensionless number that characterizes its operation at optimum conditions. Such a number is termed the **specific speed**, and it is determined in the following manner. The specific speed $\Omega_P$ of a pump is a dimensionless parameter associated with its operation at maximum efficiency, with known $\omega$, $Q$, and $H_P$. It is obtained by eliminating $D$ between Eqs. 12.3.4 and 12.3.6 and expressing the rotational speed $\omega$ to the first power:

$$\Omega_P = \frac{C_Q^{1/2}}{C_H^{3/4}} = \frac{\omega Q^{1/2}}{(gH_P)^{3/4}} \qquad (12.3.13)$$

In Eq. 12.3.13, $\omega$ is usually based on motor requirements, and the values of $Q$ and $H_P$ are those at maximum efficiency.

A preliminary pump selection can be based on the specific speed. Fig. 12.15 shows how the maximum efficiency and the discharge vary with $\Omega_P$ for radial-flow pumps. The specific speed can be correlated with the type of impeller shown in Fig. 12.2; it is given as follows:

$$
\begin{array}{ll}
\Omega_P < 1 & \text{radial-flow pump} \\[4pt]
1 < \Omega_P < 4 & \text{mixed-flow pump} \\[4pt]
\Omega_P > 4 & \text{axial-flow pump}
\end{array}
\qquad (12.3.14)
$$

For a turbine, the specific speed $\Omega_T$ is a dimensionless parameter associated with a given family of turbines operating at maximum efficiency, with known $\omega$, $H_T$, and $\dot{W}_T$. The diameter $D$ is eliminated in Eqs. 12.3.2 and 12.3.6, and $\omega$ is expressed to the first power to obtain

$$\Omega_T = \frac{C_W^{1/2}}{C_H^{5/4}} = \frac{\omega(\dot{W}_T/\rho)^{1/2}}{(gH_T)^{5/4}} \qquad (12.3.15)$$

A comparison of $\Omega_T$ for different turbines will be made in Section 12.5.[4]

---

[4] In place of Eqs. 12.3.13 and 12.3.15, it has been common industrial practice to use specific speeds in dimensional form. Hence, for a pump, the dimensional, or *homologous specific speed* is $(N_s)_P = NQ^{1/2}/H^{3/4}$, and for a turbine the corresponding homologous specific speed is $(N_s)_T = N\dot{W}^{1/2}/H^{5/4}$. For these expressions, $Q$ is in m³/s, $H$ is in ft or m, $\dot{W}$ is in hp or kW, and $N$ is in rpm.

**Fig. 12.15**   Maximum efficiency as a function of specific speed and discharge for radial-flow pumps. (PUMP HAND-BOOK, 2ND EDITION (DURO) by Karassik. Copyright 1986 by McGraw-Hill Companies, Inc. -Books. Reproduced with permission of McGraw-Hill Companies, Inc. -Books in the format Textbook via Copyright Clearance Center.)

Another measure of cavitation is *suction specific speed*, $S$. Analogous to the formulation of specific speed of a pump, Eq. 12.3.13, suction specific speed for either a pump or turbine is given by

$$S = \frac{\omega Q^{1/2}}{(gNPSH)^{3/4}} \qquad (12.3.16)$$

Here, two geometrically similar units will have the same $S$ when operating at the same flow coefficient $C_Q$. Conversely, equal values of $S$ indicate similar cavitation characteristics when the units are operating differently. Design values of $S$ are determined by experiment; when cavitation is not present, Eq. 12.3.16 is no longer valid.

## Example 12.6

Select a pump to deliver 30 L/s of water with a pressure rise of 450 kPa. Assume a rotational speed not to exceed 3600 rpm.

**Solution**

To estimate the specific speed we need the following:

$$\omega = 3600 \times \frac{\pi}{30} = 377 \text{ rad/sec}$$

$$H_P = \frac{\Delta p}{\rho g} = \frac{450}{1000} \times \frac{1000}{9.81} = 45.9 \text{ m}$$

$$Q = 30 \times 10^{-3} \text{ m}^3/\text{s} = 0.03 \text{ m}^3/\text{s}$$

Use Eq. 12.3.13 to find the specific speed to be

$$\Omega_P = \frac{\omega\sqrt{Q}}{(gH_P)^{3/4}}$$

$$= \frac{377\sqrt{0.03}}{(9.81 \times 45.9)^{3/4}} = 0.67$$

From Eq. 12.3.14, this would indicate a radial-flow pump. The pump of Fig. 12.12 could be used, even though the specific speed (0.61) would result in a lower efficiency. With $\Omega_P \approx 0.61$ (Fig. 12.12), the speed is estimated with Eq. 12.3.13 as

$$\omega = \frac{\Omega_P(gH_P)^{3/4}}{\sqrt{Q}}$$

$$= \frac{0.61 \times (9.81 \times 45.9)^{3/4}}{\sqrt{0.03}} = 344 \text{ rad/s}$$

Hence the required speed is $344 \times 30/\pi = 3300$ rpm, which does not exceed 3600 rpm. The required diameter is determined with use of Fig. 12.12, where at maximum efficiency, $C_Q = 0.0165$. This is substituted into Eq. 12.3.4 to determine the diameter:

$$D = \left(\frac{Q}{C_Q\omega}\right)^{1/3}$$

$$= \left(\frac{0.03}{0.0165 \times 344}\right)^{1/3} = 17.4 \text{ cm}$$

## 12.4  USE OF TURBOPUMPS IN PIPING SYSTEMS

The appropriate selection of one or more pumps to meet the flow demands of a piping system requires, in addition to a fundamental understanding of turbopumps, a hydraulic analysis of the pumps integrated into the piping system. A single pipe and pump arrangement is analyzed in Section 7.6.7. In Example 7.18 a system demand curve is obtained, and a trial solution to find the discharge and required pump head is shown. The technique is extended to simple piping systems in Section 11.3, wherein it is shown that one can employ either a pump performance curve or assume a constant power input to the pump. Section 11.4 deals with piping networks. In those systems, the pump performance curve can be represented by a polynomial relation (Eq. 11.4.14) and incorporated into the linearized energy equations, or alternatively, a constant pump power requirement can be employed (Eq. 11.4.15).

In this section we look more closely at the manner in which pumps are selected. The procedure involves not only satisfying the required flow and pressure requirements for a piping system, but also making sure that cavitation is avoided and that the most efficient pump is selected.

### 12.4.1  Matching Pumps to System Demand

Consider a single pipeline that contains a pump to deliver fluid between two reservoirs. The *system demand curve* is defined as

$$H_P = (z_2 - z_1) + \left(f\frac{L}{D} + \Sigma K\right)\frac{Q^2}{2gA^2} \qquad (12.4.1)$$

where atmospheric pressure is assumed at each reservoir, and the upstream and downstream reservoirs are at elevations $z_1$ and $z_2$, respectively. Note that $z_2$ need not be greater than $z_1$, and that the friction factor $f$ can vary as the discharge (i.e., Reynolds number) varies. The term $\Sigma K$ represents the minor losses in the pipe, see Section 7.6.4. Equation 12.4.1 is represented as curve (a) in Fig. 12.16. The first term on the right-hand side is the static head and the second term is the head loss due to pipe friction and minor losses. The steepness of the demand curve is

**KEY CONCEPT** *The intersection of the system demand curve with the pump head-discharge curve specifies the operating head and discharge.*

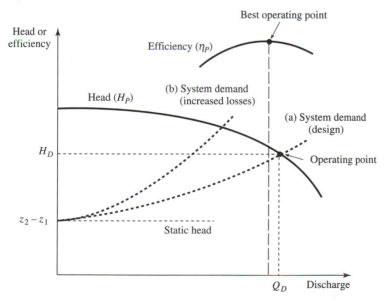

**Fig. 12.16**  Pump characteristic curve and system demand curve.

dependent on the sum of the loss coefficients in the system; as the loss coefficients increase, signified by curve (b) in Fig. 12.16, the pumping head required for a given discharge is increased. Piping systems may experience short-term changes in the demand curve such as throttling of valves, and over the long term, aging of pipes may cause the loss coefficients to increase permanently. In either case the system demand curve could change from (a) to (b) as shown.

For a given design discharge, Eq. 12.3.14 allows the selection of a pump based on the specific speed. Once the type of pump is determined, an appropriate size is selected from a manufacturer's characteristic curve; a representative curve is shown in Fig. 12.16. The intersection of the characteristic curve with the desired system demand curve will provide the *design head* $H_D$ and *design discharge* $Q_D$. It is desirable to have the intersection occur at or close to the point of maximum efficiency of the pump, designated as the *best operating point*.

## 12.4.2 Pumps in Parallel and in Series

In some instances, pumping installations may have a wide range of head or discharge requirements, so that a single pump may not meet the required range of demands. In these situations, pumps may be staged either in series or in parallel to provide operation in a more efficient manner. In this discussion, it is assumed that the pumps are placed at a single location with short lines connecting the separate units.

Where a large variation in flow demand is required, two or more pumps are placed in a parallel configuration (Fig. 12.17). Pumps are turned on individually to meet the required flow demand; in this way operation at higher efficiency can be attained. It is not necessary to have identical pumps, but individual pumps, when running in parallel, should not be operating in undesirable zones. For parallel pumping the combined characteristic curve is generated by recognizing that the head across each pump is identical, and the total discharge through the pumping system is $\Sigma Q$, the sum of the individual discharges through each pump. Note

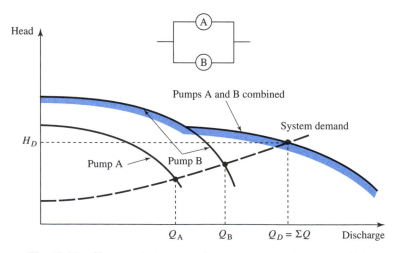

**Fig. 12.17** Characteristic curves for pumps operating in parallel.

the existence of three operating points in Fig. 12.17, in which pump A or pump B is used separately, or in which pumps A and B are combined. Other design operating points could be obtained by throttling the flow or by changing the pump speeds. The overall efficiency of pumps in parallel is

$$\eta_P = \frac{\gamma H_D \Sigma Q}{\Sigma \dot{W}_P} \tag{12.4.2}$$

in which $\Sigma \dot{W}_P$ is the sum of the individual power required by each pump.

For high head demands, pumps placed in series will produce a head rise greater than those of the individual pumps (Fig. 12.18). Since the discharge through each pump is identical, the characteristic curve is found by summing the head across each pump. Note that it is not necessary that the two pumps be identical. In Fig. 12.18 the system demand curve is such that pump A operating alone cannot deliver any liquid because its shutoff head is lower than the static system head. There are two operating points, either with pump B alone or with pumps A and B combined. The overall efficiency is

$$\eta_P = \frac{\gamma (\Sigma H_P) Q_D}{\Sigma \dot{W}_P} \tag{12.4.3}$$

in which $\Sigma H_P$ is the sum of the individual heads across each pump.

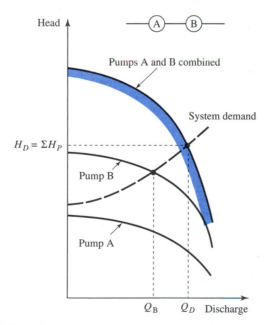

**Fig. 12.18**   Characteristic curves for pumps operating in series.

## Example 12.7

Water is pumped between two reservoirs in a pipeline with the following characteristics: $D = 300$ mm, $L = 70$ m, $f = 0.025$, $\Sigma K = 2.5$. The radial-flow pump characteristic curve is approximated by the formula

$$H_P = 22.9 + 10.7Q - 111Q^2$$

where $H_P$ is in meters and $Q$ is in m³/s.

    Determine the discharge $Q_D$ and pump head $H_D$ for the following situations: (a) $z_2 - z_1 = 15$ m, one pump placed in operation; (b) $z_2 - z_1 = 15$ m, with two identical pumps operating in parallel; and (c) the pump layout, discharge, and head for $z_2 - z_1 = 25$ m.

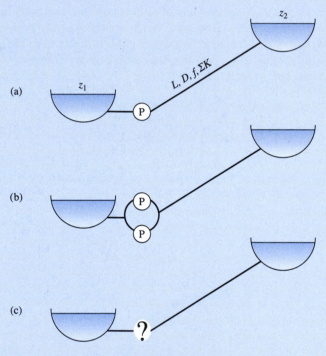

**Fig. E12.7**

### Solution

(a)   The system demand curve (Eq. 12.4.1) is developed first:

$$H_P = (z_2 - z_1) + \left(f\frac{L}{D} + \Sigma K\right)\frac{Q^2}{2gA^2}$$

$$= 15 + \left(\frac{0.025 \times 70}{0.3} + 2.5\right)\frac{Q^2}{2 \times 9.81(\pi/4 \times 0.3^2)^2}$$

$$= 15 + 85Q^2$$

To find the operating point, equate the pump characteristic curve to the system demand curve,

$$15 + 85Q_D^2 = 22.9 + 10.7Q_D - 111Q_D^2$$

Reduce and solve for $Q_D$:

$$195Q_D^2 - 10.7Q_D - 7.9 = 0$$

$$\therefore Q_D = \frac{1}{2 \times 195}[10.7 + \sqrt{10.7^2 + 4 \times 195 \times 7.9}] = 0.23 \text{ m}^3/\text{s}$$

Using the system demand curve, $H_D$ is computed as

$$H_D = 15 + 85 \times 0.23^2 = 19.5 \text{ m}$$

(b)  For two pumps in parallel, the characteristic curve is

$$H_P = 22.9 + 10.7\left(\frac{Q}{2}\right) - 111\left(\frac{Q}{2}\right)^2$$

$$= 22.9 + 5.35Q - 27.75Q^2$$

Equate this to the system demand curve and solve for $Q_D$:

$$15 + 85Q_D^2 = 22.9 + 5.35Q_D - 27.75Q_D^2$$

$$112.8Q_D^2 - 5.35Q_D - 7.9 = 0$$

$$\therefore Q_D = \frac{1}{2 \times 112.8}[5.35 + \sqrt{5.35^2 + 4 \times 112.8 \times 7.9}] = 0.29 \text{ m}^3/\text{s}$$

The design head is calculated to be

$$H_D = 15 + 85 \times 0.29^2 = 22.2 \text{ m}$$

(c)  Since z2 2 z1 is greater than the single pump shutoff head (i.e., 25 > 22.9 m), it is necessary to operate with two pumps in series. The combined pump curve is

$$H_D = 2(22.9 + 10.7Q - 111Q^2)$$

$$= 45.8 + 21.4Q - 222Q^2$$

The system demand curve is changed since $z_2 - z_1 = 25$ m. It becomes

$$H_D = 25 + 85Q^2$$

Equating the two relations above and solving for $Q_D$ and $H_P$ results in

$$25 + 85Q_D^2 = 45.8 + 21.4Q_D - 222Q_D^2$$

or

$$307Q_D^2 - 21.4Q_D = 20.8 = 0$$

$$\therefore Q_D = \frac{1}{2 \times 307}[21.4 + \sqrt{21.4^2 + 4 \times 307 \times 20.8}] = 0.30 \text{ m}^3/\text{s}$$

and

$$H_D = 25 + 85 \times 0.30^2 = 32.7 \text{ m}$$

**Fig. 12.19**   Multistage centrifugal pump. (Courtesy of Sulzer Pumps Ltd.)

### 12.4.3   Multistage Pumps

Instead of placing several pumps in series, multistage pumps are available (Fig. 12.19). Basically, the impellers are all housed in a single casing and the outlet from one impeller stage ejects into the eye of the next. Such pumps can provide extremely high heads. For the pump shown in Fig. 12.19, the pressure head range is 1500 to 3400 m, and the discharge can vary from 4500 $m^3$/h down to 260 $m^3$/h. Up to eight stages can be selected and the maximum speed is about 8000 rpm.

## 12.5  TURBINES

In many parts of the world, where sufficient head and large flow rates are possible, hydroturbines are used to produce electrical power. In contrast to pumps, turbines extract useful energy from the water flowing in a piping system. The moving component of a turbine is called a **runner**, which consists of vanes or buckets that are attached to a rotating shaft. The energy available in the liquid is transferred to the shaft by means of the rotating runner, and the resulting torque transferred by the rotating shaft can drive an electric generator. Hydroturbines vary widely in size and capacity, ranging from microunits generating 5 kW to those in large hydroelectric installations that produce over 400 MW.

   There are two types of turbines. The **reaction turbine** utilizes both flow energy and kinetic energy of the liquid; energy conversion takes place in an enclosed space at pressures above atmospheric conditions. Reaction turbines can be subdivided further, according to the available head, as either Francis or propeller type. The **impulse turbine** requires that the flow energy in the liquid be converted into kinetic energy by means of a nozzle before the liquid impacts on the runner; the energy is in the form of a high-velocity jet at or near atmospheric pressure. Turbines can be classified according to turbine specific speed, as shown in Fig. 12.20; the Pelton wheel is a particular type of impulse turbine (see Section 12.5.2).

Axis of
rotation
$\Omega_T$

| Impulse | Francis | Mixed flow | Axial flow |
| --- | --- | --- | --- |
| 0 – 1.0 | 1.0 – 3.5 | 3.5 – 7.0 | 7.0 – 14.0 |

**Fig. 12.20**   Various types of turbine runners.

### 12.5.1   Reaction Turbines

In reaction turbines, the flow is contained in a *volute* that channels the liquid into the runner (Fig. 12.21). Adjustable *guide vanes* (also called *wicket gates*) are situated upstream of the runner; their function is to control the tangential component of velocity at the runner inlet. As a result, the fluid leaves the guide vane exit and enters the runner with an acquired angular momentum. As the fluid travels through the runner region, its angular momentum is reduced and it imparts a torque on the runner, which in turn drives the shaft to produce power. The flow exits from the runner into a diffuser, called the *draft tube*, which acts to convert the kinetic energy remaining in the liquid into flow energy.

   **Francis Turbine.**   In the Francis turbine, the incoming flow through the guide vanes is radial, with a significant tangential velocity component at the entrance to the runner vanes (Fig. 12.22). As the fluid traverses the runner the velocity takes on an axial component while the tangential component is reduced. As it leaves the runner, the fluid velocity is primarily axial with little or no tangential component. The pressure at the runner exit is below atmospheric.

   The theoretical torque delivered to the runner is developed by applying Eq. 12.2.1 to the control volume shown in Fig. 12.22; the same assumptions leading to the pump torque relation, Eq. 12.2.2, apply. The resulting relation is

$$T = \rho Q (r_1 V_{t1} - r_2 V_{t2}) \tag{12.5.1}$$

Multiplying the torque by the angular speed $\omega$ gives the power delivered to the shaft:

$$\dot{W}_T = \omega T$$
$$= \rho Q (u_1 V_1 \cos \alpha_1 - u_2 V_2 \cos \alpha_2) \tag{12.5.2}$$

The fluid power input to the turbine is given by

$$\dot{W}_f = \gamma Q H_T \tag{12.5.3}$$

in which $H_T$ is the actual head drop across the turbine. Thus the overall efficiency is given by

Fig. 12.21 Reaction turbine (Francis type): (a) schematic; (b) Cross section of Francis turbine, Xingo hydro power plant, Brazil (Courtesy of Voith Hydro); (c) Turbine runner and housing, Amlach, Austria. (Courtesy of Voith Hydro.)

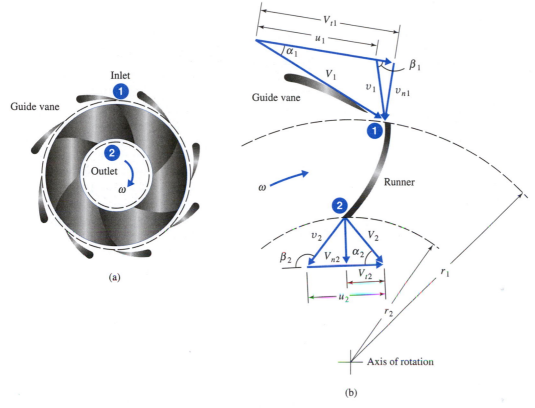

**Fig. 12.22**   Idealized Francis turbine runner: (a) runner control volume; (b) velocity diagrams at control surfaces.

$$\eta_T = \frac{\dot{W}_T}{\dot{W}_f} = \frac{\omega T}{\gamma Q H_T} \qquad (12.5.4)$$

The action of the guide vanes can be described by considering the velocity vector diagrams in Fig. 12.22b. Assume perfect guidance of the fluid along the guide vane; then the tangential velocity at the entrance to the runner is

$$V_{t1} = V_{n1} \cot \alpha_1 \qquad (12.5.5)$$

From the velocity vector diagram the tangential velocity is also given by

$$V_{t1} = u_1 + V_{n1} \cot \beta_1 \qquad (12.5.6)$$

The radial velocity component can be expressed in terms of the discharge $Q$ and the width of the runner $b_1$:

$$V_{n1} = \frac{Q}{2\pi r_1 b_1} \qquad (12.5.7)$$

Equations 12.5.5 to 12.5.7 are combined to eliminate $V_{t1}$ and $V_{n1}$. There results

$$\cot \alpha_1 = \frac{2\pi r_1^2 b_1 \omega}{Q} + \cot \beta_1 \qquad (12.5.8)$$

For a constant angular speed, in order to maintain the appropriate runner entry angle $\alpha_1$, the guide vane angle is adjusted as $Q$ changes.

Under normal operation, a turbine operates under a nearly constant head $H_T$ so that the performance characteristics are viewed differently from those of a pump. For turbine operation under constant head, the important quantities are the variations of discharge, speed, and efficiency. The interrelationships among the three parameters are shown in the isoefficiency curve of Fig. 12.23. Note the drop in discharge as the speed increases at a given guide vane setting. The efficiency is reduced by the following effects: (1) frictional head losses and draft tube head losses; (2) separation due to mismatch of flow entry angle with blade angle; (3) need to attain a certain turbine speed before useful power output is achieved; and (4) mechanical losses attributed to bearings, seals, and the like.

A representative dimensionless performance curve of a Francis turbine is shown in Fig. 12.24. The speed and head are kept constant, and the guide vanes are automatically adjusted as the discharge varies in order to attain peak efficiency.

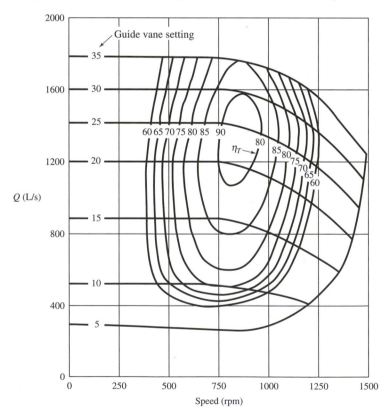

**Fig. 12.23** Isoefficiency curve for a Francis turbine: $D = 500$ mm, $H_T = 50$ m. (Courtesy of Gilbert Gilkes and Gordon, Ltd.)

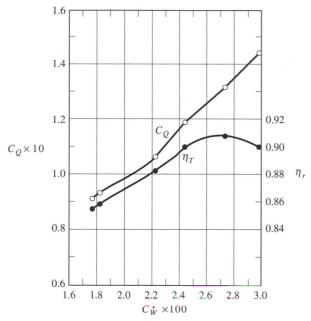

**Fig. 12.24**  Performance curve for a prototype Francis turbine: $D = 1000$ mm, $\omega = 37.7$ rad/s, $\Omega_T = 1.063$, $C_H = 0.23$. (Courtesy of Gilbert Gilkes and Gordon, Ltd.)

## Example 12.8

A reaction turbine, whose runner radii are $r_1 = 300$ mm and $r_2 = 150$ mm, operates under the following conditions: $Q = 0.057$ m³/s, $\omega = 25$ rad/s, $\alpha_1 = 30°$, $V_1 = 6$ m/s, $\alpha_2 = 80°$, and $V_2 = 3$ m/s. Assuming ideal conditions, find the torque applied to the runner, the head on the turbine, and the fluid power. Use $\rho = 1000$ kg/m³.

### Solution
The applied torque is computed using Eq. 12.5.1:

$$T = \rho Q(r_1 V_{t1} - r_2 V_{t2})$$
$$= \rho Q(r_1 V_1 \cos \alpha_1 - r_2 V_2 \cos \alpha_2)$$
$$= 1000 \times 0.057(0.3 \times 6 \times \cos 30° - 0.15 \times 3 \times \cos 80°)$$
$$= 84.4 \text{ N} \cdot \text{m}$$

Under ideal conditions, the power delivered to the shaft is the same as the fluid power input to the turbine (i.e., $\eta_T = 1$). Thus

$$\dot{W}_f = \dot{W}_T$$
$$= \omega T$$
$$= 25 \times 84.4 = 2110 \text{ W} \qquad \text{or} \qquad 2.11 \text{ kW}$$

The head on the turbine is found with the use of Eq. 12.5.3:

$$H_T = \frac{\dot{W}_f}{\gamma Q}$$
$$= \frac{2110}{9810 \times 0.057} = 3.77 \text{ m}$$

**Axial-Flow Turbine.** In an axial-flow turbine the flow is parallel to the axis of rotation (Fig. 12.25). Unlike the Francis turbine, the angular momentum of the liquid remains nearly constant and the tangential component of velocity is reduced across the blade. Both fixed-blade and pivoting-blade turbines are in use; the latter type, termed a Kaplan turbine, permits the blade angle to be adjusted to accommodate changes in head. Axial-flow turbines can be installed either vertically or horizontally. They are well suited for low-head installations.

**Cavitation Considerations.** The net positive suction head (*NPSH*) and the cavitation number, defined in Section 12.2.3, are applicable for turbines. In

Guide vanes

Runner

(a)

(b)

(c)

**Fig. 12.25** Axial-flow turbine (Kaplan type): (a) schematic; (b) Kaplan runner for the Yacyretá hydro power plant, Argentina (Courtesy of Voith Hydro); (c) Fish-friendly Kaplan turbine, Wanapum, USA. (Courtesy of Voith Hydro.)

Eqs. 12.2.21 to 12.2.23 the sign of the loss term becomes positive. The Thoma cavitation number, defined by Eq. 12.2.24 for a pump, is given in the following form for a turbine:

$$\sigma = \frac{1}{H_T}\left(\frac{p_{\text{atm}} - p_v}{\gamma} - \Delta z + h_L\right) \qquad (12.5.9)$$

Typically, location 2 is defined at the outlet of the runner, and location 1 refers to the liquid surface at the draft tube outlet (Fig. 12.26a).

   The cavitation number is commonly used to characterize the cavitation behavior of a turbine. Fig. 12.26b shows a representative plot of the cavitation number versus turbine efficiency, which is obtained experimentally by the turbine manufacturer. From such a plot operational procedures related to settings of the headwater and tailwater elevations can be established, and in addition, admissible levels of the cavitation number can be determined.

### 12.5.2  Impulse Turbines

The Pelton wheel (Fig. 12.27 on the next page) is an impulse turbine that consists of three basic components: one or more stationary inlet nozzles, a runner, and a

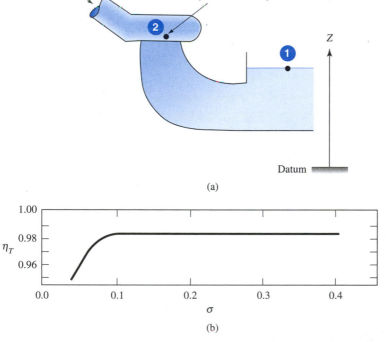

(a)

(b)

**Fig. 12.26**  Cavitation considerations: (a) schematic; (b) representative cavitation number curve.

(a)

(b)

(c)

**Fig. 12.27** Impulse turbine (Pelton type): (a) typical arrangement of twin-jet machine (Courtesy of Gilbert Gilkes and Gordon, Ltd.); (b) jet striking a bucket; (c) Pelton runner, Sedrun, Switzerland. (Courtesy of Voith Hydro.)

casing. The runner consists of multiple buckets mounted on a rotating wheel. The pressure head upstream of the nozzle is converted into kinetic energy contained in the water jet leaving the nozzle. As the jet impacts the rotating buckets, the kinetic energy is converted into a rotating torque. The buckets are shaped in a manner to divide the flow in half and turn its relative velocity vector in the horizontal plane nearly 180°; the ideal angle of 180° is not attainable since the exiting liquid must stay free of the trailing buckets.

The moment of momentum equation, illustrated in the preceding sections, can be applied to the control volume shown in Fig. 12.28. Neglecting friction, the torque delivered to the wheel by the liquid jet is

$$T = \rho Q r (V_1 - u)(1 - \cos \beta_2) \tag{12.5.10}$$

in which $Q$ is the discharge from all jets, and $u = r\omega$, $r$ is the wheel radius as shown in Fig. 12.27b. The power delivered by the fluid to the turbine runner is

$$\dot{W}_T = \rho Q u (V_1 - u)(1 - \cos \beta_2) \tag{12.5.11}$$

Usually, $\beta_2$ varies between 160 and 168°. Differentiation of Eq. 12.5.11 with respect to $u$ and setting it equal to zero shows that maximum power occurs when $u = V_1/2$.

The jet velocity can be given in terms of the available head $H_T$:

$$V_1 = C_v \sqrt{2gH_T} \tag{12.5.12}$$

The *velocity coefficient $C_v$* accounts for the nozzle losses; typically, $0.92 \leq C_v \leq 0.98$.

The efficiency is

$$\eta_T = \frac{\dot{W}_T}{\gamma Q H_T} \tag{12.5.13}$$

Substituting Eqs. 12.5.11 and 12.5.12 into this relation and rearranging produces

$$\eta_T = 2\phi(C_v - \phi)(1 - \cos \beta_2) \tag{12.5.14}$$

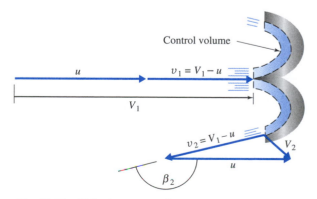

**Fig. 12.28**  Velocity vector diagram for Pelton bucket.

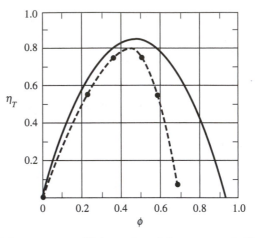

**Fig. 12.29** Speed factor versus efficiency for a laboratory-scale Pelton turbine: solid line, Eq. 12.5.14 ($C_v = 0.94$, $\beta_2 = 168°$); dashed line, experimental data.

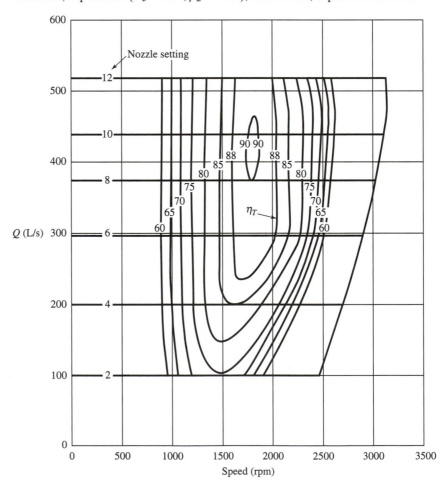

**Fig. 12.30** Isoefficiency curves for a Pelton turbine: $D = 500$ mm, $H_T = 500$ m. (Courtesy of Gilbert Gilkes and Gordon, Ltd.)

in which the *speed factor* $\phi$ is defined as

$$\phi = \frac{r\omega}{\sqrt{2gH_T}} \tag{12.5.15}$$

Maximum efficiency occurs at $\phi = C_v/2$. Fig. 12.29 shows Eq. 12.5.14 plotted for $C_v = 0.94$ and $\beta_2 = 168°$. In addition, data are plotted for a laboratory-scale Pelton wheel. The theoretical maximum efficiency is 0.874, while the experimental maximum efficiency is approximately 0.80. The efficiency is reduced because (1) the liquid jet does not strike the bucket with a uniform velocity, (2) losses occur when the jet strikes the bucket and splitter, (3) frictional loss is present due to the rotation of the bucket and splitter, and (4) there is frictional loss at the nozzle.

The interrelationships among $Q$, $\omega$, and $\eta_T$ are shown in the isoefficiency curves of Fig. 12.30. For a constant head and nozzle setting, the discharge remains constant, since the nozzle outlet is at atmospheric pressure. A representative dimensionless performance curve for a prototype Pelton turbine is shown in Fig. 12.31.

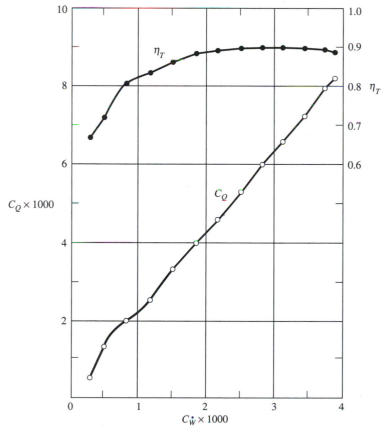

**Fig. 12.31** Performance and efficiency curves for a prototype Pelton turbine: $D = 1000$ mm, $\omega = 75.4$ rad/s, $\Omega_T = 0.135$, $C_H = 0.52$. (Courtesy of Gilbert Gilkes and Gordon, Ltd.)

## Example 12.9

A Pelton turbine rotates at an angular speed of 400 rpm, developing 67.5 kW under a head of 60 m of water. The inlet pipe diameter at the base of the single nozzle is 200 mm. The operating conditions are $C_v = 0.97$, $\phi = 0.46$, and $\eta_T = 0.83$. Determine (a) the volumetric flow rate, (b) the diameter of the jet, (c) the wheel diameter, and (d) the pressure in the inlet pipe at the nozzle base.

### Solution

(a)  The discharge is computed from Eq. 12.5.13 to be

$$Q = \frac{\dot{W}_T}{\gamma H_T \eta_T} = \frac{67\,500}{9810 \times 60 \times 0.83} = 0.138 \text{ m}^3/\text{s}$$

(b)  From Eq. 12.5.12, the velocity of the jet is

$$V_1 = C_v \sqrt{2gH_T} = 0.97 \sqrt{2 \times 9.81 \times 60} = 33.3 \text{ m/s}$$

The area of the jet is the discharge divided by $V_1$, or

$$A_1 = \frac{Q}{V_1} = \frac{0.138}{33.3} = 4.14 \times 10^{-3} \text{ m}^2$$

Hence the jet diameter $D_1$ is

$$D_1 = \sqrt{\frac{4}{\pi} A_1} = \sqrt{\frac{4}{\pi} \times 4.14 \times 10^{-3}} = 0.0726 \text{ m} \qquad \text{or} \qquad 73 \text{ mm}$$

(c)  Use Eq. 12.5.15 to compute the wheel diameter D to be

$$D = 2r = \frac{2\phi}{\omega} \sqrt{2gH_T}$$

$$= \frac{2 \times 0.46}{400 \times \pi/30} \sqrt{2 \times 9.81 \times 60} = 0.754 \text{ m} \qquad \text{or} \qquad 754 \text{ mm}$$

(d)  The area of the inlet pipe is

$$A = \frac{\pi}{4} \times 0.20^2 = 0.0314 \text{ m}^2$$

The piezometric head just upstream of the nozzle is equal to $H_T$, so that the pressure at that location is

$$p = \gamma \left( H_T - \frac{Q^2}{2gA^2} \right)$$

$$= 9810 \left( 60 - \frac{0.138^2}{2 \times 9.81 \times 0.0314^2} \right)$$

$$= 5.79 \times 10^5 \text{ Pa} \qquad \text{or} \qquad \text{approximately 580 kPa}$$

### 12.5.3   Selection and Operation of Turbines

A preliminary selection of the appropriate type of turbine for a given installation is based on the specific speed. Fig. 12.20 shows how the turbine runner varies with $\Omega_T$. Impulse turbines normally operate most economically at heads above 300 m, but small units can be used for heads as low as 60 m. Heads up to 300 m are possible for Francis units, and propeller turbines are normally used for heads lower than 30 m. Fig. 12.32 illustrates the ranges of application for a variety of hydraulic turbines.

So-called minihydro and microhydro installations have recently been subject to renewed interest, not only in developing countries, but in industrially developed countries such as the United States, where the cost of energy has increased and incentive programs have been made available to promote their use.

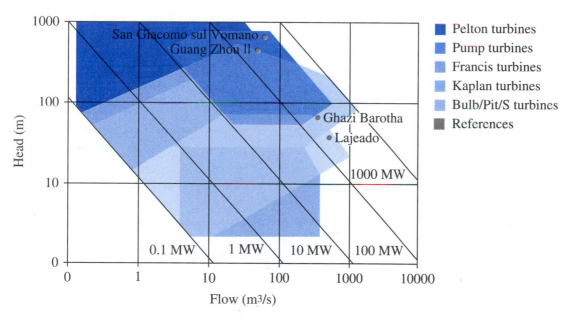

**Fig. 12.32**   Application ranges for hydraulic turbines. (Courtesy of Voith Hydro.)

*Microhydro* has been classified as units that possess a capacity of less than 100 kW, and *minihydro* refers to system capacity from 100 to 1000 kW (Warnick, 1984). It is becoming common practice for manufacturers to develop a standardized turbine unit. In addition to specialized small Pelton, Francis, and propeller units, the *cross-flow turbine* is employed; basically, it is an impulse turbine that possesses a higher rotational speed than that of other impulse turbines.

Another alternative is to use a commercially available pump and operate it in the reverse direction; in the turbine mode, the best efficiency operating point requires a larger head and flow than when the pump operates at best efficiency in the pumping mode.

Reversible pump/turbine units are used in pumped/storage hydropower installations. Water is pumped from a lower reservoir to a higher storage reservoir during periods of low demand for conventional power plants. Subsequently, water is released from the upper reservoir, and the pumps are rotated in reverse to generate power during periods of high electric power demand. Fig. 12.33 shows a representative installation consisting of a Francis reversible pump/turbine unit combined with a generator/motor. Such units require special design considerations to operate efficiently in either mode. The problem of converting the flow from the pump mode to the turbine mode, or vice versa, is a very complicated transient problem that is beyond the scope of this book. Water with enormous mass moving through the entire system must be reversed in direction; a special technique must be developed to accomplish this.

**Fig. 12.33**   Runner and top cover for pump turbine, Waldeck, Germany. (Courtesy of Voith Hydro.)

## Example 12.10

A discharge of 2100 m³/s and a head of 113 m are available for a proposed pumped-storage hydroelectric scheme. Reversible Francis pump/turbines are to be installed; in the turbine mode of operation, $\Omega_T = 2.19$, the rotational speed is 240 rpm, and the efficiency is 80%. Determine the power produced by each unit and the number of units required.

### Solution

The power produced by each unit is found using the definition of specific speed, Eq. 12.3.15. Solving for the power, we have

$$\dot{W}_T = \rho \left[ \frac{\Omega_T}{\omega} (gH_T)^{5/4} \right]^2$$

$$= 1000 \left[ \frac{2.19}{240 \times \pi/30} (9.81 \times 113)^{5/4} \right]^2 = 3.11 \times 10^8 \text{ W}$$

From Eq. 12.5.13, the discharge in each unit is

$$Q = \frac{\dot{W}_T}{\gamma H_T \eta_T}$$

$$= \frac{3.11 \times 10^8}{9800 \times 113 \times 0.8} = 351 \text{ m}^3/\text{s}$$

The required number of units is equal to the available discharge divided by the discharge in each unit, or 2100/351 = 5.98. Hence, six units are required.

## 12.6 SUMMARY

This chapter focuses on pumps or turbines that supply or extract energy by means of rotating impellers or vanes. First we emphasized the radial-flow, or centrifugal, pump and made use of the moment-of-momentum principle to derive an idealized head-discharge relation. We then contrasted that with the actual head-discharge performance curve, one that is obtained experimentally and is required for use in pump analysis and design. Axial-flow and mixed-flow pumps were introduced, and the differences between these and the radial-flow pump were discussed. The significance of pump efficiency and net positive suction head (relating to cavitation) diagrams were shown for the proper design and selection of pumps.

Dimensional analysis and similitude were applied to both pumps and turbines; the three significant dimensionless numbers that were developed are the power coefficient, flow coefficient, and head coefficient. These coefficients were combined to yield pump or turbine efficiency, or to produce the specific speed for a pump or turbine. It was shown that a family of pumps or turbines could be represented in dimensionless form by a single curve. The grouping of pumps—either singly, in parallel, or in series—to meet appropriate head and discharge requirements for piping was illustrated.

Finally, a brief introduction to turbines was provided, by relating fundamental theory along with the application of reaction, axial-flow, and impulse turbines to various hydraulic design situations.

# PROBLEMS

## Elementary Theory

**12.1** Determine the torque, power, and head input or output for each turbomachine shown in Fig. P12.1. Is a pump or turbine implied? The following data are in common: outer radius 300 mm, inner radius 150 mm, $Q = 0.057$ m$^3$/s, $\rho = 1000$ kg/m$^3$, and $\omega = 25$ rad/s.

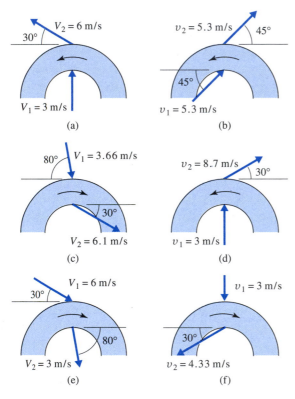

(a)

(b)

(c)

(d)

(e)

(f)

**Fig. P12.1**

**12.2** A centrifugal water pump rotates at 800 rpm. The impeller has uniform blade widths $b_1 = 50$ mm, $b_2 = 25$ mm, and radii $r_1 = 40$ mm, $r_2 = 125$ mm. The blade angles are $\beta_1 = 45°$, $\beta_2 = 30°$. Assuming no angular momentum of fluid at the blade entrance, determine the ideal flow rate, pressure head rise across the impeller, and the theoretical torque and power requirements.

**12.3** The suction and discharge pipes of a pump are connected to a differential mercury manometer. The diameter of the suction pipe is 200 mm and the diameter of the discharge pipe is 150 mm. The

specific gravity of the pumped liquid is $S = 0.81$, and the manometer reading is $H = 600$ mm. Determine the pressure head rise across the pump if the discharge is $Q = 115$ L/s. The centerlines of the suction and discharge pipes are at the same elevation.

**12.4** A centrifugal pump with the dimensions $r_2 = 62.5$ mm, $b_2 = 10$ mm, $\beta_2 = 60°$, is pumping kerosene ($S = 0.80$) at a rate of 7.6 L/s and rotating at 2000 rpm. For "best design" entry conditions, determine the theoretical pressure head rise and fluid power.

**12.5** An axial-flow pump has a stator blade positioned upstream of the impeller, and it provides an angle $\alpha_1 = 60°$ to the flow. The radius of the impeller tip is 285 mm, and the hub radius is 135 mm. Determine the required average blade angle at the exit of the impeller if the pump is to deliver 0.57 m$^3$/s of water with a theoretical head rise of 2.85 m. The rotational speed is 1500 rpm.

**12.6** For the system shown in Fig. P12.6, water flows through the pump at a rate of 50 L/s. The allowable *NPSH* provided by the manufacturer at that flow is 3 m. Determine the maximum height $\Delta z$ above the water surface at which the pump can be located to operate without cavitating. Include all losses in the suction pipe.

**Fig. P12.6**

**12.7** During a test on a water pump, no cavitation is detected when the gage pressure in the pipe at the pump inlet is $-68$ kPa, and the water temperature is 27°C. The inlet pipe diameter is 10 cm, the head across the pump is 35 m, and the discharge is 50 L/s.

(a) Compute the *NPSH* and the cavitation number if the atmospheric pressure is 101.3 kPa.

(b) If the pump is to produce the same head and discharge at a location where the atmospheric pressure is 83 kPa, what is the necessary change in elevation of the pump relative to the inlet reservoir to avoid cavitation?

---

### Dimensional Analysis and Similitude

**12.8** Consider the 371-mm-diameter pump curve of Fig. 12.9. If the pump is run at 1200 rpm to deliver 0.8 m$^3$/s of water, find the resulting head rise and fluid power.

**12.9** In Problem 12.8, how high above the suction reservoir can the pump running at 1200 rpm be located if the water temperature is 50°C and the atmospheric pressure is 101 kPa?

**12.10** Prepare dimensionless performance curves for the 205-mm-diameter radial flow pump of Fig. 12.6. Compare these with the curves shown in Fig. 12.12. Can you explain why there may be differences between the two sets of curves?

**12.11** If the 220-mm-diameter pump whose characteristics are shown Fig. 12.6 is operating at 3300 rpm, what will be the head and the *NPSH* when the discharge is 200 m$^3$/h?

**12.12** A pump is needed to transport 150 L/s of oil ($S = 0.86$) with an increase in head across the pump of 22 m when running at 1800 rpm. What type of pump is best suited for this operation?

**12.13** A pump is required to deliver 0.17 m$^3$/s of water with a head increase of 104 m when operating at maximum efficiency at 2000 rpm. Determine the type of pump best suited for this duty.

**12.14** Refer to Fig. 12.13, the dimensionless performance curves for an axial-flow pump. It is desired to deliver water at a rate of 1240 L/s and a speed of 750 rpm. Determine:

(a) The available head, diameter, power, and *NPSH* requirements.

(b) The actual head-discharge and power-discharge curves.

**12.15** A pump running at 400 rpm discharges water at a rate of 85 L/s. The total head rise across the pump is 7.6 m, and the efficiency is 0.7. A second pump, whose linear dimensions are two-thirds of the former, runs at 400 rpm under dynamically similar conditions; it is to be placed on a space satellite where simulated gravity produces a gravitational field which is 50% that of Earth's. What are the discharge, head rise, and power requirements for the second pump?

**12.16** A pump is designed to operate under optimum conditions at 600 rpm when delivering water at 22.7 m$^3$/min against a head of 19.5 m. What type of pump is recommended?

**12.17** A pump operating under the conditions stated in Problem 12.16 has a maximum efficiency of 0.70. If the same pump is now required to deliver water at a head of 30.5 m at maximum efficiency, determine the rotational speed, the discharge, and the required power.

**12.18** A manufacturer desires to double both the discharge and head on a geometrically similar pump. Determine the change in rotational speed and diameter under the new conditions.

**12.19** A pump is required to move water at 180 L/s against a head of 60 m. The approximate rotational speed is 1400 rpm. Using the appropriate dimensionless performance curve in the text, determine the actual speed and size of the pump to operate at peak efficiency.

**12.20** Select the type, size, and speed of pump to deliver a liquid ($\gamma = 8830$ N/m$^3$) at 0.66 m$^3$/s with fluid power requirements of 200 kW. (*Hint:* Assume that the pump speed can vary between 1000 and 2000 rpm.)

## Use of Turbopumps

**12.21** For the system shown in Fig. P12.21, recommend the appropriate machine to pump 1 m³/s of water if the impeller speed is 600 rpm.

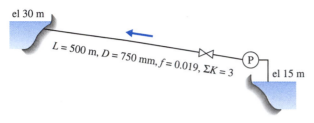

el 30 m

$L = 500$ m, $D = 750$ mm, $f = 0.019$, $\Sigma K = 3$

P

el 15 m

**Fig. P12.21**

**12.22** The following performance data were obtained from a test on a 216-mm double-entry centrifugal pump moving water at a constant speed of 1350 rpm:

| $Q$ (m³/min) | $H$ (m) | $\eta_P$ |
|---|---|---|
| 0 | 12.2 | 0 |
| 0.454 | 12.8 | 0.26 |
| 0.905 | 13.1 | 0.46 |
| 1.36 | 13.4 | 0.59 |
| 1.81 | 13.4 | 0.70 |
| 2.27 | 13.1 | 0.78 |
| 2.72 | 12.2 | 0.78 |
| 3.80 | 11.0 | 0.74 |

Plot $H$ versus $Q$, $\eta_P$ versus $Q$, and $\dot{W}_P$ versus $Q$. If the pump operates in a system whose demand curve is given by $5 + Q^2$, find the discharge, head, and power required. In the demand curve, $Q$ is given in cubic meters per minute.

**12.23** An oil ($S = 0.85$) is to be pumped through a pipe as shown in Fig. P12.23. Using the characteristic curves of Fig. 12.12 determine the required impeller diameter, speed, and input power to deliver the oil at a rate of 14.2 L/s. Would you expect the efficiency of the pump to be the value shown on the curve?

el 12.2 m

el 0 m

P

$L = 61$ m, $D = 75$ mm, $f = 0.010$, $\Sigma K = 15$

**Fig. P12.23**

**12.24** With reference to the pump data in Problem 12.22, if the pump is run at 1200 rpm, find the discharge, head, and required power. Is the operating efficiency under these conditions the same as the pump running at 1350 rpm?

**12.25** For the pump data presented in Problem 12.22, plot the dimensionless pump curves $C_H$ versus $C_Q$, $C_{\dot{W}}$ versus $C_Q$, and $\eta_P$ versus $C_Q$. What is the specific speed of the pump? (*Note:* For double-entry pumps, use one-half the discharge when computing the specific speed.)

**12.26** A closed-loop flow facility (Fig. P12.26) is to be used to study water flow in a hydraulics laboratory. It is constructed of 300-mm-diameter hydraulically smooth pipe. The loss coefficient at each of the four bends is 0.1, and the total length of the loop is 14 m. The design velocity is $V = 3$ m/s. It is proposed that an axial flow pump running at about 300 rpm be used. Verify that an axial-flow pump is the correct type to employ.

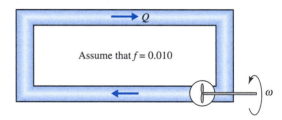

$Q$

Assume that $f = 0.010$

$\omega$

**Fig. P12.26**

**12.27** In Problem 12.26, assume that the only axial pump available is one from the family shown in Fig. 12.13. If the diameter of the impeller were restricted to the pipe diameter, what would be the rotational speed and resulting head rise across the pump?

**12.28** The 240-mm-diameter pump represented in Fig. 12.6 is used to move water in a piping system whose demand curve is $62 + 270Q^2$, where $Q$ is in cubic meters per second. Find the discharge, required input power, and the *NPSH* requirement for:

(a) One pump  (b) Two pumps in parallel

**12.29** It is desired to pump 600 m³/h of liquid ammonia ($\rho = 607$ kg/m³) at a speed of 2500 rpm and a pressure rise of 1000 kPa.

(a) Determine the type of pump best suited for this duty.

**(b)** Using the appropriate dimensionless pump characteristic curve, find the diameter and number of stages required to meet the pumping requirements.

**12.30** Crude oil ($S = 0.86$) is to be pumped through 5 km of 450-mm-diameter cast iron pipe. The elevation rise from the upstream to the downstream end is 192 m. If an available pump is the 240-mm-diameter radial-flow machine of Fig. 12.6, how many pumps in series are required to provide the most efficient operation? Find the required power.

**12.31** A water supply piping network designed for a community receives water from a 25-cm-diameter well (Fig. P12.31). The pump is situated 45 m below the 25-cm-diameter primary water main that delivers the flow to the network, and due to the available ground water, the hydraulic grade line on the suction side of the pump is 22.5 m above the pump. The design discharge is 32 L/s, and a pressure of 550 kPa is required in the primary water main. Making use of Fig. 12.12, determine the size, speed, and number of stages of a pump to satisfy the design requirements, and the required horsepower. Assume that the pump impeller diameter is 20 cm and that the friction factor in the well delivery pipe is constant, $f = 0.02$.

**Fig. P12.31**

**12.32** A model pump (Fig. P12.32) delivers 80°C water at a speed of 2400 rpm. It begins to cavitate when the inlet pressure $p_i = 83$ kPa absolute and inlet velocity $V_i = 6$ m/s.
**(a)** Compute the *NPSH* of the model pump. Neglect losses and elevation changes between the inlet section where $p_i$ is recorded and the cavitation region in the pump.

**(b)** What is the required *NPSH* of a prototype pump that is four times larger and runs at 1000 rpm?

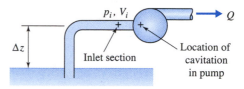

**Fig. P12.32**

**12.33** It is required to pump water at 1.25 m³/s from a lower to an upper reservoir. The water surface in the lower reservoir is at an elevation of 20 m, and the upper reservoir elevation is 23 m. The length of the 750-mm-diameter pipe is 60 m. The friction factor can be assumed constant, $f = 0.02$, and the entrance and exit loss coefficients are 0.5 and 1.0, respectively.

**(a)** What type of pump is best suited for this system if the impeller speed is to be 600 rpm?

**(b)** Assuming a reasonable efficiency from an appropriate pump curve, estimate the power in kilowatts required to operate the pump.

**12.34** Carbon tetrachloride at 20°C is pumped from a tank opened to the atmosphere through 200 m of 50-mm-diameter smooth horizontal pipe and discharges into the atmosphere at a velocity of 3 m/s. The elevation of liquid in the tank is 3 m above the pipe. A pump is situated at the pipe inlet immediately downstream of the reservoir.

**(a)** Determine the speed and size of a pump to meet the system demand.

**(b)** What is the available net positive suction head at the pump inlet? Neglect losses in the short suction pipe connecting the pump to the tank. Assume that the atmospheric pressure is 101 kPa.

**(c)** Sketch the hydraulic grade line for the system.

**12.35** A wind tunnel with a test section 1 m × 2 m is powered by an axial fan 1.5 m in diameter. The velocity in the test section is 20 m/s, and the rotational speed of the fan is 2500 rpm. Assuming negligible losses in the tunnel and a performance curve for the fan defined by Fig. 12.13:
**(a)** Calculate the power required to operate the fan.
**(b)** What is the pressure rise across the fan?

## Turbines

**12.36**  A Francis turbine has the following dimensions: $r_1 = 4.5$ m, $r_2 = 2.5$ m, $b_1 = b_2 = 0.85$ m, $\beta_1 = 75°$, $\beta_2 = 100°$. The angular speed is 120 rpm, and the discharge is 150 m³/s. For no separation at the entrance to the runner, determine the guide vane angle, theoretical torque, head, and power.

**12.37**  A model Francis turbine at 1/5 full scale develops 3 kW at 360 rpm under a head of 1.8 m. Find the speed and power of the full-size turbine when operating under a head of 5.8 m. Assume that both units are operating at maximum efficiency.

**12.38**  A model is to be built for studying the performance of a turbine having a runner diameter of 90 cm, maximum output of 2200 kW under a head of 45 m and a speed of 240 rpm. Determine the diameter of the model runner and its speed if the corresponding model power is 9 kW and the head is 7.5 m.

**12.39**  A hydroelectric site has available $H_T = 80$ m and $Q = 3$ m³/s. It is proposed to use a Francis turbine unit with the operating characteristics at optimum efficiency shown in Fig. 12.24. Determine the required speed and diameter of the machine.

**12.40**  Determine the power output, the type of turbine, and the approximate speed for the installation shown in Fig. P12.40. Neglect all minor losses except those existing at the valve.

**Fig. P12.40**

**12.41**  A water turbine is required to operate at 420 rpm under a net head of 3 m, a discharge of 0.312 m³/s, and with an efficiency of 0.9. Tests on a 1/6-scale model running at 2000 rpm are to be carried out using water. For the model turbine, determine the head drop, flow rate, output power, and expected efficiency.

**12.42**  A Pelton wheel develops 4.5 MW under a head of 120 m at a speed of 200 rpm. The wheel diameter is eight times the jet diameter. Use the experimental data of Fig. 12.29 at maximum efficiency to determine the required flow, wheel diameter, diameter of each jet, the number of jets required, and the specific speed.

**12.43**  A reaction turbine installation typically includes a draft tube (Fig. 12.26a), whose function is to convert kinetic energy of the fluid exiting the runner into flow energy. Consider a turbine operating at $Q = 85$ m³/s and $H_T = 31.8$ m. The radius of the draft tube at the outlet of the runner is 2.5 m, and the diameter at the end of the tube is 5.0 m. Losses in the draft tube can be neglected.

(a)  For water at 20°C, what is the pressure at the runner outlet if in Fig. P12.43, $z_2 - z_1 = 2.5$ m?

(b)  What is the permissible $z_2 - z_1$ if the allowable cavitation number is $\sigma = 0.14$?

**12.44**  The Ludington pumped-storage system (Fig. P12.44) on the eastern shore of Lake Michigan consists of six penstocks delivering a combined discharge of 2082 m³/s in the power-producing mode of operation. Each turbine generates 427,300 hp at an efficiency of 0.85. What diameter penstock is required for the system to operate under design conditions? Classify the type of turbine employed.

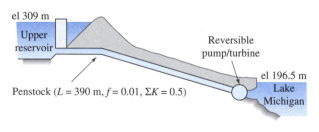

**Fig. P12.44**

**12.45**  A Francis turbine of the same family as that represented in Fig. 12.24 is running at a speed of 480 rpm under a head of 9.5 m. Determine the runner diameter, discharge, and developed power if the turbine is operating at maximum efficiency.

**12.46**  The available head from reservoir level to the four nozzles of a twin-runner Pelton wheel is 305 m. The length of the supply pipe, or penstock, is 3 km with a friction factor $f = 0.02$ and $\Sigma K = 2.0$. The turbine develops 10.4 MW with an efficiency of 0.85.

**(a)** If the head across the turbine is 95% of the available head, what is the diameter of the penstock?

**(b)** If $C_v = 0.98$ for each nozzle, calculate the diameter of each jet.

**12.47** In a projected low-head hydroelectric scheme, 282 m³/s of water is available under a head of 3.7 m. It is proposed to use Francis turbines with specific speed 2.42, for which the rotational speed is 50 rpm. Determine the number of units required and the power to be developed by each machine. Assume an efficiency of 0.9.

**12.48** Repeat Problem 12.47 if propeller turbines are substituted for the Francis units. Assume a specific speed of 4.15 and an efficiency of 0.9.

**12.49** A landowner has constructed an elevated reservoir as shown in Fig. P12.49. She estimates that 1200 L/min is available as continuous flow and desires to install a small Francis turbine to generate electricity for her own use.

**(a)** What fluid power is expected to be delivered to the turbine?

**(b)** If a turbine similar to the one whose characteristics are shown in Fig. 12.24 is to be installed, what are the speed and size required for this installation?

**(c)** Using an appropriate empirical equation, estimate the efficiency for this installation. What power is expected to be delivered to the generator?

**12.50** A proposed design for a hydroelectric project is based on a discharge of 0.25 m³/s through the penstock and turbine (Fig. P12.50). The friction factor can be assumed constant, $f = 0.015$, and the minor losses are negligible.

**(a)** Determine the power in kilowatts that can be expected from the facility, assuming that the turbine efficiency is 0.85.

**(b)** Show that the type of machine to be installed is a Francis turbine if the desired rotational speed is 1200 rpm.

**(c)** Determine the size, actual speed, and actual power output of the selected turbine.

**Fig. P12.50**

**Fig. P12.49**

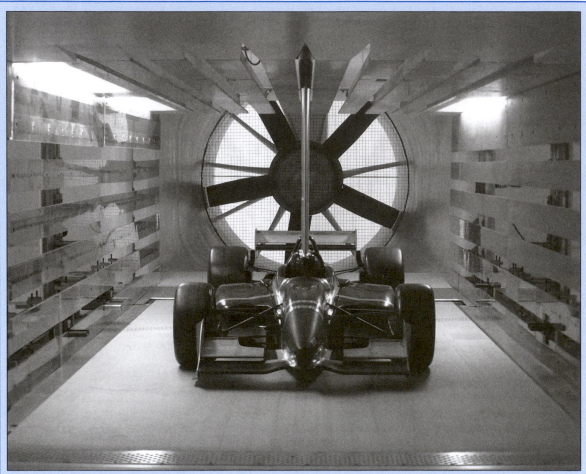

Scale model of a Formula One race car in a wind tunnel. Measurements are made of flow patterns, velocity distributions, and pressures on the surface of the vehicle. Such tests are employed to reduce the drag, resulting in increased speed of the prototype vehicle. (All American Racers archives)

# 13

# Measurements in Fluid Mechanics

## Outline

## Chapter Objectives

The objectives of this chapter are to:

▲ Describe how various flow parameters are measured; these include measurements taken at a local flow position—for example, pressure or velocity—as well as integrated measurements such as discharge or pressure averaged over a cross section

▲ Present techniques and equipment employed to visualize flow fields

▲ Introduce the concepts of uncertainty analysis and error propagation resulting from collecting laboratory and field data, and demonstrate the use of regression analysis in nonlinear flow measurement

## 13.1 INTRODUCTION

In the engineering laboratory and in many industrial situations, the need to measure fluid properties and various flow parameters, such as pressure, velocity, and discharge, are exceedingly important. In many instances these needs are obvious. Examples include the flow rate in a pipe or irrigation channel; the contaminant or sediment load in a river; the peak pressures on the surface of a high-rise building or the flow patterns around that building; the drag on an automobile or truck traveling at high speeds; the velocity field about a commercial aircraft; the wind velocity profile above a hilly terrain; and the size and distribution of ocean or lake waves.

Fluid mechanics measurements are not exclusive to the engineer. Medical diagnosticians are interested in monitoring the functions of the cardiovascular and pulmonary systems in the human body. The movement of fluids in the agricultural, petroleum, gas, chemical, beverage, and water supply and wastewater industries require large investments annually; the uncertainties present in the measurement of these flows can have a significant impact on material and monetary considerations.

Many devices have been developed to measure flow parameters; obviously, each was designed to serve a specific purpose. Before undertaking the task of measurement, it is important to define clearly the need for measuring a particular parameter. Knowledge of the underlying fluid mechanics and the physical principles involved are essential to selecting an appropriate measuring instrument and conducting a successful measurement. In addition, we must recognize that there are differences in various types of measurements. Consider, for example, discharge readings resulting from a measurement of a velocity profile integrated over a cross section; the fluctuating velocity can be time-averaged at a specific location to produce a mean value. On the other hand, instantaneous measurements of velocity in a very small sampling volume may be required to observe the transient nature of an unsteady flow. The degree of sophistication in recording a parameter can range from a simple visual reading of a manometer or dial to high-speed digital sampling of voltages representing a number of variables.

The purpose of this chapter is to provide the reader with a basic introduction to the concepts and techniques applied by engineers who measure flow parameters either in the laboratory or in an industrial environment. Included is a general overview of experimental methods commonly used in fluid mechanics teaching and research laboratories; the treatment is not exhaustive, but references are provided for additional information. In the following two sections, methods and instrumentation employed to measure pressure, velocity, and discharge are presented. Next, a discussion of flow visualization is given, and the final section is devoted to data acquisition and methods of analyzing data.

## 13.2 MEASUREMENT OF LOCAL FLOW PARAMETERS

A local flow measurement means that a quantity is measured over a relatively small sample volume of the fluid. Usually, the volume is sufficiently small so that we can say that the measurement represents the magnitude of the quantity at a point in the flow field. Two significant local flow quantities are pressure and velocity; others are temperature, density, and viscosity. Only the first two are discussed in this section; temperature measurement is not considered in this book. Density and viscosity would be determined from temperature and pressure measurements.

### 13.2.1  Dynamic Response and Averaging

Flow measurements may be classified according to whether the flow is steady or unsteady. If the magnitude of a physical quantity remains constant with time, this value is referred to as the steady-state value. Conversely, if the quantity changes with time, the measurement is transient, or unsteady. Transient measurement demands a more highly specialized measuring apparatus. Most measuring instruments require a certain amount of time to respond to the physical quantity sensed. In transient measurements this response time should be much less than the time for a significant change in the physical quantity to occur. In steady-state measurements the only concern is to take a reading after the measuring system has time to respond to the flow conditions. In many situations, for example, turbulent flow, it is desired to obtain a time-average measurement of a quantity, even though the local flow conditions are unsteady; examples are Reynolds stress and turbulence intensity. In this situation, a time-averaging mechanism must be included in the measurement. In addition, an instrument must be sufficiently small to measure the spatial variation of the measured parameter, such as the wavelength of a pressure disturbance. Generally, the issue of spatial and temporal resolution must be resolved by selecting instrumentation compatible with the flow scale and with the scale of any flow disturbance; see Goldstein (1996) for further explanation.

### 13.2.2  Pressure

Fluid pressures are measured in many different ways. The type of instrument utilized depends on the levels of precision and detail required for the particular application. Virtually all pressure measurements are based either on the manometer principle or on the concept of pressure deforming a solid material such as a crystal, membrane, tube, or plate, and then converting that deformation to an electrical signal or mechanical readout. Pressure measurements are in either the static or dynamic mode. Time-dependent pressures result from unsteadiness in the flow and from pressure disturbances; the pressure disturbances are a result of either hydrodynamic or acoustic perturbations. Often, static pressure measurements are based on time-averaged mean values, since unsteadiness is always present when the flow is turbulent. Several representative pressure gages and transducers are described below.

**Manometer.**  The manometer was analyzed in Section 2.4.3. It is a simple, inexpensive instrument for measuring static pressure, and it has no moving mechanical parts. High accuracy can be attained by using an inclined tube manometer (Fig. 13.1a) or a micromanometer (Fig. 2.6). If the pressure tap is properly machined normal to the interior wall in a duct or pipe with no burrs, and the diameter is sufficiently small (usually about 1 mm diameter), the static pressure is measured accurately (Fig. 13.1b). Often the taps are placed circumferentially around the duct in the form of a piezometer ring to average out any type of flow irregularities.

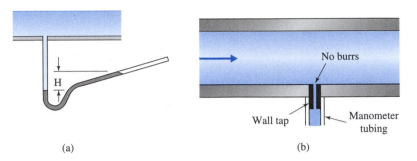

(a)                                              (b)

**Fig. 13.1**   Manometer used to measure pressure: (a) inclined tube manometer; (b) piezometer opening.

**Bourdon Gage.**   An inexpensive pressure gage for reading relatively large static pressures is the Bourdon tube (Fig. 13.2). It is a curved, flattened tube, which will expand outward when subjected to internal pressure. A mechanical linkage attached to a dial gage will provide a direct reading of the pressure. When properly designed, these gages will provide good accuracy; however, they should not be used in situations where large pressure pulsations are present, since damage to the tube may result.

**Pressure Transducer.**   Pressure transducers are electromechanical devices that are designed to measure either steady or unsteady hydrodynamic or acoustic pressures. They consist of flexible membranes or piezoelectric elements that respond to pressure changes (Fig. 13.3). All of them require accompanying electronic circuitry, which adds to their costs. Since their output is an electric signal, they can readily be interfaced with data acquisition systems for automatic storage and retrieval of data.

The condenser microphone (Fig. 13.3a) is commonly utilized in aeroacoustics. Basically, it is a capacitor whose capacitance varies as the membrane is deflected due to applied pressure. These transducers are quite delicate, since they measure very small pressures (100 μbar) and hence are not suitable for use in liquids. The piezoelectric transducer (Fig. 13.3b), commonly used with liquids, has a crystalline material (e.g., quartz) that produces an electric field when deformed. These rugged units can be made very small (diameters down to 2.5 mm are possible) and

**Fig. 13.2**   Bourdon tube pressure gage.

**Fig. 13.3**  Pressure transducers: (a) condenser microphone; (b) piezoelectric pressure transducer; (c) strain-gage transducer.

can be mounted flush with the wall of a pipe. They respond very rapidly and can measure a wide range of dynamic pressures (full vacuum to 20 MPa). A strain gage transducer (Fig. 13.3c) is another type of transducer employed with liquids; it makes use of a strain gage attached to the membrane that translates its deflection into an electrical signal. It is subject to DC drift and sensitivity changes; however, it is rugged, reliable, and insensitive to vibration.

## 13.2.3  Velocity

The fluid velocity is of primary interest in a fluid flow. By measuring the velocity we can calculate the flow rate and perhaps even piece together an image of the streamline pattern, identifying regions of separated flow, regions of stagnated flow, and other flow characteristics. There are three categories of velocity measurements: local (at a point), spatially averaged, and velocity field measurements. The discussion in this section pertains to the first two; the third is covered in Section 13.4.

Local velocities are actually measured over a small region of flow. The required precision varies, depending on the desired application. For example, time-averaged velocities measured at various points in a flow cross section may be recorded so that the velocity distribution can be integrated to provide the discharge, or flow rate. On the other hand, it may be necessary to determine the history of the time-dependent turbulent velocity components at a particular location in a region of the flow.

Instruments for velocity measurements vary considerably in their complexity and cost, depending on the type of measurement required. The desire to measure unsteady turbulent velocity components on a relatively small local scale has popularized the use of the thermal anemometer as well as the laser-doppler velocimeter. Less complicated instruments that usually measure velocity over a large spatial region are, for example, the pitot-static probe and the propeller anemometer; they are more suitable for measuring velocities that either are steady or are slowly varying with time.

**Particle Velocity Measurement.** Neutrally buoyant particles can be seeded in the flow to provide a visual image of the flow field; two of these, hydrogen bubbles in water and particles in air, are discussed in Section 13.4. Other tracer particles used in water are neutrally buoyant plastic beads, immiscible dyed droplets, and floats placed on a liquid surface or at a desired depth. It is important to ascertain if the particle is actually tracking the fluid motion. Very small particles that are natural impurities in the flow can be observed with appropriate optical equipment to measure velocities in many flows, especially microscopic flow fields.

**Pitot-Static Probe.** The pitot-static probe is analyzed in Section 3.4. It measures the local mean velocity by using the Bernoulli equation. Assuming frictionless, steady flow the velocity $u$ is given by

$$u = \sqrt{\frac{2}{\rho}(p_T - p)} \qquad (13.2.1)$$

in which $p_T$ is the total or stagnation pressure and $p$ is the static pressure. Equation 13.2.1 is valid only if the probe does not greatly disturb the flow, and if it is aligned with the flow direction so that the velocity $u$ is parallel to the pitot probe. A particular design, the so-called *Prandtl tube*, is shown in Fig. 13.4. It has the static pressure holes positioned along the horizontal tube so that the reduced pressure resulting from flow past the nose is balanced by the increased pressure due to the vertical stem. Typically, the Reynolds number based on the pitot-tube diameter should be in the neighborhood of 1000, so that viscous effects are not

**Fig. 13.4**   Pitot-static probe.

significant. If the flow is not parallel to the probe head, the measurement error is about 1% at a yaw angle of 5°, and if the yaw angle is greater than 5°, the measurement error may be substantial. When the probe is placed close to a wall, the streamlines are deflected by the interaction of the probe with the wall; errors occur when the distance between the probe axis and the wall is less than two tube diameters. In a turbulent flow, the actual pressure is less than the sensed value; generally, the results are reliable if the fluctuation intensity is less than 10%.

Other probes of a similar design are the *pitot* or *impact tube*, *Preston tube*, *static tube*, and *yaw meter*. The pitot tube measures the total pressure, the static tube the static pressure, and the yaw meter, the direction of flow. The Preston tube is a special form of pitot tube that has been used for the measurement of wall shear stress in boundary layers over smooth walls; basically, it is a hypodermic needle that senses the mean velocity adjacent to the wall. See Bean (1971) for details related to calibration and design.

**Cup or Propeller Anemometer.** Cup or propeller anemometers are employed to measure the velocity of either gases or liquids. They are relatively large, and as a result, the measurement is averaged spatially over a relatively large area. These anemometers are used for many applications where extreme precision is not required. One type of cup anemometer, the *current meter* (Fig. 13.5a), measures the velocity in water; similar devices can measure air velocity. Cup anemometers always rotate in one sense, and thus cannot provide the flow direction. On the other hand, the *propeller anemometer* will reverse rotation

**Fig. 13.5**  Anemometers: (a) current meter; (b) propeller anemometer; (c) ducted vane anemometer; (d) wind vane anemometer.

when the flow reverses (Fig. 13.5b); however, in contrast to cup anemometers, they must be aligned with the flow direction. A ducted propeller anemometer for measuring air velocities is shown in Fig. 13.5c. Unducted propellers are also employed; an example is the wind vane (Fig. 13.5d). Propellers are also placed inside pipes and used as flow meters. For all of these devices, counters or electro-mechanical readouts attached to sensors provide the rotational frequency, which is correlated to the velocity by means of a calibration test.

**Thermal Anemometer.** The *thermal* (hot wire) *anemometer* is employed to measure velocity in air. It consists of a small wire, typically made of tungsten, plat-inum, or platinum-iridium, which is insulated and mounted on supports (Fig. 13.6). The wire sensor is typically 2 mm in length and 5 μm in diameter, although smaller wires (0.2 mm long, 1 μm in diameter) have been used. When the wire is heated, the anemometer senses the changes in heat transfer as the flow speed varies; a representative wire temperature is 250°C.

The most common circuitry is one that operates in the constant-temperature mode (Fig. 13.6); earlier versions employed a constant-current arrangement. As the velocity past the probe is increased, the hot wire tends to cool, thereby attempting to lower the temperature and the resistance across the wire. The cir-cuitry then increases the current to adjust the resistance and the probe tempera-ture to their initial values, a process that appears to be instantaneous. Either the voltage of the amplifier or the sensor current is correlated with the velocity past the probe. The output is nonlinear with respect to velocity, and the sensitivity decreases as the velocity increases; often the signal is linearized electronically. The sensitivity as a percent of reading stays nearly constant so that a very wide range of velocities can be measured. When the fluctuation intensity is less than 20%, if the flow is incompressible and if the fluid temperature and density remain constant, the hot-wire anemometer can be used to measure a velocity compo-nent. One can then calculate the mean velocity, fluctuation intensity, and turbu-lence spectrum. The single-wire anemometer senses only the normal component of velocity; if the wire is properly oriented, both the mean component and the fluctuations in the mean flow direction can be measured. Two-component meas-urements are made using a probe with wires crossed in an "×" fashion, to pro-vide the correlation $\overline{u'v'}$ needed in turbulence measurements ($u'$ and $v'$ are the small perturbations). Three components may be measured either by rotating the ×-probe or adding a third wire.

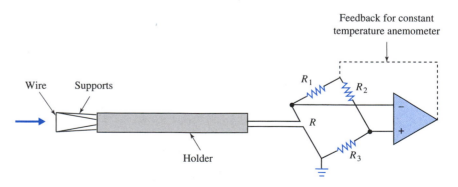

**Fig. 13.6** Thermal anemometer.

In liquids and gases that contain solid particles, the wire probe is replaced by a more rugged cylindrical film sensor. It operates on the same principle as the hot-wire probe, but it generates turbulence in its wake because of its larger size. This self-generated turbulence can limit the probe's capabilities for measuring actual fluctuation intensities in the flow field. In addition to measuring velocities, thermal anemometers have been adapted to monitor pressure and discharge.

**Laser-Doppler Velocimeter.**    A device that can be used advantageously when it is desired not to have the probe immersed in the flow is the *laser-doppler velocimeter* (LDV). The most common LDV is the dual-beam type, shown in Fig. 13.7. The frequency of the scattered light is measured by the photodetector, which converts the receiving light to an electrical signal. A simplified way to understand the principle is as follows. At the intersection of the two beams a fringe pattern is formed with fringe spacing $\Delta x$. Light scattered from particles passing through the fringe pattern is modulated with a frequency $f$, which is directly proportional to the velocity component $u$ normal to the fringes:

$$u = \frac{\Delta x}{\Delta t} = \frac{\lambda f}{2 \sin(\theta/2)} \tag{13.2.2}$$

in which $\lambda$ is the wavelength of the laser beam and $\theta$ is the beam angle. Detection of flow reversal is possible by using a moving fringe pattern. The ellipsoidal measurement volume is quite small, being on the order of 0.1 mm in diameter and 1.0 mm in length. A complete analysis of the phenomenon requires a detailed description of wave optics, which is beyond the scope of this book.

Velocities in both liquids and gases can be measured with an LDV. Liquids normally contain sufficient impurities which act as natural seeding agents, but gases usually are artificially seeded with extremely small liquid droplets or solid particles. Typically, an LDV can measure velocities ranging from 2 mm/s up to the supersonic range. The frequency response extends to 30 kHz, which enables

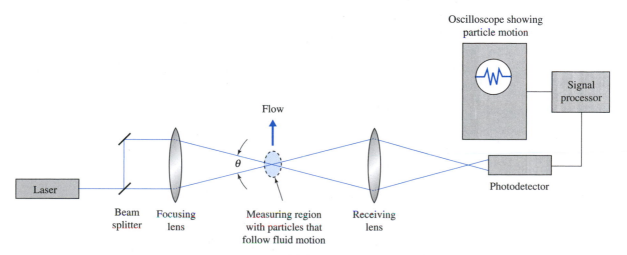

**Fig. 13.7**    Dual-beam laser-doppler velocimeter.

turbulent velocity components to be measured. In contrast to the thermal anemometer, the LDV requires no calibration and measures flow reversal with relative ease. The initial cost of the LDV is high, and extensive electronic instrumentation is required to process the output signal: a counter-type signal processor is most commonly used. The LDV has been employed to measure supersonic flow, flow in pumps, natural free convection, steam and gas turbine flows, combusting flows, blood flows, and open-channel flows.

## 13.3 FLOW RATE MEASUREMENT

The measurement of flow rate—alternately termed discharge or volumetric flow—is one of the most common measurements made in flowing fluids. Numerous devices have been invented or adapted for the purpose of flow metering, ranging widely in sophistication, size, and accuracy. Basically, instruments for flow rate measurements can be divided into those that employ direct or "quantity" means of measurement, and those that are indirect, that is, the so-called rate meters. Quantity meters either weigh or measure a volume of fluid over a known time increment; examples are liquid weigh tanks, reciprocating pistons, and for gases, bellows and the liquid-sealed drum. These direct devices usually are relatively large and possess poor frequency-response characteristics. However, they do provide high precision and accuracy, and as a result, are used most often as "primary" standards for the calibration of the indirect metering devices.

Indirect or rate meters consist of two components: the primary part, which is in contact with the fluid, and the secondary part, which converts the reaction of the primary part to a measurable quantity. They can be classified according to a characteristic operating principle: velocity–area measurement, pressure drop correlations, hydrodynamic drag, and the like. Rate meters are relatively low in cost, take up little space, and hence are commonly found in industrial and research laboratories. Several of the more common types are discussed below.

### 13.3.1 Velocity–Area Method

A number of the velocity measuring devices discussed in Section 13.2.3 may be inserted in a cross section of flow and used to measure the flow rate. Thus a thermal anemometer, a laser-doppler velocimeter, a pitot-static tube, or an anemometer can be inserted in a pipe or duct of known area and calibrated to measure the flow rate. In a manner that does not require calibration and for steady flow, a velocity distribution can be measured by systematically moving a velocity probe throughout the flow cross section, or by using a so-called "rake" where local velocities are measured simultaneously. Subsequently, a numerical integration of the velocity profile over the cross-sectional area will yield the discharge. This technique has been used in both closed conduit and open-channel flows.

### 13.3.2 Differential Pressure Meters

Differential pressure meters are widely used in industrial applications and laboratories because of their simplicity, reliability, ruggedness, and low cost. Three commonly used types are discussed in this section: the orifice meter, venturi meter, and flow nozzle. Their operation is based on the principle of an obstruction to flow being present in a duct or pipe, and consequently, a pressure differential will exist across the obstruction. This pressure drop can be correlated to the discharge by means of a calibration, and subsequently, the pressure-discharge curve can be used to determine the discharge by reading the differential pressure. In this section we deal only with the discharge of incompressible fluids in circular pipes.

The fundamental discharge relation for the differential pressure meter can be described in the following manner. Fig. 13.8 represents a thin-plate orifice meter, which can be considered as a representative differential pressure meter. A steady flow occurs in a circular duct, encounters the restrictive orifice with area $A_0$, and issues as a downstream jet. Downstream of the restriction, the streamlines converge to form a minimal flow area $A_c$, termed the *vena contracta*. Pressure taps are located at two positions: upstream of the restriction in the undisturbed flow region (location 1) and downstream at some location in the vicinity of the vena contracta (location 2). Assuming a frictionless, incompressible fluid, the Bernoulli equation applied along the center streamline from the upstream location to the vena contracta is

$$\frac{V_1^2}{2g} + \frac{p_1}{\gamma} + z_1 = \frac{V_c^2}{2g} + \frac{p_c}{\gamma} + z_c \tag{13.3.1}$$

Similarly, the continuity equation is

$$V_1 A_1 = V_c A_c \tag{13.3.2}$$

Combining Eqs. 13.3.1 and 13.3.2, and solving for $V_c$ yields

$$V_c = \sqrt{\frac{2g(h_1 - h_c)}{1 - (A_c/A_1)^2}} \tag{13.3.3}$$

**Fig. 13.8**   Flow through an orifice meter.

in which

$$h_1 = \frac{p_1}{\gamma} + z_1 \qquad h_c = \frac{p_c}{\gamma} + z_c \qquad\qquad (13.3.4)$$

The ideal discharge $Q_i$ is equal to the area multiplied by the average velocity at the vena contracta:

$$Q_i = A_c V_c$$

$$= \frac{A_c}{\sqrt{1 - (A_c/A_1)^2}} \sqrt{2g(h_1 - h_c)} \qquad\qquad (13.3.5)$$

The actual discharge differs from the ideal for two primary reasons. Because of real fluid flow, friction causes the velocity at the centerline to be greater than the average velocity at each cross section. Second, the piezometric head $h_c$, evaluated at the vena contracta in the relation, is substituted with $h_2$, the known reading at the downstream pressure tap. Also, since the area of the vena contracta is unknown, it is convenient in Eq. 13.3.5 to replace $A_c$ by $C_c A_0$, where $C_c$ is the contraction coefficient and $A_0$ is the area of the orifice opening. These anomalies are accounted for by introducing a discharge coefficient $C_d$, which is the product of the contraction coefficient and a velocity coefficient, so that the actual discharge $Q$ is given by the relation

$$Q = \frac{C_d A_0}{\sqrt{1 - (C_c A_0/A_1)^2}} \sqrt{2g(h_1 - h_2)} \qquad\qquad (13.3.6)$$

For a circular cross section, which is typical of most differential pressure meters, it is convenient to introduce the diameter ratio

$$\beta = \sqrt{\frac{A_0}{A_1}} = \frac{D_0}{D} \qquad\qquad (13.3.7)$$

where $D$ is the pipe diameter. A convenient way to express Eq. 13.3.6 is

$$Q = K A_0 \sqrt{2g(h_1 - h_2)} \qquad\qquad (13.3.8)$$

in which $K$ is the flow coefficient

$$K = \frac{C_d}{\sqrt{1 - C_c^2 \beta^4}} \qquad\qquad (13.3.9)$$

A dimensional analysis would reveal that $C_d$ and $K$ are dependent on the Reynolds number. It is convenient to evaluate the Reynolds number either at the approach region or at the obstruction.

**Orifice Meter.**    A thin plate orifice meter (Fig. 13.9) is typically manufactured in the range $0.2 \leq \beta \leq 0.8$. In the figure, two means of locating the pressure taps are shown: (1) flange taps, positioned 25 mm upstream and downstream of the orifice plate, and (2) taps placed one diameter upstream and one-half diameter downstream of the plate. The second arrangement is preferred, since it is capable of sensing a larger differential pressure, and it conforms better to geometric similarity laws. A third arrangement, not shown in the figure, has the pressure taps located in the pipe wall immediately upstream and downstream of the orifice; taps placed at this location have been termed corner taps.

Fig. 13.10 shows experimentally determined values of the flow coefficient $K$ for orifices as a function of $\beta$ and $\mathrm{Re}_0$. These data were obtained using corner taps; however, data taken with flange taps or with $D : D/2$ taps would be indistinguishable when plotted on Fig. 13.10. If greater precision is desired, numerical data for the different taps are provided in Bean (1971). Notice that for a given $\beta$, $K$ becomes nearly constant at high Reynolds numbers, but as the Reynolds number becomes lower, $K$ first increases to a maximum and then decreases. Maximum values of $K$ occur at Reynolds numbers between 100 and 1000, depending on the value of $\beta$; here $K$ is dominated by the reduced area of the vena contracta.

If the discharge is known, the Reynolds number at the orifice is known. Hence, with Fig. 13.10 one can read $K$ directly, and subsequently determine $(h_1 - h_2)$ from Eq. 13.3.8. However, one would more likely use the figure in conjunction with Eq. 13.3.8 to determine the flow rate, given that $(h_1 - h_2)$ has been read from an attached manometer or pressure transducer. In that situation, $K$ is not known a priori, since it depends on $\mathrm{Re}_0$. One can initially estimate $K$ based on an assumed $\mathrm{Re}_0$ (usually, assume $\mathrm{Re}_0$ to be large), and subsequently by trial and error improve on that estimate by successive substitution into Eq. 13.3.8.

**Fig. 13.9**    Details of a thin-plate orifice meter. (FLUID MECHANICS MEASURE-MENTS by G. E. Mattingly. Copyright 1996 by Taylor & Francis Group LLC-Books. Reproduced with permission of Taylor & Francis Group LLC-Books in the format Textbook via Copyright Clearance Center.)

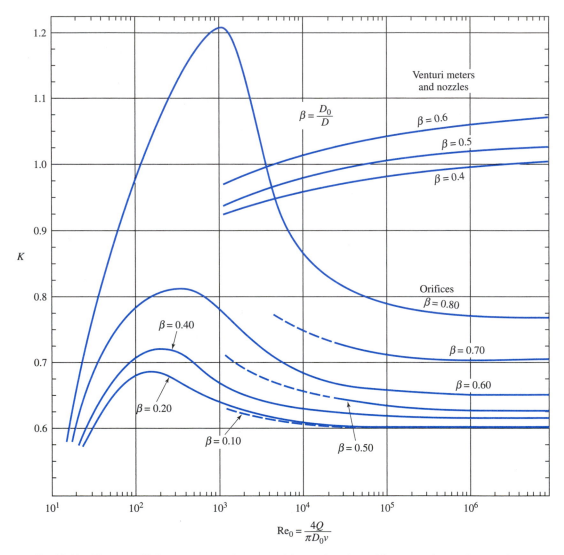

**Fig. 13.10**  Flow coefficient $K$ versus the Reynolds number for orifices, nozzles, and venturi meters. (Adapted from Engineering Fluid Mechanics, Robertson and Crowe, © 1990 John Wiley & Sons, Inc., New York. Reproduced with permission of John Wiley & Sons, Inc.)

However, if repeated readings are to be taken, it is more convenient to determine a calibration formula of the form

$$Q = C(h_1 - h_2)^m \qquad (13.3.10)$$

in which $C$ and $m$ are constants determined by a "best fit" criterion. See Section 13.5.3 for details of one well-known criterion, the method of least squares.

The orifice has become the most widely used differential meter for measuring flow rates in liquids. This is understandable, since it is relatively inexpensive and can easily be placed in an existing pipe. One disadvantage is that it produces a large head loss that cannot be recovered downstream of the orifice.

**Venturi Meter.**  The venturi meter has a shape that attempts to mimic the flow patterns through a streamlined obstruction in a pipe. The classical, or Herschel, type of venturi meter is rarely used today, since its dimensions are rather large, making it cumbersome to install and expensive to fabricate. It is made up of a 21° conical inlet contraction, followed by a short cylindrical throat, leading to a 7° or 8° conical exit expansion. The discharge coefficient is nearly unity. By contrast, the contemporary venturi tube, shown in Fig. 13.11, consists of a standard flow nozzle inlet section [ISA 1932 standard (Bean, 1971)] and a conical exit expansion no greater than 30°. Its recommended range of Reynolds numbers is limited from $1.5 \times 10^5$ to $2 \times 10^6$.

Equation 13.3.8 is valid for the venturi as well as for the orifice; representative values of the flow coefficient $K$ are shown in Fig. 13.10. Because of streamlining of the flow passage, the head loss in the venturi meter is much less than in the orifice. The vena contracta is not present, and as a result the discharge coefficient $C_d$ remains close to unity.

**Flow Nozzle.**  The flow nozzle is illustrated in Fig. 13.12. It consists of a standardized shape with pressure taps typically located one diameter upstream of the inlet and one-half diameter downstream. There are two standard shapes, either the long radius or the short radius. Due to the lack of an expansion section downstream of the nozzle, the total head loss is similar to that of an orifice, except that the vena contracta is nearly eliminated and the discharge coefficient is nearly unity. Similarly, when the pressure taps are located at $D:D/2$, the flow coefficient $K$ varies with the Reynolds number in a manner nearly identical to that for the venturi meter, as shown in Fig. 13.10. The flow nozzle has an advantage over the orifice plate in that it is less susceptible to erosion and wear, and relative to the venturi meter, it is less expensive and simpler to install.

**Elbow Meter.**  A relatively simple meter can be fabricated by locating pressure taps at the outside and inside of a pipe elbow (Fig. 13.13). By applying the linear momentum equation to a control volume encompassing the fluid in the elbow (or Euler's equation normal to the streamlines), one can derive the flow equation

$$Q = KA \sqrt{\frac{R}{D} \frac{\Delta p}{\rho}} \qquad (13.3.11)$$

$p_1$  $p_2$  15 to 30°

Pipe with diameter $D$    Throat with diameter $D_0$    Pipe with diameter $D$

**Fig. 13.11**  Venturi meter.

**Fig. 13.12** Flow nozzle.

in which $R$ is the radius of curvature of the elbow and $\Delta p$ is the pressure difference generated by centrifugal force across the bend. The flow coefficient $K$ can be most accurately determined by in-place calibration of the meter. However, a relation for $K$ has been reported (Bean, 1971) for 90° elbows with pressure taps located in a radial plane 45° from the inlet:

$$K = 1 - \frac{6.5}{\sqrt{Re}} \qquad (13.3.12)$$

which is valid when $10^4 \le Re = VD/\nu \le 10^6$, and $R/D \ge 1.5$. The elbow meter costs much less than any of the differential pressure meters, and it usually does not add to the overall head loss of the piping system since elbows are often present in any case.

### 13.3.3 Other Types of Flow Meters

**Turbine Meter.** The turbine meter consists of a propeller mounted inside a duct that is spun by the flowing fluid. The propeller's angular speed is correlated to the discharge; the angular rotation is measured in a manner identical to the anemometer discussed in Section 13.2.3. Accuracies up to $\pm 0.25\%$ of the flow rate are attainable over a relatively large discharge range (Goldstein, 1996). Viscous effects become a limiting factor at the low end, whereas at the high end, accuracy is limited by the interaction between the blade tips and the electronic sensors. Each type of turbine meter must be calibrated individually. They are commonly employed to monitor flow rates in fuel supply lines.

**Fig. 13.13** Elbow meter.

**Fig. 13.14**  Rotameter.

**Rotameter.**   The rotameter consists of a tapered tube in which the flow is directed vertically upward (Fig. 13.14). A float moves upward or downward in response to the flow rate until a position is reached where the drag force on the float is in equilibrium with its submerged weight. Calibration consists of correlating the vertical elevation of the float with the discharge. With appropriate design, the float position can be made linearly proportional to the discharge, or if desired, another relation, such as position logarithmically proportional to discharge, can be formulated. The head loss depends on the friction loss of the tube plus the loss across the floating element. The rotameter does not provide accuracy as good as the differential pressure meters; typically it is in the range of 5% full scale.

**Target Meter.**   Fig. 13.15 shows a target meter, which consists of a disk suspended on a support strut immersed in the flow. The strut is connected to a lever arm, or alternatively, has a strain gage bonded to its surface. Fluid drag on the disk will cause the strut to flex slightly, and the recorded drag force can be related to the discharge. The assembly must be sufficiently stiff so that the drag can be measured without movement or rotation of the disk; otherwise, the drag characteristics would be altered. One advantage of this device is that it is possible to record flow reversal by measuring the sense of the drag force; hence it can be used as a bidirectional flow meter. Drag meters are fairly rugged and can be used to measure discharges in sediment-laden fluids.

**Electromagnetic Flowmeter.**   This is a nonintrusive device that consists of an arrangement of magnetic coils and electrodes encircling the pipe (Fig. 13.16). The coils are insulated from the fluid and the electrodes make contact with the fluid. Sufficient electrolytes are dissolved in the fluid so that it becomes capable of conducting an electric current. When it passes through the magnetic field generated by the coils, the liquid will create an induced voltage proportional to the flow. Electromagnetic flowmeters have been used for many applications, including blood flow and seawater measurements. Commercial models are

**Fig. 13.15**   Target meter.

available in a wide range of diameters; they are costly and can be used only with liquids.

**Acoustic Flowmeter.**   The ultrasonic, or acoustic, flowmeter is based on one of two principles. The first one uses ultrasonic transmitters/receivers beamed across a flow path (Fig. 13.17). The measured differences in the travel times (or inversely, the frequencies) are directly proportional to the average velocity, and consequently, the discharge. The second type is based on the Doppler effect, in that acoustic waves are transmitted into the flow field and subsequently scattered by seeded particles or contaminants. The Doppler shift recorded between the transmitter and the receiver is then related to the discharge. Like the electromagnetic flowmeter, acoustic flowmeters are nonintrusive and have been used in both piping and free-surface flow systems.

**Vortex-Shedding Meter.**   This flow-metering device consists of a single strut or series of struts placed chord-like inside a pipe normal to the flow direction. If the Reynolds number is above a threshold value, vortices appear in the near-wake regions behind the struts (see Section 8.3.2). The periodic vortex shedding that occurs behind the struts can be correlated to the Reynolds number as shown in Fig. 8.9. Different devices are employed to detect the vortex-shedding frequency. Examples include a pressure transducer mounted on the downstream surface of the strut, a thermal sensor that detects changes in heat flux on the strut surface, or a strain gage that senses oscillation of the strut. Vortex-shedding meters are can be specified to operate in flows that range up to two orders of magnitude.

**Fig. 13.16**   Electromagnetic flowmeter.

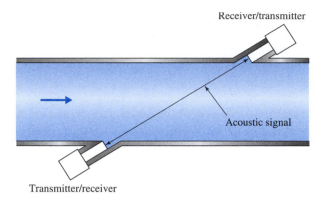

**Fig. 13.17**    Acoustic flowmeter.

**Coriolis-Acceleration Flowmeter.**    These are meters that are based on the Coriolis acceleration principle (see Eq. 3.2.15). Typically they are used to measure liquid flows in piping. A U-shaped tube filled with the liquid to be measured is vibrated normal to the plane of the U and with the tips of the U stationary. The liquid flowing in the tube produces a torsional motion about the axis of symmetry of the tube. The amplitude of the motion is proportional to the flow rate, and its frequency can be correlated to the density of the liquid. The meter provides both flow rate and density measurements in a noninvasive manner.

**Open-Channel Flowmeters.**    In Section 10.4.3, flow measurement in open-channel systems is presented. Discharge relations for the broad-crested weir and sharp-crested weir are developed, and additional methods of flow measurement are discussed. The reader is referred to that section for detailed information.

## 13.4 FLOW VISUALIZATION

Transparent flow fields have been visualized in many different ways. This is well illustrated by the hundreds of photographs presented in the publication *An Album of Fluid Motion* (Van Dyke, 1982). The book shows flows visualized with smoke, dye, bubbles, particles, shadowgraphs, schlieren images, interferometry, and other techniques. Usually, a method is selected that best shows the flow features of interest. Examples include particles used to visualize pathlines in liquid flows around submerged objects, dye released to study the mixing process in a stream, smoke released at the upwind end of a wind tunnel to study the development of a boundary layer, and airflow patterns above a solid surface visualized by coating that surface with a viscous liquid (streaky lines are formed that coincide with the streamlines near the surface). Several of these techniques are discussed in this section.

Pathlines, streaklines, and streamlines have been defined in Chapter 3. The purpose of defining them was to provide assistance in describing the motion of a flow field. A pathline is physically generated by following the motion of an individual particle such as a bubble or small neutrally buoyant sphere over a period of time; this could be achieved by time-lapse photography or video recording. On the other hand, if a trace from smoke, a train of extremely small bubbles, or dye continuously emanating from a stationary source is photographed or recorded, a streakline is observed. When the generation of a streakline is interrupted periodically, time streaklines are produced.

### 13.4.1 Tracers

Tracers are fluid additives that permit the observation of flow patterns. An effective tracer does not alter the flow pattern but is transported with the flow and is readily observable. It is important that tracers are not affected by gravitational or centrifugal forces resulting from density differences. Furthermore, their size should be at least one order of magnitude smaller than the length scale of the flow field. Several of them are discussed below.

**Hydrogen Bubbles.** A very thin metallic wire can be placed in water to serve as the cathode of a direct-current circuit, with any suitable conductive material acting as an anode. When a voltage is supplied to the circuit, hydrogen bubbles will be released at the cathode and oxygen bubbles at the anode. The primary reaction is the electrolysis of water in the manner $2H_2O \rightarrow 2H_2 + O_2$. Usually, the hydrogen bubbles are used as the tracer since they are smaller than the oxygen bubbles and more of them are formed. The bubbles are transported by the flow field away from the cathode wire as a continuous sheet. If the voltage is pulsed, discrete streaklines are formed; Fig. 13.18 shows such a pattern. Standard water supplies usually contain sufficient electrolytes to sustain a current, although addition of an electrolyte such as sodium sulfate greatly increases bubble generation. Use of extremely fine wire (0.025 to 0.05 mm in diameter) with adequate velocities (Reynolds numbers based on the wire diameter should be less than 20) may create bubbles sufficiently small to negate effects of buoyancy.

**Chemical Indicators.** Certain organic chemicals change color when the pH of water changes. For example, thymol blue solution is yellow at pH 8.0 and

**Fig. 13.18** Hydrogen bubble streaklines showing separated flow around a rotating airfoil. (Courtesy of M. Koochesfahani.)

**Fig. 13.19**   Turbulent wake behind a circular cylinder at Re = 1760. (Courtesy of R. E. Falco.)

blue at pH 9.2; phenol red solution is yellow at pH 6.8 and red at pH 8.2. Thus a change in pH caused by the injection of a base solution changes the color of the liquid. The lifetime of the colored water depends on the molecular diffusion of the hydrogen ions and the turbulence. An advantage of this technique is that color residues can be eliminated by changing the pH. One application is in studies of solid–liquid phase transport phenomena; if the chemical indicator is in contact with a metal surface and the metal surface is given a negative charge, the solution in immediate contact with the plate will change color.

**Particles in Air.**   Particles in air may be introduced as helium bubbles. By properly selecting the mixture of soap and water, one can produce helium-filled soap bubbles that are nonbuoyant; it is possible to produce bubble diameters on the order of 4 mm. Solid particles or liquid droplets could also be utilized, but they must be extremely small to avoid gravitational effects. With these small sizes one must employ an extremely strong light source to visualize the flow.

Smoke has been used successfully to study the detailed structure of complex flow phenomena. It is the most popular agent used for flow visualization in wind tunnels. One injection technique is the so-called smoke-wire method, where the smoke is generated by vaporizing oil from a fine electrically heated wire. The method can be applied to flows where the Reynolds number based on the wire diameter is less than 20. An example of such a flow structure is given in Fig. 13.19. Smoke can also be released from a small-diameter tube or "rake" to create one or more streaklines.

**Velocimetry.**   Simultaneous measurement of two- and three-dimensional flow domains has led to the definition of a generic pulsed light velocimeter (Fig. 13.20). It consists of a pulsed light source that illuminates small fluid-entrained particles over short exposure times, and a camera that is synchronized with the light to record the location of the particles. The velocity of each marker is then given by $\Delta s/\Delta t$, where $\Delta s$ is the displacement of the marker and $\Delta t$ is the time between exposures. The technique is called *Particle Image Velocimetry* (PIV). The markers are usually particles that vary in size from 1 to 20 $\mu$m. An example of PIV is shown in Fig. 13.21 on page 677. In addition to PIV, progress has been made in the use of molecular markers such as photochromic dyes in liquids, photochromic aerosols in gases, and molecular phosphorescence in both liquids and gases. A sample of *Molecular Tagging Velocimetry* (MTV) is shown in Fig. 13.22 on page 678, in which a noninvasive grid is tagged and subsequently displaced during

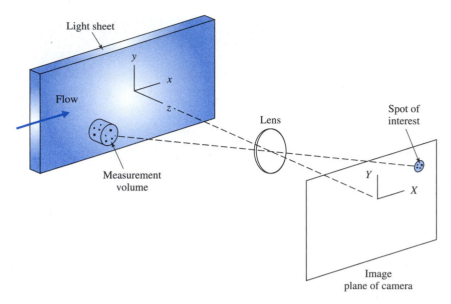

**Fig. 13.20** Pulsed light velocimeter. (Courtesy of R. Adrian.)

a brief interval. The two grids are captured with a CCD camera and analyzed to produce the velocity field shown in Fig. 13.22c. In that figure, only the left half of the velocity field is shown. The axis of symmetry is indicated by the dashed line, and the box delimits the region containing vectors derived from Figs. 13.22a and b. Velocimetry methods are particularly useful in the studies of explosions, transients in channels and ducts, flow in internal combustion engines, bubble growth and collapse, fluid-solid interaction, and the structure of turbulent flow.

**Tufts and Oil Films.** Flow visualization along a surface can be accomplished by attaching tufts to the surface, or if flow away from the surface is to be observed, they may be supported on wires. Surface tufts may be used to observe the transition from laminar to turbulent motion. They may also be used to study flow separation qualitatively, where the violent motion of the tufts or their tendency to point in the upstream direction identifies that separation is taking place. In air, surface flow patterns can also be visualized by distributing oil over the solid surface in such a manner that clear streaky patterns develop. Interpretation of the patterns is not always simple, particularly when flow reversals occur; the reason is that streak patterns do not indicate the flow direction.

**Photography and Lighting.** Many kinds of cameras are used for flow visualization, including still, stereo, and cinematic (Adrian, 1986). The video camera has become increasingly popular due to advances in resolution, and it is possible to interface certain models with a computer to obtain quantitative data. Stroboscopes, as well as laser, tungsten, and xenon lighting, are commonly employed to visualize tracers. Depending on the application, images can be produced using steady light to generate streak lines; alternatively, images can be produced using light pulses which are triggered by the camera shutter or a stroboscope to freeze the motion. If two-dimensional visualization is desired, the flow field can be illuminated by creating a narrow sheet of light directed through a window in the test section. Fig. 13.23 on page 679 demonstrates the technique

(a)

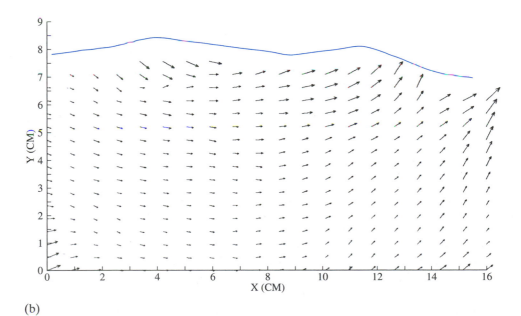

(b)

**Fig. 13.21**  Particle Image Velocimetry (PIV): (a) photograph of particle pathlines;
(b) scaled velocity vectors. (Courtesy of R. Bouwmeester.)

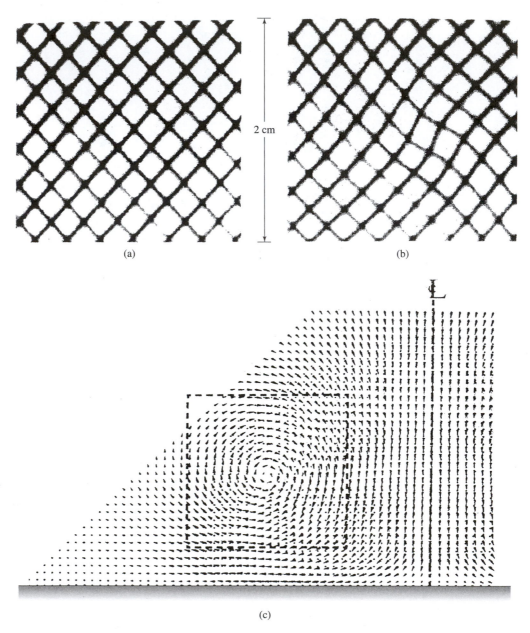

**Fig. 13.22** Molecular Tagging Velocimetry (MTV): (a) sample grid within a vortex ring just after laser tagging; (b) displaced grid 6 ms later; (c) velocity field of the vortex ring approaching a wall. (Courtesy of C. MacKinnon and M. Koochesfahani.)

of Laser Induced Fluorescence (LIF) to visualize a spanwise flow structure and mixing field in an unsteady wake flow field. Three-dimensional imaging can be accomplished by stereoscopic pictures or by use of holography.

Laser sheet    $U_O = 10$ cm/s
$Re_\theta = 100$

Spanwise flow structure
and mixing field are
examined using LIF

$U_O$

$U_O$

y

x

4 cm

z

8 cm

Upper free-stream oscillates
sinusoidally at F = 6 Hz and
rms amplitude 10% of free-
stream speed

CCD camera

(a)

(b)

**Fig. 13.23**    Laser Induced Fluorescence (LIF): (a) experimental layout; (b) detail of spanwise flow structure measured by LIF. (Courtesy of C. MacKinnon and M. Koochesfahani.)

## 13.4.2   Index of Refraction Methods

The three index of refraction techniques are schlieren, shadowgraph, and interferometry. They are used in attempting to visualize flows when the density differences are either naturally or artificially altered; both gas and liquid mediums can be employed. The methods depend on the variation of the index of refraction in a transparent fluid medium and the resulting deflection of a light beam directed through the medium. Most often the flows to be studied are two-dimensional, and the light beam is directed normal to the flow field; the result is a measurement integrated over the test section. The index of refraction is a function of the thermodynamic state of the fluid and typically depends only on the density. Examples of flows that are investigated with these techniques are high-speed flows with shock waves, convection flows, flame and combustion

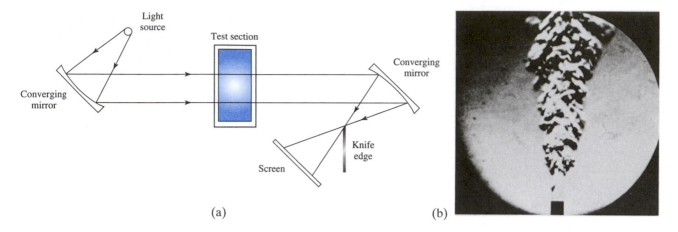

(a)                                                                        (b)

**Fig. 13.24** Schlieren system: (a) schematic; (b) helium jet entering atmospheric air. (Courtesy of R. Goldstein.)

flow fields, mixing of fluids of different densities, and forced-convection flows with heat or mass transfer.

With a *schlieren* system, one measures the angle of the light beam after it has passed through the test section and emerges into the surrounding air (Fig. 13.24a); the angle is proportional to the first spatial derivative of the index of refraction. The light beam is turned in the direction of increasing index of refraction, or for most media, toward the region of higher density. A representative schlieren image is shown in Fig. 13.24b. A *shadowgraph* system measures the linear displacement of perturbed light (Fig. 13.25). In contrast to the schlieren imaging, the light pattern in a shadowgraph is determined by the second derivative of the index of refraction.

Shadowgraph and schlieren systems are usually employed for qualitative flow visualization, whereas interferometers are often used in quantitative studies.

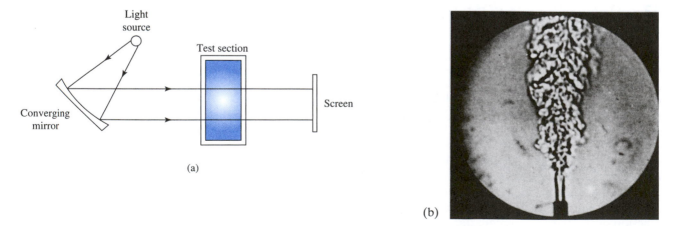

(a)

(b)

**Fig. 13.25** Shadowgraph system: (a) schematic; (b) helium jet entering atmospheric air. (Courtesy of R. Goldstein.)

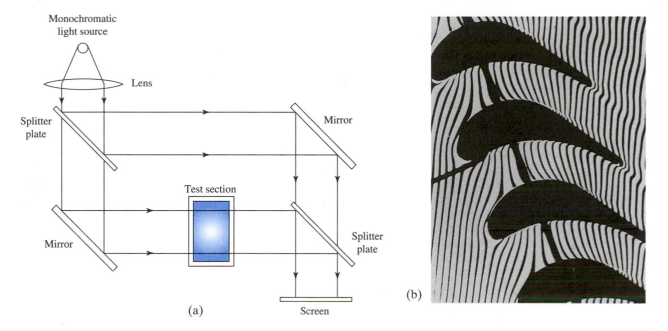

**Fig. 13.26**   Mach–Zehnder interferometer: (a) schematic; (b) flow over heated gas-turbine blades. (Courtesy of R. Goldstein.)

An interferometer responds to differences in optical path length; since a flow field with a variable density represents an optical disturbance, the phase of transmitted light waves is changed relative to undisturbed waves. A two-beam interferometer is designed to measure that phase difference (Fig. 13.26a). When the disturbed and undisturbed light beams are recombined, interference fringes appear on the screen that are a measure of the density variation in the flow field. An example of an interferogram is shown in Fig. 13.26b.

## 13.5   DATA ACQUISITION AND ANALYSIS

This section begins with a brief introduction to automated acquisition of flow data using personal computers. Next, methodology to deal with the uncertainty of data is presented, including the manner in which those uncertainties are propagated when one flow parameter is represented indirectly as the result of measuring others. Finally, since a number of flow relations involve exponential formulations, a means of curve fitting exponential data is presented.

### 13.5.1   Digital Recording of Data

Some of the measurements that have been described in previous sections are made simply by the direct reading of an instrument; those readings are probably steady-state, time-averaged, or perhaps root-mean-square values. Examples include reading a manometer attached to an orifice flowmeter, monitoring a digital rpm readout of a vane anemometer, and recording the pressure indicated on a Bourdon gage. The data acquired in this manner are usually sufficiently accurate for the desired goal; indeed, any further sophistication used to collect the data could be costly and unnecessary. On the other hand, there are situations that require more complex means of data acquisition. For instance, monitoring transient pressures in a pipeline using a strain-gage transducer, or observing turbulent velocity fluctuations in a free-stream flow, require that data be recorded continuously and accurately. Another illustration is the capture of the movement of a large number of moving particles in a two-dimensional plane in order to determine an unsteady flow field. Such measurements necessitate that data be recorded automatically, rapidly, and accurately.

The advent of the personal computer has revolutionized automated acquisition of flow data. Many of the transducers in use provide a voltage output signal (an analog signal) that must be converted into a digital signal before the computer can store and process the data. Readily available today are a number of plug-in analog input/output (I/O) boards specifically made for personal computers, which can be inserted into the central processing unit bus. These I/O boards consist of input multiplexers, an input amplifier, a sample-and-hold circuit, one or more output digital-to-analog converters, and interfacing and timing circuitry. In addition, some include built-in random-access memory, read-only memory, and real-time clocks. Compatible software is available so that programming either is unnecessary or requires minimal effort.

### 13.5.2   Uncertainty Analysis

When flow measurements are made, it is important to realize that no recorded value of a given parameter is perfectly accurate. Instruments do not measure the so-called "true value" of the parameter, but they will provide an estimate of that value. Along with each measurement, one must attempt to answer the important question: How accurate is that measurement? The inaccuracies associated with measurements can be considered as uncertainties rather than errors. An *error* is the fixed, unchangeable difference between the true value and the recorded value, whereas an *uncertainty* is the statistical value that an error may assume for any given measurement.

The purpose of this section is to describe a technique for propagating uncertainties of flow measurements into results. It has a variety of uses, among them helping us to decide whether calculations agree with collected data or lie beyond acceptable limits. Perhaps more importantly, it can help us to improve a given experiment or measuring procedure. It is not meant to be a substitute for accurate calibrations and sound experimental techniques; rather, it should be viewed as an additional aid for obtaining good, reliable results. Only a brief introduction is presented here; the reading of Coleman and Steele (1989) and Kline (1985) is recommended for a more thorough treatment of the subject.

Uncertainty can be divided into three categories: (1) uncertainty due to calibration of instruments, (2) uncertainty due to acquisition of data, and (3) uncertainty as a result of data reduction. The total uncertainty is the sum of bias and precision. *Bias* is the systematic uncertainty that is present during a test; it is considered to remain constant during repeated measurements of a certain set of parameters. There is no statistical formulation that can be applied to estimate bias, hence its value must be based on estimates. Calibrations and independent measurements will aid in its estimation. *Precision* is alternatively termed *repeatability*; it is found by taking repeated measurements from the parameter population and using the standard deviation as a precision index. For the situation in which repeated measurements are not taken (which occurs quite often for flow measurements), a single value must be used; however, less precision will be obtained. Fig. 13.27 illustrates how uncertainties are defined. A sampling of data (solid line) is shown that is assumed to be distributed in a Gaussian manner. The parameter $x_i$ is one particular measured variable, or measurand, $<x_i>$ is the average of all the measured variables, and $\delta x_i$ is the uncertainty interval, or precision, associated with $x_i$. The uncertainty for the measurand can be given as

$$x_i = <x_i> \pm \delta x_i \qquad (13.5.1)$$

Furthermore, it is necessary to state the bias and the probability $P$ of recording the measurand with the stated precision.

In addition to defining uncertainties in the data, it is necessary to propagate those uncertainties into the results. A two-step process is described as follows.

1. *Describe the uncertainty for the measurands.* Each measurand is assumed to be an independent observation, and it must come from a Gaussian population. Also, the precision for each measurand must be quoted at the same probability $P$. For example, consider an experiment to calibrate a weir, in which the measurands are the water level upstream of the weir and the discharge. The water level is measured using a water-level gage, and the discharge is recorded by measuring the time it takes to fill a known volume. Thus the two measurands are independent. In addition, based on experience and perhaps some repeated measurements, one can reasonably state that (1) there is no bias and (2) the recorded precision for each measurand has a probability of $P = 0.95$; that is, 20 out of every 21 readings of that measurand would yield the recorded value plus or minus the stated precision.

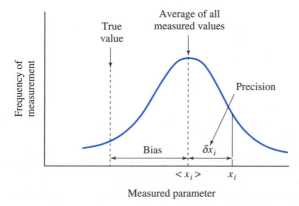

**Fig. 13.27**  Uncertainties associated with a sampling of data.

2. *Compute the overall uncertainty interval.* Let the parameter $R$ represent the result computed from $n$ variables $x_1, x_2, \ldots, x_n$; that is, $R = R(x_1, x_2, \ldots, x_n)$. Here we assume that $R$ is continuous and has continuous derivatives in the desired domain. In addition we assume that the variables are uncorrelated and the uncertainty intervals are uncorrelated; that is, they are independent of one another. Then the overall uncertainty in the result $\delta R$ can be obtained by expanding $R$ in a Taylor series about the measurands $x_1, x_2, \ldots, x_n$ and obtaining the root-sum-square result

$$\delta R = \left[ \left( \frac{\partial R}{\partial x_1} \delta x_1 \right)^2 + \left( \frac{\partial R}{\partial x_2} \delta x_2 \right)^2 + \cdots + \left( \frac{\partial R}{\partial x_n} \delta x_n \right)^2 \right]^{1/2} \qquad (13.5.2)$$

Situations arise when the expression for $R$ becomes so complicated that it is difficult to obtain the partial derivatives in Eq. 13.5.2. In such a case, one can employ finite difference approximations of the partial derivatives to simplify the analysis. For example, consider a forward finite difference approximation for the first partial derivative:

$$\frac{\partial R}{\partial x_1} \approx \frac{R(x_1 + \delta x_1, x_2, \ldots, x_n) - R(x_1, x_2, \ldots, x_n)}{(x_1 + \delta x_1) - x_1} \qquad (13.5.3)$$

Here we have assumed that in the denominator the finite difference interval is equivalent to the uncertainty interval $\delta x_1$. The numerator of Eq. 13.5.3 is the difference between $R$ evaluated at $x_1 + \delta x_1$ and $x_1$ with the remaining variables fixed. Expressing the other partial derivatives in Eq. 13.5.2 in a similar manner and substituting the results back into the relation we have the desired result:

$$\delta R \approx \left\{ \begin{array}{l} [R(x_1 + \delta x_1, x_2, \ldots, x_n) - R(x_1, x_2, \ldots, x_n)]^2 \\ + [R(x_1, x_2 + \delta x_2, \ldots, x_n) - R(x_1, x_2, \ldots, x_n)]^2 \\ + \cdots + [R(x_1, x_2, \ldots, x_n + \delta x_n) - R(x_1, x_2, \ldots, x_n)]^2 \end{array} \right\}^{1/2} \qquad (13.5.4)$$

Equation 13.5.4 can easily be evaluated using a spreadsheet algorithm.

Use of either Eq. 13.5.2 or 13.5.4 provides an estimate of $\delta R$. The overall uncertainty interval is an estimate of the standard deviation of the population of all possible experiments like the one being conducted.

## Example 13.1

Discharge and pressure data are collected for water flow through an orifice meter in the pipe of Fig. E13.1a. The orifice diameter is 35.4 mm, and the diameter of the pipe is 50.8 mm. The original data are reduced to the form shown in the accompanying table. The second column is the discharge $Q$; the precision $\delta Q$ (zero bias, $P = 0.95$) is given in the third column; and the pressure head data $\Delta h = h_1 - h_2$ are presented in the fourth column. In addition, the precision for each of the pressure measurements is determined to be the same, namely, $\delta(\Delta h) = 0.025$ m (zero bias, $P = 0.95$). The fifth and sixth columns show the Reynolds numbers and the flow coefficient $K$ computed using Eq. 13.3.8. Determine the uncertainties in the result $K$ as a result of the estimated uncertainties in the measurands $Q$ and $\Delta h$.

$$\Delta h = h_1 - h_2 = (p_1 - p_2)/\gamma$$

**Fig. E13.1a**

## Solution

Equation 13.3.8 is used in conjunction with Eq. 13.5.4 to propagate the uncertainties from $Q$ and $\Delta h$ into $K$. For example, consider the data for $i = 6$. Calculation of $\delta K$ proceeds as follows. The orifice area is

$$A_0 = \frac{\pi}{4} 0.0354^2 = 9.84 \times 10^{-4}\, \text{m}^2$$

| Data no. $i$ | $Q$ (m³/s) | $\delta Q$ (m³/s) | $\Delta h$ (m) | Re | $K$ | $\delta K$ |
|---|---|---|---|---|---|---|
| 1 | 0.00119 | 0.00010 | 0.139 | 42 800 | 0.732 | 0.085 |
| 2 | 0.00162 | 0.00012 | 0.265 | 58 300 | 0.722 | 0.062 |
| 3 | 0.00200 | 0.00015 | 0.403 | 71 900 | 0.723 | 0.058 |
| 4 | 0.00223 | 0.00015 | 0.517 | 80 200 | 0.711 | 0.051 |
| 5 | 0.00251 | 0.00015 | 0.655 | 90 300 | 0.711 | 0.045 |
| 6 | 0.00276 | 0.00017 | 0.781 | 99 300 | 0.716 | 0.046 |
| 7 | 0.00294 | 0.00018 | 0.907 | 105 700 | 0.708 | 0.045 |
| 8 | 0.00314 | 0.00019 | 1.008 | 112 900 | 0.717 | 0.045 |
| 9 | 0.00330 | 0.00016 | 1.134 | 118 700 | 0.711 | 0.035 |
| 10 | 0.00346 | 0.00017 | 1.247 | 124 400 | 0.711 | 0.036 |
| 11 | 0.00373 | 0.00017 | 1.386 | 134 200 | 0.727 | 0.034 |
| 12 | 0.00382 | 0.00017 | 1.512 | 137 400 | 0.713 | 0.032 |
| 13 | 0.00408 | 0.00018 | 1.663 | 146 700 | 0.726 | 0.033 |
| 14 | 0.00424 | 0.00019 | 1.852 | 152 500 | 0.715 | 0.032 |
| 15 | 0.00480 | 0.00021 | 2.293 | 172 600 | 0.727 | 0.032 |

and the diameter ratio is

$$\beta = \frac{35.4}{50.8} = 0.697 \approx 0.7$$

With Eq. 13.3.8, values of $K$ are determined for $(Q, \Delta h)$, $(Q + \delta Q, \Delta h)$, and $(Q, \Delta h + \delta(\Delta h))$:

*(continued)*

$$K(Q, \Delta h) = \frac{0.00276}{9.84 \times 10^{-4} \sqrt{2 \times 9.81 \times 0.781}} = 0.716$$

$$K(Q + \delta Q, \Delta h) = \frac{0.00276 + 0.00017}{9.84 \times 10^{-4} \sqrt{2 \times 9.81 \times 0.781}} = 0.761$$

$$K(Q, \Delta h + \delta(\Delta h)) = \frac{0.00276}{9.84 \times 10^{-4} \sqrt{2 \times 9.81 \times (0.781 + 0.025)}} = 0.705$$

These are then substituted into Eq. 13.5.4, with the variable $K$ replacing the result $R$ in that relation:

$$\delta K = \{[K(Q + \delta Q, \Delta h) - K(Q, \Delta h)]^2 + [K(Q, \Delta h + \delta(\Delta h)) - K(Q, \Delta h)]^2\}^{1/2}$$
$$= [(0.761 - 0.716)^2 + (0.705 - 0.716)^2]^{1/2}$$
$$= 0.046$$

The results for all measurands are similarly evaluated and shown in the last column in the table. In addition, they are plotted in Fig. E13.1b, where the vertical lines, or uncertainty bands, associated with each value of $K$ have the magnitude of $2\delta K$. The bands can be interpreted as the estimated precision for $K$, based on zero bias and $P = 0.95$ for the two measurands. Note that the precision $\delta K$ stabilizes at approximately 0.032 at the higher Reynolds numbers and that the magnitude of $\delta K$ is greater at lower Reynolds numbers. An alternative way to express this is that the uncertainty of $K$ decreases with increasing Reynolds number. The data in the figure should be compared with the orifice curve labeled $\beta = 0.7$ in Fig. 13.10.

**Fig. E13.1b**

A Mathcad solution of Example 13.1 is shown in Fig. 13.28.

**Input data:**

$D := 0.0354$       $ORIGIN := 1$     $N := 15$     $v := 1.006 \times 10^{-6}$       $g := 9.81$

$$
Q := \begin{pmatrix} 0.00119 \\ 0.00162 \\ 0.002 \\ 0.00223 \\ 0.00251 \\ 0.00276 \\ 0.00294 \\ 0.00314 \\ 0.0033 \\ 0.00346 \\ 0.00373 \\ 0.00382 \\ 0.00408 \\ 0.00424 \\ 0.0048 \end{pmatrix}
\quad \delta Q := \begin{pmatrix} 0.00010 \\ 0.00012 \\ 0.00015 \\ 0.00015 \\ 0.00015 \\ 0.00017 \\ 0.00018 \\ 0.00019 \\ 0.00016 \\ 0.00017 \\ 0.00017 \\ 0.00017 \\ 0.00018 \\ 0.00019 \\ 0.00021 \end{pmatrix}
\quad \Delta h := \begin{pmatrix} 0.139 \\ 0.265 \\ 0.403 \\ 0.517 \\ 0.655 \\ 0.781 \\ 0.907 \\ 1.008 \\ 1.134 \\ 1.247 \\ 1.386 \\ 1.512 \\ 1.663 \\ 1.852 \\ 2.293 \end{pmatrix}
\quad \delta \Delta h := \begin{pmatrix} 0.025 \\ 0.025 \\ 0.025 \\ 0.025 \\ 0.025 \\ 0.025 \\ 0.025 \\ 0.025 \\ 0.025 \\ 0.025 \\ 0.025 \\ 0.025 \\ 0.025 \\ 0.025 \\ 0.025 \end{pmatrix}
$$

Define function to compute K:    $K(q, \Delta H) := \dfrac{q}{\left(\dfrac{\pi \cdot D^2}{4}\right)} \cdot (2 \cdot g \cdot \Delta H)^{-0.5}$

Compute uncertainty in K:

$i := 1 .. N$

$KK_i := K(Q_i, \Delta h_i)$

$\delta K_i := \sqrt{(K(Q_i + \delta Q_i, \Delta h_i) - K(Q_i, \Delta h_i))^2 + (K(Q_i, \Delta h_i + \delta \Delta h_i) - K(Q_i, \Delta h_i))^2}$

$Re_i := \dfrac{4 \cdot Q_i}{\pi \cdot v \cdot D}$

**Fig. 13.28**   Mathcad Solution to Example 13.1.

Plot of K versus Re:

**Fig. 13.28 (*continued*)**

### 13.5.3 Regression Analysis

A significant number of flow-measuring devices correlate the discharge with pressure drop or head according to the general formula

$$Y = CX^m \tag{13.5.5}$$

in which $Y$ is the derived result, $X$ is the measurand, and $C$ and $m$ are constants. By taking a number of independent measurements of $X$ and $Y$, estimates of $C$ and $m$ can be found. A systematic approach to determine $C$ and $m$ is to find the best estimates based on the method of least squares. If the logarithm of each side of Eq. 13.5.5 is taken, it becomes

$$\ln Y = \ln C + m \ln X \tag{13.5.6}$$

which is of the form

$$y = b + mx \tag{13.5.7}$$

By association we observe that Eq. 13.5.6 is linear provided that $y = \ln Y$, $b = \ln C$, and $x = \ln X$.

Consider the data set $X_i, Y_i, i = 1, \ldots, n$. The objective is to generate a straight line through the logarithms of the data $(x_i, y_i)$ such that the errors of estimation are small. That error is defined as the difference between the observed value $y_i$ and the corresponding value on the line; in Fig. 13.29, $s_i$ is the error associated with the data point $(x_i, y_i)$. The method of least squares requires that to minimize the error, the sum of the squares of the errors of all the data be as small as possible. From Eq. 13.5.7,

$$S = \sum_{i=1}^{n} s_i^2$$
$$= \sum_{i=1}^{n} [y_i - (b + mx_i)]^2 \tag{13.5.8}$$

The parameter $S$ is minimized by forming the partial derivatives of $S$ with respect to $b$ and $m$, setting the resulting algebraic equations to zero, and solving for $b$ and $m$. It is left as an exercise for the reader to derive the results:

$$m = \frac{\sum x_i y_i - (\sum x_i \sum y_i)/n}{\sum x_i^2 - (\sum x_i)^2/n} \tag{13.5.9}$$

$$b = \frac{\sum y_i - m \sum x_i}{n} \tag{13.5.10}$$

$$C = e^b \tag{13.5.11}$$

A simple computer algorithm can be prepared to evaluate $m$ and $C$ based on a set of input data $(X_i, Y_i)$, or alternatively, a spreadsheet solution can be used.

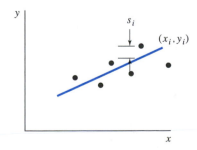

**Fig. 13.29** Method of least squares.

## Example 13.2

With the data given in Example 13.1, perform a least squares regression to determine the coefficient $C$ and exponent $m$ in Eq. 13.5.5. Compare the result with Eq. 13.3.8.

**Solution:**

Equations 13.5.9 to 13.5.11 are employed to evaluate $C$, $b$, and $m$. In Eqs. 13.5.9 and 13.5.10, $x$ is replaced by $\ln \Delta h = \ln(h_1 - h_2)$, and $y$ is replaced by $\ln Q$. The calculations are tabulated in the accompanying table.

| $i$ | $x_i = \ln \Delta h$ | $y_i = \ln Q$ | $x_i^2$ | $x_i y_i$ |
|---|---|---|---|---|
| 1 | −1.9733 | −6.7338 | 3.8939 | 13.2878 |
| 2 | −1.3280 | −6.4253 | 1.7636 | 8.5328 |
| 3 | −0.9088 | −6.2146 | 0.8259 | 5.6478 |
| 4 | −0.6597 | −6.1058 | 0.4352 | 4.0280 |
| 5 | −0.4231 | −5.9875 | 0.1790 | 2.5333 |
| 6 | −0.2472 | −5.8925 | 0.06111 | 1.4566 |
| 7 | −0.09761 | −5.8293 | 0.009528 | 0.5690 |
| 8 | 0.007968 | −5.7635 | 0.00006349 | −0.0459 |
| 9 | 0.1258 | −5.7138 | 0.01583 | −0.7188 |
| 10 | 0.2207 | −5.6665 | 0.04871 | −1.2506 |
| 11 | 0.3264 | −5.5913 | 0.1065 | −1.8250 |
| 12 | 0.4134 | −5.5675 | 0.1709 | −2.3016 |
| 13 | 0.5086 | −5.5017 | 0.2587 | −2.7982 |
| 14 | 0.6163 | −5.4632 | 0.3798 | −3.3670 |
| 15 | 0.8299 | −5.3391 | 0.6887 | −4.4309 |
| $\Sigma$ | −2.5886 | −87.7954 | 8.8374 | 19.3173 |

Substitute the summed values into Eqs. 13.5.9 to 13.5.11:

$$m = \frac{19.3173 - (-2.5886)(-87.7954)/15}{8.8374 - (-2.5886)^2/15} = 0.497$$

$$b = \frac{-87.7954 - 0.497(-2.5886)}{15} = -5.7673$$

$$C = \exp(-5.7673) = 0.00313$$

Hence the discharge is correlated to $\Delta h$ by the regression curve

$$Q = 0.00313 \, \Delta h^{0.497}$$

Based on the analysis performed in Example 13.1, the average value of $K$ is 0.718. The area of the orifice is $A_0 = 9.84 \times 10^{-4} \text{m}^2$. Substituting these into Eq. 13.3.8, we find that

$$Q = 0.00313 \, \Delta h^{0.500}$$

Thus the two results agree quite closely. The data and regression line are plotted in Fig. E13.2.

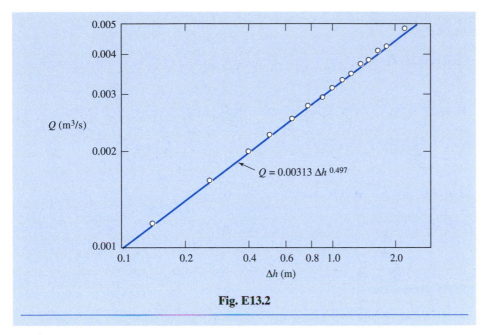

$Q = 0.00313 \, \Delta h^{0.497}$

**Fig. E13.2**

A Mathcad solution for Example 13.2 is provided in Fig. 13.30.

**Input data:**

$n := 15$

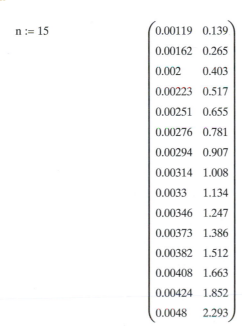

$$\begin{pmatrix} 0.00119 & 0.139 \\ 0.00162 & 0.265 \\ 0.002 & 0.403 \\ 0.00223 & 0.517 \\ 0.00251 & 0.655 \\ 0.00276 & 0.781 \\ 0.00294 & 0.907 \\ 0.00314 & 1.008 \\ 0.0033 & 1.134 \\ 0.00346 & 1.247 \\ 0.00373 & 1.386 \\ 0.00382 & 1.512 \\ 0.00408 & 1.663 \\ 0.00424 & 1.852 \\ 0.0048 & 2.293 \end{pmatrix}$$

**Fig. 13.30**   Mathcad solution to Example 13.2.

Convert to logarithmic form:    $X := \ln(M^{<1>})$   $Y := \ln(M^{<0>})$

Compute m, b, and C:

$$m := \frac{\left(\sum_{i=0}^{n-1} X_i \cdot Y_i\right) - \frac{1}{n} \cdot \sum_{i=0}^{n-1} X_i \cdot \sum_{i=0}^{n-1} Y_i}{\sum_{i=0}^{n-1} (X_i)^2 - \left(\sum_{i=0}^{n-1} X_i\right)^2 \cdot \frac{1}{n}} \qquad b := \frac{\sum_{i=0}^{n-1} Y_i - m \cdot \sum_{i=0}^{n-1} X_i}{n} \qquad C := \exp(b)$$

$m = 0.496 \qquad b = -5.767 \qquad C = 3.128 \times 10^{-3}$

**Plot Q versus Δh:**

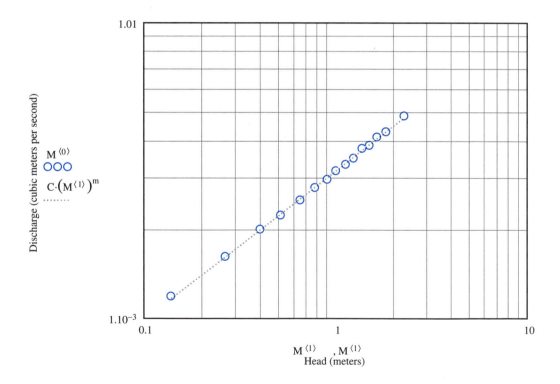

**Fig. 13.30 (*continued*)**

## 13.6 SUMMARY

There are many devices available to measure flow parameters; we have presented some of the more common ones. The monitoring of pressure and velocity at discrete locations in flow fields was first discussed, followed by descriptions of integrated flow rate or discharge measurements. Various devices are used for measurement of these parameters, ranging from purely mechanical to electromechanical.

Flow visualization is an important technique that is employed in both industrial and research laboratories to study complex flow fields, especially turbulent ones. Uses of tracers and index of refraction methods are presented. Of these methods, velocimetry techniques are rapidly gaining favor due to computer imaging and the development of laser-activated dyes.

Today, a significant number of measurements are automated and may require analog or digital interfacing with data loggers or personal computers. Regardless of the manner in which data is acquired and stored, the data taker must be aware of uncertainties that are present and unavoidable in the process of acquiring data. We have shown how the uncertainties present in measured parameters can be propagated into results using spreadsheet or computational software analyses. These techniques are useful, for example, to determine how accurate and reliable a flow meter may be, or to design an experiment to calibrate a velocity meter that will insure a desired degree of accuracy. Finally, regression analysis is applied to determine the parameters that describe an exponential relationship that correlates discharge with pressure drop or head.

## PROBLEMS

**13.1**  A 20° inclined manometer attached to a piezometer opening is used to measure the pressure at the wall of airflow. If the reading is measured to be 4 cm of mercury, calculate the pressure at the wall.

**13.2**  A pitot-static probe is to be used for repeated measurements of the velocity of an atmospheric airstream. It is desired to relate the velocity $V$ to the manometer reading $h$ in centimeters of mercury (i.e., $V = C \sqrt{h}$). Calculate the value of the constant $C$ for a laboratory at 20°C situated at:
**(a)**  Sea level
**(b)**  2000 m elevation

**13.3**  A pitot-static probe attached to a manometer that has water as the manometer fluid is proposed as the measuring device for an airflow with a velocity of 8 m/s. What manometer reading is expected? Comment as to the use of the proposed device.

**13.4**  Water at 24°C flows through a 2.5-mm-diameter tube into a calibrated tank. If 1.9 liters are collected in 10 minutes, calculate the flow rate ($m^3/s$), the

mass flux (kg/s), and the average velocity (m/s). Is the flow laminar, turbulent, or you can't tell?

**13.5**  A velocity traverse in a rectangular channel that measures 10 cm by 100 cm is represented by the following data. Estimate the flow rate and the average velocity assuming a symmetric plane flow.

| $y$ (cm) | 0 | 0.5 | 1 | 2 | 3 | 4 | 5 |
|---|---|---|---|---|---|---|---|
| $u$ (m/s) | 0 | 5.2 | 8.1 | 9.2 | 9.8 | 10 | 10 |

**13.6**  A velocity traverse in a 10-cm-diameter pipe is represented by the following data. Estimate the flow rate and the average velocity assuming a symmetric flow.

| $r$ (cm) | 0 | 1 | 2 | 3 | 4 | 4.5 |
|---|---|---|---|---|---|---|
| $u$ (m/s) | 10 | 10 | 9.8 | 9.2 | 8.1 | 5.2 |

**13.7**  The flow rate of water in a 12-cm-diameter pipe is measured with a 6-cm-diameter venturi meter to be 0.09 m³/s. What is the expected deflection on a water-mercury manometer? Assume a water temperature of 20°C.

**13.8**  The flow rate of 20°C water flow in a 24-cm-diameter pipe is desired. If a water-mercury manometer reads 12 cm, calculate the discharge if the manometer is connected to:
  (a)  A 15-cm-diameter orifice
  (b)  A 15-cm-diameter nozzle

**13.9**  Calculate the flow rate of 40°C water in the pipes shown in Fig. P13.9.

(a)

(b)

**Fig. P13.9**

**13.10**  A pressure difference across an elbow in a water pipe is measured. Estimate the flow rate in the pipe. The pressure difference, pipe diameter, and radius of curvature are, respectively:
  (a)  80 kPa, 10 cm, 20 cm
  (b)  80 kPa, 5 cm, 20 cm

Use a water temperature of 20°C.

**13.11**  A pitot tube is used to measure the velocity of gasoline flowing in a pipe. The estimate of the velocity is given by the equation

$$v_1 = \sqrt{\frac{2}{\rho}(p_T - p_1)}$$

in which $v_1$ is the local velocity at location 1, $\rho$ the density of the liquid, $p_T$ the total pressure measured by the tube, and $p_1$ the static pressure at location 1. The density is estimated to be $680 \pm 50$ kg/m³ ($P = 0.95$, 0 bias).
  (a)  Assume that the two pressures are measured separately using individual pressure gages: $p_T = 102 \pm 1$ kPa ($P = 0.95$, 0 bias), $p_1 = 95 \pm 1$ kPa ($P = 0.95$, 0 bias). Estimate the uncertainty for $v_1$.
  (b)  Assume that the pressure difference $p_T - p_1$ is measured using a single differential pressure gage: $p_T - p_1 = 7 \pm 1$ kPa ($P = 0.95$, 0 bias). Estimate the uncertainty for $v_1$.
  (c)  Which arrangement, (a) or (b), is preferable?

**13.12**  Experimental data used to calibrate discharge of water in a venturi meter are tabulated below. The diameter of the meter throat is 33.3 mm and the diameter at the upstream tap location of the differential manometer is 54.0 mm. The precision $\delta Q$ of the discharge measurements has zero bias and $P = 0.95$, and the precision for all the manometer measurements is $\delta(\Delta h) = 2$ mm Hg, with zero bias and $P = 0.95$. Note that the manometer readings are given in mm Hg. With use of Eq. 13.3.8, determine the uncertainties in $K$ as a result of the uncertainties in the two measurands. Plot the results in a fashion similar to those in Example 13.1, and compare the curve with Fig. P13.10.

| Data no.<br>$i$ | $Q$<br>(m³/s) | $\delta Q$<br>(m³/s) | $\Delta h$<br>(mm Hg) | Re |
|---|---|---|---|---|
| 1 | 0.00168 | 0.00011 | 13 | 64 200 |
| 2 | 0.00208 | 0.00014 | 22 | 79 500 |
| 3 | 0.00249 | 0.00016 | 32 | 95 100 |
| 4 | 0.00281 | 0.00019 | 40 | 107 400 |
| 5 | 0.00328 | 0.00022 | 54 | 125 300 |
| 6 | 0.00355 | 0.00024 | 62 | 135 600 |
| 7 | 0.00372 | 0.00025 | 70 | 142 100 |
| 8 | 0.00402 | 0.00023 | 82 | 153 600 |
| 9 | 0.00415 | 0.00024 | 91 | 158 600 |
| 10 | 0.00444 | 0.00025 | 100 | 169 700 |
| 11 | 0.00479 | 0.00027 | 113 | 183 000 |
| 12 | 0.00493 | 0.00028 | 123 | 188 400 |
| 13 | 0.00523 | 0.00030 | 134 | 199 800 |
| 14 | 0.00533 | 0.00035 | 140 | 203 700 |
| 15 | 0.00509 | 0.00029 | 140 | 194 500 |

**13.13**  With the numerical results obtained from Example 13.1, plot the relative uncertainty $\delta K/K$ versus Re on a linear scale. What conclusions can you draw?

**13.14**  Using the data from Problem 13.12, perform a least squares regression and determine the coefficient $C$ and exponent $m$ in Eq. 13.5.5. Compare the result with Eq. 13.3.8. Note that the manometer readings are given in mm Hg.

**13.15**  Establish a depth–discharge relation of the form $Q = CY^m$ for a 60° V-notch sharp-crested weir using the data in the following table. In the equation, $Y$ is the vertical distance from the weir notch to the undisturbed upstream water surface and $Q$ is the discharge (see Fig. 10.11). Compare the result with Eq. 10.4.27.

| Data no. $i$ | $Q$ $(m^3/s)$ | $Y$ (m) |
|---|---|---|
| 1 | 3.83 | 11.68 |
| 2 | 3.58 | 11.41 |
| 3 | 3.52 | 11.29 |
| 4 | 3.14 | 10.78 |
| 5 | 2.65 | 10.15 |
| 6 | 2.58 | 9.66 |
| 7 | 2.23 | 9.42 |
| 8 | 2.15 | 9.24 |
| 9 | 2.07 | 9.12 |
| 10 | 1.92 | 8.96 |
| 11 | 1.43 | 7.94 |
| 12 | 1.20 | 7.37 |
| 13 | 1.15 | 7.19 |
| 14 | 0.88 | 6.22 |

A CFD analysis of the cylinder of a proposed engine design. Left is the grid with areas of refined grid in critical flow areas. Right is a plot of the velocity vectors showing the flow field in a plane cross section. (Courtesy of Bassem Ramadan.)

# 14

# Computational Fluid Dynamics

## Outline

## Chapter Objectives

The objectives of this chapter are to:

▲ Present an introduction to finite-difference methods.

▲ Present ideas on consistency, numerical stability, convergence, and errors.

▲ Present methods of finite differences.

▲ Present a finite-difference solution to transient couette flow.

▲ Present a finite-difference solution to steady-state potential flow.

## 14.1 INTRODUCTION

CFD solutions pages
669–672, 807, 821

The equations governing fluid-flow problems are the *continuity*, the *Navier–Stokes*, and the *energy equations*. These equations, derived in Chapter 5, form a system of coupled quasi-linear *partial differential equations* (PDEs). Because of the nonlinear terms in these PDEs, analytical methods can yield very few solutions. In general, analytical solutions are possible only if these PDEs can be made linear, either because nonlinear terms naturally drop out (e.g., fully developed flows in ducts and flows that are irrotational everywhere) or because nonlinear terms are small when compared to other terms so that they can be neglected (e.g., flows where the Reynolds number is less than unity). Yih (1969) and Schlichting & Gersten (2000) describe most of the better-known analytical solutions. If the nonlinearities in the governing PDEs cannot be neglected, which is the situation for most engineering flows, then numerical methods are needed to obtain solutions.

*Computational fluid dynamics*, or simply CFD, is concerned with obtaining numerical solutions to fluid-flow problems by using the computer. The advent of high-speed and large-memory computers has enabled CFD to obtain solutions to many flow problems, including those that are compressible or incompressible, laminar or turbulent, chemically reacting or nonreacting, single- or multi-phase. Of the numerical methods developed to address equations governing fluid-flow problems, *finite-difference methods* (FDMs) and *finite-volume methods* (FVMs) are the most widely used. In this chapter, we give an introduction to the FDMS and present the solutions to two classical fluid-flow. Problems using finite differences (FD).

Since computers are used to obtain solutions, it is important to understand the constraints that they impose. Of these constraints, four are critical. The first of these is that computers can perform only arithmetic (i.e., $+$, $-$, $\times$, and $\div$) and logic (i.e., true and false) operations. This means non-arithmetic operations, such as derivatives and integrals, must be represented in terms of arithmetic and logic operations. The second constraint is that computers represent numbers using a finite number of digits. This means that there are round-off errors and that these errors must be controlled. The third constraint is that computers have limited storage memories. This means solutions can be obtained only at a finite number of points in space and time. Finally, computers perform a finite number of operations per unit time. This means solution procedures should minimize the computer time needed to achieve a computational task by fully utilizing all available processors on a computer and minimizing the number of operations.

With these constraints, FD methods generate solutions to PDEs through the following three major steps:

1. **Discretize the domain.** The continuous spatial and temporal domain of the problem must be replaced by a discrete one made up of grid points or cells and time levels. The ideal discretization uses the fewest number of grid points/cells and time levels to obtain solutions of desired accuracy.

2. **Discretize the PDEs.** The PDEs governing the problem must be replaced by a set of algebraic equations with the grid points/cells and the time levels as their domain. Ideally, the algebraic equations—referred to as finite-difference equations (FDEs) should describe the same physics as those described by the governing PDEs.

3. **Specify the algorithm.** The step-by-step procedure by which solutions at each grid points/cells are obtained from the FD equations when advancing from one time level to the next must be described in detail. Ideally, the algorithm should ensure not only accurate solutions but also efficiency in utilizing the computer.

These steps are illustrated through example problems in Sections 14.4 and 14.5.

## 14.2 EXAMPLES OF FINITE-DIFFERENCE METHODS

To illustrate FDMs, consider the unsteady, incompressible, laminar flow of a fluid with constant kinematic viscosity $\nu$ between two parallel plates separated by a distance $H$, as shown in Fig. 14.1. Initially, both plates are stationary, and the fluid between them is stagnant. Suddenly at time $t = 0$, the lower plate moves horizontally to the right (in the positive $x$-direction) at a constant speed of $V_0$.

The equations governing this flow are the continuity equation and the $x$- and $y$-component Navier–Stokes equations. Since the flow is parallel (i.e., $v = 0$) and the pressure is the same everywhere, these equations reduce to the following single linear PDE:

$$\frac{\partial u}{\partial t} = \nu \frac{\partial^2 u}{\partial y^2} \qquad (14.2.1)$$

The initial and boundary conditions for Eq. 14.2.1 are, assuming $u = u(y, t)$,

$$u(y, t = 0) = 0, \qquad u(0, t) = V_0 \qquad u(y = H, t) = 0$$

$$u(y, 0) = 0 \qquad u(0, t) = V_0 \qquad u(H, t) = 0 \qquad (14.2.2)$$

**Fig. 14.1**   Flow between stationary and moving parallel plates.

Since Eq. 14.2.1 is linear and the boundary conditions given by Eq. 14.2.2 are homogeneous, separation of variables is able to provide the exact analytical solution

$$u/V_0 = 1 - y/H - 2 \sum_{n=1}^{\infty} \frac{1}{n\pi} \sin(n\pi y/H) \exp(-n^2\pi^2\nu t/H^2) \quad (14.2.3)$$

This exact solution can be used to evaluate the accuracy of FDMs to be presented.

As mentioned, three steps are involved in generating solutions to PDEs by using an FDM. These three steps are illustrated below, one at a time for Eqs. 14.2.1 and 14.2.2.

### 14.2.1 Discretization of the Domain

To discretize the domain, we note that the spatial domain is a line segment between 0 and $H$, and that the temporal domain is a ray emanating from $t = 0$. Though the temporal domain is of infinite extent, the duration of interest is finite, say from $t = 0$ to $t = T$, where $T$ is the time when steady state is achieved. Here, the spatial domain, $0 \leq y \leq H$, is discretized by replacing it with $J$ equally distributed stationary grid points, and the temporal domain, $0 \leq t \leq T$ is discretized by replacing it with equally incremented time levels (see Fig. 14.2). This discretization of the domain is one of many instances. For example, the grid points do not have to be uniformly distributed nor do they have to be stationary.

Each point in the discretized domain, shown in Fig. 14.2, has coordinates $(y_j, t^n)$, given by

$$y_j = (j - 1)\,\Delta y, \qquad j = 1, 2, 3, \ldots, J \quad (14.2.4)$$

$$t^n = n\,\Delta t, \qquad n = 0, 1, 2, \ldots \quad (14.2.5)$$

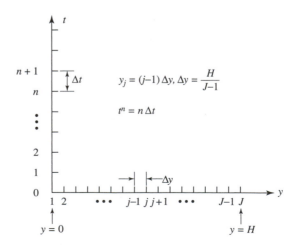

**Fig. 14.2** Grid system and time levels for problem depicted in Figure 14.1.

where $\Delta y = H/(J-1)$ is the distance between two adjacent grid points (or the grid spacing), and $\Delta t$ is the time-step size. The sizes of the grid spacing and the time step depend on the length and time scales that need to be resolved and on the properties of the discretized PDEs used to obtain solutions.

The solution sought—$u(y, t)$ in Eq. 14.2.1—will be obtained only at the grid points and the time levels. That solution is denoted as

$$u_j^n = u(y_j, t^n) \tag{14.2.6}$$

where subscripts denote the locations of the grid points and superscripts denote time levels.

## 14.2.2   Discretization of the Governing Equations

With the domain discretized, the next step is to replace the PDEs governing the problem by a set of algebraic or FDEs that use the grid points and time levels as the domain. With FDMs, PDEs are discretized by replacing derivatives with difference operators. Thus, discretization involves two parts. First, derive *difference operators*. Then, select difference operators.

**Derive Difference Operators.**   There are many different ways to derive difference operators. For problems with smooth solutions (i.e., solutions without discontinuities such as shock waves), a method based on the following theorem is often used: *If a function, u, and its derivatives are continuous, single-valued, and finite, then the value of that function at any point can be expressed in terms of u and its derivatives at any other point by using a Taylor series expansion, provided the other point is within the radius of convergence of the series.* With this theorem, $u_{j+1}$ and $u_{j-1}$ can be expressed in terms of $u_j$ and its derivatives as

$$u_{j+1} = u_j + \left(\frac{\partial u}{\partial y}\right)_j \Delta y + \left(\frac{\partial^2 u}{\partial y^2}\right)_j \frac{\Delta y^2}{2!} + \left(\frac{\partial^3 u}{\partial y^3}\right)_j \frac{\Delta y^3}{3!} + \cdots \tag{14.2.7}$$

$$u_{j-1} = u_j - \left(\frac{\partial u}{\partial y}\right)_j \Delta y + \left(\frac{\partial^2 u}{\partial y^2}\right)_j \frac{\Delta y^2}{2!} - \left(\frac{\partial^3 u}{\partial y^3}\right)_j \frac{\Delta y^3}{3!} + \cdots \tag{14.2.8}$$

From the above two Taylor series, we can readily derive the following four difference operators by solving for $(\partial u/\partial y)_j$ and $(\partial^2 u/\partial y^2)_j$ either directly or by adding or subtracting the two equations:

$$\left(\frac{\partial u}{\partial y}\right)_j = \frac{u_{j+1} - u_j}{\Delta y} + O(\Delta y) \tag{14.2.9}$$

$$\left(\frac{\partial u}{\partial y}\right)_j = \frac{u_j - u_{j-1}}{\Delta y} + O(\Delta y) \tag{14.2.10}$$

$$\left(\frac{\partial u}{\partial y}\right)_j = \frac{u_{j+1} - u_{j-1}}{2\Delta y} + O(\Delta y^2) \tag{14.2.11}$$

$$\left(\frac{\partial^2 u}{\partial y^2}\right)_j = \frac{u_{j+1} - 2u_j + u_{j-1}}{\Delta y^2} + O(\Delta y^2) \tag{14.2.12}$$

---

In the above difference operators, $O(\Delta y)$ and $O(\Delta y^2)$ denote the truncation errors (i.e., terms in the Taylor series that have been truncated). In Eqs. 14.2.9 and 14.2.10, the power of $\Delta y$ in $O(\Delta y)$ is one because the leading term of the truncation errors for these two difference operators is multiplied by $\Delta y$ raised to the first power. Such difference operators are said to be first-order accurate. In Eqs. 14.2.11 and 14.2.12, that power is two because the leading term in the truncation errors is multiplied by $\Delta y$ raised to the second power. Such difference operators are said to be second-order accurate. The higher the order of accuracy, the greater is the number of terms retained in the truncated Taylor series. For sufficiently small $\Delta y$, a higher-order accurate difference operator is a more accurate representation of a smooth function than a lower-order accurate difference operator. In general, second-order accuracy is adequate.

The grid points used to construct a difference operator form the stencil of that difference operator. If the stencil for a difference operator at $y_j$ involves only grid points with indices greater than or equal to $j$ such as Eq. 14.2.9, then that difference operator is said to be a forward-difference operator. If grid points all have indices less than or equal to $j$ such as Eq. 14.2.10, then that difference operator is said to be a backward-difference operator. If the number of grid points behind and ahead of $j$ are exactly the same such as Eqs. 14.2.11 and 14.2.12, that difference operator is said to be a central-difference operator. If the number of grid points ahead of and after $j$ are not the same, that difference operator is said to be either biased backward or biased forward.

Taylor series can be used to derive difference operators for any derivative to any order of accuracy and using any stencil. To illustrate, consider the derivation of a difference operator for a first-order derivative, $(\partial u/\partial y)_j$, with the following specifications: third-order accurate, $O(\Delta y^3)$, with a stencil that includes only one downstream grid point. The derivation of this difference operator involves the following four steps:

Step 1: Determine the number of terms to be kept in each Taylor series. That number, denoted as $N$, is equal to the order of the derivative for which a difference operator is sought plus the order of accuracy desired. For this case, the order of the derivative is 1, and the order of accuracy desired is 3. Thus, $N$ is 4.

Step 2: Decide on a stencil based on $N$ points for the difference operator at point $j$. For this case with $N = 4$ and $u > 0$, the 4 points of the stencil are $j - 2, j - 1, j, j + 1$. Other possible stencils include: $(j - 3, j - 2, j - 1, j)$ or completely backward, $(j - 1, j, j + 1, j + 2)$ or biased forward, and $(j, j + 1, j + 2, j + 3)$ or completely forward.

Step 3: Construct $N - 1$ truncated Taylor series about $u_j$ from $u$ at all points of the stencil except point $j$. Performing this operation gives

$$u_{j+1} = u_j + \left(\frac{\partial u}{\partial y}\right)_j \Delta y + \left(\frac{\partial^2 u}{\partial y^2}\right)_j \frac{\Delta y^2}{2!} + \left(\frac{\partial^3 u}{\partial y^3}\right)_j \frac{\Delta y^3}{3!} + O(\Delta y^4) \qquad (14.2.12)$$

$$u_{j-1} = u_j - \left(\frac{\partial u}{\partial y}\right)_j \Delta y + \left(\frac{\partial^2 u}{\partial y^2}\right)_j \frac{\Delta y^2}{2!} - \left(\frac{\partial^3 u}{\partial y^3}\right)_j \frac{\Delta y^3}{3!} + O(\Delta y^4) \qquad (14.2.13)$$

$$u_{j-2} = u_j - \left(\frac{\partial u}{\partial y}\right)_j 2\Delta y + \left(\frac{\partial^2 u}{\partial y^2}\right)_j \frac{(2\Delta y)^2}{2!} - \left(\frac{\partial^3 u}{\partial y^3}\right)_j \frac{(2\Delta y)^3}{3!} + O(\Delta y^4) \qquad (14.2.14)$$

In the above three equations, the unknowns are the first, second, and third derivatives, and they must be expressed in terms of $u$ at points in the stencil.

Step 4: Solve $N - 1$ coupled linear equations for the difference operator sought. For the first derivative, solving the above three equations gives

$$\left(\frac{\partial u}{\partial y}\right)_j = \frac{2u_{j+1} + 3u_j - 6u_{j-1} + u_{j-2}}{6\Delta y} + O(\Delta y^3) \qquad (14.2.15)$$

Difference operators for the second and third derivatives are also derived in the process of solving for the above difference operator. The orders of accuracy of these difference operators are second order for the second derivative and first order for the third derivative.

In Table 14.1, some commonly used difference operators are summarized. Note that if $j$ and $\Delta y$ are replaced by $n$ and $\Delta t$, then the difference operators can also be used for time derivatives.

**TABLE 14.1**  Summary of Commonly Used Difference Operators

| Description | Finite-Difference | Stencil |
|---|---|---|
| **First Derivative:** $(\partial u/\partial y)_j$ | | $j$-1 $\quad j \quad$ $j$+1 |
| central, $O(\Delta y^2)$ | $(u_{j+1} - u_{j-1})/2\Delta y$ | |
| backward, $O(\Delta y)$ | $(u_j - u_{j-1})/\Delta y$ | |
| forward, $O(\Delta y)$ | $(u_{j+1} - u_j)/\Delta y$ | |
| backward, $O(\Delta y^2)$ | $(3u_j - 4u_{j-1} + u_{j-2})/2\Delta y$ | |
| forward, $O(\Delta y^2)$ | $(-3u_j + 4u_{j+1} - u_{j+2})/2\Delta y$ | |
| **Second Derivative:** $(\partial^2 f/\partial y^2) = (\partial u/\partial y)_j, u = \partial f/\partial y$ | | |
| central, $O(\Delta y^2)$ | $(f_{j+1} - 2f_j + f_{j-1})/\Delta y^2$ | |
| backward, $O(\Delta y)$ | $(f_j - 2f_{j-1} + f_{j-2})/\Delta y^2$ | |
| forward, $O(\Delta y)$ | $(f_j - 2f_{j+1} + f_{j+2})/\Delta y^2$ | |

**Select Difference Operators.**    To select difference operators, we apply the governing PDE to be discretized (Eq. 14.2.1) at an arbitrary interior grid point (a grid point not on the boundary) and at a time between $t^n$ and $t^{n+1}$:

$$\left( \frac{\partial u}{\partial t} = \nu \frac{\partial^2 u}{\partial y^2} \right)^{n'}_{j} \qquad (14.2.16)$$

where $n'$ is $n$, $n + 1$, or some value between $n$ and $n + 1$. In this equation, it is assumed that the solution at time $t^n$ is known and that the solution at $t^{n+1}$ is sought. This assumption is generally applicable because the solution is always known at the previous time level. For example, at the zeroth time level ($n = 0$), the solution is the initial condition.

**An Explicit Method.**    If $n' = n$ in Eq. 14.2.16, then all spatial derivatives are evaluated at time $t^n$, the previous time level where the solution is known. If we choose the forward-difference operator given by Eq. 14.2.9 for $(\partial u / \partial t)^n_j$ (except replace $y$ and $j$ with $t$ and $n$) and the central-difference operator given by Eq. 14.2.12 for $(\partial^2 u / \partial y^2)^n_j$, then Eq. 14.2.16 becomes

$$\frac{u^{n+1}_j - u^n_j}{\Delta t} = \nu \frac{u^n_{j+1} - 2u^n_j + u^n_{j-1}}{\Delta y^2} + \mathrm{O}(\Delta t, \Delta y^2) \qquad (14.2.17)$$

The above FDE is first-order accurate in time and second-order accurate in space as indicated by $\mathrm{O}(\Delta t, \Delta y^2)$. In this equation, $u^{n+1}_j$ is the unknown sought, and solving for it gives

$$u^{n+1}_j = \beta u^n_{j-1} + (1 - 2\beta)u^n_j + \beta u^n_{j+1}, \qquad \beta = \nu \, \Delta t / \Delta y^2 \quad (14.2.18)$$

The FDE in the form of Eq. 14.2.17 or 14.2.18 can be applied at any interior grid point ($j = 2, 3, \ldots, J - 1$). The FDEs at the boundary grid points ($j = 1$ and $J$) are obtained by using *boundary conditions* (BCs). For this simple problem, the BCs, given by Eq. 14.2.2, readily yield the following FDEs:

$$u^n_1 = u^{n+1}_1 = V_0 \qquad (14.2.19)$$

$$u^n_J = u^{n+1}_J = 0 \qquad (14.2.20)$$

Note that the FDEs given by Eqs. 14.2.17 and 14.2.20 have only one unknown in them, namely, $u^{n+1}_j$, a consequence of setting $n' = n$ in Eq. 14.2.16. A method made exclusively of such FDEs is said to be explicit. This particular explicit method with a first-order accurate forward-time differencing is known as the *Euler explicit scheme*.

**An Implicit Method.**    If $n' = n + 1$ in Eq. 14.2.16, then all spatial derivatives are evaluated at $t^{n+1}$, the new time level where the solution is unknown. If we choose the backward-difference operator given by Eq. 14.2.10 for $(\partial u / \partial t)^{n+1}_j$ to couple $t^{n+1}$ to $t^n$ and the central-difference operator given by Eq. 14.2.11 for $(\partial^2 u / \partial y^2)^{n+1}_j$, then Eq. 14.2.16 becomes

$$\frac{u_j^{n+1} - u_j^n}{\Delta t} = \nu \frac{u_{j+1}^{n+1} - 2u_j^{n+1} + u_{j-1}^{n+1}}{\Delta y^2} + O(\Delta t, \Delta y^2) \qquad (14.2.21)$$

which can be rewritten as

$$-\beta u_{j-1}^{n+1} + (1 + 2\beta)u_j^{n+1} - \beta u_{j+1}^{n+1} = u_j^n, \quad \beta = \nu \, \Delta t/\Delta y^2 \qquad (14.2.22)$$

Similar to Eq. 14.2.18, the above equation is first-order accurate in time and second-order accurate in space. But it differs in that there are three unknowns, $u_{j-1}^{n+1}$, $u_j^{n+1}$, and $u_{j+1}^{n+1}$, instead of one in the FDE. This difference is a consequence of evaluating spatial derivatives at $t^{n+1}$ instead of $t^n$. By applying Eq. 14.2.22 at every interior grid point and by using Eqs. 14.2.19 and 14.2.20 for the boundary grid points, we obtain the following system of equations:

$$\mathbf{Ax} = \mathbf{b} \qquad (14.2.23)$$

where

$$\mathbf{A} = \begin{bmatrix} 1+2\beta & -\beta & & & & \\ -\beta & 1+2\beta & -\beta & & & \\ & -\beta & 1+2\beta & -\beta & & \\ & & \cdot & \cdot & \cdot & \\ & & & \cdot & \cdot & \cdot \\ & & & -\beta & 1+2\beta & -\beta \\ & & & & -\beta & 1+2\beta \end{bmatrix}, \mathbf{x} = \begin{bmatrix} u_2^{n+1} \\ u_3^{n+1} \\ u_4^{n+1} \\ \vdots \\ u_{J-2}^{n+1} \\ u_{J-1}^{n+1} \end{bmatrix}, \mathbf{b} = \begin{bmatrix} u_2^n + \beta V_0 \\ u_3^n \\ u_4^n \\ \vdots \\ u_{J-2}^n \\ u_{J-1}^n \end{bmatrix}$$

$$(14.2.24)$$

In the above system of equations, the FDEs at $j = 1$ and 2 and at $j = J - 1$ and $J$ have been combined. A method in which each FDE contains more than one unknown—so that solutions of simultaneous equations are needed—is said to be implicit. This particular implicit method with a first-order accurate backward-time differencing is known as the *Euler implicit scheme*.

**A Generalized Method.** If $n'$ is a value between $n$ and $n + 1$, then Eq. 14.2.16 can be written as

$$\frac{u_j^{n+1} - u_j^n}{\Delta t} = \theta \left(\frac{\partial u}{\partial t}\right)_j^{n+1} + (1 - \theta)\left(\frac{\partial u}{\partial t}\right)_j^n, \quad 0 < \theta < 1 \qquad (14.2.25)$$

where the time derivatives are set equal to the spatial derivatives according to the PDE being solved (e.g., $\partial u/\partial t = \nu \partial^2 u/\partial y^2$). If we choose the central-difference operator given by Eq. 14.2.11 for $(\partial^2 u/\partial y^2)$, then the above equation becomes

$$\frac{u_j^{n+1} - u_j^n}{\Delta t} = \theta\nu \left(\frac{u_{j+1}^{n+1} - 2u_j^{n+1} + u_{j-1}^{n+1}}{\Delta y^2}\right) + (1 - \theta)\nu\left(\frac{u_{j+1}^n - 2u_j^n + u_{j-1}^n}{\Delta y^2}\right)$$

$$(14.2.26)$$

It can readily be seen that the above equation reduces to the explicit scheme given by Eq. 14.2.18 and the implicit scheme given by Eq. 14.2.22 when $\theta = 0$ and 1, respectively. When $\theta = \frac{1}{2}$, the above equation is second-order accurate in space and in time. The resulting time-differencing formula is known as the *trapezoidal* or *Crank-Nicolson* method.

### 14.2.3 Define Solution Algorithm

With the domain and the governing equations discretized, the final step is to specify the solution algorithm (i.e., the step-by-step process of obtaining a solution). Since three sets of FDEs have been derived for Eqs. 14.2.1 and 14.2.2 there are three different solution algorithms.

For the explicit method given by Eqs. 14.2.18 and 14.2.20, the solution algorithm is as follows:

1. Specify the problem by inputting values for $H$, $V_0$, and $\nu$.
2. Specify the number of grid points desired by inputting $J$. As noted, the more the number of grid points, the higher the accuracy.
3. Specify the time-step size $\Delta t$ and the duration of interest $T$. Later, we will find that specification of $\Delta t$ depends not only on the temporal accuracy sought but also on $\Delta y$.
4. Calculate the grid spacing with $\Delta y = H/(J - 1)$.
5. Calculate constants: $\beta = \nu \, \Delta t/\Delta y^2$ and $N = T/\Delta t$.
6. Set the time level counter $n$ to 0.
7. Specify the solution at every grid point at time level $n$ by using the initial condition given by Eq. 14.2.2 ($u_j^n = 0$ for all $j$).
8. Calculate $u_j^{n+1}$ at every interior grid point ($j = 2, 3, \ldots, J - 1$) by using Eq. 14.2.18.
9. Calculate $u_j^{n+1}$ at every boundary grid point ($j = 1$ and $J$) by using Eq. 14.2.18 and 14.2.20.
10. Write the solution at time level $n$ to disk for data analysis if desired.
11. Increment the time level counter by 1; i.e., set $n$ to $n + 1$.
12. If $n < N$, repeat Steps 7 to 12.

For the implicit method given by Eqs. 14.2.22, 14.2.19, and 14.2.20, the solution algorithm is identical to the one described above for the explicit method except that Step 8 is replaced by the following:

8. Calculate $u_j^{n+1}$ at every interior grid point by solving the linear system of equations given by Eq. 14.2.24.

For the FDEs given by Eqs. 14.2.26 and 14.2.19, and 14.2.20, the solution algorithm depends on the value of $\theta$. If $\theta = 0$ or 1, then the solution algorithms are identical to those just summarized. If $\theta$ greater than 0 but less than 1, then the solution algorithm is identical to one given for the implicit method except that the elements in the matrix **A** and the vector $b$ in Eq. 14.2.24 must be replaced by those corresponding to Eq. 14.2.26 instead of Eq. 14.2.22.

The exact solution given by Eq. 14.2.3 can be used to evaluate the accuracy of the Euler explicit, the Euler implicit, and the Crank-Nicolson methods as well as give guidelines on grid spacing and time-step size needed to obtain solutions of the desired accuracy.

Before leaving this section, we present the *Thomas algorithm*, which is a highly efficient method for solving linear systems of equations with coefficient matrices that are tridiagonal such as Eq. 14.2.24. Such systems of equations are often encountered in CFD. To illustrate the Thomas algorithm, consider the following system of $N$ independent linear equations:

$$\mathbf{A}\,x = b \tag{14.2.27}$$

where

$$\mathbf{A} = \begin{bmatrix} A_1 & B_1 & & & & \\ C_2 & A_2 & B_2 & & & \\ & C_3 & A_3 & B_3 & & \\ & & \ddots & \ddots & \ddots & \\ & & & C_{N-1} & A_{N-1} & B_{N-1} \\ & & & & C_N & A_N \end{bmatrix}, \; x = \begin{bmatrix} x_1 \\ x_2 \\ x_3 \\ \vdots \\ x_{N-1} \\ x_N \end{bmatrix}, \; b = \begin{bmatrix} b_1 \\ b_2 \\ b_3 \\ \vdots \\ b_{N-1} \\ b_N \end{bmatrix} \tag{14.2.28}$$

The elements of $\mathbf{A}$ and $b$ are known, and the elements of $x$ are sought. The first step in the Thomas algorithm is to factor the coefficient matrix $\mathbf{A}$ into two bidiagonal matrices $\mathbf{L}$ and $\mathbf{U}$:

$$\mathbf{A} = \mathbf{LU} = \begin{bmatrix} L_1 & & & & & \\ P_2 & L_2 & & & & \\ & P_3 & L_3 & & & \\ & & \ddots & \ddots & & \\ & & & P_{N-1} & L_{N-1} & \\ & & & & P_N & L_N \end{bmatrix} \begin{bmatrix} 1 & Q_1 & & & & \\ & 1 & Q_2 & & & \\ & & 1 & Q_3 & & \\ & & & \ddots & \ddots & \\ & & & & 1 & Q_{N-1} \\ & & & & & 1 \end{bmatrix} \tag{14.2.29}$$

$$= \begin{bmatrix} L_1 & L_1 Q_1 & & & & \\ P_2 & P_2 Q_1 + L_2 & L_2 Q_2 & & & \\ & P_3 & P_3 Q_2 + L_3 & L_3 Q_3 & & \\ & & \ddots & \ddots & \ddots & \\ & & & P_{N-1} & P_{N-1}Q_{N-2} + L_{N-1} & L_{N-1}Q_{N-1} \\ & & & & P_N & P_N Q_{N-1} + L_N \end{bmatrix} \tag{14.2.30}$$

From the above, it can be seen that the product of the two bidiagonal matrices is a tridiagonal matrix. The next step is to determine the elements of $\mathbf{L}$ and $\mathbf{U}$. This can be accomplished by equating Eq. 14.2.30 to $\mathbf{A}$ of Eq. 14.2.28 and comparing term by term. This gives the following recursive formula for computing the elements in $\mathbf{L}$ and $\mathbf{U}$, provided $L_i \neq 0$ for $i = 1, 2, \ldots, N$:

$$P_i = C_i, \ i = 2, 3, \ldots, N; \ L_1 = A_1; \ Q_1 = B_1/L_1$$
$$L_i = A_i - P_i Q_{i-1} \text{ and } Q_i = B_i/L_i, \ i = 2, 3, \ldots, N \tag{14.2.31}$$

The third step is to substitute the **LU** factorization of **A** into Eq. 14.2.27:

$$\mathbf{A}\,x = \mathbf{L}\,\mathbf{U}\,x = b \tag{14.2.32}$$

and then split as

$$\mathbf{L}\,z = b \tag{14.2.33}$$

$$\mathbf{U}\,x = z \tag{14.2.34}$$

The fourth and final step is to solve for $z$ in Eq. 14.2.33 by forward substitution:

$$z_1 = b_1/L_1; \ z_i = (b_i - P_i z_{i-1})/L_i, \ i = 2, 3, \ldots, N \tag{14.2.35}$$

and to solve for **x** in Eq. 14.2.34 by backward substitution; i.e.,

$$x_N = z_N; \ x_i = z_i - Q_i x_{i+1}, \ i = N - 1, N - 2, \ldots, 1 \tag{14.2.36}$$

This completes the Thomas algorithm, which requires only $(5N - 4)$ arithmetic operations to obtain a solution. If **A** given by Eq. 14.2.28 is diagonally dominant (i.e., $|A_i| > |B_i| + |C_i|$ for $i = 1, 2, \ldots, N$), then it can be shown that round-off errors will not grow, which implies that $N$ can be very large (up to 100 000 or more) and still generate very accurate solutions.

### 14.2.4   Comments on Choice of Difference Operators

From Section 14.2.2, it is clear that the FDE for a given PDE is not unique because a variety of difference operators can be used for each derivative. If the highest-order time derivative is one (which is the case for the continuity, Navier–Stokes, and energy equations), then second-order spatial derivatives represent diffusion, and first-order spatial derivatives represent convection or advection. Since diffusion spreads a disturbance in all directions, second-order spatial derivatives are replaced by central-difference operators. Since convection is directional, upwind and biased-upwind operators are preferred for the first-order spatial derivatives. Upwind and biased-upwind operators are operators whose stencils have more grid points on one side than the other. The side with more grid points is the one from which the flow is coming. For example, if $u > 0$, then a backward or biased-backward difference operator should be used. If $u < 0$, then a forward or biased-forward difference operator should be used. See Hirsch (1991) and Tannehill, et al. (1997) for further discussion.

The difference operator used to replace the time derivative is very important, because it determines whether the FDEs for a PDE at different grid points will be coupled to each other or not (i.e., implicit or explicit). In Section 14.2.2, a generalized time-difference operator involving two time levels is given by Eq. 14.2.25. A generalized time-difference operator involving three time levels is as follows (Beam & Warming (1978)):

$$\frac{u_j^{n+1} - u_j^n}{\Delta t} = \frac{\theta}{1 + \gamma} \left(\frac{\partial u}{\partial t}\right)_j^{n+1} + \frac{(1 - \theta)}{1 + \gamma} \left(\frac{\partial u}{\partial t}\right)_j^n + \frac{\gamma}{1 + \gamma} \left(\frac{u_j^n - u_j^{n-1}}{\Delta t}\right) \quad (14.2.37)$$

where $\theta$ and $\gamma$ are constants specified by the user. The above formula includes Eq. 14.2.25 as a special case ($\gamma = 0$). Similar to Eq. 14.2.25, Eq. 14.2.37 is explicit when $\theta$ equals 0 and implicit otherwise. Also, it is second-order accurate in time when $\theta = \gamma + \frac{1}{2}$ and first-order accurate otherwise. Some commonly used time-difference operators that can be obtained from Eq. 14.2.37 by choosing appropriate values of $\theta$ and $\gamma$ are as follows: Euler explicit ($\theta = \gamma = 0$), Euler implicit ($\theta = 1$ and $\gamma = 0$), Crank-Nicolson ($\theta = 1/2$ and $\gamma = 0$), and three-points backward ($\theta = 1$ and $\gamma = 1/2$).

The time-difference operators in Eqs. 14.2.25 and 14.2.37 belong to the class of "single-stage" operators used to integrate ordinary differential equations (ODEs) that are initial-value problems (IVPs). In fact, any method for integrating ODEs that are IVPs, whether single-stage or multi-stage, can be used or generalized for use as difference operators for the time derivatives in PDEs. Thus, one could use any of the single-stage ODE integration operators given by the *Adams-Bashforth* or the *Adams-Moulton formulas*. The Adams-Bashforth formulas are explicit and can be written as

$$\frac{u_j^{n+1} - u_j^n}{\Delta t} = \sum_{k=0}^{m} \alpha_{mk} \left(\frac{\partial u}{\partial t}\right)_j^{n-k} + O(\Delta t^{m+1}) \quad (14.2.38)$$

The coefficients $\alpha_{mk}$ in the above equation for up to fourth-order accuracy are as follows: For first-order accuracy, $m = 0$ and $\alpha_{00} = 1$, which gives the Euler explicit formula. For second-order accuracy, $m = 1$, $\alpha_{10} = 3/2$, and $\alpha_{11} = -1/2$. For third-order accuracy, $m = 2$, $\alpha_{20} = 23/12$, $\alpha_{21} = -16/12$, and $\alpha_{22} = 5/12$. For fourth-order accuracy, $m = 3$, $\alpha_{30} = 55/24$, $\alpha_{31} = -59/24$, $\alpha_{32} = 37/24$, and $\alpha_{33} = -9/24$.

The Adams-Moulton formulas are implicit and can be written as

$$\frac{u_j^{n+1} - u_j^n}{\Delta t} = \sum_{k=0}^{m} \beta_{mk} \left(\frac{\partial u}{\partial t}\right)_j^{(n+1)-k} + O(\Delta t^{m+1}) \quad (14.2.39)$$

The coefficients $\beta_{mk}$ in the above equation for up to fourth-order accuracy are as follows: For first-order accuracy, $m = 0$ and $\beta_{00} = 1$, which gives the Euler implicit formula. For second-order accuracy, $m = 1$, $\beta_{10} = 1/2$, and $\beta_{11} = 1/2$, which is the *Crank-Nicholson formula*. For third-order accuracy, $m = 2$, $\beta_{20} = 5/12$, $\beta_{21} = 8/12$, and $\beta_{22} = -1/12$. For fourth-order accuracy, $m = 3$, $\beta_{30} = 9/24$, $\beta_{31} = 19/24$, $\beta_{32} = -5/24$, and $\beta_{33} = 1/24$.

When using the Adams-Bashforth and Adams-Moulton formulas, note that the higher-order formulas are not "self-starting" so that lower-order formulas must be used to start the computations. Also, note that the higher the order of accuracy, the greater is the number of time levels involved, which increases the amount of computer memory required since solutions at each of those time levels must be stored. Typically, we want to store solutions at no more than two to three time levels.

As an alternative to the single-stage operators, one could use "multi-stage" IVP, ODE integration operators such as Runge-Kutta. The *Runge-Kutta formulas* are constructed by interpolating and extrapolating previously computed derivatives. All Runge-Kutta methods are explicit and self-starting, regardless of the order of accuracy. One second-order Runge-Kutta operator is as follows:

$$\frac{u_j^{n+1/2} - u_j^n}{\Delta t/2} = \left(\frac{\partial u}{\partial t}\right)_j^n \tag{14.2.40}$$

$$\frac{u_j^{n+1} - u_j^n}{\Delta t} = \left(\frac{\partial u}{\partial t}\right)_j^{n+12} \tag{14.2.41}$$

Note that the $\partial u/\partial t$ terms in Eqs. 14.2.25 and 14.2.37 to 14.2.41 are to be replaced by spatial derivatives according to the PDE as illustrated by Eqs. 14.2.25 and 14.2.26.

In summary, a wide variety of single- and multi-stage operators can be used to approximate the time derivatives in PDEs. Which one to use depends on whether steady or unsteady solutions are sought. When only steady-state solutions are of interest, temporal accuracy is meaningless, and the time-difference operator is chosen to accelerate convergence to steady state. Typically, the lowest-order accurate time implicit operator is used, which is the Euler implicit operator. When unsteady solutions are of interest, either explicit or implicit operators could be used. The order of accuracy in time is typically two for single-stage difference operators to avoid self-starting problems and increased memory requirements. For multi-stage operators such as Runge-Kutta, order of accuracy can be higher than two without increasing memory requirements, though requirements in CPU time do increase somewhat.

## 14.3 STABILITY, CONVERGENCE, AND ERROR

In the previous section, a number of FDEs were obtained for the PDE given by Eq. 14.2.1. The question now is how can we judge whether a FDE is an algebraic analog of a PDE? Also, how can we be sure that round-off errors do not grow when the solution is advanced from time level to time level? Finally, how can we assess the errors in the computed solutions?

The answers to these questions can be found by examining consistency, numerical stability, convergence, and numerical errors, which are explained in the next four subsections. To facilitate the presentation of these concepts, we define the following:

$u_j^n$ = exact solution of PDE at grid point $y_j$ and time level $t^n$

$U_j^n$ = exact solution of the FDE at grid point $y_j$ and time level $t^n$ (i.e., round-off errors do not exist)

$N_j^n$ = numerical solution of the FDE at grid point $y_j$ and time level $t^n$ (i.e., round-off errors exist)

Based on the above definitions, the total error $E_j^n$ of a solution at grid point $y_j$ and time level $t^n$ is given by

$$E_j^n = u_j^n - N_j^n = (u_j^n - U_j^n) + (U_j^n - N_j^n) \qquad (14.3.1)$$

From the above equation, it can be seen that the total error is made up of two parts. The first part, $u_j^n - U_j^n$, is the error from discretizing the domain and the PDE. Hence, it is referred to as the discretization error:

$$u_j^n - U_j^n = \text{discretization error} \qquad (14.3.2)$$

The second part, $U_j^n - N_j^n$, is concerned with the propagation of round-off errors. This error is referred to as the stability error:

$$U_j^n - N_j^n = \text{stability error} \qquad (14.3.3)$$

Note that the stability error depends only on the FDE, and is independent of the PDE that it is supposed to represent.

The truncation error, TE, of a FDE is defined as

$$\text{TE} = \text{FDE} - \text{PDE} \qquad (14.3.4)$$

## 14.3.1   Consistency

An FDE is said to be consistent if for every $j$ and $n$, the following is true:

$$\lim_{\Delta y, \Delta t \to 0} \text{TE}_j^n = 0 \quad \text{or} \quad \lim_{\Delta y, \Delta t \to 0} \text{FDE}_j^n = \text{PDE}_j^n \qquad (14.3.5)$$

Thus, consistency measures how well an FDE approximates a PDE in the limit of zero grid spacing and time-step size.

To illustrate how consistency is analyzed, consider the FDE given by Eq. 14.2.17 for the PDE given by Eq. 14.2.1. For convenience, that FDE is repeated:

$$\frac{u_j^{n+1} - u_j^n}{\Delta t} = \nu \frac{u_{j+1}^n - 2u_j^n + u_{j-1}^n}{\Delta y^2} \qquad (14.3.6)$$

Since the FDE given above is algebraic, it is necessary to convert it into a PDE before it can be substituted into Eq. 14.3.5. This can be accomplished by expanding all terms in the FDE about a common grid point and time level by using Taylor series. For Eq. 14.3.6, we choose to expand $u_j^{n+1}$, $u_{j+1}^n$, and $u_{j-1}^n$ about $u_j^n$:

$$u_j^{n+1} = u_j^n + \left(\frac{\partial u}{\partial t}\right)_j^n \Delta t + \left(\frac{\partial^2 u}{\partial t^2}\right)_j^n \frac{\Delta t^2}{2!} + \left(\frac{\partial^3 u}{\partial t^3}\right)_j^n \frac{\Delta t^3}{3!} + \cdots \qquad (14.3.7)$$

$$u_{j+1}^n = u_j^n + \left(\frac{\partial u}{\partial y}\right)_j^n \Delta y + \left(\frac{\partial^2 u}{\partial y^2}\right)_j^n \frac{\Delta y^2}{2!} + \left(\frac{\partial^3 u}{\partial y^3}\right)_j^n \frac{\Delta y^3}{3!} + \cdots \qquad (14.3.8)$$

$$u_{j-1}^n = u_j^n - \left(\frac{\partial u}{\partial y}\right)_j^n \Delta y + \left(\frac{\partial^2 u}{\partial y^2}\right)_j^n \frac{\Delta y^2}{2!} - \left(\frac{\partial^3 u}{\partial y^3}\right)_j^n \frac{\Delta y^3}{3!} + \cdots \qquad (14.3.9)$$

Substituting the above three equations into Eq. 14.3.6 yields

$$\left(\frac{\partial u}{\partial t}\right)_j^n + \left(\frac{\partial^2 u}{\partial t^2}\right)_j^n \frac{\Delta t}{2!} + \left(\frac{\partial^3 u}{\partial t^3}\right)_j^n \frac{\Delta t^2}{3!} + \cdots = \nu\left[\left(\frac{\partial^2 u}{\partial y^2}\right)_j^n + 2\left(\frac{\partial^4 u}{\partial y^4}\right)_j^n \frac{\Delta y^2}{4!} + \cdots\right]$$
$$(14.3.10)$$

which becomes in the limit as $\Delta t$ and $\Delta y$ approach zero

$$\left(\frac{\partial u}{\partial t} = \nu \frac{\partial^2 u}{\partial y^2}\right)_j^n \qquad (14.3.11)$$

Since the above equation is identical to the PDE given by Eq. 14.3.1 at $(y_j, t^n)$, the FDE given by Eq. (14.3.6) is said to be consistent.

On consistency, it is important to differentiate between unconditional and conditional consistency. Unconditional consistency implies that the FDE approaches the PDE regardless of how $\Delta t$ and $\Delta y$ approach zero, which is the case for Eq. 14.3.10. Conditional consistency refers to FDEs that will approach the PDE only if $\Delta t$ and $\Delta y$ approach zero in a certain prescribed fashion (e.g., $\Delta t$ must approach zero first before $\Delta y$ can approach zero). A conditionally consistent FDE may approach a PDE that is fundamentally different from the PDE that it is intended to represent as the grid spacing and time-step size are reduced (see Problem 14.10 for an example). Thus, for a conditionally consistent FDE, reducing the grid spacing or the time-step size could increase instead of decrease errors because the FDE is approaching a different PDE. Also, note that any term multiplied by $\Delta x^p$ or $\Delta t^p$ (e.g., $\Delta y^4 u_i^n$, $\Delta t^2 u_i^n$) can be added to any FDE without violating consistency. Finally, since $\Delta y$ and $\Delta t$ are never zero, a consistent FDE is very different from the original PDE (contrast Eqs. 14.3.10 and 14.3.11). Thus, consistency, though important, is clearly inadequate in ensuring accuracy.

## 14.3.2 Numerical Stability

An FDE is said to be stable if the stability error given by Eq. 14.3.3 approaches zero or is bounded as $n$ approaches infinity. That is, the round-off error either decays or does not grow as the solution is advanced from time level to time level. Note that this definition is only valid for PDEs with a time-like coordinate. Also, note that stability depends only on the FDE and is independent of the PDE that the FDE is intended to represent.

It turns out that the stability of a FDE depends on whether some parameter $\phi$ relating grid spacing and time-step size is satisfied or not:

$$\phi \le \phi_{cr} \tag{14.3.12}$$

As an example, for the explicit Euler method given by Eq. 14.2.18, stability is ensured if $\Delta y$ and $\Delta t$ satisfy the following criteria:

$$\frac{\nu \Delta t}{\Delta y^2} \le \frac{1}{2} \tag{14.3.13}$$

Thus, $\phi = \nu \, \Delta t / \Delta y^2$ and $\phi_{cr} = \frac{1}{2}$. For this equation, if the grid spacing is $\Delta y_1$, then the largest time-step size permitted is $\Delta t_{max} = (\Delta y_1)^2 / 2\nu$. If $\Delta t$ used is greater than $\Delta t_{max}$, then the solution will oscillate wildly from grid point to grid point with the amplitude of the oscillations increasing from time level to time level so that "blow-up" will eventually take place (the numbers get so large that the computer can no longer represent them). Such solutions in which the round-off errors keep growing are said to be unstable. If $\Delta t$ used is less than or equal to $\Delta t_{max}$, then the solution obtained will either have round-off errors that approach zero or stay bounded. Such solutions are said to be stable. Note, however, though stable solutions are free of round-off errors, they may still be inaccurate from discretization errors (see Eqs. 14.3.1 and 14.3.2), a topic to be addressed in Section 14.3.4.

For the stability criterion given by Eq. 14.3.13, $\phi_{cr} = 1/2$, which is finite and greater than zero. FDEs, whose stability criteria have finite and nonzero $\phi_{cr}$, are said to be conditionally stable. FDEs of most explicit methods are conditionally stable. If $\phi_{cr}$ is infinite, then the FDE is said to be unconditionally stable. For such FDEs, there are no restrictions on the time-step size as far as stability is concerned. For accuracy, however, the time-step size should be small enough to resolve the temporal physics. FDEs of most implicit methods are unconditionally stable (this can only be proven if the PDE is linear). If $\phi_{cr}$ equals zero, then the FDE is said to be unconditionally unstable since for a given grid spacing, there is no time-step size, other than zero, that can be used. Clearly, unconditionally unstable FDEs are useless. Problem 14.14 shows a method that appears reasonable but is unconditionally unstable.

There are many ways to analyze the numerical stability of FDEs. One popular and versatile method is the *Fourier method*. The Fourier method analyzes numerical stability via the following three steps. First, introduce an arbitrary disturbance at every grid point at some arbitrary time level $t^n$. This disturbance represents a random round-off error. Next, expand that disturbance into a Fourier series. Finally, follow the evolution of each Fourier component separately. If any Fourier component grows, then the FDE is said to be unstable. If none grow, then the FDE is said to be stable.

To illustrate the Fourier method for analyzing stability, consider the FDE given by Eq. 14.2.17, which is repeated for convenience:

$$\frac{u_j^{n+1} - u_j^n}{\Delta t} = \nu \frac{u_{j+1}^n - 2u_j^n + u_{j-1}^n}{\Delta y^2} \tag{14.3.14}$$

As mentioned, the first step is to introduce an arbitrary disturbance into the solution at every grid point at $t^n$. Doing so yields

$$\frac{(u + \varepsilon)_j^{n+1} - (u + \varepsilon)_j^n}{\Delta t} = \nu \frac{(u + \varepsilon)_{j+1}^n - 2(u + \varepsilon)_j^n + (u + \varepsilon)_{j-1}^n}{\Delta y^2} \qquad (14.3.15)$$

Subtracting Eq. 14.3.14 from Eq. 14.3.15 gives an evolution equation for the disturbance, $\varepsilon$:

$$\frac{\varepsilon_j^{n+1} - \varepsilon_j^n}{\Delta t} = \nu \frac{\varepsilon_{j+1}^n - 2\varepsilon_j^n + \varepsilon_{j-1}^n}{\Delta y^2} \qquad (14.3.16)$$

The next step is to represent the disturbance in a Fourier series. Since the domain has a finite number of equally distributed grid points (see Fig. 14.2), the most general Fourier series is finite and is given by

$$\varepsilon_j^n = \sum_{m=-M}^{M} A_m^n \exp\left(Ik_m x_j\right) \qquad (14.3.17)$$

where $I = \sqrt{-1}$ is the imaginary number; $M = (IL - 1)/2$ is the maximum number of Fourier components that can be resolved on a grid with $IL$ equally spaced grid points; and $k_m = \pi m/(M \Delta y)$ is the wave number, which is related to the wavelength by $\lambda_m = 2\pi/k_m$. By substituting Eq. (14.3.17) into Eq. (14.3.16) and by noting that $y_{j \pm 1/2} = y_j \pm \Delta y$ and $\exp(y_{j \pm 1/2}) = \exp(y_j)\exp(\pm \Delta y)$, we obtain

$$\sum_{m=-M}^{M} \left[ \frac{(A_m^{n+1} - A_m^n)e^{Ik_m y_j}}{\Delta t} - \nu \frac{(A_m^n e^{Ik_m \Delta y} - 2A_m^n + A_m^n e^{-Ik_m \Delta y})e^{Ik_m y_j}}{\Delta y^2} \right] = 0 \qquad (14.3.18)$$

Since there are no interactions between the Fourier components (i.e., no multiplication of two different Fourier components), Eq. (14.3.18) requires every Fourier component to be zero. This coupled with $\exp(\pm I\phi) = \cos \phi \pm I \sin \phi$ gives

$$\begin{aligned}
A_m^{n+1} &= A_m^n + \frac{\nu \Delta t}{\Delta y^2} A_m^n \left(e^{Ik_m \Delta y} - 2 + e^{-Ik_m \Delta y}\right) \\
&= A_m^n \left[ 1 - 2\frac{\nu \Delta t}{\Delta y^2}(1 - \cos(k_m \Delta y)) \right] \\
&= A_m^n \left[ 1 - 4\frac{\nu \Delta t}{\Delta y^2} \sin^2\left(\frac{k_m \Delta y}{2}\right) \right]
\end{aligned} \qquad (14.3.19)$$

Now, define the amplification factor as

$$G_m = A_m^{n+1}/A_m^n \qquad (14.3.20)$$

The third and final step is to follow the evolution of the amplification factor as $n$ approaches infinity. For numerical stability, it is clear that the following condition must be satisfied for every Fourier component:

$$|G_m| \leq 1 \qquad (14.3.21)$$

From Eqs. (14.3.19) and (14.3.20), the above equation implies the following for every $m$:

$$-1 \leq 1 - 4\frac{\nu \Delta t}{\Delta y^2} \sin^2 \left(\frac{k_m \Delta y}{2}\right) \leq 1, \qquad (14.3.22)$$

This inequality becomes critical when $\sin^2(k_m \Delta y / 2)$ reaches a maximum of unity. At this critical point, we get the correct stability criteria, which is (see Eq. (14.3.13))

$$\frac{\nu \Delta t}{\Delta y^2} \leq \frac{1}{2} \qquad (14.3.23)$$

From the above example, we make a number of observations. First, the most unstable Fourier component is $m = M$, which corresponds to the largest wave number, but shortest wavelength. This turns out to be generally true for most FDEs. Second, this analysis assumes periodic boundary conditions because of the use of Fourier series. Thus, it cannot account for FDEs on the boundaries. If FDEs at boundary grid points can impose a more stringent stability criteria, then the matrix stability method can be used. Third, this analysis can be applied only to linear FDEs since only then can each Fourier component be analyzed separately. Nonlinear FDEs must be linearized before they can be analyzed. Finally, $A_m^n$ in the Fourier series expansion of $\varepsilon_m^n$ in Eq. (14.3.17) can be replaced by $(G_m)^{(n)}A_m^0$ because $A_m^{n+1} = G_m A_m^n = G_m G_m A_m^{n-1} = \cdots = (G_m)^{(n+1)} A_m^0$, where $(n)$ and $(n+1)$ denote to the power of (see Eq. (14.3.20)). For methods involving more than two time levels, this will be more convenient. For FDEs that involve two dimensions, replace Eq. (14.3.17) by

$$\varepsilon_{ij}^n = \sum_{m_x = -M_x}^{M_x} \sum_{m_y = -M_y}^{M_y} A_{m_x m_y}^n \exp\left[I(k_{m_x} x_i + k_{m_y} y_j)\right] \qquad (14.3.24)$$

The three-dimensional form of the above equation is a straightforward extension.

### 14.3.3   Convergence

A FDE is said to be convergent if its solutions approach the exact solution of the PDE in the limit of infinitesimal $\Delta y$ and $\Delta t$:

$$\lim_{\Delta y, \Delta t \to 0} N_j^n = u_j^n \qquad (14.3.25)$$

Equation (14.3.25) indicates that for a convergent FDE, one can get a numerical solution to any desired accuracy by simply reducing the grid spacing and time-step size. For FDEs of linear PDEs, convergence can always be analyzed analytically though the process can be tedious. Fortunately, there is a general theory known as the *Lax equivalence theorem*, which states that *an FDE of a well-posed, linear PDE is convergent, if that FDE is consistent and stable.* With this theorem, an FDE of a well-posed, linear PDE is guaranteed to be convergent if we can show that it is consistent and stable, as explained in Sections 14.3.1 and 14.3.2. Since consistency and stability are much easier to analyze than convergence, the Lax equivalence theorem is quite useful.

For FDEs of PDEs that are not linear (e.g., quasi-linear or nonlinear PDEs), we do not know how to prove convergence. But, this inability to prove convergence does not imply that these FDEs are not convergent. For these FDEs, the Lax equivalence theorem is still useful, except it becomes a necessary but not a sufficient condition for convergence. For these FDEs, a comparison between numerical solutions and experimental measured values may be needed.

### 14.3.4 Numerical Errors

From the discussions on consistency, numerical stability, and convergence given in the previous three subsections, we note the following. First, if is stable, then the total error given by Eq. (14.3.1) reduces to

$$E_j^n = u_j^n - N_j^n = (u_j^n - U_j^n) + (U_j^n - N_j^n) \approx u_j^n - U_j^n \quad (14.3.26)$$

That is, discretization error dominates with stability error being negligible. Second, when grid spacing and time-step size are finite, which is always the case, then consistency analyses show the FDEs to differ considerably from the original PDEs that they are supposed to represent (compare Eqs. (14.3.10) and (14.3.11)). It turns out that the FDEs are, in general, much more complex. They contain spurious modes (solutions not contained in the original PDEs). Thus, even if an FDE is convergent, we still need to ask the following questions to ensure that physically meaningful solutions are obtained:

- What are some of the most important properties that an FDE must possess when grid spacing and time-step sizes are finite?
- How does one choose the grid spacing and the time-step size to ensure accurate solutions?

The answers to some of these questions are briefly described below.

**Conservative Property.** One of the most important properties of PDEs governing fluid-flow problems is the conservation principle. For example, mass cannot be created or destroyed; momentum must be balanced; and total energy is conserved. Thus, it is important for the FDEs to represent this principle correctly. Fortunately, this is easy to do. Just make sure that at every common face between two abutting cells, what leaves one cell must enter the other. But, to do this requires the PDE to be written in what is referred to as the conservative form. A PDE is written in conservative form if all coefficients outside of the outermost derivatives are constants. If any coefficient is a variable, then that PDE is written in non-conservative form. Thus, for example, $\partial u/\partial t + \frac{1}{2}\partial u^2/\partial x = 0$ is written in conservative form, but $\partial u/\partial t + u\partial u/\partial x = 0$ is not. Problems 14.16 and 14.17 illustrate difficulties associated with using PDEs cast in non-conservative form.

**Transportive Property.**  For a fluid, a disturbance at any location in the flow field can spread to other locations by convection due to fluid motion, by diffusion due to random molecular motion, and by pressure waves. Convection can transport a disturbance only in the direction of fluid velocity. Diffusion and pressure waves can spread a disturbance in every direction. FDEs should possess the same transportive properties as the PDEs that they are intended to represent. Otherwise, there is transportive error. It turns out that this is not easy to do correctly. As noted in Section 14.2.4, derivatives representing convection are upwind or biased-upwind differenced, and derivatives representing diffusion and pressure are centrally differenced. For compressible flows, the convective and pressure terms should be treated collectively, and differencing is based on characteristic theory (see Hirsch (1991)). Unfortunately, the theory on upwind differencing is rigorous only for one-dimensional flows.

**Dissipative Property.**  One major error in many FDEs is excessive artificial or numerical diffusion (i.e., the "effective" diffusion coefficients such as viscosity or conductivity in the FDEs are much higher than the ones in the PDEs). This can cause shear layers, viscous forces, surface heat transfer, and conversion of mechanical to thermal energy to be computed incorrectly. There are two main culprits of high numerical diffusion. The first is how convective terms are differenced. If upwind differencing is used, then first-order formulas can produce excessive numerical diffusion. If second-order upwind differencing formulas are used, then reasonable results can be obtained. High-resolution differencing schemes such as flux-difference splitting with limiters are aimed at getting the transportive properties correct with minimal artificial diffusion (see Hirsch (1991)). The second culprit is having cells with high aspect ratios and not aligning the grid lines with the flow directions.

**Grid-Independent Solution.**  Even for a convergent set of FDEs, many different solutions can be obtained, some totally wrong, some qualitatively correct but quantitatively wrong, and some sufficiently accurate. Which solution is obtained depends on the grid spacing and time-step size used. Thus, once generating a solution on a grid, it is important to generate another solution on a finer grid and correspondingly smaller time-step size. Ideally, this process should be repeated until the relative difference between successive solutions obtained on finer and finer grids approach the level of the round-off errors. Unfortunately, for complicated engineering problems, it is often not feasible to obtain grid-independent solutions because of resource constraints. If the solution generated is not grid-independent, then its interpretation must be handled with extreme care.

## 14.4  SOLUTION OF COUETTE FLOW

Let us consider the Couette flow problem of Section 14.2. The governing differential equation is

$$\frac{\partial u}{\partial t} = \nu \frac{\partial^2 u}{\partial y^2}$$

$$(14.4.1)$$

with boundary conditions

$$u = V_0 \quad \text{at} \quad y = 0 \tag{14.4.2}$$
$$u = 0 \quad \text{at} \quad y = H \tag{14.4.3}$$

and initial condition

$$u = 0 \quad \text{at} \quad t = 0 \tag{14.4.4}$$

First, non-dimensionalize the problem by defining the following dimensionless variables:

$$Y = \frac{y}{H} \quad \theta = v\frac{t}{H^2} \quad U = \frac{u}{V_0} \tag{14.4.5}$$

The differential equation is then written as

$$\frac{\partial U}{\partial \theta} = \frac{\partial^2 U}{\partial Y^2} \tag{14.4.6}$$

Similarly, the boundary and initial conditions become

$$U = 1 \quad \text{at} \quad Y = 0 \tag{14.4.7}$$
$$U = 0 \quad \text{at} \quad Y = 1 \tag{14.4.8}$$
$$U = 0 \quad \text{at} \quad \theta = 0 \tag{14.4.9}$$

Using finite differences (see Table 14.1) and the explicit method described in Section 14.2, Eq. 14.4.6 becomes (see Eq. 14.2.17)

$$\frac{U_j^{n+1} - U_j^n}{\Delta \theta} = \frac{U_{j+1}^n - 2U_j^n + U_{j-1}^n}{\Delta Y^2} \tag{14.4.10}$$

Rearranging the above equation for $U_j^{n+1}$, the PDE (Eq. 14.4.1) takes the finite-difference form

$$U_j^{n+1} = RU_{j+1}^n + (1 - 2R)U_j^n + RU_{j-1}^n \tag{14.4.11}$$

where $R = \Delta\theta/\Delta Y^2$. The only unknown in the above equation is $U_j^{n+1}$, since the terms on the right-side of the equation are known from the previous time-step, where the initial values of $U^n$ (for $n = 0$) are known from the initial condition. Hence, the solution is obtained by solving Eq. 14.4.11 at every interior node at every time step. The next example demonstrates the use of finite differences to obtain the solution of a Couette flow.

## Example 14.1

Stagnant water at 20°C is contained between two parallel plates 2 cm apart. The lower plate is suddenly moved with a constant velocity at 0.5 m/s. Solve this Couette flow problem using finite differences and the explicit method. Use 10 space increments and a time interval of 1.6 s. Plot the velocity profile at various values of time and compare with the exact solution given in Eq. 14.2.3.

## Solution

First, calculate $\Delta Y$ and $\Delta \theta$. They are

$$\Delta Y = \frac{\Delta y}{H} = \frac{0.002}{0.02} = 0.1 \quad \Delta \theta = \frac{\nu \Delta t}{H^2} = \frac{10^{-6} \times 1.6}{0.02^2} = 0.004$$

Then $R = 0.004 / 0.01^2 = 0.4$. Note that $R$ must be less than 0.5 (see Eq. 14.3.23) in order for the method to be stable. Since $\Delta Y = 0.1$, as shown in Fig. E14.1, and the domain length (spacing between the plates) is 1, using dimensionless quantities, the number of grid points is $J = 1 + 1 / 0.1 = 11$. However, since the velocity is specified at the boundaries, then $U_1 = 1$ and $U_{11} = 0$. Hence, we only need to solve for the velocity at the interior points ($j = 2, 3, \ldots, 10$). This is done by solving Eq. 14.4.11 for $j = 2, 3, \ldots, 10$.

**Fig. E14.1**

The solution is started by using the initial condition $U_j^0 = 0$ at the grid points. For example, for grid point 2, Eq. 14.4.11 provides

$$U_2^{n+1} = 0.4U_3^n + (1 - 0.8)U_2^n + 0.4U_1^n$$

Starting with $n = 0$ ($\theta = 0$), the above equation yields

$$U_2^1 = 0.4U_3^0 + 0.2U_2^0 + 0.4U_1^0 = 0.4U_1^0 = 0.4$$

The procedure is repeated for the other grid points. After the velocity has been calculated at all grid points, the procedure is repeated for $n = 1$ ($\theta = 0.004$) and then for $n = 2$ and so on until, for example, steady state is achieved. For this example steady state is achieved at $n = 180$. Note that this solution is consistent with the steady-state analytical solution given by $U = Y$ or $u(y) = V_0 (1 - y/H)$. The table that follows lists the solution at the grid points at various values of $\theta$.

| $\theta$ | $j = 1$ | $j = 2$ | $j = 3$ | $j = 4$ | $j = 5$ | $j = 6$ |
|---|---|---|---|---|---|---|
| | $Y = 0$ | $Y = 0.1$ | $Y = 0.2$ | $Y = 0.3$ | $Y = 0.4$ | $Y = 0.5$ |
| 0.000 ($n = 0$) | 1.000 | 0.000 | 0.000 | 0.000 | 0.000 | 0.000 |
| 0.064 ($n = 16$) | 1.000 | 0.784 | 0.583 | 0.410 | 0.271 | 0.167 |
| 0.140 ($n = 35$) | 1.000 | 0.851 | 0.707 | 0.573 | 0.451 | 0.344 |
| 0.280 ($n = 70$) | 1.000 | 0.888 | 0.777 | 0.669 | 0.563 | 0.461 |
| 0.716 ($n = 180$) | 1.000 | 0.900 | 0.800 | 0.700 | 0.600 | 0.500 |

| $\theta$ | $j = 7$ $Y = 0.6$ | $j = 8$ $Y = 0.7$ | $j = 9$ $Y = 0.8$ | $j = 10$ $Y = 0.9$ | $j = 11$ $Y = 0.10$ |
|---|---|---|---|---|---|
| 0.000 ($n = 0$) | 0.000 | 0.000 | 0.000 | 0.000 | 0.000 |
| 0.064 ($n = 16$) | 0.096 | 0.051 | 0.024 | 0.010 | 0.000 |
| 0.140 ($n = 35$) | 0.252 | 0.175 | 0.109 | 0.052 | 0.000 |
| 0.280 ($n = 70$) | 0.363 | 0.269 | 0.109 | 0.088 | 0.000 |
| 0.716 ($n = 180$) | 0.400 | 0.300 | 0.200 | 0.100 | 0.000 |

The analytical solution given in Eq. 14.2.3 is written in dimensionless form as

$$U = 1 - Y - \sum_{n=1}^{\infty} \frac{2}{n\pi} \sin(n\pi Y) \exp(-n^2\pi^2\theta)$$

The table shows a comparison between values of the velocity at $Y = 0.3$ obtained from the exact solution and the finite difference solution for different values of $\theta$. The numbers show that the finite difference solution is reasonably accurate. The percent error is the difference between the two solutions expressed as a percentage of the exact solution.

| $\theta$ | $Y = 0.3$ Finite Difference | $Y = 0.3$ Exact Solution | Difference | Percent Error |
|---|---|---|---|---|
| 0.064 | 0.410 | 0.402 | 0.008 | 1.99 |
| 0.140 | 0.573 | 0.569 | 0.004 | 0.70 |
| 0.280 | 0.669 | 0.668 | 0.001 | 0.15 |
| 0.716 | 0.700 | 0.700 | 0.000 | 0.00 |

Figure E14.2 shows additional comparisons between the exact and the finite difference solutions for the entire domain and the same number of time steps. The solid lines are for the exact solution, and the dots are for the finite difference solution.

**Fig. E14.2**

## 14.5  SOLUTION OF TWO-DIMENSIONAL STEADY-STATE POTENTIAL FLOW

Two-dimensional steady flow of an incompressible non-viscous fluid satisfies Laplace's Equation; using the stream function $\psi$ (see Eq. 8.5.10), it is

$$\frac{\partial^2 \psi}{\partial x^2} + \frac{\partial^2 \psi}{\partial y^2} = 0 \qquad (14.5.1)$$

Using the notation $\psi_{i,j} = \psi(x_i, y_j)$ and central differencing (refer to Table 14.1) for the second derivative, Eq. 14.5.1 is represented in discretized form by

$$\frac{\psi_{i+1,j} - 2\psi_{i,j} + \psi_{i-1,j}}{\Delta x^2} + \frac{\psi_{i,j+1} - 2\psi_{i,j} + \psi_{i,j-1}}{\Delta y^2} = 0 \qquad (14.5.2)$$

Rearranging the above equation yields

$$2\psi_{i,j}\left(\frac{1}{\Delta x^2} + \frac{1}{\Delta y^2}\right) = \frac{\psi_{i+1,j} + \psi_{i-1,j}}{\Delta x^2} + \frac{\psi_{i,j+1} + \psi_{i,j-1}}{\Delta y^2} \qquad (14.5.3)$$

Eq. 14.5.3 shows that the value of $\psi_{i,j}$ is related to the values of the stream function at the surrounding points, as shown in Figure 14.5.

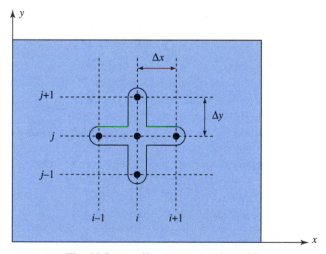

**Fig. 14.5**  A uniform rectangular grid.

Equation (14.5.3) is written for each of the $N$ grid points selected. This results in a system of $N$ linear algebraic equations that can be represented in matrix form as

$$\mathbf{A}\boldsymbol{\psi} = \mathbf{b} \qquad (14.5.4)$$

where

$$
\mathbf{A} = \begin{bmatrix} a_{11} & a_{12} & \cdots & a_{1N} \\ a_{21} & a_{22} & \cdots & a_{2N} \\ \vdots & \vdots & \vdots & \vdots \\ a_{N1} & a_{N2} & \cdots & a_{NN} \end{bmatrix}, \quad \boldsymbol{\psi} = \begin{bmatrix} \psi_1 \\ \psi_2 \\ \vdots \\ \psi_N \end{bmatrix}, \quad \mathbf{b} = \begin{bmatrix} b_1 \\ b_2 \\ \vdots \\ b_N \end{bmatrix}
$$

Note that in general, each row will have at most five nonzero elements. Hence, matrix $\mathbf{A}$ will be a sparse matrix with many zero elements. The system can be solved using either direct or iterative methods. An example of a direct method is matrix inversion, where the solution vector is expressed as

$$
\boldsymbol{\psi} = \mathbf{A}^{-1}\mathbf{b} \tag{14.5.5}
$$

However, this method is suitable for systems with a small number of equations and is computationally inefficient for large numbers of equations since it requires excessive computer resources. For large systems, iterative methods are more suitable to use since they are more computationally efficient than direct methods.

Consider the *Gauss-Seidel iterative method*. It is one of the most efficient methods for solving a system of linear algebraic equations resulting from the finite difference approximation of an elliptic partial differential equation. The rate of convergence of the method is maximized when the coefficient matrix $\mathbf{A}$ is diagonally dominant; that is, if

$$
|a_{ii}| \geq \sum_{\substack{j=1 \\ j \neq 1}}^{N} |a_{ij}| \quad \text{for all } i, \text{ and if} \quad |a_{ij}| > \sum_{\substack{j=1 \\ j \neq 1}}^{N} |a_{ij}| \quad \text{for at least one } i
$$

This means that a sufficient condition for convergence is to require for each equation that the coefficient on the diagonal be greater than or equal to the sum of the other coefficients in the equation. The "greater than" condition must hold at least for one equation (node point near a boundary).

The method may still converge even if the system is not diagonally dominant, but the rate of convergence will be slower. Fortunately, the differencing of Eq. 14.5.1 produces a diagonally dominant coefficient matrix, such that in Eq. 14.5.3 the coefficient of $\psi_{i,j}$ is always larger than the coefficients of the other unknowns in the equation.

In order to demonstrate the iterative procedure, Eq. 14.5.3 is rewritten as

$$
\psi_{i,j}^{k+1} = \frac{\psi_{i+1,j}^{k} + \psi_{i-1,j}^{k+1} + \beta^2(\psi_{i,j+1}^{k} + \psi_{i,j-1}^{k+1})}{2(1 + \beta^2)} \tag{14.5.6}
$$

where $\beta = \Delta x/\Delta y$ and the superscript $k$ represents the iteration level. The procedure for solving the system consists of the following steps:

1. An initial guess is made for all unknown values of $\psi_{i,j}^{k}$ at all grid points. This corresponds to iteration level $k = 0$.

2.   Using Eq. 14.5.6, calculate new values of $\psi_{i,j}^k$ based on the initial guess ($k = 0$) and most recently computed values ($k + 1 = 1$).

3.   Repeat Step 2 until convergence is achieved; that is, the newly computed values are not much different from the values of the previous iteration. This convergence criterion can be expressed as

$$\left| \frac{\psi_{i,j}^{k+1} - \psi_{i,j}^k}{\psi_{i,j}^k} \right| \le \varepsilon \tag{14.5.7}$$

where $\varepsilon$ is a very small number, usually less than $10^{-3}$.

In summary, the Gauss-Seidel procedure is often written as

$$\psi_i^{k+1} = \frac{1}{a_{ii}} \left( b_i - \sum_{j=1}^{i-1} a_{ij} \psi_j^{k+1} - \sum_{j=i+1}^{N} a_{ij} \psi_j^k \right) \tag{14.5.8}$$

where $i = 1, 2, ..., N$ is the node number. Eq. 14.5.8 shows that new values of $\psi_i^{k+1}$ are calculated from the values of $\psi_j^{k+1}$ from the current iteration for $1 \le j \le i - 1$, and values of $\psi_j^k$ from the previous iteration for $i + 1 \le j \le N$.

## Example 14.2

Consider a steady potential flow of air over a square block, as shown in Figure E14.2a. Estimate the streamline pattern and the velocity field around the block.

### Solution

The stream function is described by Eq. 14.5.1. Since the flow is symmetrical we will consider only one quarter of the domain, as shown in Figure E14.2b. This domain extends, far upstream of the block where the flow is uniform and the free stream velocity $U_\infty = 10$ m/s, and far above the block where the flow is expected to be uniform as well. The exact distances where the left and top boundaries need to be located depend on the block dimensions and fluid velocity. For the purpose of demonstrating the method of using finite differences for solving two-dimensional, steady potential flows we will select a domain with a reasonable number of grid points with $\Delta x = \Delta y$, as shown in Figure E14.2c. For this domain, we will specify $\psi = 0$ on the lower boundary; $\psi = U_\infty H$ on the top boundary; $\psi = U_\infty y$ on the left boundary, since the flow is assumed to be uniform; and $\partial \psi / \partial x = 0$, due to symmetry, on the right boundary. The problem is non-dimensionalized by defining the following dimensionless variables:

$$Y = \frac{y}{H}, \quad X = \frac{x}{H}, \quad \Psi = \frac{\psi}{U_\infty H}$$

The dimensionless form of Eq. 14.5.1 is $\partial^2 \Psi / \partial X^2 + \partial^2 \Psi / \partial Y^2 = 0$ with the boundary conditions shown in Figure E14.2.c. Similarly, Eq. 14.5.3 is rewritten, with $\Delta x = \Delta y$, using the dimensionless stream function $\Psi$ as

$$\Psi_{i,j} = \frac{1}{4} \left[ \Psi_{i-1,j} + \Psi_{i,j-1} + \Psi_{i+1,j} + \Psi_{i,j+1} \right]$$

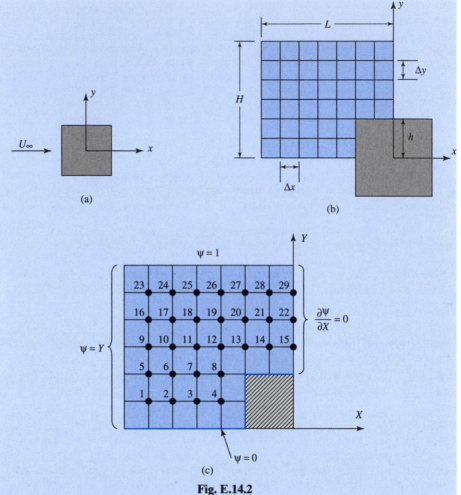

Fig. E.14.2

The grid shown in Figure E14.2c has 29 grid points. In order to demonstrate the method, the difference equation will be written for nodes 1, 9, 11, and 22 only. Similarly, difference equations can be written for the other grid points. Hence,

$$\text{Node 1: } \Psi_2 + \Psi_5 + \Delta Y - 4\Psi_1 = 0$$

$$\text{Node 9: } \Psi_5 + \Psi_{10} + \Psi_{16} + 3\Delta Y - 4\Psi_9 = 0$$

$$\text{Node 11: } \Psi_7 + \Psi_{10} + \Psi_{12} + \Psi_{18} - 4\Psi_{11} = 0$$

$$\text{Node 22: } \Psi_{15} + 2\Psi_{21} + \Psi_{29} - 4\Psi_{22} = 0$$

where we have used central differences to represent the right boundary condition as

$$\frac{\partial \Psi_i}{\partial X} = \frac{\Psi_{i+1} - \Psi_{i-1}}{2\Delta X} = 0 \qquad \therefore \; \Psi_{i+1} = \Psi_{i-1}$$

After writing the difference equations at all 29 grid points, a system of 29 linear algebraic equations results. The system can be represented in matrix form as

$$\mathbf{A}\Psi = \mathbf{b}$$

$$\mathbf{A} = \begin{bmatrix} -4 & 1 & 0 & 0 & 1 & 0 & 0 & \cdots & 0 \\ 1 & -4 & 1 & 0 & 0 & 1 & 0 & \cdots & 0 \\ \vdots & \vdots & \vdots & \vdots & \vdots & \vdots & \vdots & \vdots & \vdots \\ a_{N1} & a_{N2} & a_{N3} & a_{N4} & a_{N5} & a_{N6} & a_{N7} & \cdots & -4 \end{bmatrix} \quad \Psi = \begin{bmatrix} \Psi_1 \\ \Psi_2 \\ \vdots \\ \Psi_N \end{bmatrix} \quad \mathbf{b} = \begin{bmatrix} b_1 \\ b_2 \\ \vdots \\ b_N \end{bmatrix}$$

For the given example the values of the $b_i \cdot$ s are as follows:

$$b_1 = -\Delta Y, b_5 = -2\Delta Y, b_9 = -3\Delta Y, b_{16} = -4\Delta Y, b_{23} = -5\Delta Y - 1, b_{24} \dots b_{29} = -1$$
$$b_2 = b_3 = b_4 = b_6 = b_7 = b_8 = b_{10} \dots b_{15} = b_{17} \dots b_{22} = 0$$

Starting with an initial guess of $\Psi_i = 0$ at all grid points and using the Gauss-Seidel iterative method, the following solution results. The grid point numbers are listed in the shaded columns, and the values of $\Psi$ at the corresponding grid points are listed in the adjacent columns.

| 23 | 0.825 | 24 | 0.815 | 25 | 0.803 | 26 | 0.788 | 27 | 0.774 | 28 | 0.765 | 29 | 0.762 |
|---|---|---|---|---|---|---|---|---|---|---|---|---|---|
| 16 | 0.651 | 17 | 0.632 | 18 | 0.607 | 19 | 0.576 | 20 | 0.542 | 21 | 0.524 | 22 | 0.519 |
| 9 | 0.480 | 10 | 0.455 | 11 | 0.419 | 12 | 0.365 | 13 | 0.295 | 14 | 0.271 | 15 | 0.265 |
| 5 | 0.315 | 6 | 0.290 | 7 | 0.249 | 8 | 0.171 | | | | | | |
| 1 | 0.155 | 2 | 0.140 | 3 | 0.115 | 4 | 0.072 | | | | | | |

The solution is also shown in Fig. E14.2-1 as a plot of the $\Psi$ contours in the domain selected. The plot shows the variation in the values of $\Psi$ as a function of $X$ and $Y$.

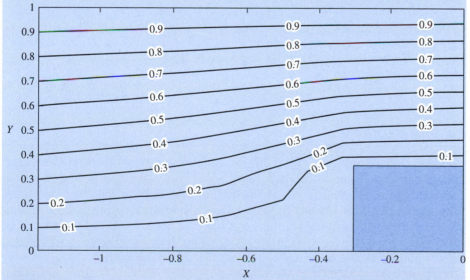

**Fig. E14.2-1.** Contour plot of $\Psi$ in the solution domain. The contour numbers are the values of $\Psi$.

The velocity filed can be obtained by calculating the $u$ and $v$ velocity components using the stream function and Eq. 8.5.11 as follows:

$$u = \frac{\partial \psi}{\partial y} \quad \text{and} \quad v = -\frac{\partial \psi}{\partial x}$$

In terms of dimensionless quantities, the velocity components can be calculated as

$$u = U_\infty \frac{\partial \Psi}{\partial Y} \quad \text{and} \quad v = -U_\infty \frac{\partial \Psi}{\partial X}$$

Using finite differences, the derivatives are approximated as

$$\left. \frac{\partial \Psi}{\partial Y} \right|_{i,j} = \frac{\Psi_{i,j+1} - \Psi_{i,j}}{\Delta Y} \quad \text{and} \quad \left. \frac{\partial \Psi}{\partial X} \right|_{i,j} = \frac{\Psi_{i+1,j} - \Psi_{i,j}}{\Delta X}$$

For example to determine the velocity at grid point 11 we write:

$$u_{11} = U_\infty \left( \frac{\Psi_{18} - \Psi_{11}}{\Delta Y} \right) = 10 \left( \frac{0.607 - 0.419}{1/6} \right) = 11.28 \text{ m/s}$$

$$v_{11} = -U_\infty \left( \frac{\Psi_{12} - \Psi_{11}}{\Delta X} \right) = -10 \left( \frac{0.365 - 0.419}{1/6} \right) = 3.24 \text{ m/s}$$

Similarly, we could have used central differences to calculate the velocity components at grid point 11 as follows:

$$u_{11} = U_\infty \left( \frac{\Psi_{18} - \Psi_{7}}{2\Delta Y} \right) = 10 \left( \frac{0.607 - 0.249}{2/6} \right) = 10.74 \text{ m/s}$$

$$v_{11} = -U_\infty \left( \frac{\Psi_{12} - \Psi_{10}}{2\Delta X} \right) = -10 \left( \frac{0.365 - 0.455}{2/6} \right) = 2.7 \text{ m/s}$$

Note that these calculated values are slightly different from the previous ones, since central difference approximations are more accurate than forward difference approximations. Values of $u$ and $v$ at all grid points can be determined in a similar way using the calculated values of $\Psi$.

## 14.6 SUMMARY

Computational fluid dynamics (CFD) in general involves a three-step process: grid-generation, flow calculation, and post-processing. The grid-generation process is the first step involved in solving fluid-flow problems. It can be a time-consuming and tedious process especially when creating grids for complicated three-dimensional geometries. However, it is worthwhile to spend the time and effort that is needed to generate a well-constructed grid since it greatly improves the accuracy of the results. A poorly constructed grid very often produces poor results. Moreover, many numerical difficulties in obtaining a solution or achieving convergence to a solution are very often due to a poorly constructed grid.

Grids can be of the structured, unstructured, or hybrid form. In a structured grid system, the geometry must be segmented into blocks and each block is meshed in a way so the grid at its boundaries is consistent with the grid at the boundaries of the attaching blocks. The grid elements in structured grids have a rectangular shape in two-dimensional space, such as shown in Figure 14.3a, and are hexahedral in three-dimensional space. Unstructured grids are formed by using triangular elements in 2D, as shown in Figure 14.3b, and tetrahedrals in 3D. Hybrid grids use a combination of structured and unstructured grids. The advantages of using a unstructured grids is that the process can be automated using a computer and hence, can be less time-consuming than generating a structured grid, which is more labor intensive. However, unstructured grids will require more computer memory and CPU time than structured grids. But, the increase in CPU time for unstructured grids may still be less than the work hours needed to generate structured grids. That depends on the solver efficiency though, since in unstructured grids grid point locations and connectivity are more random than in structured grids. Other factors to consider include the flow geometry and the ability to generate dense grids where required, such as in boundary layers. In general, it is easier to use structured grids near solid boundaries and unstructured grids elsewhere in the flow. Readers interested in learning more about these types of grids will find more information on the subject in Thompson et al. (1998).

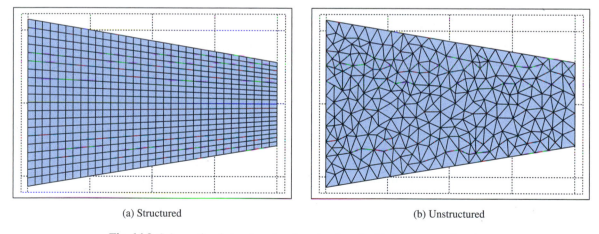

(a) Structured                    (b) Unstructured

**Fig. 14.3** Schematic of structured and unstructured grids in a converging duct.

When used correctly CFD can be a very useful and powerful tool for solving complex fluid-flow problems, especially when experimental methods cannot be used or are not possible. For example, Figure 14.4 shows how CFD was used to determine the effect of adding a shroud to an intake valve to control air flow into an engine. Figure 14.4a shows the velocity vectors around an unshrouded valve, and Figure 14.4b shows velocity vectors for a shrouded valve.

CFD allows solving transient, three-dimensional, compressible or incompressible, laminar or turbulent, reacting or non-reacting flows. CFD can help identify the controlling variables in an engineering system to provide guidelines

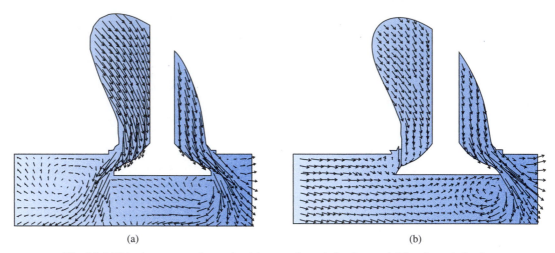

(a)                                                    (b)

**Fig. 14.4**  Velocity vectors shown for (a) an unshrouded valve, and (b) a shrouded valve.

for more rational and hence less costly experimental development efforts. CFD simulations may predict a system behavior over a wide range of design variables and operating conditions and to screen concepts prior to major hardware construction to determine trends, trade-offs, and optimize design and control. Recent developments of computer systems have made it more possible to solve complex fluid-flow problems on today's computers. Computer processing power and memory storage capabilities have improved significantly and continue to do so, which will improve accuracy and reduce computational time. However, the solution obtained using CFD may not be the right answer, and hence it is always important for the CFD user to understand the physics of the problem being solved, using the appropriate governing equations and boundary and initial conditions. Moreover, when solving transient three-dimensional flows, the amount of data generated by CFD can be overwhelming and interpreting the results can be challenging and time-consuming. In addition, the CFD results should always be verified with theoretical or experimental solutions if possible.

## REFERENCES

Beam, R.M. and Warming, R.F., "An Implicit Factored Scheme for Compressible Navier–Stokes Equations," *AIAA Journal*, Vol. 16, 1978, pp. 393–402.

Hirsch, C., *Numerical Computation of Internal and External Flows—Vol 2: Computational Methods for Inviscid and Viscous Flows*, John Wiley, Chichester, Utc, 1991.

Schlichting, H. and Gersten, K., *Boundary Layer Theory*, 8th Revised and Enlarged Ed., Springer, Berlin, 2000.

Tannehill, J.C., Anderson, A.A., and Pletcher, R.H., *Computational Fluid Mechanics and Heat Transfer*, 2nd Ed., Taylor & Francis, Washington, D.C., 1997.

Thompson, J.F., Soni, B., and Weatherill, N., Ed., *Handbook on Grid Generation*, CRC Press, Boca Raton, Fierida, 1998.

Yih, C.-S., *Fluid Mechanics, A Concise Introduction*, McGraw-Hill, New York, 1969.

# PROBLEMS

**14.1**  Expand $y = \exp(x)$ into a Taylor series about $x = 0$. What is the radius of convergence of this series? Suppose that only the first three terms of the series are added. Compare that sum against the exact solution from a calculator evaluation of $\exp(x)$ by computing the relative error for $x = 0.1, 1,$ and $2$. Explain why the error increases as $x$ increases. If you want a 1% relative error in the evaluation of $\exp(x)$, where $x = 1$, how many terms in the series must be included?

**14.2**  Write a computer program to evaluate $\exp(x)$, using single precision (use seven digits to represent a number). Afterwards, use the program to evaluate $\exp(+20)$ and $\exp(-20)$. Explain why the results for $\exp(+20)$ are correct and the results for $\exp(-20)$ are incorrect, when compared to the corresponding solution from the calculator. Note that programming is not trivial. *Hint:* If $n$ is large, $x^{n+1}$ and $(n+1)!$ are huge numbers, but $x^{n+1}/(n+1)!$ may be small. Thus: $x^{n+1}/(n+1)! = [x/(n+1)] [x/n!]$ or $[x/(n+1)] \times$ [previous term].

**14.3**  For $(\partial u/\partial y)$, derive a forward-difference and a backward-difference operator that is second-order accurate in space.

**14.4**  Program the Thomas algorithm given by Eqs. (14.2.28 to 14.2.36). Use the program to compute $\mathbf{A}x = \mathbf{b}$, where $\mathbf{A}$ and $\mathbf{b}$ are

$$\mathbf{A} = \begin{bmatrix} 20 & 5 & \\ 8 & 15 & 4 \\ & 12 & 14 \end{bmatrix} \quad \mathbf{b} = \begin{bmatrix} 1 \\ 2 \\ 3 \end{bmatrix}$$

**14.5**  Solve the following system of equations using the Gauss elimination method or the Thomas algorithm.

$$\begin{aligned} 2x_1 + 2x_2 \qquad\qquad &= 16 \\ 4x_1 - x_2 + 2x_3 \qquad &= 3 \\ 5x_2 - 3x_3 - 6x_4 &= -10 \\ -4x_3 + 3x_4 &= 6 \end{aligned}$$

**14.6**  Derive $\mathbf{A}x = \mathbf{b}$ for Eq. (14.2.26) applied at $j = 2, 3, \ldots, J - 1$ along with Eq. (14.2.19 and 14.2.20).

**14.7**  Derive the FDEs for Eqs. (14.2.1 and 14.2.2) with a second-order Runge-Kutta time differencing formula.

**14.8**  Program the solution algorithm for the explicit method in Section 14.2.3. Generate solutions with $H = 0.1$ m, $V_0 = 0.5$ m/s, $\nu = 10^{-3}$ m²/s. Try several different values of $J$. In all cases, set $\Delta t = \Delta y^2/2\nu$. Compare the solutions generated with Eq. (14.2.3).

**14.9**  Modify the FDEs and the program developed in Problem 14.8 by making the bottom plate oscillate with time; i.e., $u(0, t) = V_0 \sin(2\pi\sigma t)$. Obtain solutions with different values of $\sigma$. Examine the limiting cases of high and low $\sigma$.

**14.10**  Analyze the consistency of Eqs. (14.2.21).

**14.11**  Analyze the consistency of the following FDE:

$$\frac{u_j^{n+1} - u_j^{n-1}}{2\Delta t} = \nu \frac{u_{j+1}^n - (u_j^{n+1} + u_j^{n-1}) + u_{j-1}^n}{\Delta y^2}$$

$$+ O\left(\Delta t^2, \Delta y^2, \frac{\Delta t^4}{\Delta y^2}\right)$$

Show that it can converge to the following PDE as grid spacing and time-step size approach zero:

$$\frac{\partial u}{\partial t} + \alpha \nu \frac{\partial^2 u}{\partial t^2} = \nu \frac{\partial^2 u}{\partial y^2}$$

**14.12**  For the program developed for Problem 14.7, generate solutions for different time-step sizes to examine stability and accuracy (i.e., remove the requirement that $\Delta t = \Delta y^2/2\nu$).

**14.13**  Program the solution algorithm for the implicit method in Section 14.2.3. Generate solutions with $H = 0.1$ m, $V_0 = 0.5$ m/s, $\nu = 10^{-3}$ m²/s. Try several different values of $J$ and $\Delta t$. Compare the solution generated with Eq. (14.3.3). If only the steady-state solution is of interest, then what is the optimum $\Delta t$ (the time-step size that will enable steady-state to be achieved with the fewest time steps)?

**14.14**  Analyze the numerical stability of the FDEs given by Eqs. (14.2.21).

**14.15**  Analyze the numerical stability of the following FDE:

$$\frac{u_j^{n+1} - u_j^{n-1}}{2\Delta t} = \nu \frac{u_{j+1}^n - 2u_j^n + u_{j-1}^n}{\Delta y^2} + O(\Delta t^2, \Delta y^2)$$

$$\frac{u_j^{n+1} - u_j^{n-1}}{2\Delta t} = \nu \frac{u_{j+1}^n - (u_j^{n+1} + u_j^{n-1}) + u_{j-1}^n}{\Delta y^2}$$

$$+ O\left(\Delta t^2, \Delta y^2, \frac{\Delta t^4}{\Delta y^2}\right)$$

**14.16** For the FDE given by Eq. (14.2.26), show that it is unconditionally stable if $\theta \in (0.5, 1)$ and conditionally stable if $\theta \in [0, 0.5]$.

**14.17** Though the PDEs $\partial u/\partial t + \frac{1}{2}\partial u^2/\partial x = 0$ and $\partial u/\partial t + u \partial u/\partial x = 0$ are identical analytically, they differ when finite-differenced. If Euler explicit is used for the time derivative, and first-order backward differencing is used for the spatial derivative, which FDE will maintain the conservative property?

**14.18** Program the FDEs derived in Problem 15.16. Obtain solutions for the following initial conditions: (a) $u = 0$ when $x < -1$ and $> 1$; $u = 1, u = x + 1$ when $x$ is between $-1$ and $0$; $u = -x + 1$ when $x$ is between 0 and 1. (b) $u + 0$, when $x < -1$ and $> 1$; $u = 1$, when $x$ is between $-1$ and 1.

**14.19** Consider a two-dimensional transient fluid flow governed by the equation

$$\frac{\partial u}{\partial t} = \nu \left[ \frac{\partial^2 u}{\partial x^2} + \frac{\partial^2 u}{\partial y^2} \right]$$

Using finite differences, write the discretized form of the above PDE using (a) the explicit

method, (b) the implicit method, and (c) the Crank-Nicolson method. Adopt the notation $U_{i,j}^{(n)} = u(x_i, y_j, t^n)$.

**14.20** The steady-state values associated with selected nodal points of a two-dimensional potential flow are shown on the sketch below. Use $\Delta x = \Delta y = 0.1$ m.
   (a) Write the finite differences equations for nodes 1, 2, and 3.
   (b) Use your result in part (a) and calculate the values of $\psi$ at nodes 1, 2, and 3.

**14.21** Consider a one-dimensional transient fluid-flow system. The table that follows shows the values of the velocity at selected nodes as a function of time. Determine $V_3$ at $t = 20$ s using (a) the explicit method, (b) the implicit method, and (c) the Crank-Nicolson method. Velocities are given in m/s. (Use $R = 0.3$)

| $t$ | $V_1$ | $V_2$ | $V_3$ | $V_4$ |
|---|---|---|---|---|
| 0 | 25 | 25 | 25 | 25 |
| 10 | 100 | 58 | 34 | 27 |
| 20 | 120 | | | 30 |

**14.22** In a two-dimensional steady-state flow system the values of the stream function at selected nodes are listed in the table that follows. An iterative method was used to determine these values up to iteration number 2. Determine $\psi_5$ at the current iteration using the Gauss-Seidel method and ($\Delta x = \Delta y$).

| Iteration Number | $\psi_1$ | $\psi_2$ | $\psi_3$ | $\psi_4$ | $\psi_5$ | $\psi_6$ | $\psi_7$ | $\psi_8$ | $\psi_9$ |
|---|---|---|---|---|---|---|---|---|---|
| 0 | 0.450 | 0.440 | 0.430 | 0.400 | 0.390 | 0.370 | 0.350 | 0.320 | 0.300 |
| 1 | 0.457 | 0.443 | 0.435 | 0.431 | 0.412 | 0.387 | 0.366 | 0.307 | 0.289 |
| 2 | 0.459 | 0.445 | 0.437 | 0.434 | | | | | |

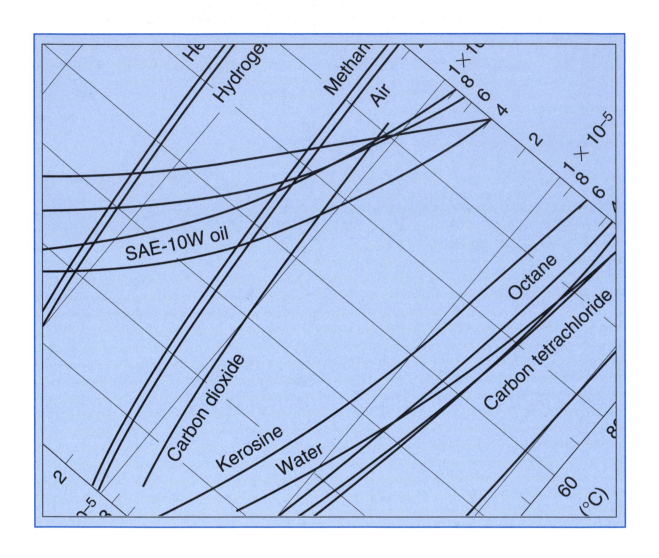

He... Hydrogen ... Methan... Air

$1 \times 10$
8
6
4
2
$1 \times 10^{-5}$
8
6

SAE-10W oil

Octane

Carbon tetrachloride

Carbon dioxide

Kerosine

Water

2

$0^{-5}$

8

60

(°C)

# Appendix

## A.  UNITS AND CONVERSIONS AND VECTOR RELATIONSHIPS

**TABLE A.1**   U.S. Engineering Units, SI Units, and Their Conversion Factors

| Quantity | English units | International system[a] SI | Conversion factor |
|---|---|---|---|
| Length | inch | millimeter | 1 in. = 25.4 mm |
| | foot | meter | 1 ft = 0.3048 m |
| | mile | kilometer | 1 mile = 1.609 km |
| Area | square inch | square centimeter | 1 in$^2$ = 6.452 cm$^2$ |
| | square foot | square meter | 1 ft$^2$ = 0.09290 m$^2$ |
| Volume | cubic inch | cubic centimeter | 1 in$^3$ = 16.39 cm$^3$ |
| | cubic foot | cubic meter | 1 ft$^3$ = 0.02832 m$^3$ |
| | gallon | | 1 gal = 0.003789 m$^3$ |
| Mass | pound-mass | kilogram | 1 lb$_m$ = 0.4536 kg |
| | slug | | 1 slug = 14.59 kg |
| Density | slug/cubic foot | kilogram/cubic meter | 1 slug/ft$^3$ = 515.4 kg/m$^3$ |
| Force | pound-force | newton | 1 lb = 4.448 N |
| Work or torque | foot-pound | newton-meter | 1 ft-lb = 1.356 N·m |
| Pressure | pound/square inch | newton/square meter (pascal) | 1 psi = 6895 Pa |
| | pound/square foot | | 1 psf = 47.88 Pa |
| Temperature | degree Fahrenheit | degree Celsius | $\_\_\_$ °F = $\frac{9}{5}$ $\_\_\_$ °C + 32 |
| | degree Rankine | kelvin | $\_\_\_$ °R = $\frac{9}{5}$ $\_\_\_$ K |
| Energy | British thermal unit | joule | 1 Btu = 1055 J |
| | calorie | | 1 cal = 4.186 J |
| | foot-pound | | 1 ft-lb = 1.356 J |
| Power | horsepower | watt | 1 hp = 745.7 W |
| | foot-pound/second | | 1 ft-lb/sec = 1.356 W |
| Velocity | foot/second | meter/second | 1 ft/sec = 0.3048 m/s |
| Acceleration | foot/second squared | meter/second squared | 1 ft/sec$^2$ = 0.3048 m/s$^2$ |
| Frequency | cycle/second | hertz | 1 cps = 1.000 Hz |
| Viscosity | pound-second/square foot | newton-second/square meter | 1 lb-sec/ft$^2$ = 47.88 N·s/m$^2$ |

[a]The reversed initials in this abbreviation come from the French form of the name: Systéme International.

**TABLE A.2**  Conversions

| Length | Force | Mass | Velocity |
|---|---|---|---|
| 1 cm = 0.3937 in. | 1 lb = 0.4536 kg | 1 oz = 28.35 g | 1 mph = 1.467 ft/sec |
| 1 m = 3.281 ft | 1 lb = 0.4448 × $10^6$ dyn | 1 lb = 0.4536 kg | 1 mph = 0.8684 knot |
| 1 km = 0.6214 mile | 1 lb = 32.17 pdl | 1 slug = 32.17 lb | 1 ft/sec = 0.3048 m/s |
| 1 in. = 2.54 cm | 1 kg = 2.205 lb | 1 slug = 14.59 kg | 1 m/s = 3.281 ft/sec |
| 1 ft = 0.3048 m | 1 pdl = 0.03108 lb | 1 kg = 2.205 lb | 1 km/h = 0.278 m/s |
| 1 mile = 1.609 km | 1 dyn = 2.248 × $10^{-6}$ lb | 1 kg = 0.06852 slug | |
| 1 mile = 5280 ft | 1 lb = 4.448 N | | |
| 1 mile = 1760 yd | 1 N = 0.2248 lb | | |

| Work, energy, and power | Pressure | Volume | Flow rate |
|---|---|---|---|
| 1 Btu = 778.2 ft-lb | 1 psi = 2.036 in. Hg | 1 $ft^3$ = 28.32 L | 1 $ft^3$/min = 4.719 × $10^{-4}$ $m^3$/s |
| 1 J = $10^7$ ergs  1 J = 0.7376 ft-lb | 1 psi = 27.7 in. $H_2O$ | 1 $ft^3$ = 7.481 gal (U.S.) | 1 $ft^3$/sec = 0.02832 $m^3$/s |
| 1 cal = 3.088 ft-lb | 14.7 psi = 22.92 in. Hg | 1 gal (U.S.) = 231 $in^3$ | 1 $m^3$/s = 35.31 $ft^3$/sec |
| 1 cal = 0.003968 Btu | 14.7 psi = 33.93 ft $H_2O$ | 1 gal (Brit.) = 1.2 gal (U.S.) | 1 gal/min = 0.002228 $ft^3$/sec |
| 1 kWh = 3413 Btu | 14.7 psi = 1.0332 kg/$cm^2$ | 1 $m^3$ = 1000 L | 1 $ft^3$/sec = 448.9 gal/min |
| 1 Btu = 1.055 kJ | 14.7 psi = 1.0133 bar | 1 $ft^3$ = 0.02832 $m^3$ | |
| 1 ft-lb = 1.356 J | 1 kg/$cm^2$ = 14.22 psi | 1 $m^3$ = 35.31 $ft^3$ | *Viscosity* |
| 1 hp = 550 ft-lb/sec | 1 in. Hg = 0.4912 psi | | |
| 1 hp = 0.7067 Btu/sec | 1 ft $H_2O$ = 0.4331 psi | | 1 stoke = $10^{-4}$ $m^2$/s |
| 1 hp = 0.7455 kW | 1 psi = 6895 Pa | | 1 poise = 0.1 N·s/$m^2$ |
| 1 W = 1 J/s | 1 psf = 47.88 Pa | | 1 lb-sec/$ft^2$ = 47.88 N·s/$m^2$ |
| 1 W = 1.0 × $10^7$ dyn-cm/s | $10^5$ Pa = 1 bar | | 1 $ft^2$/sec = 0.0929 $m^2$/s |
| 1 erg = $10^{-7}$ J | 1 kPa = 0.145 psi | | |
| 1 quad = $10^{15}$ Btu | | | |
| 1 therm = $10^5$ Btu | | | |

**TABLE A.3**  Vector Relationships

$\mathbf{A} \cdot \mathbf{B} = A_x B_x + A_y B_y + A_z B_z$

$\mathbf{A} \times \mathbf{B} = (A_y B_z - A_z B_y)\hat{\mathbf{i}} + (A_z B_x - A_x B_z)\hat{\mathbf{j}} + (A_x B_y - A_y B_x)\hat{\mathbf{k}}$

If $\mathbf{A} \perp \mathbf{B}, \mathbf{A} \cdot \mathbf{B} = 0$

If $\mathbf{A} \parallel \mathbf{B}, \mathbf{A} \times \mathbf{B} = 0$

gradient operator = $\nabla = \dfrac{\partial}{\partial x}\hat{\mathbf{i}} + \dfrac{\partial}{\partial y}\hat{\mathbf{j}} + \dfrac{\partial}{\partial z}\hat{\mathbf{k}}$

divergence of $\mathbf{V} = \nabla \cdot \mathbf{V} = \dfrac{\partial u}{\partial x} + \dfrac{\partial v}{\partial y} + \dfrac{\partial w}{\partial z}$

curl of $\mathbf{V} = \nabla \times \mathbf{V} = \left(\dfrac{\partial w}{\partial y} - \dfrac{\partial v}{\partial z}\right)\hat{\mathbf{i}} + \left(\dfrac{\partial u}{\partial z} - \dfrac{\partial w}{\partial x}\right)\hat{\mathbf{j}} + \left(\dfrac{\partial v}{\partial x} - \dfrac{\partial u}{\partial y}\right)\hat{\mathbf{k}}$

Laplace's equation = $\nabla^2 \phi = 0$

Irrotational (conservative) vector field: $\nabla \times \mathbf{V} = 0$

Stokes' theorem: $\oint_C \mathbf{V} \cdot d\mathbf{l} = \iint_A (\nabla \times \mathbf{V}) \cdot \hat{\mathbf{n}}\, dA$

Gauss' Theorem: $\oiint_A \mathbf{V} \cdot \hat{\mathbf{n}}\, dA = \iiint_V \nabla \cdot \mathbf{V}\, d\mathcal{V}$

# B.  FLUID PROPERTIES

**TABLE B.1**   Properties of Water

| Temperature $T$ (°C) | Density $\rho$ (kg/m$^3$) | Viscosity $\mu$ (N·s/m$^2$) | Kinematic viscosity $\nu$ (m$^2$/s) | Surface tension $\sigma$ (N/m) | Vapor pressure $p_v$ (kPa) | Bulk modulus $B$ (Pa) |
|---|---|---|---|---|---|---|
| 0 | 999.9 | $1.792 \times 10^{-3}$ | $1.792 \times 10^{-6}$ | 0.0762 | 0.610 | $204 \times 10^7$ |
| 5 | 1000.0 | 1.519 | 1.519 | 0.0754 | 0.872 | 206 |
| 10 | 999.7 | 1.308 | 1.308 | 0.0748 | 1.13 | 211 |
| 15 | 999.1 | 1.140 | 1.141 | 0.0741 | 1.60 | 214 |
| 20 | 998.2 | 1.005 | 1.007 | 0.0736 | 2.34 | 220 |
| 30 | 995.7 | 0.801 | 0.804 | 0.0718 | 4.24 | 223 |
| 40 | 992.2 | 0.656 | 0.661 | 0.0701 | 3.38 | 227 |
| 50 | 988.1 | 0.549 | 0.556 | 0.0682 | 12.3 | 230 |
| 60 | 983.2 | 0.469 | 0.477 | 0.0668 | 19.9 | 228 |
| 70 | 977.8 | 0.406 | 0.415 | 0.0650 | 31.2 | 225 |
| 80 | 971.8 | 0.357 | 0.367 | 0.0630 | 47.3 | 221 |
| 90 | 965.3 | 0.317 | 0.328 | 0.0612 | 70.1 | 216 |
| 100 | 958.4 | $0.284 \times 10^{-3}$ | $0.296 \times 10^{-6}$ | 0.0594 | 101.3 | $207 \times 10^7$ |

**TABLE B.2**   Properties of Air at Atmospheric Pressure

| Temperature $T$ (°C) | Density $\rho$ (kg/m$^3$) | Viscosity $\mu$ (N·s/m$^2$) | Kinematic viscosity $\nu$ (m$^2$/s) | Velocity of sound $c$ (m/s) |
|---|---|---|---|---|
| −50 | 1.582 | $1.46 \times 10^{-5}$ | $0.921 \times 10^{-5}$ | 299 |
| −30 | 1.452 | 1.56 | $1.08 \times 10^{-5}$ | 312 |
| −20 | 1.394 | 1.61 | 1.16 | 319 |
| −10 | 1.342 | 1.67 | 1.24 | 325 |
| 0 | 1.292 | 1.72 | 1.33 | 331 |
| 10 | 1.247 | 1.76 | 1.42 | 337 |
| 20 | 1.204 | 1.81 | 1.51 | 343 |
| 30 | 1.164 | 1.86 | 1.60 | 349 |
| 40 | 1.127 | 1.91 | 1.69 | 355 |
| 50 | 1.092 | 1.95 | 1.79 | 360 |
| 60 | 1.060 | 2.00 | 1.89 | 366 |
| 70 | 1.030 | 2.05 | 1.99 | 371 |
| 80 | 1.000 | 2.09 | 2.09 | 377 |
| 90 | 0.973 | 2.13 | 2.19 | 382 |
| 100 | 0.946 | 2.17 | 2.30 | 387 |
| 200 | 0.746 | 2.57 | 3.45 | 436 |
| 300 | 0.616 | $2.93 \times 10^{-5}$ | $4.75 \times 10^{-5}$ | 480 |

**TABLE B.3** Properties of the Standard Atmosphere

| Altitude (m) | Temperature (K) | Pressure (kPa) | Density (kg/m$^3$) | Velocity of sound (m/s) |
|---|---|---|---|---|
| 0 | 288.2 | 101.3 | 1.225 | 340 |
| 500 | 284.9 | 95.43 | 1.167 | 338 |
| 1 000 | 281.7 | 89.85 | 1.112 | 336 |
| 2 000 | 275.2 | 79.48 | 1.007 | 333 |
| 4 000 | 262.2 | 61.64 | 0.8194 | 325 |
| 6 000 | 249.2 | 47.21 | 0.6602 | 316 |
| 8 000 | 236.2 | 35.65 | 0.5258 | 308 |
| 10 000 | 223.3 | 26.49 | 0.4136 | 300 |
| 12 000 | 216.7 | 19.40 | 0.3119 | 295 |
| 14 000 | 216.7 | 14.17 | 0.2278 | 295 |
| 16 000 | 216.7 | 10.35 | 0.1665 | 295 |
| 18 000 | 216.7 | 7.563 | 0.1216 | 295 |
| 20 000 | 216.7 | 5.528 | 0.0889 | 295 |
| 30 000 | 226.5 | 1.196 | 0.0184 | 302 |
| 40 000 | 250.4 | 0.287 | $4.00 \times 10^{-3}$ | 317 |
| 50 000 | 270.7 | 0.0798 | $1.03 \times 10^{-3}$ | 330 |
| 60 000 | 255.8 | 0.0225 | $3.06 \times 10^{-4}$ | 321 |
| 70 000 | 219.7 | 0.00551 | $8.75 \times 10^{-5}$ | 297 |
| 80 000 | 180.7 | 0.00103 | $2.00 \times 10^{-5}$ | 269 |

**TABLE B.4** Properties of Ideal Gases at 300 K

| Gas | Chemical formula | Molar mass | $R$ $\dfrac{\text{kJ}}{\text{kg} \cdot \text{K}}$ | $c_p$ $\dfrac{\text{kJ}}{\text{kg} \cdot \text{K}}$ | k |
|---|---|---|---|---|---|
| Air | — | 28.97 | 0.287 | 1.004 | 1.40 |
| Argon | Ar | 39.94 | 0.2081 | 0.5203 | 1.667 |
| Carbon dioxide | $CO_2$ | 44.01 | 0.1889 | 0.8418 | 1.287 |
| Carbon monoxide | CO | 28.01 | 0.2968 | 1.041 | 1.40 |
| Ethane | $C_2H_6$ | 30.07 | 0.2765 | 1.766 | 1.184 |
| Helium | He | 4.003 | 2.077 | 5.193 | 1.667 |
| Hydrogen | $H_2$ | 2.016 | 4.124 | 14.21 | 1.40 |
| Methane | $CH_4$ | 16.04 | 0.5184 | 2.254 | 1.30 |
| Nitrogen | $N_2$ | 28.021 | 0.2968 | 1.042 | 1.40 |
| Oxygen | $O_2$ | 32.00 | 0.2598 | 0.9216 | 1.394 |
| Propane | $C_3H_8$ | 44.10 | 0.1886 | 1.679 | 1.12 |
| Steam | $H_2O$ | 18.02 | 0.4615 | 1.872 | 1.33 |

$c_v = c_p - R, k = c_p/c_v$

**TABLE B.5**   Properties of Common Liquids at Atmospheric Pressure and Approximately 16 to 21°C

| Liquid | Specific weight $\gamma$ N/m³ | Density $\rho$ kg/m³ | Surface tension[a] $\sigma$ N/m | Vapor pressure $p_v$ kPa (abs.) |
|---|---|---|---|---|
| Alcohol, ethyl | 7 744 | 789 | 0.022 | — |
| Benzene | 8 828 | 902 | 0.029 | 10.3 |
| Carbon tetrachloride | 15 629 | 1 593 | 0.026 | 86.2 |
| Gasoline | 6 660 | 680 | — | — |
| Glycerin | 12 346 | 1 258 | 0.063 | $1.4 \times 10^{-5}$ |
| Kerosene | 7 933 | 809 | 0.025 | — |
| Mercury | 132 800 | 13 550 | 0.467 | $1.59 \times 10$ |
| SAE 10 oil | 9 016 | 917 | 0.036 | — |
| SAE 30 oil | 9 016 | 917 | 0.035 | — |
| Turpentine | 8 529 | 871 | 0.026 | $5.31 \times 10^{-2}$ |
| Water | 9 810 | 1000 | 0.073 | 2.34 |

[a]In contact with air.

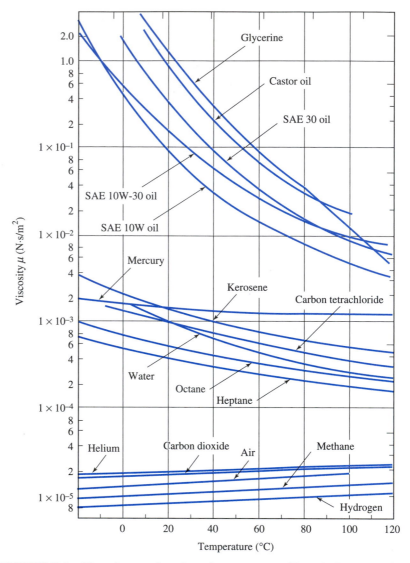

**FIGURE B.1** Viscosity as a function of temperature. (From R. W. Fox and T. A. McDonald, *Introduction to Fluid Mechanics*, 2nd ed., John Wiley & Sons, Inc., New York, 1978.)

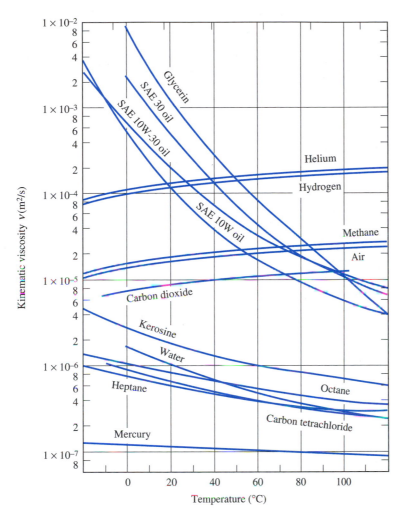

**FIGURE B.2**  Kinematic viscosity (at atmospheric pressure) as a function of temperature. (From R. W. Fox and A. T. McDonald, *Introduction to Fluid Mechanics*, 2nd ed., John Wiley & Sons, Inc., New York, 1978.)

# C. PROPERTIES OF AREAS AND VOLUMES

**TABLE C.1**  Areas

| | Sketch | Area | Centroid | Second moment |
|---|---|---|---|---|
| Rectangle | | $bh$ | $\bar{y} = h/2$ | $\bar{I} = bh^3/12$ <br> $\bar{I}_{xy} = 0$ |
| Triangle | | $bh/2$ | $\bar{y} = h/3$ | $\bar{I} = bh^3/36$ <br> $\bar{I}_{xy} = (b - 2d)bh^3/72$ |
| Circle | | $\pi D^2/4$ | $\bar{y} = r$ | $\bar{I} = \pi D^4/64$ |
| Semicircle | | $\pi D^2/8$ | $\bar{y} = 4r/3\pi$ | $I_x = \pi D^4/128$ |
| Ellipse | | $\pi ab$ | $\bar{y} = b$ | $\bar{I} = \pi ab^3/4$ |
| Semiellipse | | $\pi ab/2$ | $\bar{y} = 4b/3\pi$ | $I_x = \pi ab^3/8$ |

**TABLE C.2**  Volumes

| | Sketch | Surface area | Volume | Centroid |
|---|---|---|---|---|
| Cylinder | | $\pi Dh + \pi D^2/2$ | $\pi D^2 h/4$ | $\bar{y} = h/2$ |
| Sphere | | $\pi D^2$ | $\pi D^3/6$ | $\bar{y} = r$ |
| Cone | | $\pi(r^2 + r\sqrt{r^2 + h^2})$ | $\pi D^2 h/12$ | $\bar{y} = h/4$ |
| Hemisphere | | $3\pi D^2/4$ | $\pi D^2/12$ | $\bar{y} = 3r/8$ |

# D.  COMPRESSIBLE-FLOW TABLES FOR AIR

**TABLE D.1**   Isentropic Flow

| M | $p/p_0$ | $T/T_0$ | $A/A^*$ | M | $p/p_0$ | $T/T_0$ | $A/A^*$ |
|---|---------|---------|---------|---|---------|---------|---------|
| 0 | 1.0000 | 1.0000 | $\infty$ | .96 | .5532 | .8444 | 1.0014 |
| .02 | .9997 | .9999 | 28.9421 | .98 | .5407 | .8389 | 1.0003 |
| .04 | .9989 | .9997 | 14.4815 | 1.00 | .5283 | .8333 | 1.000 |
| .06 | .9975 | .9993 | 9.6659 | 1.02 | .5160 | .8278 | 1.000 |
| .08 | .9955 | .9987 | 7.2616 | 1.04 | .5039 | .8222 | 1.001 |
| .10 | .9930 | .9980 | 5.8218 | 1.06 | .4919 | .8165 | 1.003 |
| .12 | .9900 | .9971 | 4.8643 | 1.08 | .4800 | .8108 | 1.005 |
| .14 | .9864 | .9961 | 4.1824 | 1.10 | .4684 | .8052 | 1.008 |
| .16 | .9823 | .9949 | 3.6727 | 1.12 | .4568 | .7994 | 1.011 |
| .18 | .9776 | .9936 | 3.2779 | 1.14 | .4455 | .7937 | 1.015 |
| .20 | .9725 | .9921 | 2.9635 | 1.16 | .4343 | .7879 | 1.020 |
| .22 | .9668 | .9904 | 2.7076 | 1.18 | .4232 | .7822 | 1.025 |
| .24 | .9607 | .9886 | 2.4956 | 1.20 | .4124 | .7764 | 1.030 |
| .26 | .9541 | .9867 | 2.3173 | 1.22 | .4017 | .7706 | 1.037 |
| .28 | .9470 | .9846 | 2.1656 | 1.24 | .3912 | .7648 | 1.043 |
| .30 | .9395 | .9823 | 2.0351 | 1.26 | .3809 | .7590 | 1.050 |
| .32 | .9315 | .9799 | 1.9219 | 1.28 | .3708 | .7532 | 1.058 |
| .34 | .9231 | .9774 | 1.8229 | 1.30 | .3609 | .7474 | 1.066 |
| .36 | .9143 | .9747 | 1.7358 | 1.32 | .3512 | .7416 | 1.075 |
| .38 | .9052 | .9719 | 1.6587 | 1.34 | .3417 | .7358 | 1.084 |
| .40 | .8956 | .9690 | 1.5901 | 1.36 | .3323 | .7300 | 1.094 |
| .42 | .8857 | .9659 | 1.5289 | 1.38 | .3232 | .7242 | 1.104 |
| .44 | .8755 | .9627 | 1.4740 | 1.40 | .3142 | .7184 | 1.115 |
| .46 | .8650 | .9594 | 1.4246 | 1.42 | .3055 | .7126 | 1.126 |
| .48 | .8541 | .9560 | 1.3801 | 1.44 | .2969 | .7069 | 1.138 |
| .50 | .8430 | .9524 | 1.3398 | 1.46 | .2886 | .7011 | 1.150 |
| .52 | .8317 | .9487 | 1.3034 | 1.48 | .2804 | .6954 | 1.163 |
| .54 | .8201 | .9449 | 1.2703 | 1.50 | .2724 | .6897 | 1.176 |
| .56 | .8082 | .9410 | 1.2403 | 1.52 | .2646 | .6840 | 1.190 |
| .58 | .7962 | .9370 | 1.2130 | 1.54 | .2570 | .6783 | 1.204 |
| .60 | .7840 | .9328 | 1.1882 | 1.56 | .2496 | .6726 | 1.219 |
| .62 | .7716 | .9286 | 1.1657 | 1.58 | .2423 | .6670 | 1.234 |
| .64 | .7591 | .9243 | 1.1452 | 1.60 | .2353 | .6614 | 1.250 |
| .66 | .7465 | .9199 | 1.1265 | 1.62 | .2284 | .6558 | 1.267 |
| .68 | .7338 | .9153 | 1.1097 | 1.64 | .2217 | .6502 | 1.284 |
| .70 | .7209 | .9107 | 1.0944 | 1.66 | .2151 | .6447 | 1.301 |
| .72 | .7080 | .9061 | 1.0806 | 1.68 | .2088 | .6392 | 1.319 |
| .74 | .6951 | .9013 | 1.0681 | 1.70 | .2026 | .6337 | 1.338 |
| .76 | .6821 | .8964 | 1.0570 | 1.72 | .1966 | .6283 | 1.357 |
| .78 | .6691 | .8915 | 1.0471 | 1.74 | .1907 | .6229 | 1.376 |
| .80 | .6560 | .8865 | 1.0382 | 1.76 | .1850 | .6175 | 1.397 |
| .82 | .6430 | .8815 | 1.0305 | 1.78 | .1794 | .6121 | 1.418 |
| .84 | .6300 | .8763 | 1.0237 | 1.80 | .1740 | .6068 | 1.439 |
| .86 | .6170 | .8711 | 1.0179 | 1.82 | .1688 | .6015 | 1.461 |
| .88 | .6041 | .8659 | 1.0129 | 1.84 | .1637 | .5963 | 1.484 |
| .90 | .5913 | .8606 | 1.0089 | 1.86 | .1587 | .5910 | 1.507 |
| .92 | .5785 | .8552 | 1.0056 | 1.88 | .1539 | .5859 | 1.531 |
| .94 | .5658 | .8498 | 1.0031 | 1.90 | .1492 | .5807 | 1.555 |

**TABLE D.1**    Isentropic Flow (*continued*)

| M | $p/p_0$ | $T/T_0$ | $A/A^*$ | M | $p/p_0$ | $T/T_0$ | $A/A^*$ |
|---|---------|---------|---------|---|---------|---------|---------|
| 1.92 | .1447 | .5756 | 1.580 | 2.92 | $.3071^{-1}$ | .3696 | 3.924 |
| 1.94 | .1403 | .5705 | 1.606 | 2.94 | $.2980^{-1}$ | .3665 | 3.999 |
| 1.96 | .1360 | .5655 | 1.633 | 2.96 | $.2891^{-1}$ | .3633 | 4.076 |
| 1.98 | .1318 | .5605 | 1.660 | 2.98 | $.2805^{-1}$ | .3602 | 4.155 |
| 2.00 | .1278 | .5556 | 1.688 | 3.00 | $.2722^{-1}$ | .3571 | 4.235 |
| 2.02 | .1239 | .5506 | 1.716 | 3.02 | $.2642^{-1}$ | .3541 | 4.316 |
| 2.04 | .1201 | .5458 | 1.745 | 3.04 | $.2564^{-1}$ | .3511 | 4.399 |
| 2.06 | .1164 | .5409 | 1.775 | 3.06 | $.2489^{-1}$ | .3481 | 4.483 |
| 2.08 | .1128 | .5361 | 1.806 | 3.08 | $.2416^{-1}$ | .3452 | 4.570 |
| 2.10 | .1094 | .5313 | 1.837 | 3.10 | $.2345^{-1}$ | .3422 | 4.657 |
| 2.12 | .1060 | .5266 | 1.869 | 3.12 | $.2276^{-1}$ | .3393 | 4.747 |
| 2.14 | .1027 | .5219 | 1.902 | 3.14 | $.2210^{-1}$ | .3365 | 4.838 |
| 2.16 | $.9956^{-1}$ | .5173 | 1.935 | 3.16 | $.2146^{-1}$ | .3337 | 4.930 |
| 2.18 | $.9649^{-1}$ | .5127 | 1.970 | 3.18 | $.2083^{-1}$ | .3309 | 5.025 |
| 2.20 | $.9352^{-1}$ | .5081 | 2.005 | 3.20 | $.2023^{-1}$ | .3281 | 5.121 |
| 2.22 | $.9064^{-1}$ | .5036 | 2.041 | 3.22 | $.1964^{-1}$ | .3253 | 5.219 |
| 2.24 | $.8785^{-1}$ | .4991 | 2.078 | 3.24 | $.1908^{-1}$ | .3226 | 5.319 |
| 2.26 | $.8514^{-1}$ | .4947 | 2.115 | 3.26 | $.1853^{-1}$ | .3199 | 5.420 |
| 2.28 | $.8251^{-1}$ | .4903 | 2.154 | 3.28 | $.1799^{-1}$ | .3173 | 5.523 |
| 2.30 | $.7997^{-1}$ | .4859 | 2.193 | 3.30 | $.1748^{-1}$ | .3147 | 5.629 |
| 2.32 | $.7751^{-1}$ | .4816 | 2.233 | 3.32 | $.1698^{-1}$ | .3121 | 5.736 |
| 2.34 | $.7512^{-1}$ | .4773 | 2.274 | 3.34 | $.1649^{-1}$ | .3095 | 5.845 |
| 2.36 | $.7281^{-1}$ | .4731 | 2.316 | 3.36 | $.1602^{-1}$ | .3069 | 5.956 |
| 2.38 | $.7057^{-1}$ | .4688 | 2.359 | 3.38 | $.1557^{-1}$ | .3044 | 6.069 |
| 2.40 | $.6840^{-1}$ | .4647 | 2.403 | 3.40 | $.1512^{-1}$ | .3019 | 6.184 |
| 2.42 | $.6630^{-1}$ | .4606 | 2.448 | 3.42 | $.1470^{-1}$ | .2995 | 6.301 |
| 2.44 | $.6426^{-1}$ | .4565 | 2.494 | 3.44 | $.1428^{-1}$ | .2970 | 6.420 |
| 2.46 | $.6229^{-1}$ | .4524 | 2.540 | 3.46 | $.1388^{-1}$ | .2946 | 6.541 |
| 2.48 | $.6038^{-1}$ | .4484 | 2.588 | 3.48 | $.1349^{-1}$ | .2922 | 6.664 |
| 2.50 | $.5853^{-1}$ | .4444 | 2.637 | 3.50 | $.1311^{-1}$ | .2899 | 6.790 |
| 2.52 | $.5674^{-1}$ | .4405 | 2.686 | 3.52 | $.1274^{-1}$ | .2875 | 6.917 |
| 2.54 | $.5500^{-1}$ | .4366 | 2.737 | 3.54 | $.1239^{-1}$ | .2852 | 7.047 |
| 2.56 | $.5332^{-1}$ | .4328 | 2.789 | 3.56 | $.1204^{-1}$ | .2829 | 7.179 |
| 2.58 | $.5169^{-1}$ | .4289 | 2.842 | 3.58 | $.1171^{-1}$ | .2806 | 7.313 |
| 2.60 | $.5012^{-1}$ | .4252 | 2.896 | 3.60 | $.1138^{-1}$ | .2784 | 7.450 |
| 2.62 | $.4859^{-1}$ | .4214 | 2.951 | 3.62 | $.1107^{-1}$ | .2762 | 7.589 |
| 2.64 | $.4711^{-1}$ | .4177 | 3.007 | 3.64 | $.1076^{-1}$ | .2740 | 7.730 |
| 2.66 | $.4568^{-1}$ | .4141 | 3.065 | 3.66 | $.1047^{-1}$ | .2718 | 7.874 |
| 2.68 | $.4429^{-1}$ | .4104 | 3.123 | 3.68 | $.1018^{-1}$ | .2697 | 8.020 |
| 2.70 | $.4295^{-1}$ | .4068 | 3.183 | 3.70 | $.9903^{-2}$ | .2675 | 8.169 |
| 2.72 | $.4165^{-1}$ | .4033 | 3.244 | 3.72 | $.9633^{-2}$ | .2654 | 8.320 |
| 2.74 | $.4039^{-1}$ | .3998 | 3.306 | 3.74 | $.9370^{-2}$ | .2633 | 8.474 |
| 2.76 | $.3917^{-1}$ | .3963 | 3.370 | 3.76 | $.9116^{-2}$ | .2613 | 8.630 |
| 2.78 | $.3799^{-1}$ | .3928 | 3.434 | 3.78 | $.8869^{-2}$ | .2592 | 8.789 |
| 2.80 | $.3685^{-1}$ | .3894 | 3.500 | 3.80 | $.8629^{-2}$ | .2572 | 8.951 |
| 2.82 | $.3574^{-1}$ | .3860 | 3.567 | 3.82 | $.8396^{-2}$ | .2552 | 9.115 |
| 2.84 | $.3467^{-1}$ | .3827 | 3.636 | 3.84 | $.8171^{-2}$ | .2532 | 9.282 |
| 2.86 | $.3363^{-1}$ | .3794 | 3.706 | 3.86 | $.7951^{-2}$ | .2513 | 9.451 |
| 2.88 | $.3263^{-1}$ | .3761 | 3.777 | 3.88 | $.7739^{-2}$ | .2493 | 9.624 |
| 2.90 | $.3165^{-1}$ | .3729 | 3.850 | 3.90 | $.7532^{-2}$ | .2474 | 9.799 |

**TABLE D.1**  Isentropic Flow (*continued*)

| M | $p/p_0$ | $T/T_0$ | $A/A^*$ | M | $p/p_0$ | $T/T_0$ | $A/A^*$ |
|---|---|---|---|---|---|---|---|
| 3.92 | $.7332^{-2}$ | .2455 | 9.977 | 4.54 | $.3288^{-2}$ | .1952 | 17.13 |
| 3.94 | $.7137^{-2}$ | .2436 | 10.16 | 4.56 | $.3207^{-2}$ | .1938 | 17.42 |
| 3.96 | $.6948^{-2}$ | .2418 | 10.34 | 4.58 | $.3129^{-2}$ | .1925 | 17.72 |
| 3.98 | $.6764^{-2}$ | .2399 | 10.53 | 4.60 | $.3053^{-2}$ | .1911 | 18.02 |
| 4.00 | $.6586^{-2}$ | .2381 | 10.72 | 4.62 | $.2978^{-2}$ | .1898 | 18.32 |
| 4.02 | $.6413^{-2}$ | .2363 | 10.91 | 4.64 | $.2906^{-2}$ | .1885 | 18.63 |
| 4.04 | $.6245^{-2}$ | .2345 | 11.11 | 4.66 | $.2836^{-2}$ | .1872 | 18.94 |
| 4.06 | $.6082^{-2}$ | .2327 | 11.31 | 4.68 | $.2768^{-2}$ | .1859 | 19.26 |
| 4.08 | $.5923^{-2}$ | .2310 | 11.51 | 4.70 | $.2701^{-2}$ | .1846 | 19.58 |
| 4.10 | $.5769^{-2}$ | .2293 | 11.71 | 4.72 | $.2637^{-2}$ | .1833 | 19.91 |
| 4.12 | $.5619^{-2}$ | .2275 | 11.92 | 4.74 | $.2573^{-2}$ | .1820 | 20.24 |
| 4.14 | $.5474^{-2}$ | .2258 | 12.14 | 4.76 | $.2512^{-2}$ | .1808 | 20.58 |
| 4.16 | $.5333^{-2}$ | .2242 | 12.35 | 4.78 | $.2452^{-2}$ | .1795 | 20.92 |
| 4.18 | $.5195^{-2}$ | .2225 | 12.57 | 4.80 | $.2394^{-2}$ | .1783 | 21.26 |
| 4.20 | $.5062^{-2}$ | .2208 | 12.79 | 4.82 | $.2338^{-2}$ | .1771 | 21.61 |
| 4.22 | $.4932^{-2}$ | .2192 | 13.02 | 4.84 | $.2283^{-2}$ | .1759 | 21.97 |
| 4.24 | $.4806^{-2}$ | .2176 | 13.25 | 4.86 | $.2229^{-2}$ | .1747 | 22.33 |
| 4.26 | $.4684^{-2}$ | .2160 | 13.48 | 4.88 | $.2177^{-2}$ | .1735 | 22.70 |
| 4.28 | $.4565^{-2}$ | .2144 | 13.72 | 4.90 | $.2126^{-2}$ | .1724 | 23.07 |
| 4.30 | $.4449^{-2}$ | .2129 | 13.95 | 4.92 | $.2076^{-2}$ | .1712 | 23.44 |
| 4.32 | $.4337^{-2}$ | .2113 | 14.20 | 4.94 | $.2028^{-2}$ | .1700 | 23.82 |
| 4.34 | $.4228^{-2}$ | .2098 | 14.45 | 4.96 | $.1981^{-2}$ | .1689 | 24.21 |
| 4.36 | $.4121^{-2}$ | .2083 | 14.70 | 4.98 | $.1935^{-2}$ | .1678 | 24.60 |
| 4.38 | $.4018^{-2}$ | .2067 | 14.95 | 5.00 | $.1890^{-2}$ | .1667 | 25.00 |
| 4.40 | $.3918^{-2}$ | .2053 | 15.21 | 6.00 | $.0633^{-2}$ | .1219 | 53.19 |
| 4.42 | $.3820^{-2}$ | .2038 | 15.47 | 7.00 | $.0242^{-2}$ | .0926 | 104.14 |
| 4.44 | $.3725^{-2}$ | .2023 | 15.74 | 8.00 | $.0102^{-2}$ | .0725 | 109.11 |
| 4.46 | $.3633^{-2}$ | .2009 | 16.01 | 9.00 | $.0474^{-3}$ | .0582 | 327.19 |
| 4.48 | $.3543^{-2}$ | .1994 | 16.28 | 10.00 | $.0236^{-3}$ | .0476 | 535.94 |
| 4.50 | $.3455^{-2}$ | .1980 | 16.56 | $\infty$ | 0 | 0 | $\infty$ |
| 4.52 | $.3370^{-2}$ | .1966 | 16.84 | | | | |

**TABLE D.2**  Normal-Shock Flow

| $M_1$ | $M_2$ | $p_2/p_1$ | $T_2/T_1$ | $p_{02}/p_{01}$ |
|---|---|---|---|---|
| 1.00 | 1.000 | 1.000 | 1.000 | 1.000 |
| 1.02 | .9805 | 1.047 | 1.013 | 1.000 |
| 1.04 | .9620 | 1.095 | 1.026 | .9999 |
| 1.06 | .9444 | 1.144 | 1.039 | .9997 |
| 1.08 | .9277 | 1.194 | 1.052 | .9994 |
| 1.10 | .9118 | 1.245 | 1.065 | .9989 |
| 1.12 | .8966 | 1.297 | 1.078 | .9982 |
| 1.14 | .8820 | 1.350 | 1.090 | .9973 |
| 1.16 | .8682 | 1.403 | 1.103 | .9961 |
| 1.18 | .8549 | 1.458 | 1.115 | .9946 |
| 1.20 | .8422 | 1.513 | 1.128 | .9928 |
| 1.22 | .8300 | 1.570 | 1.141 | .9907 |
| 1.24 | .8183 | 1.627 | 1.153 | .9884 |
| 1.26 | .8071 | 1.686 | 1.166 | .9857 |
| 1.28 | .7963 | 1.745 | 1.178 | .9827 |

**TABLE D.2**  Normal-Shock Flow (*continued*)

| $M_1$ | $M_2$ | $p_2/p_1$ | $T_2/T_1$ | $p_{02}/p_{01}$ |
|-------|-------|-----------|-----------|-----------------|
| 1.30 | .7860 | 1.805 | 1.191 | .9794 |
| 1.32 | .7760 | 1.866 | 1.204 | .9758 |
| 1.34 | .7664 | 1.928 | 1.216 | .9718 |
| 1.36 | .7572 | 1.991 | 1.229 | .9676 |
| 1.38 | .7483 | 2.055 | 1.242 | .9630 |
| 1.40 | .7397 | 2.120 | 1.255 | .9582 |
| 1.42 | .7314 | 2.186 | 1.268 | .9531 |
| 1.44 | .7235 | 2.253 | 1.281 | .9476 |
| 1.46 | .7157 | 2.320 | 1.294 | .9420 |
| 1.48 | .7083 | 2.389 | 1.307 | .9360 |
| 1.50 | .7011 | 2.458 | 1.320 | .9298 |
| 1.52 | .6941 | 2.529 | 1.334 | .9233 |
| 1.54 | .6874 | 2.600 | 1.347 | .9166 |
| 1.56 | .6809 | 2.673 | 1.361 | .9097 |
| 1.58 | .6746 | 2.746 | 1.374 | .9026 |
| 1.60 | .6684 | 2.820 | 1.388 | .8952 |
| 1.62 | .6625 | 2.895 | 1.402 | .8877 |
| 1.64 | .6568 | 2.971 | 1.416 | .8799 |
| 1.66 | .6512 | 3.048 | 1.430 | .8720 |
| 1.68 | .6458 | 3.126 | 1.444 | .8640 |
| 1.70 | .6405 | 3.205 | 1.458 | .8557 |
| 1.72 | .6355 | 3.285 | 1.473 | .8474 |
| 1.74 | .6305 | 3.366 | 1.487 | .8389 |
| 1.76 | .6257 | 3.447 | 1.502 | .8302 |
| 1.78 | .6210 | 3.530 | 1.517 | .8215 |
| 1.80 | .6165 | 3.613 | 1.532 | .8127 |
| 1.82 | .6121 | 3.698 | 1.547 | .8038 |
| 1.84 | .6078 | 3.783 | 1.562 | .7948 |
| 1.86 | .6036 | 3.870 | 1.577 | .7857 |
| 1.88 | .5996 | 3.957 | 1.592 | .7765 |
| 1.90 | .5956 | 4.045 | 1.608 | .7674 |
| 1.92 | .5918 | 4.134 | 1.624 | .7581 |
| 1.94 | .5880 | 4.224 | 1.639 | .7488 |
| 1.96 | .5844 | 4.315 | 1.655 | .7395 |
| 1.98 | .5808 | 4.407 | 1.671 | .7302 |
| 2.00 | .5774 | 4.500 | 1.688 | .7209 |
| 2.02 | .5740 | 4.594 | 1.704 | .7115 |
| 2.04 | .5707 | 4.689 | 1.720 | .7022 |
| 2.06 | .5675 | 4.784 | 1.737 | .6928 |
| 2.08 | .5643 | 4.881 | 1.754 | .6835 |
| 2.10 | .5613 | 4.978 | 1.770 | .6742 |
| 2.12 | .5583 | 5.077 | 1.787 | .6649 |
| 2.14 | .5554 | 5.176 | 1.805 | .6557 |
| 2.16 | .5525 | 5.277 | 1.822 | .6464 |
| 2.18 | .5498 | 5.378 | 1.839 | .6373 |
| 2.20 | .5471 | 5.480 | 1.857 | .6281 |
| 2.22 | .5444 | 5.583 | 1.875 | .6191 |
| 2.24 | .5418 | 5.687 | 1.892 | .6100 |
| 2.26 | .5393 | 5.792 | 1.910 | .6011 |
| 2.28 | .5368 | 5.898 | 1.929 | .5921 |

**TABLE D.2** Normal-Shock Flow (*continued*)

| $M_1$ | $M_2$ | $p_2/p_1$ | $T_2/T_1$ | $p_{02}/p_{01}$ |
|-------|-------|-----------|-----------|-----------------|
| 2.30 | .5344 | 6.005 | 1.947 | .5833 |
| 2.32 | .5321 | 6.113 | 1.965 | .5745 |
| 2.34 | .5297 | 6.222 | 1.984 | .5658 |
| 2.36 | .5275 | 6.331 | 2.002 | .5572 |
| 2.38 | .5253 | 6.442 | 2.021 | .5486 |
| 2.40 | .5231 | 6.553 | 2.040 | .5401 |
| 2.42 | .5210 | 6.666 | 2.059 | .5317 |
| 2.44 | .5189 | 6.779 | 2.079 | .5234 |
| 2.46 | .5169 | 6.894 | 2.098 | .5152 |
| 2.48 | .5149 | 7.009 | 2.118 | .5071 |
| 2.50 | .5130 | 7.125 | 2.138 | .4990 |
| 2.52 | .5111 | 7.242 | 2.157 | .4991 |
| 2.54 | .5092 | 7.360 | 2.177 | .4832 |
| 2.56 | .5074 | 7.479 | 2.198 | .4754 |
| 2.58 | .5056 | 7.599 | 2.218 | .4677 |
| 2.60 | .5039 | 7.720 | 2.238 | .4601 |
| 2.62 | .5022 | 7.842 | 2.259 | .4526 |
| 2.64 | .5005 | 7.965 | 2.280 | .4452 |
| 2.66 | .4988 | 8.088 | 2.301 | .4379 |
| 2.68 | .4972 | 8.213 | 2.322 | .4307 |
| 2.70 | .4956 | 8.338 | 2.343 | .4236 |
| 2.72 | .4941 | 8.465 | 2.364 | .4166 |
| 2.74 | .4926 | 8.592 | 2.386 | .4097 |
| 2.76 | .4911 | 8.721 | 2.407 | .4028 |
| 2.78 | .4896 | 8.850 | 2.429 | .3961 |
| 2.80 | .4882 | 8.980 | 2.451 | .3895 |
| 2.82 | .4868 | 9.111 | 2.473 | .3829 |
| 2.84 | .4854 | 9.243 | 2.496 | .3765 |
| 2.86 | .4840 | 9.376 | 2.518 | .3701 |
| 2.88 | .4827 | 9.510 | 2.540 | .3639 |
| 2.90 | .4814 | 9.645 | 2.563 | .3577 |
| 2.92 | .4801 | 9.781 | 2.586 | .3517 |
| 2.94 | .4788 | 9.918 | 2.609 | .3457 |
| 2.96 | .4776 | 10.06 | 2.632 | .3398 |
| 2.98 | .4764 | 10.19 | 2.656 | .3340 |
| 3.00 | .4752 | 10.33 | 2.679 | .3283 |
| 3.02 | .4740 | 10.47 | 2.703 | .3327 |
| 3.04 | .4729 | 10.62 | 2.726 | .3172 |
| 3.06 | .4717 | 10.76 | 2.750 | .3118 |
| 3.08 | .4706 | 10.90 | 2.774 | .3065 |
| 3.10 | .4695 | 11.05 | 2.799 | .3012 |
| 3.12 | .4685 | 11.19 | 2.823 | .2960 |
| 3.14 | .4674 | 11.34 | 2.848 | .2910 |
| 3.16 | .4664 | 11.48 | 2.872 | .2860 |
| 3.18 | .4654 | 11.63 | 2.897 | .2811 |
| 3.20 | .4643 | 11.78 | 2.922 | .2762 |
| 3.22 | .4634 | 11.93 | 2.947 | .2715 |
| 3.24 | .4624 | 12.08 | 2.972 | .2668 |
| 3.26 | .4614 | 12.23 | 2.998 | .2622 |
| 3.28 | .4605 | 12.38 | 3.023 | .2577 |

**TABLE D.2**  Normal-Shock Flow (*continued*)

| $M_1$ | $M_2$ | $p_2/p_1$ | $T_2/T_1$ | $p_{02}/p_{01}$ |
|---|---|---|---|---|
| 3.30 | .4596 | 12.54 | 3.049 | .2533 |
| 3.32 | .4587 | 12.69 | 3.075 | .2489 |
| 3.34 | .4578 | 12.85 | 3.101 | .2446 |
| 3.36 | .4569 | 13.00 | 3.127 | .2404 |
| 3.38 | .4560 | 13.16 | 3.154 | .2363 |
| 3.40 | .4552 | 13.32 | 3.180 | .2322 |
| 3.42 | .4544 | 13.48 | 3.207 | .2382 |
| 3.44 | .4535 | 13.64 | 3.234 | .2243 |
| 3.46 | .4527 | 13.80 | 3.261 | .2205 |
| 3.48 | .4519 | 13.96 | 3.288 | .2167 |
| 3.50 | .4512 | 14.13 | 3.315 | .2129 |
| 3.52 | .4504 | 14.29 | 3.343 | .2093 |
| 3.54 | .4496 | 14.45 | 3.370 | .2057 |
| 3.56 | .4489 | 14.62 | 3.398 | .2022 |
| 3.58 | .4481 | 14.79 | 3.426 | .1987 |
| 3.60 | .4474 | 14.95 | 3.454 | .1953 |
| 3.62 | .4467 | 15.12 | 3.482 | .1920 |
| 3.64 | .4460 | 15.29 | 3.510 | .1887 |
| 3.66 | .4453 | 15.46 | 3.539 | .1855 |
| 3.68 | .4446 | 15.63 | 3.568 | .1823 |
| 3.70 | .4439 | 15.81 | 3.596 | .1792 |
| 3.72 | .4433 | 15.98 | 3.625 | .1761 |
| 3.74 | .4426 | 16.15 | 3.654 | .1731 |
| 3.76 | .4420 | 16.33 | 3.684 | .1702 |
| 3.78 | .4414 | 16.50 | 3.713 | .1673 |
| 3.80 | .4407 | 16.68 | 3.743 | .1645 |
| 3.82 | .4401 | 16.86 | 3.772 | .1617 |
| 3.84 | .4395 | 17.04 | 3.802 | .1589 |
| 3.86 | .4389 | 17.22 | 3.832 | .1563 |
| 3.88 | .4383 | 17.40 | 3.863 | .1536 |
| 3.90 | .4377 | 17.58 | 3.893 | .1510 |
| 3.92 | .4372 | 17.76 | 3.923 | .1485 |
| 3.94 | .4366 | 17.94 | 3.954 | .1460 |
| 3.96 | .4360 | 18.13 | 3.985 | .1435 |
| 3.98 | .4355 | 18.31 | 4.016 | .1411 |
| 4.00 | .4350 | 18.50 | 4.047 | .1388 |
| 4.02 | .4344 | 18.69 | 4.078 | .1364 |
| 4.04 | .4339 | 18.88 | 4.110 | .1342 |
| 4.06 | .4334 | 19.06 | 4.141 | .1319 |
| 4.08 | .4329 | 19.25 | 4.173 | .1297 |
| 4.10 | .4324 | 19.45 | 4.205 | .1276 |
| 4.12 | .4319 | 19.64 | 4.237 | .1254 |
| 4.14 | .4314 | 19.83 | 4.269 | .1234 |
| 4.16 | .4309 | 20.02 | 4.301 | .1213 |
| 4.18 | .4304 | 20.22 | 4.334 | .1193 |
| 4.20 | .4299 | 20.41 | 4.367 | .1173 |
| 4.22 | .4295 | 20.61 | 4.399 | .1154 |
| 4.24 | .4290 | 20.81 | 4.432 | .1135 |
| 4.26 | .4286 | 21.01 | 4.466 | .1116 |
| 4.28 | .4281 | 21.20 | 4.499 | .1098 |

**TABLE D.2**    Normal-Shock Flow (*continued*)

| $M_1$ | $M_2$ | $p_2/p_1$ | $T_2/T_1$ | $p_{02}/p_{01}$ |
|---|---|---|---|---|
| 4.30 | .4277 | 21.41 | 4.532 | .1080 |
| 4.32 | .4272 | 21.61 | 4.566 | .1062 |
| 4.34 | .4268 | 21.81 | 4.600 | .1045 |
| 4.36 | .4264 | 22.01 | 4.633 | .1028 |
| 4.38 | .4260 | 22.22 | 4.668 | .1011 |
| 4.40 | .4255 | 22.42 | 4.702 | $.9948^{-1}$ |
| 4.42 | .4251 | 22.63 | 4.736 | $.9787^{-1}$ |
| 4.44 | .4247 | 22.83 | 4.771 | $.9628^{-1}$ |
| 4.46 | .4243 | 23.04 | 4.805 | $.9473^{-1}$ |
| 4.48 | .4239 | 23.25 | 4.840 | $.9320^{-1}$ |
| 4.50 | .4236 | 23.46 | 4.875 | $.9170^{-1}$ |
| 4.52 | .4232 | 23.67 | 4.910 | $.9022^{-1}$ |
| 4.54 | .4228 | 23.88 | 4.946 | $.8878^{-1}$ |
| 4.56 | .4224 | 24.09 | 4.981 | $.8735^{-1}$ |
| 4.58 | .4220 | 24.31 | 5.017 | $.8596^{-1}$ |
| 4.60 | .4217 | 24.52 | 5.052 | $.8459^{-1}$ |
| 4.62 | .4213 | 24.74 | 5.088 | $.8324^{-1}$ |
| 4.64 | .4210 | 24.95 | 5.124 | $.8192^{-1}$ |
| 4.66 | .4206 | 25.17 | 5.160 | $.8062^{-1}$ |
| 4.68 | .4203 | 25.39 | 5.197 | $.7934^{-1}$ |
| 4.70 | .4199 | 25.61 | 5.233 | $.7809^{-1}$ |
| 4.72 | .4196 | 25.82 | 5.270 | $.7685^{-1}$ |
| 4.74 | .4192 | 26.05 | 5.307 | $.7564^{-1}$ |
| 4.76 | .4189 | 26.27 | 5.344 | $.7445^{-1}$ |
| 4.78 | .4186 | 26.49 | 5.381 | $.7329^{-1}$ |
| 4.80 | .4183 | 26.71 | 5.418 | $.7214^{-1}$ |
| 4.82 | .4179 | 26.94 | 5.456 | $.7101^{-1}$ |
| 4.84 | .4176 | 27.16 | 5.494 | $.6991^{-1}$ |
| 4.86 | .4173 | 27.39 | 5.531 | $.6882^{-1}$ |
| 4.88 | .4170 | 27.62 | 5.569 | $.6775^{-1}$ |
| 4.90 | .4167 | 27.85 | 5.607 | $.6670^{-1}$ |
| 4.92 | .4164 | 28.07 | 5.646 | $.6567^{-1}$ |
| 4.94 | .4161 | 28.30 | 5.684 | $.6465^{-1}$ |
| 4.96 | .4158 | 28.54 | 5.723 | $.6366^{-1}$ |
| 4.98 | .4155 | 28.77 | 5.761 | $.6268^{-1}$ |
| 5.00 | .4152 | 29.00 | 5.800 | $.6172^{-1}$ |
| 6.00 | .4042 | 41.83 | 7.941 | $.2965^{-1}$ |
| 7.00 | .3974 | 57.00 | 10.469 | $.1535^{-1}$ |
| 8.00 | .3929 | 74.50 | 13.387 | $.0849^{-1}$ |
| 9.00 | .3898 | 94.33 | 16.693 | $.0496^{-1}$ |
| 10.00 | .3875 | 116.50 | 20.388 | $.0304^{-1}$ |
| $\infty$ | .3780 | $\infty$ | $\infty$ | 0 |

**TABLE D.3**  Prandtl-Meyer Function

| M | $\theta$ | $\mu$ | M | $\theta$ | $\mu$ |
|---|---|---|---|---|---|
| 1.00 | 0 | 90.00 | 2.02 | 26.929 | 29.67 |
| 1.02 | .1257 | 78.64 | 2.04 | 27.476 | 29.35 |
| 1.04 | .3510 | 74.06 | 2.06 | 28.020 | 29.04 |
| 1.06 | .6367 | 70.63 | 2.08 | 28.560 | 28.74 |
| 1.08 | .9680 | 67.81 | 2.10 | 29.097 | 28.44 |
| 1.10 | 1.336 | 65.38 | 2.12 | 29.631 | 28.14 |
| 1.12 | 1.735 | 63.23 | 2.14 | 30.161 | 27.86 |
| 1.14 | 2.160 | 61.31 | 2.16 | 30.689 | 27.58 |
| 1.16 | 2.607 | 59.55 | 2.18 | 31.212 | 27.30 |
| 1.18 | 3.074 | 57.94 | 2.20 | 31.732 | 27.04 |
| 1.20 | 3.558 | 56.44 | 2.22 | 32.250 | 26.77 |
| 1.22 | 4.057 | 55.05 | 2.24 | 32.763 | 26.51 |
| 1.24 | 4.569 | 53.75 | 2.26 | 33.273 | 26.26 |
| 1.26 | 5.093 | 52.53 | 2.28 | 33.780 | 26.01 |
| 1.28 | 5.627 | 51.38 | 2.30 | 34.283 | 25.77 |
| 1.30 | 6.170 | 50.28 | 2.32 | 34.783 | 25.53 |
| 1.32 | 6.721 | 49.25 | 2.34 | 35.279 | 25.30 |
| 1.34 | 7.280 | 48.27 | 2.36 | 35.771 | 25.07 |
| 1.36 | 7.844 | 47.33 | 2.38 | 36.261 | 24.85 |
| 1.38 | 8.413 | 46.44 | 2.40 | 36.746 | 24.62 |
| 1.40 | 8.987 | 45.58 | 2.42 | 37.229 | 24.41 |
| 1.42 | 9.565 | 44.77 | 2.44 | 37.708 | 24.19 |
| 1.44 | 10.146 | 43.98 | 2.46 | 38.183 | 23.99 |
| 1.46 | 10.731 | 43.23 | 2.48 | 38.655 | 23.78 |
| 1.48 | 11.317 | 42.51 | 2.50 | 39.124 | 23.58 |
| 1.50 | 11.905 | 41.81 | 2.52 | 39.589 | 23.38 |
| 1.52 | 12.495 | 41.14 | 2.54 | 40.050 | 23.18 |
| 1.54 | 13.086 | 40.49 | 2.56 | 40.509 | 22.99 |
| 1.56 | 13.677 | 39.87 | 2.58 | 40.963 | 22.81 |
| 1.58 | 14.269 | 39.27 | 2.60 | 41.415 | 22.62 |
| 1.60 | 14.861 | 38.68 | 2.62 | 41.863 | 22.44 |
| 1.62 | 15.452 | 38.12 | 2.64 | 42.307 | 22.26 |
| 1.64 | 16.043 | 37.57 | 2.66 | 42.749 | 22.08 |
| 1.66 | 16.633 | 37.04 | 2.68 | 43.187 | 21.91 |
| 1.68 | 17.222 | 36.53 | 2.70 | 43.621 | 21.74 |
| 1.70 | 17.810 | 36.03 | 2.72 | 44.053 | 21.57 |
| 1.72 | 18.397 | 35.55 | 2.74 | 44.481 | 21.41 |
| 1.74 | 18.981 | 35.08 | 2.76 | 44.906 | 21.24 |
| 1.76 | 19.565 | 34.62 | 2.78 | 45.327 | 21.08 |
| 1.78 | 20.146 | 34.18 | 2.80 | 45.746 | 20.92 |
| 1.80 | 20.725 | 33.75 | 2.82 | 46.161 | 20.77 |
| 1.82 | 21.302 | 33.33 | 2.84 | 46.573 | 20.62 |
| 1.84 | 21.877 | 32.92 | 2.86 | 46.982 | 20.47 |
| 1.86 | 22.449 | 32.52 | 2.88 | 47.388 | 20.32 |
| 1.88 | 23.019 | 32.13 | 2.90 | 47.790 | 20.17 |
| 1.90 | 23.586 | 31.76 | 2.92 | 48.190 | 20.03 |
| 1.92 | 24.151 | 31.39 | 2.94 | 48.586 | 19.89 |
| 1.94 | 24.712 | 31.03 | 2.96 | 48.980 | 19.75 |
| 1.96 | 25.271 | 30.68 | 2.98 | 49.370 | 19.61 |
| 1.98 | 25.827 | 30.33 | 3.00 | 49.757 | 19.47 |
| 2.00 | 26.380 | 30.00 | 3.02 | 50.142 | 19.34 |

**TABLE D.3** Prandtl-Meyer Function (*continued*)

| M | $\theta$ | $\mu$ | M | $\theta$ | $\mu$ |
|---|---|---|---|---|---|
| 3.04 | 50.523 | 19.20 | 4.04 | 66.309 | 14.33 |
| 3.06 | 50.902 | 19.07 | 4.06 | 66.569 | 14.26 |
| 3.08 | 51.277 | 18.95 | 4.08 | 66.826 | 14.19 |
| 3.10 | 51.560 | 18.82 | 4.10 | 67.082 | 14.12 |
| 3.12 | 52.020 | 18.69 | 4.12 | 67.336 | 14.05 |
| 3.14 | 52.386 | 18.57 | 4.14 | 67.588 | 13.98 |
| 3.16 | 52.751 | 18.45 | 4.16 | 67.838 | 13.91 |
| 3.18 | 53.112 | 18.33 | 4.18 | 68.087 | 13.84 |
| 3.20 | 53.470 | 18.21 | 4.20 | 68.333 | 13.77 |
| 3.22 | 53.826 | 18.09 | 4.22 | 68.578 | 13.71 |
| 3.24 | 54.179 | 17.98 | 4.24 | 68.821 | 13.64 |
| 3.26 | 54.529 | 17.86 | 4.26 | 69.053 | 13.58 |
| 3.28 | 54.877 | 17.75 | 4.28 | 69.302 | 13.51 |
| 3.30 | 55.222 | 17.64 | 4.30 | 69.541 | 13.45 |
| 3.32 | 55.564 | 17.53 | 4.32 | 69.777 | 13.38 |
| 3.34 | 55.904 | 17.42 | 4.34 | 70.012 | 13.32 |
| 3.36 | 56.241 | 17.31 | 4.36 | 70.245 | 13.26 |
| 3.38 | 56.576 | 17.21 | 4.38 | 70.476 | 13.20 |
| 3.40 | 56.907 | 17.10 | 4.40 | 70.706 | 13.14 |
| 3.42 | 57.237 | 17.00 | 4.42 | 70.934 | 13.08 |
| 3.44 | 57.564 | 16.90 | 4.44 | 71.161 | 13.02 |
| 3.46 | 57.888 | 16.80 | 4.46 | 71.386 | 12.96 |
| 3.48 | 58.210 | 16.70 | 4.48 | 71.610 | 12.90 |
| 3.50 | 58.530 | 16.60 | 4.50 | 71.832 | 12.84 |
| 3.52 | 58.847 | 16.51 | 4.52 | 72.052 | 12.78 |
| 3.54 | 59.162 | 16.41 | 4.54 | 72.271 | 12.73 |
| 3.56 | 59.474 | 16.31 | 4.56 | 72.489 | 12.67 |
| 3.58 | 59.784 | 16.22 | 4.58 | 72.705 | 12.61 |
| 3.60 | 60.091 | 16.13 | 4.60 | 72.919 | 12.56 |
| 3.62 | 60.397 | 16.04 | 4.62 | 73.132 | 12.50 |
| 3.64 | 60.700 | 15.95 | 4.64 | 73.344 | 12.45 |
| 3.66 | 61.000 | 15.86 | 4.66 | 73.554 | 12.39 |
| 3.68 | 61.299 | 15.77 | 4.68 | 73.763 | 12.34 |
| 3.70 | 61.595 | 15.68 | 4.70 | 73.970 | 12.28 |
| 3.72 | 61.899 | 15.59 | 4.72 | 74.176 | 12.23 |
| 3.74 | 62.181 | 15.51 | 4.74 | 74.381 | 12.18 |
| 3.76 | 62.471 | 15.42 | 4.76 | 74.584 | 12.13 |
| 3.78 | 62.758 | 15.34 | 4.78 | 74.786 | 12.08 |
| 3.80 | 63.044 | 15.26 | 4.80 | 74.986 | 12.03 |
| 3.82 | 63.327 | 15.18 | 4.82 | 75.186 | 11.97 |
| 3.84 | 63.608 | 15.10 | 4.84 | 75.383 | 11.92 |
| 3.86 | 63.887 | 15.02 | 4.86 | 75.580 | 11.87 |
| 3.88 | 64.164 | 14.94 | 4.88 | 75.775 | 11.83 |
| 3.90 | 64.440 | 14.86 | 4.90 | 75.969 | 11.78 |
| 3.92 | 64.713 | 14.78 | 4.92 | 76.162 | 11.73 |
| 3.94 | 64.983 | 14.70 | 4.94 | 76.353 | 11.68 |
| 3.96 | 65.253 | 14.63 | 4.96 | 76.544 | 11.63 |
| 3.98 | 65.520 | 14.55 | 4.98 | 76.732 | 11.58 |
| 4.00 | 65.785 | 14.48 | 5.00 | 76.920 | 11.54 |
| 4.02 | 66.048 | 14.40 | | | |

# E. NUMERICAL SOLUTIONS FOR CHAPTER 10

```
clear;
% input diameter, Manning coefficient, and channel slope
d=5;
n=0.013;
so=0.0005;
c1-1.0;

%Set range and increments of y
y=0.01:.01:d;

%Define geometric functions
alpha=acos(1-2*y/d);
A=d^2/4*(alpha-sin(alpha).*cos(alpha));
R=A./P;

%Define discharge function, i.e., Manning's equation
Q=c1/n*A.*R.^(2/3)*sqrt(so);

%Plot Y verses Q
Plot(Q, y);
xlabel('Q(y)');
ylabel('y');
```

**(a) Algorithms**

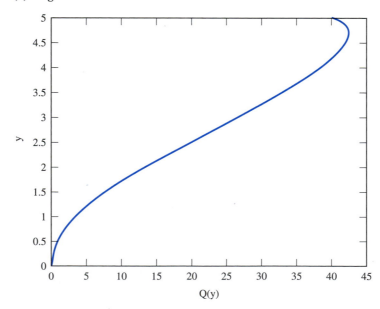

**(b) Solutions**

**FIGURE E.1**   Example 10.2 using MATLAB®

**Input side slope, bottom width, discharge, critical depth, upstream depth, and gravitational constant:**

$$m := 2.5 \qquad b := 0 \qquad Q := 20$$

$$y_c := 1.67 \qquad y_1 := 0.75 \qquad g := 9.81$$

**Define the momentum function**

$$M(y) := \frac{y^2}{6} \cdot (2 \cdot m \cdot y + 3 \cdot b) + \frac{Q^2}{g \cdot (b \cdot y + m \cdot y^2)}$$

**Find the root of M(y) within the limits $y_c < y < 3y_c$ :**

$$y_2 := root(M(y)) - M(y_1), y, y_c, 3 \cdot y_c)$$

**Hence the solution is** $y_2 = 3.218$

**FIGURE E.2**   Example 10.11 Mathcad solution

```
clear;
%Input side slope, bottom width, discharge, critical depth,
%upstream depth, and gravitational constant
m=2.5;
b=0;
q=20;
yc=1.67;
y1=0.75;
g=9.81;

%Set appropriate upper and lower limits
y=yc:.001:3*yc;

%Reducing Eq.10.5.16 with F=0 yields
f=y.^3+19.58./y.^2-35.23;

%Find value of y where f crosses zero
[s,t]=min(abs(f));
y2=y(t)
```

**FIGURE E.3**   Example 10.11 MATLAB® algorithm

|  | A | B | C | D | E | F | G | H | I |
|---|---|---|---|---|---|---|---|---|---|
| 1 |  |  |  |  |  |  |  |  |  |
| 2 |  |  | Depth [m] | Residual |  |  |  |  |  |
| 3 |  | Critical | 0.865 | 1.08665E–06 |  |  |  |  |  |
| 4 |  |  |  |  |  |  |  |  |  |
| 5 |  | Normal | 1.292 | 1.81248E–06 |  |  |  |  |  |
| 6 |  |  |  |  |  |  |  |  |  |
| 7 |  |  |  |  |  |  |  |  |  |
| 8 | Station | y [m] | A [m²] | V [m/s] | E [m] | $y_m$ [m] | $S(y_m)$ | Δx [m] | x [m] |
| 9 | 1 | 0.865 | 8.358 | 2.632 | 1.218 |  |  |  | 2000 |
| 10 | 2 | 0.950 | 9.381 | 2.345 | 1.230 | 0.908 | 2.165E–03 | –8 | 1992 |
| 11 | 3 | 1.050 | 10.631 | 2.069 | 1.268 | 1.000 | 1.527E–03 | –41 | 1951 |
| 12 | 4 | 1.150 | 11.931 | 1.844 | 1.323 | 1.100 | 1.081E–03 | –114 | 1837 |
| 13 | 5 | 1.250 | 13.281 | 1.656 | 1.390 | 1.200 | 7.866E–04 | –357 | 1480 |
| 14 | 6 | 1.270 | 13.557 | 1.623 | 1.404 | 1.260 | 6.574E–04 | –250 | 1230 |
| 15 |  |  |  |  |  |  |  |  |  |

**(a) Solution**

|  | B | C | D |
|---|---|---|---|
| 2 |  | Depth [m] | Residual |
| 3 | Critical | 0.86479527789122 | =22^2*(7.5+2*2.5*C3)/(9.81*(7.5*C3+2.5*C3^2)^3)–1 |
| 4 |  |  |  |
| 5 | Normal | 1.29156940192658 | =22*0.015*(7.5+2*C5*SQRT(1+2.5^2))^0.6667/((7.5*C5+2.5*C5^2)^1.6667*SQRT(0.0006))–1 |

|  | A | B | C | D | E | F |
|---|---|---|---|---|---|---|
| 8 | Station | y [m] | A [m²] | V [m/s] | E [m] | $y_m$ [m] |
| 9 | 1 | 0.865 | =7.5*B9+B9^2*2.5 | =22/C9 | =B9+D9^2/(2*9.81) |  |
| 10 | 2 | 0.95 | =7.5*B10+B10^2*2.5 | =22/C10 | =B10+D10^2/(2*9.81) | =0.5*(B9+B10) |
| 11 | 3 | 1.05 | =7.5*B11+B11^2*2.5 | =22/C11 | =B11+D11^2/(2*9.81) | =0.5*(B10+B11) |
| 12 | 4 | 1.15 | =7.5*B12+B12^2*2.5 | =22/C12 | =B12+D12^2/(2*9.81) | =0.5*(B11+B12) |
| 13 | 5 | 1.25 | =7.5*B13+B13^2*2.5 | =22/C13 | =B13+D13^2/(2*9.81) | =0.5*(B12+B13) |
| 14 | 6 | 1.27 | =7.5*B14+B14^2*2.5 | =22/C14 | =B14+D14^2/(2*9.81) | =0.5*(B13+B14) |

|  | G | H | I |
|---|---|---|---|
| 8 | $S(y_m)$ | Δx [m] | x [m] |
| 9 |  |  | 2000 |
| 10 | =22^2*0.015^2*(7.5*F10+F10^2*2.5)^–3.3333*((7.5+2*F10*SQRT(1+2.5^2))^1.3333) | =(E10–E9)/(0.0006–G10) | =I9+H10 |
| 11 | =22^2*0.015^2*(7.5*F11+F11^2*2.5)^–3.3333*((7.5+2*F11*SQRT(1+2.5^2))^1.3333) | =(E11–E10)/(0.0006–G11) | =I10+H11 |
| 12 | =22^2*0.015^2*(7.5*F12+F12^2*2.5)^–3.3333*((7.5+2*F12*SQRT(1+2.5^2))^1.3333) | =(E12–E11)/(0.0006–G12) | =I11+H12 |
| 13 | =22^2*0.015^2*(7.5*F13+F13^2*2.5)^–3.3333*((7.5+2*F13*SQRT(1+2.5^2))^1.3333) | =(E13–E12)/(0.0006–G13) | =I12+H13 |
| 14 | =22^2*0.015^2*(7.5*F14+F14^2*2.5)^–3.3333*((7.5+2*F14*SQRT(1+2.5^2))^1.3333) | =(E14–E13)/(0.0006–G14) | =I13+H14 |

**(b) Spreadsheet formulas**

**FIGURE E.4** Example 10.15 using Excel®

```
clear;
%Input data
Q=22;
M=2.5;
b=7.5;
L=2000;
So=0.0006;
n=0.015;
g=9.81;

%Set range and increments of y
y=0.01:.001:5;

%Define Functions
A=b*y+M*y.^2;
B=b+2*M*y;
P=b+2*y*sqrt(1+M^2);
S=Q^2*n^2./(A.^3.3333.*P.^-1.3333);
Fr=sqrt(Q^2*B./(g*A.^3));

%Find values of y where (s-So) and (Fr-1) cross zero
[s,t]=min(abs(S-So));
yN=y(t)
[s,t]=min(abs(Fr-1));
yc=y(t)

%Reset range and increments of y to go from critical
%to normal depths
clear y;
dy=0.1;
y=yc:dy:yN

%Calculate specific energy at new y values
A=b*y+M*y.^2;
E=y+Q^2./(2*g*A.^2)

%Calculate x locations according to Eq.10.7.6
ym=(y(2:5)+y(1:4))/2;
A_Ym=b*ym+M*ym.^2);
P_ym=b+2*ym*sqrt(1+M^2);
S_ym=Q^2*n^2./(A_ym.^(10/3).*P_ym.^(-4/3));

x(1)=L;
for i=2:5;
    x(i)=x(i-1)+(E(i)-E(i-1))/(S0-S_ym(i-1));
end;
```

**FIGURE E.5**   Example 10.15 using MATLAB® (a)  Algorithm

```
YN =

    1.2920

yc =

    0.8650

Y =

    0.8650      0.9650      1.0650      1.1650      1.2650

E =

    1.2181      1.2346      1.2756      1.3326      1.4006

x =

  1.0e+003 *

    2.0000.     1.9890      1.9406      1.8076      1.3591
```

**(b) Solution**

**FIGURE E.5**   Example 10.15 using MATLAB® (b) Solution

**TABLE E.1** Varied Flow Function[a], Equation 10.7.11

| u | N 2½ | N 3 | N 3⅓ | u | N 2½ | N 3 | N 3⅓ |
|---|---|---|---|---|---|---|---|
| 0.00 | 0.000 | 0.000 | 0.000 | 0.82 | 1.057 | 0.993 | 0.963 |
| 0.02 | 0.020 | 0.020 | 0.020 | 0.83 | 1.083 | 1.016 | 0.985 |
| 0.04 | 0.040 | 0.040 | 0.040 | 0.84 | 1.110 | 1.040 | 1.007 |
| 0.06 | 0.060 | 0.060 | 0.060 | 0.85 | 1.139 | 1.065 | 1.030 |
| 0.08 | 0.080 | 0.080 | 0.080 | 0.86 | 1.171 | 1.092 | 1.055 |
| 0.10 | 0.100 | 0.100 | 0.100 | 0.87 | 1.205 | 1.120 | 1.081 |
| 0.12 | 0.120 | 0.120 | 0.120 | 0.88 | 1.241 | 1.151 | 1.109 |
| 0.14 | 0.140 | 0.140 | 0.140 | 0.89 | 1.279 | 1.183 | 1.139 |
| 0.16 | 0.161 | 0.160 | 0.160 | 0.90 | 1.319 | 1.218 | 1.172 |
| 0.18 | 0.181 | 0.180 | 0.180 | 0.91 | 1.362 | 1.257 | 1.206 |
| 0.20 | 0.201 | 0.200 | 0.200 | 0.92 | 1.400 | 1.300 | 1.246 |
| 0.22 | 0.222 | 0.221 | 0.220 | 0.93 | 1.455 | 1.348 | 1.290 |
| 0.24 | 0.243 | 0.241 | 0.240 | 0.94 | 1.520 | 1.403 | 1.340 |
| 0.26 | 0.263 | 0.261 | 0.261 | 0.950 | 1.605 | 1.467 | 1.398 |
| 0.28 | 0.284 | 0.282 | 0.281 | 0.960 | 1.703 | 1.545 | 1.468 |
| 0.30 | 0.305 | 0.302 | 0.301 | 0.970 | 1.823 | 1.644 | 1.559 |
| 0.32 | 0.326 | 0.323 | 0.322 | 0.975 | 1.899 | 1.707 | 1.615 |
| 0.34 | 0.347 | 0.343 | 0.342 | 0.980 | 1.996 | 1.783 | 1.684 |
| 0.36 | 0.368 | 0.364 | 0.363 | 0.985 | 2.111 | 1.880 | 1.772 |
| 0.38 | 0.391 | 0.385 | 0.383 | 0.990 | 2.273 | 2.017 | 1.895 |
| 0.40 | 0.413 | 0.407 | 0.404 | 0.995 | 2.550 | 2.250 | 2.106 |
| 0.42 | 0.435 | 0.428 | 0.425 | 0.999 | 3.195 | 2.788 | 2.590 |
| 0.44 | 0.458 | 0.450 | 0.447 | 1.000 | $\infty$ | $\infty$ | $\infty$ |
| 0.46 | 0.481 | 0.472 | 0.469 | 1.001 | 2.786 | 2.184 | 1.907 |
| 0.48 | 0.504 | 0.494 | 0.490 | 1.005 | 2.144 | 1.649 | 1.425 |
| 0.50 | 0.528 | 0.517 | 0.512 | 1.010 | 1.867 | 1.419 | 1.218 |
| 0.52 | 0.553 | 0.540 | 0.535 | 1.015 | 1.705 | 1.286 | 1.099 |
| 0.54 | 0.578 | 0.563 | 0.557 | 1.020 | 1.602 | 1.191 | 1.014 |
| 0.56 | 0.604 | 0.587 | 0.580 | 1.03 | 1.436 | 1.060 | 0.896 |
| 0.58 | 0.631 | 0.612 | 0.604 | 1.04 | 1.321 | 0.967 | 0.813 |
| 0.60 | 0.658 | 0.637 | 0.628 | 1.05 | 1.242 | 0.896 | 0.749 |
| 0.61 | 0.673 | 0.650 | 0.641 | 1.06 | 1.166 | 0.838 | 0.697 |
| 0.62 | 0.686 | 0.663 | 0.653 | 1.07 | 1.111 | 0.790 | 0.651 |
| 0.63 | 0.700 | 0.676 | 0.666 | 1.08 | 1.059 | 0.749 | 0.618 |
| 0.64 | 0.716 | 0.690 | 0.679 | 1.09 | 1.012 | 0.713 | 0.586 |
| 0.65 | 0.731 | 0.703 | 0.692 | 0.10 | 0.973 | 0.681 | 0.558 |
| 0.66 | 0.746 | 0.717 | 0.705 | 1.11 | 0.939 | 0.652 | 0.532 |
| 0.67 | 0.762 | 0.731 | 0.718 | 1.12 | 0.907 | 0.626 | 0.509 |
| 0.68 | 0.777 | 0.746 | 0.732 | 1.13 | 0.878 | 0.602 | 0.488 |
| 0.69 | 0.795 | 0.761 | 0.746 | 1.14 | 0.851 | 0.581 | 0.479 |
| 0.70 | 0.811 | 0.776 | 0.760 | 1.15 | 0.824 | 0.561 | 0.452 |
| 0.71 | 0.828 | 0.791 | 0.775 | 1.16 | 0.802 | 0.542 | 0.436 |
| 0.72 | 0.845 | 0.807 | 0.790 | 1.17 | 0.782 | 0.525 | 0.421 |
| 0.73 | 0.863 | 0.823 | 0.805 | 1.18 | 0.760 | 0.509 | 0.406 |
| 0.74 | 0.881 | 0.840 | 0.821 | 1.19 | 0.740 | 0.494 | 0.393 |
| 0.75 | 0.900 | 0.857 | 0.837 | 1.20 | 0.723 | 0.480 | 0.381 |
| 0.76 | 0.919 | 0.874 | 0.853 | 1.22 | 0.692 | 0.454 | 0.358 |
| 0.77 | 0.940 | 0.892 | 0.870 | 1.24 | 0.662 | 0.431 | 0.338 |
| 0.78 | 0.962 | 0.911 | 0.887 | 1.26 | 0.633 | 0.410 | 0.320 |
| 0.79 | 0.985 | 0.930 | 0.905 | 1.28 | 0.609 | 0.391 | 0.303 |
| 0.80 | 1.008 | 0.950 | 0.924 | 1.30 | 0.587 | 0.373 | 0.289 |
| 0.81 | 1.032 | 0.971 | 0.943 | 1.32 | 0.568 | 0.357 | 0.275 |

**TABLE E.1**   Varied Flow Function[a], Equation 10.7.11 (*continued*)

| u | N 2½ | N 3 | N 3⅓ | u | N 2½ | N 3 | N 3⅓ |
|------|-------|-------|-------|------|-------|-------|-------|
| 1.34 | 0.549 | 0.342 | 0.262 | 2.20 | 0.220 | 0.107 | 0.071 |
| 1.36 | 0.531 | 0.329 | 0.251 | 2.3  | 0.204 | 0.098 | 0.064 |
| 1.38 | 0.513 | 0.316 | 0.239 | 2.4  | 0.190 | 0.089 | 0.057 |
| 1.40 | 0.496 | 0.304 | 0.229 | 2.5  | 0.179 | 0.082 | 0.052 |
| 1.42 | 0.481 | 0.293 | 0.220 | 2.6  | 0.169 | 0.076 | 0.048 |
| 1.44 | 0.467 | 0.282 | 0.211 | 2.7  | 0.160 | 0.070 | 0.043 |
| 1.46 | 0.455 | 0.272 | 0.203 | 2.8  | 0.150 | 0.065 | 0.040 |
| 1.48 | 0.444 | 0.263 | 0.196 | 2.9  | 0.142 | 0.060 | 0.037 |
| 1.50 | 0.432 | 0.255 | 0.188 | 3.0  | 0.135 | 0.056 | 0.034 |
| 1.55 | 0.405 | 0.235 | 0.172 | 3.5  | 0.106 | 0.041 | 0.024 |
| 1.60 | 0.380 | 0.218 | 0.158 | 4.0  | 0.087 | 0.031 | 0.017 |
| 1.65 | 0.359 | 0.203 | 0.145 | 4.5  | 0.072 | 0.025 | 0.013 |
| 1.70 | 0.340 | 0.189 | 0.135 | 5.0  | 0.062 | 0.019 | 0.010 |
| 1.75 | 0.322 | 0.177 | 0.125 | 6.0  | 0.048 | 0.014 | 0.007 |
| 1.80 | 0.308 | 0.166 | 0.116 | 7.0  | 0.038 | 0.010 | 0.005 |
| 1.85 | 0.293 | 0.156 | 0.108 | 8.0  | 0.031 | 0.008 | 0.004 |
| 1.90 | 0.279 | 0.147 | 0.102 | 9.0  | 0.027 | 0.006 | 0.003 |
| 1.95 | 0.268 | 0.139 | 0.095 | 10.0 | 0.022 | 0.005 | 0.002 |
| 2.00 | 0.257 | 0.132 | 0.089 | 20.0 | 0.015 | 0.002 | 0.001 |
| 2.10 | 0.238 | 0.119 | 0.079 |      |       |       |       |

[a]$F(u, N) = \int_0^u du/(1 - u^N)$ with constant of integration adjusted so that $F(0, N) = 0$ and $F(\infty, N) = 0$.
Source: After Henderson, 1966.

# F. NUMERICAL SOLUTIONS FOR CHAPTER 11

Given data:

$$L := \begin{pmatrix} 100 \\ 150 \\ 200 \end{pmatrix} \quad D := \begin{pmatrix} 0.05 \\ 0.075 \\ 0.085 \end{pmatrix} \quad e := \begin{pmatrix} 0.0001 \\ 0.0002 \\ 0.0001 \end{pmatrix} \quad K := \begin{pmatrix} 10 \\ 3 \\ 2 \end{pmatrix}$$

$$Q_T := 0.02 \quad g := 9.81 \quad v := 1.10^{-6}$$

Define function for resistance coeffcient:

$$R(e, D, Q, L, K, C) := \frac{1.07 \cdot L}{g \cdot D^5} \cdot \left[ \ln\left[ 0.27 \cdot \frac{e}{D} + C \cdot 5 \cdot 74 \cdot \left( \frac{\pi \cdot v \cdot D}{4 \cdot Q} \right)^{0.9} \right] \right]^{-2} + \frac{K}{2 \cdot g \cdot \left( \frac{\pi \cdot D^2}{4} \right)^2}$$

Estimate initial values of unknowns:

$$W := Q_T^2 \cdot \left( \sum_{i=0}^{2} \frac{1}{\sqrt{R(e_i, D_i, Q_T, L_i, K_i, 0)}} \right)^{-2}$$

$$i := 0..2 \quad Q_i := \sqrt{\frac{W}{R(e_i, D_i, Q_T, L_i, K_i, 1)}}$$

Solve system of equations:

$$\text{Given} \quad W = Q_T^2 \cdot \left( \sum_{i=0}^{2} \frac{1}{\sqrt{R(e_i, D_i, Q_T, L_i, K_i, 1)}} \right)^{-2} \quad Q_T = \sum_{i=0}^{2} Q_i$$

$$\text{Find}(W, Q_0, Q_1, Q_2) = \begin{pmatrix} 7.858 \\ 3.122 \times 10^{-3} \\ 7.274 \times 10^{-3} \\ 9.604 \times 10^{-3} \end{pmatrix}$$

**FIGURE F.1**   Example 11.3 Mathcad solution

$$L := \begin{pmatrix} 500 \\ 750 \\ 1000 \end{pmatrix} \qquad D := \begin{pmatrix} 0.10 \\ 0.15 \\ 0.13 \end{pmatrix} \qquad f := \begin{pmatrix} 0.025 \\ 0.020 \\ 0.018 \end{pmatrix}$$

$$H := \begin{pmatrix} 5 \\ 20 \\ 13 \end{pmatrix} \qquad K := \begin{pmatrix} 3 \\ 2 \\ 7 \end{pmatrix} \qquad g := 9.81$$

Resistance coefficients:

$$i := 0..2 \qquad R_i := \frac{8 \cdot (f_i \cdot L_i + D_i \cdot K_i)}{g \cdot \pi^2 \cdot (D_i)^5}$$

Estimate initial values of unknowns:

$$Q := \begin{pmatrix} 0.01 \\ 0.01 \\ 0.01 \end{pmatrix} \qquad H_J := 14$$

Solve for unknowns:

Given

$$H_J - H_0 = R_0 \cdot Q_0 \cdot |Q_0| \qquad\qquad H_J - H_2 = R_2 \cdot Q_2 \cdot |Q_2|$$

$$H_1 - H_J = R_1 \cdot Q_1 \cdot |Q_1| \qquad\qquad Q_0 - Q_1 + Q_2 = 0$$

$$\text{Find } (H_J, Q_0, Q_1, Q_2) = \begin{pmatrix} 15.182 \\ 9.812 \times 10^{-3} \\ 1,701 \times 10^{-2} \\ 7.201 \times 10^{-3} \end{pmatrix}$$

**FIGURE F.2**  Example 11.4 Mathcad solution

(a)

```
clear;
%Given Data
global H R;
L=[500 750 1000];
D=[0.10 0.15 0.13];
f=[0.025 0.020 0.018];
K=[3 2 7];
H=[5 20 13];
g=9.81;

%Evaluate equivalent lengths and resistence
coefficients
Le=D.*k./f;
R=8*f.*(L+Le)./(g*pi^2*D.^5);

%Initial estimates of unknowns x0=[HB Q1 Q2 Q3]
x0=[14 0.01 0.01 0.01];

%Call function file f.m and solve for unknows
options=optimset ('precondbandwidth',Inf);
[x, fval] = fsolve('f', x0, options);
x
```

(b)

```
%Define each function as f(x)=0

function F = f(x);
global H R;

F = [x(1) - H(1) - R(1)*x(2)*abs(x(2));
     H(2) - x(1) - R(2)*x(3)*abs(x(3));
     x(1) - H(3) - R(3)*x(4)*abs(x(4));
     x(2) - x(3) + x(4)];
```

(c)

```
>> ex11_4

Optimization terminated successfully:
 Relative function value changing by less than
OPTIONS.TolFun
x =

    15.1818  0.0098  0.0170  0.0072
```

**FIGURE F.3**   Example 11.4 MATLAB® solution: (a) main algorithm, (b) function subroutine, (c) output. Note that the Optimization Toolbox is required.

Given data:

$$L_1: = 50 \qquad D_1: = 0.15 \qquad f_1: = 0.02 \qquad K_1: = 2 \qquad H_1: = 10$$
$$L_2: = 100 \qquad D_2: = 0.10 \qquad f_2: = 0.015 \qquad K_2: = 1 \qquad H_2: = 30$$
$$L_3: = 300 \qquad D_3: = 0.10 \qquad f_3: = 0.025 \qquad K_3: = 1 \qquad H_3: = 15$$
$$g: = 9.81 \qquad P: = 20 \cdot 10^3$$

Equivalent lengths:

$$Leq_2: = \frac{D_2 \cdot K_2}{f_2} \qquad Leq_1: = \frac{D_1 \cdot K_1}{f_1} \qquad Leq_3: = \frac{D_3 \cdot K_3}{f_3}$$

Resistance coefficients:

$$R_1: = \frac{8 \cdot f_1 \cdot (L_1 + Leq1)}{g \cdot \pi^2 \cdot D_1^5} \qquad R_2: = \frac{8 \cdot f_2 \cdot (L_2 + Leq_2)}{g \cdot \pi^2 \cdot D_2^5} \qquad R_3: = \frac{8 \cdot f_3 \cdot (L_3 + Leq_3)}{g \cdot \pi^2 \cdot D_3^5}$$

Initial estimates of unknowns:

$$Q_1: = 0.05 \qquad Q_2: = 0.05 \qquad Q_3: = 0.05 \qquad H_B: = 20$$

Solve for unknowns:

Given

$$H_1 + \frac{P}{9800 \cdot Q_1} - H_B = R_1 \cdot Q_1 \cdot |Q_1| \qquad H_B - H_3 = R_3 \cdot Q_3 \cdot |Q_3|$$
$$H_B - H_2 = R_2 \cdot Q_2 \cdot |Q_2| \qquad\qquad Q_1 - Q_2 - Q_3 = 0$$

$$\text{Find } (HB, Q_1, Q_2, Q_3) = \begin{pmatrix} 43.845 \\ 0.054 \\ 0.032 \\ 0.021 \end{pmatrix}$$

**FIGURE F.4**  Example 11.5 Mathcad solution

(a)

```
clear;
%Given Data
global H R P;
L=[50 100 300];
D=[0.15 0.10 0.10];
f=[0.020 0.015 0.025];
K=[2 1 1];
H=[10 30 15];
g=9.81;
p=20000;

%Evaluate equivalent lengths and resistence
coefficients
Le=D.*k./f;
R=8*f.*(L+Le)./(g*pi^2*D.^5);

%Initial estimates of unknowns x0=[HB Q1 Q2 Q3]
x0=[20 0.05 0.05 0.05];

%Call function file g.m and solve for unknowns
options=optimset ('precondbandwidth', Inf);
[x, fval] = fsolve('g', x0, options);
x
```

(b)

```
%Define each function as g(x)=0

function G = g(x);
global H R P;

G = [H(1) + p/(9800*x(2)) - x(1) - R(1)*x(2)*abs(x(2));
     x(1) - H(2) - R(2)*x(3)*abs(x(3));
     x(1) - H(3) - R(3)*x(4)*abs(x(4));
     x(2) - x(3) - x(4)];
```

(c)

```
>> ex11_5
Optimization terminated successfully:
 Relative function value changing by less than
OPTIONS.TolFun

x =

   43.8449  0.0538  0.0324  0.0214
```

**FIGURE F.5**   Example 11.5 MATLAB® solution: (a) main algorithm, (b) function subroutine, (c) output. Note that the Optimization Toolbox is required.

**FIGURE F.6** Example 11.6 using Excel® (a) Solution

| | | | Iteration 1 | | | Iteration 2 | | | Iteration 3 | | | Iteration 4 | | | |
|---|---|---|---|---|---|---|---|---|---|---|---|---|---|---|---|
| | | R | Q | RQ\|Q\| | 2R\|Q\| | Q | RQ\|Q\| | 2R\|Q\| | Q | RQ\|Q\| | 2R\|Q\| | Q | RQ\|Q\| | 2R\|Q\| | Q |
| | ΔH | | | 20.000 | | | 20.000 | | | 20.000 | | | 20.000 | | |
| | Pipe 4 | 100 | -0.020 | -0.040 | 4.000 | -0.022 | -0.051 | 4.495 | -0.019 | -0.038 | 3.899 | -0.020 | -0.039 | 3.968 | -0.019 |
| | Pipe 3 | 200 | -0.060 | -0.720 | 24.000 | -0.064 | -0.809 | 25.440 | -0.062 | -0.770 | 24.817 | -0.063 | -0.787 | 25.098 | -0.062 |
| Loop 1 | Pipe 2 | 500 | -0.130 | -8.450 | 130.000 | -0.137 | -9.433 | 137.352 | -0.133 | -8.907 | 133.466 | -0.135 | -9.150 | 135.276 | -0.134 |
| | Pipe 1 | 100 | -0.320 | -10.240 | 64.000 | -0.322 | -10.399 | 64.495 | -0.319 | -10.208 | 63.899 | -0.320 | -10.230 | 63.968 | -0.319 |
| | | | | 0.550 | 222.000 | | -0.691 | 231.783 | | 0.078 | 226.081 | | -0.206 | 228.309 | |
| | | | | ΔQ = | -2.48E-03 | | ΔQ = | 2.98E-03 | | ΔQ = | -3.44E-04 | | ΔQ = | 9.03E-04 | |
| | Pipe 2 | 500 | 0.130 | 8.450 | 130.000 | 0.137 | 9.433 | 137.352 | 0.133 | 8.907 | 133.466 | 0.135 | 9.150 | 135.276 | 0.134 |
| | Pipe 8 | 300 | 0.070 | 1.470 | 42.000 | 0.074 | 1.632 | 44.251 | 0.071 | 1.530 | 42.853 | 0.073 | 1.578 | 43.519 | 0.072 |
| Loop 2 | Pipe 7 | 400 | -0.040 | -0.640 | 32.000 | -0.035 | -0.494 | 28.101 | -0.036 | -0.519 | 28.823 | -0.035 | -0.478 | 27.650 | -0.035 |
| | Pipe 6 | 300 | -0.190 | -10.830 | 114.000 | -0.185 | -10.281 | 111.075 | -0.186 | -10.382 | 111.617 | -0.185 | -10.219 | 110.738 | -0.185 |
| | | | | -1.550 | 318.000 | | 0.290 | 320.779 | | -0.464 | 316.759 | | 0.031 | 317.183 | |
| | | | | ΔQ = | 4.87E-03 | | ΔQ = | -9.03E-04 | | ΔQ = | 1.47E-03 | | ΔQ = | -9.83E-05 | |
| | Pipe 3 | 200 | 0.060 | 0.720 | 24.000 | 0.064 | 0.809 | 25.440 | 0.062 | 0.770 | 24.817 | 0.063 | 0.787 | 25.098 | 0.062 |
| Loop 2 | Pipe 5 | 400 | 0.040 | 0.640 | 32.000 | 0.041 | 0.676 | 32.898 | 0.043 | 0.724 | 34.039 | 0.043 | 0.736 | 34.325 | 0.043 |
| | Pipe 8 | 300 | -0.070 | -1.470 | 42.000 | -0.074 | -1.632 | 44.251 | -0.071 | -1.530 | 42.853 | -0.073 | -1.578 | 43.519 | -0.072 |
| | | | | -0.110 | 98.000 | | -0.146 | 102.589 | | -0.036 | 101.710 | | -0.054 | 102.941 | |
| | | | | ΔQ = | 1.12E-03 | | ΔQ = | 1.43E-03 | | ΔQ = | 3.57E-04 | | ΔQ = | 5.29E-04 | |

| A | B | C | D | E | F | G |
|---|---|---|---|---|---|---|
| | | | | Iteration 1 | | |
| | | R | Q | RQ\|Q\| | 2R\|Q\| | Q |
| | ΔH | | | 20 | | |
| | Pipe 4 | 100 | −0.02 | =C6*D6*ABS(D6) | =2*C6*ABS(D6) | =D6+F13 |
| Loop 1 | Pipe 3 | 200 | −0.06 | =C7*D7*ABS(D7) | =2*C7*ABS(D7) | =D7+F13−F32 |
| | Pipe 2 | 500 | −0.13 | =C8*D8*ABS(D8) | =2*C8*ABS(D8) | =D8+F13−F23 |
| | Pipe 1 | 100 | −0.32 | =C9*D9*ABS(D9) | =2*C9*ABS(D9) | =D9+F13 |
| | | | | =SUM(E5:E9) | =SUM(F6:F9) | |
| | | | | | ΔQ = =−E11/F11 | |
| | Pipe 2 | 500 | 0.13 | =C16*D16*ABS(D16) | =2*C16*ABS(D16) | =D16+F23−F13 |
| Loop 2 | Pipe 8 | 300 | 0.07 | =C17*D17*ABS(D17) | =2*C17*ABS(D17) | =D17+F23−F32 |
| | Pipe 7 | 400 | −0.04 | =C18*D18*ABS(D18) | =2*C18*ABS(D18) | =D18+F23 |
| | Pipe 6 | 300 | −0.19 | =C19*D19*ABS(D19) | =2*C19*ABS(D19) | =D19+F23 |
| | | | | =SUM(E16:E19) | =SUM(F15:F19) | |
| | | | | | ΔQ = =−E21/F21 | |
| | Pipe 3 | 200 | 0.06 | =C26*D26*ABS(D26) | =2*C26*ABS(D26) | =D26+F32−F13 |
| Loop 2 | Pipe 5 | 400 | 0.04 | =C27*D27*ABS(D27) | =2*C27*ABS(D27) | =D27+F32 |
| | Pipe 8 | 300 | −0.07 | =C28*D28*ABS(D28) | =2*C28*ABS(D28) | =D28+F32−F23 |
| | | | | =SUM(E26:E28) | =SUM(F26:F28) | |
| | | | | | ΔQ = =−E30/F30 | |

**FIGURE F.6 (cont'd)**  Example 11.6 using Excel® (b) Spreadsheet formulas

Given data:

$$R_1 := 100 \quad R_3 := 400 \quad R_5 := 400 \quad R_7 := 400 \qquad H_A := 50 \quad QF := 0.15$$
$$R_2 := 500 \quad R_4 := 100 \quad R_6 := 300 \quad R_8 := 300 \qquad H_B := 30 \quad QG := 0.15$$

Initial estimates of unknowns:

$$Q_1 := 0.50 \qquad Q_4 := 0.50 \qquad Q_7 := 0.50 \qquad H_C := 35 \qquad H_F := 35$$
$$Q_2 := 0.50 \qquad Q_5 := 0.50 \qquad Q_8 := 0.50 \qquad H_D := 35 \qquad H_G := 35$$
$$Q_3 := 0.50 \qquad Q_6 := 0.50 \qquad\qquad\qquad\qquad H_E := 35$$

Solve for unknowns:

Given

$$H_A - H_C = R_1 \cdot Q_1 \cdot |Q_1| \qquad H_E - H_G = R_5 \cdot Q_5 \cdot |Q_5| \qquad Q_1 = Q_2 + Q_6$$
$$H_C - H_D = R_2 \cdot Q_2 \cdot |Q_2| \qquad H_C - H_F = R_6 \cdot Q_6 \cdot |Q_6| \qquad Q_2 = Q_3 + Q_8$$
$$H_D - H_E = R_3 \cdot Q_3 \cdot |Q_3| \qquad H_F - H_G = R_7 \cdot Q_7 \cdot |Q_7| \qquad Q_3 = Q_4 + Q_5$$
$$H_E - H_B = R_4 \cdot Q_4 \cdot |Q_4| \qquad H_D - H_G = R_8 \cdot Q_8 \cdot |Q_8| \qquad Q_6 = Q_7 + QF$$
$$\qquad\qquad\qquad\qquad\qquad\qquad\qquad\qquad\qquad\qquad Q_7 + Q_8 + Q_5 = QG$$

$$\text{Find}(H_C, H_D, H_F, H_G, Q_1, Q_2, Q_3, Q_4, Q_5, Q_6, Q_7, Q_8,) = \begin{pmatrix} 39.862 \\ 30.814 \\ 30.034 \\ 29.717 \\ 29.257 \\ 0.318 \\ 0.135 \\ 0.062 \\ 0.018 \\ 0.044 \\ 0.184 \\ 0.034 \\ 0.072 \end{pmatrix}$$

**FIGURE F.7** Example 11.6 Mathcad solution

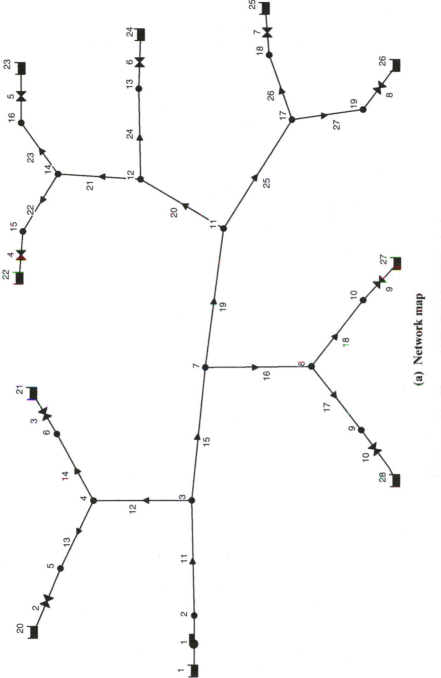

**(a) Network map**

**FIGURE F.8**   Figure 11.7a EPANET 2 solution

```
Link - Node Table:
```

| Link ID | Start Node | End Node | Length m | Diameter mm |
|---|---|---|---|---|
| 11 | 2 | 3 | 195 | 100 |
| 12 | 3 | 4 | 158 | 50 |
| 13 | 4 | 5 | 115 | 50 |
| 14 | 4 | 6 | 155 | 50 |
| 15 | 3 | 7 | 188 | 100 |
| 16 | 7 | 8 | 117 | 100 |
| 17 | 8 | 9 | 59 | 50 |
| 18 | 8 | 10 | 82 | 50 |
| 19 | 7 | 11 | 130 | 100 |
| 20 | 11 | 12 | 102 | 100 |
| 21 | 12 | 14 | 78 | 100 |
| 22 | 14 | 15 | 55 | 50 |
| 23 | 14 | 16 | 60 | 50 |
| 24 | 12 | 13 | 114 | 50 |
| 25 | 11 | 17 | 165 | 100 |
| 26 | 17 | 18 | 64 | 50 |
| 27 | 17 | 19 | 127 | 50 |
| 1 | 1 | 2 | #N/A | #N/A Pump |
| 2 | 5 | 20 | #N/A | 50 Valve |
| 3 | 6 | 21 | #N/A | 50 Valve |
| 4 | 15 | 22 | #N/A | 50 Valve |
| 5 | 16 | 23 | #N/A | 50 Valve |
| 6 | 13 | 24 | #N/A | 50 Valve |
| 7 | 18 | 25 | #N/A | 50 Valve |
| 8 | 19 | 26 | #N/A | 50 Valve |
| 9 | 10 | 27 | #N/A | 50 Valve |
| 10 | 9 | 28 | #N/A | 50 Valve |

**(b) Computed results**

```
Node Results:
--------------------------------------------------------------------------------
Node          Demand          Head            Pressure            Quality
ID            LPS             m               m
--------------------------------------------------------------------------------

2             0.00            135.85          105.85              0.00
3             0.00            89.79           57.79               0.00
4             0.00            47.59           12.59               0.00
5             0.00            40.02           3.02                0.00
6             0.00            36.36           3.36                0.00
7             0.00            58.08           28.08               0.00
8             0.00            53.37           26.37               0.00
9             0.00            37.67           13.67               0.00
10            0.00            36.41           10.41               0.00
11            0.00            49.22           22.22               0.00
12            0.00            47.26           14.26               0.00
13            0.00            38.55           3.55                0.00
14            0.00            46.56           6.56                0.00
15            0.00            42.98           2.98                0.00
16            0.00            41.74           3.74                0.00
17            0.00            46.01           20.01               0.00
18            0.00            34.75           8.75                0.00
19            0.00            30.07           6.07                0.00
1             -38.91          30.00           0.00                0.00 Reservoir
20            3.15            37.00           0.00                0.00 Reservoir
21            3.32            33.00           0.00                0.00 Reservoir
22            3.13            40.00           0.00                0.00 Reservoir
23            3.51            38.00           0.00                0.00 Reservoir
24            3.42            35.00           0.00                0.00 Reservoir
25            5.36            26.00           0.00                0.00 Reservoir
26            4.47            24.00           0.00                0.00 Reservoir
27            5.85            26.00           0.00                0.00 Reservoir
28            6.70            24.00           0.00                0.00 Reservoir
```

**FIGURE F.8   (b)  Computed results (*continued*)**

```
Link Results:
--------------------------------------------------------------------------------
Link            Flow            Velocity        Headloss        Status
  ID            LPS             m/s             m/km
--------------------------------------------------------------------------------
11              38.91           4.96            236.24          Open
12               6.47           3.30            267.05          Open
13               3.15           1.61             65.81          Open
14               3.32           1.69             72.49          Open
15              32.44           4.13            168.65          Open
16              12.55           1.60             40.21          Open
17               6.70           3.42            266.15          Open
18               5.85           2.98            206.84          Open
19              19.89           2.53             68.13          Open
20              10.06           1.28             19.27          Open
21               6.64           0.85              8.94          Open
22               3.13           1.60             65.00          Open
23               3.51           1.79             80.23          Open
24               3.42           1.74             76.36          Open
25               9.83           1.25             19.45          Open
26               5.36           2.73            176.01          Open
27               4.47           2.28            125.52          Open
 1              38.91           0.00           -105.85          Open Pump
 2               3.15           1.61              3.02          Active Valve
 3               3.32           1.69              3.36          Active Valve
 4               3.13           1.60              2.98          Active Valve
 5               3.51           1.79              3.74          Active Valve
 6               3.42           1.74              3.55          Active Valve
 7               5.36           2.73              8.75          Active Valve
 8               4.47           2.28              6.07          Active Valve
 9               5.85           2.98             10.41          Active Valve
10               6.70           3.42             13.67          Active Valve
```

**(b) Computed results (*continued*)**

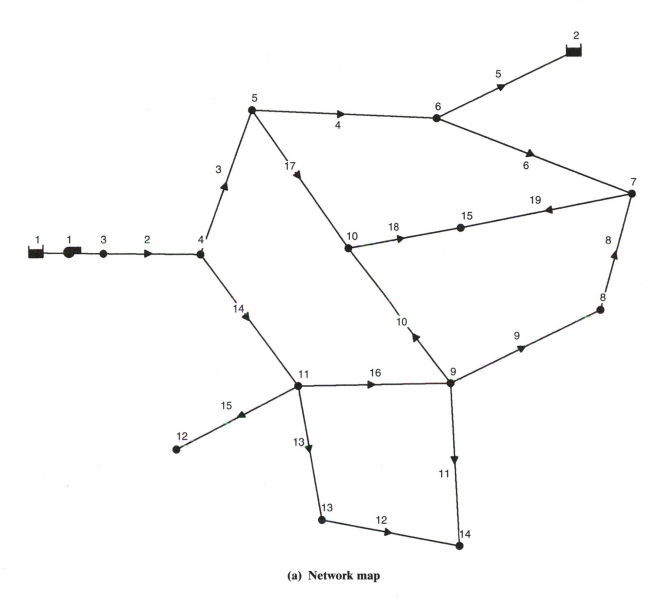

**(a)  Network map**

**FIGURE F.9**   Figure 11.7b EPANET 2 solution

```
Link - Node Table:
```

| Link ID | Start Node | End Node | Length m | Diameter mm |
|---|---|---|---|---|
| 2 | 3 | 4 | 3000 | 600 |
| 3 | 4 | 5 | 1520 | 450 |
| 4 | 5 | 6 | 1520 | 400 |
| 5 | 6 | 2 | 305 | 150 |
| 6 | 6 | 7 | 1680 | 350 |
| 8 | 7 | 8 | 1070 | 300 |
| 9 | 8 | 9 | 1680 | 350 |
| 10 | 9 | 10 | 1680 | 300 |
| 11 | 9 | 14 | 1380 | 300 |
| 12 | 14 | 13 | 760 | 150 |
| 13 | 13 | 11 | 1100 | 300 |
| 14 | 11 | 4 | 2000 | 450 |
| 15 | 11 | 12 | 1200 | 400 |
| 16 | 9 | 11 | 670 | 380 |
| 17 | 10 | 5 | 1520 | 350 |
| 18 | 10 | 15 | 900 | 350 |
| 19 | 15 | 7 | 1200 | 350 |
| 1 | 1 | 3 | #N/A | #N/A Pump |

**(b)  Computed results**

Node Results:

| Node ID | Demand LPS | Head m | Pressure m | Quality | |
|---|---|---|---|---|---|
| 3 | 0.00 | 192.15 | 177.15 | 0.00 | |
| 4 | 0.00 | 155.46 | 109.49 | 0.00 | |
| 5 | 140.00 | 132.55 | 83.55 | 0.00 | |
| 6 | 0.00 | 126.03 | 76.03 | 0.00 | |
| 7 | 100.00 | 122.64 | 73.64 | 0.00 | |
| 8 | 100.00 | 123.04 | 77.04 | 0.00 | |
| 9 | 0.00 | 130.00 | 87.00 | 0.00 | |
| 10 | 140.00 | 124.13 | 80.13 | 0.00 | |
| 11 | 0.00 | 136.56 | 92.56 | 0.00 | |
| 12 | 55.00 | 136.02 | 96.02 | 0.00 | |
| 13 | 55.00 | 133.12 | 92.12 | 0.00 | |
| 14 | 55.00 | 128.61 | 88.61 | 0.00 | |
| 15 | 85.00 | 122.59 | 76.59 | 0.00 | |
| 1 | -823.11 | 15.00 | 0.00 | 0.00 | Reservoir |
| 2 | 93.11 | 61.00 | 0.00 | 0.00 | Reservoir |

Link Results:

| Link ID | Flow LPS | Velocity m/s | Headloss m/km | Status | |
|---|---|---|---|---|---|
| 2 | 823.11 | 2.91 | 12.22 | Open | |
| 3 | 460.26 | 2.90 | 15.10 | Open | |
| 4 | 177.63 | 1.41 | 4.29 | Open | |
| 5 | 93.11 | 5.27 | 213.20 | Open | |
| 6 | 84.52 | 0.88 | 2.01 | Open | |
| 8 | -22.87 | 0.32 | 0.37 | Open | |
| 9 | -122.87 | 1.28 | 4.15 | Open | |
| 10 | 74.98 | 1.06 | 3.50 | Open | |
| 11 | 39.19 | 0.55 | 1.01 | Open | |
| 12 | -15.81 | 0.89 | 5.94 | Open | |
| 13 | -70.81 | 1.00 | 3.13 | Open | |
| 14 | -362.85 | 2.28 | 9.46 | Open | |
| 15 | 55.00 | 0.44 | 0.45 | Open | |
| 16 | -237.04 | 2.09 | 9.80 | Open | |
| 17 | -142.63 | 1.48 | 5.54 | Open | |
| 18 | 77.61 | 0.81 | 1.71 | Open | |
| 19 | -7.39 | 0.10 | 0.05 | Open | |
| 1 | 823.11 | 0.00 | -177.15 | Open | Pump |

**FIGURE F.9   (b) Computed results (*continued*)**

# Bibliography

## REFERENCES

ADRIAN, R. J., "Multi-point Optical Measurements of Simultaneous Vectors in Unsteady Flow—A Review," *Int. J. Heat Fluid Flow*, Vol. 7, No. 2, June 1986, pp. 127–145.

BAKHMETEFF, B. A., *Hydraulics of Open Channels*, McGraw-Hill Book Company, New York, 1932.

BEAN, H. S., ed., *Fluid Meters: Their Theory and Application*, 6th ed., American Society of Mechanical Engineers, New York, 1971.

BENEDICT, R. P., *Fundamentals of Pipe Flow*, John Wiley & Sons, Inc., New York, 1980.

CHAPRA, S. C., and CANALE, R. P., *Numerical Methods for Engineers*, 3rd ed., McGraw-Hill Book Company, New York, 1998.

CHOW, V. T., *Open Channel Hydraulics*, McGraw-Hill Book Company, New York, 1959.

COLEMAN, HUGH W., and STEELE, W. GLENN, JR, *Experimentation and Uncertainty Analysis for Engineers*, 2nd ed. John Wiley & Sons, New York, 1998.

CROSS, H., "Analysis of Flow in Networks of Conduits or Conductors," *University of Illinois Bulletin 286*, November 1936.

GOLDSTEIN, R. J., ed., *Fluid Mechanics Measurements*, 2nd ed., Taylor and Francis, Washington, DC, 1996.

*HEC-RAS River Analysis System*, Hydrologic Engineering Center, U.S. Army Corps of Engineers, Davis, CA, July 1995.

HENDERSON, F. M., *Open Channel Flow*, Macmillan Publishing Company, New York, 1966.

KARASSIK, I. J., MESSINA, J. P., COOPER, P., HEALD, C. C., eds., *Pump Handbook*, 3rd ed., McGraw-Hill Professional Publishing, 2000.

KING, H. W., and BRATER, E. F., *Handbook of Hydraulics*, 7th ed., McGraw-Hill Book Company, New York, 1996.

KLINE, S. J., ed., Proceedings of the Symposium on Uncertainty Analysis. *J. Fluids Eng.*, Vol. 107, No. 2, June 1985, pp. 153–178.

MCBEAN, E. A., and PERKINS, F. E., "Convergence Schemes in Water Profile Computation," *J. Hydraulics Div., ASCE*, Vol. 101, No. HY10, October 1975, pp. 1380–1384.

MARTIN, C. S., "Experimental Investigation of Column Separation with Rapid Valve Closure," *Proceedings, 4th International Conference on Pressure Surges*, BHRA Fluid Engineering, Cranfield, England, 1983, pp. 77–88.

ORMSBEE, L. E., and WOOD, D. J., "Hydraulic Design Algorithms for Pipe Networks," *J. Hydraulic Eng., ASCE*, Vol. 112, No. 12, December 1986, pp. 1195–1207.

ROBERSON, J. A., CASSIDY, J. J., and CHAUDHRY, M. H., *Hydraulic Engineering*, Houghton Mifflin Company, Boston, MA, 1988.

ROBERSON, J. A., and CROWE, C. T., *Engineering Fluid Mechanics*, 6th ed., John Wiley & Sons, 1997.

ROSSMAN, L. A., *EPANETZ Users Manual*, United States Environmental Protection Agency, June 2000.

STEPANOFF, A. J., *Centrifugal and Axial Flow Pumps*, 2nd ed., John Wiley & Sons, Inc., New York, 1957.

SWAMEE, P. K., and JAIN, A. K., "Explicit Equations for Pipe-Flow Problems," *J. Hydraulics Div., ASCE*, Vol. 102, No. HY5, May 1976, pp. 657–664.

TAYLOR, G. I., "Dispersion of Soluble Matter in Solvent Flowing Slowly Through a Tube," *Proc. Royal Soc. London*, Vol. 219A, 1953, pp. 186–203.

U. S. DEPARTMENT OF INTERIOR, BUREAU OF RECLAMATION, *Design of Small Dams*, U.S. Government Printing Office, Washington DC, 1974.

VAN DYKE, M., *An Album of Fluid Motion*, Parabolic Press, Stanford, CA, 1982.

WARNICK, C. C., *Hydropower Engineering*, Prentice-Hall, Inc., Englewood Cliffs, NJ, 1984.

WOOD, D. J., "Algorithms for Pipe Network Analysis and Their Reliability," *Research Report 127*, Water Resources Research Institute, University of Kentucky, Lexington, KY, 1981.

WYLIE, E. B., and STREETER, V. L., *Fluid Transients in Systems*, Prentice Hall, Englewood Cliffs, N.J., 1993.

## GENERAL INTEREST

DAILY, J. W., and HARLEMAN, D. R. F., *Fluid Dynamics*, Addison-Wesley Publishing Company, Inc., Reading, MA, 1968.

ESKINAZI, S., *Principles of Fluid Mechanics*, 2nd ed., Allyn and Bacon, Inc., Needham Heights, MA, 1968.

FOX, R. W., and MCDONALD, A. T., *Introduction to Fluid Mechanics*, 4th ed., John Wiley & Sons, Inc., New York, 1992.

FUNG, Y. C., *Continuum Mechanics*, 2nd ed., Prentice-Hall, Inc., Englewood Cliffs, NJ, 1977.

HANSON, A. G., *Fluid Mechanics*, John Wiley & Sons, Inc., New York, 1967.

JOHN, E. A. J., *Gas Dynamics*, Allyn and Bacon, Inc., Needham Heights, MA, 1969.

MUNSON, B. R., YOUNG, D. F., and OKIISHI, T. H., *Fundamentals of Fluid Mechanics*, 2nd ed., John Wiley & Sons, Inc., New York, 1994.

OWCZAREK, J. A., *Introduction to Fluid Mechanics*, International Textbook Company, Scranton, PA, 1968.

POTTER, M. C., and FOSS, J. F., *Fluid Mechanics*, Great Lakes Press, Okemos, MI, 1979.

ROBERSON, J. A., and CROWE, C. T., *Engineering Fluid Mechanics*, 6th ed., Houghton-Mifflin Company, Boston, MA, 1996.

SABERSKY, R. H., ACOSTA, A. J., and HAUPTMAN, E. G., *Fluid Flow*, 5th ed., Macmillan Publishing Company, New York, 2001.

SCHLICHTING, H., *Boundary Layer Theory*, 8th ed., McGraw-Hill Book Company, New York, 2000.

SHAMES., I., *Mechanics of Fluids*, 3rd ed., McGraw-Hill Book Company, New York, 1992.

SHAPIRO, A. H., *The Dynamics and Thermodynamics of Compressible Fluid Flow*, Ronald Press, New York, 1953.

STREETER, V. L., and WYLIE, E. B., *Fluid Mechanics*, 8th ed., McGraw-Hill Book Company, New York, 1985.

VAN WYLEN, G. J., and SONNTAG, R. E., *Fundamentals of Classical Thermodynamics*, 2nd ed., John Wiley & Sons, Inc., New York, 1976.

WHITE, F. M. *Fluid Mechanics*, 3rd ed., McGraw-Hill Book Company, New York, 1994.

YIH, C. S., *Fluid Mechanics*, West River Press, 1988.

YUAN, S. W., *Foundations of Fluid Mechanics*, Prentice-Hall, Inc., Englewood Cliffs, NJ, 1967.

# Answers

## Chapter 1

| | |
|---|---|
| **1.1** | C |
| **1.2** | B |
| **1.3** | A |
| **1.4** | C |
| **1.5** | C |
| **1.6** | B |
| **1.7** | D |
| **1.8** | A |
| **1.9** | D |
| **1.10** | C |
| **1.11** | C |
| **1.12** | B |
| **1.13** | D |

**1.16**   **a)** $FT^2/L^4$    **c)** $FL/T$    **e)** $FT/L$

**1.18**   **b)** $m/s^2$

**1.20**   kg/s, N/m, N

**1.22**   **a)** $1.26 \times 10^8$ N,    **c)** $6.7 \times 10^8$ Pa
      **e)** $5.2 \times 10^{-2}$ m$^2$

**1.23**   $22.5 \, m/\rho d^2$

**1.24**   **a)** $5.56 \times 10^{-5}$ m/s    **b)** 209.4 rad/s
      **c)** 373 kW       **d)** $1.67 \times 10^{-5}$ m$^3$/s
      **e)** $2 \times 10^{11}$ N/m$^2$    **f)** 5 kg/s

**1.26**   44.01 N

**1.28**   **b)** 142.2 kPa      **d)** 78.8 kPa

**1.29**   **a)** 70 kPa       **b)** 527 mm of Hg
      **c)** 7000 mN/cm$^2$    **d)** 0.07 N/mm$^2$

**1.30**   62.8 kPa, 1.95%

**1.32**   $-50.6°C$

**1.34**   2.40 N, 0.0095°

**1.35**   $\rho = 1000$ kg/m$^3$, $\gamma = 9810$ N/m$^3$

**1.36**   976 kg/m$^3$, 9560 N/m$^3$, $-0.2\%$,
      $-0.36\%$

**1.38**   **b)** 0.635 kg

**1.39**   122 liters

**1.40**   28.6 N/m$^3$, $6.86 \times 10^{-6}$ m$^2$/s

**1.42**   0.007848 N/m$^2$, 0.007848 N/m$^2$

**1.43**   $\tau_w = 0.665$ Pa

**1.44**   3.2 Pa, 6.4 Pa

**1.46**   3.10 N·m, 0.65 kW

**1.48**   $1.16 \times 10^{-3}$ N·m

**1.50**   $E(e^{-Cy} - 1)$

**1.54**   899 m

**1.56**   **a)** 1432 m/s  **b)** 1508 m/s  **c)** 1472 m/s

**1.58**   29.6 kPa, 59.2 kPa

**1.59**   395 Pa

**1.60**   8010 kPa

**1.62**   $-0.453$ mm

**1.64**   $\sqrt{8\sigma/\pi\rho g}$

**1.66**   $2\sigma\pi D$

**1.68**   50°C

**1.70**   7.4 kPa

**1.71**   61.8°C

**1.72**   16.83 km

**1.74**   1.19 kg/m$^3$

**1.75**   69.3 kg/m$^3$, 28.4 kg

**1.76**   9333 N

**1.77**   343 kPa gage

**1.78**   25.2 m

**1.79**   7.67 m/s, 10.85 m/s

**1.80**   **a)** 19.25 m/s      **c)** 16.42 m/s

**1.82**   69.2°C

**1.84**   $W = 116.98$ kJ, $Q = -116.98$ kJ

**1.86**   **b)** 243 kJ

**1.88**   804 kPa gage

**1.90**   **b)** 266.9 m/s      **d)** 1301 m/s

**1.92**   **b)** 2854 m

## Chapter 2

| | |
|---|---|
| **2.1** | C |
| **2.2** | D |
| **2.3** | C |
| **2.4** | A |
| **2.5** | B |

| | |
|---|---|
| **2.6** | A |
| **2.7** | D |
| **2.8** | A |
| **2.9** | A |

**2.12**   **a)** 25.5 m      **c)** 1.874 m

| | |
|---|---|
| **2.13** | 2.38, 2380 kg/m$^3$ |
| **2.14** | **b)** 10.2 m |
| **2.16** | −37.4 kPa |
| **2.18** | 103 kPa |
| **2.20** | 96.49 kPa, 96.44 kPa, −0.052% |
| **2.22** | **a)** 4697 kPa, −0.064% |
| | **b)** 15,657 kPa, −0.21% |
| | **c)** 46,971 kPa, −0.78% |
| **2.24** | **a)** 69.9 kPa    **c)** 30.6 kPa |
| **2.25** | 70,900 m |
| **2.26** | 33.35 kPa |
| **2.28** | **a)** 680 kg/m$^3$, gasoline  **c)** 999 kg/m$^3$, water |
| **2.29** | **a)** 30.9 m/s  **b)** 34.6 m/s  **c)** 39.9 m/s |
| | **d)** 44.6 m/s |
| **2.30** | 30.96 kPa |
| **2.32** | 17.43 cm |
| **2.34** | 15.62 kPa |
| **2.36** | 14.0 kPa |
| **2.37** | **a)** −6.28 kPa  **b)** 18.84 kPa |
| | **c)** −4.32 kPa  **d)** 20.8 kPa |
| **2.38** | 17.89 kPa |
| **2.40** | 5.87 kPa |
| **2.42** | 0.52 kPa |
| **2.43** | $H$ = 1.01 m or 3.7 m |
| **2.44** | 3.76 cm |
| **2.45** | 3.4 mm |
| **2.46** | 4.51 cm |
| **2.47** | **a)** 15 cm  **b)** 21.4 cm  **c)** 10.33 cm  **d)** 21.6 cm |
| **2.48** | 6934 N |
| **2.50** | 29 400 N vs 28 400 N so it will rise |
| **2.52** | **b)** 277.1 kN |
| **2.53** | **a)** (0.61 m, 0.76 m), $F$ = 175 kN |
| | **b)** (0.6 m, 0.8 m), 192 kN |
| | **c)** (0.61 m, 0.77 m), 180 kN |
| **2.54** | **a)** 739.7 kN at (0, −0.167) m |
| | **c)** 313.9 kN at (0.9375, −1.5) m |
| **2.55** | 74.5 kN |
| **2.56** | 523 kN |
| **2.60** | 3346 N |
| **2.62** | **b)** 0.6667 m |
| **2.63** | **a)** 3.46 m  **b)** 1.73 m  **c)** 3.12 m  **d)** 1.56 m |

| | |
|---|---|
| **2.64** | 1.55 m |
| **2.66** | **a)** won't tip    **b)** will tip |
| | **c)** will tip |
| **2.68** | 616.1 kN |
| **2.69** | 594.2 kN |
| **2.70** | **a)** $F_H$ = 706 kN, $F_V$ = 662 kN |
| **2.72** | 70.07 kN |
| **2.74** | 4580 N/m$^3$ |
| **2.75** | 8.47 kN, 0.96 |
| **2.76** | **a)** 420 kN    **b)** −23.4 kN |
| **2.77** | **a)** 17.98 kN    **b)** 14.83 kN |
| **2.78** | **a)** 1.372 m |
| **2.80** | 13.6 cm |
| **2.82** | 8.804 kN |
| **2.84** | 1.23×10$^6$ people |
| **2.86** | **a)** 5.04 m  **b)** 4.92 m  **c)** 4.875 m |
| **2.88** | **a)** 1.089    **b)** 0.01886 kg |
| **2.89** | **a)** no  **b)** no  **c)** yes |
| **2.90** | $\gamma_x > 8369$ N/m$^3$ or $\gamma_x < 1435$ N/m$^3$ |
| **2.92** | 1.1 |
| **2.93** | **a)** $G$ is below $C$, $\theta = 11.3°$ |
| | **b)** $\Delta y$ = 0.85, $\Delta x$ = 0.175, $\theta = 11.6°$ |
| **2.94** | $GM$ = 0.277 m. ∴stable |
| **2.96** | **a)** 99.6 kPa    **b)** 59.6 kPa |
| | **c)** 114.86 kPa    **d)** 50 kPa |
| **2.98** | **b)** 4.8 m/s$^2$ |
| **2.99** | **a)** 58.29 kPa    **b)** 57.46 kPa |
| **2.100** | **a)** 640 kN    **b)** 1163 kN |
| **2.102** | **a)** 0, 11.89 kPa, 11.89 kPa, |
| | **b)** −9 kPa, −3.11 kPa, 5.88 kPa |
| | **c)** −18 kPa, −6.11 kPa, 11.89 kPa |
| | **d)** 0, −5.12 kPa, −5.12 kPa |
| | **e)** −16.87 kPa, −10.74 kPa, 6.13 kPa |
| | **f)** 8.44 kPa, 20.19 kPa, 11.76 kPa |
| **2.103** | **a)** 11.1 kPa, 16.99 kPa, 5.88 kPa |
| | **b)** 4.93 kPa, 8.86 kPa, 3.92 kPa |
| **2.104** | **a)** −40.5 kPa, −34.6 kPa, 5.89 kPa |
| | **b)** −40.5 kPa, −34.6 kPa, 5.886 kPa |
| **2.105** | **a)** 7.62 rad/s    **b)** 9.34 rad/s |
| **2.106** | **a)** 8149 Pa    **c)** 17.4 kPa |
| **2.108** | **a)** 6670 N    **c)** 9920 N |

# Chapter 3

| | | | | |
|---|---|---|---|---|
| **3.1** | D | | **3.5** | B |
| **3.2** | C | | **3.6** | C |
| **3.3** | D | | **3.7** | B |
| **3.4** | C | | **3.8** | A |

**3.14** a) $x^2 - 2xy + y^2 = 4y$. A parabola

**3.16** Eulerian: Several college students would be positioned at each intersection and quantities would be recorded as a function of time.

**3.17** a) 2 m/s, 3.6 m/s  b) 0, 8.25 m/s  c) 4 m/s, 6 m/s

**3.18** a) $\hat{n} = (2\hat{i} - 3\hat{j})/\sqrt{3}$
c) $\hat{n} = (8\hat{i} + 5\hat{j})/\sqrt{89}$

**3.20** b) $8\hat{i} - 4\hat{j}$    d) $2\hat{i} - 114\hat{j} + 15\hat{k}$

**3.22** a) $-40\hat{i}$    c) $12\hat{i} - 4\hat{k}$

**3.23** a) rate-of-strain = $\begin{bmatrix} 0 & 20 & 0 \\ 20 & 0 & 0 \\ 0 & 0 & 0 \end{bmatrix}$

**3.24** a) $(9.375, 0, 0)$ m/s$^2$

**3.26** $(\partial u/\partial t)\mathbf{i}$

**3.28** b) 1.875 m/s at $t = \infty$

**3.30** $-9.11 \times 10^{-4}$ kg/m$^3$·s

**3.32** 0.04 kg/m$^3$·s

**3.34** $\mathbf{a} = \partial\mathbf{V}/\partial t + (\mathbf{V} \cdot \nabla)\mathbf{V}$

**3.35** a) $(144\hat{j} - 544\hat{i})$ m/s$^2$  b) $-465\hat{i}$ m/s$^2$

**3.36** $-52 \times 10^{-5}\hat{j} + 0.0224\hat{i}$ m/s$^2$

**3.38** Steady: a, c, e, f, h. Unsteady: b, d, g

**3.42** a) inviscid.    c) inviscid.
e) viscous inside the boundary layers and separated regions.
g) viscous.

**3.46** 9640    ∴ Turbulent

**3.48** a) Re = 795. Laminar

**3.49** 3 cm

**3.50** $x_T = 8.17$ m. Laminar

**3.51** a) compressible  b) incompressible
c) incompressible

**3.52** $u\partial\rho/\partial t + v\partial\rho/\partial y = 0$

**3.54** a) 60 m/s  b) 104 m/s  c) 33 m/s

**3.56** 57 m/s

**3.58** b) $r = r_c: p_T = \rho U_\infty^2/2$
d) $\theta = 90°: \quad p_{90} = -3\rho U_\infty^2/2$

**3.60** a) $-50\rho(2/x + 1/x^2)$  b) $50\rho$
c) $\frac{81}{2}\rho(\frac{2}{3x} + \frac{1}{9x^2})$  d) $\frac{81}{2}\rho$

**3.62** 9.39 m/s

**3.64** a) 3.14 m/s    b) 4.97 m/s

**3.66** 12.76 m/s

**3.68** b) 3.516 m/s, 193.6 kPa

**3.70** 19.82 cm

**3.71** a) 18.04 m/s    b) 36.1 m/s

**3.72** a) $-6195$ Pa    c) $-5470$ Pa

**3.74** a) 37.42 m/s, 71.4 m  b) 52.92 m/s, 142.9 m

**3.76** a) 37.4 m/s    b) 28.3 m/s

**3.78** b) $-43.3$ kPa    d) $-59$ Pa

**3.80** Negative pressure, so it will rise

**3.82** High pressure

**3.84** The higher pressure at $B$ will force the fluid toward the lower pressure at $A$, thereby causing a secondary flow normal to the pipe's axis. This results in a relatively high loss for an elbow.

# Chapter 4

**4.1** B
**4.2** D
**4.3** A
**4.4** D
**4.5** A
**4.6** C
**4.7** B
**4.8** C
**4.9** D
**4.10** A
**4.11** A
**4.12** C
**4.13** D
**4.14** A
**4.15** A

**4.16** b) $Q - W = 0$

**4.20** b) Conservation of mass
d) Energy equation

**4.24** $-0.707 (\hat{i} + \hat{j}), 0.866\hat{i} - 0.5\hat{j}, -\hat{j}, -7.07$ fps, 8.66 fps, 0

**4.26** 1559 cm$^3$

**4.32** $dm_V/dt = \dot{m} - \rho Q$

**4.34** 4.5 m/s, 51 kg/s, 0.051 m$^3$/s

**4.36** 0.1509 m$^3$/s

**4.37** 115 m/s, 1616 m/s

**4.38** 320 m/s, 20.9 kg/s, 4.712 m$^3$/s, 2.512 m$^3$/s

**4.40** b) 4.478 m

**4.42** a) 5 m/s, 320 kg/s, 0.32 m$^3$/s
c) 7.5 m/s, 480 kg/s, 0.48 m$^3$/s

**4.44** 5.08 kg/s

**4.46** 0.82 kg/s
**4.48** 63.7 kg/s, 64.9 kg/s $\therefore \overline{\rho V} \neq \overline{\rho}\,\overline{V}$
**4.50** 0.565 m
**4.51** 0.015 kg/m$^3 \cdot$s
**4.52** 12.04 m/s
**4.54** 0.244 m/s
**4.55** 3.7 m/s
**4.56** 27.3 m/s
**4.58** $\rho\pi[(d^2 - d_2^2)\dot{h}_2/4 + h_1^2 h_1 \tan^2\phi]$
**4.60** $3.99 \times 10^{-4}$ kg/s
**4.62** **b)** 8.84 mm/s
**4.64** $-14.5$ kPa/s
**4.66** 1847 W
**4.68** 0.836°C
**4.70** **b)** 524 kW
**4.72** 2.71 m, 0.61 m
**4.73** 2.47 m, 0.646 m
**4.74** 9.94 m/s
**4.76** 414.3 kPa
**4.78** 32.1 MPa
**4.80** **a)** 5.01 m$^3$/s
**4.82** **b)** 0.485 m$^3$/s
**4.84** $12.35 d_1^2 d_2^2 (H/(d_1^4 - d_2^4))^{1/2}$
**4.85** **a)** 0.0365 m$^3$/s **b)** 0.0503 m$^3$/s
**4.86** 0.663 m
**4.88** 60.4 m
**4.90** 0.131 m
**4.92** 24.8 hp
**4.94** 1304 kW
**4.95** 0.884
**4.96** **a)** 299°C
**4.97** 237.6 kW (318 hp)
**4.98** 5390 kW
**4.100** 0.5 W
**4.101** 114 hp
**4.102** 3820 W, 39.4 kPa, 416 kPa
**4.104** 1.85 m or 2.22 m
**4.106** 0.815 m
**4.108** **a)** 2.00

**4.110** $\dot{m}_f q_f = F_D V_\infty + \dot{m}[(V_2^2 - V_1^2)/2 + c_v(T_2 - T_1)]$
**4.112** $7.37 \times 10^{-5}$ m$^3$/s
**4.114** **a)** $\cong 0.32$ m$^3$/s
**4.116** **a)** 339 N **b)** 679 N **c)** 1356 N
**4.118** **a)** 692 N **c)** 2275 N
**4.119** 4.59 kN
**4.120** **b)** 2280 N, 1071 N
**4.122** 49.6 N
**4.124** 618 kN
**4.125** **a)** 2.51 m **b)** 3.23 m
**4.126** **a)** 3.2 m, 8.86 m/s **b)** 2 m, 7 m/s
**4.128** 0.77 m, 7.8 m/s
**4.130** 4420 N
**4.132** **a)** 11.46 kN **b)** 6.25 kN
**4.133** **a)** 17.85 m/s **b)** 26.65 m/s **c)** 8.85 m/s
**4.134** **b)** 202 kg/s
**4.135** **a)** 1272 N, 2203 N **b)** 318 N, 551 N **c)** 2861 N, 4956 N
**4.136** 147.3 kW
**4.138** 986 kW
**4.139** **a)** 380 hp **b)** 360 hp **c)** 283 hp
**4.140** **b)** 47°, 29.5°, 3.6 MW
**4.142** 11.08 kN, 240 kg/s, 80 kg/s
**4.144** 13.33 m/s
**4.146** 647 hp
**4.148** 110 kN
**4.149** 9.8 m/s, 2.64 m/s
**4.150** 1440 N, 69.9 m$^3$/s, 62.5%
**4.152** 395 hp, 2944 kg/s
**4.154** 141 N
**4.156** 220 m/s/m
**4.158** 3780 N
**4.160** **a)** 54.9 kW/m
**4.164** 46.9 rad/s
**4.165** 14.3 hp
**4.166** 39.6 rad/s
**4.168** $11.73(1 - e^{-31.8t})$rad/s

# Chapter 5

**5.2** See Table 5.1
**5.4** $\rho\,du/dx + u\,d\rho/dx = 0$
**5.6** $u\partial\rho/\partial x + w\partial\rho/\partial z = 0,\ \partial u/\partial x + \partial w/\partial z = 0$
**5.8** **a)** $v_r = C/r$
**5.10** **a)** $3y/(0.15x + 0.5)^2$ **b)** $-9y/(0.15x + 0.5)^4$

**5.12** $v = $ const
**5.14** $-Ay$
**5.16** $(10 - 0.4/r^2)\sin\theta$
**5.18** $(10 - 80/r^3)\cos\theta$
**5.19** $-2.32$ kg/m$^4$

**5.20** 0.541 m/s

**5.22** 0.296 m/s

**5.24** **a)** $-0.88$ m/s    **b)** 2772 m/s$^2$

**5.26** $[100\rho/(x^2 + y^2)^2]\,(x\hat{\mathbf{i}} + y\hat{\mathbf{j}})$

**5.28** $(\partial p/\partial\theta)/r = -\rho v_r v_\theta/r - \rho v_r \partial v_\theta/\partial r - \rho(v_\theta/r)\partial v_\theta/\partial\theta$

**5.30** $\rho(D\mathbf{V}/Dt + 2\Omega\times\mathbf{V} + \Omega\times(\Omega\times\mathbf{r}) + (d\Omega/dt)\times\mathbf{r}) = -\nabla p - \rho\mathbf{g}$

**5.31** $3.96\times 10^{-8}$

**5.32** $252y^2 x^{-9/5} - 5270y^3 x^{-13/5}$, $\tau_{xy} = 5.01\times 10^{-5}$ Pa

**5.34** $-\nabla p + \rho\mathbf{g} + \mu\nabla^2\mathbf{V} + \mu\nabla(\nabla\cdot\mathbf{V})/3$

**5.36** $-[(V_1 + V_2)y/h] + V_1$

**5.38** $\partial p/\partial z = \mu[\partial^2 v_z/\partial r^2 + \partial v_z/\partial r)/r]$.

**5.40** $\partial p/\partial\theta = \mu[\partial/\partial r(r^2 \partial v_\theta/\partial r)]/r - \mu v_\theta/r\sin^2\theta$

**5.44** $\rho\partial u/\partial t = \mu\partial^2 u/\partial y^2$

**5.48** $0.3 + 10y$

**5.50** $\rho D\tilde{u}/Dt = K\nabla^2 T$

**5.52** $\rho Dh/Dt = K\nabla^2 T + Dp/Dt$

**5.56** **a)** $\partial u/\partial t = v\partial^2 u/\partial y^2$, $\rho c\partial T/\partial t = K\partial^2 T/\partial y^2 + \mu(\partial u/\partial y)^2$

# Chapter 6

**6.1** A

**6.2** A

**6.3** A

**6.4** A

**6.5** C

**6.6** A

**6.8** **a)** $FT/L$    **c)** $FT^2/L^4$    **e)** $FL$

**6.10** $V\ell\rho/\mu = \text{const}$

**6.12** $V = \sqrt{gH/C}$

**6.14** $F_D/\rho\ell^2 V^2 = f_1(d/\ell, \mu/\rho\ell V)$

**6.16** $h/d = f_1(\sigma/\gamma d^2, \beta)$

**6.18** $\sigma = CMy/I$

**6.20** $V = \text{Const.}\sqrt{gH}$

**6.22** $\Delta p/\rho V^2 = f_1(\nu/Vd, L/d, e/d)$

**6.24** $Q/\sqrt{gR^5} = f_1(A/R^2, e/R, s)$

**6.26** $F_D/\rho V^2 d^2 = f_1(\mu/\rho Vd, e/d, I)$

**6.28** $T/\rho V^2 D^2 = \phi[\text{Re}]$

**6.30** $F_D/\rho V^2 d^2 = f_1(\mu/\rho Vd, e/d, r/d, Cd^2)$

**6.32** $F_L/\rho V^2\ell_c^2 = f_1(c/V, t/\ell_c, \alpha)$

**6.34** $F_D/\rho V^2 d^2 = f_1(Vd\rho/\mu, d/L, \rho/\rho_c, V/\omega d)$

**6.36** $T/\rho\omega^2 d^5 = g_1(f/\omega, H/d, \ell/d, h/d)$

**6.38** $d/D = f_1(V/V_j, \sigma/\rho V_j^2 D, \mu/\rho V_j D, \rho_a/\rho)$

**6.40** $\mu/\rho VD = f_1(H/D, \ell/D, gD/V^2)$

**6.42** $y_2/y_1 = f(gy_1/V_1^2)$

**6.44** $Q_m/Q_p = V_m\ell_m^2/V_p\ell_p^2$, $(F_p)_m/(F_p)_p = \rho_m V_m^2\ell_m^2/\rho_p V_p^2\ell_p^2$, $T_m/T_p = \rho_m V_m^2\ell_m^3/\rho_p V_p^2\ell_p^3$

**6.46** **a)** 160 kg/s, 15 MPa

**6.47** **a)** 45 N,    **b)** 90.5 N

**6.48** Full-scale is recommended.

**6.50** 1184 N

**6.52** 0.15 cm, 1.11

**6.54** $\nu_m = 6.1\times 10^{-9}\text{m}^2/\text{s}$. Impossible!

**6.56** **a)** $0.00632\ \text{m}^3/\text{s}$    **b)** 12 kN

**6.57** **a)** 18.97 m/s **b)** 4000 N

**6.58** 120 kN·m

**6.60** 1633 kW

**6.62** 4.16 N, 95 hp

**6.63** yes

**6.64** **b)** 186 m/s, 2080 N

**6.66** **a)** 6320 rpm    **b)** 2000 rpm

**6.68** 5 motions/sec

**6.70** $f\ell/V$

**6.72** $g\ell/U^2$

**6.74** **b)** $\dfrac{\partial u^*}{\partial t^*} = -\text{Re}\dfrac{\partial p^*}{\partial x^*} + \dfrac{\partial^2 u^*}{\partial r^{*2}} + \dfrac{1}{r^*}\dfrac{\partial u^*}{\partial r^*}$

**6.76** $(K/\mu c_p)(\mu/\rho U\ell) = \text{Pr}^{-1}\text{Re}^{-1}$

# Chapter 7

**7.1** D

**7.2** A

**7.3** D

**7.4** D

**7.5** A

**7.6** B

**7.7** D

**7.8** B

| | |
|---|---|
| **7.9** | A |
| **7.10** | B |
| **7.11** | A |
| **7.12** | C |
| **7.14** | **b)** 0.0015 m/s |
| **7.15** | Re = 1730 |
| **7.16** | Re = 700 000. Turbulent |
| **7.18** | **a)** 12.6 m     **c)** 25.0 m |
| **7.20** | $L_E$ = 7.2 m. Developed |
| **7.22** | 3.7 m vs 0.72 m |
| **7.24** | **a)** 20.8 m, 5.2 m |
| **7.29** | $\tau_0$ = 0.032 Pa, $f$ = 0.04 |
| **7.30** | $0.123°, 7.79 \times 10^{-6} \, \text{m}^3/\text{s}$ |
| **7.32** | **b)** $1.04 \times 10^{-4} \, \text{m}^3/\text{s}$. Laminar |
| **7.33** | 8.3 Pa |
| **7.34** | $1.15 \times 10^{-5} \, \text{m}^3/\text{s}$, 1.2 m |
| **7.36** | 23 Pa |
| **7.38** | 12.1 m/s, 3.5 Pa, 260 m |
| **7.40** | $3.85 \times 10^{-6} \, \text{m}^3/\text{s}$ |
| **7.42** | **a)** $0.707r_0$     **b)** $0.5r_0$ |
| **7.44** | 0.0063 m, 0.08 Pa, 65 m |
| **7.46** | 0.462 m/s, 0.0054 Pa |
| **7.49** | **a)** $Q$ = 48 L/min **b)** $Q$ = 70 L/min |
| **7.50** | **a)** $0.0274 \, \text{N·s/m}^2$ |
| **7.52** | $12.1 \, \text{m}^3/\text{s}$, 241 000, 60.4 m/s, 20.1 Pa |
| **7.54** | $0.028 \, \text{m}^3/\text{s}$, 2030 |
| **7.56** | 276 L/min |
| **7.58** | **a)** 18.5 m/s     **c)** −6.15 m/s |
| **7.60** | 32 kPa/m |
| **7.61** | 14.7 N |
| **7.62** | 151 N·m |
| **7.64** | 7.9 N·m |
| **7.66** | 18.9 W |
| **7.67** | 1.07 mN·m |
| **7.68** | 0.040 N·m, 12.6 W, 1650 |
| **7.70** | $4\pi\mu r_1^2 r_2^2 L\omega_2 / (r_2^2 - r_1^2)$ |
| **7.76** | $\overline{u'^2} = 51.2 \, \text{m}^2/\text{s}^2, \overline{u'v'} = 9.7 \, \text{m}^2/\text{s}^2$ |
| **7.77** | $\overline{u'v'} = -2.52 \, \text{m}^2/\text{s}^2$ |
| **7.78** | $0.016 \, \text{m}^2/\text{s}$, 0.125, 0.01 m |
| **7.80** | **b)** Rough |
| **7.82** | **b)** 0.262 m/s |

| | |
|---|---|
| **7.84** | 8.5, 7.34 m/s |
| **7.86** | −40.8 kPa/m |
| **7.88** | 24.2 m/s |
| **7.89** | **a)** 50 Pa **b)** 6.47 m/s **c)** 5.4 m/s **d)** $8.2 \times 10^5$ **e)** 42 L/s |
| **7.90** | **a)** 1.12 m/s, $0.0127 \, \text{m}^3/\text{s}$ |
| **7.92** | **a)** 0.0143     **b)** 0.0146 |
| **7.94** | **a)** 0.064     **c)** 0.034 |
| **7.96** | **a)** 17.8 m **b)** 16.8 m **c)** 12.4 m **d)** 10.4 m |
| **7.98** | **a)** 136 kPa     **c)** 20 kPa |
| **7.100** | 550 kPa |
| **7.102** | 147 kPa |
| **7.104** | **b)** $0.0033 \, \text{m}^3/\text{s}$ |
| **7.106** | **b)** 0.069 kg/s |
| **7.107** | **a)** $13.36 \, \text{m}^3/\text{s}$     **b)** $13.12 \, \text{m}^3/\text{s}$ |
| **7.108** | **a)** 0.032 m     **c)** 0.031 m |
| **7.110** | **a)** 0.121 m     **b)** 0.127 m |
| **7.112** | $0.000143 \, \text{m}^3/\text{s}$ |
| **7.114** | **b)** $0.257 \, \text{m}^3/\text{s}$ |
| **7.118** | **a)** 52.6 kPa |
| **7.120** | 11.48 |
| **7.122** | $0.0044 \, \text{m}^3/\text{s}$ |
| **7.123** | **a)** 6.9 L/s **b)** 9.9 L/s **c)** 14.1 L/s |
| **7.124** | **b)** $0.011 \, \text{m}^3/\text{s}$ |
| **7.126** | 1.3 min. |
| **7.128** | 46.1 m/s |
| **7.130** | Oscillates between laminar and turbulent flow |
| **7.131** | **a)** 555 hp **b)** 22 hp **c)** 9.8 hp |
| **7.132** | 117 kW, 13.0 m |
| **7.134** | 1400 hp, 13 cm |
| **7.136** | $0.23 \, \text{m}^3/\text{s}$, 190 kW |
| **7.138** | **a)** 195 kW     **b)** −81 kPa     **c)** 625 kPa |
| **7.140** | 1.05 MW |
| **7.142** | 17.6 Pa |
| **7.144** | **a)** $1.64 \, \text{m}^3/\text{s}$ |
| **7.145** | **a)** $4.31 \, \text{m}^3/\text{s}$ **b)** $4.65 \, \text{m}^3/\text{s}$ **c)** $2.16 \, \text{m}^3/\text{s}$ **d)** $0.6 \, \text{m}^3/\text{s}$ **e)** $0.06 \, \text{m}^3/\text{s}$ |
| **7.146** | **a)** $0.794 \, \text{m}^3/\text{s}$     **b)** $0.86 \, \text{m}^3/\text{s}$ |
| **7.148** | 3.91 m |
| **7.150** | 0.612 m |

# Chapter 8

| | |
|---|---|
| **8.1** | C |
| **8.2** | C |
| **8.3** | B |

| | |
|---|---|
| **8.4** | B |
| **8.5** | D |
| **8.6** | C |

**8.7**   C

**8.8**   D

**8.10**   $3.78 \times 10^{-5}$ m

**8.13**   **a)** $2.85 \times 10^{-4}$ m/s   **b)** $9.2 \times 10^{-5}$ m/s
     **c)** $3.7 \times 10^{-3}$ m/s

**8.14**   **b)** $7.9 \times 10^{6}$. Separated

**8.16**   2659 N, 469 N, 2.63, 0.417

**8.18**   **b)** $4.58 \times 10^{-5}$ N

**8.20**   $C_{D,\text{high}}/C_{D,\text{low}} = 2.5$

**8.21**   **a)** 30.3 m/s   **b)** 48 m/s

**8.22**   $0.5, 1.46 \times 10^{5}$

**8.24**   **a)** 1380 N, 25.7 kN·m

**8.25**   39.8 m/s, 3.7 cm

**8.26**   36 kN

**8.28**   **a)** 8.22 m/s      **b)** 15.86 m/s
     **c)** 115.7 m/s

**8.30**   **a)** 10 hp

**8.32**   **a)** 93.6 N

**8.34**   $8800

**8.35**   15.2 hp

**8.36**   1.24 hp

**8.38**   **a)** 1.58 m/s      **c)** 2.86 m/s

**8.40**   $683

**8.42**   125 km/hr

**8.44**   $< 2.5 \times 10^{-5}$ m, $> 0.0063$ m

**8.46**   0.0191 m/s

**8.48**   2.77 hp, 0.1 hp

**8.50**   900 N, 60 N

**8.52**   3.3 N, 0.29 N

**8.54**   4800 N, no cavitation

**8.56**   52 kN

**8.58**   **b)** $1.24$ m/s$^2$

**8.60**   0.587

**8.62**   13.8 hp

**8.64**   38 m/s

**8.65**   $8°, 4.86$ kN, No

**8.66**   39.9 m/s

**8.68**   **a)** 7.77% increase

**8.74**   **a)** $5x^2 + 10y^2$      **c)** $\sqrt{x^2 + y^2}$

**8.76**   $100y - 50x, 100x + 50y$

**8.78**   $-40\tan^{-1} y/x$

**8.80**   **b)** $10y + 10\tan^{-1} y/x$
     **c)** $100 - 50(2/x + 1/x^2)$ kPa
     **e)** $-12.5$ m/s$^2$

**8.82**   **a)** $(-0.025, 0)$      **b)** 0.039 m
     **c)** $\dfrac{\pi}{20}$ m      **d)** 8.25 m/s

**8.84**   $(\sqrt{2}, 0), (-\sqrt{2}, 0), 0, 0$

**8.86**   **a)** 0.314 m

**8.88**   **b)** $(0, -104)$ m/s$^2$   **c)** 13.76 kPa

**8.90**   **b)** $-8\sin\theta$      **c)** $58 - 32\sin^2\theta$ kPa
     **d)** 42.7 kN

**8.91**   $(2, 227°), (2, 313°), p_{\max} = 48.6$ Pa
     $p_{\min} = -332.4$ Pa

**8.92**   100 rad/s, $-3660$ Pa

**8.94**   1244 kN

**8.96**   $(0.729, 0.481)$ m/s

**8.97**   **a)** 0.093 m **b)** 0.12 m **c)** 0.213 m

**8.100**   **a)** 3 cm      **c)** 3 cm      **e)** 6 mm

**8.102**   $20 - 200\sin^2(x/2)$ kPa, $20\sin(x/2)$

**8.104**   $16/(2.67 - x), 256/(2.67 - x)^3$

**8.108**   **a)** $4.79\sqrt{\nu x/U_\infty}$   **b)** $0.328\mu U_\infty \sqrt{U_\infty/\nu x}$
     **c)** $\displaystyle\int_0^\delta y\cos\left[(0.189\sqrt{U_\infty/\nu})y\right]dy$

**8.110**   $6.65\sqrt{\nu x/U_\infty}, 0.451\rho U_\infty^2 \, \text{Re}_x^{-1/2}, 33\%, 36\%$

**8.112**   2.1 cm, 0.001 m, 0.001 m

**8.114**   **a)** $1.74\sqrt{\nu x/U_\infty}, 0.648\sqrt{\nu x/U_\infty}, 1.2\%,$
     0.62%

**8.115**   **a)** 0.036 m      **b)** $4.24 \times 10^{-3}$ Pa
     **c)** 0.23 N      **d)** 0.036 m/s

**8.116**   **a)** 0.0221 m      **b)** 0.00498 Pa
     **c)** 0.299 N      **d)** 0.00391 m/s

**8.118**   **a)** 0.0949 m, 0.6 Pa

**8.120**   **a)** 4.3 N   **b)** 3.77 N    **c)** 3.46 N

**8.122**   **a)** 235 m, 0.0618 Pa    **b)** 0.151 Pa, 585 m

**8.124**   **a)** 0.00213      **b)** 10.52 Pa
     **c)** $2.51 \times 10^{-5}$ m    **d)** 0.071 m

**8.126**   163 kN, 0.89 m

**8.132**   **a)** 0.0124 Pa      **b)** 0.0122 m
     **c)** 0.00527 m/s    **d)** 0.04 m$^3$/s/m

**8.133**   **a)** 0.011 Pa   **b)** 0.012 m    **c)** 0.0052 m/s
     **d)** 0.036 m$^3$/s/m

**8.134**   **a)** 32.1 Pa      **b)** 0.0248 m
     **c)** 0.109 m$^3$/s/m

**8.136**   3.85 mm, 0.00127 m/s, 0.011 Pa

**8.140**   negative, zero, positive, positive, negative

# Chapter 9

| | |
|---|---|
| **9.10** | 0.0069 s |
| **9.11** | **a)** 0.588 **b)** 0.56 **c)** 0.668 **d)** 0.61 **e)** 0.66 |
| **9.12** | 393 m |
| **9.13** | 686 m |
| **9.14** | 3.776 s |
| **9.15** | 908 m/s |
| **9.16** | $-0.035$ m/s, 0.012 K |
| **9.18** | **a)** 77.3 m/s |
| **9.20** | **a)** 0.1473 kg/s    **b)** 0.1393 kg/s |
| **9.21** | **a)** 0.232 kg/s    **b)** 0.21 kg/s |
| **9.22** | **a)** 0.1472 kg/s    **b)** 0.1423 kg/s |
| **9.24** | 211.3 kPa abs, 7.29 kg/s |
| **9.26** | 192 kPa, 1.31 kg/s, 2.62 kg/s |
| **9.28** | 97.45 kPa abs, 199.4 kPa abs |
| **9.30** | 168 m/s, 476 m/s |
| **9.32** | 1.42 kg/s |
| **9.34** | 494.2 kPa abs, 4.29 kPa abs |
| **9.36** | 0.097 m, 0.127 m, 253 m/s |
| **9.38** | 129 kPa abs, 1004 m/s |
| **9.40** | 8.16 cm |
| **9.42** | 22.1 cm$^2$, 772°C, 3670 kPa abs |
| **9.43** | $At = 6.27 \times 10^{-4}$ m$^2$ |
| | $pe = 3.78$ kPa, $Te = 100$ K |

| | |
|---|---|
| **9.44** | 1260 m/s |
| **9.46** | 412 kN |
| **9.48** | **a)** 3.774 kg/m$^3$, 808 kPa abs, 2.966, 473°C, 0.477 |
| **9.49** | 2.69, 0.498, 687 kPa, 652 K, 3.67 kg/m$^3$ |
| **9.50** | $(k + 1)/(k - 1)$ |
| **9.52** | 908 m/s, 1600 kPa abs, 8.33 kg/m$^3$ |
| **9.53** | 901 m/s, 1600 kPa, 8.44 kg/m$^3$ |
| **9.54** | 391 kPa abs, 448°C |
| **9.56** | 102.5 kPa abs, 0.471 kg/s, 0.302 kg/s |
| **9.58** | 339.3 kPa, 312 m/s, 179.2 kPa, 607.6 m/s, 10.11 kPa, 160 m/s, 100.3 kPa |
| **9.60** | 6 cm, 9.2 cm |
| **9.62** | 6.17 cm |
| **9.64** | **a)** 1.49, 120.4 kPa abs, 620 m/s |
| **9.66** | 780 m/s |
| **9.68** | 555 kPa abs, 413 kPa abs |
| **9.70** | 39.4°, 867 m/s, $-156$ °C |
| **9.71** | 39.4°, $-157$ K, 863 m/s |
| **9.72** | **a)** 14.24 kPa abs, 26.4 kPa abs |
| **9.74** | 0.0854, 0.010 |

# Chapter 10

| | |
|---|---|
| **10.2** | 2.86 m |
| **10.6** | **a)** 2.15 m, 23.5 m$^2$, 18.6 m |
| | **c)** 1.22 m, 1.85 m$^2$, 3.5 m |
| | **e)** 3.1 m, 23.25 m$^2$, 13.7 m |
| **10.8** | **a)** 0.0121 m$^3$/s    **b)** 0.244 m |
| **10.12** | **a)** $-0.17$ m    **c)** $-0.43$ m |
| **10.14** | 3.46 m |
| **10.16** | **a)** 0.3 m |
| **10.18** | **a)** 5.75 m **b)** 3.89 m |
| **10.20** | **a)** 1.00 m **b)** 0.46 m **c)** 1.81 m **d)** 1.00 m |
| **10.22** | **a)** 1.75 m    **b)** 0.83 m |
| **10.26** | $\cong 0.95$ m |
| **10.28** | **a)** 0.47 m$^3$/s |
| **10.30** | 1.03 m $\times$ 28.8 m |
| **10.31** | $Q = 8.70$ m$^3$/s, $y_c = 1.07$m |
| **10.32** | $M/by_c^2 = y^2/2y_c^2 + y_c/y$ |
| **10.34** | 3.87 m, 363 N |

| | |
|---|---|
| **10.36** | 7.03 m$^3$/s |
| **10.38** | 1.77 m, 0.339 m/s, 3.33 m/s |
| **10.40** | **a)** 3.43 m$^3$/s    **b)** 0.932 m |
| | **c)** 52 kW |
| **10.42** | 2.55 m, 3.38 kW |
| **10.44** | 11.51 m$^3$/s, 0.346 m |
| **10.46** | $y_0 = 0.094$ m, $y_1 = 0.316$ m, $y_3 = 0.088$ m, $y_{cj} = 0.243$m |
| **10.48** | $y_0 = 2.98$ m, M$_3$ followed by hydraulic jump to M$_2$ |
| **10.50** | M$_3$ followed by hydraulic jump to M$_1$ |
| **10.52** | $y_c = 1.13$ m. $y_{cj} = 1.35$ m. Hydraulic jump occurs upstream, followed by an S$_1$ curve. |
| **10.54** | **a)** $y_c = 1.50$ m. A hydraulic jump between $A$ and $C$. |
| **10.56** | $y_c = 0.7$ m. $y_0 = 0.99$ m. M$_3$ followed by a hydraulic jump to an M$_2$ |

**10.58 a)** 20.2 m      **b)** $A_2$ profile
**10.60** 15.5 m$^3$/s, $S_3$
**10.61** $y_{cj}$ = 0.94 m
**10.62** $Q$ = 7.72 m$^3$/s, $y_c$ = 0.72 m, $y_0$ = 1.17 m, $y_{cj}$ = 0.41 m. $M_1$ upstream, $M_3$ downstream

**10.64** $S_3$ followed by hydraulic jump to $S_1$
**10.66 a)** 0.87 m      **b)** 48 m
**10.68 a)** 721 kW      **b)** 59.3 m
**10.70** At $y$ = 2 m, $x \cong$ 55 km
**10.72** 3.2 m

# Chapter 11

**11.2** 750 W
**11.4 a)** 0.39 m
**11.6** 0.265 m$^3$/s
**11.8 a)** 0.32 m$^3$/s      **b)** 0.133 m$^3$/s
**11.10** 1.78 m$^3$/s, 0.285 m$^3$/s, 0.935 m$^3$/s, 1.07 MW
**11.11 a)** $Q_1$ = 0.20 m$^3$/s, $Q_2$ = 0.14 m$^3$/s, $Q_3$ = 0.06 m$^3$/s
     **b)** $Q_1$ = 0.17 m$^3$/s, $Q_2$ = 0.115 m$^3$/s, $Q_3$ = 0.051 m$^3$/s
**11.12** 0.198 m$^3$/s, 0.138 m$^3$/s, 0.06 m$^3$/s
**11.14** 16 L/s, 7.1 L/s, 5.1 L/s, 3.8 L/s
**11.16 c)** 28.2 L/s, 10.7 L/s, 8.6 L/s, 9.1 L/s
**11.18 a)** 143 L/min, 290 L/min, 166 L/min
     **b)** 114 L/min, 233 L/min, 133 L/min
**11.20** 1.05 m$^3$/s, 90.8 m, 73.6 m, 810 kW
**11.22 a)** 2.27 m$^3$/s, 0.811 m$^3$/s, 0.580 m$^3$/s, 0.880 m$^3$/s

**11.24 a)** 1.38 m   **b)** 450 L/min   **c)** 300 L/min
**11.25 a)** 22.5 hp   **b)** 368 kPa   **c)** 5 kPa
**11.26 a)** 220 L/min   **b)** 139.8 m
**11.28** 2.30, 0.43, 2.73 flow units
**11.30 a)** 96, 92, 88, 82 m   **b)** 843, 706, 813, 804 kPa
**11.31 a)** $H_J$ = 187.83 m
     **b)** $Q_1$ = 5280 L/min, $Q_2$ = 640 L/min, $Q_3$ = 600 L/min, $Q_4$ = 670 L/min, $Q_5$ = 3450 L/min
**11.32** 21.6 L/s, 6.6 L/s, 28.4 L/s
**11.34** 2.77 m$^3$/s
**11.36** 9.3, 1.6, 7.7, 0.6, 7.1 L/s
**11.38 d)** 2.06 m$^3$/s
**11.44** 69 s
**11.46** 1.57 L/s, 8.03 s
**11.48 a)** 1090 m/s   **b)** 590 kPa
**11.49 a)** 77%   **b)** 3.54 s

# Chapter 12

**12.2** 42.1 L/s, 35.6 N, 2980 W, 7.22 m
**12.4** 15.9 m, 1.27 hp
**12.6** 1.23 m
**12.7 a)** 5.04 m, 0.144   **b)** 3 m
**12.8** 17.6 m, 172 kW
**12.12** $\Omega_P$ = 1.30, so mixed-flow pump
**12.14 a)** 5.3 m, 0.685 m, 80.25 kW, 6.78 m
**12.16** $\Omega_P$ = 0.751, so radial-flow pump
**12.18** Speed and diameter are 1.19 times greater
**12.19** 1640 rpm, 0.4 m
**12.20** Mixed flow, 0.251 m, 282 rad/s
**12.22** 2.75 m$^3$/min, 12.6 m, 7.2 kW
**12.24** 2.44 m$^3$/min, 10.0 m, 5.1 kW
**12.26** $\Omega_P$ = 5.19 so axial flow

**12.28 a)** 280 m$^3$/h, 64 kW, 8.3 m
**12.30** Three pumps, 197 hp
**12.31** $D$ = 0.25 m, $N$ = 2310 rpm, 3, 50 hp
**12.32 a)** 5.67 m      **b)** 15.8 m
**12.34 a)** radial flow, 9 cm, 4680 rpm   **b)** 3.95 m
**12.36** 6.1°, 28.3 MN·m, 243 m, 360 MW
**12.38** 0.221 m, 399 rpm
**12.40** 189 kW, Francis, 2920 rpm
**12.42** 4.78 m$^3$/s, 1.95 m, 244 mm, 3 jets, 0.204
**12.44** 8 m, Francis or pump/turbine
**12.46 a)** 1.45 m      **b)** 13.6 cm
**12.48** 9.2 MW total, two units
**12.50 a)** 24.6 kW      **b)** $\Omega_T$ = 1.65, Francis
     **c)** 0.3 m, 680 rpm, 26 kW

# Chapter 13

**13.2 a)** 46.8      **b)** 51.6
**13.4** 3.2× 10$^{-6}$ m$^3$/s, 3.19 × 10$^{-3}$ kg/s, laminar
**13.6** 0.0592 m$^3$/s, 7.54 m/s
**13.8 a)** 0.064 m$^3$/s

**13.10 a)** 0.0974 m$^3$/s,   **b)** 0.035 m$^3$/s
**13.14** 0.00396, 0.50
**13.15** $Q$ = 0.7 $Y^{2.43}$

# Index

## PRINCIPAL UNITS USED IN MECHANICS

| Quantity | International System (SI) | | | U.S. Customary System (USCS) | | |
|---|---|---|---|---|---|---|
| | Unit | Symbol | Formula | Unit | Symbol | Formula |
| Acceleration (angular) | radian per second squared | | $rad/s^2$ | radian per second squared | | $rad/s^2$ |
| Acceleration (linear) | meter per second squared | | $m/s^2$ | foot per second squared | | $ft/s^2$ |
| Area | square meter | | $m^2$ | square foot | | $ft^2$ |
| Density (mass) (Specific mass) | kilogram per cubic meter | | $kg/m^3$ | slug per cubic foot | | $slug/ft^3$ |
| Density (weight) (Specific weight) | newton per cubic meter | | $N/m^3$ | pound per cubic foot | pcf | $lb/ft^3$ |
| Energy; work | joule | J | $N \cdot m$ | foot-pound | | ft-lb |
| Force | newton | N | $kg \cdot m/s^2$ | pound | lb | (base unit) |
| Force per unit length (Intensity of force) | newton per meter | | $N/m$ | pound per foot | | lb/ft |
| Frequency | hertz | Hz | $s^{-1}$ | hertz | Hz | $s^{-1}$ |
| Length | meter | m | (base unit) | foot | ft | (base unit) |
| Mass | kilogram | kg | (base unit) | slug | | $lb\text{-}s^2/ft$ |
| Moment of a force; torque | newton meter | | $N \cdot m$ | pound-foot | | lb-ft |
| Moment of inertia (area) | meter to fourth power | | $m^4$ | inch to fourth power | | $in.^4$ |
| Moment of inertia (mass) | kilogram meter squared | | $kg \cdot m^2$ | slug foot squared | | $slug\text{-}ft^2$ |
| Power | watt | W | $J/s$ $(N \cdot m/s)$ | foot-pound per second | | ft-lb/s |
| Pressure | pascal | Pa | $N/m^2$ | pound per square foot | psf | $lb/ft^2$ |
| Section modulus | meter to third power | | $m^3$ | inch to third power | | $in.^3$ |
| Stress | pascal | Pa | $N/m^2$ | pound per square inch | psi | $lb/in.^2$ |
| Time | second | s | (base unit) | second | s | (base unit) |
| Velocity (angular) | radian per second | | rad/s | radian per second | | rad/s |
| Velocity (linear) | meter per second | | m/s | foot per second | fps | ft/s |
| Volume (liquids) | liter | L | $10^{-3} m^3$ | gallon | gal. | $231\ in.^3$ |
| Volume (solids) | cubic meter | | $m^3$ | cubic foot | cf | $ft^3$ |